FREQUENTLY USED NOTATION

l_{db} = basic development length

l_{dh} = development length of a bar with a standard hook

l_{dt} = development length in tension of headed deformed bar

l_{hb} = basic development length of a standard hook in tension

l_n = clear span measured face to face of supports

l_u = unsupported length of a compression member

M_{cr} = cracking moment of concrete

M_1 = smaller end factored moment in a compression member, negative if double curvature

M_2 = larger end factored moment in a compression member

M_{1ns} = smaller factored end moment in a compression member due to loads that result in no appreciable sidesway

M_{2ns} = larger factored end moment in a compression member due to loads that result in no appreciable sidesway

M_d = moment due to dead load

M_o = total factored static moment

n = modular ratio (ratio of modulus of elasticity of steel to that of concrete)

P_c = Euler buckling load of column

P_{no} = pure axial load capacity of column

P_0 = nominal axial load strength of a member with no eccentricity

q_a = allowable soil pressure

q_e = effective soil pressure

R_n = a term used in required percentage of steel expression for flexural members ($M_u/\phi bd^2$)

s = spacing of shear or torsional reinforcing parallel to longitudinal reinforcing

V_{tu} = torsional stress

w = crack width

w_c = unit weight of concrete

y_t = distance from centroidal axis of gross section to extreme fiber in tension

z = a term used to estimate crack sizes and specify distribution of reinforcing

β = ratio of long to short dimensions: clear spans for two-way slabs; sides of column, concentrated load or reaction area; or sides of a footing

β_{dns} = ratio used to account for reduction of stiffness of columns due to sustained axial loads

β_{ds} = ratio used to account for reduction of stiffness of columns due to sustained lateral loads

β_1 = a factor to be multiplied by the depth d of a member to obtain the depth of the equivalent rectangular stress block

δ = moment magnification factor to reflect effects of member curvature between ends of compression member

δ_s = moment magnification factor for slender columns in frames not braced against sidesway

ϵ_c = strain in compression concrete

ϵ_s = strain in tension reinforcement

ϵ_s' = strain in compression reinforcement

λ_Δ = a multiplier used in computing long-term deflections

μ = coefficient of friction

ξ = a time-dependent factor for sustained loads used in computing long-term deflections

ρ = ratio of nonprestressed reinforcement in a section

ρ' = ratio of compression reinforcing in a section

ρ_b = ratio of tensile reinforcing producing balanced strain condition

ϕ = capacity reduction factor

Design of
Reinforced Concrete

Design of Reinforced Concrete

Eighth Edition

ACI 318-08 Code Edition

Jack C. McCormac
Clemson University

Russell H. Brown
Clemson University

WILEY

John Wiley & Sons, Inc.

PUBLISHER	Don Fowley
ACQUISITIONS EDITOR	Jennifer Welter
SENIOR PRODUCTION EDITOR	Valerie A. Vargas
MARKETING MANAGER	Christopher Ruel
COVER DESIGNER	Jeof Vita
PRODUCTION MANAGEMENT SERVICES	Thomson Digital
EDITORIAL ASSISTANT	Mark Owens
MEDIA EDITOR	Lauren Sapira
COVER PHOTO	Courtesy of Skidmore, Owings & Merrill LLP

This book was set in 10/12 Times New Roman by Thomson Digital and printed and bound by Hamilton Printing. The cover was printed by Lehigh Phoenix.

To order books or for customer service please, call 1-800-CALL WILEY (225-5945).

ISBN-13   978-0-470-27927-4

Printed in the United States of America
10 9 8 7 6 5 4 3 2 1

Preface

AUDIENCE

This textbook presents an introduction to reinforced concrete design. We authors hope the material is written in such a manner as to interest students in the subject and to encourage them to continue its study in the years to come. The text was prepared with an introductory three-credit course in mind, but sufficient material is included for an additional three-credit course.

NEW TO THIS EDITION

Updated Code

With the eighth edition of this text the contents have been updated to conform to the 2008 building code of the American Concrete Institute (ACI 318-08). Changes to this edition of the Code include:

- Numerous changes in notations and section numbers.
- A change in the treatment of the design of lightweight aggregate concrete throughout the Code.
- The strength reduction factor for spiral columns was increased and headed deformed bars were introduced as an alternative to hooks for providing development length.
- Clarifications for development length of galvanized, stainless steel and bundled bars.
- Use of small concrete cylinders was introduced, allowing 4×8 in. cylinders instead of 6×12 in.
- Earthquake-resistant design requirements are now related to the Seismic Design Category (SDC) to be consistent with other documents that prescribe design loads.

Updated Material

Most of the chapters have been modified reflecting the viewpoints of the new coauthor, with the concurrence of the original author.

- Discussion of the variable ϕ factor in Chapters 3, 4 and 10 has be revised and expanded. It now includes a generalized equation for the ϕ factor for any yield strength.
- A graph showing the effect of the ϕ factor on the design moment strength is added to Chapter 3 to show the impact of the variable ϕ factor on design.
- Bundled bar development lengths have been clarified by the new code and changed accordingly in Chapter 7.
- An interaction diagram showing the effect of the variable ϕ factor has been added to Chapter 10.
- Inelastic redistribution of moment section has been revised and updated in Chapter 14.
- Equations for distribution of moment to column and middle strip have been added to supplement the tables in Chapter 16.
- Design of walls loaded out-of-plane has been expanded, and an example of the rational design of a foundation wall has been added in Chapter 18.

Excel Spreadsheets

The new spreadsheets included with the text were created to provide the student and the instructor with tools to analyze and design reinforced concrete elements quickly to compare alternative solutions. Examples are given at the end of the chapters to illustrate how to use the software on Examples previously worked.

Seismic Design

A new Chapter 21 on seismic design was added. This chapter is intended only as an introduction to the topic. An entire textbook could be written on this subject alone. It does, however, familiarize the student with issues related to design of reinforced concrete structures to resist earthquakes. Chapter 21 includes two new example problems illustrating some of the new code requirements.

Shear Wall Design

The section on shear wall design in Chapter 18 has been expanded. The new material gives details and examples on how to design shear walls for combined axial load and bending moment. Interaction diagrams developed for columns in Chapter 10 are applied to design of shear walls.

INSTRUCTOR AND STUDENT RESOURCES

The website for the book is located at www.wiley.com/college/mccormac and contains the following resources.

For Instructors

Solutions Manual.   A password-protected Solutions Manual is available for download, which contains complete solutions for all homework problems in the text.

Figures in PPT format.   Also available are the figures from the text in PowerPoint format, for easy creation of lecture slides.

Lecture Presentation Slides in PPT format.   Presentation slides developed by the author.

- Sample Exams
- Course Syllabi

Visit the Instructor Companion Site portion of the books website at www.wiley.com/college/mccormac to register for a password. These resources are available for instructors who have adopted the book for their course. The website may be updated periodically with additional material.

For Students and Instructors

Excel spreadsheets are provided for most chapters of the text. Use of the spreadsheets is self-explanatory. Many of the cells contain comments to assist the new user. The spreadsheets can be modified by the student or instructor to suit their more specific needs. In most cases, calculations contained within the spreadsheets mirror those shown in the example problems in the text.

The many uses of these spreadsheets are illustrated throughout the text. At the end of most chapters are example problems demonstrating the use of the spreadsheet for that particular chapter. Space does not permit examples for all of the spreadsheet capabilities. The examples chosen were thought by the authors to be the most relevant. Visit the Student Companion Site portion of the book's Website at www.wiley.com/college/mccormac to download this software.

ACKNOWLEDGMENTS

We wish to thank the following persons who reviewed this edition:
Roger H. L. Chen (West Virginia University), Hector Estrada (University of the Pacific), and Wael Zatar (Marshall University). We also want the thank Richard Bennett (University of Tennessee), Max Porter (Iowa State University), and Mark McGinley (University of Lousiville) for their review and helpful comments on Chapter 21. Special thanks go to Richard Klingner (University of Texas at Austin) for extensive review and comment of Chapter 21.

Finally, we are also grateful to the reviewers and users of the previous editions of this book for their suggestions, corrections, and criticisms. We are always grateful to anyone who takes the time to contact us concerning any part of the book.

Jack C. McCormac
Russell H. Brown

Contents

1

Introduction

1.1 CONCRETE AND REINFORCED CONCRETE

Concrete is a mixture of sand, gravel, crushed rock, or other aggregates held together in a rocklike mass with a paste of cement and water. Sometimes one or more admixtures are added to change certain characteristics of the concrete such as its workability, durability, and time of hardening.

As with most rocklike substances, concrete has a high compressive strength and a very low tensile strength. *Reinforced concrete* is a combination of concrete and steel wherein the steel reinforcement provides the tensile strength lacking in the concrete. Steel reinforcing is also capable of resisting compression forces and is used in columns as well as in other situations to be described later.

1.2 ADVANTAGES OF REINFORCED CONCRETE AS A STRUCTURAL MATERIAL

Reinforced concrete may be the most important material available for construction. It is used in one form or another for almost all structures, great or small—buildings, bridges, pavements, dams, retaining walls, tunnels, drainage and irrigation facilities, tanks, and so on.

The tremendous success of this universal construction material can be understood quite easily if its numerous advantages are considered. These include the following:

1. It has considerable compressive strength per unit cost compared with most other materials.
2. Reinforced concrete has great resistance to the actions of fire and water and, in fact, is the best structural material available for situations where water is present. During fires of average intensity, members with a satisfactory cover of concrete over the reinforcing bars suffer only surface damage without failure.
3. Reinforced concrete structures are very rigid.
4. It is a low-maintenance material.
5. As compared with other materials, it has a very long service life. Under proper conditions, reinforced concrete structures can be used indefinitely without reduction of their load-carrying abilities. This can be explained by the fact that the strength of concrete does not decrease with time but actually increases over a very long period, measured in years, due to the lengthy process of the solidification of the cement paste.
6. It is usually the only economical material available for footings, floor slabs, basement walls, piers, and similar applications.

NCNB Tower in Charlotte, N.C. completed 1991. (Courtesy Portland Cement Association.)

7. A special feature of concrete is its ability to be cast into an extraordinary variety of shapes from simple slabs, beams, and columns to great arches and shells.

8. In most areas, concrete takes advantage of inexpensive local materials (sand, gravel, and water) and requires relatively small amounts of cement and reinforcing steel, which may have to be shipped in from other parts of the country.

9. A lower grade of skilled labor is required for erection as compared with other materials such as structural steel.

1.3 DISADVANTAGES OF REINFORCED CONCRETE AS A STRUCTURAL MATERIAL

To use concrete successfully, the designer must be completely familiar with its weak points as well as its strong ones. Among its disadvantages are the following:

1. Concrete has a very low tensile strength, requiring the use of tensile reinforcing.

2. Forms are required to hold the concrete in place until it hardens sufficiently. In addition, falsework or shoring may be necessary to keep the forms in place for roofs, walls, floors,

The 320-ft-high Pyramid Sports Arena, Memphis, Tennessee. (Courtesy of EFCO Corp.)

and similar structures until the concrete members gain sufficient strength to support themselves. Formwork is very expensive. In the United States its costs run from one-third to two-thirds of the total cost of a reinforced concrete structure, with average values of about 50%. *It should be obvious that when efforts are made to improve the economy of reinforced concrete structures the major emphasis is on reducing formwork costs.*

3. The low strength per unit of weight of concrete leads to heavy members. This becomes an increasingly important matter for long-span structures where concrete's large dead weight has a great effect on bending moments. Lightweight aggregates can be used to reduce concrete weight, but the cost of the concrete is increased.

4. Similarly, the low strength per unit of volume of concrete means members will be relatively large, an important consideration for tall buildings and long-span structures.

5. The properties of concrete vary widely due to variations in its proportioning and mixing. Furthermore, the placing and curing of concrete is not as carefully controlled as is the production of other materials such as structural steel and laminated wood.

Two other characteristics that can cause problems are concrete's shrinkage and creep. These characteristics are discussed in Section 1.11 of this chapter.

1.4   HISTORICAL BACKGROUND

Most people believe that concrete has been in common use for many centuries, but this is not the case. The Romans did make use of a cement called *pozzolana* before the birth of Christ. They found large deposits of a sandy volcanic ash near Mt. Vesuvius and in other places in Italy. When they mixed this material with quicklime and water as well as sand and gravel, it hardened into a rocklike substance and was used as a building material. One might expect that a relatively poor grade of concrete would result, as compared with today's standards, but some Roman concrete structures are still in existence today. One example is the Pantheon (a building dedicated to all gods) which is located in Rome and was completed in A.D. 126.

The art of making pozzolanic concrete was lost during the Dark Ages and was not revived until the eighteenth and nineteenth centuries. A deposit of natural cement rock was discovered in England in 1796 and was sold as "Roman cement." Various other deposits of natural cement were discovered in both Europe and America and were used for several decades.

The real breakthrough for concrete occurred in 1824 when an English bricklayer named Joseph Aspdin, after long and laborious experiments, obtained a patent for a cement which he called portland cement because its color was quite similar to that of the stone quarried on the Isle of Portland off the English coast. He made his cement by taking certain quantities of clay and limestone, pulverizing them, burning them in his kitchen stove, and grinding the resulting clinker into a fine powder. During the early years after its development, his cement was used primarily in stuccos.[1] This wonderful product was adopted very slowly by the building industry and was not even introduced into the United States until 1868; the first portland cement was not manufactured in the United States until the 1870s.

The first uses of concrete are not very well known. Much of the early work was done by the Frenchmen François Le Brun, Joseph Lambot, and Joseph Monier. In 1832 Le Brun built a concrete house and followed it with the construction of a school and a church with the same material. In about 1850, Lambot built a concrete boat reinforced with a network of parallel wires or bars. Credit is usually given to Monier, however, for the invention of reinforced concrete. In 1867 he received a patent for the construction of concrete basins or tubs and reservoirs reinforced with a mesh of iron wire. His stated goal in working with this material was to obtain lightness without sacrificing strength.[2]

From 1867 to 1881 Monier received patents for reinforced concrete railroad ties, floor slabs, arches, footbridges, buildings, and other items in both France and Germany. Another Frenchman, François Coignet, built simple reinforced concrete structures and developed basic methods of design. In 1861 he published a book in which he presented quite a few applications. He was the first person to realize that the addition of too much water in the mix greatly reduced concrete strength. Other Europeans who were early experimenters with reinforced concrete included the Englishmen William Fairbairn and William B. Wilkinson, the German G. A. Wayss, and another Frenchman, François Hennebique.[3,4]

William E. Ward built the first reinforced concrete building in the United States in Port Chester, New York, in 1875. In 1883 he presented a paper before the American Society of Mechanical Engineers in which he claimed that he got the idea of reinforced concrete by watching English laborers in 1867 trying to remove hardened cement from their iron tools.[5]

Thaddeus Hyatt, an American, was probably the first person to correctly analyze the stresses in a reinforced concrete beam, and in 1877 he published a 28-page book on the subject, entitled *An Account of Some Experiments with Portland Cement Concrete, Combined with Iron as a Building Material*. In this book he praised the use of reinforced concrete and said that "rolled

[1]Kirby, R. S., and Laurson, P. G., 1932, *The Early Years of Modern Civil Engineering* (New Haven: Yale University Press), p. 266.

[2]Kirby and Laurson, *The Early Years of Modern Civil Engineering*, pp. 273–275.

[3]Straub, H., 1964, *A History of Civil Engineering* (Cambridge: MIT Press), pp. 205–215. Translated from the German *Die Geschichte der Bauingenieurkunst* (Basel: Verlag Birkhauser), 1949.

[4]Kirby, R. S., and Laurson, P. G., 1932, *The Early Years of Modern Civil Engineering* (New Haven: Yale University Press), pp. 273–275.

[5]Ward, W. E., 1883, "Béton in Combination with Iron as a Building Material," *Transactions ASME*, 4, pp. 388–403.

Two reinforced concrete hollow cylinders each 65 ft in diameter and 185 ft high being towed to North Sea location as part of oil drilling platform.(Courtesy of United Nations, J. Moss.)

beams (steel) have to be taken largely on faith." Hyatt put a great deal of emphasis on the high fire resistance of concrete.[6]

E. L. Ransome of San Francisco reportedly used reinforced concrete in the early 1870s and was the originator of deformed (or twisted) bars, for which he received a patent in 1884. These bars, which were square in cross section, were cold-twisted with one complete turn in a length of not more than 12 times the bar diameter.[7] (The purpose of the twisting was to provide better bonding or adhesion of the concrete and the steel.) In 1890 in San Francisco, Ransome built the Leland Stanford Jr. Museum. It is a reinforced concrete building 312 feet long and 2 stories high in which discarded wire rope from a cable-car system was used as tensile reinforcing. This building experienced little damage in the 1906 earthquake and the fire that ensued. The limited damage to this building and other concrete structures that withstood the great 1906 fire led to the widespread acceptance of this form of construction on the West Coast. Since 1900–1910, the development and use of reinforced concrete in the United States has been very rapid.[8,9]

1.5   COMPARISON OF REINFORCED CONCRETE AND STRUCTURAL STEEL FOR BUILDINGS AND BRIDGES

When a particular type of structure is being considered, the student may be puzzled by the question, "Should reinforced concrete or structural steel be used?" There is much joking on this point, with the proponents of reinforced concrete referring to steel as that material which rusts and

[6]Kirby, R. S., and Laurson, P. G., 1932, *The Early Years of Modern Civil Engineering* (New Haven: Yale University Press), p. 275.
[7]American Society for Testing Materials, 1911, *Proceedings*, 11, pp. 66–68.
[8]Wang, C. K., and Salmon, C. G., 1998, *Reinforced Concrete Design*, 6th ed. (New York: HarperCollins), pp. 3–5.
[9]"The Story of Cement, Concrete and Reinforced Concrete," *Civil Engineering*, November 1977, pp. 63–65.

those favoring structural steel referring to concrete as that material which when overstressed tends to return to its natural state—that is, sand and gravel.

There is no simple answer to this question, inasmuch as both of these materials have many excellent characteristics that can be utilized successfully for so many types of structures. In fact, they are often used together in the same structures with wonderful results.

The selection of the structural material to be used for a particular building depends on the height and span of the structure, the material market, foundation conditions, local building codes, and architectural considerations. For buildings of less than 4 stories, reinforced concrete, structural steel, and wall-bearing construction are competitive. From 4 to about 20 stories, reinforced concrete and structural steel are economically competitive, with steel having taken most of the jobs above 20 stories in the past. Today, however, reinforced concrete is becoming increasingly competitive above 20 stories, and there are a number of reinforced concrete buildings of greater height around the world. The 74-story, 859-ft-high Water Tower Place in Chicago is the tallest reinforced concrete building in the world. The 1465-ft CN tower (not a building) in Toronto, Canada, is the tallest reinforced concrete structure in the world.

Although we would all like to be involved in the design of tall prestigious reinforced concrete buildings, there are just not enough of them to go around. As a result, nearly all of our work involves much smaller structures. Perhaps 9 out of 10 buildings in the United States are 3 stories or less in height, and more than two-thirds of them contain 15,000 sq ft or less of floor space.

Foundation conditions can often affect the selection of the material to be used for the structural frame. If foundation conditions are poor, a lighter structural steel frame may be desirable. The building code in a particular city may be favorable to one material over the other. For instance, many cities have fire zones in which only fireproof structures can be erected—a very favorable situation for reinforced concrete. Finally, the time element favors structural steel frames, as they can be erected more quickly than reinforced concrete ones. The time advantage, however, is not as great as it might seem at first because if the structure is to have any type of fire rating, the builder will have to cover the steel with some kind of fireproofing material after it is erected.

To make decisions about using concrete or steel for a bridge will involve several factors, such as span, foundation conditions, loads, architectural considerations, and others. In general, concrete is an excellent compression material and normally will be favored for short-span bridges and for cases where rigidity is required (as, perhaps, for railway bridges).

1.6 COMPATIBILITY OF CONCRETE AND STEEL

Concrete and steel reinforcing work together beautifully in reinforced concrete structures. The advantages of each material seem to compensate for the disadvantages of the other. For instance, the great shortcoming of concrete is its lack of tensile strength; but tensile strength is one of the great advantages of steel. Reinforcing bars have tensile strengths equal to approximately 100 times that of the usual concretes used.

The two materials bond together very well so there is little chance of slippage between the two, and thus they will act together as a unit in resisting forces. The excellent bond obtained is due to the chemical adhesion between the two materials, the natural roughness of the bars, and the closely spaced rib-shaped deformations rolled on the bar surfaces.

Reinforcing bars are subject to corrosion, but the concrete surrounding them provides them with excellent protection. The strength of exposed steel subject to the temperatures reached in fires of ordinary intensity is nil, but the enclosure of the reinforcement in concrete produces very satisfactory fire ratings. Finally, concrete and steel work well together in relation to temperature changes because their coefficients of thermal expansion are quite close to each other. For steel the coefficient is 0.0000065 per unit length per degree Fahrenheit, while it varies for concrete from about 0.000004 to 0.000007 (average value, 0.0000055).

1.7   DESIGN CODES

The most important code in the United States for reinforced concrete design is the American Concrete Institute's *Building Code Requirements for Structural Concrete* (ACI 318-08).[10] This code, which is used primarily for the design of buildings, is followed for the majority of the numerical examples given in this text. Frequent references are made to this document, and section numbers are provided. Design requirements for various types of reinforced concrete members are presented in the Code along with a "Commentary" on those requirements. The Commentary provides explanations, suggestions, and additional information concerning the design requirements. As a result, users will obtain a better background and understanding of the Code.

The ACI Code is not in itself a legally enforceable document. It is merely a statement of current good practice in reinforced concrete design. It is, however, written in the form of a code or law so that various public bodies such as city councils can easily vote it into their local building codes, and as such it becomes legally enforceable in that area. In this manner the ACI Code has been incorporated into law by countless government organizations throughout the United States. The International Building Code (IBC), which was first published in 2000 by the International Code Council, has consolidated the three regional building codes (Building Officials and Code Administrators, International Conference of Building Officials, and Southern Building Code Congress International) into one national document. The IBC Code is updated every 3 years and refers to the most recent edition of ACI 318 for most of its provisions related to reinforced concrete design, with only a few modifications. It is expected that IBC 2009 will refer to ACI 318-08 for most of its reinforced concrete provisions. The ACI 318 Code is also widely accepted in Canada and Mexico and has had tremendous influence on the concrete codes of all countries throughout the world.

As more knowledge is obtained pertaining to the behavior of reinforced concrete, the ACI revises its code. The present objective is to make yearly changes in the code in the form of supplements and to provide major revisions of the entire code every 3 years.

Other well-known reinforced concrete specifications are those of the American Association of State Highway and Transportation Officials (AASHTO) and the American Railway Engineering Association (AREA).

1.8   SI UNITS AND SHADED AREAS

Most of this book is devoted to the design of reinforced concrete structures using U.S. customary units. The authors, however, feel that it is absolutely necessary for today's engineer to be able to design in either customary or SI units. Thus SI equations, where different from those in customary units, are presented herein, along with quite a few numerical examples using SI units. The equations are taken from the American Concrete Institute's metric version of *Building Code Requirements for Structural Concrete* (ACI 318M-08).[11]

---

[10]American Concrete Institute, 2008, *Building Code Requirements for Structural Concrete* (ACI 318-08), Farmington Hills, Michigan.
[11]American Concrete Institute, 2008, *Building Code Requirements for Structural Concrete* (ACI 318M-08), Farmington Hills, Michigan.

For many people it is rather distracting to read a book in which numbers, equations, and so on are presented in two sets of units. To try to reduce this annoyance, the authors have placed a shaded area around any items pertaining to SI units throughout the text.

If readers are working at a particular time with customary units, they can completely ignore the shaded areas. On the other hand, it is hoped that the same shaded areas will enable a person working with SI units to easily find appropriate equations, examples, and so on.

1.9 TYPES OF PORTLAND CEMENT

Concretes made with normal portland cement require about two weeks to achieve a sufficient strength to permit the removal of forms and the application of moderate loads. Such concretes reach their design strengths after about 28 days and continue to gain strength at a slower rate thereafter.

On many occasions it is desirable to speed up construction by using *high-early-strength cements*, which, though more expensive, enable us to obtain desired strengths in 3 to 7 days rather than the normal 28 days. These cements are particularly useful for the fabrication of precast members in which the concrete is placed in forms where it quickly gains desired strengths and is then removed from the forms and the forms are used to produce more members. Obviously, the quicker the desired strength is obtained, the more efficient the operation. A similar case can be

One Peachtree Center in Atlanta, GA is 854 ft. high built for the 1996 Olympics. (Courtesy Portland Cement Association.)

made for the forming of concrete buildings floor by floor. High-early-strength cements can also be used advantageously for emergency repairs of concrete and for *shotcreting* (where a mortar or concrete is blown through a hose at a high velocity onto a prepared surface).

There are other special types of portland cements available. The chemical process that occurs during the setting or hardening of concrete produces heat. For very massive concrete structures such as dams, mat foundations and piers, the heat will dissipate very slowly and can cause serious problems. It will cause the concrete to expand during hydration. When cooling, the concrete will shrink and severe cracking will often occur.

Concrete may be used where it is exposed to various chlorides and/or sulfates. Such situations occur in seawater construction and for structures exposed to various types of soil. Some portland cements are manufactured that have lower heat of hydration, and others are manufactured with greater resistance to attack by chlorides and sulfates.

In the United States, the American Society for Testing and Materials (ASTM) recognizes five types of portland cement. These different cements are manufactured from just about the same raw materials, but their properties are changed by using various blends of those materials. Type I cement is the normal cement used for most construction, but four other types are useful for special situations in which high early strength or low heat or sulfate resistance is needed:

Type I—the common all-purpose cement used for general construction work.

Type II—a modified cement that has a lower heat of hydration than does Type I cement and that can withstand some exposure to sulfate attack.

Type III—a high-early-strength cement that will produce in the first 24 hours a concrete with a strength about twice that of Type I cement. This cement does have a much higher heat of hydration.

Type IV—a low-heat cement that produces a concrete which generates heat very slowly. It is used for very large concrete structures.

Type V—a cement used for concretes that are to be exposed to high concentrations of sulfate.

Should the desired type of cement not be available, various admixtures may be purchased with which the properties of Type I cement can be modified to produce the desired effect.

1.10 ADMIXTURES

Materials added to concrete during or before mixing are referred to as admixtures. They are used to improve the performance of concrete in certain situations as well as to lower its cost. There is a rather well-known saying regarding admixtures, to the effect that "they are to concrete as beauty aids are to the populace." Several of the most common types of admixtures are listed and briefly described here.

1. *Air-entraining admixtures*, conforming to the requirements of ASTM C260 and C618, are used primarily to increase concrete's resistance to freezing and thawing and provide better resistance to the deteriorating action of de-icing salts. The air-entraining agents cause the mixing water to foam, with the result that billions of closely spaced air bubbles are incorporated into the concrete. When concrete freezes, water moves into the air bubbles, relieving the pressure in the concrete. When the concrete thaws, the water can move out of the bubbles, with the result that there is less cracking than if air-entrainment had not been used.

2. The addition of *accelerating admixtures* such as calcium chloride to concrete will accelerate its early-strength development. The results of such additions (particularly useful in cold climates) are reduced times required for curing and protection of the concrete and the earlier removal of forms. (Section 3.6.3 of the ACI Code states that because of corrosion

problems, calcium chloride may not be added to concretes with embedded aluminum, concretes cast against stay-in-place galvanized steel forms, or prestressed concretes.) Other accelerating admixtures that may be used include various soluble salts as well as some other organic compounds.

3. *Retarding admixtures* are used to slow the setting of the concrete and to retard temperature increases. They consist of various acids or sugars or sugar derivatives. Some concrete truck drivers keep sacks of sugar on hand to throw into the concrete in case they get caught in traffic jams or are otherwise delayed. Retarding admixtures are particularly useful for large pours where significant temperature increases may occur. They also prolong the plasticity of the concrete, enabling better blending or bonding together of successive pours. Retarders can also slow the hydration of cement on exposed concrete surfaces or formed surfaces to produce attractive exposed aggregate finishes.

4. *Superplasticizers* are admixtures made from organic sulfonates. Their use enables engineers to reduce the water content in concretes substantially while at the same time increasing their slumps. Although superplasticizers can also be used to keep constant water-cement ratios while using less cement, they are more commonly used to produce workable concretes with considerably higher strengths while using the same amount of cement. (See Section 1.13.) A relatively new product, self-consolidating concrete, utilizes superplasticizers and modifications in mix designs to produce an extremely workable mix that requires no vibration, even for the most congested placement situations.

5. Usually, *waterproofing materials* are applied to hardened concrete surfaces, but they may be added to concrete mixes. These admixtures generally consist of some type of soap or petroleum products, as perhaps asphalt emulsions. They may help retard the penetration of water into porous concretes but probably don't help dense, well-cured concretes very much.

1.11 PROPERTIES OF REINFORCED CONCRETE

A thorough knowledge of the properties of concrete is necessary for the student before he or she begins to design reinforced concrete structures. An introduction to several of these properties is presented in this section.

Compressive Strength

The compressive strength of concrete f_c' is determined by testing to failure 28-day-old 6-in. diameter by 12-in. concrete cylinders at a specified rate of loading (4 in. diameter by 8 in. cylinders were first permitted in the 2008 Code in lieu of the larger cylinders). For the 28-day period the cylinders are usually kept under water or in a room with constant temperature and 100% humidity. Although concretes are available with 28-day ultimate strengths from 2500 psi up to as high as 10,000 to 20,000 psi, most of the concretes used fall into the 3000- to 7000-psi range. For ordinary applications, 3000- and 4000-psi concretes are used, whereas for prestressed construction, 5000- and 6000-psi strengths are common. For some applications, such as for the columns of the lower stories of high-rise buildings, concretes with strengths up to 9000 or 10,000 psi have been used and can be furnished by ready-mix companies. As a result, the use of such high-strength concretes is becoming increasingly common. At Two Union Square in Seattle, concrete with strengths up to 19,000 psi was used.

The values obtained for the compressive strength of concretes as determined by testing are to a considerable degree dependent on the sizes and shapes of the test units and the manner in which they are loaded. In many countries the test specimens are cubes 200 mm (7.87 in.) on each side. For the same batches of concrete, the testing of 6-in. by 12-in. cylinders provides compressive strengths only equal to about 80% of the values in psi determined with the cubes.

It is quite feasible to move from 3000-psi concrete to 5000-psi concrete without requiring excessive amounts of labor or cement. The approximate increase in material cost for such a strength increase is 15% to 20%. To move above 5000- or 6000-psi concrete, however, requires very careful mix designs and considerable attention to such details as mixing, placing, and curing. These requirements cause relatively larger increases in cost.

Several comments are made throughout the text regarding the relative economy of using different strength concretes for different applications, such as for beams, columns, footings, and prestressed members.

To ensure that the compressive strength of concrete in the structure is at least as strong as the specified value, f_c', the design of the concrete mix must target a higher value, f_{cr}'. Section 5.3 of the ACI Code requires that the concrete compressive strengths used as a basis for selecting the concrete proportions must exceed the specified 28-day strengths by fairly large values. For concrete production facilities that have sufficient field strength test records to enable them to calculate satisfactory standard deviations (as described in ACI Section 5.3.1.1), a set of required average compressive strengths (f_{cr}') to be used as the basis for selecting concrete properties is specified in ACI Table 5.3.2.1. For facilities that do not have sufficient records to calculate satisfactory standard deviations, ACI Table 5.3.2.2 provides increases in required average design compressive strength (f_{cr}') of 1000 psi for specified concrete strength (f_c') of less than 3000 psi and appreciably higher increases for higher f_c' concretes.

The stress–strain curves of Figure 1.1 represent the results obtained from compression tests of sets of 28-day-old standard cylinders of varying strengths. You should carefully study these curves because they bring out several significant points:

(a) The curves are roughly straight while the load is increased from zero to about one-third to one-half the concrete's ultimate strength.

(b) Beyond this range the behavior of concrete is nonlinear. This lack of linearity of concrete stress–strain curves at higher stresses causes some problems in the structural analysis of concrete structures because their behavior is also nonlinear at higher stresses.

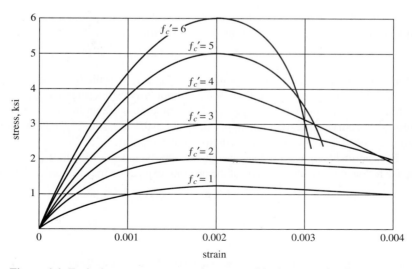

Figure 1.1 Typical concrete stress–strain curve, with short-term loading.

(c) Of particular importance is the fact that regardless of strengths, all the concretes reach their ultimate strengths at strains of about 0.002.

(d) Concrete does not have a definite yield strength; rather, the curves run smoothly on to the point of rupture at strains of from 0.003 to 0.004. It will be assumed for the purpose of future calculations in this text that concrete fails at 0.003 (ACI 10.2.3). *The reader should note that this value, which is conservative for normal-strength concretes, may not be conservative for higher-strength concretes in the 8000-psi-and-above range.* The European code uses a different value for ultimate compressive strain for columns (0.002) than for beams and eccentrically loaded columns (0.0035).[12]

(e) Many tests have clearly shown that stress–strain curves of concrete cylinders are almost identical to those for the compression sides of beams.

(f) It should be further noticed that the weaker grades of concrete are less brittle than the stronger ones—that is, they will take larger strains before breaking.

Static Modulus of Elasticity

Concrete has no clear-cut modulus of elasticity. Its value varies with different concrete strengths, concrete age, type of loading, and the characteristics and proportions of the cement and aggregates. Furthermore, there are several different definitions of the modulus:

(a) The *initial modulus* is the slope of the stress–strain diagram at the origin of the curve.

(b) The *tangent modulus* is the slope of a tangent to the curve at some point along the curve—for instance, at 50% of the ultimate strength of the concrete.

(c) The slope of a line drawn from the origin to a point on the curve somewhere between 25% and 50% of its ultimate compressive strength is referred to as a *secant modulus*.

(d) Another modulus, called the *apparent modulus* or the *long-term modulus*, is determined by using the stresses and strains obtained after the load has been applied for a certain length of time.

Section 8.5.1 of the ACI Code states that the following expression can be used for calculating the modulus of elasticity of concretes weighing from 90 to 155 lb/ft^3:

$$E_c = w_c^{1.5} 33 \sqrt{f_c'}$$

In this expression, E_c is the modulus of elasticity in psi, w_c is the weight of the concrete in pounds per cubic foot, and f_c' is its specified 28-day compressive strength in psi. This is actually a secant modulus with the line (whose slope equals the modulus) drawn from the origin to a point on the stress–strain curve corresponding approximately to the stress $(0.45 f_c')$ that would occur under the estimated dead and live loads the structure must support.

For normal-weight concrete weighing approximately 145 lb/ft^3, the ACI Code states that the following simplified version of the previous expression may be used to determine the modulus:

$$E_c = 57,000 \sqrt{f_c'}$$

---

[12]MacGregor, J. G., and Wight, J. K., *Reinforced Concrete Mechanics and Design*, 4th ed., (Upper Saddle River, NJ: Pearson Prentice Hall) p. 111.

Table A.1 (see the Appendix at the end of the book) shows values of E_c for different strength concretes having normal-weight aggregate. These values were calculated with the first of the preceding formulas assuming 145 lb/ft^3 concrete.

In SI units, $E_c = w_c^{1.5}(0.043)\sqrt{f_c'}$ with w_c varying from 1500 to 2500 kg/m^3 and with f_c' in N/mm^2 or MPa (megapascals). Should normal crushed stone or gravel concrete (with a mass of approximately 2320 kg/m^3) be used, $E_c = 4700\sqrt{f_c'}$. Table B.1 of Appendix B of this text provides moduli values for several different strength concretes.

The term *unit weight* is constantly used by structural engineers working with U.S. customary units. When using the SI system, however, this term should be replaced with the term *mass density*. A kilogram is not a force unit and only indicates the amount of matter in an object. The mass of a particular object is the same anywhere on Earth, whereas the weight of an object in our customary units varies depending on altitude because of the change in gravitational acceleration.

Concretes with strength above 6000 psi are referred to as high-strength concretes. Tests have indicated that the usual ACI equations for E_c when applied to high-strength concretes result in values that are too large. Based on studies at Cornell University, the expression to follow has been recommended for normal-weight concretes with f_c' values greater than 6000 psi and up to 12,000 psi and for lightweight concretes with f_c' greater than 6000 psi and up to 9000 psi.[13,14]

$$E_c(\text{psi}) = [40,000\sqrt{f_c'} + 10^6]\left(\frac{w_c}{145}\right)^{1.5}$$

In SI units with f_c' in MPa and w_c in kg/m^3 the expression is

$$E_c(\text{MPa}) = [3.32\sqrt{f_c'} + 6895]\left(\frac{w_c}{2320}\right)^{1.5}$$

Dynamic Modulus of Elasticity

The dynamic modulus of elasticity, which corresponds to very small instantaneous strains, is usually obtained by sonic tests. It is generally 20% to 40% higher than the static modulus and is approximately equal to the initial modulus. When structures are being analyzed for seismic or impact loads, the use of the dynamic modulus seems appropriate.

Poisson's Ratio

As a concrete cylinder is subjected to compressive loads, it not only shortens in length but also expands laterally. The ratio of this lateral expansion to the longitudinal shortening is referred to as *Poisson's ratio*. Its value varies from about 0.11 for the higher-strength concretes to as high as 0.21 for the weaker-grade concretes, with average values of about 0.16. There does not seem to be any direct relationship between the value of the ratio and the values of items such as the water–cement ratio, amount of curing, aggregate size, and so on.

For most reinforced concrete designs, no consideration is given to the so-called Poisson effect. It may very well have to be considered, however, in the analysis and design of arch dams,

[13]Nawy, E. G., 2006, *Prestressed Concrete: A Fundamental Approach*, 5th ed. (Upper Saddle River, NJ: Prentice-Hall), p. 38.

[14]Carrasquillol, R., Nilson, A., and Slate, F., 1981, "Properties of High-strength Concrete Subject to Short-term Loads." *J. ACI Proceedings*, 78(3), May–June.

Reinforced concrete bandshell in Portage, Michigan. (Courtesy of Veneklasen Concrete Construction Co.)

tunnels, and some other statically indeterminate structures. Spiral reinforcing in columns takes advantage of Poisson's ratio and will be discussed in Chapter 9.

Shrinkage

When the materials for concrete are mixed together, the paste consisting of cement and water fills the voids between the aggregate and bonds the aggregate together. This mixture needs to be sufficiently workable or fluid so that it can be made to flow in between the reinforcing bars and all through the forms. To achieve this desired workability, considerably more water (perhaps twice as much) is used than is necessary for the cement and water to react together (called *hydration*).

After the concrete has been cured and begins to dry, the extra mixing water that was used begins to work its way out of the concrete to the surface, where it evaporates. As a result, the concrete shrinks and cracks. The resulting cracks may reduce the shear strength of the members and be detrimental to the appearance of the structure. In addition, the cracks may permit the reinforcing to be exposed to the atmosphere, or chemicals such as deicers thereby increasing the possibility of corrosion. Shrinkage continues for many years, but under ordinary conditions probably about 90% of it occurs during the first year. The amount of moisture that is lost varies with the distance from the surface. Furthermore, the larger the surface area of a member in proportion to its volume, the larger the rate of shrinkage; that is, members with small cross sections shrink more proportionately than do those with large ones.

The amount of shrinkage is heavily dependent on the type of exposure. For instance, if concrete is subjected to a considerable amount of wind during curing, its shrinkage will be greater. In a related fashion a humid atmosphere means less shrinkage, whereas a dry one means more.

It should also be realized that it is desirable to use low absorptive aggregates such as those from granite and many limestones. When certain absorptive slates and sandstone aggregates are used, the result may be $1\frac{1}{2}$ or even 2 times the shrinkage with other aggregates.

To minimize shrinkage it is desirable to: (1) keep the amount of mixing water to a minimum; (2) cure the concrete well; (3) place the concrete for walls, floors, and other large items in small sections (thus allowing some of the shrinkage to take place before the next section is placed); (4) use construction joints to control the position of cracks; (5) use shrinkage reinforcement; and (6) use appropriate dense and nonporous aggregates.[15]

Creep

Under sustained compressive loads, concrete will continue to deform for long periods of time. After the initial deformation occurs, the additional deformation is called *creep*, or *plastic flow*. If a compressive load is applied to a concrete member, an immediate or instantaneous elastic shortening occurs. If the load is left in place for a long time, the member will continue to shorten over a period of several years and the final deformation will usually be two to three times the initial deformation. We will find in Chapter 6 that this means that long-term deflections may also be as much as two or three times initial deflections. Perhaps 75% of the total creep will occur during the first year.

Should the long-term load be removed, the member will recover most of its elastic strain and a little of its creep strain. If the load is replaced, both the elastic and creep strains will again be developed.

The amount of creep is largely dependent on the amount of stress. It is almost directly proportional to stress as long as the sustained stress is not greater than about one-half of f'_c. Beyond this level, creep will increase rapidly.

Long-term loads not only cause creep but also can adversely affect the strength of the concrete. For loads maintained on concentrically loaded specimens for a year or longer, there may be a strength reduction of perhaps 15% to 25%. *Thus a member loaded with a sustained load of, say, 85% of its ultimate compression strength, f'_c, may very well be satisfactory for a while, but may fail later.*[16]

Several other items affecting the amount of creep are as follows.

1. The longer the concrete cures before loads are applied, the smaller will be the creep. Steam curing, which causes quicker strengthening, will also reduce creep.

2. Higher-strength concretes have less creep than do lower-strength concretes stressed at the same values. However, applied stresses for higher-strength concretes are in all probability higher than those for lower-strength concretes, and this fact tends to cause increasing creep.

3. Creep increases with higher temperatures. It is highest when the concrete is at about 150° to 160°F.

4. The higher the humidity, the smaller will be the free pore water which can escape from the concrete. Creep is almost twice as large at 50% humidity than at 100% humidity. It is obviously quite difficult to distinguish between shrinkage and creep.

5. Concretes with the highest percentage of cement–water paste have the highest creep because the paste, not the aggregate, does the creeping. This is particularly true if a limestone aggregate is used.

6. Obviously, the addition of reinforcing to the compression areas of concrete will greatly reduce creep because steel exhibits very little creep at ordinary stresses. As creep tends to occur in the concrete, the reinforcing will block it and pick up more and more of the load.

7. Large concrete members (that is, those with large volume-to-surface area ratios) will creep proportionately less than smaller thin members where the free water has smaller distances to travel to escape.

---

[15]Leet, K., 1991, *Reinforced Concrete Design*, 2nd ed. (New York: McGraw-Hill), p. 35.
[16]Rüsch, H., 1960, "Researches Toward a General Flexure Theory for Structural Concrete," *Journal ACI*, 57(1), pp. 1–28.

Tensile Strength

The tensile strength of concrete varies from about 8% to 15% of its compressive strength. A major reason for this small strength is the fact that concrete is filled with fine cracks. The cracks have little effect when concrete is subjected to compression loads because the loads cause the cracks to close and permit compression transfer. Obviously, this is not the case for tensile loads.

Although tensile strength is normally neglected in design calculations, it is nevertheless an important property that affects the sizes and extent of the cracks that occur. Furthermore, the tensile strength of concrete members has a definite reduction effect on their deflections. (Due to the small tensile strength of concrete, little effort has been made to determine its tensile modulus of elasticity. Based on this limited information, however, it seems that its value is equal to its compression modulus.)

You might wonder why concrete is not assumed to resist a portion of the tension in a flexural member and the steel the remainder. The reason is that concrete cracks at such small tensile strains that the low stresses in the steel up to that time would make its use uneconomical. Once tensile cracking has occurred concrete has no more tensile strength remaining.

The tensile strength of concrete doesn't vary in direct proportion to its ultimate compression strength f'_c. It does, however, vary approximately in proportion to the square root of f'_c. This strength is quite difficult to measure with direct axial tension loads because of problems in gripping test specimens so as to avoid stress concentrations and because of difficulties in aligning the loads. As a result of these problems, two rather indirect tests have been developed to measure concrete's tensile strength. These are the *modulus of rupture* and the *split-cylinder tests*.

The tensile strength of concrete in flexure is quite important when considering beam cracks and deflections. For these considerations the tensile strengths obtained with the modulus of rupture test have long been used. The modulus of rupture (which is defined as the flexural tensile strength of concrete) is usually measured by loading a 6-in.× 6-in.×30-in. plain (i.e., unreinforced) rectangular beam (with simple supports placed 24 in. on center) to failure with equal concentrated loads at its one-third points as per ASTM C78-2002.[17] The load is increased until failure occurs by cracking on the tensile face of the beam. The modulus of rupture f_r is then determined from the flexure formula. In the following expressions, b is the beam width, h its depth, and M is $PL/6$ which is the maximum computed moment:

$$f_r = \frac{Mc}{I} = \frac{M\left(\frac{h}{2}\right)}{\frac{1}{12}bh^3}$$

$$f_r = \text{modulus of rupture} = \frac{6M}{bh^2} = PL/bh^2$$

The stress determined in this manner is not very accurate because in using the flexure formula we are assuming the concrete stresses vary in direct proportion to distances from the neutral axis. This assumptions is not very good.

Based on hundreds of tests, the Code (Section 9.5.2.3) provides a modulus of rupture f_r equal to $7.5\lambda\sqrt{f'_c}$, where f_r and f'_c are in units of psi. The λ term reduces the modulus of rupture when lightweight aggregates are used (see Section 1.12).

The tensile strength of concrete may also be measured with the split-cylinder test.[18] A cylinder is placed on its side in the testing machine, and a compressive load is applied

---

[17]American Society for Testing and Materials, 2002, *Standard Test Method for Flexural Strength of Concrete (Using Simple Beam with Third-Point Loading)* ASTM C78 2002, West Conshohocken, PA.

[18]American Society for Testing and Materials, 2004, *Standard Method of Test for Splitting Tensile Strength of Cylindrical Concrete Specimens* (ASTM C496–2004), West Conshohocken, PA.

*In SI units, $f_r = 0.7\sqrt{f_c}$ MPa.

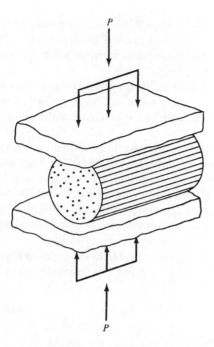

Figure 1.2 Split-cylinder test.

uniformly along the length of the cylinder, with support supplied along the bottom for the cylinder's full length (see Figure 1.2). The cylinder will split in half from end to end when its tensile strength is reached. The tensile strength at which splitting occurs is referred to as the *split-cylinder strength* and can be calculated with the following expression, in which P is the maximum compressive force, L is the length, and D is the diameter of the cylinder:

$$f_t = \frac{2P}{\pi LD}$$

Even though pads are used under the loads, some local stress concentrations occur during the tests. In addition, some stresses develop at right angles to the tension stresses. As a result, the tensile strengths obtained are not very accurate.

Shear Strength

It is extremely difficult in laboratory testing to obtain pure shear failures unaffected by other stresses. As a result, the tests of concrete shearing strengths through the years have yielded values all the way from one-third to four-fifths of the ultimate compressive strengths. You will learn in Chapter 8 that you do not have to worry about these inconsistent shear strength tests because design approaches are based on such very conservative assumptions of that strength.

1.12   AGGREGATES

The aggregates used in concrete occupy about three-fourths of the concrete volume. Since they are less expensive than the cement, it is desirable to use as much of them as possible. Both fine aggregates (usually sand) and coarse aggregates (usually gravel or crushed stone) are used. Any aggregate that passes a No. 4 sieve (which has wires spaced $\frac{1}{4}$ in. on centers in each direction) is said to be fine aggregate. Material of a larger size is coarse aggregate.

The maximum-size aggregates that can be used in reinforced concrete are specified in Section 3.3.2 of the ACI Code. *These limiting values are as follows: one-fifth of the narrowest dimensions between the sides of the forms, one-third of the depth of slabs, or three-quarters of the minimum clear spacing between reinforcing.* Larger sizes may be used if, in the judgment of the engineer, the workability of the concrete and its method of consolidation are such that the aggregate used will not cause the development of honeycomb or voids.

Aggregates must be strong, durable, and clean. Should dust or other particles be present, they may interfere with the bond between the cement paste and the aggregate. The strength of the aggregate has an important effect on the strength of the concrete, and the aggregate properties greatly affect the concrete's durability.

Concretes that have 28-day strengths equal to or greater than 2500 psi and air dry weights equal to or less than 115 lb/ft^3 are said to be structural lightweight concretes. The aggregates used for these concretes are made from expanded shales of volcanic origin, fired clays, or slag. When lightweight aggregates are used for both fine and coarse aggregate, the result is called *all-lightweight* concrete. If sand is used for fine aggregate and if the coarse aggregate is replaced with lightweight aggregate, the result is referred to as *sand-lightweight* concrete. Concretes made with lightweight aggregates may not be as durable or tough as those made with normal-weight aggregates.

Some of the structural properties of concrete are affected by the use of lightweight aggregates. ACI 318-08, Section 8.4 requires that the modulus of rupture be reduced by the introduction of the term λ in the equation

$$f_r = 7.5\lambda\sqrt{f_c'}$$ (ACI Equation 9–10)

or, in SI units with f_c' in N/mm^2, $f_r = 0.7\lambda\sqrt{f_c'}$

The value of λ depends on the aggregate that is replaced with lightweight material. If only the coarse aggregate is replaced (sand-lightweight concrete), λ is 0.85. If the sand is also replaced with lightweight material (all-lightweight concrete), λ is 0.75. Linear interpolation is permitted between the values of 0.85 and 1.0 as well as from 0.75 to 0.85 when partial replacement with lightweight material is used. Alternatively, if the average splitting tensile strength of lightweight concrete, f_{ct}, is specified, ACI 318-08 Section 8.6.1 defines λ as

$$\lambda = \frac{f_{ct}}{6.7\sqrt{f_c'}} \leq 1.0$$

For normal weight concrete, and concrete having normal weight fine aggregate and a blend of lightweight and normal weight coarse aggregate, $\lambda = 1.0$. Use of lightweight aggregate concrete can affect beam deflections, shear strength, coefficient of friction, development lengths of reinforcing bars and hooked bars, and prestressed concrete design.

1.13   HIGH-STRENGTH CONCRETES

Concretes with compression strengths exceeding 6000 psi are referred to as *high-strength concretes*. Another name sometimes given to them is *high-performance concretes* because they have other excellent characteristics besides just high strengths. For instance, the low permeability of such concretes causes them to be quite durable as regards the various physical and chemical agents acting on them that may cause the material to deteriorate.

Up until a few decades ago, structural designers felt that ready-mix companies could not deliver concretes with compressive strengths much higher than 4000 or 5000 psi. This situation, however, is no longer the case as these same companies can today deliver concretes with compressive strengths up to at least 9000 psi. Even stronger concretes than these have been used. At Two Union Square in Seattle 19,000 psi concrete was obtained using ready-mix concrete delivered to the site. Furthermore, concretes have been produced in laboratories with strengths higher than 20,000 psi. Perhaps these latter concretes should be called *super-high-strength concretes* or *super-high-performance concretes.*

If we are going to use a very high-strength cement paste, we must not forget to use a coarse aggregate that is equally as strong. If the planned concrete strength is, say, 15,000 to 20,000 psi, equally strong aggregate must be used, and such aggregate may very well not be available within reasonable distances. In addition to the strengths needed for the coarse aggregate, their sizes should be well graded, and their surfaces should be rough so that better bonding to the cement paste will be obtained. The rough surfaces of aggregates, however, may decrease the concrete's workability.

From an economical standpoint you should realize that though concretes with 12,000 to 15,000 psi strengths cost approximately 3 times as much to produce as do 3000-psi concretes, their compressive strengths are 4 to 5 times as large.

High-strength concretes are sometimes used for both precast and prestressed members. They are particularly useful in the precast industry where their strength enables us to produce smaller and lighter members, with consequent savings in storage, handling, shipping, and erection costs. In addition, they have sometimes been used for offshore structures, but their common use has been for columns of tall reinforced concrete buildings, probably over 25 to 30 stories in height where the column loads are very large, say, 1000 kips or more. Actually, for such buildings the columns for the upper floors, where the loads are relatively small, are probably constructed with conventional 4000- or 5000-psi concretes, while high-strength concretes are used for the lower heavily loaded columns. If conventional concretes were used for these lower columns, the columns could very well become so large that they would occupy excessive amounts of rentable floor space. High-strength concretes are also of advantage in constructing shear walls. (Shear walls are discussed in Chapter 18.)

To produce concretes with strengths above 6000 psi it is first necessary to use more stringent quality control of the work and to exercise special care in the selection of the materials to be used. Strength increases can be made by using lower water-cement ratios, adding admixtures, and selecting good clean and solid aggregates. The actual concrete strengths used by the designer for a particular job will depend on the size of the loads and the quality of the aggregate available.

In recent years appreciable improvements have been made in the placing, vibrating, and finishing of concrete. These improvements have resulted in lower water-cement ratios and thus higher strengths. The most important factor affecting the strength of concrete is its porosity, which is controlled primarily by the water-cement ratio. This ratio should be kept as small as possible as long as adequate workability is maintained. In this regard there are various water-reducing admixtures with which the ratios can be appreciably reduced while at the same time maintaining suitable workability.

Concretes with strengths from 6000 to 10,000 or 12,000 psi can easily be obtained if admixtures such as silica fume and superplasticizers are used. Silica fume, which is more than 90% silicon dioxide, is an extraordinarily fine powder that varies in color from light to dark gray and can even be blue-green-gray. It is obtained from electric arc furnaces as a byproduct during the production of metallic silicon and various other silicon alloys. It is available in both powder and liquid form. The amount of silica fume used in a mix varies from 5% to 30% of the weight of the cement.

Silica fume particles have diameters approximately 100 times smaller than the average cement particle, and their surface areas per unit of weight are roughly 40 to 60 times those of portland cement. As a result, they hold more water. (By the way, this increase of surface area causes the generation of more heat of hydration.) The water-cement ratios are smaller, and strengths are higher. Silica fume is a *pozzolan*: a siliceous material that by itself has no cementing quality, but when used in concrete mixes its extraordinarily fine particles react with the calcium hydroxide in the cement to produce a cementious compound. Quite a few pozzolans are available that can be used satisfactorily in concrete. Two of the most common ones are fly ash and silica fume. Here only silica fume is discussed.

When silica fume is used, it causes increases in the density and strength of the concrete. These improvements are due to the fact that the ultrafine silica fume particles are dispersed between the cement particles. Unfortunately, this causes a reduction in the workability of the concrete, and it is necessary to add *superplasticizers* to the mix. Superplasticizers, also called *high-range water reducers*, are added to concretes to increase their workability. They are made by treating formaldehyde or napthaline with sulphuric acid. Such materials used as admixtures lower the viscosity or resistance to flow of the concrete. As a result, less water can be used, thus yielding lower water-cement ratios and higher strengths.

The addition of organic polymers can be used to replace a part of the cement as the binder. An organic polymer is composed of molecules that have been formed by the union of thousands of molecules. The most commonly used polymers in concrete are latexes. Such additives improve concrete's strength, durability, and adhesion. In addition, the resulting concretes have excellent resistance to abrasion, freezing, thawing, and impact.

Another procedure that can increase the strength of concrete is *consolidation*. When precast concrete products are consolidated, excess water and air are squeezed out, thus producing concretes with optimum air contents. In a similar manner, the centrifugal forces caused by the spinning of concrete pipes during their manufacture consolidate the concrete and reduce the water and air contents. Not much work has been done in the consolidation area for cast-in-place concrete due to the difficulty of applying the squeezing forces. To squeeze such concretes it is necessary to apply pressure to the forms. You can see that one major difficulty in doing this is that very special care must be used to prevent distortion of the wet concrete members.

1.14   FIBER-REINFORCED CONCRETES

In recent years a great deal of interest has been shown in fiber-reinforced concrete, and today there is much ongoing research on the subject. The fibers used are made from steel, plastics, glass, and other materials. Various experiments have shown that the addition of such fibers in convenient quantities (normally up to about 1% or 2% by volume) to conventional concretes can appreciably improve their characteristics.

The compressive strengths of fiber-reinforced concretes are not significantly greater than they would be if the same mixes were used without the fibers. The resulting concretes, however, are substantially tougher and have greater resistance to cracking and higher impact resistance. The use of fibers has increased the versatility of concrete by reducing its brittleness. The reader should note that a reinforcing bar provides reinforcing only in the direction of the bar, while randomly distributed fibers provide additional strength in all directions.

Steel is the most commonly used material for the fibers. The resulting concretes seem to be quite durable, at least as long as the fibers are covered and protected by the cement mortar. Concretes reinforced with steel fibers are most often used in pavements, thin shells, and precast products as well as in various patches and overlays. Glass fibers are more often used for spray-on applications as in shotcrete. It is necessary to realize that ordinary glass will deteriorate when in contact with cement paste. As a result, alkali-resistant glass fibers are necessary.

The fibers used vary in length from about $\frac{1}{4}$ in. up to about 3 in. while their diameters run from approximately 0.01 in. up to 0.03 in. For improving the bond with the cement paste the fibers may be hooked or crimped. In addition, the surface characteristics of the fibers may be chemically modified in order to increase bonding.

The improvement obtained in the toughness of the concrete (the total energy absorbed in breaking a member in flexure) by adding fibers is dependent on the fibers' *aspect ratio* (length/diameter). Typically the aspect ratios used vary from about 25 up to as much as 150, with 100 being about an average value. Other factors affecting toughness are the shape and texture of the fibers. ASTM C1018[19] is the test method for determining the toughness of fiber-reinforced concrete using the third-point beam loading method described earlier.

When a crack opens up in a fiber-reinforced concrete member, the few fibers bridging the crack do not appreciably increase the strength of the concrete. They will, however, provide resistance to the opening up of the crack because of the considerable work that would be necessary to pull them out. As a result, the ductility and toughness of the concrete is increased. The use of fibers has been shown to increase the fatigue life of beams and lessen the widths of cracks when members are subject to fatigue loadings.

The use of fibers does significantly increase costs. It is probably for this reason that fiber-reinforced concretes have been used for overlays as for highway pavements and airport runways rather than for whole concrete projects. Actually in the long run, if the increased service lives of fiber-reinforced concretes are considered, they may very well prove to be quite cost-effective. For instance, many residential contractors use fiber-reinforced concrete to construct driveways instead of regular reinforced concrete.

Some people have the feeling that the addition of fibers to concrete reduces its slump and workability as well as its strength. Apparently, they feel this way because the concrete looks stiffer to them. Actually, the fibers do not reduce the slump unless the quantity is too great—that is, much above about 1 pound per cubic yard. The fibers only appear to cause a reduction in workability, but as a result concrete finishers will often add more water so that water/cement ratios are increased and strengths decreased. ASTM C1018 utilizes the third-point beam loading method described earlier to measure the toughness and first-crack strength of fiber-reinforced concrete.

1.15   CONCRETE DURABILITY

The compressive strength of concrete may be dictated by exposure to freeze-thaw conditions or chemicals such as de-icers or sulfates. These conditions may require a greater compressive strength or lower water–cement ratio than those required to carry the calculated loads. Chapter 4 of the 2008 Code imposes limits on water–cement ratio, f_c' and entrained-air for elements exposed to freeze-thaw cycles. For concrete exposed to de-icing chemicals, the amount of fly ash or other pozzolans is limited in this chapter. Finally, the water–cement ratio is limited by exposure to sulfates as well. The designer is required to determine whether structural load carrying requirements or durability requirements are more stringent and to specify the more restrictive requirements for f_c' water–cement ratio and air content.

1.16   REINFORCING STEEL

The reinforcing used for concrete structures may be in the form of bars or welded wire fabric. Reinforcing bars are referred to as *plain* or *deformed*. The deformed bars, which have ribbed projections rolled onto their surfaces (patterns differing with different manufacturers) to provide

---

[19]American Society for Testing and Materials, 1997, *Standard Test Method for Flexural Toughness and First-Crack Strength of Fiber-Reinforced Concrete* (Using Simple Beam with Third-Point Loading) (ASTM C1018-1997), West Conshohocken, PA.

Round forms for grandstand support columns at The Texas Motor Speedway, Fort Worth, Texas. (Courtesy of EFCO Corp.)

better bonding between the concrete and the steel, are used for almost all applications. Instead of rolled-on deformations, deformed wire has indentations pressed into it. Plain bars are not used very often except for wrapping around longitudinal bars, primarily in columns.

Plain round bars are indicated by their diameters in fractions of an inch as $\frac{3}{8}''\ \phi$, $\frac{1}{2}''\ \phi$, and $\frac{5}{8}''\ \phi$. Deformed bars are round and vary in sizes from #3 to #11, with two very large sizes, #14 and #18, also available. For bars up to and including #8, the number of the bar coincides with the bar diameter in eighths of an inch. For example, a #7 bar has a diameter of 7/8 in. and a cross-sectional area of 0.60 in^2 (which is the area of a circle with a 7/8 in. diameter). Bars were formerly manufactured in both round and square cross sections, but today all bars are round.

The #9, #10, and #11 bars have diameters that provide areas equal to the areas of the old 1-in. × 1-in. square bars, $1\frac{1}{8}$-in. × $1\frac{1}{8}$-in. square bars, and $1\frac{1}{4}$-in. × $1\frac{1}{4}$-in. square bars, respectively. Similarly, the #14 and #18 bars correspond to the old $1\frac{1}{2}$-in. × $1\frac{1}{2}$-in. square bars and 2-in. × 2-in. square bars, respectively. Table A.2 (see Appendix) provides details as to areas, diameters, and weights of reinforcing bars. Although #14 and #18 bars are shown in this table, the designer should check his or her suppliers to see if they have these very large sizes in stock. Reinforcing bars may be purchased in lengths up to 60 ft. Longer bars have to be specially ordered. Normally they are too flexible and difficult to handle.

Welded wire fabric is also frequently used for reinforcing slabs, pavements and shells, and places where there is normally not sufficient room for providing the necessary concrete cover required for regular reinforcing bars. The mesh is made of cold-drawn wires running in both directions and welded together at the points of intersection. The sizes and spacings of the wire may be the same in both directions or may be different, depending on design requirements. Wire mesh is easily placed, has excellent bond with the concrete, and the spacing of the wires is well controlled.

Table A.3(A) in the Appendix provides information concerning certain styles of welded wire fabric that have been recommended by the Wire Reinforcement Institute as common stock styles (normally carried in stock at the mills or at warehousing points and thus usually immediately available). Table A.3(B) in the Appendix provides detailed information about diameters, areas, weights, and spacings of quite a few wire sizes normally used to manufacture welded wire fabric. Smooth and deformed wire fabric is made from wires whose diameters range from 0.134 in. up to 0.628 in. for plain wire and from 0.225 in. up to 0.628 in. for deformed wires.

Smooth wire is denoted by the letter W followed by a number that equals the cross-sectional area of the wire in hundredths of a square inch. Deformed wire is denoted by the letter D followed by a number giving the area. For instance, a D4 wire is a deformed wire with a cross-sectional area equal to 0.04 in. Smooth wire fabric is actually included within the ACI Code's definition of deformed reinforcement because of its mechanical bonding to the concrete caused by the wire intersections. Wire fabric that actually has deformations on the wire surfaces bonds even more to the concrete because of the deformations as well as the wire intersections. According to the Code, deformed wire is not permitted to be larger than D31 or smaller than D4.

The fabric is usually indicated on drawings by the letters WWF followed by the spacings of the longitudinal wires and the transverse wires and then the total wire areas in hundredths of a square inch per foot of length. For instance, WWF6 × 12-W16 × 8 represents smooth welded wire fabric with a 6-in. longitudinal and a 12-in. transverse spacing with cross-sectional areas of 0.32 in.^{2}/ft and 0.08 in.^{2}/ft, respectively.

Headed Steel Bars for Concrete Reinforcement (ASTM A970/970M-06) were added to the ACI 318 Code in 2008. Headed bars can be used instead of straight or hooked bars, with considerably less congestion in crowded areas such as beam–column intersections. The specification covers plain and deformed bars cut to lengths and having heads either forged or welded to one or both ends. Alternatively, heads may be connected to the bars by internal threads in the head mating to threads on the bar end or by a separate threaded nut to secure the head to the bar. Heads are forge formed, machined from bar stock, or cut from plate. Figure 1.3 illustrates a headed bar detail. The International Code Council has published acceptance criteria for headed ends of concrete reinforcement (ACC 347).

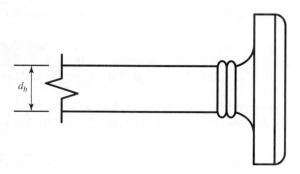

Figure 1.3 Headed deformed reinforcing bar.

1.17 GRADES OF REINFORCING STEEL

Reinforcing bars may be rolled from billet steel, axle steel, or rail steel. Only occasionally, however, are they rolled from old train rails or locomotive axles. These latter steels have been cold-worked for many years and are not as ductile as the billet steels.

There are several types of reinforcing bars, designated by the ASTM, which are listed after this paragraph. These steels are available in different grades as Grade 50, Grade 60, and so on, where Grade 50 means the steel has a specified yield point of 50,000 psi, Grade 60 means 60,000 psi, and so on.

1. ASTM A615: Deformed and plain billet steel bars. These bars, which must be marked with the letter S (for type of steel), are the most widely used reinforcing bars in the United States.

2. ASTM A706: Low alloy deformed and plain bars. These bars, which must be marked with the letter W (for type of steel), are to be used where controlled tensile properties and/or specially controlled chemical composition is required for welding purposes.

3. ASTM A996: Deformed rail steel or axle steel bars. They must be marked with the letter R (for type of steel).

4. When deformed bars are produced to meet both the A615 and A706 specifications, they must be marked with both the letters S and W.

Designers in almost all parts of the United States will probably never encounter rail or axle steel bars (A996) because they are available in such limited areas of the country. Of the 23 U.S. manufacturers of reinforcing bars listed by the Concrete Reinforcing Steel Institute,[20] only five manufacture rail steel bars and not one manufactures axle bars.

Almost all reinforcing bars conform to the A615 specification, and a large proportion of the material used to make them is not new steel but is melted reclaimed steel, such as from old car bodies. Bars conforming to the A706 specification are intended for certain uses when welding and/or bending are of particular importance. Bars conforming to this specification may not always be available from local suppliers.

There is only a small difference between the prices of reinforcing steel with yield strengths of 40 ksi and 60 ksi. As a result, the 60-ksi bars are the most commonly used in reinforced concrete design.

When bars are made from steels with yield stresses higher than 60 ksi, the ACI (Section 3.5.3.2) states that the specified yield strength must be the stress corresponding to a strain of 0.35%. The ACI (Section 9.4) has established an upper limit of 80 ksi on yield strengths permitted for design calculations for reinforced concrete. If the ACI were to permit the use of steels with yield strengths greater than 80 ksi, it would have to provide other design restrictions, since the yield strain of 80 ksi steel is almost equal to the ultimate concrete strain in compression. (This last sentence will make sense to the reader after he or she has studied Chapter 2.)

There has been gradually increasing demand through the years for Grade 75 steel, particularly for use in high-rise buildings where it is used in combination with high-strength concretes. The results are smaller columns, more rentable floor space, and smaller foundations for the lighter buildings that result.

Grade 75 steel is an appreciably higher cost steel, and the #14 and #18 bars are often unavailable from stock and will probably have to be specially ordered from the steel mills. This

---

[20]Concrete Reinforcing Steel Institute, 2001, *Manual of Standard Practice*, 27th ed., Chicago. Appendix A, pp. A-1–A-5.

means that there may have to be a special rolling to supply the steel. As a result, its use may not be economically justified unless at least 50 or 60 tons are ordered.

Yield stresses above 60 ksi are also available in welded wire fabric, but the specified stresses must correspond to strains of 0.35%. Smooth fabric must conform to ASTM A185, whereas deformed fabric cannot be smaller than size D4 and must conform to ASTM A496.

The modulus of elasticity for nonprestressed steels is considered to be equal to 29×10^6 psi. For prestressed steels it varies somewhat from manufacturer to manufacturer, with a value of 27×10^6 psi being fairly common.

Stainless steel reinforcing (ASTM A955) was introduced in the 2008 Code. It is highly resistant to corrosion, especially pitting and crevice corrosion due to exposure to chloride-containing solutions such as de-icing salts. While it is more expensive than normal carbon steel reinforcement, its life cycle cost may be less when the costs of maintenance and repairs are considered.

1.18   SI BAR SIZES AND MATERIAL STRENGTHS

The metric version of the ACI Code 318M-08 makes use of the same reinforcing bars as those made for designs using U.S. customary units. The metric bar dimensions are merely soft conversions (that is, almost equivalent) of the customary sizes. The SI concrete strengths (f'_c) and the minimum steel yield strengths (f_y) are converted from the customary values into metric units and rounded off a bit. A brief summary of metric bar sizes and material strengths is presented in the following paragraphs. These values are used for the SI examples and homework problems throughout the text.

1. The bar sizes used in the metric version of the Code correspond to our #3 through #18 bars. They are numbered 10, 13, 16, 19, 22, 25, 29, 32, 36, 43, and 57. These numbers represent the U.S. customary bar diameters rounded to the nearest millimeter (mm). For instance, the metric #10 bar has a diameter equal to 9.5 mm, the metric #13 bar has a diameter equal to 12.7 mm, and so on. Detailed information concerning metric reinforcing bar diameters, cross-sectional areas, masses, and ASTM classifications is provided in Appendix Tables B.2 and B.3.

2. The steel reinforcing grades, or minimum steel yield strengths, referred to in the Code are 300, 350, 420, and 520 MPa. These correspond, respectively, to 43,511, 50,763, 60,916, and 75,420 psi and thus correspond approximately to Grades 40, 50, 60, and 75 bars. Appendix Table B.3 provides ASTM numbers, steel grades, and bar sizes available in each grade.

3. The concrete strengths in metric units referred to in the Code are 17, 21, 24, 28, 35, and 42 MPa. These correspond respectively to 2466, 3046, 3481, 4061, 5076, and 6092 psi, that is, to 2500, 3000, 3500, 4000, 5000, and 6000 psi concretes.

In 1997 the producers of steel reinforcing bars in the United States began to produce soft metric bars. These are the same bars we have long called standard inch-pound bars, but they are marked with metric units. Today the large proportion of metric bars manufactured in the United States are soft metric. By producing the exact same bars, the industry does not have to keep two different inventories (one set of inch-pound bar sizes and another set of different bar sizes in metric units). Table 1.1 shows the bar sizes given in both sets of units.

Table 1.1 Reinforcement Bar Sizes and Areas

Standard inch-pound bars			Soft metric bars		
Bar no.	Diameter (in.)	Area (in.^{2})	Bar no.	Diameter (mm)	Area (mm^2)
3	0.375	0.11	10	9.5	71
4	0.500	0.20	13	12.7	129
5	0.625	0.31	16	15.9	199
6	0.750	0.44	19	19.1	284
7	0.875	0.60	22	22.2	387
8	1.000	0.79	25	25.4	510
9	1.128	1.00	29	28.7	645
10	1.270	1.27	32	32.3	819
11	1.410	1.41	36	35.8	1006
14	1.693	2.25	43	43.0	1452
18	2.257	4.00	57	57.3	2581

1.19   CORROSIVE ENVIRONMENTS

When reinforced concrete is subjected to de-icing salts, seawater, or spray from these substances, it is necessary to provide special corrosion protection for the reinforcing. The structures usually involved are bridge decks, parking garages, wastewater treatment plants, and various coastal structures. We must also consider structures subjected to occasional chemical spills that involve chlorides.

Should the reinforcement be insufficiently protected, it will corrode; as it corrodes, the resulting oxides occupy a volume far greater than that of the original metal. The results are large outward pressures that can lead to severe cracking and spalling of the concrete. This reduces the concrete protection or *cover* for the steel, and corrosion accelerates. Also, the *bond* or sticking of the concrete to the steel is reduced. The result of all of these factors is a decided reduction in the life of the structure.

Section 7.7.6 of the Code requires that for corrosive environments, more concrete cover must be provided for the reinforcing; it also requires that special concrete proportions or mixes be used.

The lives of such structures can be greatly increased if *epoxy-coated reinforcing* bars are used. Such bars need to be handled very carefully so as not to break off any of the coating. Furthermore, they do not bond as well to the concrete, and their embedment lengths will have to be increased somewhat for that reason, as we will learn in Chapter 7. Use of stainless steel reinforcing, as described in Section 1.14, can also significantly increase the service life of structures exposed to corrosive environments.

1.20   IDENTIFYING MARKS ON REINFORCING BARS

It is essential for people in the shop and the field to be able to identify at a glance the sizes and grades of reinforcing bars. If they are not able to do this, smaller and lower grade bars other than those intended by the designer may be used. To prevent such mistakes, deformed bars have rolled-in identification markings on their surfaces. These markings are described below and are illustrated in Figure 1.4.

1. The producing company is identified with a letter.
2. The bar size number (3 to 18) is given next.
3. Another letter is shown to identify the type of steel (S for billet, R in addition to a rail sign for rail steel, A for axle, and W for low alloy).

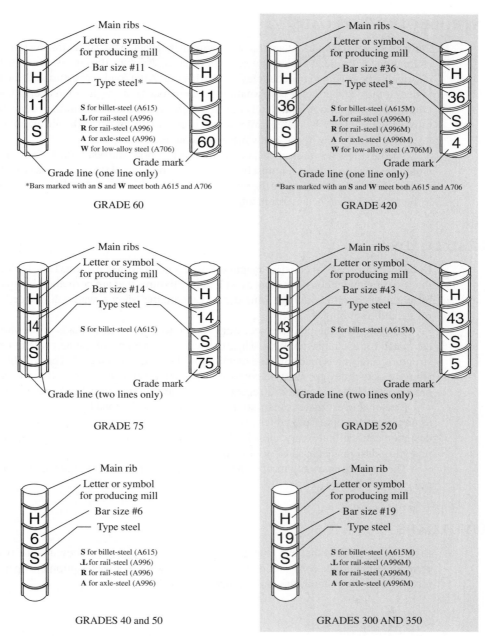

Figure 1.4 Identification marks for ASTM standard bars. (Courtesy of Concrete Reinforcing Steel Institute.)

4. Finally, the grade of the bars is shown either with numbers or with continuous lines. A Grade 60 bar has either the number 60 on it or a continuous longitudinal line in addition to its main ribs. A Grade 75 bar will have the number 75 on it or two continuous lines in addition to the main ribs.

1.21 INTRODUCTION TO LOADS

Perhaps the most important and most difficult task faced by the structural designer is the accurate estimation of the loads that may be applied to a structure during its life. No loads that may reasonably be expected to occur may be overlooked. After loads are estimated, the next problem is to decide the worst possible combinations of these loads that might occur at one time. For instance, would a highway bridge completely covered with ice and snow be simultaneously subjected to fast moving lines of heavily loaded trailer trucks in every lane and to a 90-mile lateral wind, or is some lesser combination of these loads more reasonable?

The next few sections of this chapter provide a brief introduction to the types of loads with which the structural designer must be familiar. The purpose of these sections is not to discuss loads in great detail but rather to give the reader a "feel" for the subject. As will be seen, loads are classed as being dead, live, or environmental.

1.22 DEAD LOADS

Dead loads are loads of constant magnitude that remain in one position. They include the weight of the structure under consideration, as well as any fixtures that are permanently attached to it. For a reinforced concrete building, some dead loads are the frames, walls, floors, ceilings, stairways, roofs, and plumbing.

To design a structure, it is necessary for the weights or dead loads of the various parts to be estimated for use in the analysis. The exact sizes and weights of the parts are not known until the structural analysis is made and the members of the structure selected. The weights, as determined from the actual design, must be compared with the estimated weights. If large discrepancies are present, it will be necessary to repeat the analysis and design using better estimated weights.

Reasonable estimates of structure weights may be obtained by referring to similar structures or to various formulas and tables available in most civil engineering handbooks. An experienced designer can estimate very closely the weights of most structures and will spend little time repeating designs because of poor estimates.

The approximate weights of some common materials used for floors, walls, roofs, and the like are given in Table 1.2.

1.23 LIVE LOADS

Live loads are loads that can change in magnitude and position. They include occupancy loads, warehouse materials, construction loads, overhead service cranes, equipment operating loads, and many others. In general, they are induced by gravity.

Table 1.2 Weights of Some Common Building Materials

Reinforced concrete—12 in.	150 psf	2×12 @ 16-in. double wood floor	7 psf
Acoustical ceiling tile	1 psf	Linoleum or asphalt tile	1 psf
Suspended ceiling	2 psf	Hardwood flooring ($\frac{7}{8}$-in.)	4 psf
Plaster on concrete	5 psf	1-in. cement on stone-concrete fill	32 psf
Asphalt shingles	2 psf	Movable steel partitions	4 psf
3-ply ready roofing	1 psf	Wood studs w/$\frac{1}{2}$-in. gypsum	8 psf
Mechanical duct allowance	4 psf	Clay brick wythes—4 in.	39 psf

Table 1.3 Some Typical Uniformly Distributed Live Loads

Lobbies of assembly areas	100 psf	Classrooms in schools	40 psf
Dance hall and ballrooms	100 psf	Upper-floor corridors in schools	80 psf
Library reading rooms	60 psf	Stairs and exitways	100 psf
Library stack rooms	150 psf	Heavy storage warehouse	250 psf
Light manufacturing	125 psf	Retail stores—first floor	100 psf
Offices in office buildings	50 psf	Retail stores—upper floors	75 psf
Residential dwelling areas	40 psf	Walkways and elevated platforms	60 psf

psf, pounds per square feet.

Some typical floor live loads that act on building structures are presented in Table 1.3. These loads, which are taken from Table 4–1 in ASCE 7–05,[21] act downward and are distributed uniformly over an entire floor. By contrast, roof live loads are 20 psf (pounds per square feet) maximum distributed uniformly over the entire roof.

Among the many other types of live loads are:

Traffic loads for bridges: Bridges are subjected to series of concentrated loads of varying magnitude caused by groups of truck or train wheels.

Impact loads: Impact loads are caused by the vibration of moving or movable loads. It is obvious that a crate dropped on the floor of a warehouse or a truck bouncing on uneven pavement of a bridge causes greater forces than would occur if the loads were applied gently and gradually. Impact loads are equal to the difference between the magnitude of the loads actually caused and the magnitude of the loads had they been dead loads.

Longitudinal loads: Longitudinal loads also need to be considered in designing some structures. Stopping a train on a railroad bridge or a truck on a highway bridge causes longitudinal forces to be applied. It is not difficult to imagine the tremendous longitudinal force developed when the driver of a 40-ton trailer truck traveling at 60 mph suddenly has to apply the brakes while crossing a highway bridge. There are other longitudinal load situations, such as ships running into docks and the movement of traveling cranes that are supported by building frames.

Miscellaneous loads: Among the other types of live loads with which the structural designer will have to contend are *soil pressures* (such as the exertion of lateral earth pressures on walls or upward pressures on foundations), *hydrostatic pressures* (as water pressure on dams, inertia forces of large bodies of water during earthquakes, and uplift pressures on tanks and basement structures), *blast loads* (caused by explosions, sonic booms, and military weapons), and *centrifugal forces* (such as those caused on curved bridges by trucks and trains or similar effects on roller coasters).

Live load reductions are permitted, according to Section 4.8 of ASCE 7, because is it unlikely that the entire structure will be subjected to its full design live load over its entire floor area all at one time. This reduction can significantly reduce the total design live load on a structure, resulting in much lower column loads at lower floors and footing loads.

---

[21]American Society of Civil Engineers, 2005, *Minimum Design Loads for Buildings and Other Structures*, ASCE 7-05 (Reston, VA: American Society of Civil Engineers), pp. 9–13.

Sewage treatment plant,
Redwood City, California.
(Courtesy of The Burke
Company.)

1.24   ENVIRONMENTAL LOADS

Environmental loads are loads caused by the environment in which the structure is located. For buildings, they are caused by rain, snow, wind, temperature change, and earthquake. Strictly speaking, these are also live loads, but they are the result of the environment in which the structure is located. Although they do vary with time, they are not all caused by gravity or operating conditions, as is typical with other live loads. In the next few paragraphs a few comments are made about the various kinds of environmental loads.

1. *Snow and ice.* In the colder states, snow and ice loads are often quite important. One inch of snow is equivalent to approximately 0.5 psf, but it may be higher at lower elevations where snow is denser. For roof designs, snow loads of from 10 to 40 psf are used, the magnitude depending primarily on the slope of the roof and to a lesser degree on the character of the roof surface. The larger values are used for flat roofs, the smaller ones for sloped roofs. Snow tends to slide off sloped roofs, particularly those with metal or slate surfaces. A load of approximately 10 psf might be used for 45° slopes, and a 40-psf load might be used for flat roofs. Studies of snowfall records in areas with severe winters may indicate the occurrence of snow loads much greater than 40 psf, with values as high as 80 psf in northern Maine.

   Snow is a variable load, which may cover an entire roof or only part of it. There may be drifts against walls or buildup in valleys or between parapets. Snow may slide off one roof and onto a lower one. The wind may blow it off one side of a sloping roof, or the snow may crust over and remain in position even during very heavy winds. The snow loads that are applied to a structure are dependent upon many factors, including geographic location, the pitch of the roof, sheltering, and the shape of the roof.

2. *Rain.* Although snow loads are a more severe problem than rain loads for the usual roof, the situation may be reversed for flat roofs—particularly those in warmer climates. If water on a flat roof accumulates faster than it runs off, the result is called *ponding* because the increased load causes the roof to deflect into a dish shape that can hold more water, which causes greater deflections, and so on. This process continues until equilibrium is reached or until collapse occurs. Ponding is a serious matter, as illustrated by the large number of flat-roof failures that occur due to ponding every year in the United States. It has been claimed that almost 50% of the lawsuits faced by building designers are concerned with roofing systems.[22] Ponding is one of the common subjects of such litigation.

3. *Wind.* A survey of engineering literature for the past 150 years reveals many references to structural failures caused by wind. Perhaps the most infamous of these have been bridge failures such as those of the Tay Bridge in Scotland in 1879 (which caused the deaths of 75 persons) and the Tacoma Narrows Bridge (Tacoma, Washington) in 1940. But there have also been some disastrous building failures due to wind during the same period, such as that of the Union Carbide Building in Toronto in 1958. It is important to realize that a large percentage of building failures due to wind have occurred during their erection.[23]

A great deal of research has been conducted in recent years on the subject of wind loads. Nevertheless, more study is needed as the estimation of wind forces can by no means be classified as an exact science. The magnitude and duration of wind loads vary with geographical locations, the heights of structures above ground, the types of terrain around the structures, the proximity of other buildings, the location within the structure, and the character of the wind itself.

Chapter 6 of the ASCE 7-05 specification provides a rather lengthy procedure for estimating the wind pressures applied to buildings. The procedure involves several factors with which we attempt to account for the terrain around the building, the importance of the building regarding human life and welfare, and of course the wind speed at the building site. Although use of the equations is rather complex, the work can be greatly simplified with the tables presented in the specification. The reader is cautioned, however, that the tables presented are for buildings of regular shapes. If a building having an irregular or unusual geometry is being considered, wind tunnel studies may be necessary.

The basic form of the equation presented in the specification is

$$p = qCG$$

In this equation p is the estimated wind load (in psf) acting on the structure. This wind load will vary with height above the ground and with the location on the structure. The quantity q is the reference velocity pressure. It varies with height and with exposure to the wind. The aerodynamic shape factor C is dependent upon the shape and orientation of the building with respect to the direction from which the wind is blowing. Lastly, the gust response

[22]Van Ryzin, Gary, 1980, "Roof Design: Avoid Ponding by Sloping to Drain," *Civil Engineering* (New York: ASCE, January), pp. 77–81.
[23]Task Committee on Wind Forces, Committee on Loads and Stresses, Structural Division, ASCE, 1961, "Wind Forces on Structures," Final Report, *Transactions ASCE* 126, Part II, pp. 1124–1125.

factor *G* is dependent upon the nature of the wind and the location of the building. Other considerations in determining design wind pressure include importance factor and surface roughness.

4. *Seismic loads.* Many areas of the world are in "earthquake territory," and in those areas it is necessary to consider seismic forces in design for all types of structures. Through the centuries there have been catastrophic failures of buildings, bridges, and other structures during earthquakes. It has been estimated that as many as 50,000 people lost their lives in the 1988 earthquake in Armenia.[24] The 1989 Loma Prieta and 1994 Northridge earthquakes in California caused many billions of dollars of property damage as well as considerable loss of life. The 2008 earthquake in Sichuan Province, China caused 69,000 fatalities and another 18,000 missing.

Recent earthquakes have clearly shown that the average building or bridge that has not been designed for earthquake forces can be destroyed by an earthquake that is not particularly severe. Most structures can be economically designed and constructed to withstand the forces caused during most earthquakes. On the other hand, the cost of providing seismic resistance to existing structures (called *retrofitting*) can be extremely high.

Some engineers seem to think that the seismic loads to be used in design are merely percentage increases of the wind loads. This assumption is incorrect, however, as seismic loads are different in their action and are not proportional to the exposed area of the building, but rather are proportional to the distribution of the mass of the building above the particular level being considered.

Another factor to be considered in seismic design is the soil condition. Almost all of the structural damage and loss of life in the Loma Prieta earthquake occurred in areas that have soft clay soils. Apparently these soils amplified the motions of the underlying rock.[25]

It is well to understand that earthquakes load structures in an indirect fashion. The ground is displaced, and because the structures are connected to the ground, they are also displaced and vibrated. As a result, various deformations and stresses are caused throughout the structures.

From the above information you can understand that no external forces are applied above ground by earthquakes to structures. Procedures for estimating seismic forces such as the ones presented in Chapter 12 of ASCE 7-05 are very complicated. As a result, they usually are addressed in advanced structural analysis courses such as structural dynamics or earthquake resistance design courses.

1.25 SELECTION OF DESIGN LOADS

To assist the designer in estimating the magnitudes of live loads with which he or she should proportion structures, various records have been assembled through the years in the form of building codes and specifications. These publications provide conservative estimates of live-load magnitudes for various situations. One of the most widely used design-load specifications for buildings is that published by the American Society of Civil Engineers (ASCE).[26]

The designer is usually fairly well controlled in the design of live loads by the building code requirements in his or her particular area. The values given in these various codes unfortunately vary from city to city, and the designer must be sure to meet the requirements of a particular locality. In the absence of a governing code, the ASCE code is an excellent one to follow.

[24]Fairweather, V., 1990, "The Next Earthquake," *Civil Engineering* (New York: ASCE, March), pp. 54–57.
[25]Ibid.
[26]American Society of Civil Engineers, 2005, *Minimum Design Loads for Buildings and Other Structures*, ASCE 7-05, (Reston, VA: American Society of Civil Engineers), 376 pp.

Croke Park Stadium, Dublin, Ireland. (Courtesy of EFCO Corp.)

Some other commonly used specifications are:

1. For railroad bridges, American Railway Engineering Association (AREA).[27]
2. For highway bridges, American Association of State Highway and Transportation Officials (AASHTO).[28]
3. For buildings, the International Building Code (IBC).[29]

These specifications will on many occasions clearly prescribe the loads for which structures are to be designed. Despite the availability of this information, the designer's ingenuity and knowledge of the situation are often needed to predict what loads a particular structure will have to support in years to come. Over the past several decades, insufficient estimates of future traffic loads by bridge designers have resulted in a great amount of replacement with wider and stronger structures.

1.26   CALCULATION ACCURACY

A most important point, which many students with their amazing computers and pocket calculators have difficulty in understanding, is that reinforced concrete design is not an exact science for which answers can be confidently calculated to six or eight places. The reasons for this statement should be quite obvious: The analysis of structures is based on partly true assumptions; the strengths of materials used vary widely; structures are not built to the exact dimensions shown on the plans and maximum loadings can only be approximated. With respect to this last sentence, how many users of this book could estimate to within 10% the maximum live load in pounds per square foot that will ever occur on the building floor they are now occupying? Calculations to more

[27]American Railway Engineering Association (AREA), 2003, *Manual for Railway Engineering* (Chicago, IL).
[28]*Standard Specifications for Highway Bridges*, 2002, 17th ed. (Washington, DC: American Association of State Highway and Transportation Officials [AASHTO]).
[29]International Building Code, 2006, International Code Council, Inc.

than two or three significant figures are obviously of little value and may actually mislead students into a false sense of accuracy.

1.27 IMPACT OF COMPUTERS ON REINFORCED CONCRETE DESIGN

The availability of personal computers has drastically changed the way in which reinforced concrete structures are analyzed and designed. In nearly every engineering school and office, computers are routinely used to handle structural design problems.

Many calculations are involved in reinforced concrete design, and many of these calculations are quite time consuming. With a computer, the designer can reduce the time required for these calculations tremendously and thus supposedly have time to consider alternative designs.

Although computers do increase design productivity, they do undoubtedly tend at the same time to reduce the designer's "feel" for structures. This can be a special problem for young engineers with little previous design experience. Unless designers have this "feel," computer usage, though expediting the work and reducing many errors, may occasionally result in large mistakes.

It is interesting to note that up to the present time, the feeling at most engineering schools has been that the best way to teach reinforced concrete design is with chalk and blackboard, supplemented with some computer examples.

Accompanying this text are several Excel spreadsheets that can be downloaded from this book's Website at: www.wiley.com/college/mccormac

These spreadsheets are intended to allow the student to consider multiple alternative designs and not as a tool to work basic homework problems.

PROBLEMS

1.1 Name several of the admixtures that are used in concrete mixes. What is the purpose of each?

1.2 What is Poisson's ratio, and where can it be of significance in concrete work?

1.3 What factors influence the creep of concrete?

1.4 What steps can be taken to reduce creep?

1.5 What is the effect of creep in reinforced concrete columns that are subjected to axial compression loads?

1.6 Why is silica fume used in high-strength concrete? What does it do?

1.7 Why do the surfaces of reinforcing bars have rolled-on deformations?

1.8 What are "soft metric" reinforcing bars?

1.9 What are three factors that influence the magnitude of the earthquake load on a structure?

1.10 Why are epoxy-coated bars sometimes used in the construction of reinforced concrete?

1.11 What is the diameter and cross-sectional area of a #5 reinforcing bar?

2

Flexural Analysis of Beams

2.1 INTRODUCTION

In this section it is assumed that a small transverse load is placed on a concrete beam with tensile reinforcing and that the load is gradually increased in magnitude until the beam fails. As this takes place we will find that the beam will go through three distinct stages before collapse occurs. These are: (1) the uncracked concrete stage, (2) the concrete cracked–elastic stresses stage, and (3) the ultimate-strength stage. A relatively long beam is considered for this discussion so that shear will not have a large effect on its behavior.

Uncracked Concrete Stage

At small loads when the tensile stresses are less than the *modulus of rupture* (the bending tensile stress at which the concrete begins to crack), the entire cross section of the beam resists bending, with compression on one side and tension on the other. Figure 2.1 shows the variation of stresses and strains for these small loads; a numerical example of this type is presented in Section 2.2.

Concrete Cracked–Elastic Stresses Stage

As the load is increased after the modulus of rupture of the concrete is exceeded, cracks begin to develop in the bottom of the beam. The moment at which these cracks begin to form—that is, when the tensile stress in the bottom of the beam equals the modulus of rupture—is referred to as the *cracking moment*, M_{cr}. As the load is further increased, these cracks quickly spread up to the vicinity of the neutral axis, and then the neutral axis begins to move upward. The cracks occur at those places along the beam where the actual moment is greater than the cracking moment, as shown in Figure 2.2(a).

Now that the bottom has cracked, another stage is present because the concrete in the cracked zone obviously cannot resist tensile stresses—the steel must do it. This stage will continue as long as the compression stress in the top fibers is less than about one-half of the concrete's compression strength f_c' and as long as the steel stress is less than its yield stress. The stresses and strains for this range are shown in Figure 2.2(b). In this stage the compressive stresses vary linearly with the distance from the neutral axis or as a straight line.

The straight-line stress–strain variation normally occurs in reinforced concrete beams under normal service-load conditions because at those loads the concrete stresses are generally less than $0.50f_c'$. To compute the concrete and steel stresses in this range, the transformed-area method (to be presented in Section 2.3) is used. The *service* or *working* loads are the loads that are assumed to actually occur when a structure is in use or service. Under these loads, moments develop which are

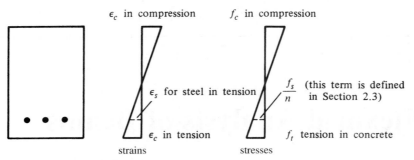

Figure 2.1 Uncracked concrete stage.

considerably larger than the cracking moments. Obviously the tensile side of the beam will be cracked. We will learn to estimate crack widths and methods of limiting their widths in Chapter 6.

Beam Failure—Ultimate-Strength Stage

As the load is increased further so that the compressive stresses are greater than one-half of f_c', the tensile cracks move further upward, as does the neutral axis, and the concrete compression stresses begin to change appreciably from a straight line. For this initial discussion it is assumed that the reinforcing bars have yielded. The stress variation is much like that shown in Figure 2.3. You should relate the information shown in this figure with that given in Figure 1.1 as to the changing ratio of stress to strain at different stress levels.

Construction of Kingdome, Seattle, Washington.(Courtesy of EFCO Corp.)

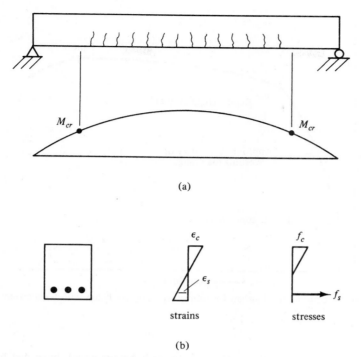

(a)

Figure 2.2 Concrete cracked–elastic stresses stage.

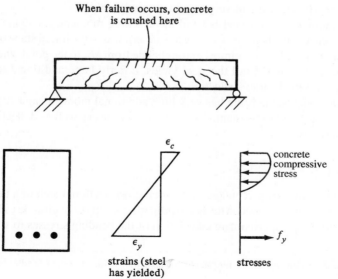

Figure 2.3 Ultimate-strength stage.

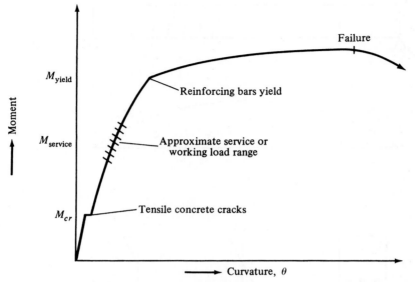

Figure 2.4 Moment–curvature diagram for reinforced concrete beam with tensile reinforcing only.

To further illustrate the three stages of beam behavior which have just been described, a moment–curvature diagram is shown in Figure 2.4.[1] For this diagram, θ is defined as the angle change of the beam section over a certain length and is computed by the following expression in which ϵ is the strain in a beam fiber at some distance y from the neutral axis of the beam:

$$\theta = \frac{\epsilon}{y}$$

The first stage of the diagram is for small moments less than the cracking moment M_{cr} where the entire beam cross section is available to resist bending. In this range the strains are small, and the diagram is nearly vertical and very close to a straight line.

When the moment is increased beyond the cracking moment, the slope of the curve will decrease a little because the beam is not quite as stiff as it was in the initial stage before the concrete cracked. The diagram will follow almost a straight line from M_{cr} to the point where the reinforcing is stressed to its yield point. Until the steel yields, a fairly large additional load is required to appreciably increase the beam's deflection.

After the steel yields, the beam has very little additional moment capacity, and only a small additional load is required to substantially increase rotations as well as deflections. The slope of the diagram is now very flat.

2.2 CRACKING MOMENT

The area of reinforcing as a percentage of the total cross-sectional area of a beam is quite small (usually 2% or less), and its effect on the beam properties is almost negligible as long as the beam is uncracked. Therefore an approximate calculation of the bending stresses in such a beam can be

---

[1]MacGregor, J. G., 2005, *Reinforced Concrete Mechanics and Design*, 4th ed. (Upper Saddle River, NJ: Prentice Hall), p. 109.

obtained based on the gross properties of the beam's cross section. The stress in the concrete at any point a distance y from the neutral axis of the cross section can be determined from the following flexure formula in which M is the bending moment, which is equal to or less than the cracking moment of the section, and I_g is the gross moment of inertia of the cross section:

$$f = \frac{My}{I_g}$$

Section 9.5.2.3 of the ACI Code states that the cracking moment of a section may be determined with ACI Equation 9-9, in which f_r is the modulus of rupture of the concrete and y_t is the distance from the centroidal axis of the section to its extreme fiber in tension. In this section, with its equation 9-10, the Code states that f_r may be taken equal to $7.5\lambda\sqrt{f_c'}$ with f_c' in psi.

Or in SI units with f_c' in $\dfrac{N}{mm^2}$ or   MPa, $f_r = 0.7\lambda\sqrt{f_c'}$.

The "lambda" term is 1.0 for normal weight concrete and is less than 1.0 for lightweight concrete as described in Section 1.12. The cracking moment is as follows:

$$M_{cr} = \frac{f_r I_g}{y_t}$$ (ACI Equation 9-9)

Example 2.1 presents calculations for a reinforced concrete beam where tensile stresses are less than its modulus of rupture. As a result, no tensile cracks are assumed to be present, and the stresses are similar to those occurring in a beam constructed with a homogeneous material.

EXAMPLE 2.1

(**a**) Assuming the concrete is uncracked, compute the bending stresses in the extreme fibers of the beam of Figure 2.5 for a bending moment of 25 ft-k. The normal weight concrete has an f_c' of 4000 psi and a modulus of rupture $f_r = 7.5(1.0)\sqrt{4000} = 474$ psi.

(**b**) Determine the cracking moment of the section.

SOLUTION   (**a**) Bending stresses:

$$I_g = \frac{1}{12}\,bh^3 \text{ with } b = 12 \text{ in. and } h = 18 \text{ in.}$$

$$I_g = \left(\frac{1}{12}\right)(12)(18)^3 = 5832 \text{ in.}^4$$

$$f = \frac{My}{I_g} \text{ with } M = 25 \text{ ft-k} = 25{,}000 \text{ ft-lb}$$

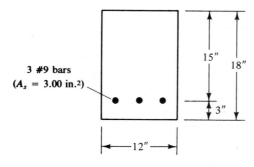

3 #9 bars
$(A_s = 3.00 \text{ in.}^2)$

15"

18"

3"

12"

Figure 2.5

Then multiply the 25,000 by 12 to obtain in.-lbs as shown below:

$$f = \frac{(12 \times 25,000)(9.00)}{5832} = 463 \text{ psi}$$

Since this stress is less than the tensile strength or modulus of rupture of the concrete of 474 psi, the section is assumed not to have cracked.

(b) Cracking moment:

$$M_{cr} = \frac{f_r I_g}{y_t} = \frac{(474)(5832)}{9.00} = 307,152 \text{ in.-lb} = \underline{\underline{25.6 \text{ ft-k}}}$$

2.3 ELASTIC STRESSES—CONCRETE CRACKED

When the bending moment is sufficiently large to cause the tensile stress in the extreme fibers to be greater than the modulus of rupture, it is assumed that all of the concrete on the tensile side of the beam is cracked and must be neglected in the flexure calculations.

The cracking moment of a beam is normally quite small compared to the service load moment. Thus when the service loads are applied, the bottom of the beam cracks. The cracking of the beam does not necessarily mean that the beam is going to fail. The reinforcing bars on the tensile side begin to pick up the tension caused by the applied moment.

On the tensile side of the beam an assumption of perfect bond is made between the reinforcing bars and the concrete. Thus the strain in the concrete and in the steel will be equal at equal distances from the neutral axis. But if the strains in the two materials at a particular point are the same, their stresses cannot be the same since they have different moduli of elasticity. Thus their stresses are in proportion to the ratio of their moduli of elasticity. The ratio of the steel modulus to the concrete modulus is called the *modular ratio n*:

$$n = \frac{E_s}{E_c}$$

If the modular ratio for a particular beam is 10, the stress in the steel will be 10 times the stress in the concrete at the same distance from the neutral axis. Another way of saying this is that when $n = 10$, 1 sq. in. of steel will carry the same total force as 10 in.^{2} of concrete.

For the beam of Figure 2.6, the steel bars are replaced with an equivalent area of fictitious concrete (nA_s), which supposedly can resist tension. This area is referred to as the *transformed area*. The resulting revised cross section or transformed section is handled by the usual methods

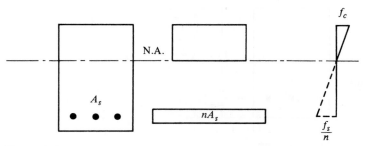

Figure 2.6

Bridge piers and deck. (PhotoDisc, Inc./ Getty Images.)

for elastic homogeneous beams. Also shown in the figure is a diagram showing the stress variation in the beam. On the tensile side a dashed line is shown because the diagram is discontinuous. There the concrete is assumed to be cracked and unable to resist tension. The value shown opposite the steel is the fictitious stress in the concrete if it could carry tension. This value is shown as f_s/n because it must be multiplied by n to give the steel stress f_s.

Examples 2.2, 2.3, and 2.4 are transformed-area problems that illustrate the calculations necessary for determining the stresses and resisting moments for reinforced concrete beams. The first step to be taken in each of these problems is to locate the neutral axis, which is assumed to be located a distance x from the compression surface of the beam. The first moment of the compression area of the beam cross section about the neutral axis must equal the first moment of the tensile area about the neutral axis. The resulting quadratic equation can be solved by completing the squares or by using the quadratic formula.

After the neutral axis is located, the moment of inertia of the transformed section is calculated, and the stresses in the concrete and the steel are computed with the flexure formula.

EXAMPLE 2.2

Calculate the bending stresses in the beam shown in Figure 2.7 by using the transformed area method; $f_c' = 3000$ psi, $n = 9$ and $M = 70$ ft-k.

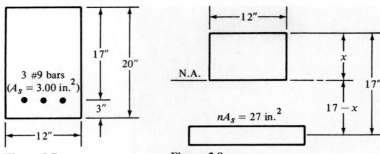

Figure 2.7   **Figure 2.8**

SOLUTION

Taking Moments about Neutral Axis (Referring to Figure 2.8)

$$(12x)\left(\frac{x}{2}\right) = (9)(3.00)(17-x)$$

$$6x^2 = 459-27.00x$$

Solving by Completing the Square

$$6x^2 + 27.00x = 459$$

$$x^2 + 4.50x = 76.5$$

$$(x+2.25)(x+2.25) = 76.5+(2.25)^2$$

$$x = 2.25 + \sqrt{76.5+(2.25)^2} = 9.03''$$

$$x = 6.780''$$

Moment of Inertia

$$I = \left(\frac{1}{3}\right)(12)(6.78)^3 + (9)(3.00)(10.22)^2 = 4067 \text{ in.}^4$$

Bending Stresses

$$f_c = \frac{My}{I} = \frac{(12)(70,000)(6.78)}{4067} = \underline{\underline{1400 \text{ psi}}}$$

$$f_s = n\frac{My}{I} = (9)\frac{(12)(70,000)(10.22)}{4067} = \underline{\underline{18,998 \text{ psi}}}$$

EXAMPLE 2.3

Determine the allowable resisting moment of the beam of Example 2.2, if the allowable stresses are $f_c = 1350$ psi and $f_S = 20{,}000$ psi.

SOLUTION

$$M_c = \frac{f_c I}{y} = \frac{(1350)(4067)}{6.78} = 809{,}800 \text{ in.-lb} = \underline{67.5 \text{ ft-k}} \leftarrow$$

$$M_s = \frac{f_s I}{ny} = \frac{(20{,}000)(4067)}{(9)(10.22)} = 884{,}323 \text{ in.-lb} = 73.7 \text{ ft-k}$$

Discussion

For a given beam, the concrete and steel will not usually reach their maximum allowable stresses at exactly the same bending moments. Such is the case for this example beam, where the concrete reaches its maximum permissible stress at 67.5 ft-k, while the steel does not reach its maximum value until 73.7 ft-k is applied. The resisting moment of the section is 67.5 ft-k because if that value is exceeded, the concrete becomes overstressed even though the steel stress is less than its allowable stress.

EXAMPLE 2.4

Compute the bending stresses in the beam shown in Figure 2.9 by using the transformed-area method; $n = 8$ and $M = 110$ ft-k.

SOLUTION   **Locating Neutral Axis (Assuming Neutral Axis below Hole)**

$$(18x)\left(\frac{x}{2}\right) - (6)(6)(x - 3) = (8)(5.06)(23 - x)$$

$$9x^2 - 36x + 108 = 931 - 40.48x$$

$$9x^2 + 4.48x = 823$$

$$x^2 + 0.50x = 91.44$$

$$(x + 0.25)(x + 0.25) = 91.44 + (0.25)^2 = 91.50$$

$$x + 0.25 = \sqrt{91.50} = 9.57$$

$$x = 9.32''>'' \quad \therefore \text{N.A. below hole as assumed}$$

Moment of Inertia

$$I = \left(\tfrac{1}{3}\right)(6)(9.32)^3(2) + \left(\tfrac{1}{3}\right)(6)(3.32)^3 + (8)(5.06)(13.68)^2 = 10{,}887 \text{ in.}^4$$

$$x^2 + 4.50x - 76.5$$

$$\frac{-4.5 + \sqrt{4.5^2 - 4(1)(-76.5)}}{2}$$

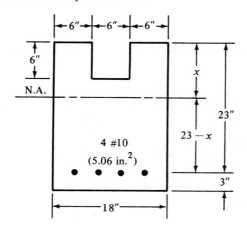

Figure 2.9

Computing Stresses

$$f_c = \frac{(12)(110{,}000)(9.32)}{10{,}887} = \underline{\underline{1130 \, \text{psi}}}$$

$$f_s = (8)\frac{(12)(110{,}000)(13.68)}{10{,}887} = \underline{\underline{13{,}269 \, \text{psi}}}$$

Example 2.5 illustrates the analysis of a doubly reinforced concrete beam—that is, one that has compression steel as well as tensile steel. Compression steel is generally thought to be uneconomical, but occasionally its use is quite advantageous.

Compression steel will permit the use of appreciably smaller beams than those that make use of tensile steel only. Reduced sizes can be very important where space or architectural requirements limit the sizes of beams. Compression steel is quite helpful in reducing long-term deflections, and such steel is useful for positioning stirrups or shear reinforcing, a subject to be discussed in Chapter 8. A detailed discussion of doubly reinforced beams is presented in Chapter 5.

The creep or plastic flow of concrete was described in Section 1.11. Should the compression side of a beam be reinforced, the long-term stresses in that reinforcing will be greatly affected by the creep in the concrete. As time goes by, the compression concrete will compact more tightly, leaving the reinforcing bars (which themselves have negligible creep) to carry more and more of the load.

As a consequence of this creep in the concrete, the stresses in the compression bars computed by the transformed-area method are assumed to double as time goes by. In Example 2.5 the transformed area of the compression bars is assumed to equal $2n$ times their area A'_s.

On the subject of "hair splitting," it will be noted in the example that the compression steel area is really multiplied by $2n - 1$. The transformed area of the compression side equals the gross compression area of the concrete plus $2nA'_s$ minus the area of the holes in the concrete $(1A'_s)$, which theoretically should not have been included in the concrete part. This equals the compression concrete area plus $(2n - 1)A'_s$. Similarly, $2n - 1$ is used in the moment of inertia calculations. The stresses in the compression bars are determined by multiplying $2n$ times the stresses in the concrete located at the same distance from the neutral axis.

EXAMPLE 2.5

Compute the bending stresses in the beam shown in Figure 2.10; $n = 10$ and $M = 118$ ft-k.

SOLUTION   **Locating Neutral Axis**

$$(14x)\left(\frac{x}{2}\right) + (20 - 1)(2.00)(x - 2.5) = (10)(4.00)(17.5 - x)$$

$$7x^2 + 38x - 95 = 700 - 40x$$

$$7x^2 + 78x = 795$$

$$x^2 + 11.14x = 113.57$$

$$x + 5.57 = \sqrt{113.57 + (5.57)^2} = 12.02$$

$$x = 6.45 \text{ in.}$$

Moment of Inertia

$$I = \left(\tfrac{1}{3}\right)(14)(6.45)^3 + (20-1)(2.00)(3.95)^2 + (10)(4.00)(11.05)^2 = 6729 \text{ in.}^4$$

Bending Stresses

$$f_c = \frac{(12)(118{,}000)(6.45)}{6729} = 1357 \text{ psi}$$

$$f_s' = 2n\frac{My}{I} = (2)(10)\frac{(12)(118{,}000)(3.95)}{6729} = 16{,}624 \text{ psi}$$

$$f_s = (10)\frac{(12)(118{,}000)(11.05)}{6729} = 23{,}253 \text{ psi}$$

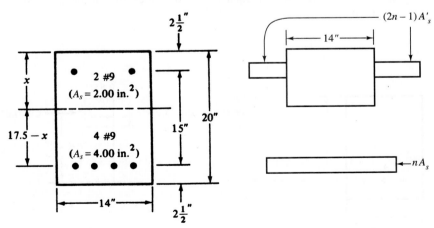

(a) Actual section   (b) Transformed section

Figure 2.10

2.4 ULTIMATE OR NOMINAL FLEXURAL MOMENTS

In this section a very brief introduction to the calculation of the ultimate or nominal flexural strength of beams is presented. This topic is continued at considerable length in the next chapter, where formulas, limitations, designs, and other matters are presented. For this discussion it is assumed that the tensile reinforcing bars are stressed to their yield point before the concrete on the compressive side of the beam crushes. We will learn in Chapter 3 that the ACI Code requires all our beam designs to fall into this category.

After the concrete compression stresses exceed about $0.50f'_c$ they no longer vary directly as the distance from the neutral axis or as a straight line. Rather, they vary much as shown in Figure 2.11(b). It is assumed for the purpose of this discussion that the curved compression diagram is replaced with a rectangular one with a constant stress of $0.85f'_c$, as shown in part (c) of the figure. The rectangular diagram of depth a is assumed to have the same c.g. (center of gravity) and total magnitude as the curved diagram. (In Section 3.4 of Chapter 3 of this text we will learn that this distance a is set equal to $\beta_1 c$, where β_1 is a value determined by testing and specified by the Code.) These assumptions will enable us easily to calculate the theoretical or nominal flexural strength of reinforced concrete beams. Experimental tests show that with the assumptions used here, accurate flexural strengths are determined.

To obtain the nominal or theoretical moment strength of a beam, the simple steps to follow are used as is illustrated in Example 2.6.

1. Compute total tensile force $T = A_s f_y$.

2. Equate total compression force $C = 0.85f'_c\,ab$ to $A_s f_y$ and solve for a. In this expression ab is the assumed area stressed in compression at $0.85f'_c$. The compression force C and the tensile force T must be equal to maintain equilibrium at the section.

3. Calculate the distance between the centers of gravity of T and C. (For a rectangular beam cross-section it equals $d-a/2$.)

4. Determine M_n, which equals T or C times the distance between their centers of gravity.

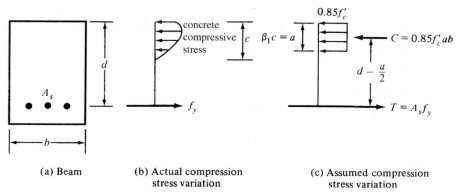

(a) Beam   (b) Actual compression   (c) Assumed compression
                       stress variation   stress variation

Figure 2.11 Compression and tension couple at nominal moment.

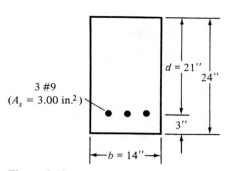
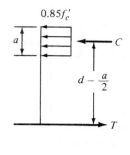

Figure 2.12

EXAMPLE 2.6

Determine M_n the nominal or theoretical ultimate moment strength of the beam section shown in Figure 2.12, if $f_y = 60{,}000$ psi and $f'_c = 3000$ psi.

SOLUTION

Computing Tensile and Compressive Forces T and C

$$T = A_s f_y = (3.00)(60) = 180 \text{ k}$$
$$C = 0.85 f'_c ab = (0.85)(3)(a)(14) = 35.7a$$

Equating T and C and Solving for a

$$T = C \text{ for equilibrium}$$
$$180 = 35.7a$$
$$a = 5.04 \text{ in.}$$

Computing $d - \frac{a}{2}$ and M_n

$$d - \frac{a}{2} = 21 - \frac{5.04}{2} = 18.48 \text{ in.}$$
$$M_n = (180)(18.48) = 3326.4 \text{ in.-k} = \underline{\underline{277.2 \text{ ft-k}}}$$

In Example 2.7, the nominal moment capacity of another beam is determined much as it was for Example 2.6. The only difference is that the cross section of the compression area (A_c) stressed at $0.85f'_c$ is not rectangular. As a result, once this area is determined we need to locate its center of gravity. The c.g. of the concrete compression area for the beam of Figure 2.13 is shown as being a distance $\bar{y}$ from the top of the beam in Figure 2.14. Then the lever arm from C to T is equal to $d - \bar{y}$ (which corresponds to $d - a/2$ in Example 2.6) and M_n equals $A_s f_y (d - \bar{y})$.

With this very simple procedure, values of M_n can be computed for tensilely reinforced beams of any cross section.

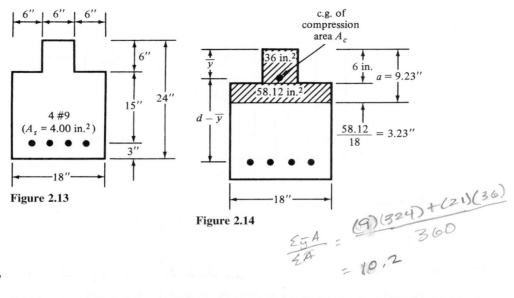

Figure 2.13

Figure 2.14

$$\frac{\Sigma \bar{y} A}{\Sigma A} = \frac{(9)(324) + (21)(36)}{360}$$

$$= 10.2$$

EXAMPLE 2.7

Calculate the nominal or theoretical ultimate moment strength of the beam section shown in Figure 2.13, if $f_y = 60{,}000$ psi and $f'_c = 3000$ psi. The 6-in.-wide ledges on top are needed for the support of precast concrete slabs.

SOLUTION

$$T = A_s f_y = (4.00)(60) = 240 \text{ k}$$

$$C = (0.85f'_c)(\text{area of concrete } A_c \text{ stressed to } 0.85f'_c)$$

$$= 0.85f'_c A_c$$

Σ

$240 = 45.9 a$

Finger piers for U.S. Coast Guard base, Boston. (Courtesy of EFCO Corp.)

Equating T and C and solving for A_c

$$A_c = \frac{T}{0.85f'_c} = \frac{240}{(0.85)(3)} = 94.12 \text{ in.}^2$$

The top 94.12 in.^{2} of the beam of Figure 2.14 is stressed in compression to $0.85f'_c$. This area can be shown to extend 9.23 in. down from the top of the beam. Its c.g. is located by taking moments at the top of the beam as follows:

$$\bar{y} = \frac{(36)(3) + (58.12)\left(6 + \frac{3.23}{2}\right)}{94.12} = 5.85 \text{ in.}$$

$$d - \bar{y} = 21 - 5.85 = 15.15 \text{ in.}$$

$$M_n = (240)(15.15) = 3636 \text{ in.-k} = \underline{\underline{303 \text{ ft-k}}}$$

2.5   EXAMPLE PROBLEM USING SI UNITS

In Example 2.8, the nominal moment strength of a beam is computed using SI units. Appendix Tables B.1 to B.9 provide information concerning various concrete and steel grades, as well as bar diameters, areas, and so on, all given in SI units.

EXAMPLE 2.8

Determine the nominal moment strength of the beam shown in Figure 2.15, if $f'_c = 28$ MPa and $f_y = 420$ MPa.

SOLUTION

$$T = C$$

$$A_s f_y = 0.85 f'_c ab$$

$$a = \frac{A_s f_y}{0.85 f'_c b} = \frac{(1530)(420)}{(0.85)(28)(300)} = 90 \text{ mm}$$

$$M_n = T\left(d - \frac{a}{2}\right) = C\left(d - \frac{a}{2}\right) = A_s f_y \left(d - \frac{a}{2}\right)$$

$$= (1530)(420)\left(430 - \frac{90}{2}\right)$$

$$= 2.474 \times 10^8 \text{ N} \cdot \text{mm} = \underline{\underline{247.4 \text{ kN} \cdot \text{m}}}$$

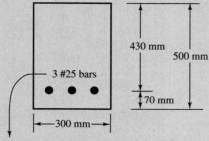

3 #25 bars

430 mm

500 mm

70 mm

300 mm

($A_s = 1530$ mm^2 from Appendix Table B.4.)

Figure 2.15

2.6 COMPUTER SPREADSHEETS

On the John Wiley website for this textbook, several spreadsheets have been provided for the student to use in assisting in the solution of problems. They are categorized by chapter. Note that most of the spreadsheets have multiple worksheets indicated by tabs at the bottom. The three worksheets available for Chapter 2 include (1) calculation of cracking moment, (2) stresses in singly reinforced rectangular beams, (3) T Beam Analysis ASD and (4) nominal strength of singly reinforced rectangular beams.

EXAMPLE 2.9

Repeat Example 2.1 using the spreadsheet provided for Chapter 2.

SOLUTION   Open the Chapter 2 spreadsheet and select the worksheet called Cracking Moment. Input only the cells highlighted in yellow[2] (the first six values below).

$f'_c =$	4000	psi
$M =$	25	kip-ft
$b =$	12	in.
$h =$	18	in.
$\gamma_c =$	145	pcf
$\lambda =$	1.00	
$I_g = bh^3/12 =$	5832	in.^{4}
$f_r = 7.5\,\lambda\,SQRT(f'_c) =$	474	psi
$f =$	463	psi
$M_{cr} =$	307,373	in.lb
$M_{cr} =$	25.6	kip-ft

The last five values are the same as calculated in Example 2.1.

EXAMPLE 2.10

Repeat Example 2.2 using the spreadsheet provided for Chapter 2.

SOLUTION   Open the Chapter 2 spreadsheet and select the worksheet called Elastic Stresses. Input only the cells highlighted in yellow[2] (the first seven values below).

$b =$	12	in.
$d =$	17	in.
$n =$	9	
$A_s =$	3	in.^{2}
$M =$	70	kip-ft
$f'_c =$	3000	psi
$\gamma_c =$	145	pcf
$E_c =$	3,155,924	psi
$n =$	9.19	
$n\rho =$	0.132	

---

[2]Yellow highlighting is in the Excel Spreadsheets, not in the printed example.

$x =$	**6.78**	**in.**
$I_{cr} =$	**4067**	**in.**^{4}
$f_c = Mx/I =$	**1401**	**psi**
$f_s = nM(d - x)/I =$	**18,996**	**psi**

The last four values are the same (within a small roundoff) as calculated in Example 2.2.

EXAMPLE 2.11

Repeat Example 2.6 using the spreadsheet provided for Chapter 2.

SOLUTION   Open the Chapter 2 spreadsheet and select the worksheet called Nominal Moment Strength. Input only the cells highlighted yellow^2 (the first five values below).

$f'_c =$	3000	psi
$b =$	14	in.
$d =$	21	in.
$A_s =$	3	in.^{2}
$f_y =$	60	ksi
$a =$	5.04	
$M_n =$	3326.2	in.-k
$=$	277.2	ft-k

The third worksheet called Nominal Moment Strength can be used to easily work Example 2.6. In this case, enter the first five values and the results are the same as in the example. The process can be reversed if "goal seek" is used. Suppose you would like to know how much reinforcing steel, A_s, is needed to resist a moment, M_n, of 320 ft-k for the beam shown in Example 2.6. Highlight the cell where M_n is calculated in ft-k (cell C11). Then go to "Data" at the top of the Excel window and select "What-If Analysis" and "Goal seek..." The Goal Seek window shown will open. In "Set cell," C11 appears because it was highlighted when you selected Goal Seek. In "To value," type 320 because that is the moment you are seeking. Finally, for "By changing cell" insert C7 because the area of reinforcing steel is what you want to vary to produce a moment of 320 ft-k. Click OK and the value of A_s will change to 3.55. This means that a steel area of 3.55 in.^{2} is required to produce a moment capacity M_n of 320 ft-k. The Goal Seek feature can be used in a similar manner for most of the spreadsheets provided in this text.

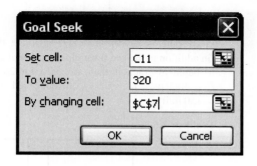

PROBLEMS

Cracking Moments

For Problems 2.1 to 2.5, determine the cracking moments for the sections shown, if $f'_c = 4000$ psi and if the modulus of rupture $f_r = 7.5\sqrt{f'_c}$.

Problem 2.1 (*Ans.* 40.7 ft-k)

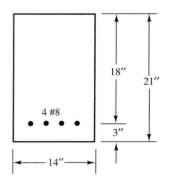

$\rightarrow$ **Problem 2.4**

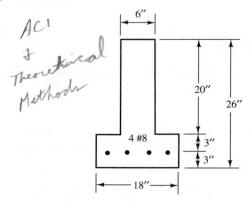

Problem 2.5 (*Ans.* 59.8 ft-k)

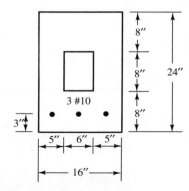

Problem 2.2

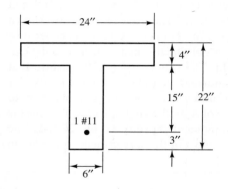

For Problems 2.6 through 2.7, calculate the uniform load (in addition to the beam weight) which will cause the sections to begin to crack if they are used for 28-ft simple spans. $f'_c = 4000$ psi, $f_r = 7.5\sqrt{f'_c}$. Reinforced concrete weight = 150 lb/ft^3.

Problem 2.6

$\rightarrow$ **Problem 2.3** (*Ans.* 25.6 ft-k)

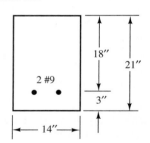

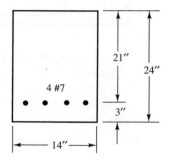

Problem 2.7 (*Ans.* 0.343 k/ft)

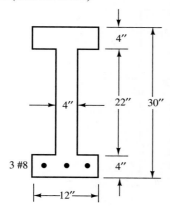

Problem 2.10

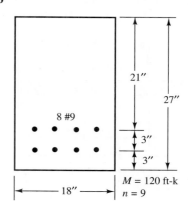

$M = 120$ ft-k
$n = 9$

Transformed-Area Method

For Problems 2.8 to 2.14, assume the sections have cracked and use the transformed-area method to compute their flexural stresses for the loads or moments given.

Problem 2.8

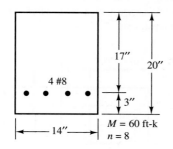

$M = 60$ ft-k
$n = 8$

Problem 2.11 (*Ans.* $f_c = 1187$ psi, $f_s = 12{,}289$ psi)

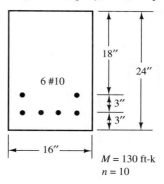

$M = 130$ ft-k
$n = 10$

2.9 Repeat Problem 2.8 if four #7 bars are used.
(*Ans.* $f_c = 1214$ psi, $f_s = 19{,}826$ psi)

Problem 2.12

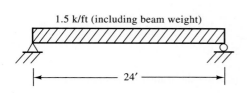

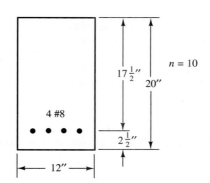

$n = 10$

Problem 2.13 (*Ans.* $f_c = 1762$ psi, $f_s = 28{,}635$ psi)

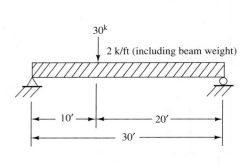

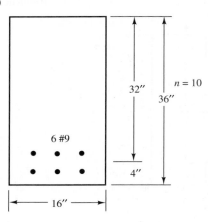

Problem 2.14

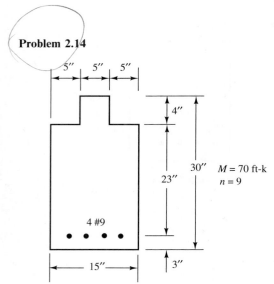

2.17 Using transformed area, what allowable uniform load can this beam support in addition to its own weight for a 28-ft simple span? Concrete weight = 150 lb/ft³, $f_s = 24{,}000$ psi, and $f_c = 1800$ psi. (*Ans.* 3.11 k/ft)

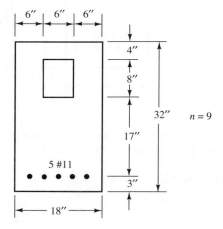

2.15 Using the transformed-area method, compute the resisting moment of the beam of Problem 2.10 if $f_s = 24{,}000$ psi and $f_c = 1800$ psi. (*Ans.* 258.8 ft-k)

2.16 Compute the resisting moment of the beam of Problem 2.13, if eight #10 bars are used and if $n = 10$, $f_s = 20{,}000$ psi, and $f_c = 1125$ psi. Use the transformed-area method.

For Problems 2.18 to 2.21 determine the flexural stresses in these members using the transformed-area method.

Problem 2.18

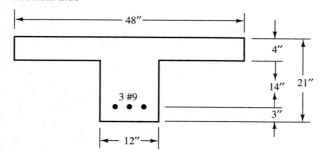

$M = 100$ ft-k

$n = 10$

Problem 2.19 (*Ans.* $f_c = 1223$ psi, $f_s = 30,185$ psi)

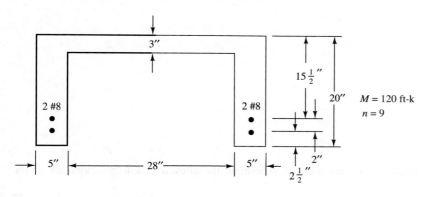

$M = 120$ ft-k

$n = 9$

Problem 2.20

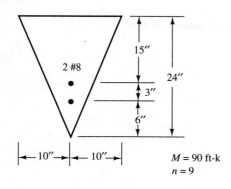

$M = 90$ ft-k

$n = 9$

Problem 2.21 (*Ans.* $f_c = 1280$ psi; $f_s' = 15,463$ psi; $f_s = 26,378$ psi)

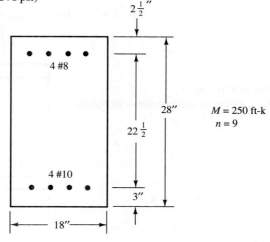

$M = 250$ ft-k

$n = 9$

2.22 Compute the allowable resisting moment
of the section shown using transformed area, if $f_c = 1800$ psi,
$f_s = f_s' = 24,000$ psi, and $n = 8$.

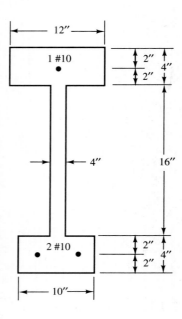

For Problems 2.23 to 2.25 using the transformed-area method, determine the allowable resisting moments of the
sections shown.

Problem 2.23 (*Ans.* 140.18 ft-k)

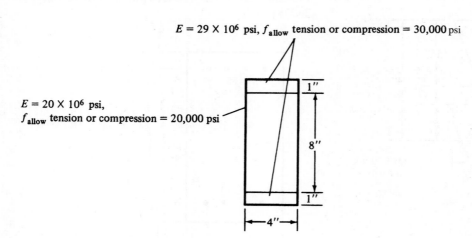

$E = 29 \times 10^6$ psi, f_{allow} tension or compression $= 30,000$ psi

$E = 20 \times 10^6$ psi,
f_{allow} tension or compression $= 20,000$ psi

Problem 2.24

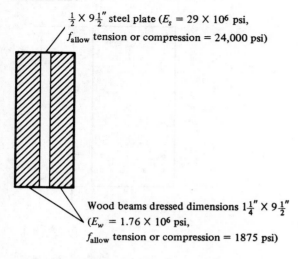

$\frac{1}{2} \times 9\frac{1}{2}''$ steel plate ($E_s = 29 \times 10^6$ psi,

f_{allow} tension or compression = 24,000 psi)

Wood beams dressed dimensions $1\frac{1}{4}'' \times 9\frac{1}{2}''$

($E_w = 1.76 \times 10^6$ psi,

f_{allow} tension or compression = 1875 psi)

Problem 2.25 (*Ans.* 124.4 ft-k)

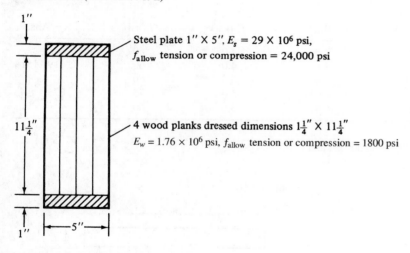

Steel plate $1'' \times 5''$, $E_s = 29 \times 10^6$ psi,

f_{allow} tension or compression = 24,000 psi

4 wood planks dressed dimensions $1\frac{1}{4}'' \times 11\frac{1}{4}''$

$E_w = 1.76 \times 10^6$ psi, f_{allow} tension or compression = 1800 psi

Nominal Strength Analysis

For Problems 2.26 to 2.29 determine the nominal or theoretical moment capacity M_n of each beam, if $f_y = 60,000$ psi and $f_c' = 4000$ psi.

Problem 2.26   **Problem 2.27** (*Ans.* 837.3 ft-k)

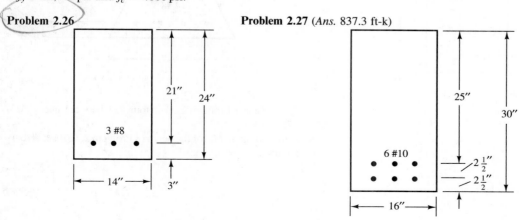

Problem 2.28

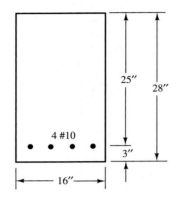

Problem 2.29 (*Ans.* 680.7 ft-k)

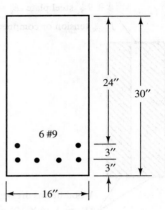

For Problems 2.30 to 2.34 determine the nominal moment capacity M_n for each of the rectangular beams.

Prob. No.	b (in.)	d (in.)	Bars	f'_c (ksi)	f_y(ksi)	Ans.
2.30	14	21	3 #9	4.0	60	—
2.31	18	28.5	8 #10	4.0	60	1191.1 ft-k
2.32	14	20.5	4 #10	5.0	60	—
2.33	18	25.5	4 #11	5.0	75	876.4 ft-k
2.34	22	36	6 #11	3.0	60	—

For Problems 2.35 to 2.39, determine M_n if $f_y = 60,000$ psi and $f'_c = 4000$ psi.

Problem 2.35 (*Ans.* 502.2 ft-k)

Problem 2.36

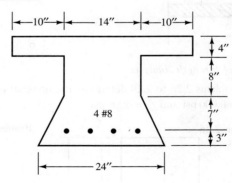

2.37 Repeat Problem 2.35 if four #11 bars are used. (*Ans.* 771.3 ft-k)

2.38 Compute M_n for the beam of Problem 2.36 if six #8 bars are used.

Problem 2.39 (*Ans.* 456 ft-k)

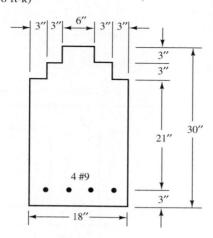

2.40 Determine the nominal uniform load, w_n (including beam weight), which will cause a bending moment equal to $M_n \cdot f_y = 60{,}000$ psi and $f_c' = 4000$ psi.

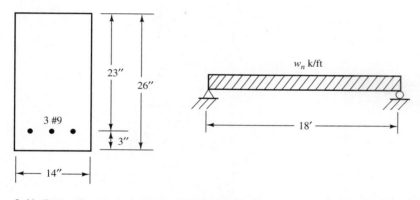

2.41 Determine the nominal uniform load, w_n (including beam weight) which will cause a bending moment equal to $M_n \cdot f_c' = 3000$ psi and $f_y = 60{,}000$ psi (Ans. 5.74 k/ft)

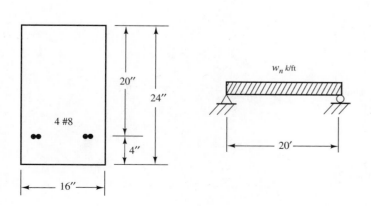

Problems in SI Units

For Problems 2.42 to 2.44, determine the cracking moments for the sections shown if, $f_c' = 28$ MPa and if the modulus of rupture is $f_r = 0.7\sqrt{f_c'}$ with f_c' in MPa.

Problem 2.42

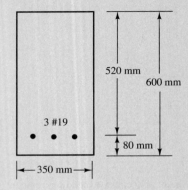

Problem 2.44

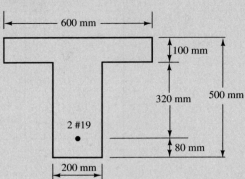

Problem 2.43 (*Ans.* 54.0 kN-m)

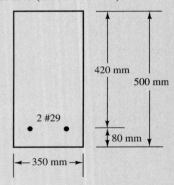

In Problems 2.45 to 2.47 compute the flexural stresses in the concrete and steel for the beams shown using the transformed-area method.

Problem 2.45 (*Ans.* $f_c' = 10.07$ MPa, $f_s = 155.9$ MPa)

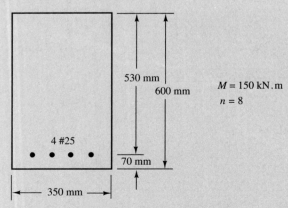

Problem 2.46

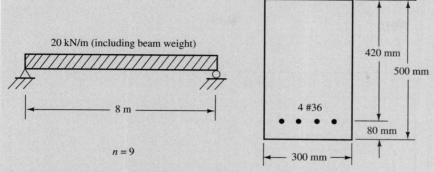

20 kN/m (including beam weight)

8 m

n = 9

420 mm

500 mm

80 mm

4 #36

300 mm

Problem 2.47 (*Ans.* $f_c = 9.68$ MPa, $f_s' = 118.4$ MPa, $f_s = 164.1$ MPa.

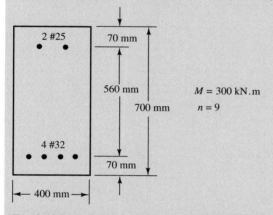

2 #25

70 mm

560 mm

700 mm

4 #32

70 mm

400 mm

$M = 300$ kN.m

$n = 9$

For Problems 2.48 to 2.55, compute M_n values.

Prob. No.	b (mm)	d (mm)	Bars	f_c'(MPa)	f_y (MPa)	Ans.
2.48	300	600	3 #36	35	350	—
2.49	320	600	3 #36	28	350	560.5 kN-m
2.50	350	530	3 #25	24	420	—
2.51	400	660	3 #32	42	420	644 kN-m

Problem 2.52

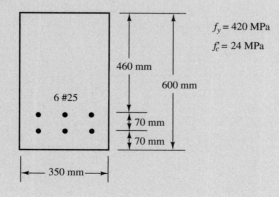

$f_y = 420$ MPa

$f_c' = 24$ MPa

460 mm

600 mm

6 #25

70 mm

70 mm

350 mm

2.53   Repeat Problem 2.48 if four #36 bars are used. (*Ans.* 734 kN · m)

Problem 2.54

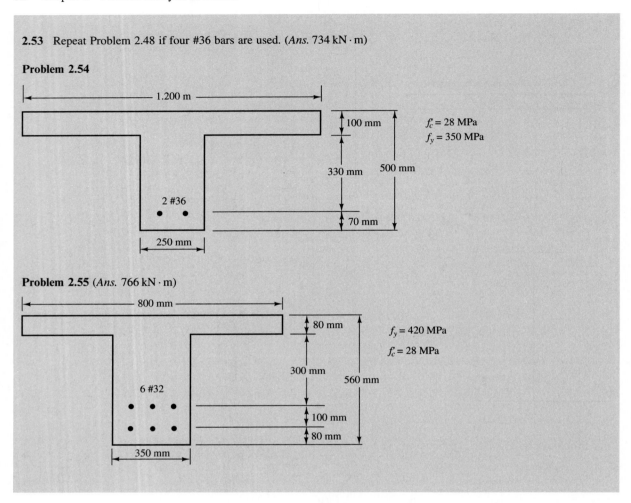

Problem 2.55 (*Ans.* 766 kN · m)

2.56 Repeat Problem 2.27 using Chapter 2 spreadsheets.
2.57 Repeat Problem 2.28 using Chapter 2 spreadsheets.
(*Ans.* 561.9 ft-k)

2.58 Prepare a flowchart for the determination of M_n for a rectangular tensilely reinforced concrete beam.

3
Strength Analysis of Beams According to ACI Code

3.1 DESIGN METHODS

From the early 1900s until the early 1960s, nearly all reinforced concrete design in the United States was performed by the working-stress design method (also called *allowable-stress design* or *straight-line design*). In this method, frequently referred to as WSD, the dead and live loads to be supported, called *working loads* or *service loads*, were first estimated. Then the members of the structure were proportioned so that stresses calculated by a transformed area did not exceed certain permissible or allowable values.

After 1963 the ultimate-strength design method rapidly gained popularity because (1) it makes use of a more rational approach than does WSD; (2) a more realistic consideration of safety is used; and (3) it provides more economical designs. With this method (now called *strength design*) the working dead and live loads are multiplied by certain load factors (equivalent to safety factors) and the resulting values are called *factored loads*. The members are then selected so they will theoretically just fail under the factored loads.

Even though almost all of the reinforced concrete structures the reader will encounter will be designed by the strength design method, it is still rather desirable to be familiar with WSD for several reasons. These include the following:

1. Some designers use WSD for proportioning fluid-containing structures (such as water tanks and various sanitary structures). When these structures are designed by WSD, stresses are kept at fairly low levels, with the result that there is appreciably less cracking and less consequent leakage. (If the designer uses strength design and makes use of proper crack control methods as described in Chapter 6, there should be few leakage problems.)

2. The ACI method for calculating the moments of inertia to be used for deflection calculations requires some knowledge of the working-stress procedure.

3. The design of prestressed concrete members is based not only on the strength method but also on elastic stress calculations at service load conditions.

The reader should realize that working-stress design has several disadvantages. When using the method, the designer has little knowledge about the magnitudes of safety factors against collapse; no consideration is given to the fact that different safety factors are desirable for dead and live loads; the method does not account for variations in resistances and loads, nor does it account for the possibility that as loads are increased, some increase at different rates than others.

In 1956, as an appendix, the ACI Code for the first time included ultimate-strength design, although the concrete codes of several other countries had been based on such considerations for several decades. In 1963 the Code gave ultimate-strength design equal status with working-stress design, the 1971 Code made the method the predominant method and only briefly mentioned the working-stress method. From 1971 until 1999 each issue of the Code permitted designers to use working-stress design and set out certain provisions for its application. *Beginning with the 2002 Code, however, permission is not included for using the method.*

Today's design method was called *ultimate-strength design* for several decades; but, as mentioned, the Code now uses the term *strength design*. The strength for a particular reinforced concrete member is a value given by the Code and is not necessarily the true ultimate strength of the member. Therefore, the more general term *strength design* is used whether beam strength, column strength, shear strength, or others are being considered.

3.2　ADVANTAGES OF STRENGTH DESIGN

Among the several advantages of the strength design method as compared to the no longer permitted working-stress design method are the following:

1. The derivation of the strength design expressions takes into account the nonlinear shape of the stress–strain diagram. When the resulting equations are applied, decidedly better estimates of load-carrying ability are obtained.

2. With strength design, a more consistent theory is used throughout the designs of reinforced concrete structures. For instance, with working-stress design the transformed-area or straight-line method was used for beam design, and a strength design procedure was used for columns.

3. A more realistic factor of safety is used in strength design. The designer can certainly estimate the magnitudes of the dead loads that a structure will have to support more accurately than he or she can estimate the live and environmental loads. With working-stress design the same safety factor was used for dead, live, and environmental loads. This is not the case for strength design. For this reason, use of different load or safety factors in strength design for the different types of loads is a definite improvement.

4. A structure designed by the strength method will have a more uniform safety factor against collapse throughout. The strength method takes considerable advantage of higher-strength steels, whereas working-stress design did only partly so. The result is better economy for strength design.

5. The strength method permits more flexible designs than did the working-stress method. For instance, the percentage of steel may be varied quite a bit. As a result, large sections may be used with small percentages of steel, or small sections may be used with large percentages of steel. Such variations were not the case in the relatively fixed working-stress method. If the same amount of steel is used in strength design for a particular beam as would have been used with WSD, a smaller section will result. If the same size section is used as required by WSD, a smaller amount of steel will be required.

3.3　STRUCTURAL SAFETY

The structural safety of a reinforced concrete structure can be calculated with two methods. The first method involves calculations of the stresses caused by the working or service loads and their comparison with certain allowable stresses. Usually the safety factor against collapse when the working-stress method was used was said to equal the smaller of f_c'/f_c or f_y/f_s.

Water Tower Place, Chicago, Illinois, tallest reinforced concrete building in the United States (74 stories, 859 ft). (Courtesy of Symons Corporation.)

The second approach to structural safety is the one used in strength design in which uncertainty is considered. The working loads are multiplied by certain load factors that are larger than one. The resulting larger or factored loads are used for designing the structure. The values of the load factors vary depending on the type and combination of the loads.

To accurately estimate the ultimate strength of a structure, it is necessary to take into account the uncertainties in material strengths, dimensions, and workmanship. This is done by multiplying the theoretical ultimate strength (called the *nominal strength* herein) of each member by the *strength reduction factor* ϕ, which is less than one. These values generally vary from 0.90 for bending down to 0.65 for some columns.

In summary, the strength design approach to safety is to select a member whose computed ultimate load capacity multiplied by its strength reduction factor will at least equal the sum of the service loads multiplied by their respective load factors.

Member capacities obtained with the strength method are appreciably more accurate than member capacities predicted with the working-stress method.

3.4   DERIVATION OF BEAM EXPRESSIONS

Tests of reinforced concrete beams confirm that strains vary in proportion to distances from the neutral axis even on the tension sides and even near ultimate loads. Compression stresses vary approximately in a straight line until the maximum stress equals about $0.50 f_c'$. This is not the case, however, after stresses go higher. When the ultimate load is reached, the strain and stress variations are approximately as shown in Figure 3.1.

The compressive stresses vary from zero at the neutral axis to a maximum value at or near the extreme fiber. The actual stress variation and the actual location of the neutral axis vary somewhat

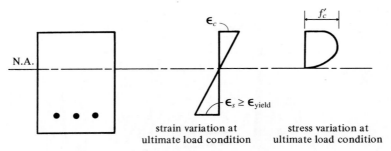

Figure 3.1 Ultimate load.

from beam to beam depending on such variables as the magnitude and history of past loadings, shrinkage and creep of the concrete, size and spacing of tension cracks, speed of loading, and so on.

If the shape of the stress diagram were the same for every beam, it would easily be possible to derive a single rational set of expressions for flexural behavior. Because of these stress variations, however, it is necessary to base the strength design on a combination of theory and test results.

Although the actual stress distribution given in Figure 3.2(b) may seem to be important, any assumed shape (rectangular, parabolic, trapezoidal, etc.) can be used practically if the resulting equations compare favorably with test results. The most common shapes proposed are the rectangle, parabola, and trapezoid, with the rectangular shape used in this text as shown in Figure 3.2(c) being the most common one.

If the concrete is assumed to crush at a strain of about 0.003 (which is a little conservative for most concretes) and the steel to yield at f_y, it is possible to make a reasonable derivation of beam formulas without knowing the exact stress distribution. However, it is necessary to know the value of the total compression force and its centroid.

Whitney[1] replaced the curved stress block with an equivalent rectangular block of intensity $0.85f'_c$ and depth $\alpha = \beta_1 c$, as shown in Figure 3.2(c). The area of this rectangular block should equal that of the curved stress block, and the centroids of the two blocks should coincide. Sufficient

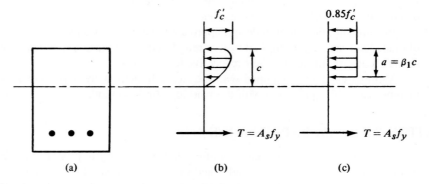

Figure 3.2 Some possible stress distribution shapes.

---

[1]Whitney, C. S., 1942, "Plastic Theory of Reinforced Concrete Design," *Transactions ASCE*, 107, pp. 251–326.

test results are available for concrete beams to provide the depths of the equivalent rectangular stress blocks. The values of β_1 given by the Code (10.2.7.3) are intended to give this result. For f'_c values of 4000 psi or less, $\beta_1 = 0.85$, and it is to be reduced continuously at a rate of 0.05 for each 1000-psi increase in f'_c above 4000 psi. Their value may not be less than 0.65. The values of β_1 are reduced for high-strength concretes primarily because of the shapes of their stress–strain curves (see Figure 1.1).

For concretes with $f'_c > 4000$ psi, β_1 can be determined with the following formula:

$$\beta_1 = 0.85 - \left(\frac{f'_c - 4000}{1000}\right)(0.05) \geq 0.65$$

In SI units, β_1 is to be taken equal to 0.85 for concrete strengths up to and including 30 MPa. For strengths above 30 MPa, β_1 is to be reduced continuously at a rate of 0.05 for each 7 MPa of strength in excess of 30 MPa but shall not be taken less than 0.65.

For concretes with $f'_c > 30$ MPa, β_1 can be determined with the following expression:

$$\beta_1 = 0.85 - 0.008(f'_c - 30) \geq 0.65$$

Based on these assumptions regarding the stress block, statics equations can easily be written for the sum of the horizontal forces and for the resisting moment produced by the internal couple. These expressions can then be solved separately for a and for the moment M_n.

A very clear statement should be made here regarding the term M_n because it otherwise can be confusing to the reader. M_n is defined as the theoretical or nominal resisting moment of a section. In Section 3.3 it was stated that the usable strength of a member equals its theoretical strength times the strength reduction factor, or, in this case, ϕM_n. The usable flexural strength of a member, ϕM_n, must at least be equal to the calculated factored moment, M_u, caused by the factored loads

$$\phi M_n \geq M_u$$

For writing the beam expressions, reference is made to Figure 3.3. Equating the horizontal forces C and T and solving for a, we obtain

$$0.85f'_c ab = A_s f_y$$

$$a = \frac{A_s f_y}{0.85f'_c b} = \frac{\rho f_y d}{0.85f'_c}, \quad \text{where } \rho = \frac{A_s}{bd} = \text{percentage of tensile steel}$$

Because the reinforcing steel is limited to an amount such that it will yield well before the concrete reaches its ultimate strength, the value of the nominal moment M_n can be written as

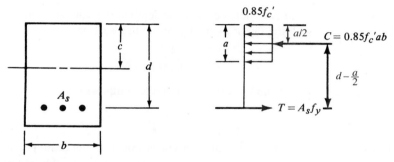

Figure 3.3

$$M_n = T\left(d - \frac{a}{2}\right) = A_s f_y\left(d - \frac{a}{2}\right)$$

and the usable flexural strength is

$$\phi M_n = \phi A_s f_y\left(d - \frac{a}{2}\right) \tag{Eq. 3-1}$$

If we substitute into this expression the value previously obtained for a (it was $\rho f_y d/0.85 f_c'$), replace A_s with $\rho b d$ and equate ϕM_n to M_u, we obtain the following expression:

$$\phi M_n = M_u = \phi b d^2 f_y \rho\left(1 - \frac{\rho f_y}{1.7 f_c'}\right) \tag{Eq. 3-2}$$

Replacing A_s with $\rho b d$ and letting $R_n = M_u/\phi b d^2$, we can solve this expression for ρ (the percentage of steel required for a particular beam) with the following results:

$$\rho = \frac{0.85 f_c'}{f_y}\left(1 - \sqrt{1 - \frac{2R_n}{0.85 f_c'}}\right) \tag{Eq. 3-3}$$

Instead of substituting into this equation for ρ when rectangular sections are involved, the reader will find Tables A.8 to A.13 in Appendix A of this text to be quite convenient. (For SI units refer to Tables B.8 and B.9 in Appendix B.) Another way to obtain the same information is to refer to Graph 1 which is also located in Appendix A. The user, however, will have some difficulty in reading this small-scale graph accurately. This expression for ρ is also very useful for tensilely reinforced rectangular sections that do not fall into the tables. An iterative technique for determination of reinforcing steel area is also presented later in this chapter.

3.5   STRAINS IN FLEXURAL MEMBERS

As previously mentioned, Section 10.2.2 of the Code states that the strains in concrete members and their reinforcement are to be assumed to vary directly with distances from their neutral axes. (This assumption is not applicable to deep flexural members whose depths over their clear spans are greater than 0.25.) Furthermore, in Section 10.2.3 the Code states that the maximum usable strain in the extreme compression fibers of a flexural member is to be 0.003. Finally, Section 10.3.3 states that for Grade 60 reinforcement and for all prestressed reinforcement we may set the strain in the steel equal to 0.002 at the balanced condition. (Theoretically, for 60,000-psi steel it equals $\dfrac{f_y}{E_s} = \dfrac{60,000}{29 \times 10^6} = 0.00207$.)

In Section 3.4 a value was derived for a, the depth of the equivalent stress block of a beam. It can be related to c with the factor β_1 also given in that section.

$$a = \frac{A_s f_y}{0.85 f_c' b} = \beta_1 c$$

Then the distance c from the extreme concrete compression fibers to the neutral axis is

$$c = \frac{a}{\beta_1}$$

In Example 3.1 the values of a and c are determined for the beam previously considered in Example 2.6, and by straight-line proportions the strain in the reinforcing ϵ_t is computed.

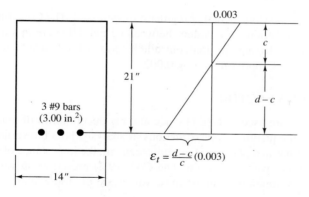

Figure 3.4

EXAMPLE 3.1

Determine the values of a, c, and ϵ_t for the beam shown in Figure 3.4. $f_y = 60{,}000$ psi and $f'_c = 3000$ psi.

SOLUTION

$$a = \frac{A_s f_y}{0.85 f'_C B} = \frac{(3.00)(60)}{(0.85)(3)(14)} = \underline{\underline{5.04''}}$$

$\beta_1 = 0.85$ for 3000 psi concrete

$$c = \frac{a}{\beta_1} = \frac{5.04}{0.85} = \underline{\underline{5.93''}}$$

$$\epsilon_t = \frac{d-c}{c}(0.003) = \left(\frac{21 - 5.93}{5.93}\right)(0.003) = \underline{0.00762}$$

This value of strain is much greater than the yield strain of 0.002. This is an indication of ductile behavior of the beam, because the steel is well into its yield plateau before concrete crushes.

3.6   BALANCED SECTIONS, TENSION-CONTROLLED SECTIONS, AND COMPRESSION-CONTROLLED OR BRITTLE SECTIONS

A beam that has a *balanced steel ratio* is one for which the tensile steel will theoretically just reach its yield point at the same time the extreme compression concrete fibers attain a strain equal to 0.003. Should a flexural member be so designed that it has a balanced steel ratio or be a member whose compression side controls (that is, if its compression strain reaches 0.003 before the steel yields), the member can suddenly fail without warning. As the load on such a member is increased, its deflections will usually not be particularly noticeable, even though the concrete is highly stressed in compression and failure will probably occur without warning to users of the structure. These members are *compression controlled* and are referred to as *brittle members*. Obviously, such members must be avoided.

The Code, in Section 10.3.4, states that members whose computed tensile strains are equal to or greater than 0.0050 at the same time the concrete strain is 0.003 are to be referred to as *tension-controlled sections*. For such members the steel will yield before the compression side crushes and deflections will be large, giving users warning of impending failure. Furthermore, members with $\epsilon_t \geq 0.005$ are considered to be fully ductile. The ACI chose the 0.005 value for ϵ_t to apply to all types of steel permitted by the Code, whether regular or prestressed. The Code further states that members that have net steel strains or ϵ_t values between ϵ_y and 0.005 are in a transition region between compression-controlled and tension-controlled sections. For Grade 60 reinforcing steel, which is quite common, ϵ_y is approximated by 0.002.

3.7   STRENGTH REDUCTION OR ϕ FACTORS

Strength reduction factors are used to take into account the uncertainties of material strengths, inaccuracies in the design equations, approximations in analysis, possible variations in dimensions of the concrete sections and placement of reinforcement, the importance of members in the structures of which they are part, and so on. The Code (9.3) prescribes ϕ values or strength reduction factors for most situations. Among these values are the following:

0.90 for tension-controlled beams and slabs

0.75 for shear and torsion in beams

0.65 or 0.75 for columns

0.65 or 0.75 to 0.9 for columns supporting very small axial loads

0.65 for bearing on concrete

The sizes of these factors are rather good indications of our knowledge of the subject in question. For instance, calculated nominal moment capacities in reinforced concrete members seem to be quite accurate, whereas computed bearing capacities are more questionable.

For ductile or tension-controlled beams and slabs where $\epsilon_t \geq 0.005$, the value of ϕ for bending is 0.90. Should ϵ_t be less than 0.005 it is still possible to use the sections if ϵ_t is not less than certain values. This situation is shown in Figure 3.5, which is similar to Figure R.9.3.2 in the ACI Commentary to the 2008 Code. One significant change in the 2008 Code is illustrated by comparing the ϕ factors for spiral members for the 2005 and 2008 Codes in Figure 3.5. This liberalization was "to recognize the superior performance of spirally reinforced columns when subjected to extraordinary loads."[2]

Members subject to axial loads equal to or less than $0.10f_c'A_g$ may be used only when ϵ_t is no lower than 0.004 (ACI Section 10.3.5). An important implication of this limit is that reinforced concrete beams must have a tension strain of at least 0.004. Should the members be subject to axial loads $\geq 0.10f_c'A_g$ they may be used when ϵ_t is as small as 0.002. When ϵ_t values fall between 0.002 and 0.005, they are said to be in the transition range between tension-controlled and compression-

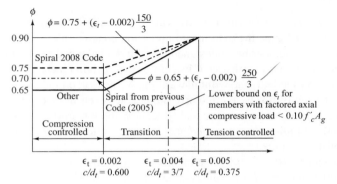

Figure 3.5 Variation of ϕ with net tensile strain ϵ_t and c/d_t for Grade 60 reinforcement and for prestressing steel.

---

[2]Wight, James K., 2007, *Concrete International*, July 2007, Farmington Hills, Michigan.

controlled sections. In this range ϕ values will fall between 0.65 or 0.70 and 0.90 as shown in the figure.

The procedure for determining ϕ values in the transition range is described later in this section. *You must clearly understand that the use of flexural members in this range is usually uneconomical, and it is probably better, if the situation permits, to increase member depths and/ or decrease steel percentages until ϵ_t is equal to or larger than 0.005.* If this is done, not only will ϕ values equal 0.9 but also steel percentages will not be so large as to cause crowding of reinforcing bars. The net result will be slightly larger concrete sections, with consequent smaller deflections. Furthermore, as we will learn in subsequent chapters, the bond of the reinforcing to the concrete will be increased as compared to cases where higher percentages of steel are used.

We have computed values of steel percentages for different grades of concrete and steel for which ϵ_t will exactly equal 0.005 and present them in Appendix Tables A.7 and B.7 of this textbook. It is therefore desirable, under ordinary conditions, to design beams with steel percentages that are no larger than these values, and we have shown them as suggested maximum percentages to be used.

The bottom half of Figure 3.5 gives values for c/d ratios. If c/d for a particular flexural member is ≤ 0.375, the beam will be ductile, and if > 0.600 it will be brittle. In between is the transition range. You may prefer to compute c/d for a particular beam to check its ductility rather than computing ρ or ϵ_t. In the transition region, interpolation to determine ϕ using c/d_t instead of ϵ_t can be performed using the equations

$$\phi = 0.75 + 0.15 \left[\frac{1}{c/d_t} - \frac{5}{3} \right] \ \textit{for spiral members}$$

$$\phi = 0.65 + 0.25 \left[\frac{1}{c/d_t} - \frac{5}{3} \right] \ \textit{for other members}$$

The equations for ϕ above and in Figure 3.5 are for the special case where $f_y = 60$ ksi and for prestressed concrete. For other cases, replace 0.002 with $\epsilon_y = f_y/E_s$. Figure 10.25 shows Figure 3.5 for the general case where ϵ_y is not assumed to be 0.002.

The resulting general equations are

$$\phi = 0.75 + (\epsilon_t - \epsilon_y) \frac{0.15}{(0.005 - \epsilon_y)} \ \textit{for spiral members}$$

and

$$\phi = 0.65 + (\epsilon_t - \epsilon_y) \frac{0.25}{(0.005 - \epsilon_y)} \ \textit{for other members}$$

The impact of the variable ϕ factor on moment capacity is shown in Figure 3.6. The two curves show the moment capacity with and without the application of the ϕ factor. Point A corresponds to a tensile strain, ϵ_t, of 0.005 and $\rho = 0.0181$ (Table A7). This is the largest value of ρ for $\phi = 0.9$. Above this value of ρ, ϕ decreases to as low as 0.65 as shown by point B which corresponds to ϵ_t of ϵ_y. ACI 10.3.5 requires ϵ_t not to exceed 0.004 for flexural members with low axial loads. This situation corresponds to point C in Figure 3.6. The only allowable range for ρ is below point C. From the figure, it is clear that little moment capacity is gained in adding steel area above point A. The variable ϕ factor provisions essentially permit a constant value of ϕM_n when ϵ_t is less than 0.005. It is important for the designer to know this because often actual bar selections result in more steel area than theoretically required. If the slope between points A and C were negative, the designer could not use a larger area. Knowing the slope is slightly positive, the designer can use the larger bar area with confidence that the design capacity is not reduced.

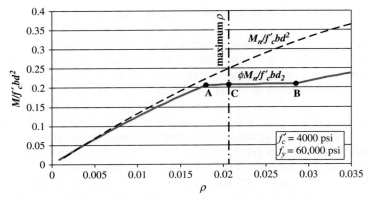

Figure 3.6 Moment capacity vs. ρ.

For values of f_y of 75 ksi and higher, the slope between point A and B in Figure 3.6 is actually negative. It is therefore especially important when using high strength reinforcing steel to verify your final design to be sure the bars you have selected do not result in a moment capacity less than the design value.

Continuing our consideration of Figure 3.5, we can see that when ϵ_t is less than 0.005, the values of ϕ will vary along a straight line from their 0.90 value for ductile sections to 0.65 at balanced conditions where ϵ_t is 0.002. Later, we will learn that ϕ can equal 0.75 rather than 0.65 at this latter strain situation if spirally reinforced sections are being considered.

3.8 MINIMUM PERCENTAGE OF STEEL

A brief discussion of the modes of failure that occur for various reinforced beams was presented in Section 3.6. Sometimes because of architectural or functional requirements, beam dimensions are selected that are much larger than are required for bending alone. Such members theoretically require very small amounts of reinforcing.

Actually, another mode of failure can occur in very lightly reinforced beams. If the ultimate resisting moment of the section is less than its cracking moment, the section will fail immediately when a crack occurs. This type of failure may occur without warning. To prevent such a possibility, the ACI (10.5.1) specifies a certain minimum amount of reinforcing that must be used at every section of flexural members where tensile reinforcing is required by analysis, whether for positive or negative moments. In the following equations, b_w represents the web width of beams.

$$A_{s,\min} = \frac{3\sqrt{f_c'}}{f_y} b_w d$$

(ACI Equation 10-3)

$$\text{nor less than } \frac{200 b_w d}{f_y}$$

[In SI units, these expressions are $\left(\dfrac{\sqrt{f_c'}}{4 f_y}\right) b_w d$ and $\left(\dfrac{1.4 b_w d}{f_y}\right)$, respectively.]

The $\left(\dfrac{200 b_w d}{f_y}\right)$ value was obtained by calculating the cracking moment of a plain concrete section and equating it to the strength of a reinforced concrete section of the same size, applying a safety factor of 2.5 and solving for the steel required. It has been found, however, that when f_c'

Fountain Hills, Arizona Wastewater Treatment Plant. (Courtesy of Economy Forms Corporation.)

is higher than about 5000 psi, this value may not be sufficient. Thus the $\left(\dfrac{3\sqrt{f'_c}}{f_y}\right) b_w d$ value is also required to be met, and it will actually control when f'_c is greater than 4440 psi.

This ACI equation (10-3) for the minimum amount of flexural reinforcing can be written as a percentage, as follows:

$$\rho_{\min} \text{ for flexure} = \frac{3\sqrt{f'_c}}{f_y} \geq \frac{200}{f_y}$$

Values of $\rho_{\min}$ for flexure have been calculated by the authors and are shown for several grades of concrete and steel in Appendix Table A.7 of this text. They are also included in Appendix Tables A.8 to A.13. (For SI units, the appropriate tables are B.7 to B.9.)

Section 10.5.3 of the Code states that the preceding minimums do not have to be met if the area of the tensile reinforcing furnished at every section is at least one-third greater than the area required by moment. Furthermore, ACI Section 10.5.4 states that for slabs and footings of uniform thickness, the minimum area of tensile reinforcing in the direction of the span is that specified in ACI Section 7.12 for shrinkage and temperature steel which is much lower. When slabs are overloaded in certain areas there is a tendency for those loads to be distributed laterally to other parts of the slab, thus substantially reducing the chances of sudden failure. This explains why a reduction of the minimum reinforcing percentage is permitted in slabs of uniform thickness. Supported slabs, such as slabs on grade, are not considered to be structural slabs in this section unless they transmit vertical loads from other parts of the structure to the underlying soil.

3.9 BALANCED STEEL PERCENTAGE

In this section an expression is derived for ρ_b, the percentage of steel required for a balanced design. At ultimate load for such a beam, the concrete will theoretically fail (at a strain of 0.00300), and the steel will simultaneously yield (see Figure 3.7).

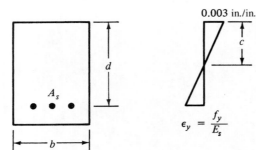

Figure 3.7 Balanced conditions.

The neutral axis is located by the triangular strain relationships that follow, noting that $E_s = 29 \times 10^6$ psi for the reinforcing bars:

$$\frac{c}{d} = \frac{0.00300}{0.00300 + (f_y/E_s)} = \frac{0.00300}{0.003 + (f_y/29 \times 10^6)}$$

This expression is rearranged and simplified, giving

$$c = \frac{87,000}{87,000 + f_y} d$$

In Section 3.4 of this chapter an expression was derived for a by equating the values of C and T. This value can be converted to c by dividing it by β_1

$$a = \frac{\rho f_y d}{0.85 f_c'}$$

$$c = \frac{a}{\beta_1} = \frac{\rho f_y d}{0.85 \beta_1 f_c'}$$

Two expressions are now available for c, and they are equated to each other and solved for the percentage of steel. This is the balanced percentage ρ_b.

$$\frac{\rho f_y d}{0.85 \beta_1 f_c'} = \frac{87,000}{87,000 + f_y} d$$

$$\rho_b = \left(\frac{0.85 \beta_1 f_c'}{f_y} \right) \left(\frac{87,000}{87,000 + f_y} \right)$$

or in SI units $\left(\dfrac{0.85 \beta_1 f_c'}{f_y} \right) \left(\dfrac{600}{600 + f_y} \right)$.

Values of ρ_b can easily be calculated for different values of f_c' and f_y and tabulated for U.S. customary units as shown in Appendix A.7. For SI units it's Appendix Table B.7.

Previous Codes (1963–1999) limited flexural members to 75% of the balanced steel ratio, ρ_b. However, this approach was changed in the 2002 Code to the new philosophy explained in Section 3.7, whereby the member capacity is penalized by reducing the ϕ factor when the strain in the reinforcing steel at ultimate is less than 0.005.

3.10 EXAMPLE PROBLEMS

Examples 3.2 to 3.4 present the computation of the design moment capacities of three beams using the ACI Code limitations. Remember that according to the Code (10.3.5) beams whose axial load is less than $0.10f'_c A_y$ may not, when loaded to their nominal strengths, have net tensile calculated strains less than 0.004.

EXAMPLE 3.2

Determine the ACI design moment capacity ϕM_n of the beam shown in Figure 3.8 if $f'_c = 4000$ psi and $f_y = 60,000$ psi.

SOLUTION

Checking Steel Percentage

$$\rho = \frac{A_s}{bd} = \frac{4.00}{(15)(24)} = 0.0111$$

$$\left. \begin{array}{l} > \rho_{\min} = 0.0033 \\ < \rho_{\max} = 0.0181 \end{array} \right\} \begin{array}{l} \text{both from} \\ \text{Appendix Table A.7} \end{array}$$

$$a = \frac{A_s f_y}{0.85 f'_c b} = \frac{(4.00)(60,000)}{(0.85)(4000)(15)} = 4.71 \text{ in.}$$

$$\beta_1 = 0.85 \text{ for 4000 psi concrete}$$

$$c = \frac{a}{\beta_1} = \frac{4.71}{0.85} = 5.54 \text{ in.}$$

Drawing Strain Diagram (Figure 3.9)

$$\epsilon_t = \frac{d-c}{c}(0.003) = \frac{18.46}{5.54}(0.003) = 0.0100$$

$$> 0.005 \therefore \text{tension controlled}$$

$$M_n = A_s f_y \left(d - \frac{a}{2} \right) = (4.00)(60)\left(24 - \frac{4.71}{2} \right)$$

$$= 5194.8 \text{ in.-k} = 432.9 \text{ ft-k}$$

$$\phi M_n = (0.9)(432.9) = \underline{\underline{389.6}} \text{ ft-k}$$

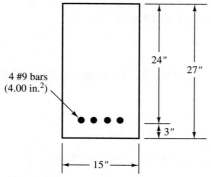

4 #9 bars
(4.00 in.²)

24″

27″

3″

15″

Figure 3.8

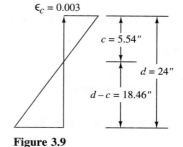

$\epsilon_c = 0.003$

$c = 5.54''$

$d = 24''$

$d - c = 18.46''$

Figure 3.9

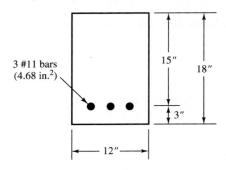

Figure 3.10

EXAMPLE 3.3

Determine the ACI design moment capacity ϕM_n of the beam shown in Figure 3.10 if $f'_c = 4000$ psi and $f_y = 60,000$ psi.

SOLUTION   **Checking Steel Percentage**

$$\rho = \frac{A_s}{bd} = \frac{4.68}{(12)(15)} = 0.026 > \rho_{min} = 0.0033 \text{ and}$$

$> \rho_{max} = 0.0181$ from Appendix Table A.7. As a result, we know that ϵ_t will be < 0.005.

Computing Value of ϵ_t

$$a = \frac{A_s f_y}{0.85 f'_c b} = \frac{(4.68)(60,000)}{(0.85)(4000)(12)} = 6.88 \text{ in.}$$

$\beta_1 = 0.85$ for 4000 psi concrete

$$c = \frac{a}{\beta_1} = \frac{6.88}{0.85} = 8.09 \text{ in.}$$

$$\epsilon_t = \frac{d - c}{c}(0.003) = \frac{15 - 8.09}{8.09}(0.003)$$

$$= 0.00256 < 0.004$$

∴ *Section is not ductile and may not be used as per ACI Section 10.3.5.*

EXAMPLE 3.4

Determine the ACI design moment capacity ϕM_n for the beam of Figure 3.11 if $f'_c = 4000$ psi and $f_y = 60,000$ psi.

SOLUTION   **Checking Steel Percentage**

$$\rho = \frac{A_s}{bd} = \frac{3.00}{(10)(15)} = 0.020 > \rho_{min} = 0.0033$$

but also $< \rho_{max} = 0.0181 (\text{for } \epsilon_t = 0.005)$

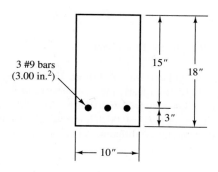

Figure 3.11

Computing Value of ϵ_t

$$a = \frac{A_s f_y}{0.85 f'_c b} = \frac{(3.00)(60,000)}{(0.85)(4000)(10)} = 5.29 \text{ in.}$$

$$\beta_1 = 0.85 \text{ for 4000 psi concrete}$$

$$c = \frac{a}{\beta_1} = \frac{5.29}{0.85} = 6.22 \text{ in.}$$

$$\epsilon_t = \frac{d-c}{c}(0.003) = \left(\frac{15 - 6.22}{6.22}\right)(0.003) = 0.00423 > 0.004 < 0.005$$

$$\therefore \text{ Beam is in transition zone and}$$

$$\phi \text{ from Figure 3.5} = 0.65 + (0.00423 - 0.002)\left(\frac{250}{3}\right) = 0.836$$

$$M_n = A_s f_y\left(d - \frac{a}{2}\right) = (3.00)(60)\left(15 - \frac{5.29}{2}\right) = 2223.9 \text{ in-k} = 185.3 \text{ ft-k}$$

$$\phi M_n = (0.836)(185.3) = \underline{154.9 \text{ ft-k}}$$

3.11   COMPUTER EXAMPLES

EXAMPLE 3.5

Repeat Example 3.2 using the Excel spreadsheet provided for Chapter 3.

SOLUTION   Open the Chapter 3 spreadsheet and open the Rectangular Beam worksheet. Enter values only in the cells highlighted yellow. The final result is $\phi M_n = 389.6$ ft-k (same answer as Example 3.2).

EXAMPLE 3.6

Repeat Example 3.3 using the Excel spreadsheet provided for Chapter 3.

SOLUTION   Open the Chapter 3 spreadsheet and the Rectangular Beam worksheet. Enter values only in the cells highlighted yellow. The spreadsheet displays a message, "code violation … too much steel." This is an indication that the beams violates ACI Section 10.3.5 and is not ductile. This beam is not allowed by the ACI Code.

.007,84 × 10⁻³
.00784

EXAMPLE 3.7

Repeat Example 3.4 using the Excel spreadsheet provided for Chapter 3.

SOLUTION   Open the Chapter 3 spreadsheet and the Rectangular Beam worksheet. Enter values only in the cells highlighted yellow. The final result is $\phi M_n = 154.5$ ft-k (nearly the same answer as Example 3.4). The ϕ factor is also nearly the same as Example 3.4 (0.0834 compared with 0.0836). The difference is the result of the spreadsheet using the more general value for ϵ_y of $f_y/E_s = 0.00207$, instead of the approximate value of 0.002 permitted by the code for Grade 60 reinforcing steel. A difference of this magnitude is not important, as discussed in Section 1.25, Calculation Accuracy.

PROBLEMS

3.1 What are the advantages of the strength design method as compared to the allowable stress or alternate design method?

3.2 What is the purpose of strength reduction factors? Why are they smaller for columns than for beams?

3.3 What are the basic assumptions of the strength design theory?

3.4 Why does the ACI Code specify that a certain minimum percentage of reinforcing be used in beams?

3.5 Distinguish between tension-controlled and compression-controlled beams.

3.6 Explain the purpose of the minimum cover requirements for reinforcing specified by the ACI Code.

3.7 For Problems 3.7 to 3.9 determine the values of ϵ_t, ϕ and ϕM_n for the sections shown.

Problem 3.7 (*Ans.* $\phi M_n = 264.1$ ft-k)

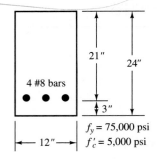

Problems 3.8

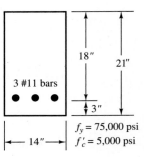

Problems 3.9 (*Ans.* Section has $\epsilon_t = 0.00368 < 0.004$. Therefore it may not be used.)

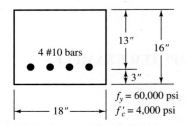

4

Design of Rectangular Beams and One-Way Slabs

4.1 LOAD FACTORS

Load factors are numbers, almost always larger than 1.0, which are used to increase the estimated loads applied to structures. They are used for loads applied to all types of members, not just beams and slabs. The loads are increased to attempt to account for the uncertainties involved in estimating their magnitudes. How close can you estimate the largest wind or seismic loads that will ever be applied to the building which you are now occupying? How much uncertainty is present in your answer?

You should note that the load factors for dead loads are much smaller than the ones used for live and environmental loads. Obviously, the reason is that we can estimate the magnitudes of dead loads so much more accurately than we can the magnitudes of those other loads. In this regard, you will notice that the magnitudes of loads that remain in place for long periods of time are much less variable than are those loads applied for brief periods such as wind and snow.

Section 9.2 of the Code presents the load factors and combinations that are to be used for reinforced concrete design. The required strength U, or the load-carrying ability of a particular reinforced concrete member, must at least equal the largest value obtained by substituting into ACI equations 9-1 to 9-7. The following equations conform to the requirements of the International Building Code (IBC)[1] as well as to the values required by ASCE Standard 7.[2]

$$U = 1.4(D + F) \qquad \text{(ACI Equation 9-1)}$$
$$U = 1.2(D + F + T) + 1.6(L + H) + 0.5(L_r \text{ or } S \text{ or } R) \qquad \text{(ACI Equation 9-2)}$$
$$U = 1.2D + 1.6(L_r \text{ or } S \text{ or } R) + (1.0L \text{ or } 0.8W) \qquad \text{(ACI Equation 9-3)}$$
$$U = 1.2D + 1.6W + 1.0L + 0.5(L_r \text{ or } S \text{ or } R) \qquad \text{(ACI Equation 9-4)}$$
$$U = 1.2D + 1.0E + 1.0L + 0.2S \qquad \text{(ACI Equation 9-5)}$$
$$U = 0.9D + 1.6W + 1.6H \qquad \text{(ACI Equation 9-6)}$$
$$U = 0.9D + 1.0E + 1.6H \qquad \text{(ACI Equation 9-7)}$$

---

[1]International Code Council, 2006 *International Building Code* Falls Church, Virginia 22041-3401, 755 pages.

[2]American Society of Civil Engineers, *Minimum Design Loads for Buildings and Other Structures*. ASCE 7-05 (Reston, VA: American Society of Civil Engineers), p. 5.

In the preceding expressions the following values are used:

U = the design or ultimate load the structure needs to be able to resist

D = dead load

F = loads due to the weight and pressure of fluids

T = total effects of temperature, creep, shrinkage, differential settlement and shrinkage–compensating concrete

L = live load

H = loads due to weight and lateral earth pressure of soils, groundwater pressure or pressure of bulk materials

L_r = roof live load

S = snow load

R = rain load

W = wind load

E = seismic or earthquake load effects

When impact effects need to be considered, they should be included with the live loads as per ACI Section 9.2.2. Such situations occur when those loads are quickly applied, as they are for parking garages, elevators, loading docks, and others.

The load combinations presented in ACI Equations 9-6 and 9-7 contain a $0.9D$ value. This 0.9 factor accounts for cases where larger dead loads tend to reduce the effects of other loads. One obvious example of such a situation may occur in tall buildings that are subject to lateral wind and seismic forces where overturning may be a possibility. As a result, the dead loads are reduced by 10% to take into account situations where they may have been overestimated.

The reader must realize that the sizes of the load factors do not vary in relation to the seriousness of failure. You may think that larger load factors should be used for hospitals or high-rise buildings than for cattle barns, but such is not the case. The load factors were developed on the assumption that designers would consider the seriousness of possible failure in specifying the magnitude of their service loads. Furthermore, the ACI load factors are minimum values, and designers are perfectly free to use larger factors as they desire. The magnitude of wind loads and seismic loads, however, reflect the importance of the structure. For example, in ASCE7[3] a hospital must be designed for an earthquake load 50% larger than a comparable building with less serious consequences of failure.

For some special situations, ACI Section 9.2 permits reductions in the specified load factors. These situations are as follows:

(a) In ACI Equations 9-3 to 9-5 the factor used for live loads may be reduced to 0.5 except for garages, for areas used for public assembly, and all areas where the live loads exceed 100 psf.

(b) If the design wind load has been obtained without using the wind directionality factor, the designer is permitted to use $1.3W$ instead of $1.6W$ in Equations 9-4 and 9-6.

(c) Frequently, building codes and design load references convert seismic loads to strength-level values (that is, in effect they have already been multiplied by a load factor). This is the situation assumed in ACI Equations 9-5 and 9-7. If, however, service load seismic forces are specified, it will be necessary to use $1.4E$ in these two equations.

---

[3]American Society of Civil Engineers, *Minimum Design Loads for Buildings and Other Structures.* ASCE 7-05 (Reston, VA: American Society of Civil Engineers). p. 116.

Example 4.1 presents the calculation of factored loads for a reinforced concrete column using the ACI load combinations. The largest value obtained is referred to as the critical or governing load combination and is the value to be used in design. *Notice that the values of the wind and seismic loads can be different depending on the direction of those forces, and it may be possible for the sign of those loads to be different (that is, compression or tension).* This is the situation assumed to exist in the column of this example. These rather tedious calculations can be easily handled with the Excel Spreadsheet entitled Load Combinations on this book's website: www.wiley.com/college/mccormac

EXAMPLE 4.1

The compression gravity axial loads for a building column have been estimated with the following results: $D = 150$k, live load from roof $L_r = 60$k and live loads from floors $L = 300$k. Compression wind $W = 70$k, tensile wind $W = 60$k, seismic compression load $= 50$k, and tensile seismic load $= 40$k. Determine the critical design load using the ACI load combinations.

SOLUTION

(9-1)		$U = (1.4)(150 + 0) = 210$k
(9-2)		$U = (1.2)(150 + 0 + 0) + (1.6)(300 + 0) + (0.5)(60) = 690$k
(9-3)(a)		$U = (1.2)(150) + (1.6)(60) + (1.0)(300) = 576$k
	(b)	$U = (1.2)(150) + (1.6)(60) + (0.8)(70) = 332$k
	(c)	$U = (1.2)(150) + (1.6)(60) + (0.8)(-60) = 228$k
(9-4)(a)		$U = (1.2)(150) + (1.6)(70) + (1.0)(300) + (0.5)(60) = 622$k
	(b)	$U = (1.2)(150) + (1.6)(-60) + (1.0)(300) + (0.5)(60) = 414$k
(9-5)(a)		$U = (1.2)(150) + (1.0)(50) + (1.0)(300) + (0.2)(0) = 530$k
	(b)	$U = (1.2)(150) + (1.0)(-40) + (1.0)(300) + (0.2)(0) = 440$k
(9-6)(a)		$U = (0.9)(150) + (1.6)(70) + (1.6)(0) = 247$k
		$U = (0.9)(150) + (1.6)(-60) + (1.6)(0) = 39$k
(9-7)(a)		$U = (0.9)(150) + (1.0)(50) + (1.6)(0) = 185$k
	(b)	$U = (0.9)(150) + (1.0)(-40) + (1.6)(0) = 95$k

<u>Answer:</u> Largest value $= 690$k. Notice that overturning is not a problem.

For most of the example problems presented in this textbook, in the interest of reducing the number of computations, only dead and live loads are specified. As a result, the only load factor combination usually applied herein is the one presented by ACI Equation 9-2. Occasionally, when the dead load is quite large compared to the live load, it is also necessary to consider Equation 9-1.

4.2 DESIGN OF RECTANGULAR BEAMS

Before the design of an actual beam is attempted, several miscellaneous topics need to be discussed. These include the following:

1. *Beam proportions.* Unless architectural or other requirements dictate the proportions of reinforced concrete beams, the most economical beam sections are usually obtained for shorter beams (up to 20 or 25 ft in length), when the ratio of d to b is in the range of $1\frac{1}{2}$ to 2. For longer spans, better economy is usually obtained if deep, narrow sections are used. The depths may be as large as 3 or 4 times the widths. However, today's reinforced concrete designer is often confronted with the need to keep members rather shallow to reduce floor heights. As a result, wider and shallower beams are used more frequently than in the past. You will notice that the overall beam dimensions are selected to whole inches. This is done for simplicity in constructing forms or for the rental of forms, which are usually available in 1- or 2-in. increments. Furthermore, beam widths are often selected in multiples of 2 or 3 in.

   2. *Deflections.* Considerable space is devoted in Chapter 6 to the topic of deflections in reinforced concrete members subjected to bending. However, the ACI Code provides minimum thicknesses of beams and one-way slabs for which such deflection calculations are not required. These values are shown in Table 4.1. The purpose of such limitations is to prevent deflections of such magnitudes as would interfere with the use of or cause injury to the structure. If deflections are computed for members of lesser thicknesses than those listed in the table and are found to be satisfactory, it is not necessary to abide by the thickness rules. For simply supported slabs, normal-weight concrete, and grade 60 steel, the minimum depth given when deflections are not computed equals $\ell/20$, where ℓ is the span length of the slab. *For concretes of other weights and for steels of different yield strengths, the minimum depths required by the ACI Code are somewhat revised as indicated in the footnotes to Table 4.1.* The ACI does not specify changes in the table for concretes weighing between 120 and 145 lb/ft because substitution into the correction expression given yields correction factors almost exactly equal to 1.0.

   The minimum thicknesses provided apply only to members that are not supporting or attached to partitions or other construction likely to be damaged by large deflections.

   3. *Estimated beam weight.* The weight of the beam to be selected must be included in the calculation of the bending moment to be resisted, because the beam must support itself as well as the external loads. The weight estimates for the beams selected in this text are generally very close because the authors were able to perform a little preliminary paperwork before making their estimates. You are not expected to be able to glance at a problem and give an exact estimate of the weight of the beam required. Following the same procedures as did the authors, however, you can do a little figuring on the side and make a very reasonable estimate. For instance, you could calculate the moment due to the external loads only, select a beam size, and calculate its weight. From this beam size, you should be able to make a very good estimate of the weight of the final beam section.

   Another practical method for estimating beam sizes is to assume a minimum overall depth h equal to the minimum depth specified by the ACI if deflections are not to be calculated. The ACI minimum for the beam in question may be determined by referring to Table 4.1. Then the beam width

Table 4.1 Minimum Thickness of Nonprestressed Beams or One-Way Slabs Unless Deflections Are Computed[1,2]

	Minimum thickness, h			
	Simply supported	One end continuous	Both ends continuous	Cantilever
Member	Members not supporting or attached to partitions or other construction likely to be damaged by large deflections			
Solid one-way slabs	$\ell/20$	$\ell/24$	$\ell/28$	$\ell/10$
Beams or ribbed one-way slabs	$\ell/16$	$\ell/18.5$	$\ell/21$	$\ell/8$

[1]Span length ℓ is in inches.

[2]Values given shall be used directly for members with normal-weight concrete and Grade 60 reinforcement. For other conditions, the values shall be modified as follows:

(a) For lightweight concrete having equilibrium density in the range 90 to 115 lb/ft³, the values shall be multiplied by $(1.65 - 0.005w_c)$ but not less than 1.09, where w_c is the unit weight in lb/ft³.

(b) For f_y other than 60,000 psi, the values shall be multiplied by $(0.4 + f_y/100,000)$.

can be roughly estimated equal to about one-half of the assumed value of h and the weight of this estimated beam calculated $= \frac{bh}{144}$ times the concrete weight per cubic foot. Because concrete weighs approximately 150 pcf (if the weight of steel is included), a "quick and dirty" calculation of self-weight is simply $b \times h$ because the concrete weight approximately cancels the 144 conversion factor.

After M_u is determined for all of the loads, including the estimated beam weight, the section is selected. If the dimensions of this section are significantly different from those initially assumed, it will be necessary to recalculate the weight and M_u and repeat the beam selection. At this point you may very logically ask, "What's a significant change?" Well, you must realize that we are not interested academically in how close our estimated weight is to the final weight, but rather we are extremely interested in how close our calculated M_u is to the actual M_u. In other words, our estimated weight may be considerably in error, but if it doesn't affect M_u by more than say 1% or $1\frac{1}{2}$%, forget it.

In Example 4.2 beam proportions are estimated as just described, and the dimensions so selected are taken as the final ones. As a result, you can see that it is not necessary to check the beam weight and recalculate M_u and repeat the design.

In Example 4.3 a beam is designed for which the total value of M_u (including the beam weight) has been provided, as well as a suggested steel percentage.

Finally, with Example 4.4, the authors have selected a beam whose weight is unknown. Without doubt many students initially have a little difficulty understanding how to make reasonable member weight estimates for cases such as this one. To show how easily, quickly, and accurately this may be done for beams, this example is included.

We dreamed up a beam weight estimated out of the blue equal to 400 lb/ft. (We could just as easily and successfully have made it 10 lb/ft or 1000 lb/ft.) With this value a beam section was selected and its weight calculated to equal 619 lb/ft. With this value a very good weight estimate was then made. The new section obviously would be a little larger than the first one. So we estimated the weight a little above the 619 lb/ft value, recalculated the moment, selected a new section, and determined its weight. The results were very satisfactory.

4. *Selection of bars.* After the required reinforcing area is calculated, Appendix Table A.4 is used to select bars that provide the necessary area. For the usual situations, bars of sizes #11 and smaller are practical. It is usually convenient to use bars of one size only in a beam, although occasionally two sizes will be used. Bars for compression steel and stirrups are usually a different size, however. Otherwise the workmen may become confused.

5. *Cover.* The reinforcing for concrete members must be protected from the surrounding environment; that is, fire and corrosion protection need to be provided. To do this the reinforcing is located at certain minimum distances from the surface of the concrete so that a protective layer of concrete, called *cover*, is provided. In addition, the cover improves the bond between the concrete and the steel. In Section 7.7 of the ACI Code, specified cover is given for reinforcing bars under different conditions. Values are given for reinforced concrete beams, columns, and slabs, for cast-in-place members, for precast members, for prestressed members, for members exposed to earth and weather, for members not so exposed, and so on. The concrete for members that are to be exposed to deicing salts, brackish water, seawater, or spray from these sources must be especially proportioned to satisfy the exposure requirements of Chapter 4 of the Code. These requirements pertain to air entrainment, water–cement ratios, cement types, concrete strength, and so on.

The beams designed in Examples 4.2, 4.3, and 4.4 are assumed to be located inside a building and thus protected from the weather. For this case the Code requires a minimum cover of $1\frac{1}{2}$ in. of concrete outside of any reinforcement.

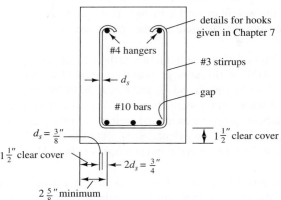

Minimum edge distance = cover + d_s + $2d_s$
= $1.50 + \frac{3}{8} + (2)\left(\frac{3}{8}\right) = 2\frac{5}{8}$

Figure 4.1 Determining minimum edge distance.

In Chapter 8 you will learn that vertical stirrups are used in most beams for shear reinforcing. A sketch of a stirrup is shown in the beam of Figure 4.1. The minimum stirrup diameter (d_s) which the Code permits us to use is $\frac{3}{8}$ in., when the longitudinal bars are #10 or smaller, while for #11 and larger bars the minimum stirrup diameter is $\frac{1}{2}$ in. The minimum inside radius of the 90-degree stirrup bent around the outside longitudinal bars is two times the stirrup diameter ($2d_s$). As a result, when the longitudinal bars are #14 or smaller there will be a gap between the bars and the stirrups, as shown in the figure. This is based on the assumption that each outside longitudinal bar is centered over the horizontal point of tangency of the stirrup corner bend. For #18 bars, however, the half-bar diameter is larger than $2d_s$ and controls.

For the beam of Figure 4.1 it is assumed that 1.50 in. clear cover, #3 stirrups, and #10 longitudinal bars are used. The minimum horizontal distance from the center of the outside longitudinal bars to the edge of the concrete can be determined as follows:

$$\text{Minimum edge distance} = \text{cover} + d_s + 2d_s = 1.50 + \frac{3}{8} + (2)\left(\frac{3}{8}\right) = 2\frac{5}{8} \text{ in.}$$

The minimum cover required for concrete cast against earth, as in a footing, is 3 in., and for concrete cast not against the earth but later exposed to it, as by backfill, 2 in. Precast and prestressed concrete or other concrete cast under plant control conditions requires less cover, as described in Sections 7.7.2 and 7.7.3 of the ACI Code.

Notice the two #4 bars called *hangers* placed in the compression side of this beam. Their purpose is to provide support for the stirrups and to hold the stirrups in position.

If concrete members are exposed to very harsh surroundings, such as deicing salts, smoke, or acid vapors, the cover should be increased above these minimums.

6. *Minimum spacing of bars.* The Code (7.6) states that the clear distance between parallel bars cannot be less than 1 in.[*] or less than the nominal bar diameter. If the bars are placed in more than one layer, those in the upper layers are required to be placed directly over the ones in the lower layers and the clear distance between the layers must be not less than 1 in.

---

[*]25 mm in SI.

Reinforcing bars. Note the supporting metal chairs. (Courtesy of Alabama Metal Industries Corporation.)

A major purpose of these requirements is to enable the concrete to pass between the bars. The ACI Code further relates the spacing of the bars to the maximum aggregate sizes for the same purpose. In Code Section 3.3.2, maximum permissible aggregate sizes are limited to the smallest of (a) one-fifth of the narrowest distance between side forms, (b) one-third of slab depths, and (c) three-fourths of the minimum clear spacing between bars.

A reinforcing bar must extend an appreciable length in both directions from its point of highest stress in order to develop its stress by bonding to the concrete. The shortest length in which a bar's stress can be increased from 0 to f_y is called its *development length*.

If the distance from the end of a bar to a point where it theoretically has a stress equal to f_y is less than its required development length, the bar may very well pull loose from the concrete. Development lengths are discussed in detail in Chapter 7. There we will learn that required development lengths for reinforcing bars vary appreciably with their spacings and their cover. *As a result, it is sometimes wise to use greater cover and larger bar spacings than the specified minimum values in order to reduce development lengths.*

When selecting the actual bar spacing, the designer will comply with the preceding code requirements and, in addition, will give spacings and other dimensions in inches and fractions, not in decimals. The workers in the field are accustomed to working with fractions and would be confused by a spacing of bars such as 3 at 1.45 in. The designer should always strive for simple spacings, for such dimensions will lead to better economy.

Each time a beam is designed, it is necessary to select the spacing and arrangement of the bars. To simplify these calculations, Appendix Table A.5 is given. Corresponding information is provided in SI units in Appendix Table B.5. These tables show the minimum beam widths required for different numbers of bars. The values given are based on the assumptions that $\frac{3}{8}$-in. stirrups and $1\frac{1}{2}$-in. cover are required except for #18 bars where the stirrup diameter is $\frac{1}{2}$ in. If three #10 bars are required, it can be seen from the table that a minimum beam width of 10.4 in. (say 11 in.) is required.

This value can be checked as follows, noting that $2d_s$ is the radius of bend of the bar and the minimum clear spacing between bars in this case is d_b:

$$\text{Minimum beam width} = \text{cover} + d_s + 2d_s + \frac{d_b}{2} + d_b + d_b + d_b + \frac{d_b}{2} + 2d_s + d_s + \text{cover}$$

$$= 1.50 + \frac{3}{8} + (2)\left(\frac{3}{8}\right) + \frac{1.27}{2} + (3)(1.27) + \frac{1.27}{2}$$

$$+ (2)\left(\frac{3}{8}\right) + \frac{3}{8} + 1.50$$

$$= 10.33 \text{ in. rounded to } 10.4 \text{ in.}$$

4.3   BEAM DESIGN EXAMPLES

Example 4.2 illustrates the design of a simple span rectangular beam. For this introductory example, approximate dimensions are assumed for the beam cross section. The depth h is assumed to equal about one-tenth of the beam span, while its width b is assumed to equal about $1/2\ h$. Next the percentage of reinforcing needed is determined with the equation derived in Section 3.4, and reinforcing bars are selected to satisfy that percentage. Finally, ϕM_n is calculated for the final design.

EXAMPLE 4.2

Design a rectangular beam for a 22-ft simple span if a dead load of 1 k/ft (not including the beam weight) and a live load of 2 k/ft are to be supported. Use $f_c' = 4000$ psi and $f_y = 60{,}000$ psi.

SOLUTION   **Estimating Beam Dimensions and Weight**

$$\text{Assume } h = (0.10)(22) = 2.2 \text{ ft} \quad \underline{\text{Say 27 in.}} \quad (d = 24.5 \text{ in.})$$

$$\text{Assume } b = \frac{1}{2}h = \frac{27}{2} \quad \underline{\text{Say 14 in.}}$$

$$\text{Beam } wt = \frac{(14)(27)}{144}(150) = 394 \text{ lb/ft}$$

Computing w_u and M_u

$$w_u = (1.2)(1 + 0.394) + (1.6)(2) = 4.873 \text{ k/ft}$$

$$M_u = \frac{w_u L^2}{8} = \frac{(4.873)(22)^2}{8} = 294.8 \text{ ft-k}$$

Assuming $\phi = 0.90$ and computing ρ with the following expression which was derived in Section 3.4.

$$\rho = \frac{0.85 f_c'}{f_y}\left(1 - \sqrt{1 - \frac{2R_n}{0.85 f_c'}}\right)$$

$$R_n = \frac{M_u}{\phi bd^2} = \frac{(12)(294,800)}{(0.90)(14)(24.5)^2} = 467.7$$

$$\rho = \frac{(0.85)(4000)}{60,000}\left(1-\sqrt{1-\frac{(2)(467.7)}{(0.85)(4000)}}\right) = 0.00842$$

Selecting Reinforcing

$$A_s = \rho bd = (0.00842)(14)(24.5) = 2.89 \text{ in.}^2$$

Use three #9 bars $(A_s = 3.00 \text{ in.}^2)$

Appendix A.5 indicates a minimum beam width of 9.8 in. for interior exposure for three #9 bars. If five #7 bars had been selected, a minimum width of 12.8 in. would be required. Either choice would be acceptable since the beam width of 14 in. exceeds either requirement. If we had selected a beam width of 12 in. earlier in the design process, we may have been limited to the larger #9 bars because of this minimum beam width requirement.

Checking Solution

$$\rho = \frac{A_s}{bd} = \frac{3.00}{(14)(24.5)} = 0.00875 > \rho_{\min} = 0.0033$$

$$< \rho_{\max} = 0.0181 (\rho \text{ values from Appendix Table A.7}). \quad \therefore \underline{\text{section is ductile and } \phi = 0.90.}$$

$$a = \frac{A_s f_y}{0.85 f'_c b} = \frac{(3.00)(60)}{(0.85)(4)(14)} = 3.78 \text{ in.}$$

$$\phi M_n = \phi A_s f_y \left(d - \frac{a}{2}\right) = (0.90)(3.00)(60)\left(24.5 - \frac{3.78}{2}\right)$$

$$= 3662 \text{ in.-k} = 305.2 \text{ ft-k} > 294.8 \text{ ft-k} \quad \underline{\underline{\text{OK}}}$$

Final Section (Figure 4.2)

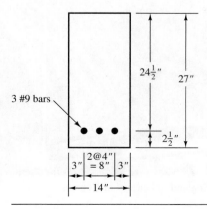

3 #9 bars

$24\frac{1}{2}''$ $27''$

$2\frac{1}{2}''$

$3''$ $\begin{array}{c}2@4''\\=8''\end{array}$ $3''$

$14''$

Figure 4.2

Use of Graphs and Tables

In Section 3.4 the following equation was derived:

$$M_u = \phi A_s f_y d \left(1 - \frac{1}{1.7}\frac{\rho f_y}{f_c'}\right)$$

If A_s in this equation is replaced with ρbd, the resulting expression can be solved for $M_u/\phi bd^2$.

$$M_u = \phi \rho bd f_y d \left(1 - \frac{1}{1.7}\frac{\rho f_y}{f_c'}\right)$$

and from this

$$\frac{M_u}{\phi bd^2} = \rho f_y \left(1 - \frac{1}{1.7}\frac{\rho f_y}{f_c'}\right)$$

For a given steel percentage ρ and for a certain concrete f_c' and certain steel f_y, the value of $M_u/\phi bd^2$ can be calculated and plotted in tables, as is illustrated in Appendix Tables A.8 through A.13 or in graphs (see Graph 1 of Appendix A). SI values are provided in Appendix Tables B.8 to B.9. It is much easier to accurately read the tables than the graphs (at least to the scale to which the graphs are shown in this text). For this reason the tables are used for the examples here. The units for $M_u/\phi bd^2$ in both the tables and the graphs of Appendix A are pounds per square inch. In Appendix B, the units are MPa.

Once $M_u/\phi bd^2$ is determined for a particular beam, the value of M_u can be calculated as illustrated in the alternate solution for Example 3.1. The same tables and graphs can be used for either the design or analysis of beams.

The value of ρ, determined in Example 4.2 by substituting into that long and tedious equation, can be directly selected from Appendix Table A.13. We enter that table with the $\frac{M_u}{\phi bd^2}$ value previously calculated in the example, and we read a value of ρ between 0.0084 and 0.0085. Interpolation can be used to find the actual value of 0.00842, but such accuracy is not really necessary. It is conservative to use the higher value (0.0085) to calculate the steel area.

Barnes Meadow Interchange, Northampton, England. (Courtesy of Cement and Concrete Association.)

In Example 4.3, which follows, a value of ρ was specified in the problem statement, and the long equation was used to determine the required dimensions of the structure as represented by bd^2. Again it is much easier to use the appropriate appendix table to determine this value. In nearly every other case herein in this textbook the tables are used for design or analysis purposes.

Once bd^2 is determined, the author takes what seem to him to be reasonable values for b (in this case 12, 14, and 16 in.) and computes the required d for each width so that the required bd^2 is satisfied. Finally, a section is selected in which b is roughly $\frac{1}{2}$ to $\frac{2}{3}$ of d. (For long spans d may be $2\frac{1}{2}$, or 3 or more times b for economical reasons.)

EXAMPLE 4.3

A beam is to be selected with $\rho = 0.0120$, $M_u = 600$ ft-k, $f_y = 60,000$ psi and $f_c' = 4000$ psi.

SOLUTION   Assuming $\phi = 0.90$ and substituting into the following equation from Section 3.4:

$$\frac{M_u}{\phi bd^2} = \rho f_y \left(1 - \frac{1}{1.7}\frac{\rho f_y}{f_c'}\right)$$

$$\frac{(12)(600,000)}{(0.9)(bd^2)} = (0.0120)(60,000)\left[1 - \left(\frac{1}{1.7}\right)\frac{(0.0120)(60,000)}{4000}\right]$$

$$bd^2 = 12,427 \begin{cases} b \times d \\ 12 \times 32.18 \\ 14 \times 29.79 \leftarrow \\ 16 \times 27.87 \end{cases} \begin{cases} \text{This one seems} \\ \text{pretty reasonable} \\ \text{to the authors.} \end{cases}$$

Note: Alternatively, we could have used tables to help calculate bd^2. Upon entering Appendix Table A.13, we find $\frac{M_u}{\phi bd^2} = 643.5$ when $\rho = 0.0120$.

$$\therefore bd^2 = \frac{(12)(600,000)}{(0.90)(643.5)} = 12,432 \quad \underline{\underline{OK}}$$

Try $14 \times 33 (d = 30.00 \text{ in.})$

$$A_s = \rho bd = (0.0120)(14)(30) = 5.04 \text{ in.}^2$$

$$\text{use } 4\#10(A_s = 5.06 \text{ in.}^2)$$

Note: Appendix A.5 indicates a minimum beam width of 12.9 in. for this bar selection. Since our width is 14 in. the bars will fit.

Checking Solution

$$\rho = \frac{A_s}{bd} = \frac{5.06}{(14)(30)} = 0.01205 > \rho_{min} = 0.0033$$

$$< \rho_{max} = 0.0181 \text{ from Appendix Table A.7}$$

Note: A value of 0.0206 is permitted by the Code, but the corresponding value of ϕ would be less than 0.9 (See Fig. 3.5 and Table A.7). Since a value of ϕ of 0.9 was used in the above calculations, it is necessary to use 0.0181.

With $\rho = 0.01205$, $\frac{M_u}{\phi bd^2}$ by interpolation from Appendix Table A.13 equals 645.85.

$$\phi M_n = 645.85 \, \phi bd^2 = (645.85)(0.9)(14)(30)^2$$

$$= 7,323,939 \text{ in.-lb} = 610.3 \text{ ft-k} > 600 \text{ ft-k}$$

Final Section (Figure 4.3)

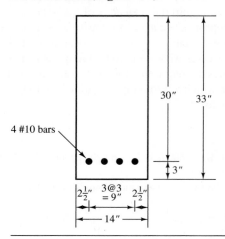

Figure 4.3

Through quite a few decades of reinforced concrete design experience, it has been found that if steel percentages are kept fairly small, say roughly $0.18f_c'/f_y$ or perhaps $0.375\rho_b$, beam cross sections will be sufficiently large so that deflections will seldom be a problem. As the areas of steel required will be fairly small, there will be little problem fitting them into beams without crowding.

If these relatively small percentages of steel are used, there will be little difficulty in placing the bars and in getting the concrete between them. Of course, from the standpoint of deflection, higher percentages of steel, and thus smaller beams, can be used for short spans where deflections present no problem. Whatever steel percentages are used, the resulting members will have to be carefully checked for deflections, particularly for long-span beams, cantilever beams, and shallow beams and slabs. Of course, such deflection checks are not required if the minimum depths specified in Table 4.1 of this chapter are met.

Another reason for using smaller percentages of steel is given in ACI Section 8.4 where a plastic redistribution of moments (a subject to be discussed in Chapter 14) is permitted in continuous members whose ϵ_t values are 0.0075 or greater. Such tensile strains will occur when smaller percentages of steel are used. For the several reasons mentioned here, structural designers believe that keeping steel percentages fairly low will result in good economy.

EXAMPLE 4.4

A rectangular beam is to be sized with $f_y = 60,000$ psi, $f_c' = 3000$ psi, and a ρ equal to $0.18f_c'/f_y$. It is to have a 25-ft simple span and to support a dead load in addition to its own weight equal to 2 k/ft and a live load equal to 3 k/ft.

SOLUTION   **Assume Beam** $wt = 400$ **lb/ft**

$$w_u = (1.2)(2 + 0.400) + (1.6)(3) = 7.68 \text{ k/ft}$$

$$M_u = \frac{(7.68)(25)^2}{8} = 600 \text{ ft-k}$$

$$\rho = \frac{(0.18)(3)}{60} = 0.009$$

$$\frac{M_u}{\phi b d^2} = 482.6 \text{ from Appendix Table A.12}$$

$$bd^2 = \frac{M_u}{\phi \, 482.6} = \frac{(12)(600,000)}{(0.9)(482.6)}$$

Solving this expression for bd^2 and trying varying values of b and d.

$$b \times d$$

$$= 16{,}577 \begin{cases} 16 \times 32.19 \\ 18 \times 30.35 \quad \leftarrow \quad \text{seems reasonable} \\ 20 \times 28.79 \end{cases}$$

Try 18 × 33 Beam ($d = 30.50$ in.)

$$\text{Bm wt} = \frac{(18)(33)}{144}(150) = 619 \text{ lb/f}$$

$$> \text{the estimated 400 lb/ft} \qquad\qquad \text{No good}$$

Assume Beam wt a Little Higher than 619 lb/ft

$$\text{Estimate wt} = 650 \text{ lbs/ft}$$

$$w_u = (1.2)(2 + 0.650) + (1.6)(3) = 7.98 \text{ k/ft}$$

$$M_u = \frac{(7.98)(25)^2}{8} = 623.4 \text{ ft-k}$$

$$bd^2 = \frac{M_u}{\phi\, 482.6} = \frac{(12)(623{,}400)}{(0.9)(482.6)}$$

$$= 17{,}223 \begin{cases} 16 \times 32.81 \\ 18 \times 30.93 \quad \leftarrow \quad \text{seems reasonable} \\ 20 \times 29.35 \end{cases}$$

Try 18 × 34 Beam ($d = 31.00$ in.)

$$\text{Bm wt} = \frac{(18)(34)}{144}(150) = 637.5 \text{ lbs/ft} < 650 \text{ lbs/ft} \qquad \text{OK}$$

$$A_s = \rho bd = (0.009)(18)(31) = 5.02 \text{ in.}^2$$

Try five #9 bars (minimum width is 14.3 in. from Appendix A.5) OK

Normally a bar selection should exceed the theoretical value of A_s. In this case, the area chosen was less than, but very close to, the theoretical area, and it will be checked to be sure it has enough capacity.

$$a = \frac{A_s f_y}{0.85 f'_c b} = \frac{5.00(60)}{(0.85)(3)(18)} = 6.54 \text{ in.}$$

$$\phi M_n = \phi A_s f_y \left(d - \frac{a}{2}\right) = 0.9(5.00)(60)\left[31 - \frac{6.54}{2}\right] = 7487.6 \text{ in.-lb} = 623.9 \text{ ft-k} > M_u$$

The reason a beam with less reinforcing steel than calculated is acceptable is because a value of d exceeding the theoretical value was selected ($d = 31$ in. > 30.93 in.) Whenever the value of b and d selected results in a bd^2 that exceeds the calculated value based on the assumed ρ, the actual value of ρ will be lower than the assumed value.

If a value of $b = 18$ in. and $d = 30$ in. had been selected, the result would have been that the actual value of ρ would be greater than the assumed value of 0.009. Using the actual values of b and d to recalculate ρ

$$\frac{M_u}{\phi bd^2} = \frac{(12)(623{,}400)}{0.9(18)(30)^2} = 513.1$$

From Appendix Table A.12, $\rho = 0.00965$ which exceeds the assumed value of 0.009. The required value of A_s will be larger than that required for $d = 31$ in.

$$A_s = \rho b d = (0.00965)(18)(30) = 5.21 \text{ in.}^2$$

Either design is acceptable. This kind of flexibility is sometimes perplexing to the student who simply wants to know the right answer. One of the best features of reinforced concrete is that there is so much flexibility in the choices that can be made.

4.4   MISCELLANEOUS BEAM CONSIDERATIONS

This section introduces two general limitations relating to beam design: lateral bracing and deep beams.

Lateral Support

It is unlikely that laterally unbraced reinforced concrete beams of any normal proportions will buckle laterally, even if they are deep and narrow, unless they are subject to appreciable lateral torsion. As a result, the ACI Code (10.4.1) states that lateral bracing for a beam is not required closer than 50 times the least width b of the compression flange or face. Should appreciable torsion be present, however, it must be considered in determining the maximum spacing for lateral support.

Skin Reinforcement for Deep Beams

Beams with web depths that exceed 3 ft have a tendency to develop excessively wide cracks in the upper parts of their tension zones. To reduce these cracks, it is necessary to add some additional longitudinal reinforcing in the zone of flexural tension near the vertical side faces of their webs, as shown in Figure 4.4. The Code (10.6.7) states that additional skin reinforcement must be uniformly distributed along both side faces of members with $d > 36$ in. for distances equal to $d/2$ nearest the flexural reinforcing.

The spacing s_{sk} between this skin reinforcement must not be greater than the least of $d/6$, 12 in., or $1000A_b/(d-30)$. In this latter expression A_b is the area of a single bar or wire. These additional bars may be used in computing the bending strengths of members only if appropriate strains for their positions relative to neutral axes are used to determine bar stresses. The total area of the skin reinforcement in both side faces of the beam does not have to exceed one-half of the

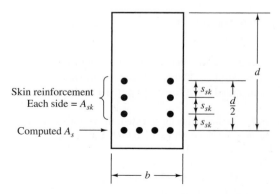

Figure 4.4 Skin reinforcement for deep beams with $d > 36$ in., as required by ACI Code Section 10.6.7.

required bending tensile reinforcement in the beam. The ACI does not specify the actual area of skin reinforcing; it merely states that some additional reinforcement should be placed near the vertical faces of the tension zone to prevent cracking in the beam webs.

Some special requirements must be considered relating to shear in deep beams, as described in the ACI Code (11.7) and in Section 8.14 of this text. Should these latter provisions require more reinforcing than required by ACI Section 10.6.7, the larger values will govern.

> For a beam designed in SI units with an effective depth > 1 m, additional skin reinforcement must be determined with the following expression in which A_{sk} is the area of skin reinforcement per meter of height on each side of the beam:
> Its maximum spacing may not exceed $d/6$ on 300 mm or $1000A_b/(d-750)$.

Other Items

The next four chapters of this book are devoted to several other important items relating to beams. These include different shaped beams, compression reinforcing, cracks, bar development lengths, and shear.

Further Notes on Beam Sizes

From the standpoints of economy and appearance, only a few different sizes of beams should be used in a particular floor system. Such a practice will save appreciable amounts of money by simplifying the formwork and at the same time will provide a floor system that has a more uniform and attractive appearance.

If a group of college students studying the subject of reinforced concrete were to design a floor system and then compare their work with a design of the same floor system made by an experienced structural designer, the odds are that the major difference between the two designs would be in the number of beam sizes. The practicing designer would probably use only a few different sizes, whereas the average student would probably use a larger number.

The designer would probably examine the building layout to decide where to place the beams and then would make the beam subject to the largest bending moment as small as practically possible (that is, with a fairly high percentage of reinforcing). Then he or she would proportion as many as possible of the other similar beams with the same outside dimensions. The reinforcing percentages of these latter beams might vary quite a bit because of their different moments.

4.5   DETERMINING STEEL AREA WHEN BEAM DIMENSIONS ARE PREDETERMINED

Sometimes the external dimensions of a beam are predetermined by factors other than moments and shears. The depth of a member may have been selected on the basis of the minimum thickness requirements discussed in Section 4.2 for deflections. The size of a whole group of beams may have been selected to simplify the formwork as discussed in Section 4.4. Finally, a specific size may have been chosen for architectural reasons. Next we briefly mention three methods for computing the reinforcing required. Example 4.5 illustrates the application of each of these methods.

Appendix Tables

The value of $M_u/\phi bd^2$ can be computed, and ρ can be selected from the tables. For most situations this is the quickest and most practical method. *The tables given in Appendices A and B of this text apply only to tensilely reinforced rectangular sections. Furthermore, we must remember to check ϕ values.*

Use of ρ Formula

The following equation was previously developed in Section 3.4 for rectangular sections.

$$\rho = \frac{0.85f_c'}{f_y}\left(1 - \sqrt{1 - \frac{2R_n}{0.85f_c'}}\right)$$

Trial-and-Error Method

A value of a can be assumed, the value of A_s computed, the value of a determined for that value of A_s, another value of a assumed, and so on. Alternatively, a value of the lever arm from C to T (it's $d - a/2$ for rectangular sections) can be estimated and used in the trial-and-error procedure. *This method is a general one that will work for all cross sections with tensile reinforcing.* It is particularly useful for T beams, as will be illustrated in the next chapter.

EXAMPLE 4.5

The dimensions of the beam shown in Figure 4.5 have been selected for architectural reasons. Determine the reinforcing steel area by each of the methods described in this section.

SOLUTION   **Using Appendix Tables**

$$\frac{M_u}{\phi bd^2} = \frac{(12)(160,000)}{(0.9)(16)(21)^2} = 302.3 \text{ psi}$$

$$\rho \text{ from Table A.12} = 0.0054$$

$$A_s = (0.0054)(16)(21) = 1.81 \text{ in.}^2$$

$$\text{Use 3 \#7 bars}(1.80 \text{ in.}^2)$$

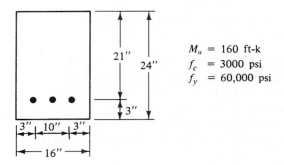

$$M_u = 160 \text{ ft-k}$$
$$f_c = 3000 \text{ psi}$$
$$f_y = 60,000 \text{ psi}$$

Figure 4.5

Using ρ Formula

$$R_R = \frac{M_u}{\phi b d^2} = 302.3$$

$$\rho = \frac{(0.85)(3000)}{60,000}\left(1 - \sqrt{1 - \frac{(2)(302.3)}{(0.85)(3000)}}\right)$$

$$= 0.0054$$

Trial-and-Error Method

Here it is necessary to estimate the value of a. The student probably has no idea of a reasonable value for this quantity, but the accuracy of the estimate is not a matter of importance. He or she can assume some value probably considerably less than $\frac{d}{2}$ and then compute $d - \frac{a}{2}$ and A_s. With this value of A_s, a new value of a can be computed and the cycle repeated. After two or three cycles, a very good value of a will be obtained.

Assume $a = 2''$:

$$A_s = \frac{M_u}{\phi f_y\left(d - \frac{a}{2}\right)} = \frac{(12)(160,000)}{(0.9)(60,000)\left(21 - \frac{2}{2}\right)} = 1.78 \text{ in.}^2$$

$$a = \frac{A_s f_y}{0.85 f'_c} = \frac{(1.78)(60,000)}{(0.85)(3000)(16)} = 2.62''$$

Assume $a = 2.6''$:

$$A_s = \frac{(12)(160,000)}{(0.9)(60,000)\left(21 - \frac{2.62}{2}\right)} = 1.81 \text{ in.}^2$$

$$a = \frac{(1.81)(60,000)}{(0.85)(3000)(16)} = 2.66 \text{ in. } \underline{\text{(close enough)}}$$

4.6   BUNDLED BARS

Sometimes when large amounts of steel reinforcing are required in a beam or column, it is very difficult to fit all the bars in the cross section. For such situations, groups of parallel bars may be bundled together. Up to four bars can be bundled, provided they are enclosed by stirrups or ties. The ACI Code (7.6.6.3) states that bars larger than #11 shall not be bundled in beams or girders. This is primarily because of crack control problems, a subject discussed in Chapter 6 of this text. That is, if the ACI crack control provisions are to be met, bars larger than #11 cannot practically be used. The AASHTO permits the use of 2-, 3-, and 4-bar bundles for bars up through the #11 size. For bars larger than #11, however, AASHTO limits the bundles to two bars (AASHTO Sections 8.21.5 ASD and 5.10.3.1.5 strength design).

Typical configurations for 2-, 3-, and 4-bar bundles are shown in Figure 4.6. When bundles of more than one bar deep vertically are used in the plane of bending, they may not practically be hooked or bent as a unit. If end hooks are required, it is preferable to stagger the hooks of the individual bars within the bundle.

Although the ACI permits the use of bundled bars, their use in the tension areas of beams may very well be counterproductive because of the other applicable code restrictions that are brought into play as a result of their use.

Figure 4.6 Bundled-bar arrangements.

When spacing limitations and cover requirements are based on bar sizes, the bundled bars may be treated as a single bar for computation purposes; the diameter of the fictitious bar is to be calculated from the total equivalent area of the group. When individual bars in a bundle are cut off within the span of beams or girders, they should terminate at different points. The Code (7.6.6.4) requires that there be a stagger of at least 40 bar diameters.

4.7 ONE-WAY SLABS

Reinforced concrete slabs are large flat plates that are supported by reinforced concrete beams, walls, or columns, by masonry walls, by structural steel beams or columns, or by the ground. If they are supported on two opposite sides only, they are referred to as *one-way slabs* because the bending is in one direction only—that is, perpendicular to the supported edges. Should the slab be supported by beams on all four edges, it is referred to as a *two-way slab* because the bending is in both directions. Actually, if a rectangular slab is supported on all four sides, but the long side is two or more times as long as the short side, the slab will, for all practical purposes, act as a one-way slab, with bending primarily occurring in the short direction. Such slabs are designed as one-way slabs. You can easily verify these bending moment ideas by supporting a sheet of paper on two opposite sides or on four sides with the support situation described. This section is concerned with one-way slabs; two-way slabs are considered in Chapters 16 and 17. It should be realized that a large percentage of reinforced concrete slabs fall into the one-way class.

A one-way slab is assumed to be a rectangular beam with a large ratio of width to depth. Normally, a 12-in.-wide piece of such a slab is designed as a beam (see Figure 4.7), the slab being assumed to consist of a series of such beams side by side. The method of analysis is somewhat conservative due to the lateral restraint provided by the adjacent parts of the slab. Normally, a beam will tend to expand laterally somewhat as it bends, but this tendency to expand by each of the 12-in. strips is resisted by the adjacent 12-in.-wide strips, which tend to expand also. In other words, Poisson's ratio is assumed to be zero. Actually, the lateral expansion tendency results in a very slight stiffening of the beam strips, which is neglected in the design procedure used here.

The 12-in.-wide beam is quite convenient when thinking of the load calculations because loads are normally specified as so many pounds per square foot, and thus the load carried per foot of length of the 12-in.-wide beam is the load supported per square foot by the slab. The load supported by the one-way slab, including its own weight, is transferred to the members supporting the edges of the slab. Obviously, the reinforcing for flexure is placed perpendicular to these supports—that is, parallel to the long direction of the 12-in.-wide beams. This flexural reinforcing may not be spaced farther on center than three times the slab thickness, or 18 in., according to the

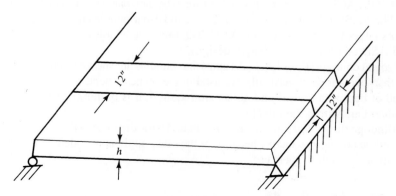

Figure 4.7 A 12-in. strip in a simply supported one-way slab.

ACI (7.6.5). Of course, there will be some reinforcing placed in the other direction to resist shrinkage and temperature stresses.

The thickness required for a particular one-way slab depends on the bending, the deflection, and shear requirements. As described in Section 4.2, the ACI Code (9.5.2.1) provides certain span/depth limitations for concrete flexural members where deflections are not calculated.

Because of the quantities of concrete involved in floor slabs, their depths are rounded off to closer values than are used for beam depths. Slab thicknesses are usually rounded off to the nearest $\frac{1}{4}$ in. on the high side for slabs of 6 in. or less in thickness and to the nearest $\frac{1}{2}$ in. on the high side for slabs thicker than 6 in.

As concrete hardens, it shrinks. In addition, temperature changes occur that cause expansion and contraction of the concrete. When cooling occurs, the shrinkage effect and the shortening due to cooling add together. The Code (7.12) states that shrinkage and temperature reinforcement must be provided in a direction perpendicular to the main reinforcement for one-way slabs. (For two-way slabs, reinforcement is provided in both directions for bending.) The Code states that for grades 40 or 50 deformed bars, the minimum percentage of this steel is 0.002 times the gross cross-sectional area of the slab. Notice that the gross cross-sectional area is bh (where h is the slab thickness). The Code (7.12.2.2) states that shrinkage and temperature reinforcement may not be spaced farther apart than five times the slab thickness, or 18 in. When grade 60 deformed bars or welded wire fabric are used, the minimum area is $0.0018bh$. For slabs with $f_y > 60,000$ psi, the minimum value is $\frac{(0.0018 \times 60,000)}{f_y} \geq 0.0014$.

> In SI units, the minimum percentages of reinforcing are 0.002 for grades 300 and 350 steels and 0.0018 for grade 420 steel. When $f_y > 420$ MPa, the minimum percent equals $\frac{(0.0018 \times 420)}{f_y}$. The reinforcing may not be spaced farther apart than five times the slab thickness, or 500 mm.

Should structural walls or large columns provide appreciable resistance to shrinkage and temperature movements, it may very well be necessary to increase the minimum amounts listed.

Shrinkage and temperature steel serves as mat steel in that it is tied perpendicular to the main flexural reinforcing and holds it firmly in place as a mat. This steel also helps to distribute concentrated loads transversely in the slab. (In a similar manner, the AASHTO gives minimum permissible amounts of reinforcing in slabs transverse to the main flexural reinforcing for lateral distribution of wheel loads.)

Areas of steel are often determined for 1-ft widths of reinforced concrete slabs, footings, and walls. A table of "areas of bars in slabs" such as Table A.6 in the Appendix is very useful in such cases for selecting the specific bars to be used. A brief explanation of the preparation of this table is provided here.

For a 1-ft width of concrete, the total steel area obviously equals the total or average number of bars in a 1-ft width times the cross-sectional area of one bar. This can be expressed as (12 in./bar spacing c. to c.)(area of one bar). Some examples follow, and the values obtained can be checked in the table. Understanding these calculations enables one to expand the table as desired.

1. #9 bars, 6-in. o.c. Total area in 1-ft width $= \left(\frac{12}{6}\right)(1.00) = 2.00$ in.^{2}
2. #9 bars, 5-in. o.c. Total area in 1-ft width $= \left(\frac{12}{5}\right)(1.00) = 2.40$ in.^{2}

Example 4.6 illustrates the design of a one-way slab. It will be noted that the Code (7.7.1.c) cover requirement for reinforcement in slabs (#11 and smaller bars) is $\frac{3}{4}$ in. clear, unless corrosion or fire protection requirements are more severe.

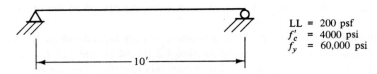

LL = 200 psf
f'_c = 4000 psi
f_y = 60,000 psi

Figure 4.8

EXAMPLE 4.6

Design a one-way slab for the inside of a building using the span, loads, and other data given in Figure 4.8. Normal-weight aggregate concrete is specified with a density of 145 pcf.

SOLUTION   **Minimum Total Slab Thickness h if Deflections Are Not Computed (See Table 4.7)**

$$h = \frac{\ell}{20} = \frac{(12)(10)}{20} = 6''$$

Assume 6-in. slab (with d = approximately $6'' - \frac{3}{4}'$ Cover $-\frac{1}{4}'$ for estimated half-diameter of bar size = 5.0''). The moment is calculated, and then the amount of steel required is determined. If this value seems unreasonable, a different thickness is tried.

Design a 12-in.-wide strip of the slab. Usually 5 pcf is added to account for the weight of reinforcement, so 150 pcf is used in calculating the weight of a normal-weight concrete member.

$$DL = \text{slab wt} = \left(\frac{6}{12}\right)(150) = 75 \text{ psf}$$

$$LL = 200 \text{ psf}$$

$$w_u = (1.2)(75) + (1.6)(200) = 410 \text{ psf}$$

$$M_u = \frac{(0.410)(10)^2}{8} = 5.125 \text{ ft-k}$$

$$\frac{M_u}{\phi bd^2} = \frac{(12)(5125)}{(0.9)(12)(5.00)^2} = 227.8 \text{ psi}$$

$$\rho = 0.00393 \text{ (from Appendix Table A.13)}$$

$$> \rho_{\min} = 0.0033$$

$$A_s = (0.00393)(12)(5.0) = 0.236 \text{ in.}^2/\text{ft}$$

Use #4 @ 10'' from Appendix Table A.6 ($A_s = 0.24$ in^2/ft)

spacing < maximum of 18 in. as per ACI 7.6.5

Transverse Direction—Shrinkage and Temperature Steel

$$A_s = 0.0018bh = (0.0018)(12)(6) = 0.1296 \text{ in.}^2/\text{ft}$$

Use #3 @ 10''(0.13 in.^{2}/ft) as selected from Appendix Table A.6

Spacing < maximum of 18 in. as per ACI 7.12.2.2   **OK**

The #4 bars are placed below the #3 bars in this case. The #4 bars are the primary flexural reinforcing, and the value of d is based on this assumption. The #3 bars are for temperature and shrinkage control, and their depth within the slab is not as critical.

The designers of reinforced concrete structures must be very careful to comply with building code requirements for fire resistance. If the applicable code requires a certain fire resistance rating for floor systems, that requirement may very well cause the designer to use thicker slabs than might otherwise be required to meet the ACI strength design requirements. *In other words, the designer of a building should study carefully the fire resistance provisions of the governing building code before proceeding with the design.* Section 7.7.8 of ACI 318-08 includes such a requirement.

4.8   CANTILEVER BEAMS AND CONTINUOUS BEAMS

Cantilever beams supporting gravity loads are subject to negative moments throughout their lengths. As a result, their reinforcement is placed in their top or tensile sides, as shown in Figures 4.9 and 4.10(a). The reader will note that for such members the maximum moments occur at the faces of the fixed supports. As a result, the largest amounts of reinforcing are required at those points. You should also note that the bars cannot be stopped at the support faces. They must be extended or anchored in the concrete beyond the support face. We will later call this development length. The *development length* does not have to be straight as shown in the figure, because the bars may be hooked at 90° or 180°. Development lengths and hooked bars are discussed in depth in Chapter 7.

Up to this point, only statically determinate members have been considered. The very common situation, however, is for beams and slabs to be continuous over several supports as shown in Figure 4.10. Because reinforcing is needed on the tensile sides of the beams, we will place it in the bottoms when we have positive moments and in the tops when we have negative moments. There are several ways in which the reinforcing bars can be arranged to resist the positive and negative moments in continuous members. One possible arrangement is shown in Figure 4.10(a). These members, including bar arrangements, are discussed in detail in Chapter 14.

Placing concrete slab. (Courtesy of Bethlehem Steel Corporation.)

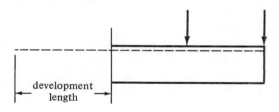

Figure 4.9 Cantilever beam.

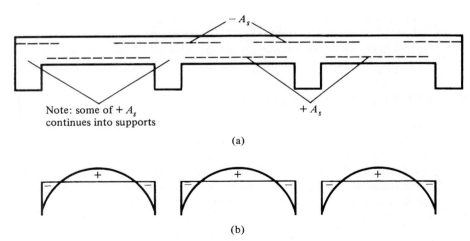

Note: some of $+A_s$
continues into supports

$+A_s$

(a)

(b)

Figure 4.10 Continuous slab showing theoretical placement of bars for given moment diagram.

4.9 SI EXAMPLE

Example 4.7 illustrates the design of a beam using SI units.

EXAMPLE 4.7

Design a rectangular beam for a 10-m simple span to support a dead load of 20 kN/m (not including beam weight) and a live load of 30 kN/m. Use $\rho = 0.5\rho_b$, $f_c' = 28$ MPa, and $f_y = 420$ MPa, and concrete weight is 23.5 kN/m³. Do not use the ACI thickness limitation.

SOLUTION   Assume that the beam weight is 10 kN/m and $\phi = 0.90$.

$$w_u = (1.2)(30) + (1.6)(30) = 84 \text{ kN/m}$$

$$M_u = \frac{(84)(10)^2}{8} = 1050 \text{ kN} \cdot \text{m}$$

$$\rho = \left(\frac{1}{2}\right)(0.0283) = 0.01415 \text{ from Appendix Table B.7}$$

$$M_u = \phi \rho f_y b d^2 \left(1 - \frac{1}{1.7}\rho\frac{f_y}{f_c'}\right)$$

$$(10^6)(1050) = (0.9)(0.01415)(420)(bd^2)\left[1 - \left(\frac{1}{1.7}\right)(0.01415)\left(\frac{420}{28}\right)\right]$$

$$bd^2 = 2.2432 \times 10^8 \text{ mm}^3 \begin{cases} 400 \times 749 \\ 450 \times 706 \\ 500 \times 670 \leftarrow \end{cases}$$

Use 500 mm × 800 mm section ($d = 680$ mm)

$$\text{Beam wt} = \frac{(500)(800)}{10^6}(23.5) = 9.4 \text{ kN/m}$$

$$< 10 \text{ kN/m assumed}$$

$$A_s = (0.01415)(500)(680) = 4811 \text{ mm}^2 \qquad \underline{\text{OK}}$$

Use six #32 bars in two rows (4914 mm²). One row could be used here.

$$a = \frac{A_s f_y}{0.85 f'_c b} = \frac{(4914)(420)}{(0.85)(28)(500)} = 173 \text{ mm}$$

$$c = \frac{a}{\beta_1} = \frac{173}{0.85} = 204 \text{ mm}$$

$$\epsilon_t = \frac{680-204}{204}(0.003) = 0.0070 > 0.005 \therefore \underline{\phi = 0.90}$$

Note: Can more easily be checked with ρ values.

$b_{min} = 267$ mm from Appendix Table B.5 for 3 bars in a layer

< 500 mm   OK

The final section is shown in Figure 4.11.

Note: This problem can be solved more quickly by making use of the Appendix tables. In Table B.9 with $f_y = 420$ MPa, $f'_c = 28$ MPa, and $\rho = 0.01415$.

$$\frac{M_u}{\phi b d^2} = 5.201$$

$$bd^2 = \frac{M_u}{\phi(5.201)} = \frac{(1050)(10)^6}{(0.9)(5.201)} = 2.2432 \times 10^8 \text{ mm}^3$$

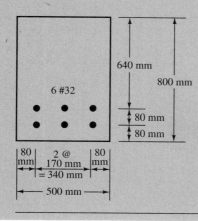

Figure 4.11

4.10   COMPUTER EXAMPLE

EXAMPLE 4.8

Repeat Example 4.4 using the Excel Spreadsheet for Chapter 4.

SOLUTION   Use the worksheet called Beam Design. Enter material properties (f'_c, f_y) and M_u (can be taken from the bottom part of the spreadsheet or just entered if you already know it). Input $\rho = 0.009$ (given in the example). The two tables with headings b and d give some choices for b and d based on the ρ value you picked. Larger assumed values of ρ result in smaller values of b and d and vice versa. Select $b = 18$ in. and $d = 31$ in. (many other choices are also correct). Add 2.5 in. or more to d to get h and enter that value (used only to find beam weight below). The spreadsheet recalculates ρ and A_s from actual values of b and d chosen, so note that ρ is not the same as originally assumed (0.00895 instead of 0.009). This results in a slightly smaller calculated steel area than in Example 4.4. You can also enter the number of bars and size to get a value for A_s. This value must exceed the theoretical value or an error message will appear. You should check to see if this bar selection will fit within the width selected.

At the bottom of the spreadsheet, the design moment M_u can be obtained if the beam is simply supported and uniformly loaded with only dead and live loads. The beam self-weight is calculated based on the input values for b and h (Cell D23 & D25). You may have to iterate a few times before these values all agree. In this example, the dead load is 2 klf plus self weight. The input value for w_D is $2.0 + 0.65$ plf, with the second term being taken from the spreadsheet. In working this problem the first time, you probably would not have these dimensions for b and h, hence the self weight would not be correct. Iteration as done in Example 4.4 is also required with the spreadsheet, although it is much faster.

Design of singly reinforced rectangular beams

$f'_c =$ | **3** ksi
$f_y =$ | **60** ksi
$\beta_1 =$ | 0.85

$M_u =$ | **623.4** ft-k

> **Instructions: Enter values only in cells that are highlighted yellow. Other values calculated from those input values.**

Assume $\rho = \boxed{0.009}$

$$bd^2 = \frac{M_u}{\phi f_y \rho \left(1 - \rho f_y/(1.7f'_c)\right)} = 17{,}215 \text{ in.}^3$$

$R = d/R$

R	b	d		b	d
1	25.82	25.82		14	35.07
1.2	22.86	27.44		15	33.88
1.4	20.63	28.88		16	32.80
1.5	19.70	29.56		17	31.82
1.6	18.87	30.20		18	30.93
1.7	18.13	30.82		19	30.10
1.8	17.45	31.41		20	29.34
1.9	16.83	31.98		21	28.63
2	16.27	32.58		22	27.97

Select b and d | $b =$ | **18** in.
| $d =$ | **31** in.
| $h =$ | **34** in.
| $R_n =$ | 480.52

> These tables give some choices for b and d that you may round up to enter here.

$\rho = 0.85f'_c/f_y \left(1 - (1 - 2R_n/(.85f'_c))^{0.5}\right) = 0.00895$

$A_s = \rho bd = 4.99 \text{ in.}^2$

No. of bars | Bar size | -
select bars | **5** | # **9** | $A_s = 5.00 \text{ in.}^2$

Calculation of M_u for simply supported beam with D & L uniformly distributed loads		
$w_D =$	2.65	plf
$w_L =$	3	plf
span =	25	ft.
$w_u =$	7.980	plf
$M_u =$	623.4	ft-k
$\gamma_c =$	145	pcf
self wt. =	0.6375	plf

PROBLEMS

4.1 The estimated service or working axial loads and bending moments for a particular column are as follows: $P_D = 100$ k, $P_L = 60$ k, $M_D = 30$ ft-k and $M_L = 20$ ft-k. Compute the axial load and moment values that must be used in the design. (*Ans.* $P_u = 216$ k, $M_u = 68$ ft-k)

4.2 Determine the required design strength for a column for which $P_D = 120$ k, $P_L = 40$ k, and wind $P_W = 60$ k compression, or 80 k tension.

4.3 A reinforced concrete slab must support a dead working floor load of 90 psf, which includes the weight of the concrete slab and a live working load of 40 psf. Determine the factored uniform load for which the slab must be designed. (*Ans.* $w_u = 172$ psf)

4.4 Using the Chapter 4 spreadsheet, repeat the following problems:
(a) Problem 4.1
(b) Problem 4.2
(c) Problem 4.3

In Problems 4.5 to 4.9 design rectangular sections for the beams, loads, and values given. Beam weights are not included in the given loads. Show sketches of beam cross sections, including bar sizes, arrangement, and spacing. Assume concrete weighs 150 lb/ft³. Use $h = d + 2.5''$

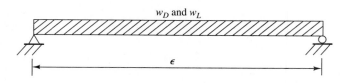

w_D and w_L

ϵ

Problem No.	$f_y(\phi)$	$f_c'(\phi)$	Span ℓ (ft)	w_D not incl. beam wt (k/ft)	w_L(k/ft)	ρ^*
4.5	60,000	4000	24	2	2.8	$\varepsilon_t = 0.0075$
4.6	60,000	4000	30	2	2	$\dfrac{0.18 f_c'}{f_y}$
4.7	50,000	3000	28	3	3	$\dfrac{1}{2}\rho_b$
4.8	60,000	4000	32	2	1.8	$\dfrac{1}{2}\rho_b$
4.9	60,000	3000	25	1.8	1.5	$\varepsilon_t = 0.005$

*See Table A.7 for ρ values that correspond to the ϵ_t values listed.
One ans. Prob. 4.5: 18 in. × 26 in. with six #9 bars.
One ans. Prob. 4.7: 22 in. × 33 in. with 6 #11 bars.
One ans. Prob. 4.9: 16 in. × 25 in. with 5 #9 bars.

In Problems 4.10 to 4.22 design rectangular sections for the beams, loads, and ρ values shown in the accompanying illustrations. Beam weights are not included in the loads shown. Show sketches of cross sections, including bar sizes, arrangement, and spacing. Assume concrete weighs $150\,\text{lb/ft}^3$, $f_y = 60{,}000\,\text{psi}$, and $f'_c = 4000\,\text{psi}$ unless given otherwise.

Problem 4.10

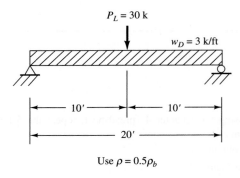

Use $\rho = 0.5\rho_b$

4.11 Repeat Problem 4.10 if $w_D = 2\,\text{k/ft}$ and if $P_L = 10\,\text{k}$. (*One ans.* 12 in. × 21 in. with 4 #8 bars.)

Problem 4.12

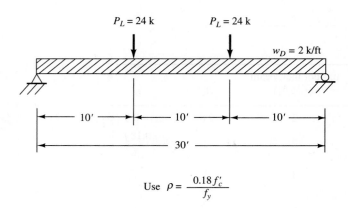

Use $\rho = \dfrac{0.18 f'_c}{f_y}$

4.13 Repeat Problem 4.12 if $w_D = 1.5\,\text{k/ft.}$ and $P_L = 30\,\text{k}$. (*One ans.* 16 in. × 34 in. with 4 #11 bars.)

Problem 4.14

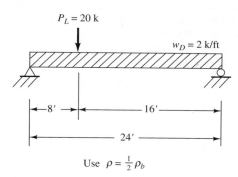

$P_L = 20$ k

$w_D = 2$ k/ft

8' 16'

24'

Use $\rho = \frac{1}{2}\rho_b$

Problem 4.20

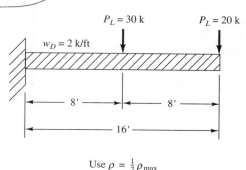

$P_L = 30$ k $P_L = 20$ k

$w_D = 2$ k/ft

8' 8'

16'

Use $\rho = \frac{1}{2}\rho_{max}$

4.15 Repeat Problem 4.14 if $w_D = 3$ k/ft and if $P_L = 40$ k, $f'_c = 3000$ psi and if $\rho = 0.5\rho_b$. (*One ans.* 16 in.× 33 in. with 5 #10 bars)

Problem 4.16

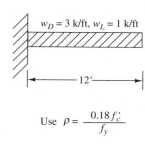

$w_D = 3$ k/ft, $w_L = 1$ k/ft

12'

Use $\rho = \dfrac{0.18 f'_c}{f_y}$

4.21 Select reinforcing bars for the beam shown if $M_u = 150$ ft-k, $f_y = 60{,}000$ psi, and $f'_c = 4000$ psi. (*Hint:* Assume that the distance from the c.g. of the tensile steel to the c.g. of the compression block equals 0.9 times the effective depth d of the beam.) After a steel area is computed, check the assumed distance and revise the steel area if necessary. Is $\epsilon_t \geq 0.005$? (*Ans.* $A_s = 1.57$ in.^{2}, $\epsilon_t > 0.005$)

4.17 Repeat Problem 4.16 if the beam span $= 14$ ft. (*One ans.* 20 in. × 27 in. with 6 #9 bars in top)

Problem 4.18

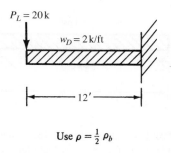

$P_L = 20$ k

$w_D = 2$ k/ft

12'

Use $\rho = \frac{1}{2}\,\rho_b$

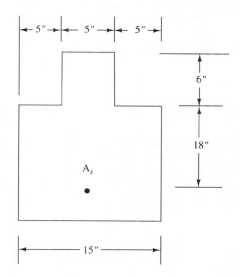

5" 5" 5"

6"

18"

A_s

15"

4.19 Repeat Problem 4.18 if $P_L = 10$ k, $w_D = 1$ k/ft, $\ell = 16$ ft and $\rho = \frac{1}{2}\rho_b$. (*One ans.* 14 in. × 27 in. with 4 #10 in top)

4.22 Repeat Problem 4.21 for $M_u = 250$ ft-k.

In Problems 4.23 and 4.24 design rectangular sections for the beams and loads shown in the accompanying illustrations. Beam weights are not included in the given loads. $f_y = 60,000$ psi and $f'_c = 4000$ psi. Live loads are to be placed where they will cause the most severe conditions at the sections being considered. Select beam size for the largest moment (positive or negative) and then select the steel required for maximum positive moment and for maximum negative moment. Finally, sketch the beam and show approximate bar locations.

Problem 4.23 (*One ans.*: 12 in. × 23 in. with 3 #9 bars negative reinforcement and 4 #7 bars positive reinforcement)

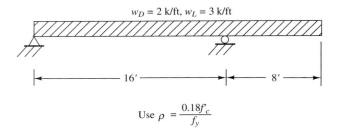

$w_D = 2$ k/ft, $w_L = 3$ k/ft

16′   8′

Use $\rho = \dfrac{0.18f'_c}{f_y}$

Problem 4.24

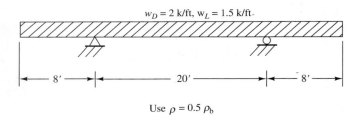

$w_D = 2$ k/ft, $w_L = 1.5$ k/ft

8′   20′   8′

Use $\rho = 0.5\,\rho_b$

In Problems 4.25 and 4.26 design interior one-way slabs for the situations shown. Concrete weight $= 150$ lb/ft^3, $f_y = 60,000$ psi and $f'_c = 4000$ psi. Do not use the ACI Code's minimum thickness for deflections (Table 4.1). Steel percentages are given in the figures. The only dead load is the weight of the slab.

Problem 4.25 (*One ans.* 5 in. slab with #6 at 10 in. main reinf.)

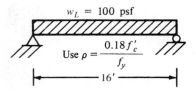

$w_L = 100$ psf

Use $\rho = \dfrac{0.18f'_c}{f_y}$

16′

Problem 4.26

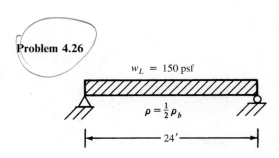

$w_L = 150$ psf

$\rho = \tfrac{1}{2}\rho_b$

24′

4.27 Repeat Problem 4.26 using the ACI Code's minimum thickness requirement for cases where deflections are not computed (Table 4.1). Do not use the ρ given in Problem 4.26. (*Ans.* $14\tfrac{1}{2}$ in. slab with #6 @ 9 in. main reinf.)

4.28 Using $f'_c = 3000$ psi, $f_y = 60,000$ psi, and ρ corresponding to $\epsilon_t = 0.005$, determine the depth of a simple beam to support itself for a 200-ft simple span.

4.29 Determine the depth required depth for a beam to support itself only for a 100-ft span. Given $f'_c = 4000$ psi, $f_y = 60,000$ psi, and $\rho = 0.5\rho_b$. (*Ans. d = 32.5 in.*)

4.30 Determine the stem thickness for maximum moment for the retaining wall shown in the accompanying illustration. Also, determine the steel area required at the bottom and mid-depth of the stem if $f'_c = 4000$ psi, and $f_y = 60,000$ psi. Assume that #8 bars are to be used and that the stem thickness is constant for the 20-ft height. Also, assume that the clear cover required is 2 in. $\rho = 0.5\rho_b$.

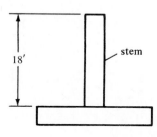
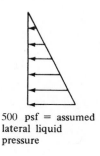

18' stem

500 psf = assumed
lateral liquid
pressure

4.31 (**a**) Design a 24-in.-wide precast concrete slab to support a 60 psf live load for a simple span of 15 ft. Assume minimum concrete cover required is $\frac{5}{8}$ in. as per section 7.7.3 of the Code. Use welded wire fabric for reinforcing, $f_y = 60,000$ psi and $f'_c = 3000$ psi, $\rho = 0.18f'_c/f_y$. (*Ans. $3\frac{1}{2}$ in. slab with 4 × 8 D12/D6*)

(**b**) Can a 380-lb football tackle walk across the center of the span when the other live load is not present? Assume 100% impact. (*Ans. yes*)

4.32 Prepare a flow chart for the design of tensilely reinforced rectangular beams.

4.33 Using the Chapter 4 spreadsheets, solve the following problems.

(**a**) Problem 4.6. (*Ans.* 16 in. × 33 in. 5 #10 bars)

(**b**) Problem 4.18. (*Ans.* 18 in. × 27 in. with 5 #10 bars)

Problems in SI Units

In Problems 4.34 to 4.39 design rectangular sections for the beams, loads, and ρ values shown in the accompanying illustrations. Beam weights are not included in the loads given. Show sketches of cross sections - including bar sizes, arrangements, and spacing. Assume concrete weights 23.5 kN/m³, $f_y = 420$ MPa, and $f'_c = 28$ MPa.

Problem 4.34

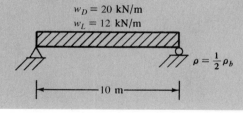

$w_D = 20$ kN/m
$w_L = 12$ kN/m

$\rho = \frac{1}{2}\rho_b$

10 m

Problem 4.35 (*One ans.:* 500 mm × 830 mm with 5 #36 bars.)

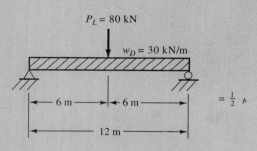

$P_L = 80$ kN

$w_D = 30$ kN/m

6 m   6 m

12 m

$= \frac{1}{2} b$

Problem 4.36

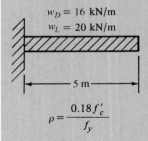

$w_D = 16$ kN/m

$w_L = 20$ kN/m

5 m

$$\rho = \frac{0.18 f_c'}{f_y}$$

Problem 4.37 Place live loads to cause maximum positive and negative moments. $\rho = 0.18 f_c'/f_y$ (*One ans.* 400 mm × 830 mm with 5 #32 bars positive reinf.)

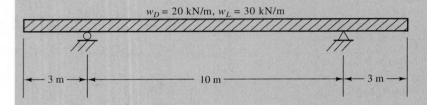

$w_D = 20$ kN/m, $w_L = 30$ kN/m

3 m   10 m   3 m

Problem 4.38

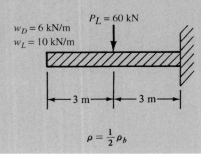

$P_L = 60$ kN

$w_D = 6$ kN/m

$w_L = 10$ kN/m

3 m   3 m

$$\rho = \frac{1}{2}\rho_b$$

4.39 Design the one-way slab shown in the accompanying figure to support a live load of 10 kN/m². Do not use the ACI thickness limitation for deflections and assume concrete weighs 23.5 kN/m3. Assume that $f_c' = 28$ MPa and $f_y = 420$ MPa. Use $\rho = \rho_{max}$ (*One ans.* 220 mm slab with #29 at 200 mm main steel.)

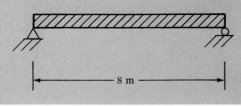

8 m

5

Analysis and Design of T Beams and Doubly Reinforced Beams

5.1 T BEAMS

Reinforced concrete floor systems normally consist of slabs and beams that are placed monolithically. As a result, the two parts act together to resist loads. In effect, the beams have extra widths at their tops, called *flanges*, and the resulting T-shaped beams are called *T beams*. The part of a T beam below the slab is referred to as the *web* or *stem*. (The beams may be L-shaped if the stem is at the end of a slab.) The stirrups (described in Chapter 8) in the webs extend up into the slabs, as perhaps do bent-up bars, with the result that they further make the beams and slabs act together.

There is a problem involved in estimating how much of the slab acts as part of the beam. Should the flanges of a T beam be rather stocky and compact in cross section, bending stresses will be fairly uniformly distributed across the compression zone. If, however, the flanges are wide and thin, bending stresses will vary quite a bit across the flange due to shear deformations. The further a particular part of the slab or flange is away from the stem, the smaller will be its bending stress.

Instead of considering a varying stress distribution across the full width of the flange, the ACI Code (8.12.2) calls for a smaller width with an assumed uniform stress distribution for design purposes. The objective is to have the same total compression force in the reduced width that actually occurs in the full width with its varying stresses.

The hatched area in Figure 5.1 shows the effective size of a T beam. For T beams with flanges on both sides of the web, the Code states that the effective flange width may not exceed one-fourth of the beam span, and the overhanging width on each side may not exceed eight times the slab thickness or one-half the clear distance to the next web. An isolated T beam must have a flange thickness no less than one-half the web width, and its effective flange width may not be larger than four times the web width (ACI 18.12.4). If there is a flange on only one side of the web, the width of the overhanging flange cannot exceed one-twelfth the span, $6h_f$ or half the clear distance to the next web (ACI 8.12.3).

The analysis of T beams is quite similar to the analysis of rectangular beams in that the specifications relating to the strains in the reinforcing are identical. To repeat briefly, it is desirable to have ϵ_t values ≥ 0.005, and they may not be less than 0.004 unless the member is subjected to an axial load $\geq 0.10f'_c A_g$. You will learn that ϵ_t values are almost always much larger than 0.005 in T beams because of their very large compression flanges. For such members the values of c are normally very small, and calculated ϵ_t values very large.

The neutral axis (N.A.) for T beams can fall either in the flange or in the stem, depending on the proportions of the slabs and stems. If it falls in the flange, and it almost always does for positive moments, the rectangular beam formulas apply, as can be seen in Figure 5.2(a). The concrete below the neutral axis is assumed to be cracked, and its shape has no effect on the flexure

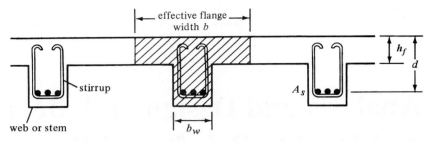

Figure 5.1 Effective width of T beams.

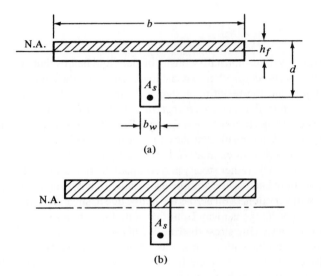

Figure 5.2 Neutral axis locations.

calculations (other than weight). The section above the neutral axis is rectangular. If the neutral axis is below the flange, however, as shown for the beam of Figure 5.2(b), the compression concrete above the neutral axis no longer consists of a single rectangle, and thus the normal rectangular beam expressions do not apply.

If the neutral axis is assumed to fall within the flange, the value of a can be computed as it was for rectangular beams:

$$a = \frac{A_s f_y}{0.85 f'_c b} = \frac{\rho f_y d}{0.85 f'_c}$$

The distance to the neutral axis c equals a/β_1. If the computed value of a is equal to or less than the flange thickness, the section for all practical purposes can be assumed to be rectangular even though the computed value of c is actually greater than the flange thickness.

A beam does not really have to look like a T beam to be one. This fact is shown by the beam cross sections shown in Figure 5.3. For these cases the compression concrete is T shaped, and the shape or size of the concrete on the tension side, which is assumed to be cracked, has no effect on the theoretical resisting moments. It is true, however, that the shapes, sizes, and weights of the tensile concrete do affect the deflections that occur (as is described in Chapter 6), and their dead weights affect the magnitudes of the moments to be resisted.

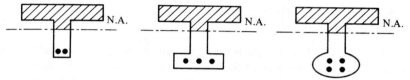

Figure 5.3 Various cross sections of T beams.

Natural History Museum, Kensington, London, England. (Courtesy of Cement and Concrete Association.)

5.2 ANALYSIS OF T BEAMS

The calculation of the design strengths of T beams is illustrated in Examples 5.1 and 5.2. In the first of these problems, the neutral axis falls in the flange, while for the second it is in the web. The procedure used for both examples involves the following steps:

1. Check A_s min as per ACI Section 10.5.1 using b_w as the web width.

2. Compute $T = A_s f_y$.

3. Determine the area of the concrete in compression (A_c) stressed to $0.85f_c'$.

$$C = T = 0.85f_c' A_c$$

$$A_c = \frac{T}{0.85f_c'}$$

4. Calculate a, c, and ϵ_r.

5. Calculate ϕM_n.

For Example 5.1, where the neutral axis falls in the flange, it would be logical to apply the normal rectangular equations of Sections 3.4 of this book, but the author has used the couple method as a background for the solution of Example 5.2, where the neutral axis falls in the web. This same procedure can be used for determining the design strengths of tensilely reinforced concrete beams of any shape (T,Γ,Π, triangular, circular, and so on).

EXAMPLE 5.1

Determine the design strength of the T beam shown in Figure 5.4, with $f'_c = 4000$ psi and $f_y = 60{,}000$ psi. The beam has a 30 ft span and is cast integrally with a floor slab that is 4 in. thick. The clear distance between webs is in 50 in.

SOLUTION **Check Effective Flange Width**

$$b \leq 16h_f + b_w = 16(4) + 10 = 74 \text{ in.}$$

$$b \leq \text{ average clear distance to adjacent webs} + b_w = 50 + 10 = 60 \text{ in.} \leftarrow$$

$$b \leq \text{span}/4 = 30/4 = 7.5 \text{ ft} = 90 \text{ in.}$$

Checking $A_{s\ min}$

$$A_{s\ min} = \frac{3\sqrt{f'_c}}{f_y} b_w d = \frac{(3\sqrt{4000})}{60{,}000}(10)(24) = 0.76 \text{ in.}^2$$

$$\text{nor less than } \frac{200 b_w d}{f_y} = \frac{(200)(10)(24)}{60{,}000} = 0.80 \text{ in.}^2 \leftarrow$$

$$< A_s = 6.00 \text{ in.}^2 \ \underline{\text{OK}}$$

Computing T

$$T = A_s f_y = (6.00)(60) = 360 \text{ k}$$

Determining A_c

$$A_c = \frac{T}{0.85 f'_c} = \frac{360}{(0.85)(4)} = 105.88 \text{ in.}^2$$

$$< \text{flange area} = (60)(4) = 240 \text{ in.}^2 \therefore \ \underline{\text{N.A. is in flange}}$$

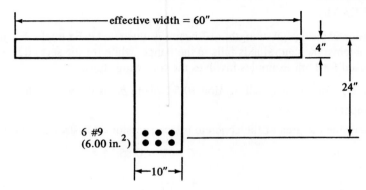

Figure 5.4

Calculating a, c, and ϵ_t

$$a = \frac{105.88}{60} = 1.76 \text{ in.}$$

$$c = \frac{a}{\beta_1} = \frac{1.76}{0.85} = 2.07 \text{ in.}$$

$$\epsilon_t = \left(\frac{d-c}{c}\right)(0.003) = \left(\frac{24-2.07}{2.07}\right)(0.003)$$

$$= 0.0318 > 0.005$$

$$\therefore \text{Section is ductile and } \phi = 0.90$$

Calculating ϕM_n

Obviously, the stress block is entirely within the flange, and the rectangular formulas apply. However, using the couple method as follows:

$$\text{Lever arm} = z = d - \frac{a}{2} = 24 - \frac{1.76}{2} = 23.12 \text{ in.}$$

$$\phi M_n = \phi Tz = (0.90)(360)(23.12)$$

$$= 7490.9 \text{ in.-k} = \underline{624.2 \text{ ft-k}}$$

EXAMPLE 5.2

Compute the design strength for the T beam shown in Figure 5.5 in which $f_c' = 4000$ psi and $f_y = 60,000$ psi.

SOLUTION   **Checking $A_{s\,min}$**

$$A_{s\,min} = \frac{3\sqrt{4000}}{60,000}(14)(30) = 1.33 \text{ in.}^2$$

$$\text{nor less than} \frac{(200)(14)(30)}{60,000} = 1.40 \text{ in.}^2 \leftarrow$$

$$< A_s = 10.12 \text{ in.}^2 \underline{\text{OK}}$$

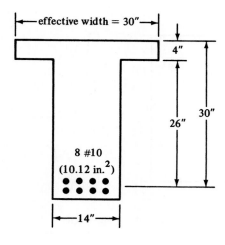

Figure 5.5

Computing T

$$T = A_s f_y = (10.12)(60) = 607.2 \text{ k}$$

Determining A_c and Its Center of Gravity

$$A_c = \frac{T}{0.85 f_c'} = \frac{607.2}{(0.85)(4)} = 178.59 \text{ in.}^2$$

$$> \text{flange area} = (30)(4) = 120 \text{ in.}^2$$

Obviously, the stress block must extend below the flange to provide the necessary compression area = $178.6 - 120 = 58.6 \text{ in.}^2$, *as shown in Figure 5.6.*

Computing the Distance $\bar{y}$ from the Top of the Flange to the Center of Gravity of A_c

$$\bar{y} = \frac{(120)(2) + (58.6)\left(4 + \dfrac{4.19}{2}\right)}{178.6} = 3.34 \text{ in.}$$

The Lever Arm Distance from T to $C = 30.00 - 3.34 = 26.66$ in. $= z$
Calculating a, c, and ϵ_t

$$a = 4 + 4.19 = 8.19 \text{ in.}$$

$$c = \frac{a}{\beta_1} = \frac{8.19}{0.85} = 9.64 \text{ in.}$$

$$\epsilon_t = \frac{d - c}{c}(0.003) = \left(\frac{30 - 9.64}{9.64}\right)(0.003) = 0.00634$$

$$> 0.005 \therefore \text{ Section is ductile and } \phi = 0.90$$

Calculating ϕM_n

$$\phi M_n = \phi T z = (0.90)(607.2)(26.66) = 14{,}569 \text{ in.-k}$$

$$= \underline{1214 \text{ ft-k}}$$

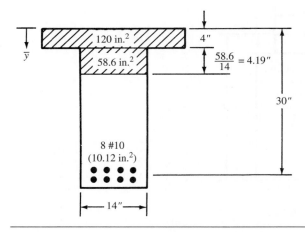

Figure 5.6

5.3   ANOTHER METHOD FOR ANALYZING T BEAMS

The preceding section presented a very important method of analyzing reinforced concrete beams. It is a general method that is applicable to tensily reinforced beams of any cross section, including T beams. T beams are so very common, however, that many designers prefer another method that is specifically designed for T beams.

First, the value of a is determined as previously described in this chapter. Should it be less than the flange thickness h_f, we will have a rectangular beam and the rectangular beam formulas will apply. Should it be greater than the flange thickness h_f (as was the case for Example 5.2), the special method to be described here will be very useful.

The beam is divided into a set of rectangular parts consisting of the overhanging parts of the flange and the compression part of the web (see Figure 5.7).

The total compression C_w in the web rectangle and the total compression in the overhanging flange C_f are computed:

$$C_w = 0.85 f_c' a b_w$$

$$C_f = 0.85 f_c' (b - b_w)(h_f)$$

Then the nominal moment M_n is determined by multiplying C_w and C_f by their respective lever arms from their centroids to the centroid of the steel:

$$M_n = C_w \left(d - \frac{a}{2}\right) + C_f \left(d - \frac{h_f}{2}\right)$$

This procedure is illustrated in Example 5.3. Although it seems to offer little advantage in computing M_n, we will learn that it does simplify the design of T beams when $a > h_f$ because it permits a direct solution of an otherwise trial-and-error problem.

EXAMPLE 5.3

Repeat Example 5.2 using the value of a (8.19 in.) previously obtained and the alternate formulas just developed. Reference is made to Figure 5.8, the dimensions of which were taken from Figure 5.5.

SOLUTION

(Noting that $a > h_f$)
Computing C_w and C_f

$$C_w = (0.85)(4)(8.19)(14) = 389.8 \text{ k}$$

$$C_f = (0.85)(4)(30 - 14)(4) = 217.6 \text{ k}$$

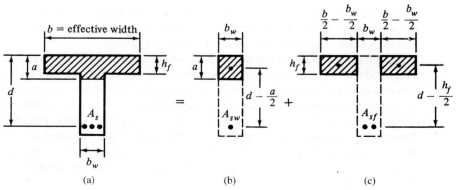

Figure 5.7 Separation of T beam into rectangular parts.

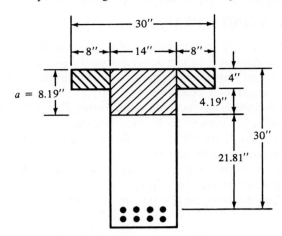

Figure 5.8

Computing c and ϵ_t

$$c = \frac{\alpha}{\beta_1} = \frac{8.19}{0.85} = 9.64 \text{ in.}$$

$$\epsilon_t = \left(\frac{d-c}{c}\right)(0.003) = \left(\frac{30 - 9.64}{9.64}\right)(0.003) = 0.00634$$

$$> 0.005 \therefore \text{ Section is ductile and } \phi = 0.90$$

Calculating M_n and ϕM_n

$$M_n = C_w\left(d - \frac{a}{2}\right) + C_f\left(d - \frac{h_f}{2}\right)$$

$$= (389.8)\left(30 - \frac{8.19}{2}\right) + (217.6)\left(30 - \frac{4}{2}\right) = 16,190 \text{ in.-k} = 1349 \text{ ft-k}$$

$$\phi M_n = (0.90)(1349) = \underline{1214 \text{ ft-k}}$$

5.4   DESIGN OF T BEAMS

For the design of T beams, the flange has normally already been selected in the slab design, as it is the slab. The size of the web is normally not selected on the basis of moment requirements but probably is given an area based on shear requirements; that is, a sufficient area is used so as to provide a certain minimum shear capacity, as will be described in Chapter 8. It is also possible that the width of the web may be selected on the basis of the width estimated to be needed to put in the reinforcing bars. Sizes may also have been preselected as previously described in Section 4.5 to simplify formwork, for architectural requirements, or for deflection reasons. For the examples that follow (5.4 and 5.5), the values of h_f, d and b_w are given.

The flanges of most T beams are usually so large that the neutral axis probably falls within the flange and thus the rectangular beam formulas apply. Should the neutral axis fall within the web, a trial-and-error process is often used for the design. In this process a lever arm from the center of gravity of the compression block to the center of gravity of the steel is estimated to equal the larger of $0.9d$ or $d - (h_f/2)$ and from this value called z, a trial steel area is calculated ($A_s = M_n/f_y z$). Then by the process used in Example 5.2, the value of the estimated lever arm is checked. If there is much difference, the estimated value of z is revised and a new A_s determined. This process is continued until the change in A_s is quite small. T beams are designed in Examples 5.4 and 5.5 by this process.

Often a T beam is part of a continuous beam that spans over interior supports such as columns. The bending moment over the support is negative so the flange is in tension. Also, the magnitude of the negative moment is usually larger than that of the positive moment near midspan. This situation will control the design of the T beam because the depth and web width will be determined for this case. Then, when the beam is designed for positive moment at midspan, the width and depth are already known. See Section 5.5 for other details on T beams with negative moments.

Example 5.6 presents a more direct approach for the case where $a > h_f$. This is the case where the beam is assumed to be divided into its rectangular parts.

EXAMPLE 5.4

Design a T beam for the floor system shown in Figure 5.9 for which b_w and d are given. $M_D = 80$ ft-k, $M_L = 100$ ft-k, $f'_c = 4000$ psi, $f_y = 60,000$ psi, and simple span $= 20$ ft.

SOLUTION

Effective Flange Width

(a) $\frac{1}{4} \times 20 = 5'0'' = 60''$

(b) $12 + (2)(8)(4) = 76''$ ✓

(c) $10'0'' = 120''$

Computing Moments Assuming $\phi = 0.90$

$$M_u = (1.2)(80) + (1.6)(100) = 256 \text{ ft-k}$$

$$M_n = \frac{M_u}{\phi} = \frac{256}{0.90} = 284.4 \text{ ft-k}$$

Assuming a Lever Arm z Equal to the Larger of $0.9d$ or $d - (h_f/2)$

$$z = (0.9)(18) = \underline{16.20 \text{ in.}}$$

$$z = 18 - \frac{4}{2} = 16.00 \text{ in.}$$

Trial Steel Area

$$A_s f_y z = M_n$$

$$A_s = \frac{(12)(284.4)}{(60)(16.20)} = 3.51 \text{ in.}^2$$

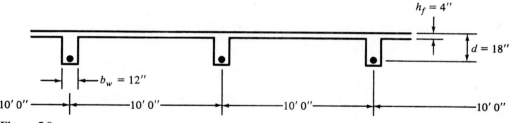

Figure 5.9

Computing Values of a and z

$$0.85f'_c A = A_s f_y$$

$$(0.85)(4)(A_c) = (3.51)(60)$$

$$A_c = 61.9 \text{ in.}^2 < (4)(60) = 240 \text{ in.}^2 \therefore \underline{\text{N.A. in flange}}$$

$$a = \frac{61.9}{60} = 1.03 \text{ in.}$$

$.9(18)=16.2 \qquad 18 - \left(\frac{1.03}{2} \right)$

Calculating A_s with This Revised z

$$A_s = \frac{(12)(284.4)}{(60)(17.48)} = 3.25 \text{ in.}^2$$

Computing Values of a and z

$$A_c = \frac{(3.25)(60)}{(0.85)(4)} = 57.4 \text{ in.}^2$$

$$a = \frac{57.4}{60} = 0.96 \text{ in.}$$

$$z = 18 - \frac{0.96}{2} = 17.52 \text{ in.}$$

Calculating A_s with This Revised z

$$A_s = \frac{(12)(284.4)}{(60)(17.52)} = 3.25 \text{ in.}^2 \quad \underline{\text{OK, close enough to previous value}}$$

Checking Minimum Reinforcing

$$A_{s\,\text{min}} = \frac{3\sqrt{f'_c}}{f_y} b_w d = \frac{3\sqrt{4000}}{60,000}(12)(18) = 0.68 \text{ in.}^2$$

but not less than

$$A_{s\,\text{min}} = \frac{200 b_w d}{f_y} = \frac{(200)(12)(18)}{60,000} = 0.72 \text{ in.}^2 < 3.25 \text{ in.}^2 \quad \underline{\text{OK}}$$

or ρ_{min} from Appendix Table A.7 = 0.0033

$$A_{s\,\text{min}} = (0.0033)(12)(18) = 0.71 \text{ in.}^2 < 3.25 \text{ in.}^2 \quad \underline{\text{OK}}$$

Computing c, ϵ_t, and ϕ

$$c = \frac{a}{\beta_1} = \frac{0.96}{0.85} = 1.13 \text{ in.}$$

$$\epsilon_t = \left(\frac{d-c}{c} \right)(0.003) = \left(\frac{18 - 1.13}{1.13} \right)(0.003)$$

$$= 0.045 > 0.005 \therefore \underline{\phi = 0.90 \text{ as assumed}}$$

$$\underline{\underline{A_{s\,\text{reqd}} = 3.25 \text{ in.}^2}}$$

EXAMPLE 5.5

Design a T beam for the floor system shown in Figure 5.10, for which b_w and d are given. $M_D = 200$ ft-k, $M_L = 425$ ft-k, $f'_c = 3000$ psi, $f_y = 60,000$ psi, and simple span $= 18$ ft.

SOLUTION

Effective Flange Width

    (a) $\frac{1}{4} \times 18' = 4'6'' = \underline{54''}$

    (b) $15'' + (2)(8)(3) = 63''$

    (c) $6'0'' = 72''$

Moments Assuming $\phi = 0.90$

$$M_u = (1.2)(200) + (1.6)(425) = 920 \text{ ft-k}$$

$$M_n = \frac{M_u}{0.90} = \frac{920}{0.90} = 1022 \text{ ft-k}$$

Assuming a Lever Arm z (Note that the compression area in the slab is very wide, and thus its required depth is very small.)

$$z = (0.90)(24) = 21.6''$$

$$z = 24 - 3/2 = \underline{22.5''}$$

Trial Steel Area

$$A_s = \frac{(12)(1022)}{(60)(22.5)} = 9.08 \text{ in.}^2$$

Checking Values of a and z

$$A_c = \frac{(60)(9.08)}{(0.85)(3)} = 213.6 \text{ in.}^2$$

The stress block extends down into the web, as shown in Figure 5.11.

Computing the Distance $\bar{y}$ from the Top of the Flange to the Center of Gravity of A_c

$$\bar{y} = \frac{(162)(1.5) + (51.6)\left(3 + \dfrac{3.44}{2}\right)}{213.6} = 2.28 \text{ in.}$$

$$z = 24 - 2.28 = 21.72 \text{ in.}$$

$$A_s = \frac{(12)(1022)}{(60)(21.72)} = 9.41 \text{ in.}^2$$

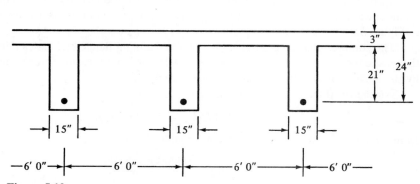

Figure 5.10

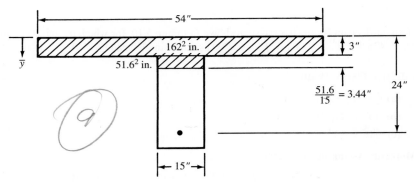

Figure 5.11

The steel area required (9.41 in.^{2}) could be refined a little by repeating the design, but space is not used to do this. (If this is done, $A_s = 9.51$ in.^{2})

Checking Minimum Reinforcing

ρ_{min} from Appendix Table A.7 = 0.00333

or

$A_{s\,min} = (0.00333)(15)(24) = 1.20$ in.$^2 < 9.58$ in.^{2} <u>OK</u>

Checking Values of ϵ_t and ϕ

$a = 3 + 3.44 = 6.44$ in.

$c = \dfrac{a}{\beta_1} = \dfrac{6.44}{0.85} = 7.58$ in.

$\epsilon_t = \left(\dfrac{d-c}{c}\right)(0.003) = \left(\dfrac{24 - 7.58}{7.58}\right)(0.003)$

$= 0.00650 > 0.005 \;\; \therefore \;\; \phi = 0.90$ as assumed

---

Our procedure for designing T beams has been to assume a value of z, compute a trial steel area of A_s, determine a for that steel area assuming a rectangular section, and so on. Should $a > h_f$, we will have a real T beam. A trial-and-error process was used for such a beam in Example 5.5. It is easily possible, however, to determine A_s directly using the method of Section 5.3 where the member was broken down into its rectangular components. For this discussion, reference is made to Figure 5.7.

The compression force provided by the overhanging flange rectangles must be balanced by the tensile force in part of the tensile steel A_{sf} while the compression force in the web is balanced by the tensile force in the remaining tensile steel A_{sw}.

For the overhanging flange we have

$$0.85f'_c(b - b_w)(h_f) = A_{sf}f_y$$

from which the required area of steel A_{sf} equals

$$A_{sf} = \frac{0.85f'_c(b - b_w)h_f}{f_y}$$

The design strength of these overhanging flanges is

$$M_{uf} = \phi A_{sf} f_y \left(d - \frac{h_f}{2} \right)$$

The remaining moment to be resisted by the web of the T beam and the steel required to balance that value are determined next.

$$M_{uw} = M_u - M_{uf}$$

The steel required to balance the moment in the rectangular web is obtained by the usual rectangular beam expression. The value $M_{uw}/\phi b_w d^2$ is computed, and ρ_w is determined from the appropriate Appendix table or the expression for ρ_w previously given in Section 3.4 of this book. Think of ρ_w as the reinforcement ratio for the beam shown in Figure 5.7(b). Then

$$A_{sw} = \rho_w b_w d$$
$$A_s = A_{sf} + A_{sw}$$

EXAMPLE 5.6

Rework Example 5.5 using the rectangular component method just described.

SOLUTION

First assume $a \leq h_f$ (which is very often the case). Then the design would proceed like that of a rectangular beam with a width equal to the effective width of the T beam flange.

$$\frac{M_u}{\phi b d^2} = \frac{920(12,000)}{(0.9)(54)(24)^2} = 394.4 \text{ psi}$$

from Table A.12, $\rho = 0.0072$

$$a = \frac{\rho f_y d}{0.85 f'_c} = \frac{0.0072(60)(24)}{(0.85)(3)} = 4.06 \text{ in.} > h_f = 3 \text{ in.}$$

The beams acts like a T beam, not a rectangular beam, and the values for ρ and a above are not correct. If the value of a had been $\leq h_f$, the value of A_s would have been simply $\rho b d = 0.0072(54)(24) = 9.33$ in.^{2}. Now break the beam up into two parts (Figure 5.7) and design it as a T beam.

Assuming $\phi = 0.90$

$$A_{sf} = \frac{(0.85)(3)(54-15)(3)}{60} = 4.97 \text{ in.}^2$$
$$M_{uf} = (0.9)(4.97)(60)(24 - \tfrac{3}{2}) = 6039 \text{ in.-k} = 503 \text{ ft-k}$$
$$M_{uw} = 920 - 503 = 417 \text{ ft-k}$$

Designing a rectangular beam with $b_w = 15$ in. and $d = 24$ in. to resist 417 ft-k

$$\frac{M_{uw}}{\phi b_w d^2} = \frac{(12)(417)(1000)}{(0.9)(15)(24)^2} = 643.5$$
$$\rho_w = 0.0126 \text{ from Appendix Table A.12}$$
$$A_{sw} = (0.0126)(15)(24) = 4.54 \text{ in.}^2$$
$$A_s = 4.97 + 4.54 = \underline{9.51 \text{ in.}^2}$$

5.5   DESIGN OF T BEAMS FOR NEGATIVE MOMENTS

When T beams are resisting negative moments, their flanges will be in tension and the bottom of their stems will be in compression, as shown in Figure 5.12. Obviously, for such situations the rectangular beam design formulas will be used. Section 10.6.6 of the ACI Code requires that part of the flexural steel in the top of the beam in the negative-moment region be distributed over the effective width of the flange or over a width equal to one-tenth of the beam span, whichever is smaller. Should the effective width be greater than one-tenth of the span length, the Code requires that some additional longitudinal steel be placed in the outer portions of the flange. The intention of this part of the Code is to minimize the sizes of the flexural cracks that will occur in the top surface of the flange perpendicular to the stem of a T beam subject to negative moments.

In Section 3.8 it was stated that if a rectangular section had a very small amount of tensile reinforcing, its design-resisting moment ϕM_n might very well be less than its cracking moment. If this were the case, the beam might fail without warning when the first crack occurred. The same situation applies to T beams with a very small amount of tensile reinforcing.

When the flange of a T beam is in tension, the amount of tensile reinforcing needed to make its ultimate resisting moment equal to its cracking moment is about twice that of a rectangular section or that of a T section with its flange in compression. As a result, ACI Section 10.5.1 states that the minimum amount of reinforcing required equals the larger of the two values that follow:

$$A_{s\,min} = \frac{3\sqrt{f'_c}}{f_y}b_w d \qquad\qquad \text{ACI Equation(10-3)}$$

or

$$A_{s\,min} = \frac{200\,b_w d}{f_y}$$

For statically determinate members with their flanges in tension, b_w in the above expression is to be replaced with either $2b_w$ or the width of the flange, whichever is smaller.

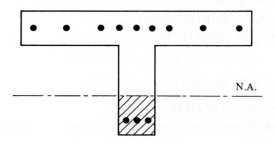

N.A.

Figure 5.12 T beam with flange in tension and bottom (hatched) in compression (a rectangular beam).

Reinforced concrete building in Calgary, Canada. (Courtesy of EFCO Corp.)

New Comiskey Park, Chicago, Illinois. (Courtesy of Economy Forms Corporation.)

5.6   L-SHAPED BEAMS

The author assumes for this discussion that L beams (that is, edge T beams with a flange on one side only) are not free to bend laterally. Thus they will bend about their horizontal axes and will be handled as symmetrical sections exactly as T beams.

For L beams the effective width of the overhanging flange may not be larger than one-twelfth the span length of the beam, six times the slab thickness, or one-half the clear distance to the next web (ACI 8.12.3).

If an L beam is assumed to be free to deflect both vertically and horizontally it will be necessary to analyze it as an unsymmetrical section with bending about both the horizontal and vertical axes. An excellent reference on this topic is given in a book by MacGregor.[1]

5.7   COMPRESSION STEEL

The steel that is occasionally used on the compression sides of beams is called *compression steel*, and beams with both tensile and compressive steel are referred to as *doubly reinforced beams*. Compression steel is not normally required in sections designed by the strength method because use of the full compressive strength of the concrete decidedly decreases the need for such reinforcement, as compared to designs made with the working-stress design method.

Occasionally, however, space or aesthetic requirements limit beams to such small sizes that compression steel is needed in addition to tensile steel. To increase the moment capacity of a beam beyond that of a tensilely reinforced beam with the maximum percentage of steel (when ($\epsilon_t = 0.005$)) it is necessary to introduce another resisting couple in the beam. This is done by adding steel in both the compression and tensile sides of the beam. Compressive steel increases not only the resisting moments of concrete sections, but also the amount of curvature that a member can take before flexural failure. This means that the ductility of such sections will be appreciably increased. Though expensive, compression steel makes beams tough and ductile, enabling them to withstand large moments, deformations, and stress reversals such as might occur during earthquakes. As a result, many building codes for earthquake zones require that certain minimum amounts of compression steel be included in flexural members.

Compression steel is very effective in reducing long-term deflections due to shrinkage and plastic flow. In this regard you should note the effect of compression steel on the long-term deflection expression in Section 9.5.2.5 of the Code (to be discussed in Chapter 6 of this text). Continuous compression bars are also helpful for positioning stirrups (by tying them to the compression bars) and keeping them in place during concrete placement and vibration.

Tests of doubly reinforced concrete beams have shown that even if the compression concrete crushes, the beam may very well not collapse if the compression steel is enclosed by stirrups. Once the compression concrete reaches its crushing strain, the concrete cover spalls or splits off the bars, much as in columns (see Chapter 9). If the compression bars are confined by closely spaced stirrups, the bars will not buckle until additional moment is applied. This additional moment cannot be considered in practice because beams are not practically useful after part of their concrete breaks off. (Would you like to use a building after some parts of the concrete beams have fallen on the floor?)

Section 7.11.1 of the ACI Code states that compression steel in beams must be enclosed by ties or stirrups, or by welded wire fabric of equivalent area. In Section 7.10.5.1 the Code states that the ties must be at least #3 in size for longitudinal bars #10 and smaller and at least #4 for larger longitudinal bars and bundled longitudinal bars. The ties may not be spaced farther apart than 16 bar diameters, 48 tie diameters, or the least dimension of the beam cross section (Code 7.10.5.2).

---

[1]MacGregor, J. G., 2005, *Reinforced Concrete Mechanics and Design*, 4th ed. (Upper Saddle River, NJ: Pearson Prentice Hall), pp. 198–201.

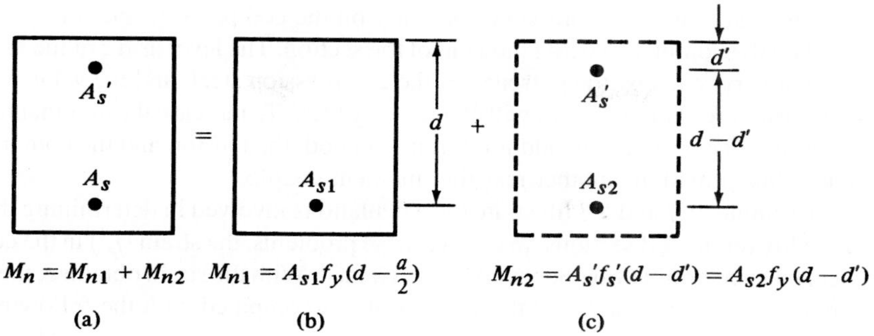

$$M_n = M_{n1} + M_{n2} \quad M_{n1} = A_{s1}f_y\left(d - \frac{a}{2}\right) \qquad M_{n2} = A_s'f_s'(d - d') = A_{s2}f_y(d - d')$$

(a)   (b)   (c)

Figure 5.13

For doubly reinforced beams an initial assumption is made that the compression steel yields as well as the tensile steel. (The tensile steel is always assumed to yield because of the ductile requirements of the ACI Code.) If the strain at the extreme fiber of the compression concrete is assumed to equal 0.00300 and the compression steel A_s' is located two-thirds of the distance from the neutral axis to the extreme concrete fiber, then the strain in the compression steel equals $\frac{2}{3} \times 0.003 = 0.002$. If this is greater than the strain in the steel at yield, as say $50,000/(29 \times 10^6) = 0.00172$ for 50,000-psi steel, the steel has yielded. It should be noted that actually the creep and shrinkage occurring in the compression concrete help the compression steel to yield.

Sometimes the neutral axis is quite close to the compression steel. As a matter of fact, in some beams with low steel percentages, the neutral axis may be right at the compression steel. For such cases the addition of compression steel adds little, if any, moment capacity to the beam. It can, however, make the beam more ductile.

When compression steel is used, the nominal resisting moment of the beam is assumed to consist of two parts: the part due to the resistance of the compression concrete and the balancing tensile reinforcing and the part due to the nominal moment capacity of the compression steel and the balancing amount of the additional tensile steel. This situation is illustrated in Figure 5.13. In the expressions developed here, the effect of the concrete in compression, which is replaced by the compressive steel A_s' is neglected. This omission will cause us to overestimate M_n by a very small and negligible amount (less than 1%). The first of the two resisting moments is illustrated in Figure 5.13(b).

$$M_{n1} = A_{s1} f_y\left(d - \frac{a}{2}\right)$$

The second resisting moment is that produced by the additional tensile and compressive steel (A_{s2} and A_s'), which is presented in Figure 5.13(c).

$$M_{n2} = A_s' f_y(d - d')$$

Up to this point it has been assumed that the compression steel has reached its yield stress. If such is the case, the values of A_{s2} and A_s' will be equal because the addition to T of $A_{s2}f_y$ must be equal to the addition to C of $A_s'f_y$ for equilibrium. If the compression steel has not yielded, A_s' must be larger than A_{s2}, as will be described later in this section.

Combining the two values, we obtain

$$M_n = A_{s1}f_y\left(d - \frac{a}{2}\right) + A_{s2}f_y(d - d')$$

$$\phi M_n = \phi\left[A_{s1}f_y\left(d - \frac{a}{2}\right) + A_{s2}f_y(d - d')\right]$$

The addition of compression steel only on the compression side of a beam will have little effect on the nominal resisting moment of the section. The lever arm z of the internal couple is not affected very much by the presence of the compression steel, and the value of T will remain the same. Thus the value $M_n = Tz$ will change very little. To increase the nominal resisting moment of a section, it is necessary to add reinforcing on both the tension and the compression sides of the beam, thus providing another resisting moment couple.

Examples 5.7 and 5.8 illustrate the calculations involved in determining the design strengths of doubly reinforced sections. In each of these problems, the strain (f_s') in the compression steel is checked to determine whether or not it has yielded. With the strain obtained, the compression steel stress (f_s') is determined, and the value of A_{s2} is computed with the following expression:

$$A_{s2}f_y = A_s'f_s'$$

In addition, it is necessary to compute the strain in the tensile steel (ϵ_t) because if it is less than 0.005, the value of the bending ϕ will have to be computed inasmuch as it will be less than its usual 0.90 value. The beam may not be used in the unlikely event that ϵ_t is less than 0.004.

To determine the value of these strains an equilibrium equation is written, which upon solution will yield the value of c and thus the location of the neutral axis. To write this equation, the nominal tensile strength of the beam is equated to its nominal compressive strength. Only one unknown appears in the equation, and that is c.

Initially the stress in the compression steel is assumed to be at yield $(f_s' = f_y)$. From Figure 5.14, summing forces horizontally in the force diagram and substituting $\beta_1 c$ for a leads to

$$A_s f_y = 0.85 f_c' \beta_1 c + A_s' f_y$$

$$c = \frac{(A_s - A_s')f_y}{0.85 f_c' \beta_1}$$

Referring to the strain diagram of Figure 5.14, from similar triangles

$$\epsilon_s' = \frac{c - d'}{c} 0.003$$

If the strain in the compression steel, $\epsilon_s' > \epsilon_y = f_y/E_s$, the assumption is valid and f_s' is at yield, f_y. If $\epsilon_s' < \epsilon_y$, the compression steel is not yielding and the value of c calculated above is not correct. A new equilibrium equation must be written that assumes $f_s' < f_y$.

$$A_s f_y = 0.85 f_c' \beta_1 cb + A_s' \left(\frac{c - d'}{c} \right)(0.003)(29,000)$$

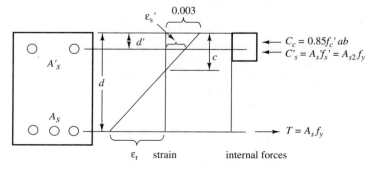

Figure 5.14 Internal strains and forces for doubly reinforced rectangular beam.

The value of c determined enables us to compute the strains in both the compression and tensile steels and thus their stresses. Even though the writing and solving of this equation are not too tedious, use of the Excel spreadsheet for beams with compression steel makes short work of the whole business.

Examples 5.7 and 5.8 illustrate the computation of the design moment strength of doubly reinforced beams. In the first of these examples the compression steel yields, while in the second it does not.

EXAMPLE 5.7

Determine the design moment capacity of the beam shown in Figure 5.15 for which $f_y = 60,000$ psi and $f_c' = 3000$ psi.

SOLUTION   Writing the equilibrium equation assuming $f_s' = f_y$

$$A_s f_y = 0.85 f_c' b \beta_1 c + A_s' f_y$$

$$(6.25)(60) = (0.85)(3)(14)(0.85c) + (2.00)(60)$$

$$c = \frac{(6.25 - 2.00)(60)}{(0.85)(3)(0.85)} = 8.40 \text{ in.}$$

$$a = \beta_1 c = (0.85)(8.40) = 7.14 \text{ in.}$$

Computing strains in compression steel to verify assumption that it is yielding

$$\epsilon_s' = \frac{c - d'}{c} 0.003 = \frac{8.40 - 2.5}{8.40} 0.003 = 0.00211$$

$$\epsilon_y = \frac{f_y}{E_s} = \frac{60,000}{29,000,000} = 0.00207 < \epsilon_s' \therefore f_s' = f_y \quad \text{as assumed.}$$

Note: Example 5.8 shows what to do if this assumption is not correct.

$$A_{s2} = \frac{A_s' f_s'}{f_y} = \frac{(2.00)(60,000)}{60,000} = 2.00 \text{ in.}^2$$

$$A_{s1} = A_s - A_{s2} = 6.25 - 2.00 = 4.25 \text{ in.}^2$$

$$\epsilon_t = \frac{d - c}{c} 0.003 = \frac{24 - 8.40}{8.40}(0.003) = 0.00557 > 0.005 \therefore \phi = 0.9$$

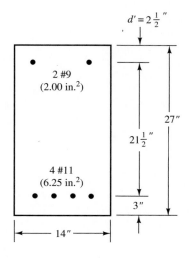

Figure 5.15

Then the design moment strength is

$$\phi M_n = \phi \left[A_{s1} f_y \left(d - \frac{a}{2} \right) + \left(A'_s f'_s (d - d') \right) \right]$$

$$= 0.9 \left[(4.25)(60) \left(24 - \frac{7.14}{2} \right) + (2.00)(60)(24 - 2.5) \right]$$

$$= 7010 \text{ in.-k} = 584.2 \text{ ft-k}$$

EXAMPLE 5.8

Compute the design moment strength of the section shown in Figure 5.16 is $f_y = 60{,}000$ psi and $f'_c = 4000$ psi.

SOLUTION   Writing the equilibrium equation assuming $f'_s = f_y$

$$A_s f_y = 0.85 f'_c b \beta_1 c + A'_s f_y$$

$$(5.06)(60) = (0.85)(4)(14)(0.85c) + (1.20)(60)$$

$$c = \frac{(5.06 - 1.20)(60)}{(0.85)(4)(0.85)} = 5.72 \text{ in.}$$

$$a = \beta_1 c = (0.85)(5.72) = 4.86 \text{ in.}$$

Computing strains in compression steel to verify assumption that it is yielding

$$\epsilon'_s = \frac{c - d'}{c} 0.003 = \frac{5.72 - 2.5}{5.72} 0.003 = 0.00169$$

$$\epsilon_y = \frac{f_y}{E_s} = \frac{60{,}000}{29{,}000{,}000} = 0.00207 > \epsilon'_s \therefore f'_s \neq f_y \quad \text{as assumed.}$$

Since the assumption is not valid, we have to use the equilibrium equation that is based on f'_s not yielding.

$$A_s f_y = 0.85 f'_c \beta_1 cb + A'_s \left[\frac{c - d'}{c} \right] (0.003) E_s$$

$$(5.06)(60) = (0.85)(4)(0.85c)(14) + (1.20) \frac{c - 2.5}{c} (0.003)(29{,}000)$$

Solving the quadratic equation for $c = 6.00$ in.² and $a = \beta_1 c = 5.10$ in.²

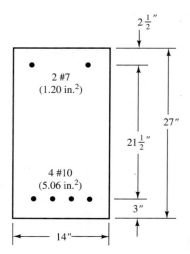

Figure 5.16

Computing strains, stresses and steel areas

$$\epsilon_s' = \frac{c-d'}{c}0.003 = \frac{6.00-2.5}{6.00}0.003 = 0.00175 < \epsilon_y$$

$$f_s' = \epsilon_s'E_s = (0.00175)(29,000) = 50.75 \text{ ksi}$$

$$A_{s2} = \frac{A_s'f_s'}{f_y} = \frac{(1.20)(50,750)}{60,000} = 1.015 \text{ in.}^2$$

$$A_{s1} = A_s - A_{s2} = 5.06 - 1.015 = 4.045 \text{ in.}^2$$

$$\epsilon_t = \frac{d-c}{c}0.003 = \frac{24-6.00}{6.00}(0.003) = 0.0090 > 0.005 \therefore \phi = 0.9$$

Then the design moment strength is

$$\phi M_n = \phi\left[A_{s1}f_y\left(d-\frac{a}{2}\right) + A_s'f_s'(d-d')\right]$$

$$= 0.9\left[(4.045)(60)\left(24-\frac{5.10}{2}\right) + (1.20)(50.75)(24-2.5)\right]$$

$$= 5863 \text{ in.-k} = 488.6 \text{ ft-k}$$

5.8   DESIGN OF DOUBLY REINFORCED BEAMS

Sufficient tensile steel can be placed in most beams so that compression steel is not needed. But if it is needed, the design is usually quite straightforward. Examples 5.9 and 5.10 illustrate the design of doubly reinforced beams. The solutions follow the theory used for analyzing doubly reinforced sections.

EXAMPLE 5.9

SOLUTION   Design a rectangular beam for $M_D = 325$ ft-k and $M_L = 400$ ft-k if $f_c' = 4000$ psi and $f_y = 60,000$ psi. The maximum permissible beam dimensions are shown in Figure 5.17.

$$M_u = (1.2)(325) + (1.6)(400) = 1030 \text{ ft-k}$$

Assuming $\phi = 0.90$

$$M_n = \frac{M_u}{\phi} = \frac{1030}{0.90} = 1144.4 \text{ ft-k}$$

Assuming maximum possible tensile steel with no compression steel and computing beams nominal moment strength

$$\rho_{max} \text{ from Appendix Table A.7} = 0.0181$$

$$A_{s1} = (0.0181)(15)(28) = 7.60 \text{ in.}^2$$

For $\rho = 0.0181 \frac{M_u}{\phi bd^2}$ from Appendix Table A.13 = 912.0

$$M_{u1} = (912.0)(0.9)(15)(28)^2 = 9,652,608 \text{ in.-lb}$$

$$= 804.4 \text{ ft-k}$$

$$M_{n1} = \frac{804.4}{0.90} = 893.8 \text{ ft-k}$$

$$M_{n2} = M_n - M_{n1} = 1144.4 - 893.8 \text{ ft-k} = 250.6 \text{ ft-k}$$

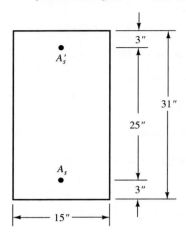

Figure 5.17

Checking to See Whether Compression Steel Has Yielded

$$a = \frac{(7.60)(60)}{(0.85)(4)(15)} = 8.94 \text{ in.}$$

$$c = \frac{8.94}{0.85} = 10.52 \text{ in.}$$

$$\epsilon'_s = \left(\frac{10.52 - 3}{10.52}\right)(0.00300) = 0.00214 > 0.00207$$

Therefore, compression steel has yielded.

$$\text{Theoretical } A'_s \text{ required} = \frac{M_{n2}}{(f_y)(d - d')} = \frac{(12)(250.6)}{(60)(28 - 3)} = 2.00 \text{ in.}^2$$

$$A'_s f'_s = A_{s2} f_y$$

$$A_{s2} = \frac{A'_s f'_s}{f_y} = \frac{(2.00)(60)}{60} = \underline{2.00 \text{ in.}^2} \quad \frac{\text{Try}}{2 \text{ \#9 } (2.00 \text{ in.}^2)}$$

$$A_s = A_{s1} + A_{s2}$$

$$A_s = 7.60 + 2.00 = 9.60 \text{ in.}^2 \quad \frac{\text{Try}}{8 \text{ \#10 } (10.12 \text{ in}^2.)}$$

If we had been able to select bars with exactly the same areas as calculated here, ϵ_t would have remained = 0.005 and $\phi = 0.90$, but such was not the case.

Writing the quadratic equation c is found to equal 11.08 in. and $a = 9.43$ in.

$$\epsilon'_s = \left(\frac{11.80 - 3}{11.80}\right)(0.003) = 0.00224 > 0.00207 \qquad \underline{\underline{\text{OK}}}$$

$$\epsilon_t = \left(\frac{28 - 11.80}{11.80}\right)(0.003) = 0.00412 < 0.827$$

$$\phi = 0.65 + (0.00412 - 0.002)\left(\frac{250}{3}\right) = 0.827$$

$$\phi M_n = 0.827\left[(10.12 - 2.00)(60)\left(28 - \frac{9.43}{2}\right) + (2.00)(60)(25)\right]$$

$$= 11,863 \text{ in. -lbs} = 989 \text{ ft-k} < 1030 \text{ ft-k} \qquad \underline{\underline{\text{No good}}}$$

The reason the beam does not have sufficient capacity is because of the variable ϕ factor. This can be avoided if you are careful in picking bars. Note that the actual value of A_s' is exactly the same as the theoretical value. The actual value of A_s, on the other hand, is higher than the theoretical value by $10.12 - 9.6 = 0.52$ in.^{2} If a new bar selection for A_s' is made whereby the actual value of A_s' exceeds the theoretical value by about this much (0.52 in.^{2}), the design will be adequate. Select three #8 bars ($A_s' = 2.36$ in.^{2}) and repeat the previous steps. Note that the actual steel areas are used below, not the theoretical ones. As a result, the value of c, a, ϵ_s' and f_s' must be recalculated.

Assume $f_s' = f_y$.

$$c = \frac{(A_s - A_s')f_y}{0.85 f_c' b \beta_1} = \frac{(10.12 - 2.36)60}{0.85(4)(15)(0.85)} = 10.74 \text{ in.}$$

$$\epsilon_s' = \frac{c - d'}{c} 0.003 = \frac{10.74 - 3}{10.74} 0.003 = 0.00216 > \epsilon_y \therefore \text{ assumption valid}$$

$$\epsilon_t = \frac{d - c}{c} 0.003 = \frac{28 - 10.74}{10.74} 0.003 = 0.00482 < 0.005 \therefore \phi \neq 0.9$$

$$\phi = 0.65 + (\epsilon_t - 0.002)\frac{250}{3} = 0.88$$

$$A_{s2} = \frac{A_s' f_s'}{f_y} = \frac{2.36(60)}{60} = 2.36 \text{ in.}^2$$

$$A_{s1} = A_s - A_{s2} = 10.12 - 2.36 = 7.76 \text{ in.}$$

$$M_{n1} = A_{s1} f_y \left(d - \frac{a}{2}\right) = 7.76(60)\left[28 - \frac{0.85(10.74)}{2}\right] = 10{,}912 \text{ ft-k}$$

$$M_{n2} = A_{s2} f_y (d - d') = 2.36(60)[28 - 3] = 3540 = 295 \text{ ft-k}$$

$$M_n = M_{n1} + M_{n2} = 909.3 + 295 = 1204.3 \text{ ft-k}$$

$$\phi M_n = 0.88(1204.3) = 1059.9 \text{ ft-k} > M_u \qquad \underline{\text{OK}}$$

Note that eight #10 bars will not fit in a single layer in this beam. If they are placed in two layers, the centroid would have to be more than 3 in. from the bottom of the section. It would be necessary to increase the beam depth, h, in order to provide for two layers or to use bundled bars (Section 7.4).

EXAMPLE 5.10

A beam is limited to the dimensions shown in Figure 5.17. If $M_D = 170$ ft-k, $M_L = 225$ ft-k, $f_c' = 4000$ psi, and $f_y = 60,000$ psi, select the reinforcing required $b = 15$ in., $d = 20$ in. and $d' = 4$ in.

SOLUTION

$$M_u = (1.2)(170) + (1.6)(225) = 564 \text{ ft-k}$$

Assuming $\phi = 0.90$

$$M_n = \frac{564}{0.90} = 626.7 \text{ ft-k}$$

$$\text{Max } A_{s1} = (0.0181)(15)(20) = 5.43 \text{ in.}^2$$

$$\text{For } \rho = 0.0181 \; \frac{M_u}{\phi b d^2} = 912.0 \text{ from Appendix Table A.13}$$

$$M_{u1} = (912)(0.90)(15)(20)^2 = 4{,}924{,}800 \text{ in.-lb} = 410.4 \text{ ft-k}$$

$$M_{n1} = \frac{410.4}{0.90} = 456.0 \text{ ft-k}$$

$$M_{n2} = 626.7 - 456.0 = 170.7 \text{ ft-k}$$

Checking to See if Compression Steel Has Yielded

$$a = \frac{A_{s1}f_y}{0.85f_c'b} = \frac{(5.43)(60)}{(0.85)(4)(15)} = 6.39 \text{ in.}$$

$$c = \frac{6.39}{0.85} = 7.52 \text{ in.}$$

$$\epsilon_s' = \left(\frac{7.52 - 4.00}{7.52}\right)(0.003) = 0.00140 < \frac{60}{29,000} = 0.00207$$

$$\therefore f_s' = (0.00140)(29000) = 40.6 \text{ ksi}$$

$$\text{Theoretical } A_s' \text{ reqd} = \frac{M_{n2}}{f_s'(d - d')}$$

$$= \frac{(12)(170.7)}{(40.6)(20 - 4)} = \underline{3.15 \text{ in.}^2} \quad \underline{\text{Try 4\#8 } (3.14 \text{ in.}^2)}$$

$$A_s'f_s^1 = A_{s2}f_y$$

$$A_{s2} = \frac{(3.15)(40.6)}{60} = 2.13 \text{ in.}^2$$

$$A_s = A_{s1} + A_{s2} = 5.43 + 2.13 = \underline{7.56 \text{ in.}^2} \quad \underline{\text{Try 6 \#10 } (7.59 \text{ in.}^2)}$$

Subsequent checks for c, ϵ_t, and so on, prove this design to be satisfactory.

5.9   SI EXAMPLES

Examples 5.11 and 5.12 illustrate the analysis of a T beam and the design of a doubly reinforced beam using SI units.

EXAMPLE 5.11

Determine the design strength of the T beam shown in Figure 5.18 if $f_y = 420$ MPa, $f_c' = 35$ MPa, and $E_s = 200,000$ MPa.

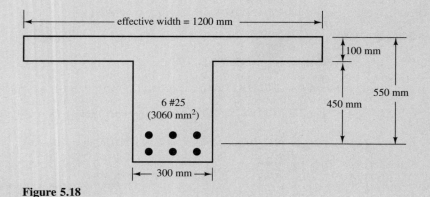

Figure 5.18

SOLUTION   **Computing T and A_c**

$$T = (3060)(420) = 1\,285\,200\,\text{N}$$

$$A_c = \frac{T}{0.85f_c'} = \frac{1\,285\,200}{(0.85)(35)} = 43\,200\,\text{mm}^2$$

$$\rho = \frac{A_s}{b_w d} = \frac{3060}{(300)(550)} = 0.0185 < \rho_{max}$$

$$= 0.0216\,\text{from Appendix Table B.7}\quad\underline{\text{OK}}$$

$$\rho_{min} = \frac{\sqrt{f_c'}}{4f_y} = \frac{\sqrt{35}}{(4)(420)} = 0.003\,52 < 0.0185\quad\underline{\text{OK}}$$

or

$$\frac{1.4}{f_y} = \frac{1.4}{420} = 0.003\,33 < 0.0185\quad\underline{\underline{\text{OK}}}$$

Calculating Design Strength

$$a = \frac{43,200}{1200} = 36\,\text{mm} < h_f = 100\,\text{mm}$$

$$\therefore\text{stress block is entirely within flange}$$

$$z = d - \frac{a}{2} = 550 - \frac{36}{2} = 532\,\text{mm}$$

$$\phi M_n = \phi Tz$$

$$= (0.9)(1\,285\,200)(532)$$

$$= 6.153 \times 10^8\,\text{N}\cdot\text{mm} = \underline{\underline{615.3\,\text{kN}\cdot\text{m}}}$$

EXAMPLE 5.12

If $M_u = 1225\,\text{kN}\cdot\text{m}$, determine the steel area required for the section shown in Figure 5.19. Should compression steel be required, assume that it will be placed 70 mm from the compression face. $f_c' = 21\,\text{MPa}$, $f_y = 420\,\text{MPa}$, and $E_s = 200,000\,\text{MPa}$.

SOLUTION

$$M_n = \frac{1225}{0.9} = 1361\,\text{kN}\cdot\text{m}$$

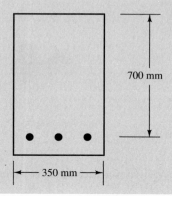

700 mm

350 mm

Figure 5.19

$$\rho_{\max} \text{ if singly reinforced} = 0.0135 \text{ from Appendix Table B.7}$$

$$A_{s1} = (0.0135)(350)(700) = 3307 \text{ mm}^2$$

$$\frac{M_{u1}}{\phi b d^2} = 4.769 \text{ from Appendix Table B.8}$$

$$M_{u1} = (\phi b d^2)(4.769) = \frac{(4.769)(0.9)(350)(700)^2}{10^6} = 736.1 \text{kN} \cdot \text{m}$$

$$M_{n1} = \frac{736.1}{0.9} = 818 \text{ kN} \cdot \text{m}$$

$$< M_n \text{ of } 1361 \text{ kN} \cdot \text{m} \quad \therefore \underline{\text{double reinf. required}}$$

$$M_{n2} = M_n - M_{n1} = 1361 - 818 = 543 \text{ kN} \cdot \text{m}$$

Checking to See if Compression Steel Yields

$$a = \frac{(3307)(420)}{(0.85)(21)(350)} = 222.32 \text{ mm}$$

$$c = \frac{222.32}{0.85} = 261.55 \text{ mm}$$

$$\epsilon_s' = \left(\frac{261.55 - 70}{261.55} \right)(0.003) = 0.00220$$

$$> \frac{420}{200\,000} = 0.002\,10 \quad \therefore \underline{\text{compression steel yields}}$$

$$A_s' \text{ reqd} = \frac{M_{n2}}{f_y(d - d')} = \frac{543 \times 10^6}{(700 - 70)(420)} = \underline{\underline{2052 \text{ mm}^2}}$$

$$\underline{\underline{\text{Use 3 #32 bars } (2457 \text{ mm}^2)}}$$

$$A_s = A_{s1} + A_{s2} = 3307 + 2052 = \underline{\underline{5359 \text{ mm}^2}}$$

$$\text{Use 6 #36 bars} \underline{\underline{(A_s = 6036 \text{ mm}^2)}}$$

5.10   COMPUTER EXAMPLES

EXAMPLE 5.13

Repeat Example 5.3 using the Excel spreadsheet.

SOLUTION   Open the Excel spreadsheet for T beams and select the Analysis worksheet tab at the bottom. Input only cells C3 through C9 (yellow highlight).

T Beam Analysis

$f'_c =$	4000	psi
$f_y =$	60,000	psi
$b_{eff} =$	30	in.
$b_w =$	14	in.
$d =$	30	in.
$h_f =$	4	in.
$A_s =$	10.12	in.^{2}

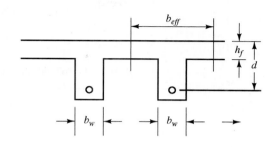

$\beta_1 =$ 0.85

$$A_{s\,min} = \frac{3\sqrt{f'_c}}{f_y} b_w d = 1.33 \text{ in.}^2$$

$$A_{s\,min} = \frac{200}{f_y} b_w d = 1.40 \text{ in.}^2 \qquad A_{s\,min} = 1.40 \text{ in.}^2 \qquad \text{minimum steel is } \underline{\underline{\text{OK}}}$$

If a $\leq h_f$: $a > h_f$, so this analysis is not valid, go to the box below

$$a = \frac{A_s f_y}{0.85 f'_c b} = 5.95 \text{ in.}$$

$$M_n = A_s f_y (d - a/2) = \qquad - \quad \text{in.-lb} \quad = \quad \begin{array}{c}\text{See}\\\text{solution}\\\text{below}\end{array} \quad \text{ft-k}$$

$c = a/\beta_1 = 7.00346$

$$\epsilon_t = \frac{d-c}{c} 0.003 = 0.009851 \qquad -$$

$\phi = 0.9$

$$\phi M_n = \qquad - \quad \text{in.-lb} \quad = \quad \begin{array}{c}\text{see}\\\text{solution}\\\text{below}\end{array} \quad \text{ft-k}$$

If a $> h_f$: $a > h_f$ so this analyse is valid-acts like a T-Beam

$$A_{sf} = \frac{0.85 f'_c (b - b_w) h_f}{f_y} = 3.627 \text{ in.}^2$$

$$A_{sw} = A_s - A_{sf} \qquad = 6.493 \text{ in.}^2$$

$$a = \frac{A_{sw} f_y}{0.85 f'_c b} \qquad = 8.185 \text{ in.}$$

$$M_n = A_{sf} f_y \left(d - \frac{h_f}{2}\right) + A_{sw} f_y \left(d - \frac{a}{2}\right) = 16{,}186{,}387 \text{ in.-lb} = 1348.9 \text{ ft-k}$$

$$c = a/\beta_1 = 9.629263 \text{ in.}$$

$$\epsilon_t = \frac{d-c}{c}0.003 = 0.006347 \qquad -$$

$$\phi = 0.9$$

$$\phi\,M_n = 14{,}567{,}748 \text{ in.-1b} = 1214.0 \text{ ft-k}$$

EXAMPLE 5.14

Repeat Example 5.6 using the Excel spreadsheet.

SOLUTION   Open the Excel spreadsheet for T beams and select the T Beam Design worksheet tab at the bottom. Input only cells C3 through C9 (yellow highlight).

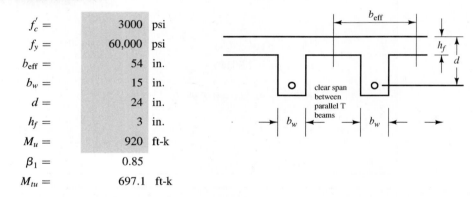

T Beam Design

$f_c' =$	3000	psi
$f_y =$	60,000	psi
$b_{eff} =$	54	in.
$b_w =$	15	in.
$d =$	24	in.
$h_f =$	3	in.
$M_u =$	920	ft-k
$\beta_1 =$	0.85	
$M_{tu} =$	697.1	ft-k

If $a \le h_f$: $a > h_f$ so this analysis is not valid, go to box below

$$R_n = \frac{M_u}{\phi b d^2} = 394.38 \text{ in.}$$

$$\rho = \frac{0.85f_c'}{f_y}\left[1 - \sqrt{1 - \frac{2R_n}{0.85f_c'}}\right] = 0.007179$$

$$a = \frac{\rho f_y d}{0.85f_c'} = 4.054201 \text{ in.}$$

$$c = a/\beta_1 = 4.769648$$

$$\epsilon_t = \frac{d-c}{c} 0.003 = 0.012095 \qquad A_{s\,min} = \frac{3\sqrt{f_c'}}{f_y} b_w d = \quad 0.985901 \text{ in.}^2$$

$$\phi = 0.9 \qquad\qquad - \qquad\qquad -$$

$$A_s = \rho bd = 9.304391 \text{ in.}^2 \qquad A_{s\,min} = \frac{200}{f_y} b_w d = 1.2 \text{ in.}^2$$

$$-$$

If $a > h_f$: $a > h_f$ so this analyse is valid-acts like a T-Beam

$$A_{sf} = \frac{0.85 f_c'(b - b_w)h_f}{f_y} = 4.97 \text{ in.}^2$$

$$M_{uf} = A_{sf} f_y (d - hf/2) = 503.5 \text{ ft-k}$$

$$M_{uw} = M_{uf} - M_u = 416.5$$

$$R_{nw} = \frac{M_{uw}}{\phi b d^2} = 642.8$$

$$\rho_w = \frac{0.85 f_c'}{f_y}\left[1 - \sqrt{1 - \frac{2R_{nw}}{0.85 f_c'}}\right] = 0.012573$$

$$a = \frac{\rho_w f_y d}{0.85 f_c'} = 7.100128 \text{ in.}$$

$$c = a/\beta_1 = 8.353092 \text{ in.}$$

$$\epsilon_t = \frac{d-c}{c} 0.003 = 0.00562 \qquad\qquad -$$

$$\phi = 0.9 \qquad\qquad -$$

$$A_{sw} = \rho_\omega b_w d = \quad 4.53 \text{ in.}^2 \qquad A_{s\,min} = \frac{3\sqrt{f_c'}}{f_y} b_w d = 0.985901 \text{ in.}^2 \quad -$$

$$\boxed{A_s = A_{sw} + A_{sf} = \quad 9.50 \text{ in.}^2} \qquad A_{s\,min} = \frac{200}{f_y} b_w d = 1.2 \text{ in.}^2$$

Note: Solution is based on $\phi = 0.9$

EXAMPLE 5.15

Repeat Example 5.7 using the Excel spreadsheet.

SOLUTION   Open the Excel spreadsheet for Beams with Compression Steel and select the Analysis worksheet tab at the bottom. Input only cells C3 through C9 (yellow highlight). Other values are calculated from those input values. See comment on cell E22 for goalseek instructions.

Analysis of Doubly Reinforced Beams by ACI 318-08

A_s, A'_s, b, d, M_u, f'_c, f_y known or specified

$b =$	14	in.
$d =$	24	in.
$d' =$	2.5	in.
$A'_s =$	2.00	in.^{2}
$A_s =$	6.25	in.^{2}
$f_c =$	3000	psi
$f_y =$	60,000	psi
$\beta_1 =$	0.85	

Determine the location of the neutral axis, c.

$c =$	8.40	in.
$\epsilon'_s =$	0.00211	
$a =$	7.14	in.
$f'_s =$	60.00	psi Compression steel is at yield
$0.85 f'_c \beta_1 cb + A'_s f'_s =$	$A_s f_y$	
$0.85 f'_c \beta_1 cb =$	255	
$A'_s f'_s =$	120	kips
$A_s f_y =$	375	kips
$0.85 f'_c \beta_1 cb + A'_s f' - A_s f_y =$	0	OK
$A_{s2} = A'_s f'_s / f_y =$	2.000	in.^{2}
$A_{s1} = A_s - A_{s2} =$	4.250	in.^{2}

You must use "Goal Seek" to set this cell $= 0$ by changing c in cell E14. Go to "Tools" at the top of the screen to find Goal Seek.

$\epsilon_t = \dfrac{d-c}{c} 0.003 =$	0.005568	OK tensile strain exceeds 0.004		
$\phi =$	0.900	–		
$M_{n1} = (A_s{-}f_y{-}A_s{'}f'_s)(d{-}a/2) =$	5209.29	in.-lb $=$	434.11	kip-ft
$\phi M_{ni} =$	4688.36	in.-lb $=$	390.70	kip-ft
$M_{n2} =$	2580.00	in.-lb $=$	215.00	kip-ft
$\phi M_{n2} =$	2322.00	in.-lb $=$	193.50	kip-ft
$M_n =$	7789.29	in.-lb $=$	649.11	kip-ft
$\phi M_n =$	7010.36	in.-lb $=$	584.20	kip-ft

EXAMPLE 5.16

Repeat Example 5.9 using the Excel spreadsheet.

SOLUTION   Open the Excel spreadsheet for Beams with Compression Steel and select the ACI 318-08 Case I worksheet tab at the bottom. Input only cells C3 through C9 (yellow highlight). Other values calculated from those input values.

Design of Doubly Reinforced Beams by ACI 318-08 when both A_s and A'_s are unknown (Case I)

Case I: A_s and A'_s are unknown; b, d, M_u, f'_u, f_y know or specified

$$
\begin{aligned}
M_u &= 1030.00 \text{ k-ft} \\
b &= 15 \text{ in.} \\
d &= 28 \text{ in.} \\
d' &= 3 \text{ in.} \\
f'c &= 4000 \text{ psi} \\
f_y &= 60{,}000 \text{ psi} \\
\beta_1 &= 0.85
\end{aligned}
$$

1. Determine the maximum ultimate moment permitted by the code for the beam if it were singly reinforced (using the maximum value of ρ associated with $\phi = 0.9$).

$$\rho = 0.375(0.85\beta_1 f'_c/f_y) = 0.018063$$

$$M_{max} = \phi\rho bd^2 f_y\left(1 - \frac{\rho f_y}{1.7 f'_m}\right) \quad \begin{aligned} &= 96{,}42313.4 \text{ in.-lb} \\ &= 803.53 \text{ k-ft} \end{aligned}$$

2. If $M_{max} \geq M_u$, compression steel is not needed. Design as singly reinforced beam.
   If $M_{max} < M_u$, continue to step 3.

3. The most economical design uses $M_{u1} = M_{max}$ which corresponds to ρ_1 = the maximum value of ρ associated with $\phi = 0.9$.

$$
\begin{aligned}
\rho_1 &= 0.018063 \\
A_{s1} = \rho_1 bd &= 7.586 \text{ in.}^2 \\
M_{u1} = \phi A_{s1} f_y (d - a/2) &= 9642313.41 \text{ in.-lb} \\
&= 803.53 \text{ k-ft}
\end{aligned}
$$

$$a = \frac{A_s f_y}{.85 f'_c b} = 8.925 \text{ in.} \qquad c = \frac{\alpha}{\beta_1} = 10.500 \text{ in.}$$

$$\epsilon_t = \frac{d-c}{c}0.003 = 0.00500$$

$$\phi = 0.65 + (\epsilon_t - .002)\cdot 250/3 = 0.900$$

4. $M_{u2} = M_u - M_{u1} \qquad = 226.47 \text{ k-ft}$

5. $A_{s2} = \dfrac{M_{u2}}{.\phi \cdot f_y(d - d')} \qquad = 2.013 \text{ in.}^2$

6. $c = a/\beta_1 \qquad = 10.500 \text{ in.}$

7. $f'_s = \dfrac{c - d'}{c}87{,}000 \qquad = 62143 \text{ psi} \qquad$ if $f'_s > f_y$, use $f_s = f_y \qquad f'_s = 60{,}000 \text{ psi}$

8. $A_s = \dfrac{A_{s2}f_y}{f'_s} \qquad = 2.01 \text{ in.}^2 \qquad$ Select bars $\underset{\text{No. of bars}}{3} \quad \underset{\text{Bar size}}{\#8} \quad A'_s = 2.36 \text{ in.}^2$

9. $A'_s = A_{s1} + A_{s2} \qquad = 9.60 \text{ in.}^2 \qquad$ Select bars $8 \quad \#10 \quad A_s = 10.13 \text{ in.}^2$

PROBLEMS

5.1 What is the effective width of a T beam? What does it represent?

5.2 What factors affect the selection of the dimensions of T beam stems?

5.3 If additional reinforcing bars are placed only in the compression side of a reinforced concrete beam, will they add significantly to the beam's flexural strength? Explain your answer.

5.4 Why is compression reinforcing particularly important in reinforced concrete flexual members located in earthquake-prone areas?

Analysis of T Beams

In Problems 5.5 to 5.15 determine the design moment strengths, ϕM_n, of the sections shown. Use $f_y = 60,000$ psi and $f'_c = 4000$ psi, except for Problem 5.9 where $f'_c = 5000$ psi. Check each section to see if it is ductile.

Problem 5.5 (*Ans.* 393.5 ft-k)

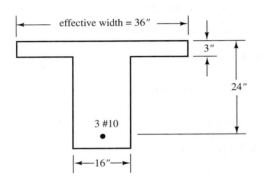

5.6 Repeat Problem 5.5 if four #10 bars are used.

5.7 Repeat Problem 5.5 if ten #8 bars are used.
(*Ans.* 775.7 ft-k)

Problem 5.8

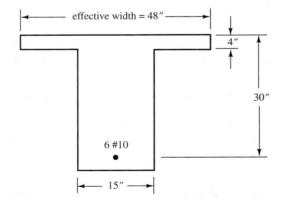

5.9 Repeat Problem 5.8 if $f'_c = 5000$ psi. (*Ans.* 986.5 ft-k)

5.10 Repeat Problem 5.8 if ten #10 bars are used and if $f'_c = 4000$ psi.

5.11 (*Ans.* 415.8 ft-k)

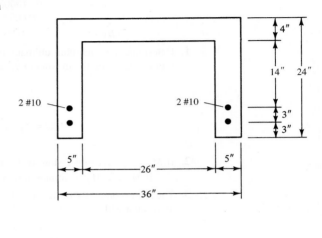

Problem 5.12

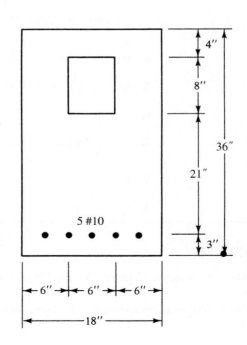

Problem 5.13 (*Ans.* 493.1 ft-k)

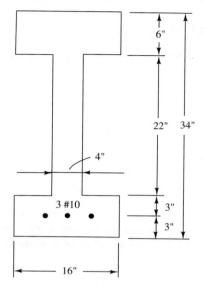

Problem 5.15   (*Ans.* 838.1 ft-k)

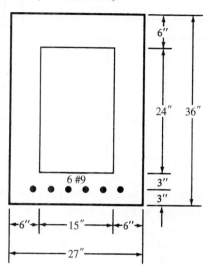

5.16 Calculate the design strength ϕM_n for one of the T beams if $f'_c = 5000$ psi, $f_y = 60,000$ psi, and the section has a 24-ft simple span. Is $\epsilon_t \geq 0.005$?

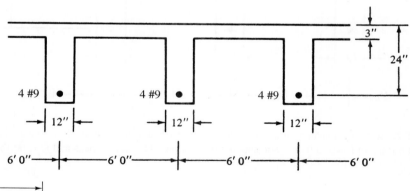

Problem 5.14

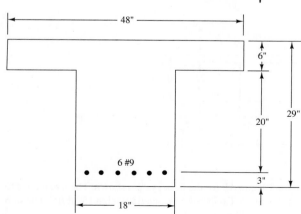

5.17 Repeat Problem 5.16 if $f'_c = 3000$ psi and 3 #10 bars are used. (*Ans.* 396.7 ft-k)

Design of T Beams

5.18 Determine the area of reinforcing steel required for the T beam shown if $f'_c = 4000$ psi, $f_y = 60,000$ psi, $M_u = 400$ ft-k, and $L = 28$ ft. Clear distance between flanges = 3 ft.

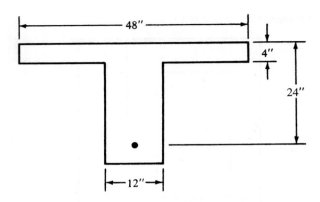

5.19 Repeat Problem 5.18 if $M_u = 500$ ft-k (*Ans.* 4.80 in.^{2})

5.20 Repeat Problem 5.18 if $f_y = 50{,}000$ psi, and $f_c' = 5000$ psi

5.21 Determine the amount of reinforcing steel required for each T beam in the accompanying illustration if $f_y = 60{,}000$ psi. $f_c' = 4000$ psi, simple span 20 ft, clear distance between stems = 3 ft, $M_D = 200$ ft-k (includes effect of concrete weight), and $M_L = 500$ ft-k (*Ans.* 8.11 in.^{2})

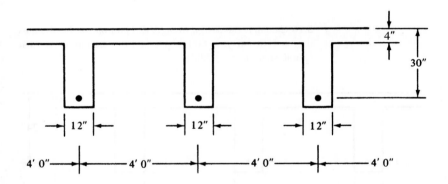

5.22 Select the tensile reinforcing needed for the T beams if the reinforced concrete is assumed to weigh 150 lb/ft^3 and if a live floor load of 140 lb/ft^2 is to be supported. Assume 40 ft simple spans and $f_y = 60{,}000$ psi and $f_c' = 3000$ psi.

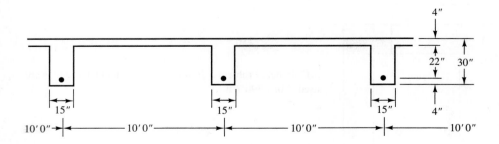

5.23 With $f_y = 60{,}000$ psi and $f_c' = 4000$ psi, select the reinforcing for T beam *AB* for the floor system shown. The live load is to be 100 psf, while the dead load in addition to the concrete's weight is to be 80 psf. Concrete is assumed to weigh 150 lb/ft^3. The slab is 4 in. thick, while d is to be 24 in. and b_w is 16 in. (*Ans.* 3.84 in.^{2} Use 4 #9 bars)

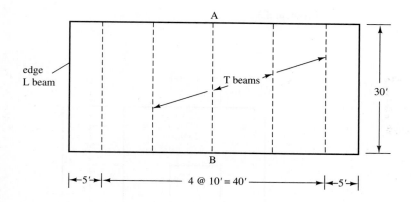

$\leftarrow$5'$\rightarrow$ | $\leftarrow$———— 4 @ 10' = 40' ————$\rightarrow$ | $\leftarrow$5'$\rightarrow$

5.24 Repeat Problem 5.23 if the span is 36 ft and the live load is 100 psf.

5.25 Prepare a flowchart for the design of tensilely reinforced T beams with $\phi = 0.9$.

Analysis of Doubly Reinforced Beams

In Problems 5.26 to 5.32 compute the design strengths of the beams shown if $f_y = 60,000$ psi and $f_c' = 4000$ psi. Check the maximum permissible A_s in each case to ensure tensile failure.

Problem 5.27 (*Ans.* 777.0 ft-k)

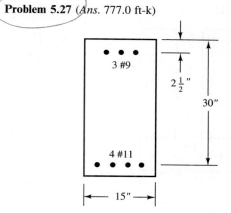

Problem 5.26

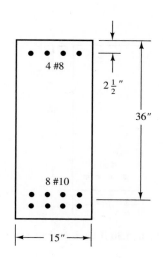

Problem 5.28

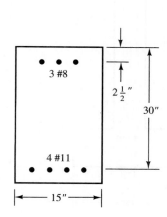

Problem 5.29 (*Ans.* 505.1 ft-k)

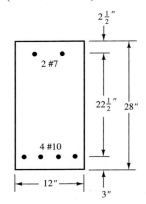

Problem 5.30

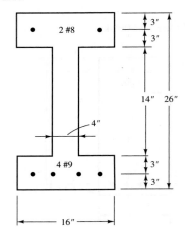

Problem 5.31 (*Ans.* 500.1 ft-k)

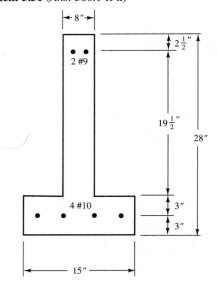

Problem 5.32

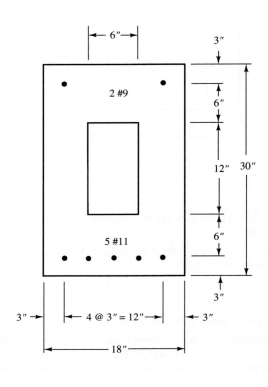

5.33 Compute the design strength of the beam shown in the accompanying illustration. How much can this permissible moment be increased if three #9 bars are added to the top $2\frac{1}{2}$ in. from the compression face if $f_c' = 4000$ psi and $f_y = 60,000$ psi? (*Ans.* 740.5 ft-k, 36.5 ft-k)

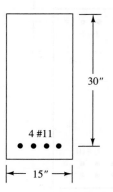

5.34 Repeat Problem 5.30 if three #8 bars are used in top.

Design of Doubly Reinforced Beams

In Problems 5.35 to 5.38 determine the steel areas required for the sections shown in the accompanying illustrations. In each case the dimensions are limited to the values shown. If compression steel is required, assume it will be placed 3 in. o.c. from the compression face. $f'_c = 4000$ psi and $f_y = 60,000$ psi.

Problem 5.35 (*Ans.* $A_s = 9.03$ in.^{2}, $A'_s = 2.44$ in.^{2})

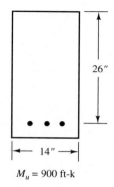

$M_u = 900$ ft-k

Problem 5.36

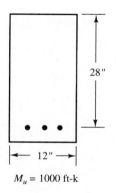

$M_u = 1000$ ft-k

Problem 5.37 (*Ans.* $A_s = 8.36$ in.^{2}, $A'_s = 2.58$ in.^{2})

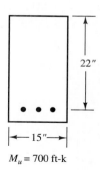

$M_u = 700$ ft-k

Problem 5.38

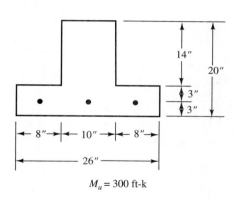

$M_u = 300$ ft-k

5.39 Prepare a flowchart for the design of doubly reinforced rectangular beams.

Computer Problems

Solve Problems 5.40 to 5.45 using the Chapter 5 spreadsheet.

5.40 Problem 5.5

5.41 Problem 5.7 (*Ans.* 775.7 ft-k)

5.42 Problem 5.14

5.43 Problem 5.21 (*Ans.* $A_s = 8.11$ in.^{2})

5.44 Problem 5.27

5.45 Problem 5.35 (*Ans.* $A_s = 9.02$ in.^{2}, $A'_s = 2.45$ in.^{2})

Problems in SI Units

In Problems 5.46 and 5.47 determine the design moment strengths of the beams shown in the accompanying illustrations if $f_c' = 28$ MPa and $f_y = 420$ MPa. Are the steel percentages in each case sufficient to ensure tensile behavior, that is, $\epsilon_t \geq 0.005$.

Problem 5.46

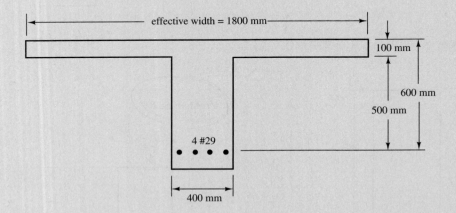

Problem 5.47 (*Ans.* 1516 kN · m)

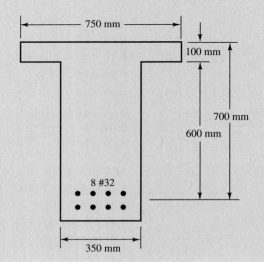

In Problems 5.48 and 5.49 determine the area of reinforcing steel required for the T beams shown if $f'_c = 28$ MPa and $f_y = 420$ MPa.

Problem 5.48

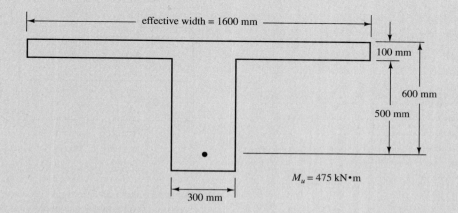

effective width = 1600 mm

100 mm

600 mm

500 mm

$M_u = 475$ kN•m

300 mm

Problem 5.49 (*Ans.* 4112 mm^2)

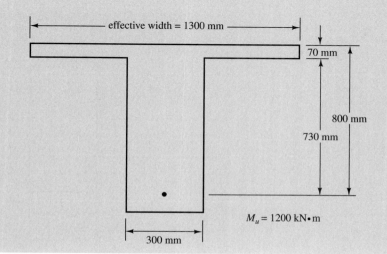

effective width = 1300 mm

70 mm

800 mm

730 mm

$M_u = 1200$ kN•m

300 mm

In Problems 5.50 to 5.52 compute the design moment strengths of the beams shown if $f_y = 420$ MPa and $f'_c = 21$ MPa. Check the maximum permissible A_s in each case to ensure ductile failure. $E_s = 200\,000$ MPa.

Problem 5.52

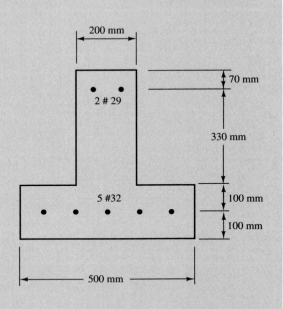

Problem 5.50

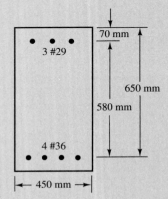

In Problems 5.53 and 5.54 determine the steel areas required for the sections shown in the accompanying illustration. In each case the dimensions are limited to the values shown. If compression steel is required, assume it will be placed 70 mm from the compression face. $f'_c = 28$ MPa, $f_y = 420$ MPa, and $E_s = 200\,000$ MPa.

Problem 5.51 (*Ans.* 995.0 kN · m)

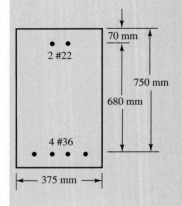

Problem 5.53 (*Ans.* $A_s = 5925$ mm², $A'_s = 2154$ mm²)

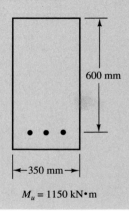

Problem 5.54

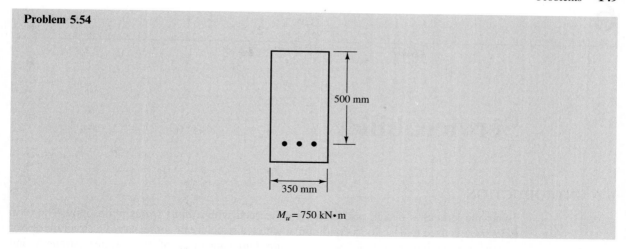

500 mm

350 mm

$M_u = 750$ kN·m

More Detailed Problems

5.55 Two-foot-wide, 4-in.-deep precast reinforced concrete slabs are to be used for a flat roof deck. The slabs are to be supported at their ends by precast rectangular beams spanning the 30-ft width of the roof (measured c. to c. of the supporting masonry walls). Select f_y, f_c', and design the slabs including their length, and design one of the supporting interior beams. Assume a 30-psf-roof live load and a 6-psf built-up roof. (*One ans.* Use 12 × 24 in. beams with three #9 bars.)

5.56 Repeat Problem 5.55 if the beams span 40 ft and the roof live load is 40 psf.

5.57 For the same building considered in Problem 5.55, a 6-in.-deep cast-in-place concrete slab has been designed. It is to be supported by T beams cast integrally with the slabs. The architect says that the 30-ft-long T beams are to be supported by columns that are to be spaced 18 ft o.c. The building is to be used for light manufacturing (see Table 1.3 in Chapter 1). Select f_y, f_c' and design one of the interior T beams. (*One ans.* Use T beam web 12 in. wide, $h = 32$ in. and four #10 bars.)

5.58 Repeat Problem 5.57 if the building is to be used for offices. The beam spans are to be 36 ft and the columns are to be placed 20 ft. o.c.

5.59 Determine the least cost design for a tensily reinforced concrete beam for the conditions that follow $f_y = 60$ ksi, $f_c' = 4$ ksi, $M_u = 400$ ft-k, $l = 24$ ft, $h = d + 2.5$ in.; concrete costs \$120 per yard and weighs 150 lb/ft^3, reinforcing bars cost \$0.95/lb and weigh 490 lbs/ft^3. Design the beam for the moment given with $d = 1.5b$ and calculate its cost per linear ft. Plot the cost per linear ft of beam (y-axis) versus steel percentage (x-axis). Then change the beam size, recalculate ρ and the new cost. Limit beam sizes to increments of 1 in. for b. Find the lowest cost design and the corresponding value of ρ. (*Ans.* Approx $\rho = 0.0139$ and Cost = \$26.03/ft.)

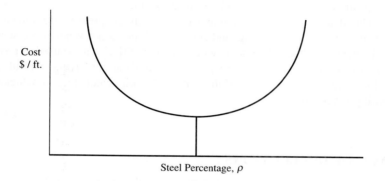

Cost $/ft.

Steel Percentage, ρ

5.60 Repeat Problem 5.59 if $d = 2b$.

6

Serviceability

6.1 INTRODUCTION

Today the structural design profession is concerned with a *limit states* philosophy. The term *limit state* is used to describe a condition at which a structure or some part of a structure ceases to perform its intended function. There are two categories of limit states: strength and serviceability.

Strength limit states are based on the safety or load-carrying capacity of structures and include buckling, fracture, fatigue, overturning, and so on. Chapters 3 to 5 have been concerned with the bending limit state of various members.

Serviceability limit states refer to the performance of structures under normal service loads and are concerned with the uses and/or occupancy of structures. Serviceability is measured by considering the magnitudes of deflections, cracks, and vibrations of structures as well as by considering the amounts of surface deterioration of the concrete and corrosion of the reinforcing. You will note that these items may disrupt the use of structures but do not usually involve collapse.

This chapter is concerned with serviceability limits for deflections and crack widths. The ACI Code contains very specific requirements relating to the strength limit states of reinforced concrete members but allows the designer some freedom of judgment in the serviceability areas. This doesn't mean that the serviceability limit states are not significant, but by far the most important consideration (as in all structural specifications) is the life and property of the public. As a result, public safety is not left up to the judgment of the individual designer.

Vertical vibration for bridge and building floors, as well as lateral and torsional vibration in tall buildings, can be quite annoying to users of these structures. Vibrations, however, are not usually a problem in the average size reinforced concrete building, but we should be on the lookout for the situations where they can be objectionable.

The deterioration of concrete surfaces can be greatly minimized by exercising good control of the mixing, placing, and curing of the concrete. When those surfaces are subjected to harsh chemicals, special cements with special additives can be used to protect the surfaces. The corrosion of reinforcing can be greatly minimized by giving careful attention to concrete quality, using good vibration of the concrete, using adequate cover thickness for the bars, and limiting crack sizes.

6.2 IMPORTANCE OF DEFLECTIONS

The adoption of the strength design method, together with the use of higher-strength concretes and steels, has permitted the use of relatively slender members. As a result, deflections and deflection cracking have become more severe problems than they were a few decades ago.

Georgia Dome, Atlanta, Georgia. (Courtesy of Economy Forms Corporation.)

The magnitudes of deflections for concrete members can be quite important. Excessive deflections of beams and slabs may cause sagging floors, ponding on flat roofs, excessive vibrations, and even interference with the proper operation of supported machinery. Such deflections may damage partitions and cause poor fitting of doors and windows. In addition, they may damage a structure's appearance or frighten the occupants of the building, even though the building may be perfectly safe. Any structure used by people should be quite rigid and relatively vibration-free so as to provide a sense of security.

Perhaps the most common type of deflection damage in reinforced concrete structures is the damage to light masonry partitions. They are particularly subject to injury due to concrete's long-term creep. When the floors above and below deflect, the relatively rigid masonry partitions do not bend easily and are often severely damaged. On the other hand, the more flexible gypsum board partitions are much more adaptable to such distortions.

6.3   CONTROL OF DEFLECTIONS

One of the best ways to reduce deflections is by increasing member depths—but designers are always under pressure to keep members as shallow as possible. (As you can see, shallower members mean thinner floors, and thinner floors mean buildings with less height, with consequent reductions in many costs such as plumbing, wiring, elevators, outside materials on buildings, and so on.) Reinforced concrete specifications usually limit deflections by specifying certain minimum depths or maximum permissible computed deflections.

Minimum Thicknesses

Table 4.1 of Chapter 4, which is Table 9.5(a) of the ACI Code, provides a set of minimum thicknesses for beams and one-way slabs to be used, unless actual deflection calculations indicate that lesser thicknesses are permissible. These minimum thickness values, which were developed primarily on the basis of experience over many years, should be used only for beams and slabs that are not supporting or attached to partitions or other members likely to be damaged by deflections.

Maximum Deflections

If the designer chooses not to meet the minimum thicknesses given in Table 4.1, he or she must compute deflections. If this is done, the values determined may not exceed the values specified in Table 6.1, which is Table 9.5(b) of the ACI Code.

Camber

The deflection of reinforced concrete members may also be controlled by cambering. The members are constructed of such a shape that they will assume their theoretical shape under some service loading condition (usually dead load and perhaps some part of the live load). A simple beam would be constructed with a slight convex bend so that under certain gravity loads it would become straight as assumed in the calculations. (See Figure 6.1.) Some designers take into account both dead and full live loads in figuring the amount of camber. Camber is generally used only for longer-span members.

Table 6.1 Maximum Permissible Computed Deflections

Type of member	Deflection to be considered	Deflection limitation
Flat roofs not supporting or attached to nonstructural elements likely to be damaged by large deflections	Immediate deflection due to live load L	$\dfrac{\ell^*}{180}$
Floors not supporting or attached to nonstructural elements likely to be damaged by large deflections	Immediate deflection due to live load L	$\dfrac{\ell}{360}$
Roof or floor construction supporting or attached to nonstructural elements likely to be damaged by large deflections	That part of the total deflection occurring after attachment of nonstructural elements (sum of the long-term deflection due to all sustained loads and the immediate deflection due to any additional live load)[†]	$\dfrac{\ell^‡}{480}$
Roof or floor construction supporting or attached to nonstructural elements not likely to be damaged by large deflections		$\dfrac{\ell^§}{240}$

[*]Limit not intended to safeguard against ponding. Ponding should be checked by suitable calculations of deflection, including added deflections due to ponded water, and considering long-term effects of all sustained loads, camber, construction tolerances, and reliability of provisions for drainage.

[†]Long-term deflection shall be determined in accordance with 9.5.2.5 or 9.5.4.3 but may be reduced by amount of deflection calculated to occur before attachment of nonstructured elements. This amount shall be determined on the basis of accepted engineering data relating to time-deflection characteristics of members similar to those being considered.

[‡]Limit may be exceeded if adequate measures are taken to prevent damage to supported or attached elements.

[§]But not greater than tolerance provided for nonstructural elements. Limit may be exceeded if camber is provided so that total deflection minus camber does not exceed limit.

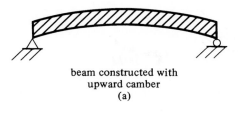

beam constructed with
upward camber
(a)

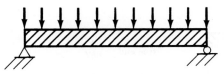

beam straight under dead load
plus some percentage of live load
(b)

Figure 6.1 Cambering.

6.4   CALCULATION OF DEFLECTIONS

Deflections for reinforced concrete members can be calculated with the usual deflection expressions, several of which are shown in Figure 6.2. A few comments should be made about the magnitudes of deflections in concrete members as determined by the expressions given in this figure. It can be seen that the ℂ deflection of a uniformly loaded simple beam [Figure 6.2(a)] is 5 times as large as the ℂ deflection of the same beam if its ends are fixed [Figure 6.2(b)]. Nearly all concrete beams and slabs are continuous, and their deflections fall somewhere between the two extremes mentioned here.

Because of the very large deflection variations that occur with different end restraints, it is essential that those restraints be considered if realistic deflection calculations are to be made. For most practical purposes it is sufficiently accurate to calculate the ℂ deflection of a member as though it is simply supported and to subtract from that value the deflection caused by the average of the negative moments at the member ends. (This can be done by using a combination of expressions taken from Figure 6.2. For instance, the deflection equation of part (a) may be used together with the one of part (g) applied at one or both ends as necessary.) Loads used in these expressions are unfactored loads. In some cases, only the live load is considered; while in others, both live and dead (sustained) loads are considered.

6.5   EFFECTIVE MOMENTS OF INERTIA

Regardless of the method used for calculating deflections, there is a problem in determining the moment of inertia to be used. The trouble lies in the amount of cracking that has occurred. If the bending moment is less than the cracking moment (that is, if the flexural stress is less than the modulus of rupture of about $7.5\lambda\sqrt{f_c'}$ the full uncracked section provides rigidity, and the moment

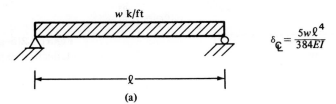

$$\delta_{\text{ℂ}} = \frac{5w\ell^4}{384EI}$$

Figure 6.2(a) Some deflection expressions.

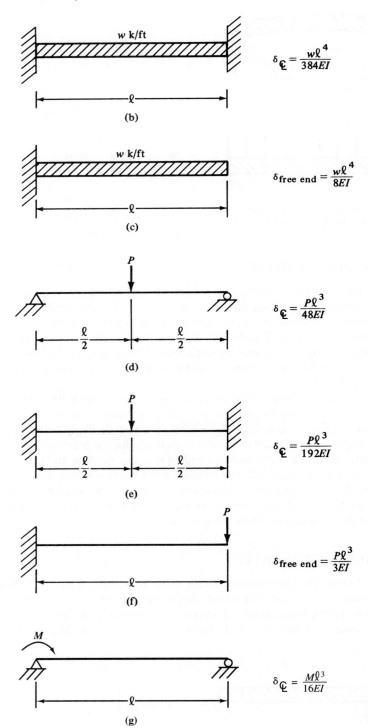

$\delta_{\mathbb{C}} = \dfrac{w\ell^4}{384EI}$

(b)

$\delta_{\text{free end}} = \dfrac{w\ell^4}{8EI}$

(c)

$\delta_{\mathbb{C}} = \dfrac{P\ell^3}{48EI}$

(d)

$\delta_{\mathbb{C}} = \dfrac{P\ell^3}{192EI}$

(e)

$\delta_{\text{free end}} = \dfrac{P\ell^3}{3EI}$

(f)

$\delta_{\mathbb{C}} = \dfrac{M\ell^3}{16EI}$

(g)

Figure 6.2(b–g) Some deflection expressions.

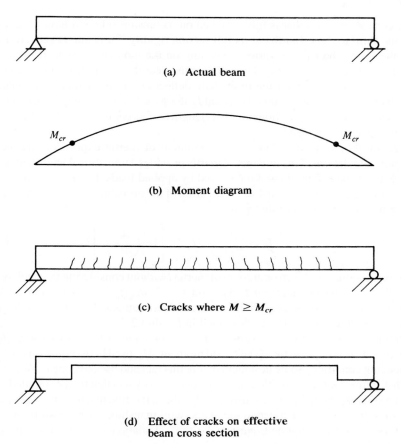

(a)   **Actual beam**

(b)   **Moment diagram**

(c)   **Cracks where** $M \geq M_{cr}$

(d)   **Effect of cracks on effective beam cross section**

Figure 6.3 Effects of cracks on deflections.

of inertia for the gross section I_g is available. When larger moments are present, different size tension cracks occur and the position of the neutral axis varies.

Figure 6.3 illustrates the problem involved in selecting the moment of inertia to be used for deflection calculations. Although a reinforced concrete beam may be of constant size (or prismatic) throughout its length, for deflection calculations it will behave as though it is composed of segments of different-size beams.[1]

For the portion of a beam where the moment is less than the cracking moment, M_{cr}, the beam can be assumed to be uncracked and the moment of inertia can be assumed to equal I_g. When the moment is greater than M_{cr}, the tensile cracks that develop in the beam will, in effect, cause the beam cross section to be reduced, and the moment of inertia may be assumed to equal the transformed value, I_{cr}. It is as though the beam consists of the segments shown in Figure 6.3(d).

The problem is even more involved than indicated by Figure 6.3. It is true that at cross sections where tension cracks are actually located, the moment of inertia is probably close to the transformed I_{cr}, but in between cracks it is perhaps closer to I_g. Furthermore, diagonal tension cracks may exist in areas of high shear, causing other variations. As a result, it is difficult to decide what value of I should be used.

---

[1]Leet, K., 1997, *Reinforced Concrete Design*, 3rd ed. (New York: McGraw-Hill), p. 155.

A concrete section that is fully cracked on its tension side will have a rigidity of anywhere from one-third to three-fourths of its full section rigidity if it is uncracked. At different sections along the beam, the rigidity varies depending on the moment present. It is easy to see that an accurate method of calculating deflections must take these variations into account.

If it is desired to obtain the immediate deflection of an uncracked prismatic member, the moment of inertia may be assumed to equal I_g along the length of the member. Should the member be cracked at one or more sections along its length or if its depth varies along the span, a more exact value of I needs to be used.

Section 9.5.2.3 of the Code gives a moment of inertia expression that is to be used for deflection calculations. This moment of inertia provides a transitional value between I_g and I_{cr} that depends upon the extent of cracking caused by applied loads. It is referred to as I_e, the effective moment of inertia, and is based on an estimation of the probable amount of cracking caused by the varying moment throughout the span:[2]

$$I_e = \left(\frac{M_{cr}}{M_a}\right)^3 (I_g) + \left[1 - \left(\frac{M_{cr}}{M_a}\right)^3\right] I_{cr} \qquad \text{(ACI Equation 9-8)}$$

In this expression, I_g is the gross amount of inertia (without considering the steel) of the section and M_{cr} is the cracking moment $= f_r I_g / y_t$, with $f_r = 7.5\lambda \sqrt{f_c'}{}^*$. M_a is the maximum service-load moment occurring for the condition under consideration and I_{cr} is the transformed moment of inertia of the cracked section, as described in Section 2.3.

You will note that the values of the effective moment of inertia vary with different loading conditions. This is because the service load moment, M_a, used in the equation for I_e, is different for each loading condition. Some designers ignore this fact and use only one I_e for each member, even though different loading conditions are considered. They feel that their computed values are just as accurate as those obtained with the different I_e values. It is true that the varying conditions involved in constructing reinforced concrete members (workmanship, curing conditions, age of members when loads were first applied, etc.) make the calculation of deflections by any present-day procedure a very approximate process.

In this chapter the authors compute I_e for each different loading condition. The work is a little tedious, but it can be greatly expedited with various tables such as the ones provided in the *ACI Design Handbook*.[3]

6.6 LONG-TERM DEFLECTIONS

With I_e and the appropriate deflection expressions, instantaneous or immediate deflections are obtained. Long-term or sustained loads, however, cause significant increases in these deflections due to shrinkage and creep. The factors affecting deflection increases include humidity, temperature, curing conditions, compression steel content, ratio of stress to strength, and the age of the concrete at the time of loading.

If concrete is loaded at an early age, its long-term deflections will be greatly increased. Excessive deflections in reinforced concrete structures can very often be traced to the early application of loads. The creep strain after about 5 years (after which creep is negligible) may be as

---

[2]Branson, D. E., 1965, "Instantaneous and Time-Dependent Deflections on Simple Continuous Reinforced Concrete Beams," HPR Report No. 7, Part 1, Alabama Highway Department, Bureau of Public Roads, August 1963, pp. 1–78.

*or $0.7\lambda \sqrt{f_c'}$ in SI.

[3]American Concrete Institute, 1997, *ACI Design Handbook* (Farmington Hills, MI), Publication SP-17 (97), pp. 110–124.

Table 6.2 Time Factor for Sustained Loads (ACI Code 9.5.2.5)

Duration of sustained load	Time-Dependent factor ξ
5 years or more	2.0
12 months	1.4
6 months	1.2
3 months	1.0

1 year = 1.4
2 year = 1.6
3 year = 1.8

high as 4 or 5 times the initial strain when loads were first applied 7 to 10 days after the concrete was placed, while the ratio may only be 2 or 3 when the loads were first applied 3 or 4 months after concrete placement.

Because of the several factors mentioned in the last two paragraphs, the magnitudes of long-term deflections can only be estimated. The Code (9.5.2.5) states that to estimate the increase in deflection due to these causes, the part of the instantaneous deflection that is due to sustained loads may be multiplied by the empirically derived factor λ at the end of this paragraph and the result added to the instantaneous deflection.[4]

$$\lambda_\Delta = \frac{\xi}{1 + 50\rho'} \qquad \text{(ACI Equation 9-11)}$$

In this expression, which is applicable to both normal and lightweight concrete, ξ is a time-dependent factor that may be determined from Table 6.2.

Should times differing from the values given in Table 6.2 be used, values of ξ may be selected from the curve of Figure 6.4.

The effect of compression steel on long-term deflections is taken into account in the λ expression with the term ρ'. It equals A_s'/bd and is to be computed at midspan for simple and continuous spans and at the supports for cantilevers.

The full dead load of a structure can be classified as a sustained load, but the type of occupancy will determine the percentage of live load that can be called sustained. For an apartment house or for an office building, perhaps only 20 to 25% of the service live load should be considered as being sustained, whereas perhaps 70 to 80% of the service live load of a warehouse might fall into this category.

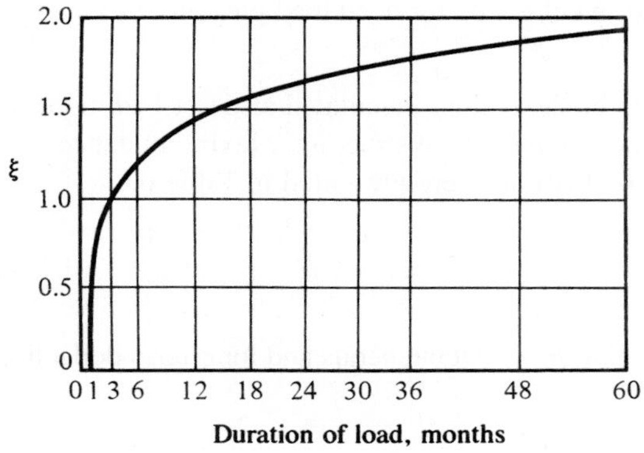

Figure 6.4 Multipliers for long-time deflections. (ACI Commentary Figure R9.5.2.5.)

[4]Branson, D. E., 1971, "Compression Steel Effect on Long-Time Deflections," *Journal ACI*, 68(8), pp. 555–559.

A study by the ACI indicates that under controlled laboratory conditions, 90% of test specimens had deflections between 20% below and 30% above the values calculated by the method described in this chapter.[5] The reader should realize, however, that field conditions are not lab conditions and deflections in actual structures will vary much more than those occurring in the lab specimens. Despite the use of plans and specifications and field inspection, it is difficult to control field work adequately. Construction crews may add a little water to the concrete to make it more workable. Further, they may not obtain satisfactory mixing and compaction of the concrete, with the result that voids and honeycomb occur. Finally, the forms may be removed before the concrete has obtained its full design strength. If this is the case, the moduli of rupture and elasticity will be low and excessive cracks may occur in beams that would not have occurred if the concrete had been stronger. All of these factors can cause reinforced concrete structures to deflect appreciably more than is indicated by the usual computations.

It is logical to assume that the live load cannot act on a structure when the dead load is not present. As a result of this fact we will compute an effective I_e and a deflection δ_D for the case where the dead load alone is acting. Then we will compute an I_e and a deflection δ_{D+L} for the case where both dead and live loads are acting. This will enable us to determine the initial live load part of the deflection as follows:

$$\delta_L = \delta_{D+L} - \delta_D$$

The long-term deflection will equal the initial live load deflection δ_L plus the infinitely long-term multiplier λ_∞ times the dead load deflection δ_D plus λ_r, the live load sustained multiplier, times the initial live load deflection δ_{SL}.

$$\delta_{LT} = \delta_L + \lambda_\infty \delta_D + \lambda_t \delta_{SL}$$

The steps involved in calculating instantaneous and long-term deflections can be summarized as follows:

(a) Compute the instantaneous or short-term deflection δ_D for dead load only.

(b) Compute instantaneous deflection δ_{D+L} for dead plus full live load.

(c) Determine instantaneous deflection δ_L for full live load only.

(d) Compute instantaneous deflection due to dead load plus the sustained part of the live load $\delta_D + \delta_{SL}$.

(e) Determine instantaneous deflection δ_L for the part of the live load that is sustained.

(f) Determine the long-term deflection for dead load plus the sustained part of the live load δ_{LT}.

As previously mentioned, the deflections calculated as described in this chapter should not exceed certain limits, depending on the type of structure. Maximum deflections permitted by the ACI for several floor and roof situations were presented in Table 6.1.

6.7   SIMPLE-BEAM DEFLECTIONS

Example 6.1 presents the calculation of instantaneous and long-term deflections for a uniformly loaded simple beam.

---

[5] ACI Committee 435, 1972, "Variability of Deflections of Simply Supported Reinforced Concrete Beams," *Journal ACI,* 69(1), p. 29.

$$n = \frac{E_s}{E_c} = \frac{29000000}{3160,000}$$

$$n = 9$$

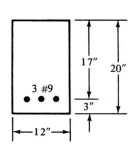

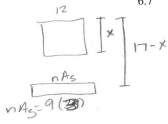

$$17 - x$$

$$nAs$$

$$nA_s = 9 \left(\frac{3}{3}\right)$$

Figure 6.5

$$\frac{-b \pm \sqrt{b^2 - 4ac}}{2a} = \frac{+27 \pm \sqrt{27^2 - 4(6)(-459)}}{2(6)}$$

$$x = 6.78$$

$$12x \left(\frac{x}{2}\right) = 9(3)(17-x)$$

$$6x^2 = 27(17-x)$$

$$6x^2 = 459 - 27x$$

$$6x^2 + 27x - 459 = 0$$

$$I_a = \frac{1}{3}(12)(6.78)^3 + (9)(3)(17-6.78)^2$$

$$I_{cr} = 1523 + 343 \, in^4$$

$$I_{cr} = 4067$$

EXAMPLE 6.1

The beam of Figure 6.5 has a simple span of 20 ft and supports a dead load including its own weight of 1 klf and a live load of 0.7 klf. $f'_c = 3000$ psi.

(a) Calculate the instantaneous deflection for $D + L$.

(b) Calculate the deflection assuming that 30% of the live load is continuously applied for three years.

SOLUTION

(a) Instantaneous or short-term dead load deflection (δ_D)

$$I_g = \left(\tfrac{1}{12}\right)(12)(20)^3 = 8000 \text{ in.}^4$$

$$M_{cr} = \frac{f_r I_g}{y_t} = \frac{(7.5\sqrt{3000})(8000)}{10} = 328{,}633 \text{ in.-lb} = 27.4 \text{ ft-k}$$

$$M_a = \frac{(1.0)(20)^2}{8} = 50 \text{ ft-k} = M_D$$

Should the dead load moment M_D be less than the cracking moment M_{cr}, we should use $M_{cr} = M_a$ and $I_e = I_g$.

By transformed-area calculations the values of x and I_{cr} can be determined as previously illustrated in Example 2.2.

$$x = 6.78''$$

$$I_{cr} = 4067 \text{ in.}^4$$

Then I_e is calculated with ACI Equation 9-8

$$I_e = \left(\frac{27.4}{50}\right)^3 (8000) + \left[1 - \left(\frac{27.4}{50}\right)^3\right] 4067 = 4714 \text{ in.}^4$$

$$E_c = 57{,}000\sqrt{3000} = 3.122 \times 10^6 \text{ psi}$$

$$\delta_D = \frac{5wl^4}{384 E_c I_e} = \frac{(5)\left(\frac{1000}{12}\right)(12 \times 20)^4}{(384)(3.122 \times 10^6)(4714)} = \underline{0.245 \text{ in.}}*$$

(b) Instantaneous or short-term deflection for dead + full live load (δ_{D+L})

$$M_a = \frac{(1.7)(20)^2}{8} = 85 \text{ ft-k}$$

Noting that the value of I_e changes when the moments change

---

*The authors really got carried away in this chapter when they calculated deflection to three digits. We cannot expect this kind of accuracy, and one digit beyond the decimal is sufficient. These instances are denoted in this chapter by *'s.

$$I_e = \left(\frac{27.4}{85}\right)^3 (8000) + \left[1 - \left(\frac{27.4}{85}\right)^3\right](4067) = 4199 \text{ in.}^4$$

$$\delta_{D+L} = \frac{(5)\left(\dfrac{1700}{12}\right)(12 \times 20)^4}{(384)(3.122 \times 10^6)(4199)} = 0.467 \text{ in.*}$$

(c) Initial deflection for full live load (δ_L)

$$\delta_L = \delta_{D+L} - \delta_D = 0.467 - 0.245 = \underline{0.222 \text{ in.*}}$$

This is the live load deflection that would be compared with the first or second row of Table 6.1. If the beam is part of a floor system that is "not supporting or attached to nonstructural elements likely to be damaged by large deflections" (left column of Table 6.1), then the deflection limit is $\ell/360 = (20)(12)/360 = 0.67$ in. This limit would easily be satisfied in this case as the calculated deflection is only 0.22 in.

(d) Initial deflection due to dead load + 30% live load ($\delta_D + \delta_{SL}$)

$$M_a = \frac{(1.0 + 0.30 \times 0.7)(20)^2}{8} = 60.5 \text{ ft-k}$$

$$I_e = \left(\frac{27.4}{60.5}\right)^3 (8000) + \left[1 - \left(\frac{27.4}{60.5}\right)^3\right](4067) = 4432 \text{ in.}^4$$

$$\delta_D + \delta_{SL} = \frac{(5)\dfrac{(1000 + 0.30 \times 700)}{12}(12 \times 20)^4}{(384)(3.122 \times 10^6)(4432)} = 0.315 \text{ in.*}$$

(e) Initial deflection due to 30% live load (δ_{SL})

$$\delta_{SL} = (\delta_D + \delta_{SL}) - \delta_D = 0.315 - 0.245 = \underline{0.070 \text{ in.*}}$$

(f) Long-term deflection for dead load plus three years of 30% sustained live load (δ_{LT})

$$\lambda_\infty = \frac{2.0}{1 + 50\rho'} = \frac{2.0}{1 + 0} = 2.0$$

$$\lambda_{3 \text{ years}} = \frac{1.80}{1 + 0} = 1.80$$

$$\delta_{LT} = \delta_L + \lambda_\infty \delta_D + \lambda_{3 \text{ years}} \delta_{SL}$$

$$= 0.222 + (2.0)(0.245) + (1.80)(0.070) = 0.838 \text{ in.*}$$

The middle column of Table 6.1 describes this deflection for the last two rows of the table. The answer is compared with either $\ell/480$ or $\ell/240$, depending on whether the structural member supports elements likely to be damaged by large deflections.

6.8   CONTINUOUS-BEAM DEFLECTIONS

The following discussion considers a continuous T beam subjected to both positive and negative moments. As shown in Figure 6.6, the effective moment of inertia used for calculating deflections varies a great deal throughout the member. For instance, at the center of the span at section 1–1

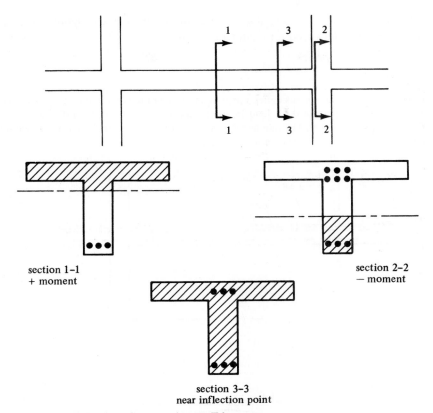

Figure 6.6 Deflections for a continuous T beam.

where the positive moment is largest, the web is cracked and the effective section consists of the hatched section plus the tensile reinforcing in the bottom of the web. At section 2–2 in the figure, where the largest negative moment occurs, the flange is cracked and the effective section consists of the hatched part of the web (including any compression steel in the bottom of the web) plus the tensile bars in the top. Finally, near the points of inflection, the moment will be so low that the beam will probably be uncracked, and thus the whole cross section is effective, as shown for section 3–3 in the figure. (For this case I is usually calculated only for the web, and the effect of the flanges is neglected as shown in Figure 6.10.)

From the preceding discussion it is obvious that to calculate the deflection in a continuous beam, theoretically it is necessary to use a deflection procedure that takes into account the varying moment of inertia along the span. Such a procedure would be very lengthy, and it is doubtful that the results so obtained would be within ±20% of the actual values. For this reason the ACI Code (9.5.2.4) permits the use of a constant moment of inertia throughout the member equal to the average of the I_e values computed at the critical positive- and negative-moment sections. *The I_e values at the critical negative-moment sections are averaged with each other, and then that average is averaged with I_e at the critical positive-moment section.* It should also be noted that the multipliers for long-term deflection at these sections should be averaged, as were the I_e values.

Example 6.2 illustrates the calculation of deflections for a continuous member. Although much of the repetitious math is omitted from the solution given herein, you can see that the calculations are still very lengthy and you will understand why approximate deflection calculations are commonly used for continuous spans.

EXAMPLE 6.2

Determine the instantaneous deflection at the midspan of the continuous T beam shown in Figure 6.7(a). The member supports a dead load including its own weight of 1.5 k/ft and a live load of 2.5 k/ft. $f_c' = 3000$ psi and $n = 9$. The moment diagram for full dead and live loads is shown in Figure 6.7(b), and the beam cross section is shown in Figure 6.7(c).

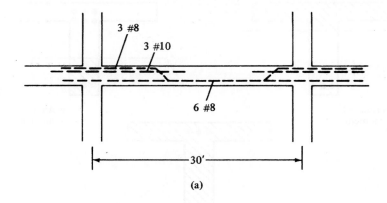

(a)

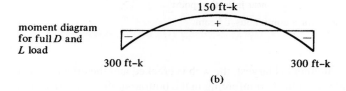

moment diagram for full D and L load

(b)

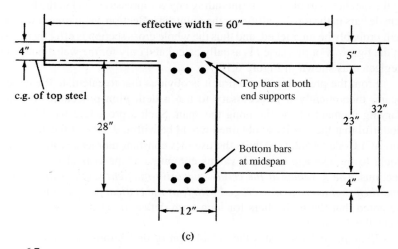

(c)

Figure 6.7

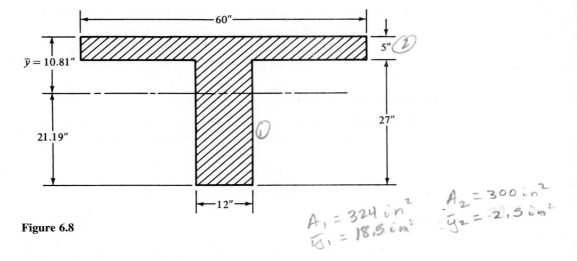

Figure 6.8

$A_1 = 324 \, in^2$ $A_2 = 300 \, in^2$
$\bar{y}_1 = 18.5 \, in^2$ $\bar{y}_2 = 2.5 \, in^2$

SOLUTION   **For Positive-Moment Region**

1. Locating centroidal axis for uncracked section and calculating gross moment of inertia I_g and cracking moment M_{cr} for the positive-moment region (Figure 6.8)

$$\bar{y} = 10.81''$$

$$I_g = 60,185 \text{ in.}^4$$

$$M_{cr} = \frac{(7.5)(\sqrt{3000})(60,185)}{21.19} = 1,166,754 \text{ in.-lb} = 97.2 \text{ ft-k}$$

2. Locating centroidal axis of cracked section and calculating transformed moment of inertia I_{cr} for the positive-moment region (Figure 6.9)

$$x = 5.65''$$

$$I_{cr} = 24,778 \text{ in.}^4$$

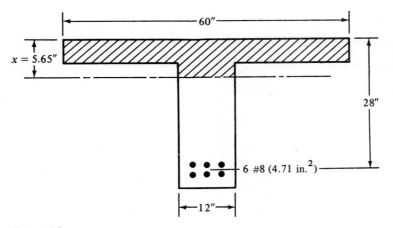

Figure 6.9

3. Calculating the effective moment of inertia in the positive-moment region

$$M_a = 150 \text{ ft-k}$$

$$I_e = \left(\frac{97.2}{150}\right)^3 (60{,}185) + \left[1 - \left(\frac{97.2}{150}\right)^3\right] 24{,}778 = 34{,}412 \text{ in.}^4$$

For Negative-Moment Region

1. Locating the centroidal axis for uncracked section and calculating gross moment of inertia I_g and cracking moment M_{cr} for the negative-moment region, considering only the hatched rectangle shown in Figure 6.10

$$\bar{y} = \left(\frac{32}{2}\right) = 16''$$

$$I_g = \left(\frac{1}{12}\right)(12)(32)^3 = 32{,}768 \text{ in.}^4$$

$$M_{cr} = \frac{(7.5)(\sqrt{3000})(32{,}768)}{16} = 841{,}302 \text{ in.-lb} = 70.1 \text{ ft-k}$$

The Code does not require that the designer ignore the flanges in tension for this calculation. The authors used this method to be conservative. If the tension flanges are considered, then the cracking moment is calculated from the section in Figure 6.8. The value of $\bar{y}$ is taken to the top of the section (10.81″) because the top is in tension for negative moment, so

$$M_{cr} = \frac{7.5\sqrt{3000}(60{,}185)}{10.81} = 2{,}287{,}096 \text{ in-lb} = 190.6 \text{ ft-k}$$

If this larger value for M_{cr} were used in step 3 below, the value of I_e would be 33,400 in⁴.

2. Locating the centroidal axis of the cracked section and calculating the transformed moment of inertia I_{tr} for the negative-moment region (Figure 6.11). See Example 2.5 for this type of calculation.

$$x = 10.43''$$

$$I_{cr} = 24{,}147 \text{ in.}^4$$

3. Calculating the effective moment of inertia in the negative-moment region

$$M_a = 300 \text{ ft-k}$$

$$I_e = \left(\frac{70.1}{300}\right)^3 (32{,}768) + \left[1 - \left(\frac{70.1}{300}\right)^3\right] 24{,}147 = 24{,}257 \text{ in.}^4$$

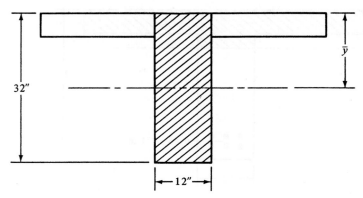

Figure 6.10

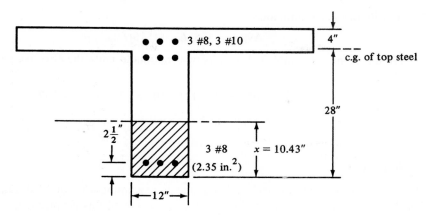

Figure 6.11

Instantaneous Deflection

The I_e to be used is obtained by averaging the I_e at the positive-moment section, with the average of I_e computed at the negative-moment sections at the ends of the span:

$$\text{Average } I_e = \frac{1}{2}\left[\left(\frac{24{,}257 + 24{,}257}{2}\right) + 34{,}412\right] = 29{,}334 \text{ in.}^4$$

$$E_c = 57{,}000\sqrt{3000} = 3.122 \times 10^6 \text{ psi}$$

Using the equation from Figure 6.2(b) and using only live loads to calculate deflections,

$$\delta_L = \frac{w_L l^4}{384 E_c I_e} = \frac{(2.5)(30)^4}{(384)(3122)(29{,}334)}(1728) = 0.10 \text{ in.}$$

In this case the authors used an approximate method to calculate δ_L. Instead of the cumbersome equation $(\delta_L = \delta_{D+L} - \delta_D)$ we used earlier in Example 6.1(c), we simply used w_L as the load in the above equation and average I_e. This approximation ignores the difference between I_e for dead load compared with I_e for dead and live load. This method gives a larger deflection, so it is conservative. Many designers have conservative approximations that they try first on many engineering calculations. If they work, there is no need to carry out the more cumbersome ones.

It has been shown that for continuous spans the Code (9.5.2.4) suggests an averaging of the I_e values at the critical positive- and negative-moment sections. The ACI Commentary (R9.5.2.4) says that for approximate deflection calculations for continuous prismatic members it is satisfactory to use the midspan section properties for simple and continuous spans and at supports for cantilevers. This is because these properties, which include the effect of cracking, have the greatest effect on deflections.

ACI Committee 435 has shown that better results for the deflections in continuous members can be obtained if an I_e is used that gives greater weight to the midspan values.[6] The committee suggests the use of the following expressions in which I_{em}, I_{e1}, and I_{e2} are the computed effective moments of inertia at the midspan and the two ends of the span, respectively.

Beams with two ends continuous

$$\text{Avg } I_e = 0.70 I_{em} + 0.15(I_{e1} + I_{e2})$$

---

[6]ACI Committee 435, 1978, "Proposed Revisions by Committee 435 to ACI Building Code and Commentary Provisions on Deflections," *Journal ACI*, 75(6), pp. 229–238.

Beams with one end continuous

$$\text{Avg } I_e = 0.85 I_{em} + 0.15(I_{\text{cont. end}})$$

For the beam of Example 6.2 with its two continuous ends, the effective moment of inertia would be

$$\text{Avg } I_e = (0.70)(34{,}412) + (0.15)(24{,}257 + 24{,}257)$$

$$= 31{,}365 \text{ in.}^4$$

6.9   TYPES OF CRACKS

This section presents a few introductory comments concerning some of the several types of cracks that occur in reinforced concrete beams. The remainder of this chapter is concerned with the estimated widths of flexural cracks and recommended maximum spacings of flexural bars to control cracks.

Flexural cracks are vertical cracks that extend from the tension sides of beams up to the region of their neutral axes. They are illustrated in Figure 6.12(a). Should beams have very deep webs (more than 3 or 4 ft), the cracks will be very closely spaced, with some of them coming together above the reinforcing and some disappearing there. These cracks may be wider up in the middle of the beam than at the bottom.

Inclined cracks due to shear can develop in the webs of reinforced concrete beams either as independent cracks or as extensions of flexural cracks. Occasionally, inclined cracks will develop independently in a beam, even though no flexural cracks are in that locality. These cracks, which are called *web-shear cracks* and which are illustrated in Figure 6.12(b), sometimes occur in the webs of prestressed sections, particularly those with large flanges and thin webs.

The usual type of inclined shear cracks are the *flexure-shear cracks*, which are illustrated in Figure 6.12(c). They commonly develop in both prestressed and nonprestressed beams.

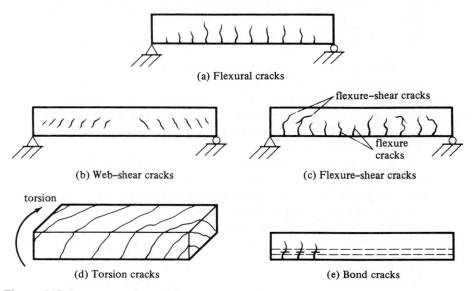

(a) Flexural cracks

(b) Web–shear cracks

(c) Flexure–shear cracks

flexure–shear cracks

flexure cracks

(d) Torsion cracks

torsion

(e) Bond cracks

Figure 6.12 Some types of cracks in concrete members.

Torsion cracks, which are illustrated in Figure 6.12(d), are quite similar to shear cracks except they spiral around the beam. Should a plain concrete member be subjected to pure torsion, it will crack and fail along 45° spiral lines due to the diagonal tension corresponding to the torsional stresses. For a very effective demonstration of this type of failure, you can take a piece of chalk in your hands and twist it until it breaks. Although torsion stresses are very similar to shear stresses, they will occur on all faces of a member. As a result, they add to the shear stresses on one side and subtract from them on the other.

Sometimes bond stresses between the concrete and the reinforcing lead to a splitting along the bars, as shown in Figure 6.12(e).

Of course, there are other types of cracks not illustrated here. Members that are loaded in axial tension will have transverse cracks through their entire cross sections. Cracks can also occur in concrete members due to shrinkage, temperature change, settlements, and so on. Considerable information concerning the development of cracks is available.[7]

6.10   CONTROL OF FLEXURAL CRACKS

Cracks are going to occur in reinforced concrete structures because of concrete's low tensile strength. For members with low steel stresses at service loads, the cracks may be very small and in fact may not be visible except upon careful examination. Such cracks, called *microcracks*, are generally initiated by bending stresses.

When steel stresses are high at service load, particularly where high-strength steels are used, visible cracks will occur. These cracks should be limited to certain maximum sizes so that the appearance of the structure is not spoiled and so that corrosion of the reinforcing does not result. The use of high-strength bars and the strength method of design have made crack control a very important item indeed. Because the yield stresses of reinforcing bars in general use have increased from 40 ksi to 60 ksi and above, it has been rather natural for designers to specify approximately the same size bars as they are accustomed to using, but fewer of them. The result has been more severe cracking of members.

Although cracks cannot be eliminated, they can be limited to acceptable sizes by spreading out or distributing the reinforcement. In other words, smaller cracks will result if several small bars are used with moderate spacings rather than a few large ones with large spacings. Such a practice will usually result in satisfactory crack control even for Grades 60 and 75 bars. An excellent rule of thumb to use as regards cracking is *"don't use a bar spacing larger than about 9 in."*

The maximum crack widths that are considered to be acceptable vary from approximately 0.004 to 0.016 in., depending on the location of the member in question, the type of structure, the surface texture of the concrete, illumination, and other factors. Somewhat smaller values may be required for members exposed to very aggressive environments such as de-icing chemicals and saltwater spray.

ACI Committee 224, in a report on cracking,[8] presented a set of approximately permissible maximum crack widths for reinforced concrete members subject to different exposure situations. These values are summarized in Table 6.3.

---

[7]MacGregor, J. G., 2005, *Reinforced Concrete Mechanics and Design*, 4th ed. (Upper Saddle River, NJ: Prentice-Hall), pp. 393–401.

[8]ACI Committee 224, 1972, "Control of Cracking in Concrete Structures," *Journal ACI*, 69(12), pp. 717–753.

Table 6.3 Permissible Crack Widths

Members subjected to	Permissible crack widths	
	(in.)	(mm)
Dry air	0.016	0.41
Moist air, soil	0.012	0.30
De-icing chemicals	0.007	0.18
Seawater and seawater spray	0.006	0.15
Use in water-retaining structures	0.004	0.10

Definite data are not available as to the sizes of cracks above which bar corrosion becomes particularly serious. As a matter of fact, tests seem to indicate that concrete quality, cover thickness, amount of concrete vibration, and other variables may be more important than crack sizes in their effect on corrosion.

Results of laboratory tests of reinforced concrete beams to determine crack sizes vary. The sizes are greatly affected by shrinkage and other time-dependent factors. The purpose of crack control calculations is not really to limit cracks to certain rigid maximum values but rather to use reasonable bar details, as determined by field and laboratory experience, that will in effect keep cracks within a reasonable range.

Lake Point Tower, Chicago, Illinois.
(Courtesy of Portland Cement Association.)

In 1968 the following equation was developed for the purpose of estimating the maximum widths of cracks that will occur in the tension faces of flexural members.[9] It is merely a simplification of the many variables affecting crack sizes.

$$w = 0.076\beta_h f_s \sqrt[3]{d_c A}$$

Where w = the estimated cracking width in thousandths of inches

β_h = ratio of the distance to the neutral axis from the extreme tension concrete fiber to the distance from the neutral axis to the centroid of the tensile steel (values to be determined by the working-stress method)

f_s = steel stress, in kips per square inch at service loads (designer is permitted to use $0.6f_y$ for normal structures)

d_c = the cover of the outermost bar measured from the extreme tension fiber to the center of the closest bar or wire. (For bundled bars d_c is measured to the c.g. of the bundles)

A = the effective tension area of concrete around the main reinforcing (having the same centroid as the reinforcing) divided by the number of bars

This expression is referred to as the Gergely-Lutz equation after its developers. In applying it to beams, reasonable results are usually obtained if β_h is set equal to 1.20. For thin one-way slabs, however, more realistic values are obtained if β_h is set equal to 1.35.

The number of reinforcing bars present in a particular member decidedly affects the value of A to be used in the equation and thus the calculated crack width. If more and smaller bars are used to provide the necessary area, the value of A will be smaller, as will the estimated crack widths.

Should all the bars in a particular group not be the same size, their number (for use in the equation) should be considered to equal the total reinforcing steel area actually provided in the group divided by the area of the largest bar size used.

Example 6.3 illustrates the determination of the estimated crack widths occurring in a tensilely reinforced rectangular beam.

EXAMPLE 6.3

Assuming $\beta_h = 1.20$ and $f_y = 60$ ksi, calculate the estimated width of flexural cracks that will occur in the beam of Figure 6.13. If the beam is to be exposed to moist air, is this width satisfactory as compared to the values given in Table 6.3 of this chapter? Should the cracks be too wide, revise the design of the reinforcing and recompute the crack width.

SOLUTION   **Substituting into the Gergely-Lutz Equation**

$$w = (0.076)(1.20)(0.6 \times 60) \sqrt[3]{(3)\left(\frac{6 \times 16}{3}\right)}$$

$$= \frac{15.03}{1000} = 0.015 \text{ in.} > 0.012 \text{ in.} \qquad \underline{\underline{\text{No good}}}$$

---

[9]Gergely, P., and Lutz, L. A., 1968, "Maximum Crack Width in Reinforced Flexural Members," *Causes, Mechanisms and Control of Cracking in Concrete*, SP-20 (Detroit: American Concrete Institute), pp. 87–117.

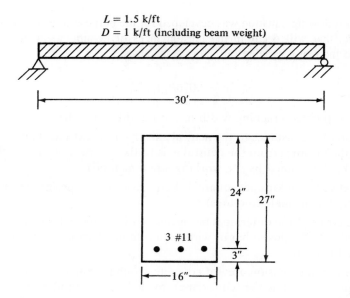

L = 1.5 k/ft
D = 1 k/ft (including beam weight)

30′

24″ 27″

3 #11

3″

16″

Figure 6.13

Replace the three #11 bars (4.68 in.^{2}) with five #9 bars (5.00 in.^{2})

$$w = (0.076)(1.20)(0.6 \times 60) \sqrt[3]{(3)\left(\frac{6 \times 16}{5}\right)} = \frac{12.68}{1000} = 0.0127 \text{ in.} > 0.012 \text{ in.} \qquad \underline{\underline{\text{No good}}}$$

Try six #8 bars (4.71 in.^{2})

$$w = (0.076)(1.20)(0.6 \times 60) \sqrt[3]{(3)\left(\frac{6 \times 16}{6}\right)} = \frac{11.93}{1000} = 0.0119 \text{ in.} < 0.012 \text{ in.} \qquad \underline{\underline{\text{OK}}}$$

<u>use 6#8 bars</u>

If reinforced concrete members are tested under carefully controlled laboratory conditions and cracks measured for certain loadings, considerable variations in crack sizes will occur. Consequently, the calculations of crack widths described in this chapter should only be used to help the designer select good details for reinforcing bars. The calculations are clearly not sufficiently accurate for comparison with field crack sizes.

The bond stress between the concrete and the reinforcing steel decidedly affects the sizes and spacings of the cracks in concrete. When bundled bars are used, there is appreciably less contact between the concrete and the steel as compared to the cases where the bars are placed separately from each other. To estimate crack widths successfully with the Gergely-Lutz equation when bundled bars are used, it is necessary to take into account this reduced contact surface.[10]

When bundled bars are present, some designers use a very conservative procedure in computing the value of A. For this calculation they assume each bundle is one bar, that bar having an area equal to the total area of the bars in that bundle. Certainly, the bond properties of a group of bundled bars are better than those of a single large "equivalent bar."

---

[10]Nawy, E. G., 2005, *Reinforced Concrete: A Fundamental Approach*, 5th ed. (Upper Saddle River, NJ: Prentice Hall), pp. 301–303.

Particular attention needs to be given to crack control for doubly reinforced beams where it is common to use small numbers of large-diameter tensile bars. Calculation of crack widths for such beams may result in rather large values, thus in effect requiring the use of a larger number of rather closely spaced smaller bars.

Special rules are given in ACI Section 10.6.6 for the spacings of reinforcing to help control the amount of cracking in T beams whose flanges are in tension.

6.11   ACI CODE PROVISIONS CONCERNING CRACKS

In the ACI Code, Sections 10.6.3 and 10.6.4 require that flexural tensile reinforcement be well distributed within the zones of maximum tension so that the center-to-center spacing of the reinforcing closest to a tension surface is not greater than the value computed with the following expression:

$$s = (15)\left(\frac{40,000}{f_s}\right) - 2.5c_c \le (12)\left(\frac{40,000}{f_s}\right) \qquad \text{(ACI Equation 10-4)}$$

In this expression f_s is the computed tensile stress at working load. It may be calculated by dividing the unfactored bending moment by the beam's internal moment arm (see Example 2.2) or it may simply be taken equal to $0.6f_y$. The term c_c represents the clear cover from the nearest surface in tension to the surface of the tensile reinforcement in inches.

For beams with Grade 60 reinforcing and with 2-in. clear cover, the maximum Code-permitted bar spacing is

$$s = (15)\left(\frac{40,000}{0.6 \times 60,000}\right) - (2.5)(2) = 11.67 \text{ in.} < (12)\left(\frac{40,000}{0.6 \times 60,000}\right) = 13.33 \text{ in.}$$

A bar spacing not more than 11.67 in. would thus be required. This limit can control the spacing of bars in one-way slabs, but is not likely to control beam bar spacings.

It is felt that these ACI maximum bar-spacing provisions are quite reasonable for one-way slabs and for beams with wide webs. For beams with normal web widths used in ordinary buildings, it is felt that estimating crack widths with the Gergely-Lutz equation and comparing the results to the values given in Table 6.3 of this chapter may be a more reasonable procedure.[11]

The ACI equation for maximum spacing does not apply to beams with extreme exposure or to structures that are supposed to be watertight. Special consideration must be given to such situations. It is probably well to use the Gergely-Lutz equation and a set of maximum crack widths such as those of Table 6.3 for such situations.

The effect of cracks and their widths on the corrosion of reinforcing is not clearly understood. There does not seem to be a direct relationship between crack widths and corrosion, at least at the reinforcing stresses occurring when members are subjected to service loads. Thus the 2005 ACI Code does not distinguish between interior and exterior exposure as did the 1995 Code. Present research seems to indicate that the total corrosion occurring in reinforcing is independent of crack widths. It is true, however, that the time required for corrosion to begin in reinforcing is inversely related to the widths of cracks.

---

[11]Nawy, 2005, *Reinforced Concrete*, p. 303.

When using the Gergely-Lutz crack width expression with SI units, the equation is $w = 0.0113\beta_h f_s \sqrt[3]{d_c A}$, with the resulting crack widths in mm.

The SI version of the ACI Code for the maximum spacing of tensile bars from the standpoint of crack widths is given here. To use this expression correctly, s and c_c must be used in mm while f_s must be in MPa.

$$s = (380)\left(\frac{280}{f_s}\right) - 2.5c_c \le (300)\left(\frac{280}{f_s}\right)$$

6.12 MISCELLANEOUS CRACKS

The beginning designer will learn that it is wise to include a few reinforcing bars in certain places in some structures, even though there seems to be no theoretical need for them. Certain spots in some structures (such as in abutments, retaining walls, building walls near openings, etc.) will develop cracks. The young designer should try to learn about such situations from more experienced people. Better structures will be the result.

6.13 SI EXAMPLE

EXAMPLE 6.4

Is the spacing of the bars shown in Figure 6.14 within the requirements of the ACI Code from the standpoint of cracking if $f_y = 420$ MPa?

SOLUTION   For $f_y = 420$ MPa and $c_c = 75 - \dfrac{28.7}{2} = 60.65$ mm,

$$s = (380)\left(\frac{280}{0.6 \times 420}\right) - (2.5)(60.65) = 271 \text{ mm} < (300)\left(\frac{280}{0.6 \times 420}\right) = 333 \text{ mm}$$

Since the actual bar spacing of 75 mm is less than 271 mm, this spacing is acceptable.

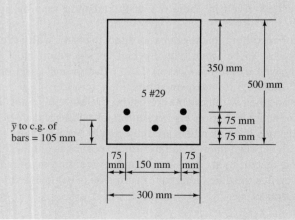

Figure 6.14

6.14 COMPUTER EXAMPLES

EXAMPLE 6.5

Repeat Example 6.1 using the Excel spreadsheet in Chapter 6.

Deflection calculator for simply supported, uniformly loaded, rectangular beam

	$b =$	12	in.
	$d =$	17	in.
	$h =$	20	in.
	$A_s =$	3.00	in.^{2}
	$A_s' =$	0.00	in.^{2}
	$f_c' =$	3	ksi
	$f_y =$	60	ksi
	$\gamma_c =$	145	pcf
	$\lambda =$	1	
	$\xi =$ (from Table 6.2 or Figure 6.4)	2.0	
	$w_D =$	1000	plf
	$w_L =$	700	plf
	$\ell =$	20	ft
	Deflection limit (denominator from Table 6.1)	180	
	% live load that is sustained	30	%
	$E_c =$	3156	ksi
	$n = E_s/E_c$	9.189	9
	$\rho =$	0.015	
	$n\rho =$	0.132	
	$k =$	0.399	
	$x =$	6.78	in.
	$I_{cr} =$	4067	in.^{4}
	$I_g =$	8000	in.^{4}
	$f_r =$	410.8	psi
	$M_{cr} =$	27.4	k-ft

Dead + full live load			
	$M_{a,D+L} =$	85	k-ft
	$(M_{cr}/M_{a,D+L})^3 =$	0.0334	
	$I_e = \left(\frac{M_{cr}}{M_a}\right)^3 I_g + \left[1 - \left(\frac{M_{cr}}{M_a}\right)^3\right] I_{cr} =$	4198.3	in.^{4}
	$\delta_{D+L} =$	0.462	in.
Dead load only			
	$M_{a,D} =$	50	k-ft
	$(M_{cr}/M_{a,D})^3$	0.1643	
	$I_e =$	4713.1	in.^{4}
	$\delta_D =$	0.242	in.
Live load only			
	$\delta_L = \delta_{D+L} - \delta_D =$	0.220	in.
Initial δ from D + %L			
	$M_{a,D+\%L} =$	60.5	
	$(M_{cr}/M_{a,D+\%L})^3 =$	0.0928	
	$I_e =$	4431.6	in.^{4}
	$\delta_{D+\%L} =$	0.311	in.
Initial δ from %L only			
	$\delta_{\%L} = (\delta_D + \delta_{\%L}) - \delta_D =$	0.069	in.
Long term δ for D + long term sustained L			
	$\rho' =$	0	
	$\lambda_\Delta = \xi/(1 + \rho')$	2	
	$\delta_{LT} = d_L + \lambda_\Delta \delta_D + \lambda_\Delta \delta_{\%L} =$	0.843	in.
	$\delta_{\text{limit}} =$	1.3333333	in.
	Deflection complies with Table 6.1		

PROBLEMS

6.1 What factors make it difficult to estimate accurately the magnitude of deflections in reinforced concrete members?

6.2 Why do deflections in concrete members increase as time goes by?

6.3 How can the deflection of concrete beams be limited?

6.4 Why is it necessary to limit the width of cracks in reinforced concrete members? How can it be done?

Deflections

In Problems 6.5 to 6.10, calculate the instantaneous deflections for the dead and live loads shown. Use $f_y = 60,000$ psi, $f_c = 4000$ psi, and $n = 8$. Beam weights are included in the w_D values.

Problem 6.5 (*Ans.* 0.418 in.)

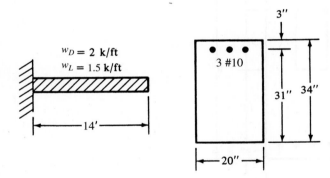

due 10/22

Problem 6.6

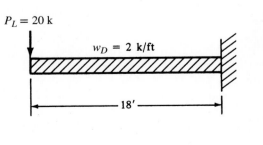

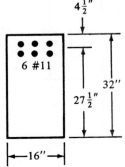

Problem 6.7 (*Ans.* 1.14 in.)

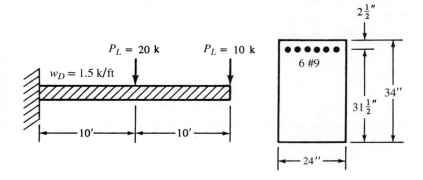

Problem 6.8

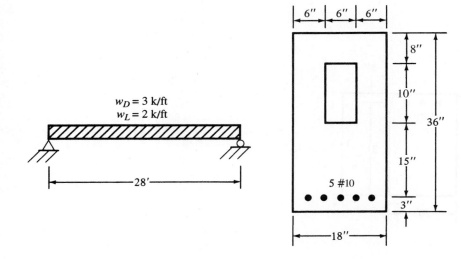

Problem 6.9 (*Ans.* 1.54 in.)

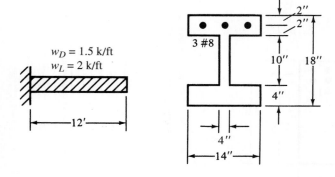

Problem 6.10 Repeat Problem 6.8 if a 25-k concentrated live load is added at the ℄ of the span.

In Problems 6.11 to 6.12, calculate the instantaneous deflections and the long-term deflections after 4 years, assuming that 30% of the live loads are continuously applied for 48 months. $f_y = 60,000$ psi, $f'_c = 400$ psi, $n = 8$.

Problem 6.11   (*Ans.* Instantaneous δ for full $w_D + w_L = 0.610$ in., long-term $\delta = 1.043$ in.)

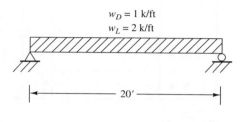
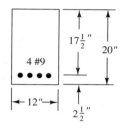

due 10/22

Problem 6.12

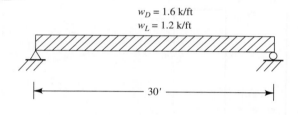
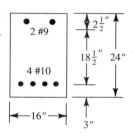

6.13   Repeat Problem 6.12 if the two top #9 compression bars are removed. (*Ans.* Instantaneous δ for full $D + L = 1.47$ in., long-term $\delta = 2.66$ in.)

6.14   Repeat Problem 6.12 using sand-lightweight concrete ($\gamma_c = 125$ pcf)

6.15   Repeat Problem 6.13 using all-lightweight concrete ($\gamma_c = 100$ pcf) (*Ans.* Instantaneous δ for full $w_D + w_L = 1.76$ in., long-term $\delta = 3.18$ in.)

Crack Widths

6.16   Select a rectangular beam section for the span and loads shown in the accompanying illustration. Use $\rho = \frac{1}{2}\rho_b$, #9 bars, $f'_c = 3000$ psi, and $f_y = 60,000$ psi. Compute the estimated maximum crack widths using the Gergely-Lutz equation. Are they less than the suggested maximum value given in Table 6.3 for dry air?

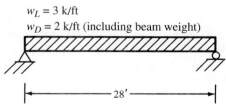

Problem 6.16

In Problems 6.17 and 6.18 estimate maximum crack widths with the Gergely-Lutz equation. Compare the results with the suggested maximums given in Table 6.3. Assume $f_y = 60$ ksi, and $\beta_h = 1.20$. Also calculate maximum permissible bar spacings as per ACI Equation 10.4. Assume moist air conditions.

Problem 6.17 (*Ans.* 0.0144 > 0.012 in.; max. ACI spacing = 10.75 in.)

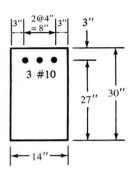

Problem 6.20

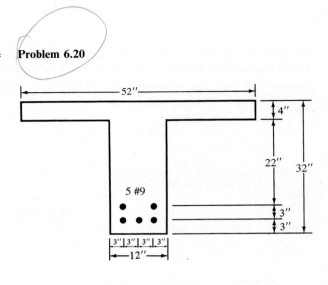

Problem 6.21 (*Ans.* 0.0165 in. > 0.016 in.; 8.25 in.)

Problem 6.18

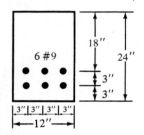

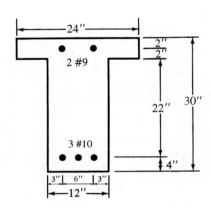

Problems 6.19, 6.20, and 6.21. Same questions as for Problems 6.17 and 6.18, but assume interior exposure.

Problem 6.19 (*Ans.* 0.0129 in. < 0.016 in.; max. ACI spacing = 10.93 in.)

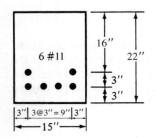

Problem 6.22 What is the maximum permissible spacing of #5 bars in the one-way slab shown which will satisfy the ACI Code crack requirements? f_y = 60,000 psi.

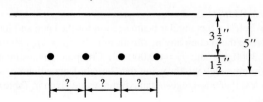

Problems in SI Units

In Problems 6.23 to 6.25 calculate the instantaneous deflections. Use normal weight concrete with $f'_c = 28$ Mpa, $f_y = 420$ Mpa. Assume that the w_D values shown include the beam weights. $E_s = 200\,000$ MPa.

Problem 6.23   (*Ans.* 9.88 mm)

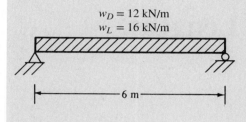

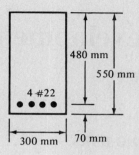

$w_D = 12$ kN/m
$w_L = 16$ kN/m

6 m

480 mm

550 mm

4 #22

300 mm   70 mm

Problem 6.24

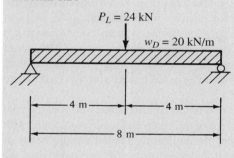

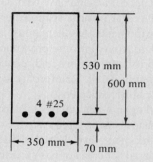

$P_L = 24$ kN

$w_D = 20$ kN/m

4 m   4 m

8 m

530 mm

600 mm

4 #25

350 mm   70 mm

Problem 6.25   (*Ans.* 10.22 mm)

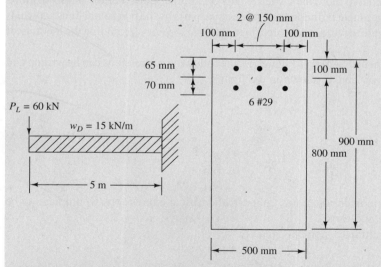

2 @ 150 mm

100 mm   100 mm

65 mm

70 mm

100 mm

6 #29

$P_L = 60$ kN

$w_D = 15$ kN/m

5 m

900 mm

800 mm

500 mm

Problems 6.26 and 6.27 Do these beams meet the maximum spacing requirements of the ACI Code if $f_y = 420$ MPa?

Problem 6.26 The beam of Problem 6.21.

Problem 6.27 The beam of Problem 6.25 ($s = 275$ mm).

Problem 6.28 Rework Problem 6.17 using the Excel spreadsheet in Chapter 6.

7

Bond, Development Lengths, and Splices

7.1 CUTTING OFF OR BENDING BARS

The beams designed up to this point have been selected on the basis of maximum moments. These moments have occurred at or near span centerlines for positive moments and at the faces of supports for negative moments. At other points in the beams, the moments were less. Although it is possible to vary beam depths in some proportion to the bending moments, it is normally more economical to use prismatic sections and reduce or cut off some reinforcing when the bending moments are sufficiently small. Reinforcing steel is quite expensive, and cutting it off where possible may appreciably reduce costs.

Should the bending moment fall off 50% from its maximum, approximately 50% of the bars can be cut off or perhaps bent up or down to the other face of the beam and made continuous with the reinforcing in the other face. For this discussion, the uniformly loaded simple beam of Figure 7.1 is considered. This beam has six bars, and it is desired to cut off two bars when the moment falls off a third and two more bars when it falls off another third. For the purpose of this discussion, the maximum moment is divided into three equal parts by the horizontal lines shown. If the moment diagram is drawn to scale, a graphical method is satisfactory for finding the theoretical cutoff points.

For the parabolic moment diagram of Figure 7.1, the following expressions can be written and solved for the bar lengths x_1 and x_2 shown in the figure:

$$\frac{x_1^2}{(\ell/2)^2} = \frac{2}{6}$$

$$\frac{x_2^2}{(\ell/2)^2} = \frac{4}{6}$$

For different-shaped moment diagrams, other mathematical expressions would have to be written or a graphical method used.

Actually, the design ultimate moment capacity

$$\phi M_n = \phi A_s f_y \left(d - \frac{a}{2} \right)$$

does not vary exactly in proportion to the area of the reinforcing bars as is illustrated in Example 7.1 because of variations in the depth of the compression block as the steel area is changed.

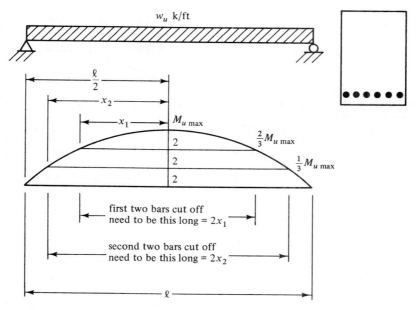

Figure 7.1

The change is so slight, however, that for all practical purposes the moment capacity of a beam can be assumed to be directly proportional to the steel area.

It will be shown in this chapter that the moment capacities calculated as illustrated in this example problem will have to be reduced if sufficient lengths are not provided beyond the theoretical cut-off points for the bars to develop their full stresses.

EXAMPLE 7.1

For the uniformly loaded simple beam of Figure 7.2, determine the theoretical points on each end of the beam where two bars can be cut off and then determine the points where two more bars can be cut off. $f_c' = 3000$ psi, $f_y = 60,000$ psi.

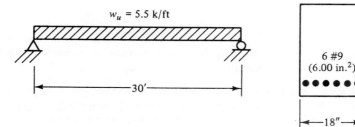

Figure 7.2

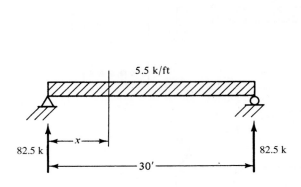

Figure 7.3

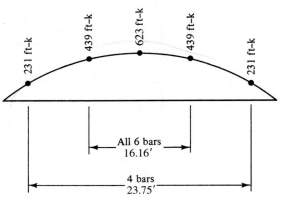

Figure 7.4 ϕM_n diagram for beam in Figure 7.2.

SOLUTION   When the beam has only four bars,

$$a = \frac{(4.00)(60)}{(0.85)(3)(18)} = 5.23''$$

$$\phi M_n = (0.9)(4.00)(60)\left(27 - \frac{5.23}{2}\right) = 5267 \text{ in.-k} = 439 \text{ ft-k}$$

When the moment falls off to 439 ft-k, two of the six bars can theoretically be cut off.
    When the beam has only two bars,

$$a = \frac{(2.00)(60)}{(0.85)(3)(18)} = 2.61''$$

$$\phi M_n = (0.9)(2.00)(60)\left(27 - \frac{2.61}{2}\right) = 2775 \text{ in.-k} = 231 \text{ ft-k}$$

When the moment falls off to 231 ft-k, two more bars can theoretically be cut off leaving two bars in the beam.

(Notice that ρ with 6 bars $= \frac{6.00}{(18)(27)} = 0.0123$, which is less than $\rho_{max} = 0.0136$ from Appendix Table A.7. Also, this ρ is $> \rho_{min}$ of $\frac{200}{60,000} = 0.00333$.)

The moment at any section in the beam at a distance x from the left support is as follows, with reference being made to Figure 7.3:

$$M = 82.5x - (5.5x)\left(\frac{x}{2}\right)$$

From this expression the location of the points in the beam where the moment is 439 ft-k and 231 ft-k can be determined. The results are shown in Figure 7.4.

Discussion If the approximate procedure had been followed (where bars are cut off purely on the basis of the ratio of the number of bars to the maximum moment, as was illustrated with the equations on page 181), the first two bars would have had lengths equal to 17.2 ft (as compared to the theoretically correct value of 16.16 ft) and the second two bar lengths equal to 24.45 ft (as compared to the theoretically correct value of 23.75 ft). It can then be seen that the approximate procedure yields fairly reasonable results.

Lab Building, Portland Cement Association, Skokie, Illinois. (Courtesy of Portland Cement Association.)

In this section the theoretical points of cutoff have been discussed. As will be seen in subsequent sections of this chapter, the bars will have to be run additional distances because of variations of moment diagrams, anchorage requirements of the bars, and so on. The discussion of cutting off or bending bars is continued in Section 7.11

7.2 BOND STRESSES

A basic assumption made for reinforced concrete design is that there must be absolutely no slippage of the bars in relation to the surrounding concrete. In other words, the steel and the concrete should stick together or *bond* so they will act as a unit. If there is no bonding between the two materials and if the bars are not anchored at their ends, they will pull loose from the concrete. As a result, the concrete beam will act as an unreinforced member and will be subject to sudden collapse as soon as the concrete cracks.

It is obvious that the magnitude of bond stresses will change in a reinforced concrete beam as the bending moments in the beam change. The greater the rate of bending moment change (occurring at locations of high shear), the greater will be the rate of change of bar tensions and thus bond stresses.

What may not be so obvious is the fact that bond stresses are also drastically affected by the development of tension cracks in the concrete. At a point where a crack occurs, all of the longitudinal tension will be resisted by the reinforcing bar. At a small distance along the bar at a point away from the crack, the longitudinal tension will be resisted by both the bar and the uncracked concrete. In this small distance there can be a large change in bar tension due to the fact that the uncracked concrete is now resisting tension. Thus the bond stress in the surrounding

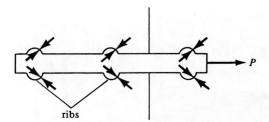

ribs

Figure 7.5 Bearing forces on bar and bearing of bar ribs on concrete.

concrete, which was zero at the crack, will drastically change within this small distance as the tension in the bar changes.

In the past it was common to compute the maximum theoretical bond stresses at points in the members and to compare them with certain allowable values obtained by tests. It is the practice today, however, to look at the problem from an ultimate standpoint, where the situation is a little different. Even if the bars are completely separated from the concrete over considerable parts of their length, the ultimate strength of the beam will not be affected if the bars are so anchored at their ends that they cannot pull loose.

The bonding of the reinforcing bars to the concrete is due to several factors, including the chemical adhesion between the two materials, the friction due to the natural roughness of the bars, and the bearing of the closely spaced rib-shaped deformations on the bar surfaces against the concrete. The application of the force P to the bar shown in Figure 7.5 is considered in the discussion that follows.

When the force is first applied to the bar, the resistance to slipping is provided by the adhesion between the bar and the concrete. If plain bars were used, it would not take much tension in the bars to break this adhesion, particularly adjacent to a crack in the concrete. If this were to happen for a smooth surface bar, only friction would remain to keep the bar from slipping. There is also some Poisson's effect due to the tension in the bars. As they are tensioned they become a little smaller, enabling them to slip more easily. If we were to use straight, plain, or smooth reinforcing bars in beams, there would be very little bond strength and the beams would only be a little stronger than if there were no bars. The introduction of deformed bars was made so that in addition to the adhesion and friction there would also be a resistance due to the bearing of the concrete on the lugs or ribs (or deformations) of the bars as well as the so-called shear-friction strength of the concrete between the lugs.

Deformed bars are used in almost all work. However, plain bars or plain wire fabrics are sometimes used for lateral reinforcement in compression members (as ties or spirals as described in Chapter 9), for members subject to torsion, and for confining reinforcing in splices (ACI R3.5.4).

As a result of these facts, reinforcing bars are made with rib-type deformations. The chemical adhesion and friction between the ribs are negligible, and thus bond is primarily supplied by bearing on the ribs. Based on testing, the crack patterns in the concrete show that the bearing stresses are inclined to the axis of the bars from about 45° to 80° (the angle being appreciably affected by the shape of the ribs.)[1]

---

[1]Goto, Y., 1971, "Cracks Formed on Concrete Around Deformed Tensioned Bar," *ACI Journal,* Proceedings 68, p. 244.

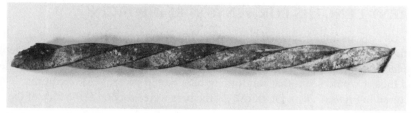

Twisted square bar, formerly used to increase bond between concrete and steel.
(Courtesy of Clemson University Communications Center.)

Equal and opposite forces develop between the reinforcing bars and the concrete as shown in Figure 7.5. These internal forces are caused by the wedging action of the ribs bearing against the concrete. They will cause tensile stresses in a cylindrical piece of concrete around each bar. It's rather like a concrete pipe filled with water which is pressing out against the pipe wall, causing it to be placed in tension. If the tension becomes too high, the pipe will split.

In a similar manner, if the bond stresses in a beam become too high, the concrete will split around the bars and eventually the splits will extend to the side and/or bottom of the beam. If either of these types of splits runs all the way to the end of the bar, the bar will slip and the beam will fail. The closer the bars are spaced together and the smaller the cover, the thinner will be the concrete cylinder around each bar and the more likely that a bond-splitting failure will occur.

Figure 7.6 shows examples of bond failures that may occur for different values of concrete cover and bar spacing. These are as shown by MacGregor.[2]

Splitting resistance along bars depends on quite a few factors, such as the thickness of the concrete cover, the spacing of the bars, the presence of coatings on the bars, the types of aggregates used, the transverse confining effect of stirrups, and so on. Because there are so many variables, it is impossible to make comprehensive bond tests that are good for a wide range of structures. Nevertheless, the ACI has attempted to do just this with its equations, as will be described in the sections to follow.

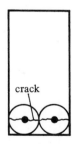

(a) Side cover
and one-half
clear spacing
between bars
< bottom cover

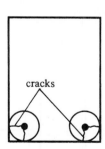

(b) Cover on sides
and bottom equal
and < one-half
clear spacing
between bars

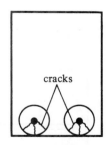

(c) Bottom cover
< side cover and
< one-half clear
spacing between
bars

Figure 7.6 Types of bond failures.

[2]MacGregor, J. G., 2005, *Reinforced Concrete Mechanics and Design*, 4th ed. (Upper Saddle River, NJ: Prentice Hall), p. 334.

7.3   DEVELOPMENT LENGTHS FOR TENSION REINFORCING

For this discussion, reference is made to the cantilever beam of Figure 7.7. It can be seen that both the maximum moment in the beam and the maximum stresses in the tensile bars occur at the face of the support. Theoretically, a small distance back into the support the moment is zero, and thus it would seem that reinforcing bars would no longer be required. This is the situation pictured in Figure 7.7(a). Obviously, if the bars were stopped at the face of the support, the beam would fail.

The bar stresses must be transferred to the concrete by bond between the steel and the concrete before the bars can be cut off. In this case the bars must be extended some distance back into the support and out into the beam to anchor them or develop their strength. This distance, called the *development length* (ℓ_d), is shown in Figure 7.7(b). It can be defined as the minimum length of embedment of bars that is necessary to permit them to be stressed to their yield point plus some extra distance to ensure member toughness. A similar case can be made for bars in other situations and in other types of beams.

As previously mentioned, the ACI for many years required designers to calculate bond stresses with a formula that was based on the change of moment in a beam. Then the computed values were compared to allowable bond stresses in the Code. Originally, bond strength was measured by means of pullout tests. A bar would be cast in a concrete cylinder, and a jack used to see how much force was required to pull it out. The problem with such a test is that the concrete is placed in compression, preventing the occurrence of cracks. In a flexural member, however, we have an entirely different situation due to the off-again/on-again nature of the bond stresses caused by the tension cracks in the concrete. In recent years, more realistic tests have been made with beams; the ACI Code development length expressions to be presented in this chapter are based primarily on such tests at the National Institute of Standards and Technology and the University of Texas.

The development lengths used for deformed bars or wires in tension may not be less than the values computed with ACI Equation 12-1 or 12 in. If the equation is written as $\left(\frac{\ell_d}{d_b}\right)$, the results obtained will be in terms of bar diameters. This form of answer is very convenient to use as, say, 30 bar diameters, 40 bar diameters, and so on.

$$\ell_d = \frac{3}{40}\frac{f_y}{\lambda\sqrt{f_c'}}\frac{\psi_t\psi_e\psi_s}{\left(\dfrac{c_b + K_{tr}}{d_b}\right)}d_b \qquad \text{(ACI Equation 12-1)}$$

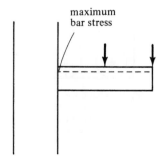

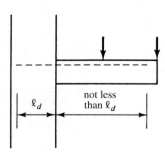

(a) No development
    length at support
    (beam will fail)

(b) Bars extended into
    the support a distance $= \ell_d$

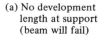

Figure 7.7

or

$$\frac{\ell_d}{d_b} = \frac{3}{40} \frac{f_y}{\lambda \sqrt{f_c'}} \frac{\psi_t \psi_e \psi_s}{\left(\dfrac{c_b + K_{tr}}{d_b} \right)}$$

Or in SI units,

$$\ell_d = \frac{9}{10} \frac{f_y}{\lambda \sqrt{f_c'}} \frac{\psi_t \psi_e \psi_s}{\left(\dfrac{c_b + K_{tr}}{d_b} \right)} d_b$$

This expression, which seems to include so many terms, is much easier to use than it might at first appear because several of the terms are usually equal to 1.0. Even if not equal to 1.0, the factors can be quickly obtained.

In the following paragraphs all of the terms in the equation that have not previously been introduced are described. Then their values for different situations are given in Table 7.1.

1. *Location of reinforcement.* Horizontal bars that have a least 12 in.[*] of fresh concrete placed beneath them do not bond as well to concrete as do bars placed nearer the bottom of the concrete. These bars are referred to as *top bars.* During the placing and vibration of the concrete, some air and excess water tend to rise toward the top of the concrete, and some portion may be caught under the higher bars. In addition, there may be some settlement of the concrete below. As a result, the reinforcement does not bond as well to the concrete underneath, and increased development lengths will be needed. To account for this effect the *reinforcement location factor* ψ_t is used.

2. *Coating of bars.* Epoxy-coated reinforcing bars are frequently used today to protect the steel from severe corrosive situations, such as where de-icing chemicals are used. Bridge decks and parking garage slabs in the colder states fit into this class. When bar coatings are used, bonding is reduced and development lengths must be increased. To account for this fact, the term ψ_e—the *coating factor*—is used in the equation.

3. *Sizes of reinforcing.* If small bars are used in a member to obtain a certain total cross-sectional area, the total surface area of the bars will be appreciably larger than if fewer but larger bars are used to obtain the same total bar area. As a result, the required development lengths for smaller bars with their larger surface bonding areas (in proportion to their cross-sectional areas) are less than those required for larger diameter bars. This factor is accounted for with the *reinforcement size factor* ψ_s.

4. *Lightweight aggregates.* The dead weight of concrete can be substantially reduced by substituting lightweight aggregate for the regular stone aggregate. The use of such aggregates (expanded clay or shale, slag, etc.) generally results in lower strength concretes. Such concretes have lower splitting strengths, and so development lengths will have to be larger. In the equation, λ is the *lightweight concrete modification factor* discussed in Section 1.12.

5. *Spacing of bars or cover dimensions.* Should the concrete cover or the clear spacing between the bars be too small, the concrete may very well split, as was previously shown in Figure 7.6. This situation is accounted for with the $(c_b + K_{tr})/d_b$ term in the development length expression. It is called the *confinement term.* In the equation c_b represents the smaller of the distance from the center of the tension bar or wire to the nearest concrete surface, or one-half the center-to-center spacing of the reinforcement.

[*]300 mm in SI.

Table 7.1 Factors for Use in the Expressions for Determining Required Development Lengths for Deformed Bars and Deformed Wires in Tension (ACI 12.2.4)

---

(1) ψ_t = *reinforcement location factor*

Horizontal reinforcement so placed that more than 12 in. of fresh concrete is cast in the member below the development length or splice. 1.3

Other reinforcement. 1.0

(2) ψ_e = *coating factor*

Epoxy-coated bars or wires with cover less than $3d_b$, or clear spacing less than $6d_b$ 1.5

All other epoxy-coated bars or wires. 1.2

Uncoated and zinc-coated reinforcement. 1.0

However, the product of $\psi_t\psi_e$ need not be taken greater than 1.7.

(3) ψ_s = *reinforcement size factor*

No. 6 and smaller bars and deformed wires 0.8

No. 7 and larger bars. 1.0

> In SI units
> No. 19 and smaller bars and deformed wires 0.8
> No. 22 and larger bars 1.0

(4) λ *(lambda)* = *lightweight aggregate concrete factor*

When lightweight aggregate concrete is used, λ shall not exceed 0.75

However, when f_{ct} is specified, λ shall be permitted to be taken as 6.7 $\sqrt{f'_c}/f_{ct}$

> It's $\dfrac{\sqrt{f'_c}}{1.8f_{ct}}$ in SI.

but not greater than 1.0

When normal weight concrete is used. 1.0

(5) c_b = spacing or cover dimension, in.

Use the smaller of either the distance from the center of the bar or wire to the nearest concrete surface, or one-half the center-to-center spacing of the bars or wires being developed.

In this expression, K_{tr} is a factor called the *transverse reinforcement index*. It is used to account for the contribution of confining reinforcing (stirrups or ties) across possible splitting planes.

$$K_{tr} = \frac{40A_{tr}}{sn}$$

$$K_{tr} = \frac{A_{tr}f_{yt}}{1500sn}$$

where

A_{tr} = the total cross-sectional area of all transverse reinforcement having the center-to-center spacing s and a yield strength f_{yt}

n = the number of bars or wires being developed along the plane of splitting. If steel is in two layers, n is the largest number of bars in a single layer.

s = center to center spacing of transverse reinforcing

The Code in Section 12.2.3 conservatively permits the use of $K_{tr} = 0$ to simplify the calculations even if transverse reinforcing is present. ACI 12.2.3 limits the value of $c_b + K_{tr}/d_b$ used in the equation to a maximum value of 2.5. (It has been found that if values larger than 2.5 are used, the shorter development lengths resulting will increase the danger of pullout-type failures.)

The calculations involved in applying ACI Equation 12-1 are quite simple, as is illustrated in Example 7.2.

$$\text{In SI units, } K_{tr} = \frac{A_{tr}f_{yt}}{10sn}$$

EXAMPLE 7.2

Determine the development length required for the #8 uncoated bottom bars shown in Figure 7.8.

(a) assume $K_{tr} = 0$ and

(b) use the computed value of K_{tr}.

SOLUTION   From Table 7.1

$\psi_t = 1.0$ for bottom bars

$\psi_e = 1.0$ for uncoated bars

$\psi_s = 1.0$ for #8 bars

$\lambda = 1.0$ for normal weight concrete

$c_b = $ side cover of bars measured from center of bars $= 2\frac{1}{2}$ in.

or

$c_b = $ one-half of c. to c. spacing of bars $= 1\frac{1}{2}$ in. ←

(a) Using ACI Equation 12-1 with $K_{tr} = 0$

$$\frac{c_b + K_{tr}}{d_b} = \frac{1.50 + 0}{1.00} = 1.50 < 2.50 \qquad \text{limit for pullout failure} \qquad \underline{\underline{OK}}$$

$$\frac{\ell_d}{d_b} = \frac{3}{40}\frac{f_y}{\lambda\sqrt{f'_c}}\frac{\psi_t\psi_e\psi_s}{\left(\dfrac{c_b + K_{tr}}{d_b}\right)}$$

$$= \left(\frac{3}{40}\right)\left(\frac{60{,}000}{(1.0)\sqrt{3000}}\right)\frac{(1.0)(1.0)(1.0)}{1.50}$$

$$= \underline{\underline{55 \text{ diameters}}} \;\rightarrow 55'' \; f$$

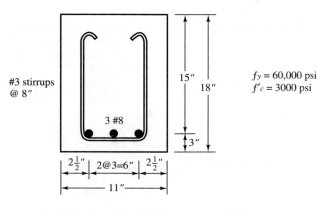

#3 stirrups @ 8"

3 #8

15"
18"
3"

$2\frac{1}{2}''$ 2@3=6" $2\frac{1}{2}''$

11"

$f_y = 60{,}000$ psi
$f'_c = 3000$ psi

Figure 7.8

(b) Using Computed Value of K_{tr} and ACI Equation 12-1

$$K_{tr} = \frac{40\,A_{tr}}{sn} = \frac{(40)(2)(0.11)}{(8)(3)} = 0.367$$

$$\frac{c_b + K_{tr}}{d_b} = \frac{1.50 + 0.367}{1.0} = 1.867 \; < \; 2.5 \qquad\qquad \text{OK}$$

$$\frac{\ell_d}{d_b} = \left(\frac{3}{40}\right)\left(\frac{60{,}000}{\sqrt{3000}}\right)\frac{(1.0)(1.0)(1.0)(1.0)}{1.867} = \underline{44 \text{ diameters}}$$

In determining required development lengths, there are two more ACI specifications to keep in mind:

1. Section 12.1.2 states that values of $\sqrt{f_c'}$ used in the equations cannot be greater than 100 psi or $\frac{25}{3}$ MPa in SI. (This limit is imposed because there has not been a sufficient amount of research on the development of bars in higher strength concretes to justify higher $\sqrt{f_c'}$ values, which would result in smaller $\frac{\ell_d}{d_b}$ values.)

2. When the amount of flexural reinforcing provided exceeds the theoretical amount required and where the specifications being used do not specifically require that the development lengths be based on f_y, the value of $\frac{\ell_d}{d_b}$ may be multiplied by $(A_{s\,\text{required}}/A_{s\,\text{provided}})$ according to ACI 12.2.5. This reduction factor may not be used for the development of reinforcement at supports for positive reinforcement, for the development of shrinkage and temperature reinforcement, and for a few other situations referenced in R12.2.5. This reduction also is not permitted in regions of high seismic risk, as described in ACI 318-08, Chapter 21.

Instead of using ACI Equation 12-1 for computing development lengths, the ACI in its Section 12.2 permits the use of a somewhat simpler and more conservative approach (as shown in Table 7.2 herein) for certain conditions. With this approach the ACI recognizes that in a very large percentage of cases, designers use spacing and cover values and confining reinforcing that result in a value of $\frac{c_b + K_{tr}}{d_b}$ equal to at least 1.5. Based on this value and the appropriate values of ψ_s, the expressions in Table 7.2 were determined.

For SI values, see Section 12.2.2 of the 318M-08 Code.

If a minimum cover equal to d_b and a minimum clear spacing between bars of $2d_b$ (or a minimum clear spacing of bars equal to d_b, along with a minimum of ties or stirrups) are used, the

Table 7.2 Simplified Development Length Equations

	#6 and smaller bars and deformed wires	#7 and larger bars
Clear spacing of bars being developed or spliced not less than d_b, clear cover not less than d_b, and stirrups or ties throughout ℓ_d not less than the code minimum or Clear spacing of bars being developed or spliced not less than $2d_b$ and clear cover not less than d_b	$\dfrac{\ell_d}{d_b} = \dfrac{f_y \psi_t \psi_e}{25\,\lambda\,\sqrt{f_c'}}$	$\dfrac{\ell_d}{d_b} = \dfrac{f_y \psi_t \psi_e}{20\,\lambda\,\sqrt{f_c'}}$
Other cases	$\dfrac{\ell_d}{d_b} = \dfrac{3 f_y \psi_t \psi_e}{50\,\lambda\,\sqrt{f_c'}}$	$\dfrac{\ell_d}{d_b} = \dfrac{3 f_y \psi_t \psi_e}{40\,\lambda\,\sqrt{f_c'}}$

expressions in Table 7.2 can be used. Otherwise it is necessary to use the more rigorous ACI Equation 12-1.

The author feels that the application of the so-called simplified equations requires almost as much effort as is needed to use the longer equation. Furthermore, the development lengths computed with the "simpler" equations are often so much larger than the ones determined with the regular equation as to be uneconomical.

For these reasons the author recommends the use of Equation 12-1 for computing development lengths. In using this long form equation, however, you may very well like to assume that $K_{tr} = 0$, as the results obtained usually are only slightly more conservative than those obtained with the full equation. *The author uses Equation 12-1 with $K_{tr} = 0$ for all applications after this chapter.*

Examples 7.3 and 7.4, which follow, present the determination of development lengths using each of the methods that have been described in this section.

EXAMPLE 7.3

The #7 bottom bars shown in Figure 7.9 are epoxy coated. Assuming normal weight concrete, $f_y = 60,000$ psi, and $f'_c = 3500$ psi, determine required development lengths

(a) Using the simplified equations of Table 7.2.

(b) Using the full ACI Equation 12-1 with the calculated value of K_{tr}.

(c) Using ACI Equation 12-1 with $K_{tr} = 0$.

SOLUTION   With reference to Table 7.1

$$\psi_t = 1.0 \text{ for bottom bars}$$

$$\psi_e = 1.5 \text{ for epoxy-coated bars with clear spacing} < 6d_b$$

$$\psi_t\psi_e = (1.0)(1.5) = 1.5 < 1.7 \qquad\qquad \underline{\underline{\text{OK}}}$$

$$\psi_s = 1.0 \text{ for \#7 and larger bars}$$

$$\lambda = 1.0 \text{ for normal weight concrete}$$

$$c_b = \text{cover} = 3 \text{ in.}$$

or

$$c_b = \text{one-half of c. to c. spacing of bars} = 1\tfrac{1}{2} \text{ in.} \leftarrow$$

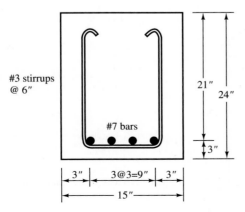

Figure 7.9

(a) Using Simplified Equation

$$\frac{\ell_d}{d_b} = \frac{f_y \psi_t \psi_e}{20 \lambda \sqrt{f_c'}} = \frac{(60,000)(1.0)(1.5)}{20(1.0)\sqrt{3500}} = 76 \text{ diameters}$$

(b) Using ACI Equation 12-1 with Computed Value of K_{tr}

$$K_{tr} = \frac{40 A_{tr}}{sn} = \frac{(40)(2)(0.11)}{(6)(4)} = 0.367$$

$$\frac{c_b + K_{tr}}{d_b} = \frac{1.5 + 0.367}{0.875} = 2.13 \ < \ 2.50 \qquad\qquad \underline{\underline{\text{OK}}}$$

$$\frac{\ell_d}{d_b} = \frac{3}{40} \frac{f_y}{\lambda \sqrt{f_c'}} \frac{\psi_t \psi_e \psi_s}{\dfrac{c_b + K_{tr}}{d_b}}$$

$$= \left(\frac{3}{40}\right) \left(\frac{60,000}{(1.0)\sqrt{3500}}\right) \frac{(1.0)(1.5)(1.0)}{2.13}$$

$$= \underline{\underline{54 \text{ diameters}}}$$

(c) Using ACI Equation 12-1 with $K_{tr} = 0$

$$\frac{c_b + K_{tr}}{d_b} = \frac{1.5 + 0}{0.875} = 1.71 \ < \ 2.50 \qquad\qquad \underline{\underline{\text{OK}}}$$

$$\frac{\ell_d}{d_b} = \left(\frac{3}{40}\right) \left(\frac{60,000}{(1.0)\sqrt{3500}}\right) \frac{(1.0)(1.5)(1.0)}{1.71}$$

$$= \underline{\underline{67 \text{ diameters}}}$$

EXAMPLE 7.4

The required reinforcing steel area for the lightweight concrete beam of Figure 7.10 is 2.88 in.[2] The #8 top bars shown are uncoated. Compute development lengths if $f_y = 60,000$ psi and $f_c' = 3500$ psi.

(a) Using simplified equations.

(b) Using the full ACI Equation 12-1.

(c) Using Equation 12-1 with $K_{tr} = 0$.

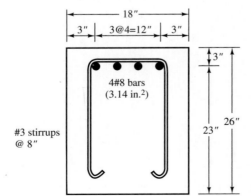

Figure 7.10 Cross section of cantilever beam.

SOLUTION   With reference to Table 7.1

$$\psi_t = 1.3 \text{ for top bars}$$

$$\psi_e = 1.0 \text{ for uncoated bars}$$

$$\psi_t\psi_e = (1.3)(1.0) \ < \ 1.7 \qquad\qquad \underline{\underline{\text{OK}}}$$

$$\psi_s = 1.0 \text{ for \#7 and larger bars}$$

$$\lambda = 0.75 \text{ for lightweight concrete}$$

$$c_b = \text{cover} = 3 \text{ in.}$$

or

$$c_b = \text{one-half of c. to c. spacing of bars} = 2 \text{ in.} \leftarrow$$

(a) Using Simplified Equations

$$\frac{\ell_d}{d_b} = \frac{f_y\psi_t\psi_e}{20\lambda\sqrt{f_c'}} = \frac{(60{,}000)(1.3)(1.0)}{20(0.75)\sqrt{3500}} = 88 \text{ diameters}$$

$\dfrac{\ell_d}{d_b}$ reduced for excess reinforcement to $\left(\dfrac{2.88}{3.14}\right)(88) = \underline{\underline{81 \text{ diameters}}}$

(b) Using ACI Equation 12-1 with Computed Value of K_{tr}

$$K_{tr} = \frac{40A_{tr}}{sn} = \frac{(40)(2)(0.11)}{(8)(4)} = 0.275$$

$$\frac{c_b + K_{tr}}{d_b} = \frac{2.0 + 0.275}{1.0} = 2.275 \ < \ 2.5 \qquad\qquad \underline{\underline{\text{OK}}}$$

$$\frac{\ell_d}{d_b} = \frac{3}{40}\frac{f_y}{\lambda\sqrt{f_c'}}\frac{\psi_t\psi_e\psi_s}{\dfrac{c_b + K_{tr}}{d_b}}$$

$$= \left(\frac{3}{40}\right)\left(\frac{60{,}000}{(0.75)\sqrt{3500}}\right)\frac{(1.3)(1.0)(1.0)}{2.275}$$

$$= 58 \text{ diameters}$$

$\dfrac{\ell_d}{d_b}$ reduced for excess reinforcement to $\left(\dfrac{2.88}{3.14}\right)(58) = \underline{\underline{53 \text{ diameters}}}$

(c) Using ACI Equation 12-1 with $K_{tr} = 0$

$$\frac{c_b + K_{tr}}{d_b} = \frac{2.0 + 0}{1.0} = 2.0 \ < \ 2.5$$

$$\frac{\ell_d}{d_b} = \left(\frac{3}{40}\right)\left(\frac{60{,}000}{(0.75)\sqrt{3500}}\right)\frac{(1.3)(1.0)(1.0)}{2.0}$$

$$= 66 \text{ diameters}$$

$\dfrac{\ell_d}{d_b}$ reduced for excess reinforcement to $\left(\dfrac{2.88}{3.14}\right)(66) = \underline{\underline{61 \text{ diameters}}}$

7.4 DEVELOPMENT LENGTHS FOR BUNDLED BARS

When bundled bars are used, greater development lengths are needed because there is not a "core" of concrete between the bars to provide resistance to slipping. The Code, Section 12.4.1, states that splice and development lengths for bundled bars are to be determined by first computing the lengths needed for the individual bars and then by increasing those values by 20% for three-bar bundles and 33% for four-bar bundles.

When the factors relating to cover and clear spacing are being computed for a particular bundle, the bars are treated as though their area is furnished by a single bar. In other words, it is necessary to replace the bundle of bars with a fictitious single bar with a diameter such that its cross-sectional area equals that of the bundle of bars. This is conservative because the bond properties of the bundled bars are actually better than for the fictitious single bar. When determining c_b, the confinement term, and the ψ_e factor, the fictitious bar is considered to have a centroid coinciding with that of the bar bundle. Example 7.5 presents the calculation of the development length needed for a three-bar bundle of #8 bars.

EXAMPLE 7.5

Compute the development length required for the uncoated bundled bars shown in Figure 7.11, if $f_y = 60,000$ psi and $f'_c = 4000$ psi with normal-weight concrete. Use ACI Equation 12-1 and assume $K_{tr} = 0$.

SOLUTION   With reference to Table 7.1

$$\psi_t = \psi_e = \psi_s = \lambda = 1.0$$

Area of 3 #8 bars $= 2.35$ in.^{2}

Diameter d_{bf} of a single bar of area 2.35 in.^{2}

$$\frac{\pi d_{bf}^2}{4} = 2.35$$
$$d_{bf} = 1.73 \text{ in.}$$

Find the lowest value for c_b (Figure 7.11(b)).

$$c_{b1} = \text{side cover of bars} = 2 + \tfrac{3}{8} + 1.00 = 3.38 \text{ in.} \leftarrow$$
$$c_{b2} = \text{bottom cover of bars} = 2 + 3/8 + 0.79 d_b = 2 + 3/8 + 0.79(1.00) = 3.16 \text{ in.}$$

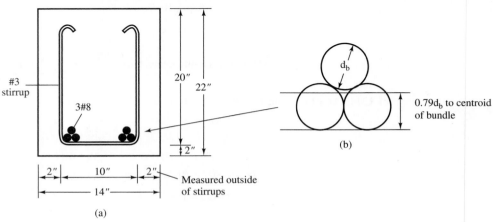

(a)

(b)

Figure 7.11

[3]See Figure 7.11(b) for this dimension.

where d_b is the actual (not the fictitious) bar diameter.

$$c_{b3} = \tfrac{1}{2}\text{c. to c. spacing of bars} = \frac{10 - (2)\left(\tfrac{3}{8}\right) - (2)(1.00)}{2} = 3.62 \text{ in.}$$

Using ACI Equation 12-1 with $K_{tr} = 0$

$$\frac{c_b + K_{tr}}{d_{bf}} = \frac{3.16 + 0}{1.73} = 1.83 < 2.5$$

$$\frac{\ell_d}{d_b} = \left(\frac{3}{40}\right)\left(\frac{60{,}000}{(1.0)\sqrt{4000}}\right)\frac{(1.0)(1.0)(1.0)}{1.83}$$

$$= 39 \text{ diameters}$$

But should be increased 20% for a 3-bar bundle according to ACI Section 12.4.1.

$$\frac{\ell_d}{d_b} = (1.20)(39) = 47 = 47d_b = 47 \text{ in.}$$

Note that the actual bar diameter is used in the last equation, not the fictitious bar.

7.5 HOOKS

When sufficient space is not available to anchor *tension* bars by running them straight for their required development lengths, as described in the last section of this text, hooks may be used. (*Hooks are considered useless for compression bars for development length purposes.*)

Figure 7.12 shows details of the standard 90° and 180° hooks specified in Sections 7.1 and 7.2 of the ACI Code. Either the 90° hook with an extension of 12 bar diameters ($12d_b$) at the free end or

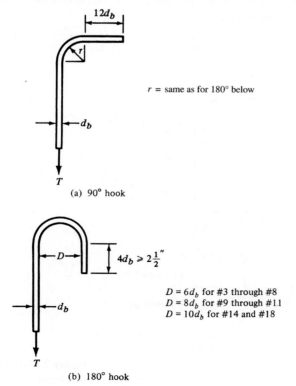

$12d_b$

r

r = same as for 180° below

d_b

T

(a) 90° hook

D

$4d_b \geqslant 2\tfrac{1}{2}''$

$D = 6d_b$ for #3 through #8
$D = 8d_b$ for #9 through #11
$D = 10d_b$ for #14 and #18

d_b

T

(b) 180° hook

Figure 7.12 Hooks.

the 180° hook with an extension of 4 bar diameters ($4d_b$), but not less than $2\frac{1}{2}$ in., may be used at the free end. The radii and diameters shown are measured on the inside of the bends.

The dimensions given for hooks were developed to protect members against splitting of the concrete or bar breakage no matter what concrete strengths, bar sizes, or bar stresses are used. Actually, hooks do not provide an appreciable increase in anchorage strength because the concrete in the plane of the hook is somewhat vulnerable to splitting. This means that adding more length (i.e., more than the specified $12d_b$, or $4d_b$ values) onto bars beyond the hooks doesn't really increase their anchorage strengths.

The development length needed for a hook is directly proportional to the bar diameter. This is because the magnitude of compressive stresses in the concrete on the inside of the hook is governed by d_b. To determine the development lengths needed for standard hooks, the ACI (12.5.2) requires the calculation of

$$\ell_{dh} = \frac{0.02\psi_e f_y d_b}{\lambda \sqrt{f_c'}}$$

The value of ℓ_{dh}, according to ACI Section 12.5.1, may not be less than 6 in. or $8\ d_b$. For deformed bars the ACI, Section 12.5.2, states that ψ_e in this expression can be taken equal to 1.2 for epoxy-coated reinforcing and the λ used equal to 0.75 for lightweight aggregate concrete. For all other cases ψ_e and λ are to be set equal to 1.0.

In SI units, $\ell_{dh} = \dfrac{0.24\psi_e f_y}{\lambda \sqrt{f_c'}} d_b$

The development length ℓ_{dh}, is measured from the critical section of the bar to the outside end or edge of the hooks, as shown in Figure 7.13.

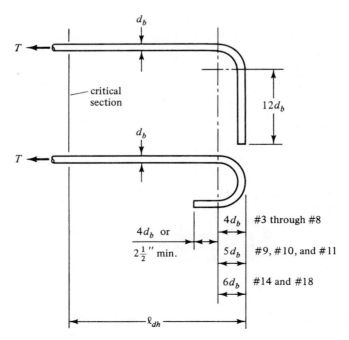

Figure 7.13 Hooked-bar details for development of standard hooks.

The modification factors that may have to be successively multiplied by ℓ_{dh}, are listed in Section 12.5.3 of the Code and are summarized in subparagraphs (a) to (d). These values apply only for cases where standard hooks are used. The effect of hooks with larger radii is not covered by the Code. For the design of hooks, no distinction is made between top bars and other bars. (It is difficult to distinguish top from bottom anyway when hooks are involved.)

(a) *Cover.* When hooks are made with #11 or smaller bars and have side cover values normal to the plane of the hooks no less than $2\frac{1}{2}$ in. and where the cover on the bar extensions beyond 90° hooks are not less than 2 in., multiply by 0.7.

(b) *Ties or stirrups.* When hooks made of #11 or smaller bars are enclosed either vertically or horizontally within ties or stirrup ties along their full development length ℓ_{dh}, and the stirrups or ties are spaced no farther apart than $3d_b$ (where d_b is the diameter of the hooked bar), multiply by 0.8. This situation is shown in Figure 7.14. (Detailed dimensions are given for stirrup and tie hooks in Section 7.1.3 of the ACI Code.)

(c) When 180° hooks consisting of #11 or smaller bars are used and are enclosed within ties or stirrups placed perpendicular to the bars being developed and spaced no further than $3d$ apart along the development length ℓ_{dh} of the hook, multiply by 0.8. If the 90° hook shown in Figure 7.14 is replaced with a 180° hook and ties or stirrups are perpendicular (not parallel) to the longitudinal bar being developed, Figure 7.14 applies to this case as well.

(d) Should anchorage or development length not be specially required for f_y of the bars, it is permissible to multiply ℓ_{dh} by A_s required/A_s provided.

The danger of a concrete splitting failure is quite high if both the side cover (perpendicular to the hook) and the top and bottom cover (in the plane of the hook) are small. The Code (12.5.4), therefore, states that when standard hooks with less than $2\frac{1}{2}$ in. side and top or bottom cover are used at discontinuous ends of members, the hooks shall be enclosed within ties or stirrups spaced no farther than $3d_b$ for the full development length ℓ_{dh}. The first tie or stirrup must enclose the bent part of the hook within a distance of $2d_{bh}$ of the outside of the bend. Furthermore, the modification factor of 0.8 of items (b) and (c) herein shall not be applicable. If the longitudinal bar being developed with the hook shown in Figure 7.14 were at a discontinuous end of a member, such as

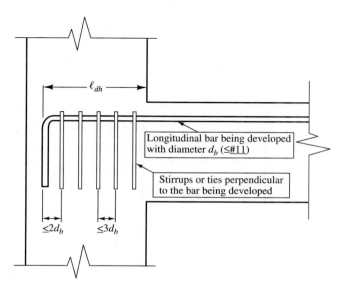

Figure 7.14 Stirrup or tie detail for 90° hooks complying with the 0.8 multiplier. The stirrups or ties shown can be either vertical (as illustrated) or horizontal.

the free end of a cantilever beam, the ties or stirrups shown in that figure are *required* unless side and top cover both are at least 2-$\frac{1}{2}$ in.

Example 7.6, which follows, illustrates the calculations necessary to determine the development lengths required at the support for the tensile bars of a cantilever beam. The lengths for straight or hooked bars are determined.

EXAMPLE 7.6

Determine the development or embedment length required for the epoxy-coated bars of the beam shown in Figure 7.15.

(a) If the bars are straight, assuming $K_{tr} = 0$.

(b) If a 180° hook is used.

(c) If a 90° hook is used.

The six #9 bars shown are considered to be top bars. $f_c' = 4000$ psi and $f_y = 60,000$ psi.

SOLUTION

(a) Straight Bars

$\psi_t = 1.3$ for top bars

$\psi_e = 1.5$ for coated bars with cover $< 3d_b$ or clear spacing $< 6d_b$

$\psi_t\psi_e = (1.3)(1.5) = 1.95 > \underline{1.7} \therefore$ Use 1.7

$\psi_s = 1.0$ for 9 bars

$\lambda = 1.0$ for normal weight concrete

$c_b =$ side cover $=$ top cover $= 2.5$ in.

$c_b =$ one-half of c. to c. spacing of bars $= 2.25$ in.$\leftarrow$

$$\frac{c_b + K_{tr}}{d_b} = \frac{2.25 + 0}{1.128} = 1.99 < 2.5 \qquad \underline{\underline{OK}}$$

$$\frac{\ell_d}{d_b} = \left(\frac{3}{40}\right)\left(\frac{60,000}{(1.0)\sqrt{4000}}\right)\frac{(1.7)(1.0)}{1.99} = 61 \text{ diameters}$$

$$\ell_d = (61)(1.128) = \underline{\underline{69 \text{ in.}}}$$

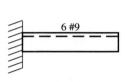

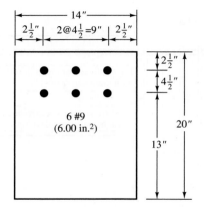

6 #9

14"

2$\frac{1}{2}$" | 2@4$\frac{1}{2}$=9" | 2$\frac{1}{2}$"

2$\frac{1}{2}$"

4$\frac{1}{2}$"

6 #9
(6.00 in.²)

13"

20"

Figure 7.15

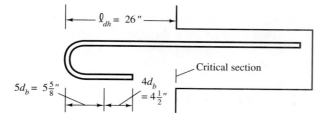

Figure 7.16 180° hook.

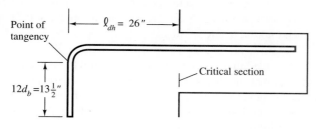

Figure 7.17

(b) Using 180° hooks (see Figure 7.16) note that $\psi_e = 1.2$ as required in ACI Section 12.5.2 for epoxy-coated hooks

$$\ell_{dh} = \frac{0.02\psi_e f_y d_b}{\lambda \sqrt{f_c'}} = \frac{(0.02)(1.2)(60,000)(1.128)}{(1.0)\sqrt{4000}}$$

$$= 25.68 \text{ in. } \underline{\text{Say 26 in.}}$$

Note: The dimensions shown in the beam cross section (Figure 7.15) indicate 2-1/2 in. from the bar center to the top and side of the beam. The cover is $2.5 - d_b/2 = 1.936'' < 2.5''$. If this hook were in the free end of a cantilever beam, ties or stirrups would be required and the 0.8 reduction factor would not be applicable. In this example, the hook is in a column, so special ties are not required. If they were provided, a reduction of 0.8 would apply. In this example, they are not provided.

(c) Using 90° hooks (see Figure 7.17)

$$\ell_{dh} = 26''$$

as the 0.8 reduction factor does not apply because ties or stirrups are not provided.

7.6   DEVELOPMENT LENGTHS FOR WELDED WIRE FABRIC IN TENSION

Section 12.7 of the ACI Code provides minimum required development lengths for deformed welded wire fabric, whereas Section 12.8 provides minimum values for plain welded wire fabric.

The minimum required development length for deformed welded wire fabric in tension measured from the critical section equals the value determined for ℓ_d as per ACI Section 12.2.2 or 12.2.3 multiplied by a wire fabric factor, ψ_w, from ACI Section 12.7.2 or 12.7.3.

This factor, which follows, contains the term s, which is the spacing of the wire to be developed. The resulting development length may not be less than 8 in. except in the computation of lap splices. You might note that epoxy coatings seem to have little effect on the lengths needed for welded wire fabric, and it is thus permissible to use $\psi_e = 1.0$.

The wire fabric factor, ψ_w, for welded wire fabric with at least one crosswire within the development length not less than 2 inches from the critical section is

$$\psi_w = \frac{f_y - 35{,}000}{f_y} \text{ not less than } \frac{5d_b}{s}$$

but need not be taken > 1.0.

In SI units for welded wire fabric with at least one crosswire within the development length and not less than 50 mm from the point of the critical section, the wire fabric, factor ψ_w, is $\frac{f_y - 240}{f_y}$ not less than $\frac{5d_b}{s}$, but need not be taken > 1.0.

The yield strength of welded plain wire fabric is considered to be adequately developed by two crosswires if the closer one is not less than 2 in. from the critical section. The Code (Section 12.8), however, says that the development length, ℓ_d, measured from the critical section to the outermost crosswire may not be less than the value computed from the following equation, in which A_w is the area of the individual wire to be developed.

$$\ell_d = 0.27 \frac{A_w}{s} \left(\frac{f_y}{\lambda \sqrt{f_c'}} \right) \text{ but not } < 6 \text{ in.}$$

Or in SI units

$$\ell_d = 3.3 \frac{A_w}{s} \left(\frac{f_y}{\lambda \sqrt{f_c'}} \right) \text{ but not } < 150 \text{ mm}$$

The development lengths obtained for either plain or deformed wire may be reduced, as were earlier development lengths by multiplying them by $(A_{s \text{ required}}/A_{s \text{ furnished}})$ (ACI 12.2.5), but the results may not be less than the minimum values given above.

7.7 DEVELOPMENT LENGTHS FOR COMPRESSION BARS

There is not a great deal of experimental information available about bond stresses and needed embedment lengths for compression steel. It is obvious, however, that embedment lengths will be smaller than those required for tension bars. For one reason, there are no tensile cracks present to encourage slipping. For another, there is some bearing of the ends of the bars on concrete, which also helps develop the load.

The Code (12.3.2) states that the minimum basic development length provided for compression bars (ℓ_{dc}) may not be less than the value computed from the following expression.

$$\ell_{dc} = \frac{0.02 f_y d_b}{\lambda \sqrt{f_c'}} \geq 0.0003 \, f_y d_b \text{ but not less than 8 in.}$$

Or in SI units

$$\ell_{dc} = \frac{0.02 f_y d_b}{\lambda \sqrt{f_c'}} \geq 0.0003 \, f_y d_b \text{ but not less than 200 mm}$$

If more compression steel is used than is required by analysis, ℓ_{dc} may be multiplied by $(A_{s\ required})/(A_{s\ provided})$ as per ACI Section 12.3.3. When bars are enclosed in spirals for any kind of concrete members, the members become decidedly stronger due to the confinement or lateral restraint of the concrete. The normal use of spirals is in spiral columns, which are discussed in Chapter 9. Should compression bars be enclosed by spirals of not less than $\frac{1}{4}$ in. diameter and with a pitch not greater than 4 in. or within #4 ties spaced at not more than 4 in. on center, the value of ℓ_{dc} may be multiplied by 0.75 (ACI 12.3.3). In no case can the development length be less than 8 in. Thus

$$\ell_d = \ell_{dc} \times \text{applicable modification factors} \geq 8.0 \text{ in.}$$

An introductory development length problem for compression bars is presented in Example 7.7. The forces in the bars at the bottom of the column of Figure 7.18 are to be transferred down into a reinforced concrete footing by means of dowels. Dowels such as these are usually bent at their bottoms (as shown in the figure) and set on the main footing reinforcing where they can be tied securely in place. The bent or hooked parts of the dowels, however, do not count as part of the required development lengths for compression bars (ACI 12.5.5), as they are ineffective.

In a similar fashion, the dowel forces must be developed up into the column. In Example 7.7 the required development lengths up into the column and down into the footing are different because the f_c' values for the footing and the column are different in this case. The topic of dowels and force transfer from walls and columns to footings is discussed in some detail in Chapter 12. (The development lengths determined in this example are for compression bars as would normally be the case at the base of columns. If uplift is possible, however, it will be necessary to consider tension development lengths, which could very well control.)

EXAMPLE 7.7

The forces in the column bars of Figure 7.18 are to be transferred into the footing with #9 dowels. Determine the development lengths needed for the dowels (a) down into the footing and (b) up into the column if $f_y = 60,000$ psi. The concrete in both the column and the footing is normal weight.

SOLUTION   **(a)** Down into the footing,

$$\ell_{dc} = \frac{0.02 d_b f_y}{\lambda\sqrt{f_c'}} = \frac{(0.02)(1.128)(60,000)}{(1.0)\sqrt{3000}} = 24.71'' \leftarrow$$

$$\ell_{dc} = (0.0003)(1.128)(60,000) = 20.30''$$

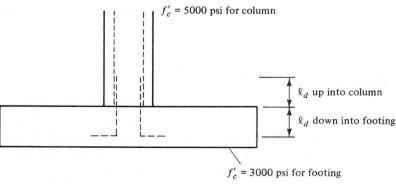

Figure 7.18

Hence $\ell_d = 24.71''$, say $25''$, as there are no applicable modification factors. Under no circumstances may ℓ_d be less than 8 in.

(b) Up into column,

$$\ell_{dc} = \frac{(0.02)(1.128)(60,000)}{(1.0)\sqrt{5000}} = 19.14''$$

$$\ell_{dc} = (0.0003)(1.128)(60,000) = 20.30'' \longleftarrow$$

Hence $\ell_d = 20.30''$, say $21''$ (can't be $< 8''$), as there are no applicable modification factors. (*Answer:* Extend the dowels $25''$ down into the footing and $21''$ up into the column.)

Note: The bar details shown in Figure 7.18 are unsatisfactory for seismic areas as the bars should be bent inward and not outward. The reason for this requirement is that the Code, Chapter 21, on seismic design, stipulates that hooks must be embedded in confined concrete.

7.8   CRITICAL SECTIONS FOR DEVELOPMENT LENGTH

Before the development length expressions can be applied in detail, it is necessary to understand clearly the critical points for tensile and compressive stresses in the bars along the beam.

First, it is obvious that the bars will be stressed to their maximum values at those points where maximum moments occur. Thus those points must be no closer in either direction to the bar ends than the ℓ_d values computed.

There are, however, other critical points for development lengths. As an illustration, a critical situation occurs whenever there is a tension bar whose neighboring bars have just been cut off or bent over to the other face of the beam. Theoretically, if the moment is reduced by a third, one-third of the bars are cut off or bent and the remaining bars would be stressed to their yield points. The full development lengths would be required for those bars.

This could bring up another matter in deciding the development length required for the remaining bars. The Code (12.10.3) requires that bars that are cut off or bent be extended a distance beyond their theoretical flexure cutoff points by d or 12 bar diameters, whichever is greater. In addition, the point where the other bars are bent or cut off must also be at least a distance ℓ_d from their points of maximum stress (ACI 12.10.4). Thus these two items might very well cause the remaining bars to have a stress less than f_y, thus permitting their development lengths to be reduced somewhat. A conservative approach is normally used, however, in which the remaining bars are assumed to be stressed to f_y.

7.9   EFFECT OF COMBINED SHEAR AND MOMENT ON DEVELOPMENT LENGTHS

The ACI Code does not specifically consider the fact that shear affects the flexural tensile stress in the reinforcing. The Code (12.10.3) does require bars to be extended a distance beyond their theoretical cutoff points by a distance no less than the effective depth of the member d or 12 bar diameters, whichever is larger. The Commentary (R12.10.3) states that this extension is required to account for the fact that the locations of maximum moments may shift due to changes in loading, support settlement, and other factors. It can be shown that a diagonal tension crack in a beam without stirrups can shift the location of the computed tensile stress a distance approximately equal to d toward the point of zero moment. When stirrups are present, the effect is still there but is somewhat less severe.

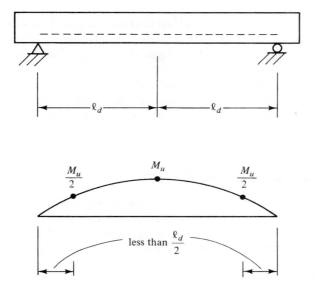

Figure 7.19

The combined effect of shear and bending acting simultaneously on a beam may produce premature failure due to overstress in the flexural reinforcing. Professor Charles Erdei[4-6] has done a great deal of work on this topic. His work demonstrates that web reinforcing participates in resisting bending moment. He shows that the presence of inclined cracks increases the force in the tensile reinforcing at all points in the shear span except in the region of maximum moment. The result is just as though we have a shifted moment diagram and leads us to the thought that we should be measuring ℓ_d from the shifted moment diagram rather than from the basic one. He clearly explains the moment shift and the relationship between development length and the shift in the moment diagram.

The late Professor P. M. Ferguson[7] stated that whether or not we decide to use the shifted moment concept, it is nevertheless desirable to stagger the cutoff points of bars (and it is better to bend them than to cut them).

7.10   EFFECT OF SHAPE OF MOMENT DIAGRAM ON DEVELOPMENT LENGTHS

A further consideration of development lengths will show the necessity of considering the shape of the moment diagram. To illustrate this point, the uniformly loaded beam of Figure 7.19 with its parabolic moment diagram is considered. *It is further assumed that the length of the reinforcing bars on each side of the beam centerline equals the computed development length* ℓ_d. The discussion to follow will prove that this distance is not sufficient to properly develop the bars for this moment diagram.[8]

[4]Erdei, C. K., 1961, "Shearing Resistance of Beams by the Load-Factor Method," *Concrete and Constructional Engineering*, 56(9), pp. 318–319.

[5]Erdei, C. K., 1962, "Design of Reinforced Concrete for Combined Bending and Shear by Ultimate Lead Theory," *Journal of the Reinforced Concrete Association*, 1(1).

[6]Erdei, C. K., 1963, "Ultimate Resistance of Reinforced Concrete Beams Subjected to Shear and Bending," *European Concrete Committees Symposium on Shear*, Wiesbaden, West Germany, pp. 102–114.

[7]Ferguson, P. M., 1979, *Reinforced Concrete Fundamentals*, 4th ed. (New York: John Wiley & Sons), p. 187.

[8]Ferguson, *Reinforced Concrete Fundamentals*, pp. 191–193.

At the centerline of the beam of Figure 7.19, the moment is assumed to equal M_u and the bars are assumed to be stressed to f_y. Thus the development length of the bars on either side of the beam centerline must be no less than ℓ_d. If one then moves along this parabolic moment diagram on either side to a point where the moment has fallen off to a value of $M_u/2$, it is correct to assume a required development length from this point equal to $\ell_d/2$.

The preceding discussion clearly shows that the bars will have to be extended farther out from the centerline than ℓ_d. For the moment to fall off 50%, one must move more than halfway toward the end of the beam.

7.11   CUTTING OFF OR BENDING BARS (CONTINUED)

This section presents a few concluding remarks concerning the cutting off of bars, a topic that was introduced in Section 7.1. The last several sections have offered considerable information that affects the points where bars may be cut off. Here we give a summary of the previously mentioned requirements, together with some additional information. First, a few comments concerning shear are in order.

When some of the tensile bars are cut off at a point in a beam, a sudden increase in the tensile stress will occur in the remaining bars. For this increase to occur there must be a rather large increase in strain in the beam. Such a strain increase quite possibly may cause large tensile cracks to develop in the concrete. If large cracks occur, there will be a reduced beam cross section left to provide shear resistance—and thus a greater possibility of shear failure.

To minimize the possibility of a shear failure, Section 12.10.5 of the ACI Code states that *at least one* of the following conditions must be met if bars are cut off in a tension zone:

1. The shear at the cutoff point must not exceed two-thirds of the design shear strength ϕV_n in the beam, including the strength of any shear reinforcing provided (ACI 12.10.5.1).

2. An area of shear reinforcing in excess of that required for shear and torsion must be provided for a distance equal to $\frac{3}{4}d$ from the cutoff point. The minimum area of this reinforcing and its maximum spacing are provided in Section 12.10.5.2 of the Code.

3. When #11 or smaller bars are used, the continuing bars should provide twice the area of steel required for flexure at the cutoff point, and the shear should not exceed three-fourths of the permissible shear (ACI 12.10.5.3).

The moment diagrams used in design are only approximate. Variations in loading, settlement of supports, the application of lateral loads, and other factors may cause changes in those diagrams. In Section 7.9 of this chapter we saw that shear forces could appreciably offset the tensile stresses in the reinforcing bars, thus in effect changing the moment diagrams. As a result of these factors, the Code (12.10.3) says that reinforcing bars should be continued for a distance of 12 bar diameters or the effective depth d of the member, whichever is greater (except at the supports of simple spans and the free ends of cantilevers), beyond their theoretical cutoff points.

Various other rules for development lengths apply specifically to positive-moment reinforcement, negative-moment reinforcement, and continuous beams. These are addressed in Chapter 14 of this text. Another item presented in that chapter, which is usually of considerable interest to students, are the rules of thumb which are frequently used in practice to establish cutoff and bend points.

Another rather brief development length example is presented in Example 7.8. A rectangular section and satisfactory reinforcing have been selected for the given span and loading condition. It is desired to determine where two of the four bars may be cut off, considering both moment and development length.

Los Angeles County water project. (Courtesy of The Burke Company.)

EXAMPLE 7.8

The rectangular beam with four #8 bars shown in Figure 7.20(b) has been selected for the span and loading shown in part (a) of the figure. Determine the cutoff point for two of the bars, considering both the actual moment diagram and the required development length.

The design moment capacity (ϕM_n) of this beam has been computed to equal 359.7 ft-k when it has four bars and 185.3 ft-k when it has two bars. (Notice that ρ with two bars $= \frac{1.57}{(18)(27)} = 0.00323 < \rho_{min} = \frac{200}{60,000} = 0.00333$, but is considered to be close enough.) In addition, ℓ_d for the bars has been determined to equal 41 in., using ACI Equation 12-1 with $\lambda = 0.75$ and $K_{tr} = 0.275$ based on #3 stirrups at $s = 8$ in. (similar to Example 7.4).

SOLUTION   The solution for this problem is shown in Figure 7.21. There are two bars beginning at the left end of the beam. As no development length is available, the design moment capacity of the member is zero. If we move a distance ℓ_d from point A at the left end of the beam to point B, the design moment capacity will increase in a straight line from 0 to 185.3 ft-k. From point B to point C it will remain equal to 185.3 ft-k.

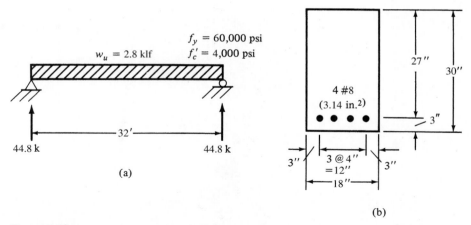

Figure 7.20

At point C we reach the cutoff point of the bars, and from C to D (a distance equal to ℓ_d) the design moment capacity will increase from 185.3 ft-k to 359.7 ft-k. (In Figure 7.21(a) the bars seem to be shown in two layers. They are actually on one level, but the author has shown them this way so that the reader can get a better picture of how many bars there are at any point along the beam.)

At no point along the span may the design strength of the beam be less than the actual bending moment caused by the loads. We can then see that point C is located where the actual bending moment equals 185.3 ft-k. The left reaction for this beam is 44.8 k, as shown in Figure 7.20(a). Using this value, an expression is written for the moment at point C (185.3 ft-k) at a distance x from the left support. The resulting expression can be solved for x.

$$M_u = (44.8)(8.25) - (2.8)(8.25)\left(\frac{8.25}{2}\right) = 274.3 \text{ ft-k}$$

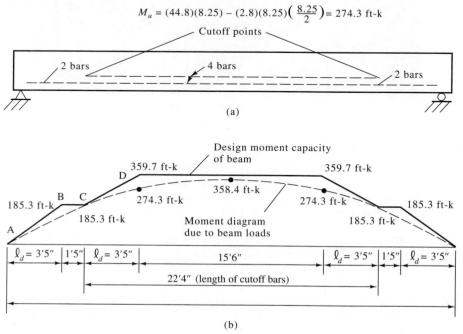

Figure 7.21

$$44.8x - (2.8x)\left(\frac{x}{2}\right) = 185.3$$
$$x = 4.88 \text{ ft} \quad \text{Say, 4 ft 10 in.}$$

By the time we reach point D (3 ft 5 in. to the right of C and 8 ft 3 in. from the left support), the required moment capacity is

$$M_u = (44.8)(8.25) - (2.8)(8.25)\left(\frac{8.25}{2}\right) = 274.3 \text{ ft-k}$$

Earlier in this section reference was made to ACI Section 12.10.5, where shear at bar cutoff points was considered. It is assumed that this beam will be properly designed for shear as described in the next chapter and will meet the ACI shear requirements.

7.12   BAR SPLICES IN FLEXURAL MEMBERS

Field splices of reinforcing bars are often necessary because of the limited bar lengths available, requirements at construction joints, and changes from larger bars to smaller bars. Although steel fabricators normally stock reinforcing bars in 60-ft lengths, it is often convenient to work in the field with bars of shorter lengths, thus necessitating the use of rather frequent splices.

The reader should carefully note that the ACI Code, Sections 1.2.1(h) and 12.14.1 clearly states that the designer is responsible for specifying the types and locations for splices for reinforcement.

The most common method of splicing #11 or smaller bars is simply to lap the bars one over the other. Lapped bars may be either separated from each other or placed in contact, with the contact splices being much preferred since the bars can be wired together. Such bars also hold their positions better during the placing of the concrete. Although lapped splices are easy to make, the complicated nature of the resulting stress transfer and the local cracks that frequently occur in the vicinity of the bar ends are disadvantageous. Obviously, bond stresses play an important part in transferring the forces from one bar to another. Thus the required splice lengths are closely related to development lengths. It is necessary to understand that the minimum specified clear distances between bars also apply to the distances between contact lap splices and adjacent splices or bars (ACI Section 7.6.4).

Lap splices are not very satisfactory for several situations. They include: (1) where they would cause congestion, (2) where the laps would be very long, as they are for #9 to #11 Grade 60 bars, (3) where #14 or #18 bars are used because the Code (12.14.2) does not permit them to be lap spliced except in a few special situations, and (4) where very long bar lengths would be left protruding from existing concrete structures for purposes of future expansion. For such situations, other types of splices such as those made by welding or by mechanical devices may be used. Welded splices, from the view of stress transfer, are the best splices, but they may be expensive and may cause metallurgical problems. The result may be particularly disastrous in high seismic zones. The ACI Code (12.14.3.4) states that welded splices must be accomplished by welding the bars together so that the connection will be able to develop at least 125% of the specified yield strength of the bars. It is considered desirable to butt the bars against each other, particularly for #7 and larger bars. Splices not meeting this strength requirement can be used at points where the bars are not stressed to their maximum tensile stresses. It should be realized that welded splices are usually the most expensive due to the high labor costs and due to the costs of proper inspection.

Mechanical connectors usually consist of some type of sleeve splice, which fits over the ends of the bars to be joined and into which a metallic grout filler is placed to interlock the grooves inside the sleeve with the bar deformations. From the standpoint of stress transfer, good mechanical connectors are next best to welded splices. They do have the disadvantage that some slippage may occur in the connections; as a result, there may be some concrete cracks in the area of the splices.

Before the specific provisions of the ACI Code are introduced, the background for these provisions should be explained briefly. The following remarks are taken from a paper by George F. Leyh of the CRSI.[9]

1. Splicing of reinforcement can never reproduce exactly the same effect as continuous reinforcing.

2. The goal of the splice provisions is to require a ductile situation where the reinforcing will yield before the splices fail. Splice failures occur suddenly without warning and with dangerous results.

3. Lap splices fail by splitting of the concrete along the bars. If some type of closed reinforcing is wrapped around the main reinforcing (such as ties and spirals described for columns in Chapter 9), the chances of splitting are reduced and smaller splice lengths are needed.

4. When stresses in reinforcement are reduced at splice locations, the chances of splice failure are correspondingly reduced. For this reason, the Code requirements are less restrictive where stresses are low.

Splices should be located away from points of maximum tensile stress. Furthermore, not all of the bars should be spliced at the same locations—that is, the splices should be staggered. Should two bars of different diameters be lap spliced, the lap length used shall be the splice length required for the smaller bar or the development length required for the larger bar, whichever is greater (ACI Code 12.15.3).

The length of lap splices for bundled bars must be equal to the required lap lengths for individual bars of the same size, but increased by 20% for three-bar bundles and 33% for four-bar bundles (ACI Code 12.4) because there is a smaller area of contact between the bars and the concrete, and thus less bond. Furthermore, individual splices within the bundles are not permitted to overlap each other.

7.13   TENSION SPLICES

The Code (12.15) divides tension lap splices into two classes, A and B. The class of splice used is dependent on the level of stress in the reinforcing and on the percentage of steel that is spliced at a particular location.

Class A splices are those where the reinforcing is lapped for a minimum distance of $1.0\,\ell_d$ (but not less than 12 in.) and where one-half or less of the reinforcing is spliced at any one location.

Class B splices are those where the reinforcing is lapped for a minimum distance of $1.3\,\ell_d$ (but not less than 12 in.) and where all the reinforcing is spliced at the same location.

The Code (12.15.2) states that lap splices for deformed bars and deformed wire in tension must be Class B unless (1) the area of reinforcing provided is equal to two or more times the area required by analysis over the entire length of the splice and (2) one-half or less of the reinforcing is spliced within the required lap length. A summary of this information is given in Table 7.3, which is Table R12.15.2 in the ACI Commentary.

---

[9]Portland Cement Association, 1972, *Proceedings of the PCA-ACI Teleconference on ACI 318-71 Building Code Requirements* (Skokie, IL: Portland Cement Association), p. 14–1.

Table 7.3 Tension Lap Splices

$\dfrac{A_{s\ \text{provided}}}{A_{s\ \text{required}}}$	Maximum percent of A_s spliced within required lap length	
	50	100
Equal to or greater than 2	Class A	Class B
Less than 2	Class B	Class B

In calculating the ℓ_d to be multiplied by 1.0 or 1.3, the reduction for excess reinforcing furnished should not be used because the class of splice (A or B) already reflects any excess reinforcing at the splice location (see ACI Commentary R12.15.1).

7.14 COMPRESSION SPLICES

Compression bars may be spliced by lapping, by end bearing, and by welding or mechanical devices. (Mechanical devices consist of bars or plates or other pieces welded or otherwise attached transversely to the flexural bars in locations where sufficient anchorage is not available.) The Code (12.16.1) says that the minimum splice length of such bars should equal $0.0005 f_y d_b$ for bars with f_y of 60,000 psi or less, $(0.0009 f_y - 24) d_b$ for bars with higher f_y values, but not less than 12 in. Should the concrete strengths be less than 3000 psi, it is necessary to increase the computed laps by one-third. Reduced values are given in the Code for cases where the bars are enclosed by ties or spirals (12.17.2.4 and 12.17.2.5).

The required length of lap splices for compression bars of different sizes is the larger of the computed compression lap splice length of the smaller bars or the compression development length, ℓ_{dc} of the larger bars. It is permissible to lap splice #14 and #18 compression bars to #11 and smaller bars (12.16.2).

The transfer of forces between bars that are always in compression can be accomplished by end bearing, according to Section 12.16.4 of the Code. For such transfer to be permitted, the bars must have their ends square cut (within $1\frac{1}{2}°$ of a right angle), must be fitted within 3° of full bearing after assembly, and they must be suitably confined (by closed ties, closed stirrups, or spirals). Section 12.17.4 further states that when end-bearing splices are used in columns, in each face of

Extruded coupler splice.
(Courtesy of Dywidag Systems
International, USA, Inc.)

the column more reinforcement has to be added that is capable of providing a tensile strength at least equal to 25% of the yield strength of the vertical reinforcement provided in that face.

The Code (12.14.2.1), with one exception, prohibits the use of lap splices for #14 or #18 bars. When column bars of those sizes are in compression, it is permissible to connect them to footings by means of dowels of smaller sizes with lap splices, as described in Section 15.8.2.3 of the Code.

7.15   HEADED AND MECHANICALLY ANCHORED BARS

Headed deformed bars (Figure 1.3) were added to the Code in the 2008 edition. Such devices transfer force from the bar to the concrete through a combination of bearing force at the head and bond forces along the bar. There are several limitations to the use of headed bars, as follows:

(a) bar f_y shall not exceed 60,000 psi

(b) bar size shall not exceed No. 11

(c) concrete shall be normal weight

(d) net bearing area of head A_{brg} shall not be less than 4 times the area of the bar A_b

(e) clear cover for bar shall not be less than $2d_b$

(f) clear spacing between bars shall not be less than $4d_b$

Clear cover and clear spacing requirements in (e) and (f) are measured to the bar, not to the head.

The development length in tension for headed deformed bars that comply with the ASTM A970 and other special requirements pertaining to obstructions (ACI Section 3.5.9) is given by

$$\ell_{dt} = \frac{0.016\psi_e f_y}{\sqrt{f'_c}} d_b$$

In applying this equation, f'_c cannot be taken greater than 6000 psi, and ψ_e is 1.2 for epoxy-coated bars and 1.0 otherwise. The calculated value of ℓ_{dt} cannot be less than $8d_b$ or 6 in., whichever is larger. The multiplier used earlier, $(A_{s\ required}/A_{s\ provided})$ is permitted except in cases where development of f_y is specifically required. There are no λ, ψ_t or ψ_s terms in this expression.

$$\text{In SI units, } \ell_{dt} = \frac{0.192\psi_e f_y}{\sqrt{f'_c}} d_b$$

The Code also permits other mechanical devices (ACI 12.6.4) shown by tests to be effective and approved by the building official.

EXAMPLE 7.9

Repeat Example 7.6 using a headed bar and compare with the results of Example 7.6.

$$\ell_{dt} = \frac{0.016\psi_e f_y}{\sqrt{f'_c}} d_b = \frac{(0.016)(1.2)(60,000)}{\sqrt{4000}} 1.128 = 20.54 \text{ in., say 21 in.}$$

This value compares with 69 in. for a straight bar and 26 in. for a 90° or 180° hooked bar.

7.16   SI EXAMPLE

EXAMPLE 7.10

Determine the development length required for the epoxy-coated bottom bars shown in Figure 7.22.

(a) assuming $K_{tr} = 0$ and

(b) computing K_{tr} with the appropriate equation, $f_y = 420$ MPa and $f'_c = 21$ MPa.

SOLUTION   From Table 6.1

$$\psi_t = 1.0 \text{ for bottom bars}$$

$$\psi_e = 1.5 \text{ for epoxy-coated bars with clear spacing } < 6d_b$$

$$\psi_t \psi_e = (1.0)(1.5) = 1.5 \ < \ 1.7 \qquad\qquad \underline{\underline{\text{OK}}}$$

$$\psi_s = \lambda = 1.0$$

$$c_b = \text{side cover of bars} = 80 \text{ mm}$$

$$c_b = \frac{1}{2} \text{ of c. to c. spacing of bars} = 40 \text{ mm}$$

(a) **Using SI Equation 12-1 with $K_{tr} = 0$**

$$\frac{c_b + K_{tr}}{d_b} = \frac{40 + 0}{25.4} = 1.575 \ < \ 2.5 \qquad\qquad \underline{\underline{\text{OK}}}$$

$$\frac{\ell_d}{d_b} = \frac{9}{10} \frac{f_y}{\lambda \sqrt{f'_c}} \frac{\psi_t \psi_e \psi_s}{\dfrac{c_b + K_{tr}}{d_b}}$$

$$= \left(\frac{9}{10}\right) \left(\frac{420}{(1.0)\sqrt{21}}\right) \frac{(1.0)(1.5)(1.0)}{1.575} = \underline{\underline{78.6 \text{ diameters}}}$$

(b) **Using Computed Value of K_{tr} and SI Equation 12-1**

$$K_{tr} = \frac{42A_{tr}}{sn} = \frac{(42)(2)(71)}{(200)(4)} = 7.45$$

$$\frac{c_b + K_{tr}}{d_b} = \frac{40 + 7.45}{25.4} = 1.87 \ < \ 2.5 \qquad\qquad \underline{\underline{\text{OK}}}$$

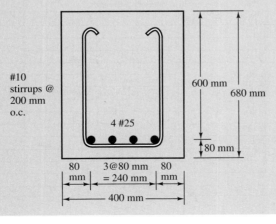

#10
stirrups @
200 mm
o.c.

4 #25

600 mm

680 mm

80 mm

80
mm

3@80 mm
= 240 mm

80
mm

400 mm

Figure 7.22

$$\frac{\ell_d}{d_b} = \left(\frac{9}{10}\right)\left(\frac{420}{(1.0)\sqrt{21}}\right)\frac{(1.0)(1.5)(1.0)}{1.87} = \underline{\underline{66.2 \text{ diameters}}}$$

7.17   COMPUTER EXAMPLE

EXAMPLE 7.11

Using the worksheet entitled Devel length tens - calc As in the spreadsheet for Chapter 7, determine the required tension development length ℓ_d of the beam shown in Figure 7.20 if lightweight aggregate concrete and #3 stirrups at 8 in. centers are used.

SOLUTION   Input the values of the cells in yellow highlight. Some cells are optional (see note on output below). Pass the cursor over cells for comments explaining what is to be input. Note that two answers are given, one with the $A_{s\,\text{reqd}}/A_{s\,\text{provided}}$ reduction and one without. In this example, there is little difference because this ratio is nearly 1.0.

Development Length, Tension

$f'_c =$	4000	psi		
$f_y =$	60,000	psi		
$f_{yt} =$	60,000	psi		
$b =$	18	in.*		
$d =$	27	in.*		
$h =$	30	in.*		
$A_s =$	3.14	in.²*	A_s	
$A_{tr} =$	0.22	in.²*		
$d_b =$	1	in.		
$n =$	4	*		
$s =$	8	in.*		
$\psi_t =$	1.00			
$\psi_e =$	1.00		*Cells indicate that this information is	
$\psi_s =$	1.00		optional. M_u, b, d, h, and A_s are needed only to calculate.	
$\lambda =$	0.75		$A_{s\,\text{reqd.}}$, A_{tr}, n, and s are needed only if the K_{tr} term is	
$c_b =$	2.00	in.	to be used. All terms with * can be omitted and	
$M_u =$	358.40	ft-k*	a conservative value of l_d will result.	
$\psi_t \psi_e =$	1			
$K_{tr} =$	0.275			
$(c_b + K_{tr})/d_b =$	2.28			

$$l_d = \frac{3}{40}\frac{f_y}{\lambda\sqrt{f_c}}\frac{\psi_t\,\psi_e\,\psi_s}{\left[\dfrac{c_b + K_{tr}}{d_b}\right]}\,d_b =$$

41.7	diameters
41.7	in. (not adjusted for $A_s/A_{s\,\text{provided}}$)

$A_{s\,\text{reqd}} =$	3.12736	in.²
$A_{s\,\text{reqd}}/A_{s\,\text{provided}} =$	0.995975	$l_d =$

	41.5 in. (adjusted for $A_s/A_{s\,\text{provided}}$)	
but not less than	12	in.

PROBLEMS

7.1 Why is it very difficult to calculate actual bond stresses?

7.2 What are top bars? Why are their required development lengths greater than they would be if they were not top bars?

7.3 Why do the cover of bars and the spacing of those bars affect required development lengths?

7.4 Why isn't the anchorage capacity of a standard hook increased by extending the bar well beyond the end of the hook?

7.5 For the cantilever beam shown, determine the point where two bars can be cut off from the standpoint of the calculated moment strength ϕM_n of the beam. $f_y = 60,000$ psi, $f'_c = 3000$ psi. (*Ans.* 9.09 ft from free end.)

Nov. 3

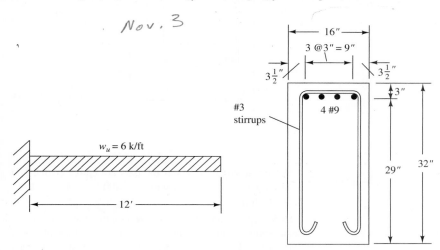

$$K_{tr} = \frac{A_{tr} f_{yt}}{1500 \, sn}$$

In Problems 7.6 to 7.9 determine the development lengths required for the tension bar situations described, (a) using ACI Equation 12-1 and assuming $K_{tr} = 0$, and (b) using ACI Equation 12-1 and the calculated value of K_{tr}.

7.6 Uncoated bars in normal-weight concrete. $A_{s \text{ required}} = 3.44$ in.^{2}

7.7 Uncoated bars in normal-weight concrete. $A_{s \text{ required}} = 3.40$ in.^{2} (*Ans.* 34.2 in., 23.5 in.)

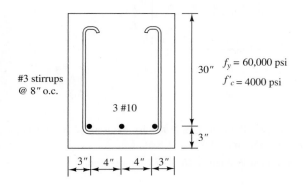

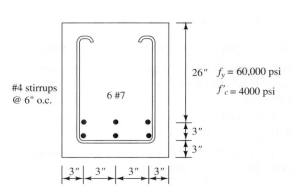

7.8 Epoxy-coated bars in lightweight concrete, $A_{s\ required} = 2.76$ in.^{2}

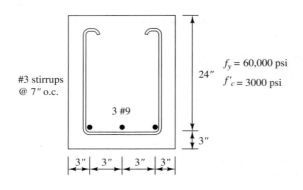

#3 stirrups
@ 7" o.c.

24"

$f_y = 60,000$ psi
$f'_c = 3000$ psi

3 #9

3"

| 3" | 3" | 3" | 3" |

7.9 Uncoated top bars in normal-weight concrete. $A_{s\ required} = 2.92$ in.^{2} (*Ans.* 35.1 in., 30.9 in.)

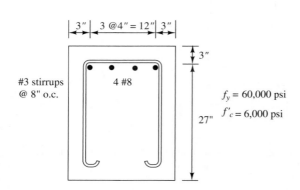

| 3" | 3 @4" = 12" | 3" |

3"

#3 stirrups
@ 8" o.c.

4 #8

27"

$f_y = 60,000$ psi
$f'_c = 6,000$ psi

7.10 Repeat Problem 7.6 if the bars are epoxy coated.

7.11 Repeat Problem 7.7 if all-lightweight concrete with $f'_c = 3000$ psi and epoxy-coated bars are used. (*Ans.* 79.0 in., 54.2 in.)

7.12 Repeat Problem 7.8 if three uncoated #6 bars are used and $A_{s\ required} = 1.20$ in.^{2}

7.13 Repeat Problem 7.9 if the bars are epoxy coated and all-lightweight concrete is used. (*Ans.* 61.2 in., 53.8 in.)

7.14 The bundled #10 bars shown are uncoated and are used in normal-weight concrete. $A_{s\ required} = 4.44$ in.^{2}

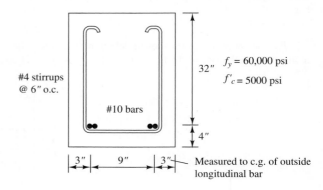

#4 stirrups
@ 6" o.c.

32"

$f_y = 60,000$ psi
$f'_c = 5000$ psi

#10 bars

4"

| 3" | 9" | 3" | Measured to c.g. of outside longitudinal bar

7.15 Repeat Problem 7.14 if the bars are epoxy coated and used in lightweight concrete with $f'_c = 4000$ psi. (*Ans.* 78.8 in., 63.4 in. etc.)

7.16 Set up a table for required development lengths for the beam shown using $f_y = 60,000$ psi, and f'_c values of 3000, 3500, 4000, 4500, 5000, 5500, and 6000 psi. Assume the bars are uncoated and normal-weight concrete is used. Use ACI Equation 12-1 and assume $K_{tr} = 0$.

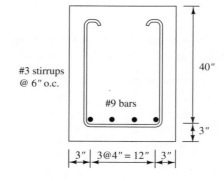

#3 stirrups
@ 6" o.c.

40"

#9 bars

3"

| 3" | 3@4" = 12" | 3" |

7.17 Repeat Problem 7.16 if #8 bars are used. (*Ans.* 41.1 in., 38.0 in., 35.6 in., 33.5 in., 31.8 in., etc.)

7.18 Repeat Problem 7.16 if #7 bars are used.

7.19 Repeat Problem 7.16 if #6 epoxy-coated bars are used in lightweight concrete.(*Ans.* 39.4 in., 36.5 in., 34.2 in., 32.2 in., 30.5 in., etc.)

7.20 (a) Determine the tensile development length required for the uncoated #8 bars shown if normal-weight concrete is used and the bars are straight. Use ACI Equation 12-1 and compute the value of K_{tr}. $f_c' = 4000$ psi, $f_y = 60,000$ psi.

(b) Repeat part (a) if 180° hooks are used.

Assume side, top, and bottom cover in all cases to be at least $2\frac{1}{2}$ in.

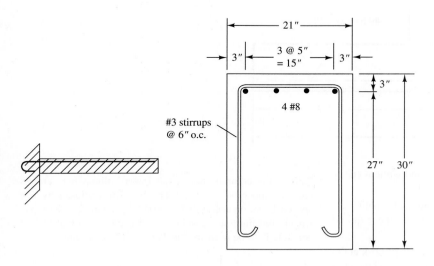

7.21 Are the uncoated #8 bars shown anchored sufficiently with their 90° hooks? $f_c' = 4000$ psi, $f_y = 60$ ksi. Side and top cover is $2\frac{1}{2}$ on bar extensions. Normal-weight concrete is used. $A_{s\ reqd} = 2.20$ in.² (*Ans.* $\ell_{dh} = 12.4$ in., sufficient)

7.22 Repeat Problem 7.21 if headed bars are used instead of 90° hooks.

7.23 Repeat Problem 7.7 if the bars are in compression. (*Ans.* 15.6 in.)

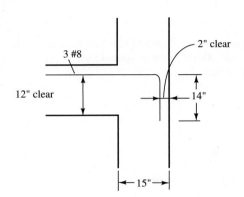

For Problems 7.24 to 7.29 use ACI Equation 12-1 and assume $K_{tr} = 0$.

7.24 The required bar area for the wall footing shown is 0.87 in.² per foot of width: #9 epoxy-coated bars 12 in. on center are used. Maximum moment is assumed to occur at the face of the wall. If $f_y = 60,000$ psi and $f'_c = 4,000$ psi, do the bars have sufficient development lengths? Assume $c_b = 3$ in.

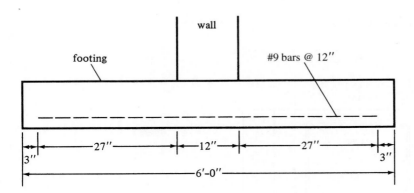

7.25 Repeat Problem 7.24 using #8 @ 9″ and without epoxy coating. (*Ans.* $\ell_d = 23.6 < 27$ in. <u>OK</u>)

7.26 Problem 7.24 has insufficient embedment length. List four design modifications that would reduce the required development length.

7.27 The beam shown is subjected to an M_u of 200 ft-k at the support. If $c_b = 1.5$, $K_{tr} = 0$, the concrete is lightweight, $f_y = 60,000$ psi, and $f'_c = 4000$ psi, do the following: (a) select #8 bars to be placed in one row, (b) determine the development lengths required if straight bars are used in the beam, and (c) determine the development lengths needed if 180° hooks are used in the support. (*Ans.* 3 #8, 74.5 in., 22.9 in.)

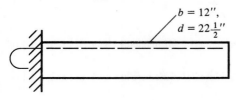

7.28 In the column shown, the lower column bars are #8 and the upper ones are #7. The bars are enclosed by ties spaced 12 in. on center. If $f_y = 60,000$ psi and $f'_c = 4000$ psi, what is the minimum lap splice length needed? Normal-weight concrete is to be used for the 12 in. × 12 in. column.

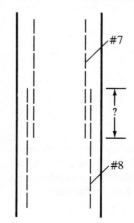

7.29 Calculations show that 2.24 in.² of top or negative steel is required for the beam shown. Three #8 bars have been selected. Are the 4 ft. 6 in. embedment lengths shown satisfactory if $f'_c = 4,000$ psi and $f_y = 60,000$ psi? Bars are spaced 3 in. o.c. with 3 in. side and top cover measured from c.g. of bars. Use $K_{tr}=0$ (*Ans.* No; $\ell_d = 58.8$ in.)

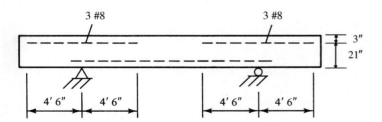

7.30 Calculations show that 4.90 in.^{2} of top or negative steel is required for the beam shown. If four uncoated #10 bars have been selected and if $f_c' = 4000$ psi and $f_y = 60,000$ psi, determine the minimum development length needed for the standard 90° hooks shown. Assume bars have 3 in. side and top cover measured from c.g. of bars and are used in normal-weight concrete. The bars are not enclosed by ties or stirrups spaced at $3d_b$ or less.

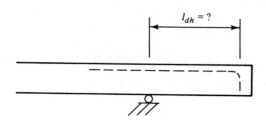

7.31 If $f_y = 75,000$ psi, $f_c' = 3,000$ psi, $w_D = 1.5$ k/ft, and $w_L = 5$ k/ft, are the development lengths of the straight bars satisfactory? Assume that the bars extend 6 in. beyond the $\mathcal{C}$ of the reactions and that $K_{tr} = 0$. $A_{s\,required} = 3.05$ in.^{2} The bars are uncoated, and the concrete is normal weight. (*Ans.* $\ell_d = 66.5$ in., embedment length is adequate)

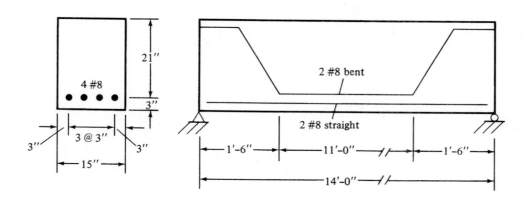

Compression Splices

7.32 Determine the compression lap splices needed for a 14 in. × 14 in. reinforced concrete column with ties (whose effective area exceeds 0.0015 hs as described in Section 12.17.2.4 of the Code) for the cases to follow. There

are eight #8 longitudinal bars equally spaced around the column.

(a) $f_c' = 4000$ psi and $f_y = 60,000$ psi
(b) $f_c' = 2000$ psi and $f_y = 50,000$ psi

Problems in SI Units

In Problems 7.33 to 7.36, determine the tensile development lengths required (a) using ACI Metric Equation 12-1, assuming $K_{tr} = 0$ and (b) using ACI Metric Equation 12-1 and the computed value of K_{tr}. Use $f_y = 420$ MPa and $f'_c = 28$ MPa.

Problem 7.33 (*Ans.* 922 mm, 769 mm)

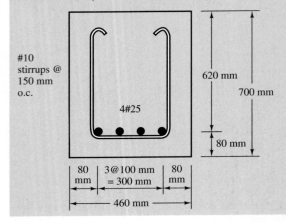

#10 stirrups @ 150 mm o.c.

4#25

620 mm

700 mm

80 mm

80 mm | 3@100 mm = 300 mm | 80 mm

460 mm

Problem 7.34

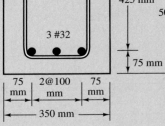

#13 stirrups @ 200 mm o.c.

3 #32

425 mm

500 mm

75 mm

75 mm | 2@100 mm | 75 mm

350 mm

7.35 Repeat Problem 7.33 if the bars are #19 and if #10 stirrups 150 mm o.c. are used. (*Ans.* 437 mm, 437 mm)

7.36 Repeat Problem 7.34 if the bars are epoxy coated.

Computer Problems

For Problems 7.37 and 7.38, use the Chapter 7 spreadsheet.

7.37 Repeat Problem 7.6. (*Ans.* 52.1 in, 44.0 in.)

7.38 Repeat Problem 7.9.

7.39 Repeat Problem 7.22 (*Ans.* $\ell_{dt} = 14.2$ in. > 13 in. available $\therefore$ No good)

8

Shear and Diagonal Tension

8.1 INTRODUCTION

As repeatedly mentioned earlier in this book, the objective of today's reinforced concrete designer is to produce ductile members that provide warning of impending failure. To achieve this goal, the Code provides design shear values that have larger safety factors against shear failures than do those provided for bending failures. The failures of reinforced concrete beams in shear are quite different from their failures in bending. Shear failures occur suddenly with little or no advance warning. Therefore beams are designed to fail in bending under loads that are appreciably smaller than those that would cause shear failures. As a result, those members will fail ductilely. They may crack and sag a great deal if overloaded, but they will not fall apart as they might if shear failures were possible.

8.2 SHEAR STRESSES IN CONCRETE BEAMS

Although no one has ever been able to accurately determine the resistance of concrete to pure shearing stress, the matter is not very important because pure shearing stress is probably never encountered in concrete structures. Furthermore, according to engineering mechanics, if pure shear is produced in a member, a principal tensile stress of equal magnitude will be produced on another plane. Because the tensile strength of concrete is less than its shearing strength, the concrete will fail in tension before its shearing strength is reached.

You have previously learned that in elastic homogeneous beams, where stresses are proportional to strains, two kinds of stresses occur (bending and shear) and they can be calculated with the following expressions:

$$f = \frac{Mc}{I}$$

$$v = \frac{VQ}{Ib}$$

An element of a beam not located at an extreme fiber or at the neutral axis is subject to both bending and shear stresses. These stresses combine into inclined compressive and tensile stresses, called *principal stresses*, which can be determined from the following expression:

$$f_p = \frac{f}{2} \pm \sqrt{\left(\frac{f}{2}\right)^2 + v^2}$$

Massive reinforced concrete structures. (Courtesy of Bethlehem Steel Corporation.)

The direction of the principal stresses can be determined with the formula to follow, in which α is the inclination of the stress to the beam's axis:

$$\tan 2\alpha = \frac{2v}{f}$$

Obviously, at different positions along the beam, the relative magnitudes of v and f change, and thus the directions of the principal stresses change. It can be seen from the preceding equation that at the neutral axis the principal stresses will be located at a 45° angle with the horizontal.

You understand by this time that tension stresses in concrete are a serious matter. Diagonal principal tensile stresses, called *diagonal tension*, occur at different places and angles in concrete beams, and they must be carefully considered. If they reach certain values, additional reinforcing, called *web reinforcing*, must be supplied.

The discussion presented up to this point relating to diagonal tension applies rather well to plain concrete beams. If, however, reinforced concrete beams are being considered, the situation is quite different because the longitudinal bending tension stresses are resisted quite satisfactorily by the longitudinal reinforcing. These bars, however, do not provide significant resistance to the diagonal tension stresses.

8.3 LIGHTWEIGHT CONCRETE

In the 2008 ACI 318 Code, the effect of lightweight aggregate concrete on shear strength was modified by the introduction of the term λ (see Section 1.12). This term was added to most equations containing $\sqrt{f_c'}$. The resulting combined term, $\lambda \sqrt{f_c'}$, appears throughout this chapter as well as in Chapter 7 on development length and Chapter 15 on torsion. If normal-weight concrete is used, then λ is simply taken as 1. This unified approach to the effects of lightweight aggregate on the strength and other properties of concrete is a logical and simplifying improvement found in the 2008 ACI Code.

8.4   SHEAR STRENGTH OF CONCRETE

A great deal of research has been done on the subject of shear and diagonal tension for nonhomogeneous reinforced concrete beams, and many theories have been developed. Despite all this work and all the resulting theories, no one has been able to provide a clear explanation of the failure mechanism involved. As a result, design procedures are based primarily on test data.

If V_u is divided by the effective beam area $b_w d$, the result is what is called an *average shearing stress*. This stress is not equal to the diagonal tension stress but merely serves as an indicator of its magnitude. Should this indicator exceed a certain value, shear or web reinforcing is considered necessary. In the ACI Code the basic shear equations are presented in terms of shear forces and not shear stresses. In other words, the average shear stresses described in this paragraph are multiplied by the effective beam areas to obtain total shear forces.

For this discussion V_n is considered to be the nominal or theoretical shear strength of a member. This strength is provided by the concrete and by the shear reinforcement.

$$V_n = V_c + V_s$$

The design shear strength of a member, ϕV_n, is equal to ϕV_c plus ϕV_s, which must at least equal the factored shear force to be taken, V_u

$$V_u = \phi V_c + \phi V_s$$

The shear strength provided by the concrete, V_c, is considered to equal an average shear stress strength (normally $2\lambda\sqrt{f'_c}$) times the effective cross-sectional area of the member, $b_w d$, where b_w is the width of a rectangular beam or of the web of a T beam or an I beam.

$$V_c = 2\lambda\sqrt{f'_c}b_w d \qquad\qquad \text{(ACI Equation 11-3)}$$

Or in SI units with f'_c in MPa

$$V_c = \left(\frac{\lambda\sqrt{f'_c}}{6}\right)b_w d$$

Beam tests have shown some interesting facts about the occurrence of cracks at different average shear stress values. For instance, where large moments occur even though appropriate longitudinal steel has been selected, extensive flexural cracks will be evident. As a result, the uncracked area of the beam cross section will be greatly reduced and the nominal shear strength V_c can be as low as $1.9\lambda\sqrt{f'_c}b_w d$. On the other hand, in regions where the moment is small, the cross section will be either uncracked or slightly cracked and a large portion of the cross section is available to resist shear. For such a case, tests show that a V_c of about $3.5\lambda\sqrt{f'_c}b_w d$ can be developed before shear failure occurs.[1]

Based on this information, the Code (11.2.1.1) suggests that, conservatively, V_c (the shear force that the concrete can resist without web reinforcing) can go as high as $2\lambda\sqrt{f'_c}b_w d$. As an alternative, the following shear force (from Section 11.2.1.2 of the Code) may be used, which takes into account the effects of the longitudinal reinforcing and the moment and shear magnitudes. This value must be calculated separately for each point being considered in the beam.

$$V_c = \left(1.9\lambda\sqrt{f'_c} + 2500\rho_w\frac{V_u d}{M_u}\right)b_w d \leq 3.5\lambda\sqrt{f'_c}b_w d \qquad\qquad \text{(ACI Equation 11-5)}$$

---

[1]ACI-ASCE Committee 326, 1962, "Shear and Diagonal Tension," part 2, *Journal ACI*, 59, p. 277.

In SI units

$$V_c = \left(\lambda\sqrt{f_c'} + 120\rho_w \frac{V_u d}{M_u}\right)\frac{b_w d}{7} \leq 0.3\lambda\sqrt{f_c'}b_w d$$

In these expressions $\rho_w = A_s/b_w d$ and M_u is the factored moment occurring simultaneously with V_u, the factored shear at the section considered. The quantity $V_u d/M_u$ cannot be taken to be greater than unity in computing V_c by means of the above expressions.

From these expressions it can be seen that V_c increases as the amount of reinforcing (represented by ρ_w) is increased. As the amount of steel is increased, the length and width of cracks will be reduced. If the cracks are kept narrower, more concrete is left to resist shear and there will be closer contact between the concrete on opposite sides of the cracks. Hence there will be more resistance to shear by friction (called *aggregate interlock*) on the two sides of cracks.

Although this more complicated expression for V_c can easily be used for computer designs, it is quite tedious to apply when hand-held calculators are used. The reason is that the values of ρ_w, V_u, and M_u are constantly changing as we move along the span, requiring the computation of V_c at numerous positions. As a result, the alternate value $2\lambda\sqrt{f_c'}b_w d$ is normally used. If the same member is to be constructed many times, the use of the more complex expression may be justified.

8.5 SHEAR CRACKING OF REINFORCED CONCRETE BEAMS

Inclined cracks can develop in the webs of reinforced concrete beams either as extensions of flexural cracks or occasionally as independent cracks. The first of these two types is the *flexure–shear crack*, an example of which is shown in Figure 8.1. These are the ordinary types of shear cracks found in both prestressed and nonprestressed beams. For them to occur, the moment must be larger than the cracking moment and the shear must be rather large. The cracks run at angles of about 45° with the beam axis and probably start at the top of a flexure crack. The approximately vertical flexure cracks shown are not dangerous unless a critical combination of shear stress and flexure stress occurs at the top of one of the flexure cracks.

Occasionally, an inclined crack will develop independently in a beam, even though no flexure cracks are in that locality. Such cracks, which are called *web–shear cracks*, will sometimes occur in the webs of prestressed sections, particularly those with large flanges and thin webs. They also sometimes occur near the points of inflection of continuous beams or near simple supports. At such locations small moments and high shear often occur. These types of cracks will form near the

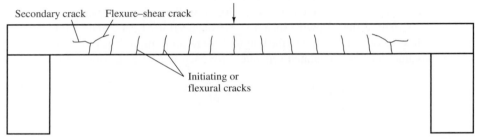

Figure 8.1 Flexure–shear crack.

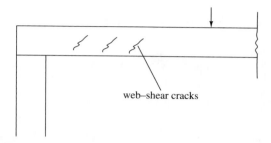

Figure 8.2 Web–shear cracks.

mid-depth of sections and will move on a diagonal path to the tension surface. Web–shear cracks are illustrated in Figure 8.2.

As a crack moves up to the neutral axis, the result will be a reduced amount of concrete left to resist shear—meaning that shear stresses will increase on the concrete above the crack.

It will be remembered that at the neutral axis the bending stresses are zero and the shear stresses are at their maximum values. The shear stresses will therefore determine what happens to the crack there.

After a crack has developed, the member will fail unless the cracked concrete section can resist the applied forces. If web reinforcing is not present, the items that are available to transfer the shear are as follows: (1) the shear resistance of the uncracked section above the crack (estimated to be 20% to 40% of the total resistance), (2) the aggregate interlock, that is, the friction developed due to the interlocking of the aggregate on the concrete surfaces on opposite sides of the crack (estimated to be 33% to 50% of the total), (3) the resistance of the longitudinal reinforcing to a frictional force, often called *dowel action* (estimated to be 15% to 25%), and (4) a tied-arch type of behavior that exists in rather deep beams produced by the longitudinal bars acting as the tie and by the uncracked concrete above and to the sides of the crack acting as the arch above.[2]

8.6   WEB REINFORCEMENT

When the factored shear V_u is high, it shows that large cracks are going to occur unless some type of additional reinforcing is provided. This reinforcing usually takes the form of stirrups that enclose the longitudinal reinforcing along the faces of the beam. The most common stirrups are ⊔ shaped, but they can be ⊔⊓⊔ shaped or perhaps have only a single vertical prong, as shown in Figure 8.3 (a) to (c). Multiple stirrups such as the ones shown in Figure 8.3(e) are considered to inhibit splitting in the plane of the longitudinal bars. As a consequence, they are generally more desirable for wide beams than the ones shown in Figure 8.3(d). Sometimes it is rather convenient to use lap-spliced stirrups such as the ones shown in Figure 8.3(g). These stirrups, which are described in ACI Section 12.13.5, are occasionally useful for deep members, particularly those with gradually varying depths. *However, they are considered to be unsatisfactory in seismic areas.*

Bars called *hangers* (usually with about the same diameter as that of the stirrups) are placed on the compression sides of beams to support the stirrups, as illustrated in Figure 8.3(a) to 8.3(j). The stirrups are passed around the tensile steel and to meet anchorage requirements are run as far into the compression side of the beam as practical and are hooked around the hangers. Bending of the stirrups around the hangers reduces the bearing stresses under the hooks. If these bearing stresses are too high, the concrete will crush and the stirrups will tear out. When a significant amount of

---

[2]Taylor, H. P. J., 1974, "The Fundamental Behavior of Reinforced Concrete Beams in Bending and Shear," *Shear in Reinforced Concrete*, Vol. 1, SP-42 (Detroit: American Concrete Institute), pp. 43–47.

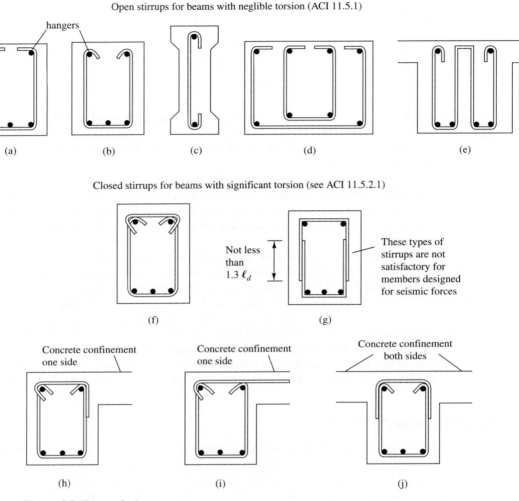

Figure 8.3 Types of stirrups.

torsion is present in a member, it will be necessary to use closed stirrups as shown in parts (f) through (j) of Figure 8.3 and as discussed in Chapter 15.

The width of diagonal cracks is directly related to the strain in the stirrups. Consequently, the ACI 11.4.2 does not permit the design yield stress of the stirrups to exceed 60 ksi. This requirement limits the width of cracks that can develop. Such a result is important from the standpoint of both appearance and aggregate interlock. When the width of cracks is limited, it enables more aggregate interlock to develop. A further advantage of a limited yield stress is that the anchorage requirements at the top of the stirrups are not quite as stringent as they would be for stirrups with greater yield strengths.

The 60,000 psi limitation does not apply to deformed welded wire fabric because recent research has shown that the use of higher strength wires has been quite satisfactory. Tests have shown that the width of inclined shear cracks at service load conditions is less for high strength wire fabric than for those occurring in beams reinforced with deformed Grade 60 stirrups. The maximum stress permitted for deformed welded wire fabric is 80,000 psi (ACI 11.4.2).

In SI units, the maximum design yield stress values that may be used are 420 MPa for regular shear reinforcing and 550 MPa for welded deformed wire fabric.

8.7   BEHAVIOR OF BEAMS WITH WEB REINFORCEMENT

The actual behavior of beams with web reinforcement is not really understood, although several theories have been presented through the years. One theory, which has been widely used for 100 years, is the so-called truss analogy, wherein a reinforced concrete beam with shear reinforcing is said to behave much as a statically determinate parallel chord truss with pinned joints. The flexural compression concrete is thought of as the top chord of the truss, whereas the tensile reinforcing is said to be the bottom chord. The truss web is made up of stirrups acting as vertical tension members and pieces of concrete between the approximately 45° diagonal tension cracks acting as diagonal compression members.[3,4] The shear reinforcing used is similar in its action to the web members of a truss. For this reason the term *web reinforcement* is used when referring to shear reinforcing. A "truss" of the type described here is shown in Figure 8.4.

Although the truss analogy has been used for many years to describe the behavior of reinforced concrete beams with web reinforcing, it does not accurately describe the manner in which shear forces are transmitted. For example, the web reinforcing does increase the shearing strength of a beam, but it has little to do with shear transfer in a beam before inclined cracks form.

The Code requires web reinforcement for all major beams. In Section 11.4.6.1, a minimum area of web reinforcing is required for all concrete flexural members except (a) footings and solid slabs, (b) certain hollow core units, (c) concrete floor joists, (d) shallow beams with h not more than 10 in., (e) beams integral with slabs with h less than 24 in., and h not greater than the larger of $2\frac{1}{2}$ times their flange thicknesses or one-half their web widths, or (f) beams constructed with steel fiber-reinforced, normal-weight concrete with f_c' not exceeding 6000 psi, h not greater than 24 in., and V_u not greater than $2\phi\sqrt{f_c'}b_w d$. Various tests have shown that shear failures do not occur before bending failures in shallow members. Shear forces are distributed across these wide sections. For joists the redistribution is via the slabs to adjacent joists. Hooked or crimped steel fibers in dosages $\geq$ 100 lb per cubic yard exhibit higher shear strengths in laboratory tests. However, use of such fibers is not recommended when the concrete is exposed to chlorides, such as de-icing salts.

Inclined or diagonal stirrups lined up approximately with the principal stress directions are more efficient in carrying the shears and preventing or delaying the formation of diagonal cracks.

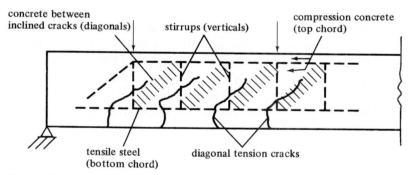

Figure 8.4 Truss analogy.

---

[3]Ritter, W., 1899, *Die Bauweise Hennebique* (Schweizerische Bauzeitung, XXXIII, no. 7).

[4]Mörsch, E., 1912, *Der Eisenbetonbau, seine Theorie und Anwendung* (Stuttgart: Verlag Konrad Witttwer).

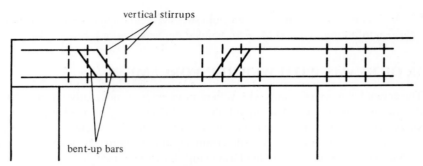

Figure 8.5 Bent-up bar web reinforcing.

Such stirrups, however, are not usually considered to be very practical in the United States because of the high labor costs required for positioning them. Actually, they can be rather practical for precast concrete beams where the bars and stirrups are preassembled into cages before being used and where the same beams are duplicated many times.

Bent-up bars (usually at 45° angles) are another satisfactory type of web reinforcing (see Figure 8.5). Although bent-up bars are commonly used in flexural members in the United States, the average designer seldom considers the fact that they can resist diagonal tension. Two reasons for not counting their contribution to diagonal tension resistance are that there are only a few, if any, bent-up bars in a beam and that they may not be conveniently located for use as web reinforcement.

Diagonal cracks will occur in beams with shear reinforcing at almost the same loads that they occur in beams of the same size without shear reinforcing. The shear reinforcing makes its presence known only after the cracks begin to form. At that time, beams must have sufficient shear reinforcing to resist the shear force not resisted by the concrete.

After a shear crack has developed in a beam, only a little shear can be transferred across the crack unless web reinforcing is used to bridge the gap. When such reinforcing is present, it keeps the pieces of concrete on the two sides of the crack from separating. Several benefits result. These include:

1. The steel reinforcing passing across the cracks carries shear directly.
2. The reinforcing keeps the cracks from becoming larger, and this enables the concrete to transfer shear across the cracks by aggregate interlock.
3. The stirrups wrapped around the core of concrete act like hoops and thus increase the beam's strength and ductility. In a related fashion, the stirrups tie the longitudinal bars into the concrete core of the beam and restrain them from prying off the covering concrete.
4. The holding together of the concrete on the two sides of the cracks helps keep the cracks from moving into the compression zone of the beam. Remember that other than for deformed wire fabric, the yield stress of the web reinforcing is limited to 60 ksi to limit the width of the cracks.

8.8 DESIGN FOR SHEAR

The maximum shear V_u in a beam must not exceed the design shear capacity of the beam cross section ϕV_n, where ϕ is 0.75 and V_n is the nominal shear strength of the concrete and the shear reinforcing.

$$V_u \leq \phi V_n$$

The value of ϕV_n can be broken down into the design shear strength of the concrete ϕV_c plus the design shear strength of the shear reinforcing ϕV_s. The value of ϕV_c is provided in the Code for different situations, and thus we are able to compute the required value of ϕV_s for each situation:

$$V_u \leq \phi V_c + \phi V_s$$

For this derivation an equal sign is used:

$$V_u = \phi V_c + \phi V_s$$

The purpose of stirrups is to minimize the size of diagonal tension cracks or to carry the diagonal tension stress from one side of the crack to the other. Very little tension is carried by the stirrups until after a crack begins to form. Before the inclined cracks begin to form, the strain in the stirrups is equal to the strain in the adjacent concrete. Because this concrete cracks at very low diagonal tensile stresses, the stresses in the stirrups at that time are very small, perhaps only 3 to 6 ksi. You can see that these stirrups do not prevent inclined cracks and that they really aren't a significant factor until the cracks begin to develop.

Tests made on reinforced concrete beams show that they will not fail by the widening of the diagonal tension cracks until the stirrups going across the cracks have been stressed to their yield stresses. For the derivation to follow, it is assumed that a diagonal tension crack has developed and has run up into the compression zone but not all the way to the top, as shown in Figure 8.6. It is further assumed that the stirrups crossing the crack have yielded.

The nominal shear strength of the stirrups V_s crossing the crack can be calculated from the following expression, where n is the number of stirrups crossing the crack and A_v is the cross-sectional area each stirrup has crossing the crack. (If a ⊔ stirrup is used, $A_v = 2$ times the cross-sectional area of the stirrup bar. If it is a ⊔⊓⊔, $A_v = 4$ times the cross-sectional area of the stirrup bar.)

$$V_s = A_v f_y n$$

If it is conservatively assumed that the horizontal projection of the crack equals the effective depth d of the section (thus a 45° crack), the number of stirrups crossing the crack can be determined from the expression to follow, in which s is the center-to-center spacing of the stirrups:

$$n = \frac{d}{s}$$

Then

$$V_s = A_v f_y \frac{d}{s} \qquad\qquad \text{(ACI Equation 11-15)}$$

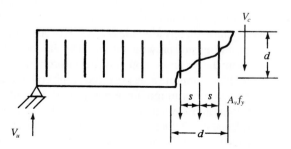

Figure 8.6

From this expression the required spacing of vertical stirrups is

$$s = \frac{A_v f_y d}{V_s}$$

and the value of V_s can be determined as follows:

$$V_u = \phi V_c + \phi V_s$$

$$V_s = \frac{V_u - \phi V_c}{\phi}$$

Going through a similar derivation, the following expression can be determined for the required area for inclined stirrups, in which α is the angle between the stirrups and the longitudinal axis of the member. Inclined stirrups should be placed so they form an angle of at least 45° with respect to the longitudinal bars, and they must be securely tied in place.

$$V_s = \frac{A_v f_y (\sin \alpha + \cos \alpha) d}{s}$$ (ACI Equation 11-16)

And for a bent-up bar or a group of bent-up bars at the same distance from the support, we have

$$V_s = A_v f_y \sin \alpha \leq 3\sqrt{f_c'} b_w d$$ (ACI Equation 11-17)

8.9 ACI CODE REQUIREMENTS

This section presents a detailed list of the Code requirements controlling the design of web reinforcing, even though some of these items have been previously mentioned in this chapter:

1. When the factored shear V_u exceeds one-half the shear design strength ϕV_c, the Code (11.4.6.1) requires the use of web reinforcing. The value of V_c is normally taken as $2\lambda\sqrt{f_c'} b_w d$, but the Code (11.2.2.1) permits the use of the following less conservative value:

$$V_c = \left(1.9\lambda\sqrt{f_c'} + 2500\rho_w \frac{V_u d}{M_u}\right) b_w d \leq 3.5\lambda\sqrt{f_c'} b_w d$$ (ACI Equation 11-5)

As previously mentioned, M_u is the moment occurring simultaneously with V_u at the section in question. The value of $V_u d / M_u$ must not be taken greater than 1.0 in calculating V_c, according to the Code.

2. When shear reinforcing is required, the Code states that the amount provided must fall between certain clearly specified lower and upper limits. If the amount of reinforcing is too low, it may yield or even snap immediately after the formation of an inclined crack. As soon as a diagonal crack develops, the tension previously carried by the concrete is transferred to the web reinforcing. To prevent the stirrups (or other web reinforcing) from snapping at that time, their area is limited to the minimum value provided at the end of the next paragraph.

ACI Section 11.4.6.3 specifies a minimum amount of web reinforcing such as to provide an ultimate shear strength no less than $0.75\lambda\sqrt{f_c'} b_w s$. Using this provision of the Code should prevent a sudden shear failure of the beam when inclined cracks occur. The shear strength calculated with this expression may not be less than $50 b_w s$. If a $0.75\lambda\sqrt{f_c'}$ psi strength is available for a web width b_w and a length of beam s equal to the stirrup spacing, we will have

$$0.75\sqrt{f_c'}b_w s = A_v f_y$$

$$A_{v\ minimum} = \frac{0.75\sqrt{f_c'}b_w s}{f_y} \qquad \text{(ACI Equation 11-13)}$$

but not less than the value obtained with a 50 psi strength $50b_w s/f_y$.

If $\sqrt{f_c'}$ is greater than 4444 psi, the minimum value of A_v is controlled by the expression $0.75\sqrt{f_c'}b_w s/f_y$. Should f_c' be less than 4444 psi, the minimum A_v value will be controlled by the $50b_w s/f_y$ expression.

In SI units

$$A_{v\ min}\ \frac{1}{16}\sqrt{f_c'}\frac{b_w s}{f_y} \geq \frac{0.33 b_w s}{f_y}$$

This expression from ACI Section 11.4.6.3 provides the minimum area of web reinforcing A_v which is to be used as long as the factored torsional moment T_u does not exceed one-fourth of the cracking torque T_{cr}. Such a torque will not cause an appreciable reduction in the flexural or shear strength of a member and may be neglected (ACI Section 11.5.1). For nonprestressed members this limiting value is

$$\phi\lambda\sqrt{f_c'}\left(\frac{A_{cp}^2}{p_{cp}}\right)$$

In SI units

$$\frac{\phi\lambda\sqrt{f_c'}}{12}\frac{A_{cp}^2}{p_{cp}}$$

In this expression $\phi = 0.75$, A_{cp} = the area enclosed by the outside perimeter of the concrete cross section, and p_{cp} is the outside perimeter of the concrete cross section. The computation of T_u and T_{cr} for various situations is presented in Chapter 15.

Although you may feel that the use of such minimum shear reinforcing is not necessary, studies of earthquake damage in recent years have shown very large amounts of shear damage occurring in reinforced concrete structures, and it is felt that the use of this minimum value will greatly improve the resistance of such structures to seismic forces. Actually, many designers believe that the minimum area of web reinforcing should be used throughout beams and not just where V_u is greater than $\phi V_c/2$.

This requirement for a minimum amount of shear reinforcing may be waived if tests have been conducted showing that the required bending and shear strengths can be met without the shear reinforcing (ACI 11.4.6.2).

3. As previously described, stirrups cannot resist appreciable shear unless they are crossed by an inclined crack. Thus to make sure that each 45° crack is intercepted by at least one stirrup, the maximum spacing of vertical stirrups permitted by the Code (11.4.5.1) is the lesser of $d/2$ or 24 in. for nonprestressed members and $\frac{3}{4}h$ for prestressed members or 24 in. where h is the overall thickness of a member. Should, however, V_s exceed $4\sqrt{f_c'}b_w d$,[*]

[*]In SI $V_s = \frac{1}{3}\sqrt{f_c'}b_w d$.

Placement of reinforcing bars for hemispherical dome of nuclear power plant reinforced concrete containment structure. (Courtesy of Bethlehem Steel Corporation.)

these maximum spacings are to be reduced by one-half (ACI 11.4.5.3). These closer spacings will lead to narrower inclined cracks.

Another advantage of limiting maximum spacing values for stirrups is that closely spaced stirrups will hold the longitudinal bars in the beam. They reduce the chance that the steel may tear or buckle through the concrete cover or possibly slip on the concrete.

Under no circumstances may V_s be allowed to exceed $8\sqrt{f'_c}b_w d$ (Code 11.4.7.9).[†] The shear strength of a beam cannot be increased indefinitely by adding more and more shear reinforcing because the concrete will eventually disintegrate no matter how much shear reinforcing is added. The reader can understand the presence of an upper limit if he or she will think for a little while about the concrete above the crack. The greater the shear in the member that is transferred by the shear reinforcing to the concrete above, the greater will be the chance of a combination shear and compression failure of that concrete.

4. Section 11.1.2 of the Code states that the values of $\sqrt{f'_c}$ used for the design of web reinforcing may not exceed 100 psi[‡] except for certain cases listed in Section 11.1.2.1. In that section, permission is given to use a larger value for members having the minimum reinforcing specified in ACI Sections 11.4.6.3, 11.4.6.4 and 11.5.5.2. Members meeting these requirements for extra shear reinforcing have sufficient post-crack capacities to prevent diagonal tension failures.

5. Section 12.13 of the Code provides requirements about dimensions, development lengths, and so forth. For stirrups to develop their design strengths they must be adequately anchored. Stirrups may be crossed by diagonal tension cracks at various points along their depths. Since these cracks may cross very close to the tension or compression edges of the

[†]It's $\frac{2}{3}\sqrt{f'_c}b_w d$ in SI units.

[‡]It's $\frac{25}{3}$ MPa in SI.

members, the stirrups must be able to develop their yield strengths along the full extent of their lengths. It can then be seen why they should be bent around longitudinal bars of greater diameters than their own and extended beyond by adequate development lengths. Should there be compression reinforcing, the hooking of the stirrups around them will help prevent them from buckling.

Stirrups should be carried as close as possible to the compression and tension faces of beams as the specified cover and longitudinal reinforcing will permit. The ends of stirrup legs should ideally have 135° or 180° hooks bent around longitudinal bars, with development lengths as specified in ACI Sections 8.1 and 12.13. Detailed information on stirrups follows:

(a) Stirrups with 90° bends and $6d_b$ extensions at their free ends may be used for #5 and smaller bars as shown in Figure 8.7(a). Tests have shown that 90° bends with $6d_b$ extensions should not be used for #6 or larger bars (unless f_y is 40,000 psi or less) because they tend to "pop out" under high loads.

(b) If f_y is greater than 40,000 psi, #6, #7, and #8 bars with 90° bends may be used if the extensions are $12d_b$ [see Figure 8.7(b)]. The reason for this specification is that it is not possible to bend these higher strength bars tightly around the longitudinal bars.

(c) Stirrups with 135° bends and $6d_b$ extensions may be used for #8 and smaller bars, as shown in Figure 8.7(c).

6. When a beam reaction causes compression in the end of a member in the same direction as the external shear, the shearing strength of that part of the member is increased. Tests of such reinforced concrete members have shown that in general, as long as a gradually varying shear is present (as with a uniformly loaded member), the first crack will occur at a distance d from the face of the support. It is therefore permissible, according to the Code (11.1.3.1), to decrease somewhat the calculated shearing force for a distance d from the face of the support. This is done by using a V_u in that range equal to the calculated V_u at a distance d from the face of the support. Should a concentrated load be applied in this region, no such shear reduction is permitted. Such loads will be transmitted directly to the

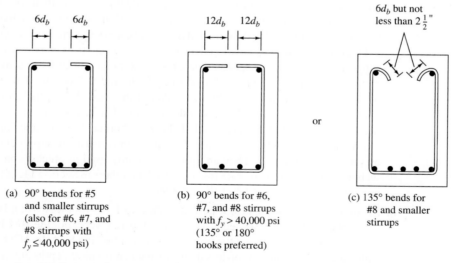

(a) 90° bends for #5
and smaller stirrups
(also for #6, #7, and
#8 stirrups with
$f_y \leq 40,000$ psi)

(b) 90° bends for #6,
#7, and #8 stirrups
with $f_y > 40,000$ psi
(135° or 180°
hooks preferred)

(c) 135° bends for
#8 and smaller
stirrups

Note: Fit stirrups as close to compression and tension surfaces as cover and other reinforcing permits.

Figure 8.7 Stirrup details.

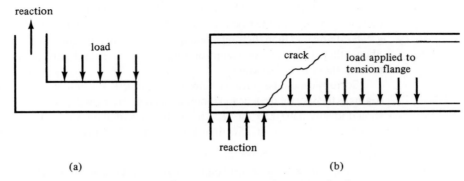

Figure 8.8 Two situations where end shear reduction is not permitted.

support above the 45° cracks, with the result that we are not permitted a reduction in the end shear for design purposes.

Should the reaction tend to produce tension in this zone, no shear stress reduction is permitted because tests have shown that cracking may occur at the face of the support or even inside it. Figure 8.8 shows two cases where the end shear reduction is not permitted. In the situation shown in Figure 8.8(a) the critical section will be at the face of the support. In Figure 8.8(b) an I-shaped section is shown, with the load applied to its tension flange. The loads have to be transferred across the inclined crack before they reach the support. Another crack problem like this one occurs in retaining wall footings and is discussed in Section 13.10 of this text.

7. Various tests of reinforced concrete beams of normal proportions with sufficient web reinforcing have shown that shearing forces have no significant effect on the flexural capacities of the beams. Experiments with deep beams, however, show that large shears will often keep those members from developing their full flexural capacities. As a result, the Code requirements given in the preceding paragraphs are not applicable to beams whose clear spans divided by their effective depths are less than four or for regions of beams that are loaded with concentrated loads within a distance from the support equal to the member depth and that are loaded on one face and supported on the opposite face. Such a situation permits the development of compression struts between the loads and the supports. For such members as these, the Code in its Appendix A provides an alternate method of design, which is referred to as "strut and tie" design. This method is briefly described in Appendix C of this text. Should the loads be applied through the sides or bottom of such members, their shear design should be handled as it is for ordinary beams. Members falling into this class include beams, short cantilevers, and corbels. Corbels are brackets that project from the sides of columns and are used to support beams and girders, as shown in Figure 8.9. They are quite commonly used in precast construction. Special web reinforcing provisions are made for such members in Section 11.7 of the Code and are considered in Section 8.12 of this chapter.

8. Section 8.11.8 of the ACI Code permits a shear of $1.1V_c$ for the ribs of joist construction, as where we have closely spaced T beams with tapered webs. For the 10% increase in V_c, the joist proportions must meet the provisions of ACI Section 8.11. In ACI Section 8.11.2 it is stated that the ribs must be no less than 4 in. wide, must have depths not more than 3.5 times the minimum width of the ribs, and may not have clear spacings between the ribs greater than 30 in.

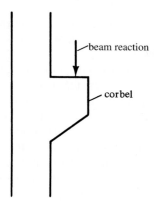

Figure 8.9 Corbel supporting beam reaction.

8.10   EXAMPLE SHEAR DESIGN PROBLEMS

Example 8.1 illustrates the selection of a beam with a sufficiently large cross section so that no web reinforcing is required. The resulting beam is unusually large. It is normally considered much better practice to use appreciably smaller sections constructed with web reinforcing. The reader should also realize that it is good construction practice to use some stirrups in all reinforced concrete beams (even though they may not be required by shear) because they enable the workers to build for each beam a cage of steel that can be conveniently handled.

EXAMPLE 8.1

Determine the minimum cross section required for a rectangular beam from a shear standpoint so that no web reinforcing is required by the ACI Code if $V_u = 38$ k and $f'_c = 4000$ psi. Use the conservative value of $V_c = 2\lambda\sqrt{f'_c}b_w d$.

SOLUTION

Shear strength provided by concrete is determined by the equation

$$\phi V_c = (0.75)\left(2(1.0)\sqrt{4000}b_w d\right) = 94.87 b_w d$$

But the ACI Code 11.4.6.1 states that a minimum area of shear reinforcement is to be provided if V_u exceeds $\frac{1}{2}\phi V_c$

$$38{,}000 = \left(\frac{1}{2}\right)(94.87 b_w d)$$

$$b_w d = 801.1 \text{ in.}$$

Use 24″ × 36″ beam ($d = 33.5″$)

The design of web reinforcing is illustrated by Examples 8.2 through 8.6. Maximum vertical stirrup spacings have been given previously, whereas no comment has been made about minimum spacings. Stirrups must be spaced far enough apart to permit the aggregate to pass through, and, in addition, they must be reasonably few in number so as to keep within reason the amount of labor involved in fabricating and placing them. Accordingly, minimum spacings of 3 or 4 in. are normally used. Usually #3 stirrups are assumed, and if the calculated design spacings are less than $d/4$, larger-diameter stirrups can be used. Another alternative is to use ⊔⊔ stirrups instead of ⊔ stirrups. Different diameter stirrups should not be used in the same beam, or confusion will result.

As is illustrated in Examples 8.3, 8.5, and 8.6, it is quite convenient to draw the V_u diagram and carefully label it with values of such items as ϕV_c, $\phi V_c/2$, and V_u at a distance d from the face of the support and to show the dimensions involved.

Some designers place their first stirrup a distance of one-half of the end-calculated spacing requirement from the face. Others put the first one 2 or 3 in. from the support.

From a practical viewpoint, stirrups are usually spaced with center-to-center dimensions that are multiples of 3 or 4 in., to simplify the fieldwork. Although this procedure may require an additional stirrup or two, total costs should be less because of reduced labor costs. A common field procedure is to place chalk marks at 2-ft intervals on the forms and to place the stirrups by eye in between those marks. This practice is combined with a somewhat violent placing of the heavy concrete in the forms followed by vigorous vibration. These field practices should clearly show the student that it is foolish to specify odd theoretical stirrup spacings such as 4 @ $6\frac{7}{16}$ in. and 6 @ $5\frac{3}{8}$ in. because such careful positioning will not be achieved in the actual members. Thus the designer will normally specify stirrup spacings in multiples of whole inches and perhaps in multiples of 3 or 4 in.

With available computer programs it is easily possible to obtain theoretical arrangements of stirrups with which the least total amounts of shear reinforcing will be required. The use of such programs is certainly useful to the designer, but he or she needs to take the resulting values and revise them into simple economical patterns with simple spacing arrangements—as in multiples of 3 in., for example.

A summary of the steps required to design vertical stirrups is presented in Table 8.1. For each step, the applicable section number of the Code is provided. The author has found this to be a very useful table for students to refer to while designing stirrups.

Table 8.1 Summary of Steps Involved in Vertical Stirrup Design

Is shear reinforcing necessary?	
1. Draw V_u diagram.	
2. Calculate V_u at a distance d from the support (with certain exceptions).	11.1.3.1 and Commentary (R11.1.3.1)
3. Calculate $\phi V_c = 2\phi\lambda\sqrt{f_c'}b_wd$ (or use the alternate method).	11.2.1.1 11.2.2.1
4. Stirrups are needed if $V_u > \frac{1}{2}\phi V_c$ (with some exceptions for slabs, footings, shallow members, hollow core units, steel fiber reinforced beams, and joists).	11.4.6.1
Design of stirrups	
1. Calculate theoretical stirrup spacing $s = A_vf_yd/V_s$ where $V_s = (V_u - \phi V_c)/\phi$.	11.4.7.2
2. Determine maximum spacing to provide minimum area of shear reinforcement $s = \dfrac{A_vf_y}{0.75\sqrt{f_c'}b_w}$ but not more than $\dfrac{A_vf_y}{50b_w}$	11.4.6.3
3. Compute maximum spacing: $d/2 \leq 24$ in., if $V_s \leq 4\sqrt{f_c'}b_wd$.	11.4.5.1
4. Compute maximum spacing: $\dfrac{d}{4} \leq 12$ in., if $V_s > 4\sqrt{f_c'}b_wd$.	11.4.5.3
5. V_s may not be $> 8\sqrt{f_c'}b_wd$ 6. Minimum practical spacing ≈ 3 or 4 in.	11.4.7.9

EXAMPLE 8.2

The beam shown in Figure 8.10 was selected using $f_y = 60{,}000$ psi and $f'_c = 3000$ psi, normal weight. Determine the theoretical spacing of #3 ⊔ stirrups for each of the following shears:

(a) $V_u = 12{,}000$ lb

(b) $V_u = 40{,}000$ lb

(c) $V_u = 60{,}000$ lb

(d) $V_u = 150{,}000$ lb

SOLUTION

(a) $V_u = 12{,}000$ lb (using $\lambda = 1.0$ for normal weight concrete)

$$\phi V_c = (0.75)\big(2(1.0)\sqrt{3000}\big)(14)(24) = 27{,}605 \text{ lb}$$

$$\frac{1}{2}\phi V_c = 13{,}803 \text{ lb} > 12{,}000 \text{ lb} \qquad \therefore \text{Stirrups not required}$$

(b) $V_u = 40{,}000$ lb

Stirrups needed because $V_u > \frac{1}{2}\phi V_c$.
Theoretical spacing

$$\phi V_c + \phi V_s = V_u$$

$$V_s = \frac{V_u - \phi V_c}{\phi} = \frac{40{,}000 - 27{,}605}{0.75} = 16{,}527 \text{ lb}$$

$$s = \frac{A_v f_y d}{V_s} = \frac{(2)(0.11)(60{,}000)(24)}{16{,}527} = 19.17'' \leftarrow.$$

Maximum spacing to provide minimum A_v

$$s = \frac{A_v f_y}{0.75\sqrt{f'_c}b_w} = \frac{(2)(0.11)(60{,}000)}{(0.75\sqrt{3000})(14)} = 22.95''$$

$$\leq s = \frac{A_v f_y}{50 b_w} = \frac{(2)(0.11)(60{,}000)}{(50)(14)} = 18.86''$$

$$V_s = 16{,}527 < (4)(\sqrt{3000})(14)(24) = 73{,}614 \text{ lb}$$

$$\therefore \text{Maximum } s = \frac{d}{2} = 12'' \qquad\qquad \underline{\underline{s = 12.0 \text{ in.}}}$$

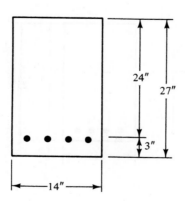

24"

27"

3"

14"

Figure 8.10

(c) $V_u = 60,000\,\text{lb}$

Theoretical spacing

$$V_s = \frac{V_u - \phi V_c}{\phi} = \frac{60,000 - 27,605}{0.75} = 43,193\,\text{lb}$$

$$s = \frac{A_v f_y d}{V_s} = \frac{(2)(0.11)(60,000)(24)}{43,193} = 7.33'' \leftarrow$$

Maximum spacing to provide minimum A_v

$$s = \frac{A_v f_y}{0.75\sqrt{f'_c}b_w} = \frac{(2)(0.11)(60,000)}{(0.75\sqrt{3000})(14)} = 22.95''$$

$$\leq s = \frac{A_v f_y}{50 b_w} = \frac{(2)(0.11)(60,000)}{(50)(14)} = 18.86''$$

$$V_s = 43,193 < (4)(\sqrt{3000})(14)(24) = 73,614\,\text{lb}$$

$$\therefore \text{Maximum } s = \frac{d}{2} = 12'' \qquad\qquad \underline{s = 7.33\,\text{in.}}$$

(d) $V_u = 150,000\,\text{lb}$

$$V_s = \frac{150,000 - 27,605}{0.75} = 163,193\,\text{lb}$$

$$163,193\,\text{lb} > (8)(\sqrt{3000})(14)(24) = 147,228\,\text{lb}$$

V_s may not be taken $> 8\sqrt{f'_c}b_w d$

$$\therefore \underline{\text{Need larger beam and/or one with larger } f'_c \text{ value}}$$

EXAMPLE 8.3

Select #3 ⊔ stirrups for the beam shown in Figure 8.11, for which $D = 4\,\text{k/ft}$ and $L = 6\,\text{k/ft}$. $f'_c = 4000\,\text{psi}$, normal weight, and $f_y = 60,000\,\text{psi}$.

SOLUTION

$$V_u \text{ at left end} = 7(1.2 \times 4 + 1.6 \times 6) = 100.8\text{k} = 100,800\,\text{lb}$$

$$V_u \text{ at a distance } d \text{ from face of support} = \left(\frac{84 - 22.5}{84}\right)(100,800) = 73,800\,\text{lb}$$

$$\phi V_c = \phi 2\lambda\sqrt{f'_c}b_w d = (0.75)(2)\big((1.0)\sqrt{4000}\big)(15)(22.5) = 32,018\,\text{lb}$$

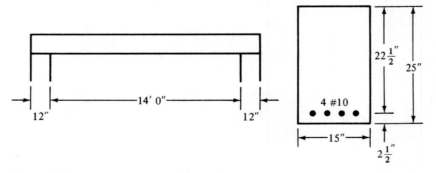

Figure 8.11

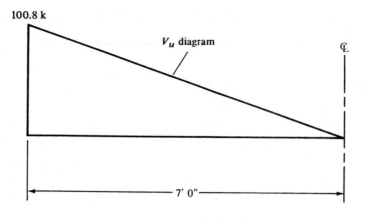

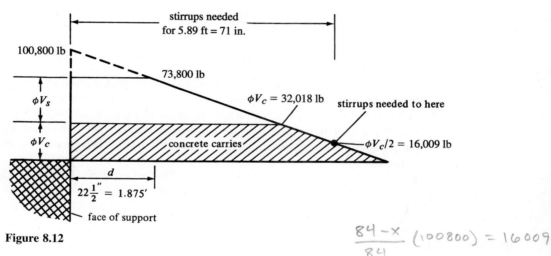

Figure 8.12

$$\frac{84-x}{84}(100800) = 16009$$

These values are shown in Figure 8.12.

$$V_u = \phi V_c + \phi V_s$$

$$\phi V_s = V_u - \phi V_c = 73{,}800 - 32{,}018 = 41{,}782 \text{ lb}$$

$$V_s = \frac{41{,}782}{0.75} = 55{,}709 \text{ lb}$$

Maximum spacing of stirrups $= d/2 = 11.25''$, since V_s is $< 4\sqrt{f_c'}b_w d = 85{,}382$ lb. Maximum theoretical spacing at left end

$$s = \frac{A_v f_y d}{V_s} = \frac{(2)(0.11)(60{,}000)(22.5)}{(55{,}709)} = 5.33 \text{ in.}$$

Maximum spacing to provide minimum A_v of stirrups

$$s = \frac{A_v f_y}{0.75\sqrt{f_c'}b_w} = \frac{(2)(0.11)(60{,}000)}{(0.75\sqrt{4000})(15)} = 18.55 \text{ in.}$$

$$\leq s = \frac{A_v f_y}{50b_w} = \frac{(2)(0.11)(60{,}000)}{(50)(15)} = 17.6 \text{ in.}$$

For convenience, theoretical spacings are calculated at different points along the span and are listed in the following table:

Distance from face of support (ft)	V_u (lb)	$V_s = \dfrac{V_u - \phi V_c}{\phi}$ (lb)	Theoretical $s = \dfrac{A_v f_y d}{V_s}$ (in.)
0 to d = 1.875	73,800	55,709	5.33
2	72,000	53,309	5.57
3	57,600	34,109	8.71
4	43,200	14,909	> Maximum of $\frac{d}{2}$ = 11.25

Spacings selected

1 @ 3 in. = 3 in.

6 @ 6 in. = 36 in.

4 @ 9 in. = 36 in.

symmetric about ℄

As previously mentioned, it is a good practice to space stirrups at multiples of 3 or 4 in. on center. As an illustration, it is quite reasonable to select for Example 8.3 the following spacings: 1 @ 3 in., 6 @ 6 in., and 4 @ 9 in. In rounding off the spacings to multiples of 3 in., it was necessary to exceed the theoretical spacings by a small amount near the end of the beam. However, the values are quite close to the required ones, and the overall number of stirrups used in the beam is more than adequate.

In Example 8.4, which follows, the value of V_c for the beam of Example 8.3 is computed by the alternate method of Section 11.2.2.1 of the Code.

EXAMPLE 8.4

Compute the value of V_c at a distance 3 ft from the left end of the beam of Example 8.3 and Figure 8.11 by using ACI Equation 11-5.

$$V_c = \left(1.9\lambda\sqrt{f_c'} + 2500\rho_w \frac{V_u d}{M_u}\right)b_w d \le 3.5\lambda\sqrt{f_c'}b_w d$$

SOLUTION

$$\lambda = 1.0 \text{ (normal weight aggregate)}$$

$$w_u = (1.2)(4) + (1.6 \times 6) = 14.4 \text{ k/ft}$$

$$V_u \text{ at } 3' = (7)(14.4) - (3)(14.4) = 57.6 \text{ k}$$

$$M_u \text{ at } 3' = (100.8 \times 3) - (14.4)(3)(1.5) = 237.6 \text{ ft-k}$$

$$\rho_w = \frac{5.06}{(15)(22.5)} = 0.0150$$

$$\frac{V_u d}{M_u} = \frac{(57.6)(22.5)}{(12)(237.6)} = 0.456 < 1.0 \quad \underline{\underline{OK}}$$

$$V_c = [1.9(1.0)\sqrt{4000} + (2500)(0.0150)(0.456)](15)(22.5)$$

$$= \underline{\underline{46,328 \text{ lb}}} < 3.5\sqrt{4000}(15)(22.5) = 74,709 \text{ lb}$$

For the uniformly loaded beams considered up to this point, it has been assumed that both dead and live loads extended from end to end of the spans. Although this practice will produce the maximum V_u at the ends of simple spans, it will not produce maximums at interior points. For such points, maximum shears will be obtained when the uniform live load is placed from the point in question to the most distant end support. For Example 8.5, shear is determined at the beam end (live load running for entire span) and then at the beam centerline (live load to one side only), and a straight-line relationship is assumed in between. Although the ACI does not specifically comment on the variable positioning of live load to produce maximum shears, it certainly is their intent for engineers to position loads so as to maximize design shear forces.

EXAMPLE 8.5

Select #3 ⊔ stirrups for the beam of Example 8.3, assuming the live load is placed to produce maximum shear at beam end and ℄.

SOLUTION   Maximum V_u at left end $= (7)(1.2 \times 4 + 1.6 \times 6) = 110.8\,\text{k} = 100,800\,\text{lb}$.

For maximum V_u at ℄, the live load is placed as shown in Figure 8.13.

$$V_u \text{ at } \text{℄} = 50,400 - (7)(1.2 \times 4) = 16.8\text{k} = 16,800\,\text{lb}$$

$$V_c = 2(1.0)\sqrt{4000}(15)(22.5) = 42,691\,\text{lb}$$

V_u at a distance d from face of support $= 78,300\,\text{lb}$ as determined by proportions from Figure 8.14.

$$V_u = \phi V_c + \phi V_s$$

$$\phi V_s = V_u - \phi V_c = 78,300 - (0.75)(42,691) = 46,282\,\text{lb at left end}$$

$$V_s = \frac{46,282}{0.75} = 61,709\,\text{lb}$$

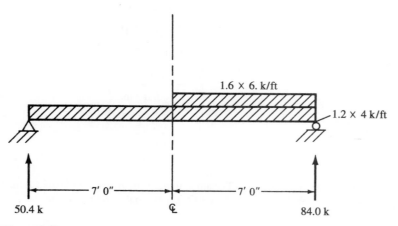

Figure 8.13

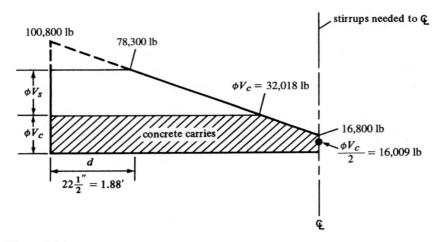

Figure 8.14

The limiting spacings are the same as in Example 8.3. The theoretical spacings are given in the following table:

Distance from face of support (ft)	V_u (lb)	$V_s = \dfrac{V_u - \phi V_c}{\phi}$ (lb)	Theoretical spacing required $s = \dfrac{A_v f_y d}{V_s}$ (in.)
0 to $d = 1.875$	78,300	61,709	4.81
2	76,800	59,709	4.97
3	64,800	43,709	6.79
4	52,800	27,709	10.71
5	40,800	11,709	>Maximum 11.25

One possible arrangement (#4 stirrups might be better)

$$1 @ 3 \text{ in.} = 3 \text{ in.}$$
$$7 @ 4 \text{ in.} = 28 \text{ in.}$$
$$2 @ 6 \text{ in.} = 12 \text{ in.}$$
$$\underline{4 @ 9 \text{ in.} = 36 \text{ in.}}$$
$$6 \text{ ft } 7 \text{ in.} \qquad\qquad \text{Symmetrical about } \text{℄}$$

EXAMPLE 8.6

Select spacings for #3 ⊔ stirrups for a T beam with $b_w = 10$ in. and $d = 20$ in. for the V_u diagram shown in Figure 8.15, with $f_y = 60,000$ psi and $f'_c = 3000$ psi, normal weight concrete

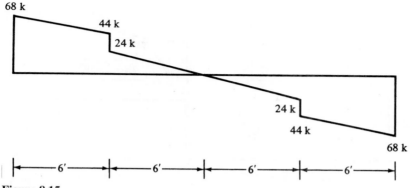

Figure 8.15

SOLUTION   **(with reference to Figure 8.16)**

V_u at a distance d from face of support

$$= 44,000 + \left(\frac{72-20}{72}\right)(68,000 - 44,000) = 61,333 \text{ lb}$$

$\lambda = 1.0$ for normal weight concrete.

$$\phi V_c = (0.75)\left(2(1.0)\sqrt{3000}\right)(10)(20) = 16,432 \text{ lb}$$

$$\frac{\phi V_c}{2} = \frac{16,432}{2} = 8216 \text{ lb}$$

Stirrups are needed for a distance $= 72 + \left(\frac{24,000 - 8216}{24,000}\right)(72) = 119.5 \text{ in.}$

V_s at left end $(V_u - \phi V_c)/\phi = (61,333 - 16,432)/0.75 = 59,868$ which is larger than $4\sqrt{f_c'}b_w d = (4\sqrt{3000})(10)(20) = 43,818$ lb but less than $8\sqrt{f_c'}b_w d$. Therefore the maximum spacing of stirrups in that range is $d/4 = 5$ in.

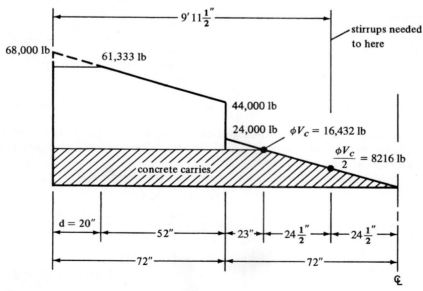

Figure 8.16

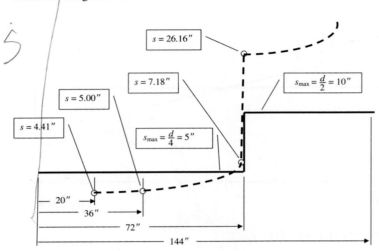

Figure 8.17

Maximum spacing of stirrups $= \frac{d}{4} = \frac{20}{4} = 5$ in. when $V_s > 4\sqrt{f'_c}b_w d = (4\sqrt{3000})(10)(20) = 43{,}818$ lb. By proportions from the V_s column in the table, V_s falls to 43,818 lb at approximately 4.66 ft or 56 in. from the left end of the beam.

Maximum spacing permitted to provide minimum A_v of stirrups is the smaller of the two following values of s.

$$s = \frac{A_v f_y}{0.75\sqrt{f'_c}b_w} = \frac{(2)(0.11)(60{,}000)}{(0.75\sqrt{3000})(10)} = 32.13 \text{ in.}$$

$$s = \frac{A_v f_y}{50 b_w} = \frac{(2)(0.11)(60{,}000)}{(50)(10)} = 26.4 \text{ in.}$$

The theoretical spacings at various points in the beam are computed in the following table:

Distance from face of support (ft.)	V_u (lb.)	$V_s = \frac{V_u - \phi V_c}{\phi}$ (lb.)	Theoretical spacing required $s = \frac{A_v f_y d}{V_s}$ (in.)	Maximum spacing (in.)
0 to $d = 1.667$	61,333	59,868	4.41"	4.41
3	56,000	52,758	5.00"	5.00
6−	44,000	36,757	7.18"	5.00
6+	24,000	10,091	26.16"	$\frac{d}{2} = 10.00$

A summary of the results of the preceding calculations is shown in Figure 8.17, where the solid dark line represents the maximum stirrup spacings permitted by the Code and the dashed line represents the calculated theoretical spacings required for $V_u - \phi V_c$.

From this information the author selected the following spacings:

$$1@3 \text{ in.} = 3 \text{ in.}$$
$$17@4 \text{ in.} = 68 \text{ in.}$$
$$\underline{5@10 \text{ in.} = 50 \text{ in.}}$$

$$121 \text{ in.}$$

<u>Symmetrical about ℄</u>

8.11   ECONOMICAL SPACING OF STIRRUPS

When stirrups are required in a reinforced concrete member, the Code specifies maximum permissible spacings varying from $d/4$ to $d/2$. On the other hand, it is usually thought that stirrup spacings less than $d/4$ are rather uneconomical. Many designers use a maximum of three different spacings in a beam. These are $d/4$, $d/3$, and $d/2$. It is easily possible to derive a value of ϕV_s for each size and style of stirrups for each of these spacings.[5]

Note that the number of stirrups is equal to d/s and that if we use spacings of $d/4$, $d/3$, and $d/2$ we can see that n equals 4, 3, or 2. Then the value of ϕV_s can be calculated for any particular spacing, size, and style of stirrup. For instance, for #3 ⊔ stirrups spaced at $d/2$ with $\phi = 0.75$ and $f_y = 60$ ksi,

$$\phi V_s = \frac{\phi A_v f_y d}{s} = \frac{(0.75)(2 \times 0.11)(60)(d)}{d/2} = 19.8 \text{ k}$$

The values shown in Table 8.2 were computed in this way for 60-ksi stirrups.

For an example using this table, reference is made to the beam and V_u diagram of Example 8.3, which was shown in Figure 8.12 where 60-ksi #3 ⊔ stirrups were selected for a beam with a d of $22\frac{1}{2}$ in. For our closest spacing, $d/4$, we can calculate $\phi V_c + 39.6 = 32.018 + 39.6 = 71.6$ k. Similar calculations are made for $d/3$ and $d/2$ spacings, and we obtain, respectively, 61.7 k and 51.8 k. The shear diagram is repeated in Figure 8.18, and the preceding values are located on the diagram by proportions or by scaling.

From this information we can see that we can use $d/4$ for the first 2.72 ft, $d/3$ for the next 0.68 ft, and $d/2$ for the remaining 2.49 ft. Then the spacings are smoothed (preferably to multiples of 3 in.). Also, for this particular beam we would probably use the $d/4$ spacing on through the 0.68-ft section and then use $d/2$ the rest of the required distance.

Table 8.2 Values for 60-ksi Stirrups

s	ϕV_s for #3 ⊔ stirrups (kips)	ϕV_s for #4 ⊔ stirrups (kips)
$d/2$	19.8	36
$d/3$	29.7	54
$d/4$	39.6	72

8.12   SHEAR FRICTION AND CORBELS

If a crack occurs in a reinforced concrete member (whether caused by shear, flexure, shrinkage, etc.) and if the concrete pieces on opposite sides of the crack are prevented from moving apart, there will be a great deal of resistance to slipping along the crack due to the rough and irregular concrete surfaces. If reinforcement is provided across the crack to prevent relative displacement along the crack, shear will be resisted by friction between the faces, by resistance to shearing off of protruding portions of the concrete, and by dowel action of the reinforcing crossing the crack. The transfer of shear under these circumstances is called *shear friction*.

Shear friction failures are most likely to occur in short, deep members subject to high shears and small bending moments. These are the situations where the most nearly vertical cracks will occur. If moment and shear are both large, diagonal tension cracks will occur at rather large angles from the vertical. This situation has been discussed in Sections 8.1 through 8.11.

---

[5]Neville, B. B., ed., 1984, *Simplified Design Reinforced Concrete Buildings of Moderate Size and Height* (Skokie, IL: Portland Cement Association), pp. 3-12–3-16.

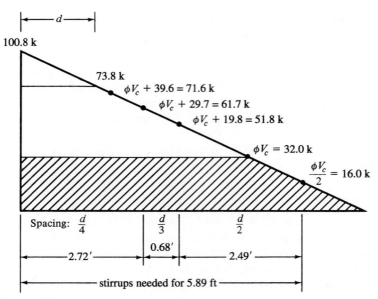

Figure 8.18

A short cantilever member having a ratio of clear span to depth (a/d) of 1.0 or less is often called a *bracket* or *corbel*. One such member is shown in Figure 8.19. The shear friction concept provides a convenient method for designing for cases where diagonal tension design is not applicable. The most common locations where shear friction design is used are for brackets, corbels, and precast connections, but it may also be applied to the interfaces between concretes cast at different times, to the interfaces between concrete and steel sections, and so on.

When brackets or corbels or short, overhanging ends or precast connections support heavy concentrated loads, they are subject to possible shear friction failures. The dashed lines in Figure 8.19 show the probable locations of these failures. It will be noted that for the end-bearing situations the cracks tend to occur at angles of about 20° from the direction of the force application.

Space is not taken in this chapter to provide an example of shear friction design, but a few general remarks about them are presented. (In Section 12.13 of this text, a numerical shear friction

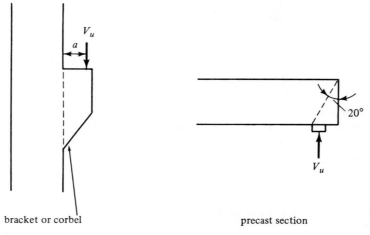

Figure 8.19 Possible shear friction failures.

example is presented in relation to the transfer of horizontal forces at the base of a column to a footing.) It is first assumed that a crack has occurred as shown by the dashed lines of Figure 8.19. As slip begins to occur along the cracked surface, the roughness of the concrete surfaces tends to cause the opposing faces of the concrete to separate.

As the concrete separates, it is resisted by the tensile reinforcement (A_{vf}) provided across the crack. It is assumed that this steel is stretched until it yields. (An opening of the crack of 0.01 in. will probably be sufficient to develop the yield strength of the bars.) The clamping force developed in the bars $A_{vf}f_y$ will provide a frictional resistance equal to $A_{vf}f_y\mu$, where μ is the coefficient of friction (values of which are provided for different situations in Section 11.6.4.3 of the Code).

Then the design shear strength of the member must at least equal the shear, to be taken as

$$\phi V_n = V_u = \phi A_{vf}f_y\mu$$

The value of f_y used in this equation cannot exceed 60 ksi and the shear friction reinforcement across or perpendicular to the shear crack may be obtained by

$$A_{vf} = \frac{V_u}{\phi f_y \mu}$$

This reinforcing should be appropriately placed along the shear plane. If there is no calculated bending moment, the bars will be uniformly spaced. If there is a calculated moment, it will be necessary to distribute the bars in the flexural tension area of the shear plane. The bars must be anchored sufficiently on both sides of the crack to develop their yield strength by means of embedment, hooks, headed bars, or other methods. Since space is often limited in these situations, it is often necessary to weld the bars to special devices such as crossbars or steel angles. The bars should be anchored in confined concrete (that is, column ties or external concrete or other reinforcing shall be used).

When beams are supported on brackets or corbels, there may be a problem with shrinkage and expansion of the beams, producing horizontal forces on the bracket or corbel. When such forces are present, the bearing plate under the concentrated load should be welded down to the tensile steel. Based on various tests, the ACI Code (11.8.3.4) says that the horizontal force used must be at least equal to $0.2V_u$ unless special provisions are made to avoid tensile forces.

The presence of direct tension across a cracked surface obviously reduces the shear-transfer strength. Thus direct compression will increase its strength. As a result, Section 11.6.7 of the Code permits the use of a permanent compressive load to increase the shear friction clamping force. A typical corbel design and its reinforcing are shown in Figure 8.20.

Enough concrete area must also be provided, and Section 11.6.5 of the Code gives the upper limits on the shear force, V_n, transferred across a shear-friction failure surface based on concrete strength and contact area. For normal-weight concrete placed monolithically or placed against intentionally roughened concrete, V_n cannot exceed the smaller of

$$V_n \leq 0.2f_c'A_c,$$
$$\leq (480 + 0.08f_c')A_c, \quad \text{and}$$
$$\leq 1600A_c$$

For all other cases,

$$V_n \leq 0.2f_c'A_c,$$
$$\leq 800A_c$$

where A_c is the concrete contact area along the shear-friction failure surface.

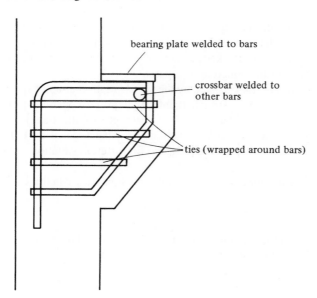

bearing plate welded to bars

crossbar welded to
other bars

ties (wrapped around bars)

Figure 8.20

8.13   SHEAR STRENGTH OF MEMBERS SUBJECTED TO AXIAL FORCES

Reinforced concrete members subjected to shear forces can at the same time be loaded with axial compression or axial tension forces due to wind, earthquake, gravity loads applied to horizontal or inclined members, shrinkage in restrained members, and so on. These forces can affect the shear design of our members. Compressive loads tend to prevent cracks from developing. As a result, they provide members with larger compressive areas and thus greater shear strengths. On the other hand, tensile forces exaggerate cracks and reduce shear resistances because they will decrease compression areas.

When we have appreciable axial compression, the following equation can be used to compute the shear carrying capacity of a concrete member:

$$V_c = 2\left(1 + \frac{N_u}{2000\,A_g}\right)\lambda\sqrt{f_c'}\,b_w d \qquad \text{(ACI Equation 11-4)}$$

For a member subjected to a significant axial tensile force, the shear capacity of the concrete may be determined from the following expression:

$$V_c = 2\left(1 + \frac{N_u}{500\,A_g}\right)\lambda\sqrt{f_c'}\,b_w d \qquad \text{(ACI Equation 11-8)}$$

In this expression N_u the axial load is minus if the load is tensile. You might note that if the computed value of N_u/A_g for use in this equation is 500 psi or more, the concrete will have lost its capacity to carry shear. (The value of V_c used need not be taken less than zero.)

The SI values for ACI Equations 11-4 and 11-8 are, respectively,

$$V_c = \left(1 + \frac{N_u}{14A_g}\right)\left(\frac{\lambda\sqrt{f_c'}}{6}\right)b_w d$$

$$V_c = \left(1 + \frac{0.3N_u}{A_g}\right)\left(\frac{\lambda\sqrt{f_c'}}{6}\right)b_w d$$

Instead of using ACI Equation 11-4 to compute the shear capacity of sections subject to axial compressive loads, ACI Equation 11-5 may be used. In this equation a revised moment, M_m, may be substituted for M_u at the section in question and $V_u d / M_u$ shall not be > 1.0. For this case V_c may not be larger than the value obtained with ACI Equation 11-7.

$$V_c = \left(1.9\lambda\sqrt{f_c'} + 2500\rho_w\frac{V_u d}{M_u}\right)b_w d \leq 3.5\lambda\sqrt{f_c'}b_w d \qquad \text{(ACI Equation 11-5)}$$

$$M_m = M_u - N_u\left(\frac{4h - d}{8}\right) \qquad \text{(ACI Equation 11-6)}$$

$$V_c \text{ may not be } > 3.5\lambda\sqrt{f_c'}b_w d\sqrt{1 + \frac{N_u}{500A_g}} \qquad \text{(ACI Equation 11-7)}$$

In SI units ACI Equation 11-5 is

$$V_c = \left(\lambda\sqrt{f_c'} + 120\rho_w\frac{V_u d}{M_u}\right)\frac{b_w d}{7} \leq 0.3\lambda\sqrt{f_c'}b_w d$$

Equation 11-7 is

$$V_c = 0.3\lambda\sqrt{f_c'}b_w d\sqrt{1 + \frac{0.3N_u}{A_g}}$$

Example 8.7, which follows, illustrates the computation of the shear strength of an axially loaded concrete member.

EXAMPLE 8.7

For the concrete section shown in Figure 8.21 for which f_c' is 3000 psi, normal weight ($\lambda = 1.0$)

(a) Determine V_c if no axial load is present using ACI Equation 11-3.

(b) Compute V_c using ACI Equation 11-4 if the member is subjected to an axial compression load of 12,000 lb.

(c) Repeat part (b) using revised ACI Equation 11-5. At the section in question, assume $M_u = 30$ ft-k and $V_u = 40$ k. Use M_m in place of M_u.

(d) Compute V_c if the 12,000-lb load is tensile.

SOLUTION

(a) $V_c = 2(1.0)\sqrt{3000}(14)(23) = \underline{35{,}273 \text{ lb}}$

(b) $V_c = 2\left(1 + \frac{20{,}000}{(2000)(14)(26)}\right)\left((1.0)\sqrt{3000}\right)(14)(23) = \underline{\underline{36{,}242 \text{ lb}}}$

$\qquad < 3.5(1.0)\sqrt{3000}(14)(23) = 61{,}728 \text{ lb}$

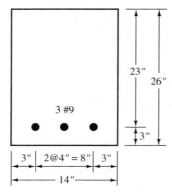

Figure 8.21

(c) $M_m = (12)(30{,}000) - 12{,}000\left(\dfrac{4 \times 26 - 23}{8}\right) = 238{,}500 \text{ in.-lb.}$

$\dfrac{V_u d}{M_m} = \dfrac{(40)(23)}{238.5} = 3.857 > 1.00 \;\therefore 1.0$

$V_c = \left[1.9(1.0)\sqrt{3000} + (2500)\left(\dfrac{3.00}{(14)(23)}\right)(1.00)\right](14)(23)$

$\underline{\underline{= 41{,}010 \text{ lb}}}$

But not $> \left(3.5(1.0)\sqrt{3000}\right)(14)(23)\sqrt{1 + \dfrac{12{,}000}{(500)(14)(26)}}$

$= 63{,}731 \text{ lb}$ OK

(d) $V_c = 2\left(1 + \dfrac{-12{,}000}{(500)(14)(26)}\right)\left((1.0)\sqrt{3000}(14)(23)\right)$

$\underline{\underline{= 32{,}950 \text{ lb}}}$

8.14   SHEAR DESIGN PROVISIONS FOR DEEP BEAMS

There are some special shear design provisions given in Section 11.7 of the Code for deep flexural members with ℓ_n/d values equal to or less than four that are loaded on one face and supported on the other face so that compression struts can develop between the loads and the supports. Such a member is shown in Figure 8.22(a). Some members falling into this class are short, deep, heavily loaded beams, wall slabs under vertical loads, shear walls, and perhaps floor slabs subjected to horizontal loads.

If the loads are applied through the sides or the bottom (as where beams are framing into its sides or bottom) of the member, as illustrated in Figure 8.22(b) and 8.22(c), the usual shear design provisions described earlier in this chapter are to be followed whether or not the member is deep.

The angles at which inclined cracks develop in deep flexural members (measured from the vertical) are usually much smaller than 45°—on some occasions being very nearly vertical. As a result, web reinforcing when needed has to be more closely spaced than for beams of regular depths. Furthermore, the web reinforcing needed is in the form of both horizontal and vertical reinforcing. These almost vertical cracks indicate that the principal tensile forces are primarily horizontal, and thus horizontal reinforcing is particularly effective in resisting them.

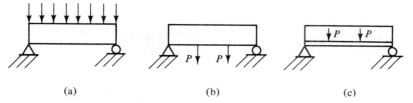

(a)   (b)   (c)

Figure 8.22

The detailed provisions of the Code relating to shear design for deep beams, together with the applicable ACI Section numbers, are as follows:

1. Deep beams are to be designed using the procedure described in Appendix A of the Code (Appendix C in this textbook) or by using a nonlinear analysis. (ACI 11.7.2)

2. The nominal shear strength V_n for deep beams shall not exceed $10\sqrt{f_c'}b_w d$. (ACI 11.7.3)

3. The area of shear reinforcing A_v perpendicular to the span must at least equal $0.0025\,b_w s$ and s may not be greater than $d/5$ or 12 in. (ACI 11.7.4) s is the spacing of the shear or torsion reinforcing measured in a direction parallel to the logitudinal reinforcing.

4. The area of shear reinforcing parallel to the span must not be less than $0.0015\,b_w s_2$ and s_2 may not be greater than $d/5$ or 12 in. (ACI 11.7.5) s_2 is the spacing of shear reinforcing measured in a direction perpendicular to the beams longitudinal reinforcement.

You will note that more vertical than horizontal shear reinforcing is required because vertical reinforcing has been shown to be more effective than horizontal reinforcing. The subject of deep beams is continued in Appendix C of this textbook.

8.15   INTRODUCTORY COMMENTS ON TORSION

Until recent years, the safety factors required by design codes for proportioning members for bending and shear were very large, and the resulting large members could almost always be depended upon to resist all but the very largest torsional moments. Today, however, with the smaller members selected using the strength design procedure, this is no longer true and torsion needs to be considered much more frequently.

Torsion may be very significant for curved beams, spiral staircases, beams that have large loads applied laterally off center, and even in spandrel beams running between exterior building columns. These latter beams support the edges of floor slabs, floor beams, curtain walls, and façades, and they are loaded laterally on one side. Several situations where torsion can be a problem are shown in Figure 8.23.

When plain concrete members are subjected to pure torsion, they will crack along 45° spiral lines when the resulting diagonal tension exceeds the design strength of the concrete. Although these diagonal tension stresses produced by twisting are very similar to those caused by shear, they will occur on all faces of a member. As a result, they add to the stresses caused by shear on one side of the beam and subtract from them on the other.

Reinforced concrete members subjected to large torsional forces may fail quite suddenly if they are not provided with torsional reinforcing. The addition of torsional reinforcing does not change the magnitude of the torsion that will cause diagonal cracks, but it does prevent the members from tearing apart. As a result, they will be able to resist substantial torsional moments without failure. Tests have shown that both longitudinal bars and closed stirrups or spirals are

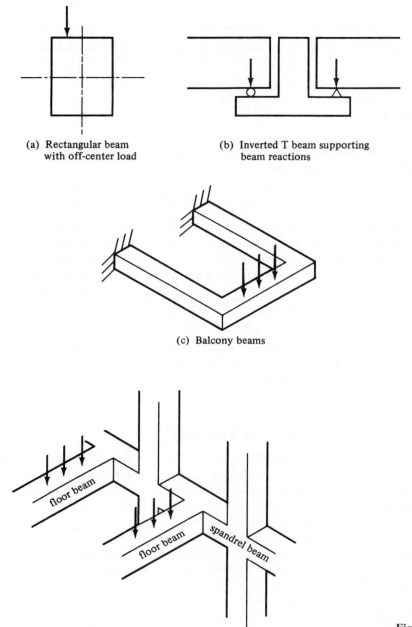

(a) Rectangular beam
with off-center load

(b) Inverted T beam supporting
beam reactions

(c) Balcony beams

(d) Spandrel beam with torsion caused
by floor beams

Figure 8.23 Some situations where torsion stresses may be significant.

necessary to intercept the numerous diagonal tension cracks that occur on all surfaces of beams subject to appreciable torsional forces. There must be a longitudinal bar in each corner of the stirrups to resist the horizontal components of the diagonal tension caused by torsion. Chapter 15 of this text is completely devoted to torsion.

8.16   SI EXAMPLE

EXAMPLE 8.8

Determine required spacing of #10 stirrups at the left end of the beam shown in Figure 8.24, if $f_c' = 21$ MPa, normal weight, and $f_y = 420$ MPa.

SOLUTION

$$V_u \text{ @ left end} = (4)(84.6) = 338.4 \text{ kN}$$
$$V_u \text{ @ a distance } d \text{ from left end}$$
$$= 338.4 - \left(\frac{750}{1000}\right)(84.6) = 274.95 \text{ kN}$$
$$\phi V_c = (\phi)\left(\frac{\lambda\sqrt{f_c'}}{6}\right)b_w d = (0.75)\left(\frac{(1.0)\sqrt{21}}{6}\right)(400)(750)$$
$$= 171\,847 \text{ kN} = 171.85 \text{ kN}$$

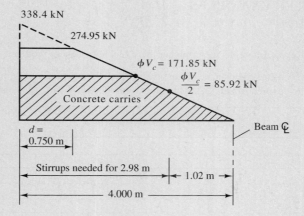

338.4 kN

274.95 kN

$\phi V_c = 171.85$ kN

$\dfrac{\phi V_c}{2} = 85.92$ kN

Concrete carries

Beam ℄

$d = 0.750$ m

Stirrups needed for 2.98 m

1.02 m

4.000 m

$$V_u = \phi V_c + \phi V_s$$
$$V_s = \frac{V_u - \phi V_c}{\phi} = \frac{274.95 - 171.85}{0.75} = 137.47 \text{ kN}$$

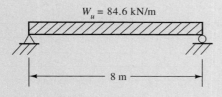

$W_u = 84.6$ kN/m

8 m

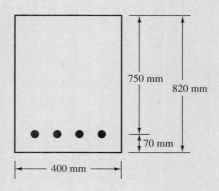

750 mm

820 mm

70 mm

400 mm

Figure 8.24

Assuming #10 stirrups

$$\text{Theoretical } s = \frac{A_v f_y d}{V_s} = \frac{(2)(71)(420)(750)}{(137.47)(10^3)} = 325 \text{ mm}$$

Maximum s to provide minimum A_v for stirrups

$$s = \frac{3A_v f_y}{b_w} = \frac{(3)(2 \times 71)(420)}{400} = 447 \text{ mm} \qquad \text{(ACI metric Equation 11-13)}$$

$$V_s = 137.47 \text{ kN} < \frac{1}{3}\sqrt{f_c'}b_w d \qquad \text{(From ACI metric Section 11.4.4.3)}$$

$$= \frac{1}{3}\sqrt{21}(400)(750) = 458,257 \text{ N} = 458.26 \text{ kN} \qquad \underline{\underline{\text{OK}}}$$

$$\therefore \text{Maximum } s = \frac{d}{2} = \frac{700}{2} = 375 \text{ mm} \leftarrow$$

Use $s = 325$ mm

8.17 COMPUTER EXAMPLES

EXAMPLE 8.9

Repeat Example 8.2 (c) using the Excel spreadsheet provided for Chapter 8.

SOLUTION   Open the spreadsheet and enter values in the cells in yellow highlight (shown in shaded cells below, but yellow in the spreadsheets). These include values for $V_u, f_c', \lambda, b_w, d, A_v,$ and f_{yt}. The required stirrup spacing s is shown

Shear Design—Beams

$V_u =$	60,000	lb				
$f_c' =$	3000	psi				
$\lambda =$	1					
$b_w =$	14	in.				
$d =$	24	in.				
$A_v =$	0.22	in.^{2}				
$f_{yt} =$	60,000	psi				
$\phi =$	0.75					
$V_c =$	36,807	lb				
$\phi V_c =$	27,605	lb	-			
$1/2\phi V_c =$	13,803	lb				
$V_s = (V_u - \phi V_c)/\phi =$	43,193	lb				
			-			
Required $\phi V_s =$	32,395	lb				
$s =$	7.33	in.				
choose s =	7.00	in.	-			
$s_{\max} =$	12	in.	Code Section 11.4.5	-		
$A_{v\,\min} =$	0.082	in.^{2}	Code Eq. 11-13	-		
$s_{\max} =$	22.95	in.	also Code Eq. 11-13			
$s_{\max} =$	18.86	in.	Code Eq. 11-13 with 50 psi limit			
Controlling $s_{\max} =$	12.00	in.				
Actual $\phi V_s =$	33,943	ℓb				
Check $\phi V_c - \phi V_s =$	61,548	lb				

in cell C19 ($s = 7.33$ in.). Use good judgement to enter an actual value for spacing in the cell below (choose s). A value of choose $s = 7.00$ in. is shown. This value must not exceed the calculated value of s as well as the "Controllling s_{max}" listed a few cells below. In the cell labeled "Check $\phi V_c + \phi V_s$" is the shear capacity of the section with the actual stirrup spacing you entered in "choose s". It will exceed the input value of V_u if the design is okay. In this case the capacity is 61548 lb which exceeds V_u Of 60,000 lb. Several warnings will appear if your "choose $s =$" value is too large.

PROBLEMS

8.1 The ACI Code provides the following limiting shear values for members subject only to shear and flexure: $2\sqrt{f_c'}$, $4\sqrt{f_c'}$ and $8\sqrt{f_c'}$. What is the significance of each of these limits?

8.2 Should the maximum shear force in a member occur at a support, the Code in the presence of a certain condition permits the designer to calculate the shear at a distance d from the face of the support. Describe the situation when this reduced shear may be used.

8.3 Why does the Code limit the maximum design yield stress that may be used in the design calculations for shear reinforcing to 60,000 psi (not including welded wire fabric)?

8.4 What is shear friction, and where is it most likely to be considered in reinforced concrete design?

Shear Analysis

8.5 What is the design shear strength of the beam shown if $f_c' = 4000$ psi and $f_y = 60,000$ psi? No shear reinforcing is provided. (*Ans.* 29,219)

Problem 8.5

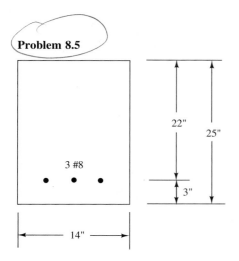

8.6 Repeat Problem 8.5 if the total depth of the beam is 32 in. and $f_c' = 3000$ psi.

In Problems 8.7 to 8.9, compute ϕV_n for the sections shown if f_y of stirrups is 60 ksi and $f_c' = 4000$ psi.

Problem 8.7 (*Ans.* 58,572)

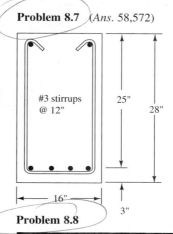

Problem 8.8

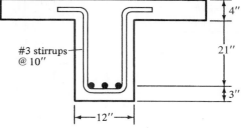

Problem 8.9 (*Ans.* 49,580 lb)

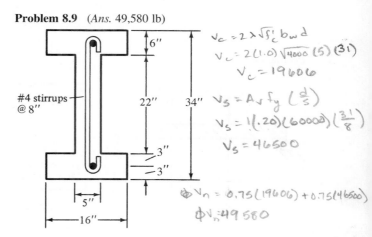

$V_c = 2 \times \sqrt{f_c'}\, b_w d$

$V_c = 2(1.0)\sqrt{4000}\,(5)(31)$

$V_c = 19606$

$V_s = A_v f_y \left(\frac{d}{s}\right)$

$V_s = 1(.20)(60000)\left(\frac{31}{8}\right)$

$V_s = 46500$

$\phi V_n = 0.75(19606) + 0.75(46500)$

$\phi V_n = 49580$

Shear Design

8.10 If $f_c' = 3000$ psi, $V_u = 60$ k, and $b_w = \frac{1}{2}d$, select a rectangular beam section if no web reinforcement is used. Use sandlightweight concrete and b_w is an integer inch.

In Problems 8.11 to 8.19 for the beams and loads given, select stirrup spacings if $f'_c = 4000$ psi normal weight concrete and $f_y = 60,000$ psi. The dead loads shown include beam weights. Do not consider movement of live loads unless specifically requested. Assume #3 ⊔ stirrups unless given otherwise.

Problem 8.11   (*One ans.* 1 @ 6 in., 10 @ 12 in.)

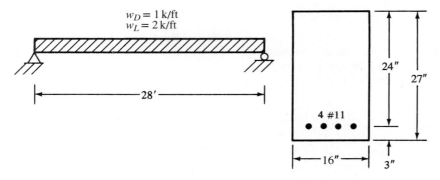

Problem 8.12

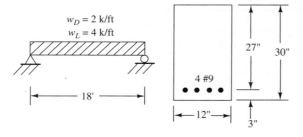

8.13 Repeat Problem 8.12 if live load positions are considered to cause maximum end shear and maximum ℄ shear. (*One ans.* 1 @ 4 in., 4 @ 8 in., 2 @ 10 in., 4 @ 13 in.)

Problem 8.14

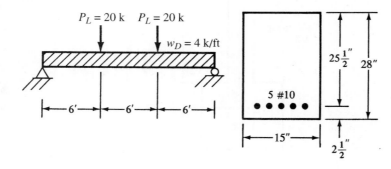

Problem 8.15   (*One ans.* 1 @ 6 in., 8 @ 12 in.)

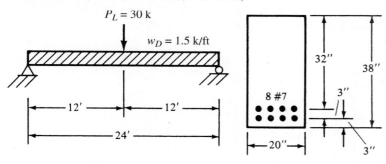

Problem 8.16 Use #4 ⊔ stirrups.

Problem 8.17 Use #4 ⊔ stirrups. (*One ans.* 1 @ 4 in., 5 @ 9 in., 10 @ 12 in.)

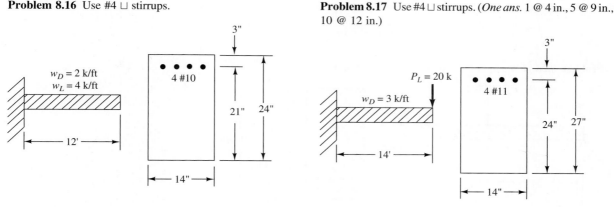

Problem 8.18

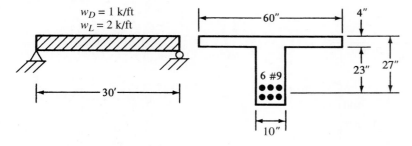

8.19 If the beam of Problem 8.14 has a factored axial compression load of 120 k in addition to other loads, calculate ϕV_c and redesign the stirrups. (*One ans.* 1 @ 4 in., 3 @ 10 in., 4 @ 12 in.)

8.20 Repeat Problem 8.19 if the axial load is tensile. Use #4 ⊔ stirrups.

In Problems 8.21 to 8.22 repeat the problems given using the Excel Spreadsheet for Chapter 8.

8.21 If $V_u = 56{,}400$ lb at a particular section determine the theoretical spacing of #3 ⊔ stirrups for the beam of Problem 8.11. (*Ans.* Theoretical $s = 11.90$ in., use maximum = 11 in.)

8.22 If $V_u =$ equals 79,600 lb at a particular section determine the spacing of #4 ⊔ stirrups for the beam of Problem 8.12.

8.23 Prepare a flowchart for the design of stirrups for rectangular, T, or I beams.

Problems in SI Units

In Problems 8.24 to 8.26 for the beams and loads given, select stirrup spacings if $f'_c = 21$ MPa and $f_y = 420$ MPa. The dead loads shown include beam weights. Do not consider movement of live loads. Use #10 $\sqcup$ stirrups.

Problem 8.24

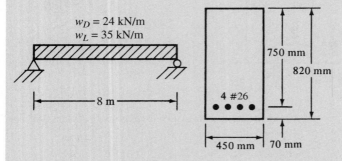

Problem 8.25 (*One ans.* 1 @ 100 mm, 13 @ 300 mm)

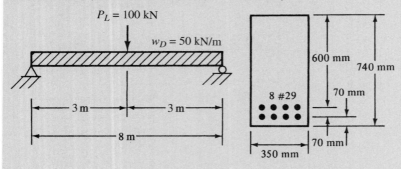

Problem 8.26

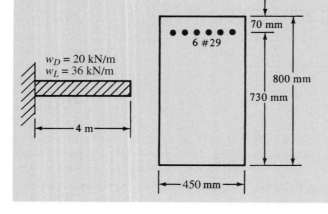

9

Introduction to Columns

9.1 GENERAL

This chapter presents an introductory discussion of reinforced concrete columns, with particular emphasis on short, stocky columns subjected to small bending moments. Such columns are often said to be "axially loaded." Short, stocky columns with large bending moments are discussed in Chapter 10, while long or slender columns are considered in Chapter 11.

Concrete columns can be roughly divided into the following three categories:

Short compression blocks or pedestals. If the height of an upright compression member is less than three times its least lateral dimensions, it may be considered to be a pedestal. The ACI (2.2 and 10.14) states that a pedestal may be designed with unreinforced or plain concrete with a maximum design compressive stress equal to $0.85\phi f_c'$, where ϕ is 0.65. Should the total load applied to the member be larger than $0.85\phi f_c' A_g$, it will be necessary either to enlarge the cross-sectional area of the pedestal or to design it as a reinforced concrete column, as described in Section 9.9 of this chapter.

Short reinforced concrete columns. Should a reinforced concrete column fail due to initial material failure, it is classified as a short column. The load that it can support is controlled by the dimensions of the cross section and the strength of the materials of which it is constructed. We think of a short column as being a rather stocky member with little flexibility.

Long or slender reinforced concrete columns. As columns become more slender, bending deformations will increase, as will the resulting secondary moments. If these moments are of such magnitude as to significantly reduce the axial load capacities of columns, those columns will be referred to as being *long* or *slender*.

When a column is subjected to *primary moments* (those moments caused by applied loads, joint rotations, etc.), the axis of the member will deflect laterally, with the result that additional moments equal to the column load times the lateral deflection will be applied to the column. These latter moments are called *secondary moments* or *PΔ moments* and are illustrated in Figure 9.1.

A column that has large secondary moments is said to be a slender column, and it is necessary to size its cross section for the sum of both the primary and secondary moments. The ACI's intent is to permit columns to be designed as short columns if the secondary or *PΔ* effect does not reduce their strength by more than 5%. Effective slenderness ratios are described and evaluated in Chapter 11 and are used to classify columns as being short or slender. When the ratios are larger than certain values (depending on whether the columns are braced or unbraced laterally), they are classified as slender columns.

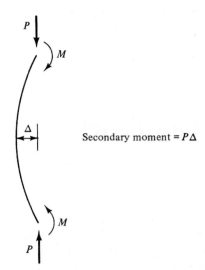

Figure 9.1 Secondary or $P\Delta$ moment.

The effects of slenderness can be neglected in about 40% of all unbraced columns and about 90% of those braced against sidesway.[1] These percentages are probably decreasing year by year, however, due to the increasing use of slenderer columns designed by the strength method using stronger materials and with a better understanding of column buckling behavior.

9.2 TYPES OF COLUMNS

A plain concrete column can support very little load, but its load-carrying capacity will be greatly increased if longitudinal bars are added. Further substantial strength increases may be made by providing lateral restraint for these longitudinal bars. Under compressive loads, columns tend not only to shorten lengthwise but also to expand laterally due to the Poisson effect. The capacity of such members can be greatly increased by providing lateral restraint in the form of closely spaced closed ties or helical spirals wrapped around the longitudinal reinforcing.

Reinforced concrete columns are referred to as *tied* or *spiral* columns, depending on the method used for laterally bracing or holding the bars in place. If the column has a series of closed ties, as shown in Figure 9.2(a), it is referred to as a *tied column*. These ties are effective in increasing the column strength. They prevent the longitudinal bars from being displaced during construction, and they resist the tendency of the same bars to buckle outward under load, which would cause the outer concrete cover to break or spall off. Tied columns are ordinarily square or rectangular, but they can be octagonal, round, L-shaped, and so forth.

The square and rectangular shapes are commonly used because of the simplicity of constructing the forms. Sometimes, however, when they are used in open spaces, circular shapes are very attractive. The forms for round columns are often made from cardboard or plastic tubes, which are peeled off and discarded once the concrete has sufficiently hardened.

If a continuous helical spiral made from bars or heavy wire is wrapped around the longitudinal bars, as shown in Figure 9.2(b), the column is referred to as a *spiral column*.

[1]Portland Cement Association, 2005, *Notes on ACI 318-05. Building Code Requirements for Structural Concrete* (Skokie, IL), p. 11-3.

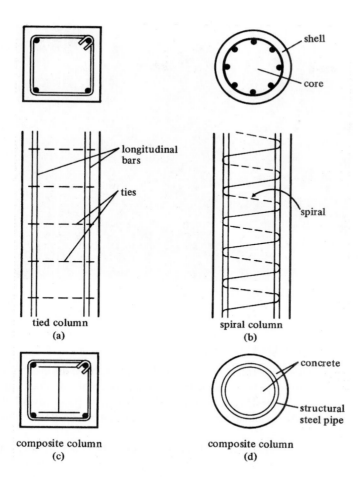

Figure 9.2 Types of columns.

Spirals are even more effective than ties in increasing a column's strength. The closely spaced spirals do a better job of holding the longitudinal bars in place, and they also confine the concrete inside and greatly increase its resistance to axial compression. As the concrete inside the spiral tends to spread out laterally under the compressive load, the spiral that restrains it is put into hoop tension, and the column will not fail until the spiral yields or breaks, permitting the bursting of the concrete inside. Spiral columns are normally round, but they also can be made into rectangular, octagonal, or other shapes. For such columns, circular arrangements of the bars are still used. Spirals, though adding to the resilience of columns, appreciably increase costs. As a result, they are usually used only for large heavily loaded columns and for columns in seismic areas due to their considerable resistance to earthquake loadings. (In nonseismic zones, probably more than 9 out of 10 existing reinforced concrete columns are tied.) Spirals very effectively increase the ductility and toughness of columns, but they are much more expensive than ties.

Composite columns, illustrated in Figures 9.2(c) and 9.2(d), are concrete columns that are reinforced longitudinally by structural steel shapes, which may or may not be surrounded by structural steel bars, or they may consist of structural steel tubing filled with concrete (commonly called *lally columns*).

Column forms. (Courtesy of Economy Forms Corporation.)

9.3 AXIAL LOAD CAPACITY OF COLUMNS

In actual practice there are no perfect axially loaded columns, but a discussion of such members provides an excellent starting point for explaining the theory involved in designing real columns with their eccentric loads. Several basic ideas can be explained for purely axially loaded columns, and the strengths obtained provide upper theoretical limits that can be clearly verified with actual tests.

It has been known for several decades that the stresses in the concrete and the reinforcing bars of a column supporting a long-term load cannot be calculated with any degree of accuracy. You might think that such stresses could be determined by multiplying the strains by the appropriate moduli of elasticity. But this idea does not work too well practically because the modulus of elasticity of the concrete is changing during loading due to creep and shrinkage. So it can be seen that the parts of the load carried by the concrete and the steel vary with the magnitude and duration of the loads. For instance, the larger the percentage of dead loads and the longer they are applied, the greater the creep in the concrete and the larger the percentage of load carried by the reinforcement.

Though stresses cannot be predicted in columns in the elastic range with any degree of accuracy, several decades of testing have shown that the ultimate strength of columns can be estimated very well. Furthermore, it has been shown that the proportions of live and dead loads, the length of loading, and other such items have little effect on the ultimate strength. It does not even matter whether the concrete or the steel approaches its ultimate strength first. If one of the two materials is stressed close to its ultimate strength, its large deformations will cause the stress to increase quicker in the other material.

For these reasons, only the ultimate strength of columns is considered here. At failure, the theoretical ultimate strength or nominal strength of a short axially loaded column is quite accurately determined by the expression that follows, in which A_g is the gross concrete area and A_{st} is the total cross-sectional area of longitudinal reinforcement, including bars and steel shapes:

$$P_n = 0.85f'_c(A_g - A_{st}) + f_y A_{st}$$

9.4   FAILURE OF TIED AND SPIRAL COLUMNS

Should a short, tied column be loaded until it fails, parts of the shell or covering concrete will spall off and, unless the ties are quite closely spaced, the longitudinal bars will buckle almost immediately, as their lateral support (the covering concrete) is gone. Such failures may often be quite sudden, and apparently they have occurred rather frequently in structures subjected to earthquake loadings.

When spiral columns are loaded to failure, the situation is quite different. The covering concrete or shell will spall off, but the core will continue to stand, and if the spiral is closely spaced, the core will be able to resist an appreciable amount of additional load beyond the load that causes spalling. The closely spaced loops of the spiral together with the longitudinal bars form a cage that very effectively confines the concrete. As a result, the spalling off of the shell of a spiral column provides a warning that failure is going to occur if the load is further increased.

American practice is to neglect any excess capacity after the shell spalls off since it is felt that once the spalling occurs the column will no longer be useful—at least from the viewpoint of the occupants of the building. For this reason the spiral is designed so that it is just a little stronger than the shell that is assumed to spall off. The spalling gives a warning of impending failure, and then the column will take a little more load before it fails. Designing the spiral so that it is just a little stronger than the shell does not increase the column's ultimate strength much, but it does result in a more gradual or ductile failure.

The strength of the shell is given by the following expression, where A_c is the area of the core, which is considered to have a diameter that extends from out to out of the spiral:

$$\text{Shell strength} = 0.85f_c'(A_g - A_c)$$

By considering the estimated hoop tension that is produced in spirals due to the lateral pressure from the core and by tests, it can be shown that spiral steel is at least twice as effective in increasing the ultimate column capacity as is longitudinal steel.[2,3] Therefore, the strength of the spiral can be computed approximately by the following expression, in which ρ_s is the percentage of spiral steel:

$$\text{Spiral strength} = 2\rho_s A_c f_y$$

Equating these expressions and solving for the required percentage of spiral steel, we obtain

$$0.85f_c'(A_g - A_c) = 2\rho_s A_c f_y$$

$$\rho_s = 0.425\frac{(A_g - A_c)f_c'}{A_c f_y} = 0.425\left(\frac{A_g}{A_c} - 1\right)\frac{f_c'}{f_y}$$

To make the spiral a little stronger than the spalled concrete, the Code (10.9.3) specifies the minimum spiral percentage with the expression to follow in which f_y is the specified yield strength of the spiral reinforcement up to 100,000 psi.

$$\rho_s = 0.45\left(\frac{A_g}{A_c} - 1\right)\frac{f_c'}{f_y} \qquad \text{(ACI Equation 10-5)}$$

[2]Park, A., and Paulay, T., 1975, *Reinforced Concrete Structures* (New York: John Wiley & Sons), pp. 25, 119–121.

[3]Considere, A., 1902, "Compressive Resistance of Concrete Steel and Hooped Concrete, Part I," *Engineering Record*, December 20, pp. 581–583; "Part II," December 27, pp. 605–606.

Round spiral columns. (Courtesy of Economy Forms Corporation.)

Once the required percentage of spiral steel is determined, the spiral may be selected with the expression to follow, in which ρ_s is written in terms of the volume of the steel in one loop:

$$
\begin{aligned}
\rho_s &= \frac{\text{volume of spiral in one loop}}{\text{volume of concrete core for a pitch } s} \\
&= \frac{V_{\text{spiral}}}{V_{\text{core}}} \\
&= \frac{a_s \pi (D_c - d_b)}{(\pi D_c^2 / 4) s} = \frac{4 a_s (D_c - d_b)}{s D_c^2}
\end{aligned}
$$

In this expression, D_c is the diameter of the core out to out of the spiral, a_s is the cross-sectional area of the spiral bar, and d_b is the diameter of the spiral bar. Here reference is made to Figure 9.3. The designer can assume a diameter for the spiral bar and solve for the pitch required. If the results do not seem reasonable, he or she can try another diameter. The pitch used must be within the limitations listed in the next section of this chapter. Actually, Table A.14 (see Appendix), which is based on this expression, permits the designer to select spirals directly.

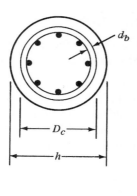

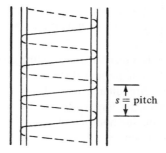

Figure 9.3

Columns and hammerhead cap forms for the Gandy Bridge, Tampa, Florida. (Courtesy of Economy Forms Corporation.)

9.5 CODE REQUIREMENTS FOR CAST-IN-PLACE COLUMNS

The ACI Code specifies quite a few limitations on the dimensions, reinforcing, lateral restraint, and other items pertaining to concrete columns. Some of the most important limitations are as follows.

1. The percentage of longitudinal reinforcement may not be less than 1% of the gross cross-sectional area of a column (ACI Code 10.9.1). It is felt that if the amount of steel is less than 1%, there will be a distinct possibility of a sudden nonductile failure, as might occur in a plain concrete column. The 1% minimum steel value will also lessen creep and shrinkage and provide some bending strength for the column. Actually, the Code (10.8.4) does permit the use of less than 1% steel if the column has been made larger than is necessary to carry the loads because of architectural or other reasons. In other words, a column can be designed with 1% longitudinal steel to support the factored load and then more concrete can be added with no increase in reinforcing and no increase in calculated load-carrying capacity. In actual practice the steel percentage for such members is kept to an absolute minimum of 0.005.

2. The maximum percentage of steel may not be greater than 8% of the gross cross-sectional area of the column (ACI Code 10.9.1). This maximum value is given to prevent too much crowding of the bars. Practically, it is rather difficult to fit more than 4% or 5% steel into the forms and still get the concrete down into the forms and around the bars. When the percentage of steel is high, the chances of having honeycomb in the concrete is decidedly increased. If this happens, there can be a substantial reduction in the column's load-carrying capacity. Usually the percentage of reinforcement should not exceed 4% when the bars are to be lap-spliced. It is to be remembered that if the percentage of steel is very high, the bars may be bundled.

3. The minimum numbers of longitudinal bars permissible for compression members (ACI 10.9.2) are as follows: 4 for bars within rectangular or circular ties, 3 for bars within triangular-shaped ties, and 6 for bars enclosed within spirals. Should there be fewer than 8 bars in a circular arrangement, the orientation of the bars will affect the moment strength of eccentrically loaded columns. This matter should be considered in design according to the ACI Commentary (R10.9.2).

4. The Code does not directly provide a minimum column cross-sectional area, but to provide the necessary cover outside of ties or spirals and to provide the necessary clearance between longitudinal bars from one face of the column to the other, it is obvious that minimum widths or diameters of about 8 to 10 in. are necessary. To use as little rentable floor space as possible, small columns are frequently desirable. In fact, thin columns may often be enclosed or "hidden" in walls.

5. When tied columns are used, the ties shall not be less than #3, provided that the longitudinal bars are #10 or smaller. The minimum size is #4 for longitudinal bars larger than #10 and for bundled bars. Deformed wire or welded wire fabric with an equivalent area may also be used (ACI 7.10.5.1).

   In SI units, ties should not be less than #10 for longitudinal bars #32 or smaller and #13 for larger longitudinal bars.

   The center-to-center spacing of ties shall not be more than 16 times the diameter of the longitudinal bars, 48 times the diameter of the ties, or the least lateral dimension of the column. The ties must be arranged so that every corner and alternate longitudinal bar will have lateral support provided by the corner of a tie having an included angle not greater

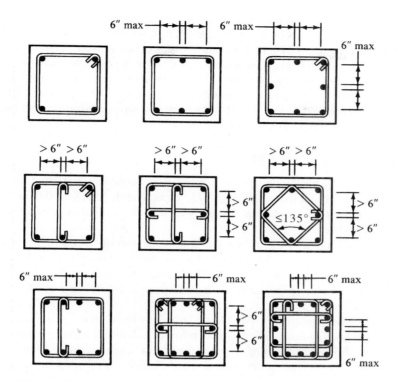

Figure 9.4 Typical tie arrangements.

than 135°. No bars can be located a greater distance than 6 in. clear* on either side from such a laterally supported bar. These requirements are given by the ACI Code in its Section 7.10.5. Figure 9.4 shows tie arrangements for several column cross sections. Some of the arrangements with interior ties, such as the ones shown in the bottom two rows of the figure, are rather expensive. Should longitudinal bars be arranged in a circle, round ties may be placed around them and the bars do not have to be individually tied or restrained otherwise (7.10.5.3). The ACI also states (7.10.3) that the requirements for lateral ties may be waived if tests and structural analysis show that the columns are sufficiently strong without them and that such construction is feasible.

There is little evidence available concerning the behavior of spliced bars and bundled bars. For this reason, Section R7.10.5 of the Commentary states that it is advisable to provide ties at each end of lap-spliced bars and provides recommendations concerning the placing of ties in the region of end-bearing splices and offset bent bars.

Ties should not be placed more than one-half a spacing above the top of a footing or slab and not more than one-half a spacing below the lowest reinforcing in a slab or drop panel (to see a drop panel, refer to Figure 16.1 of this text). Where beams frame into a column from all four directions, the last tie may be below the lowest reinforcing in any of the beams.

6. The Code (7.10.4) states that spirals may not have diameters less than $\frac{3}{8}$ in.** and that the clear spacing between them may not be less than 1 in. or greater than 3 in.† Should splices

*150 mm in SI.

**10 mm in SI.

†25 and 75 mm in SI.

be necessary in spirals, they are to be provided by welding or by lapping deformed uncoated spiral bars or wires by the larger of 48 diameters or 12 in.[*] Other lap splice lengths are also given in ACI Section 7.10.4 for plain uncoated bars and wires, for epoxy-coated deformed bars and wires, and so on. Special spacer bars may be used to hold the spirals in place and at the desired pitch until the concrete hardens. These spacers consist of vertical bars with small hooks. Spirals are supported by the spacers, not by the longitudinal bars. Section R7.10.4 of the ACI Commentary provides suggested numbers of spacers required for different-size columns.

9.6 SAFETY PROVISIONS FOR COLUMNS

The values of ϕ to be used for columns as specified in Section 9.3.2 of the Code are well below those used for flexure and shear (0.90 and 0.75, respectively). A value of 0.65 is specified for tied columns and 0.75 for spiral columns. A slightly larger ϕ is specified for spiral columns because of their greater toughness.

The failure of a column is generally a more severe matter than is the failure of a beam because a column generally supports a larger part of a structure than does a beam. In other words, if a column fails in a building, a larger part of the building will fall down than if a beam fails. This is particularly true for a lower-level column in a multistory building. As a result, lower ϕ values are desirable for columns.

There are other reasons for using lower ϕ values in columns. As an example, it is more difficult to do as good a job in placing the concrete for a column than it is for a beam. The reader can readily see the difficulty of getting concrete down into narrow column forms and between the longitudinal and lateral reinforcing. As a result, the quality of the resulting concrete columns is probably not as good as that of beams and slabs.

The failure strength of a beam is normally dependent on the yield stress of the tensile steel—a property that is quite accurately controlled in the steel mills. On the other hand, the failure strength of a column is closely related to the concrete's ultimate strength, a value that is quite variable. The length factors also drastically affect the strength of columns and thus make the use of lower ϕ factors necessary.

It seems impossible for a column to be perfectly axially loaded. Even if loads could be perfectly centered at one time, they would not stay in place. Furthermore, columns may be initially crooked or have other flaws, with the result that lateral bending will occur. Wind and other lateral loads cause columns to bend, and the columns in rigid-frame buildings are subjected to moments when the frame is supporting gravity loads alone.

9.7 DESIGN FORMULAS

In the pages that follow, the letter e is used to represent the eccentricity of the load. The reader may not understand this term because he or she has analyzed a structure and has computed an axial load P_u and a bending moment M_u, but no specific eccentricity e for a particular column. The term e represents the distance the axial load P_u would have to be off center of the column to produce M_u. Thus

$$P_u e = M_u$$

or

$$e = \frac{M_u}{P_u}$$

---

[*]300 mm in SI.

Royal Towers, Baltimore, Maryland. (Courtesy of Simpson Timber Company.)

Nonetheless, there are many situations where there are no calculated moments for the columns of a structure. For many years the Code specified that such columns had to be designed for certain minimum moments even though no calculated moments were present. This was accomplished by requiring designers to assume certain minimum eccentricities for their column loads. These minimum values were 1 in. or $0.05h$, whichever was larger, for spiral columns and 1 in. or $0.10h$ for tied columns. (The term h represents the outside diameter of round columns or the total depth of square or rectangular columns.) A moment equal to the axial load times the minimum eccentricity was used for design.

In today's Code, minimum eccentricities are not specified, but the same objective is accomplished by requiring that theoretical axial load capacities be multiplied by a factor sometimes called α, which is equal to 0.85 for spiral columns and 0.80 for tied columns. Thus, as shown in Section 10.3.6 of the Code, the axial load capacity of columns may not be greater than the following values:

For spiral columns ($\phi = 0.75$)

$$\phi P_n(\max) = 0.85\phi[0.85f_c'(A_g - A_{st}) + f_y A_{st}] \quad \text{(ACI Equation 10-1)}$$

For tied columns ($\phi = 0.65$)

$$\phi P_n(\max) = 0.80\phi[0.85f_c'(A_g - A_{st}) + f_y A_{st}] \quad \text{(ACI Equation 10-2)}$$

It is to be clearly understood that the preceding expressions are to be used only when the moment is quite small or when there is no calculated moment.

The equations presented here are applicable only for situations where the moment is sufficiently small so that e is less than $0.10h$ for tied columns or less than $0.05h$ for spiral columns. Short columns can be completely designed with these expressions as long as the e values are under the limits described. Should the e values be greater than the limiting values and/or should the columns be classified as long ones, it will be necessary to use the procedures described in the next two chapters.

9.8 COMMENTS ON ECONOMICAL COLUMN DESIGN

Reinforcing bars are quite expensive, and thus the percentage of longitudinal reinforcing used in reinforced concrete columns is a major factor in their total costs. This means that under normal circumstances a small percentage of steel should be used (perhaps in the range of 1.5% to 3%). This can be accomplished by using larger column sizes and/or higher strength concretes. Furthermore, if the percentage of bars is kept in approximately this range, it will be found that there will be sufficient room for conveniently placing them in the columns.

Higher strength concretes can be used more economically in columns than in beams. Under ordinary loads, only 30% to 40% of a beam cross section is in compression, while the remaining 60% to 70% is in tension and thus assumed to be cracked. This means that if a high-strength concrete is used for a beam, 60% to 70% of it is wasted. For the usual column, however, the situation is quite different because a much larger percentage of its cross section is in compression. As a result, it is quite economical to use high-strength concretes for columns. Although some designers have used concretes with ultimate strengths as high as 19,000 psi (as at Two Union Square in Seattle) for column design with apparent economy, the use of 5000- to 6000-psi columns is the normal rule when higher strengths are specified for columns.

Grade 60 reinforcing bars are generally used for best economy in the columns of most structures. However, Grade 75 bars may provide better economy in high-rise structures particularly when they are used in combination with higher strength concretes.

In general, tied columns are more economical than spiral columns, particularly if square or rectangular cross sections are to be used. Of course, spiral columns, high-strength concretes, and high percentages of steel save floor space.

As few different column sizes as possible should be used throughout a building. In this regard, it is completely uneconomical to vary a column size from floor to floor to satisfy the different loads it must support. This means that the designer may select a column size for the top floor of a multistory building (using as small a percentage of steel as possible) and then continue to use that same size vertically for as many stories as possible by increasing the steel percentage floor by floor as required. Furthermore, it is desirable to use the same column size as much as possible on each floor level. This consistency of sizes will provide appreciable savings in labor costs.

The usual practice for the columns of multistory reinforced concrete buildings is to use one-story-length vertical bars tied together in preassembled cages. This is the preferred procedure when the bars are #11[*] or smaller, where all the bars can be spliced at one location just above the floor line. For columns where staggered splice locations are required (as for larger size bars), the number of splices can be reduced by using preassembled two-story cages of reinforcing.

Unless the least column dimensions or longitudinal bar diameters control tie spacings, the selection of the largest practical tie sizes will increase their spacings and reduce their number. This can result in some savings. Money can also be saved by avoiding interior ties such as the ones shown in the bottom two rows of columns in Figure 9.4. With no interior ties, the concrete can be placed more easily and lower slumps used (thus lower cost concrete).

In fairly short buildings, the floor slabs are often rather thin, and thus deflections may be a problem. As a result, rather short spans and thus close column spacings may be used. As buildings become taller, the floor slabs will probably be thicker to help provide lateral stability. For such buildings, slab deflections will not be as much of a problem, and the columns may be spaced farther apart.

Even though the columns in tall buildings may be spaced at fairly large intervals, they still will occupy expensive floor space. For this reason, many designers try to place many of their columns

*#36 in SI.

on the building perimeters so they will not use up the valuable interior space. In addition, the omission of interior columns provides more flexibility for the users for placement of partitions and also makes large open spaces available.

9.9   DESIGN OF AXIALLY LOADED COLUMNS

As a brief introduction to columns, the design of three axially loaded short columns is presented in this section and the next. Moment and length effects are completely neglected. Examples 9.1 and 9.3 present the design of axially loaded square tied columns, while Example 9.2 illustrates the design of a similarly loaded round spiral column. Table A.15 of the Appendix provides several properties for circular columns that are particularly useful for designing round columns.

EXAMPLE 9.1

Design a square tied column to support an axial dead load D of 130 K and an axial live load L of 180 k. Initially assume that 2% longitudinal steel is desired, $f'_c = 4000$ psi, and $f_y = 60,000$ psi.

SOLUTION

$$P_u = (1.2)(130) + (1.6)(180) = 444\,\text{k}$$

Selecting Column Dimensions

$$\phi P_n = \phi 0.80[0.85 f'_c (A_g - A_{st}) + f_y A_{st}]$$

$$444 = (0.65)(0.80)[(0.85)(4)(A_g - 0.02A_g) + (60)(0.02A_g)] \qquad \text{(ACI Equation 10-2)}$$

$$A_g = 188.40\,\text{in.}^2 \quad \underline{\underline{\text{Use } 14 \times 14 (A_g = 196\,\text{in.}^2)}}$$

Selecting Longitudinal Bars

Substituting into column equation with known A_g and solving for A_{st}, we obtain

$$444 = (0.65)(0.80)[(0.85)(4)(196 - A_{st}) + 60A_{st}]$$

$$A_{st} = 3.31\,\text{in.}^2 \quad \underline{\underline{\text{Use 6 \#7 bars } (3.61\,\text{in.}^2)}}$$

Design of Ties (Assuming #3 Bars)

Spacing: **(a)** $48 \times \frac{3}{8} = 18''$

(b) $16 \times \frac{7}{8} = 14'' \leftarrow$

(c) Least dim. $= 14'' \leftarrow$ $\underline{\underline{\text{Use \#3 ties at } 14''}}$

A sketch of the column cross section is shown in Figure 9.5.

Check Code Requirements

Listed below are the ACI Code limitations for columns. Space is not taken in future examples to show all of these essential checks, but they must be made.

(7.6.1)   Longitudinal bar clear spaceing $= \frac{9}{2} - \frac{7}{8} = 3.625\,\text{in.} > 1$ in. and d_b of $\frac{7}{8}$ in. $\underline{\underline{\text{OK}}}$

(10.9.1)   Steel percentage $0.01 < \rho = \frac{3.61}{(14)(14)} = 0.0184 < 0.08$ $\underline{\underline{\text{OK}}}$

(10.9.2)   Number of bars $= 6 >$ min. no. of 4 $\underline{\underline{\text{OK}}}$

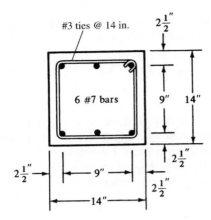

Figure 9.5

(7.10.5.1) Minimum tie size = #3 for #7 bars   <u>OK</u>

(7.10.5.2) Spacing of ties   <u>OK</u>

(7.10.5.3) Arrangement of ties   <u>OK</u>

EXAMPLE 9.2

Design a round spiral column to support an axial dead load P_D of 240 and an axial live load P_L of 300 k. Initially assume that approximately 2% longitudinal steel is desired, $f_c' = 4000$ psi, and $f_y = 60,000$ psi.

SOLUTION

$$P_u = (1.2)(240) + (1.6)(300) = 768$$

Selecting Column Dimensions and Bar Sizes

$$\phi P_n = \phi 0.85[0.85 f_c'(A_g - A_{st}) + f_y A_{st}] \qquad \text{(ACI Equation 10-1)}$$

$$768 = (0.75)(0.85)[(0.85)(4)(A_g - 0.02A_g) + (60)(0.02A_g)]$$

$$A_g = 266 \text{ in.}^2 \qquad \underline{\underline{\text{Use } 18'' \text{ diameter column } (255 \text{ in.}^2)}}$$

$$768 = (0.75)(0.85)[(0.85)(4)(255 - A_{st}) + 60A_{st}]$$

$$A_{st} = 5.97 \text{ in.}^2 \qquad \underline{\underline{\text{Use 6 \#9 bar } (6.00 \text{ in.}^2)}}$$

Check Code requirements as in Example 9.1. A sketch of the column cross section is shown in Figure 9.6.

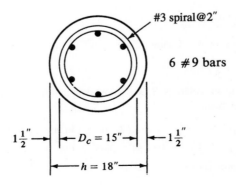

Figure 9.6

Design of Spiral

$$A_c = \frac{(\pi)(15)^2}{4} = 177 \text{ in.}^2$$

$$\text{Minimum } \rho_s = (0.45)\left(\frac{A_g}{A_c} - 1\right)\frac{f_c'}{f_y} = (0.45)\left(\frac{255}{177} - 1\right)\left(\frac{4}{60}\right) = 0.0132$$

Assume a #3 spiral, $d_b = 0.375$ in. and $a_s = 0.11$ in.^{2}.

$$\rho_s = \frac{4a_s(D_c - d_b)}{sD_c^2}$$

$$0.0132 = \frac{(4)(0.11)(15 - 0.375)}{(s)(15)^2}$$

$$s = 2.17'' \quad \underline{\underline{\text{say } 2''}}$$

(Checked with Appendix Table A.14.)

9.10   SI EXAMPLE

EXAMPLE 9.3

Design an axially loaded short square tied column for $P_u = 2600$ kN if $f_c' = 28$ MPa and $f_y = 350$ MPa. Initially assume $\rho = 0.02$.

SOLUTION   **Selecting Column Dimensions**

$$\phi P_n = \phi 0.80[0.85f_c'(A_g - A_{st}) + f_y A_{st}] \qquad \text{(ACI Equation 10-2)}$$

$$2600 \times 10^3 = (0.65)(0.80)[(0.85)(28)(A_g - 0.02A_g) + (350)(0.02A_g)]$$

$$A_g = 164\,886 \text{ mm}^2$$

$$\underline{\underline{\text{Use 400 mm} \times \text{400 mm } (A_g = 160\,000 \text{ mm}^2)}}$$

Selecting Longitudinal Bars

$$2600 \times 10^3 = (0.65)(0.80)[(0.85)(28)(160\,000 - A_{st}) + 350A_{st}]$$

$$A_{st} = 3654 \text{ mm}^2$$

$$\underline{\underline{\text{Use 6 #29 } (3870 \text{ mm}^2)}}$$

Design of Ties (Assuming #10 SI ties)

(a) $16 \times 28.7 = 459.2$ mm

(b) $48 \times 9.5 = 456$ mm

(c) Least col. dim. $= 400$ mm $\leftarrow$ <u>Use #10 ties @ 400 mm</u>

Check Code requirements as in Example 9.1. A sketch of the column cross section is shown in Figure 9.7.

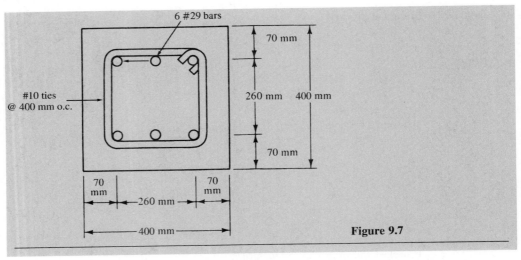

6 #29 bars

70 mm

260 mm   400 mm

70 mm

#10 ties
@ 400 mm o.c.

70
mm

70
mm

260 mm

400 mm

Figure 9.7

9.11 COMPUTER EXAMPLE

EXAMPLE 9.4

Using the Excel spreadsheets for Chapters 9 and 10, repeat Example 9.2.

SOLUTION Open the Circular Column worksheet and enter the material properties ($f'_c = 4000$ psi, $f_y = 60,000$ psi). For γ any value less than 1 is acceptable for Chapter 8 problems with no moment or eccentricity. Enter a trial value of h (cell C4) and A_{st} (cell C8). The corresponding axial load capacity will appear in cell D19 which is identified as ϕP_0. If this value is greater than or equal to 768 kips, the design is acceptable. It is a more economical design if the capacity is also close to the design value of 768 kips. As an example, start with $h = 10$ in. and $A_{st} = 1.00$ in.^{2}. The value of ϕP_0 is only 206 kips. Obviously a larger column is needed. Keep increasing h until the ϕP_0 value is close to 768 kips, keeping in mind that the value of A_{st} is still set very low. Several iterations shows that for $h = 18$ in., $\phi P_0 = 588$ kips. Now begin incrementing A_{st} and see its effect on ϕP_0. Several trials leads to $A_{st} = 6.00$ in.^{2} with a corresponding value of $\phi P_0 = 768$ kips.

It is also possible to use "goal seek" to solve this problem. Input a trial value for h (say 10 in.). Then highlight cell D19 and select goal seek from tools on the Excel toolbar. Input 768 in. the second window and

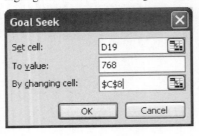

C8 in the bottom one (as shown). Click OK, and a value of $A_{st} = 16.57$ in.^{2} will appear in cell C8. This is way too much steel because the steel percentage exceeds 8%. Clearly a larger diameter column is needed. Repeat this process, increasing h until an acceptable value of A_{st} is obtained. If $h = 16$ in. is input, Goal seek indicates $A_{st} = 9.21$ in.^{2}. This may not be the best choice, but it shows how the spreadsheet can be used to get different answers, all of which may be acceptable.

Circular Column Capacity

h =	16	in.
γ =	0.7	
f'_c=	4000	psi
f_y =	60000	psi
A_{st} =	9.21	in.^{2}
A_g =	201.1	in.^{2}
ρ_t =	0.0458	
β_1 =	0.85	
ε_y =	0.00207	
E_s =	29000	ksi
c_{bal} =	8.05	in.
$c_{.005}$ =	5.1	in.

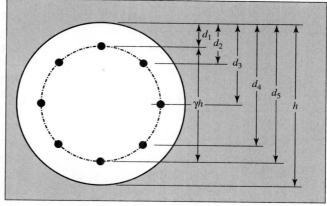

$$P_o = (0.85f'_c A_n + A_s f_y) = 1204.7 \text{ kips}$$
$$\phi P_o = 768.0 \text{ kips} \qquad \text{ACI Equation 10-1}$$

PROBLEMS

9.1 Distinguish between tied, spiral, and composite columns.

9.2 What are primary and secondary moments?

9.3 Distinguish between long and short columns.

9.4 List several design practices that may help make the construction of reinforced concrete columns more economical.

Analysis of Axially Loaded Columns

For Problems 9.5 to 9.8 compute the load-bearing capacity of the concentrically loaded short columns. $f_y = 60,000$ psi and $f'_c = 4000$ psi.

9.5 A 24 in. square column reinforced with 12 #10 bars. (*Ans.* 1465.4 k)

Problem 9.6

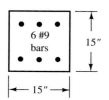

6 #9 bars

15″

15″

Problem 9.7 (*Ans.* 566.7 k)

8 #8 bars

12″

18″

Problem 9.8

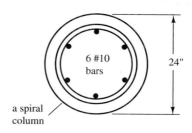

6 #10 bars

a spiral column

24″

Problem 9.9 Determine the load-bearing capacity of the concentrically loaded short column shown if $f_y = 60{,}000$ psi and $f'_c = 3000$ psi. (*Ans.* 1181.8 k)

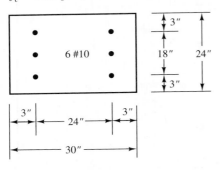

Design of Axially Loaded Columns

In Problems 9.10 to 9.15 design columns for axial load only. Include the design of ties or spirals and a sketch of the cross sections selected, including bar arrangements. All columns are assumed to be short and form sizes are available in 2-in. increments.

9.10 Square tied column: $P_D = 300$ k, $P_L = 500$ k, $f'_c = 4000$ psi, and $f_y = 60{,}000$ psi. Initially assume $\rho_g = 2\%$.

9.11 Repeat Problem 9.10 if ρ_g is to be 3% initially. (*One ans.* 22 in. × 22 in. column with 10 #10 bars)

9.12 Round spiral column; $P_D = 300$ k, $P_L = 400$ k, $f'_c = 3500$ psi, and $f_y = 60{,}000$ psi. Initially assume $\rho_g = 4\%$.

9.13 Round spiral column; $P_D = 400$ k, $P_L = 250$ k, $f'_c = 4000$ psi, $f_y = 60{,}000$ psi, and p_g initially assumed $= 2\%$. (*One ans.* 20 in. diameter column with 6 #9 bars)

9.14 Smallest possible square tied column; $P_D = 200$ k, $P_L = 300$ k, $f'_c = 4000$ psi, and $f_y = 60{,}000$ psi.

9.15 Design a rectangular tied column with the long side equal to two times the length of the short side. $P_D = 700$, $P_L = 400$ k, $f'_c = 3000$ psi, $f_y = 60{,}000$ psi, and initially assume that $p_g = 2\%$. (*One ans.* 20 in. × 40 in. column with 9 #11 bars)

Problems in SI Units

In Problems 9.16 to 9.18 design columns for axial load only for the conditions described. Include the design of ties or spirals and a sketch of the cross sections selected, including bar arrangements. All columns are assumed to be short and not exposed to the weather. Form sizes are in 50-mm increments.

9.16 Square tied column; $P_D = 600$ kN, $P_L = 800$ kN, $f'_c = 24$ MPa, and $f_y = 420$ MPa. Initially assume $\rho_g = 0.02$.

9.17 Smallest possible square tied column; $P_D = 700$ kN, $P_L = 300$ kN, $f'_c = 28$ MPa, and $f_y = 300$ MPa. (*One ans.* 250 mm × 250 mm column with six #29 bars)

9.18 Round spiral column; $P_D = 500$ kN, $P_L = 650$ kN, $f'_c = 35$ MPa, and $f_y = 420$ MPa. Initially assume $\rho_g = 0.03$.

For problem 9.19 to 9.21 use the Excel spreadsheets for Chapters 9 and 10. Assumed $d' = 2.5$ in. for each column.

9.19 Repeat Problem 9.6 (*Ans.* $\phi P_n = 574.4$ k)

9.20 Repeat Problem 9.10.

9.21 Repeat Problem 9.12. (*One ans.* 20 in. diameter column with 8 #11 bars for which $\phi P_n = 1050$ k)

10

Design of Short Columns Subject to Axial Load and Bending

10.1 AXIAL LOAD AND BENDING

All columns are subjected to some bending as well as axial forces, and they need to be proportioned to resist both. The so-called axial load formulas presented in Chapter 9 do take into account some moments because they include the effect of small eccentricities with the 0.80 and 0.85 factors. These values are approximately equivalent to the assumption of actual eccentricities of $0.10h$ for tied columns and $0.05h$ for spiral columns.

Columns will bend under the action of moments, and those moments will tend to produce compression on one side of the columns and tension on the other. Depending on the relative magnitudes of the moments and axial loads, there are several ways in which the sections might fail. Figure 10.1 shows a column supporting a load P_n. In the various parts of the figure the load is placed at greater and greater eccentricities (thus producing larger and larger moments) until finally in part (f) the column is subject to such a large bending moment that the effect of the axial load is negligible. Each of the six cases shown is briefly discussed in the paragraphs to follow, where the letters (a) through (f) correspond to those same letters in the figure. The column is assumed to reach its ultimate capacity when the compressive concrete strain reaches 0.003.

(a) *Large axial load with negligible moment:* For this situation, failure will occur by the crushing of the concrete, with all reinforcing bars in the column having reached their yield stress in compression.

(b) *Large axial load and small moment such that the entire cross section is in compression:* When a column is subject to a small bending moment (that is, when the eccentricity is small), the entire column will be in compression, but the compression will be higher on one side than on the other. The maximum compressive stress in the column will be $0.85f_c'$, and failure will occur by the crushing of the concrete with all the bars in compression.

(c) *Eccentricity larger than in case (b) such that tension begins to develop on one side of the column:* If the eccentricity is increased somewhat from the preceding case, tension will begin to develop on one side of the column and the steel on that side will be in tension but less than the yield stress. On the other side the steel will be in compression. Failure will occur by crushing of the concrete on the compression side.

(d) *A balanced loading condition:* As we continue to increase the eccentricity, a condition will be reached at which the reinforcing bars on the tension side will reach their yield stress at the same time that the concrete on the opposite side reaches its maximum compression $0.85f_c'$. This situation is called the *balanced loading condition*.

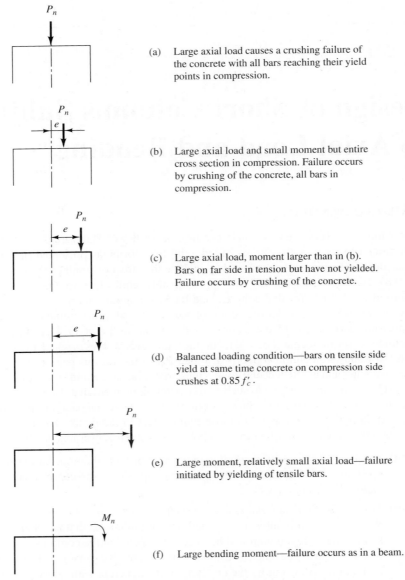

Figure 10.1 Column subject to load with larger and larger eccentricities.

(e) *Large moment with small axial load:* If the eccentricity is further increased, failure will be initiated by the yielding of the bars on the tensile side of the column prior to concrete crushing.

(f) *Large moment with no appreciable axial load:* For this condition, failure will occur as it does in a beam.

10.2 THE PLASTIC CENTROID

The eccentricity of a column load is the distance from the load to the *plastic centroid* of the column. The plastic centroid represents the location of the resultant force produced by the steel and the concrete. It is the point in the column cross section through which the resultant column

Pennsylvania Southern Expressway,
Philadelphia, Pennsylvania.
(Courtesy of Economy Forms
Corporation.)

load must pass to produce uniform strain at failure. For locating the plastic centroid, all concrete is assumed to be stressed in compression to $0.85f_c'$ and all steel to f_y in compression. For symmetrical sections, the plastic centroid coincides with the centroid of the column cross section, while for nonsymmetrical sections it can be located by taking moments.

Example 10.1 illustrates the calculations involved in locating the plastic centroid for a nonsymmetrical cross section. The ultimate load P_n is determined by computing the total compressive forces in the concrete and the steel and adding them together. Then P_n is assumed to act downward at the plastic centroid at a distance $\bar{x}$ from one side of the column, and moments are taken on that side of the column of the upward compression forces acting at their centroids and the downward P_n.

EXAMPLE 10.1

Determine the plastic centroid of the T-shaped column shown in Figure 10.2 if $f_c' = 4000$ psi and $f_y = 60{,}000$ psi.

SOLUTION   The plastic centroid falls on the x-axis as shown in Figure 10.2 due to symmetry. The column is divided into two rectangles, the left one being $16'' \times 6''$ and the right one $8'' \times 8''$. C_1 is assumed to be the total compression in the left concrete rectangle, C_2 the total compression in the right rectangle, and C_s' the total compression in the reinforcing bars.

$$C_1 = (16)(6)(0.85)(4) = 326.4\text{ k}$$
$$C_2 = (8)(8)(0.85)(4) = 217.6\text{ k}$$

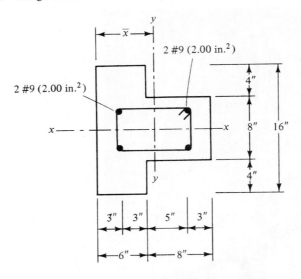

Figure 10.2

In computing C'_s, the concrete where the bars are located is subtracted; that is,

$$C'_s = (4.00)(60 - 0.85 \times 4) = 226.4\,\text{k}$$
$$\text{Total compression} = P_n = 326.4 + 217.6 + 226.4 = 770.4\,\text{k}$$

Taking Moments about Left Edge of Column

$$-(326.4)(3) - (217.6)(10) - (226.4)(7) + (770.4)(\bar{x}) = 0$$

$$\underline{\underline{x = 6.15''}}$$

10.3 DEVELOPMENT OF INTERACTION DIAGRAMS

Should an axial compressive load be applied to a short concrete member, it will be subjected to a uniform strain or shortening, as is shown in Figure 10.3(a). If a moment with zero axial load is applied to the same member, the result will be bending about the member's neutral axis such that strain is proportional to the distance from the neutral axis. This linear strain variation is shown in Figure 10.3(b). Should axial load and moment be applied at the same time, the resulting strain diagram will be a combination of two linear diagrams and will itself be linear, as illustrated in Figure 10.3(c). As a result of this linearity, we can assume certain numerical values of strain in one part of a column and determine strains at other locations by straight-line interpolation.

As the axial load applied to a column is changed, the moment that the column can resist will change. This section shows how an interaction curve of nominal axial load and moment values can be developed for a particular column.

Assuming the concrete on the compression edge of the column will fail at a strain of 0.003, a strain can be assumed on the far edge of the column and the values of P_n and M_n can be computed by statics. Then holding the compression strain at 0.003 on the far edge, we can assume a series of different strains on the other edge and calculate P_n and M_n for each.[1] Eventually a sufficient number of values will be obtained to plot an interaction curve such as the one shown in Figure 10.8. Example 10.2 illustrates the calculation of P_n and M_n for a column for one set of assumed strains.

---

[1]Leet, K., 1991, *Reinforced Concrete Design*, 2nd ed. (New York: McGraw-Hill), pp. 316–317.

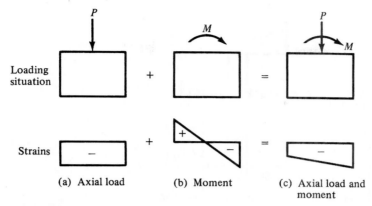

Figure 10.3 Column strains.

EXAMPLE 10.2

It is assumed that the column of Figure 10.4 has a strain on its compression edge equal to -0.00300 and has a tensile strain of $+0.00200$ on its other edge. Determine the values of P_n and M_n that cause this strain distribution if $f_y = 60$ ksi and $f'_c = 4$ ksi.

SOLUTION Determine the values of c and of the steel strains ϵ'_s, and ϵ_s by proportions with reference to the strain diagram shown in Figure 10.5.

$$c = \left(\frac{0.00300}{0.00300 + 0.00200}\right)(24) = 14.40 \text{ in.}$$

$$\epsilon'_s = \left(\frac{11.90}{14.40}\right)(0.00300) = 0.00248 > 0.00207 \quad \underline{\therefore \text{ yields}}$$

$$\epsilon_s = \left(\frac{7.10}{9.60}\right)(0.00200) = 0.00148 \quad \underline{\text{does not yield}}$$

In the following calculations, C_c is the total compression in the concrete, C'_s is the total compression in the compression steel, and T_s is the total tension in the tensile steel. Each of these values is computed below.

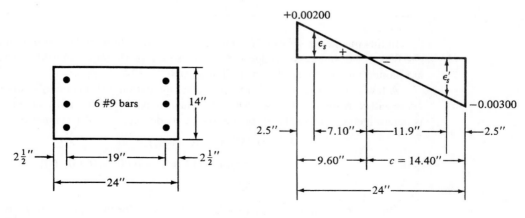

Figure 10.4   **Figure 10.5**

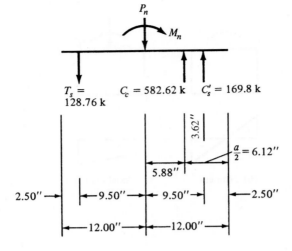

Figure 10.6

The reader should note that C'_s is reduced by $0.85f'_cA'_s$ to account for concrete displaced by the steel in compression.

$$a = (0.85)(14.40) = 12.24 \text{ in.}$$

$$C_c = (0.85)(12.24)(14)(4.0) = -582.62 \text{ k}$$

$$C'_s = (60)(3.0) - (0.85)(3.0)(4.0) = -169.8 \text{ k}$$

$$T_s = (0.00148)(29{,}000)(3.0) = +128.76 \text{ k}$$

By statics, P_n and M_n are determined with reference made to Figure 10.6, where the values of C_c, C'_s, and T_s are shown.

$\Sigma V = 0$

$$-P_n + 169.8 + 582.62 - 128.76 = 0$$

$$P_n = 623.7 \text{ k}$$

$\Sigma M = 0$ **about tensile steel**

$$(623.7)(9.50) + M_n - (582.62)(15.38) - (169.8)(19.00) = 0$$

$$M_n = 6261.3 \text{ in.-k} = 521.8 \text{ ft-k}$$

In this manner, a series of P_n and M_n values are determined to correspond with a strain of -0.003 on the compression edge and varying strains on the far column edge. The resulting values are plotted on a curve as shown in Figure 10.8.

A few remarks are made here concerning the extreme points on this curve. One end of the curve will correspond to the case where P_n is at its maximum value and M_n is zero. For this case, P_n is determined as in Chapter 9 for the axially loaded column of Example 10.2.

$$P_n = 0.85f'_c(A_g - A_s) + A_sf_y$$

$$= (0.85)(4.0)(14 \times 24 - 6.00) + (6.00)(60)$$

$$= 1482 \text{ k}$$

On the other end of the curve, M_n is determined for the case where P_n is zero. This is the procedure used for a doubly reinforced member as previously described in Chapter 5. For the column of Example 10.2, M_n is equal to 297 ft-k.

Washington Redskins Stadium. (Courtesy of EFCO.)

A column reaches its ultimate capacity when the concrete reaches a compressive strain of 0.003. If the steel closest to the extreme tension side of the column reaches yield strain or even more when the concrete reaches a strain of 0.003, the column is said to be tension controlled; otherwise, it is compression controlled. The transition point between these regions is the balance point. In Chapter 3 the term *balanced section* was used in referring to a section whose compression concrete strain reached 0.003 at the same time as the tensile steel reached its yield strain at f_y/E_s. In a beam, this situation theoretically occurred when the steel percentage equaled ρ_b. A column can undergo a balanced failure no matter how much steel it has if it has the right combination of moment and axial load.

For columns the definition of balanced loading is the same as it was for beams—that is, a column that has a strain of 0.003 on its compression side at the same time that its tensile steel on the other side has a strain of f_y/E_s. Although it is easily possible to prevent a balanced condition in beams by requiring that tensile steel strains be kept well above f_y/E_s such is not the case for columns. Thus for columns it is not possible to prevent sudden compression failures or balanced failures. For every column there is a balanced loading situation where an ultimate load P_{bn} placed at an eccentricity e_b will produce a moment M_{bn}, at which time the balanced strains will simultaneously be reached.

At the balanced condition we have a strain of -0.003 on the compression edge of the column and a strain of $f_y/29 \times 10^3 = 60/29 \times 10^3 = 0.00207$ in the tensile steel. This information is shown in Figure 10.7. The same procedure used in Example 10.2 is used to find $P_n = 504.4$ k and $M_n = 559.7$ ft-k.

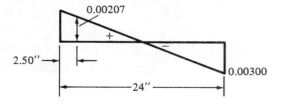

Figure 10.7

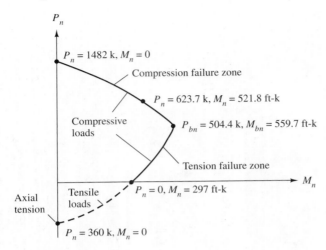

Figure 10.8 Interaction curve for the column of Figure 10.4. Notice these are nominal values.

The curve for P_n and M_n for a particular column may be extended into the range where P_n becomes a tensile load. We can proceed in exactly the same fashion as we did when P_n was compressive. A set of strains can be assumed, and the usual statics equations can be written and solved for P_n and M_n. Several different sets of strains were assumed for the column of Figure 10.4, and then the values of P_n and M_n were determined. The results were plotted at the bottom of Figure 10.8 and were connected with the dashed line which is labeled "tensile loads."

Because axial tension and bending are not very common for reinforced concrete columns, the tensile load part of the curves is not shown in subsequent figures in this chapter. You will note that the largest tensile value of P_n will occur when the moment is zero. For that situation, all of the column steel has yielded, and all of the concrete has cracked. Thus P_n will equal the total steel area A_s times the yield stress. For the column of Figure 10.4

$$P_n = A_s f_y = (6.0)(60) = 360 \text{ k}$$

Round columns. (Courtesy of Economy Forms Corporation.)

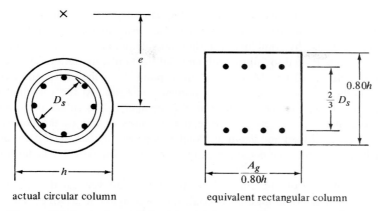

actual circular column   equivalent rectangular column

Figure 10.9 Replacing circular column with equivalent rectangular one.

On some occasions, members subject to axial load and bending have unsymmetrical arrangements of reinforcing. Should this be the case, you must remember that eccentricity is correctly measured from the plastic centroid of the section.

In this chapter P_n values were obtained only for rectangular tied columns. The same theory could be used for round columns, but the mathematics would be somewhat complicated because of the circular layout of the bars, and the calculations of distances would be rather tedious. Several approximate methods have been developed that greatly simplify the mathematics. Perhaps the best known of these is the one proposed by Charles Whitney, in which equivalent rectangular columns are used to replace the circular ones.[2] This method gives results that correspond quite closely with test results.

In Whitney's method, the area of the equivalent column is made equal to the area of the actual circular column, and its depth in the direction of bending is 0.80 times the outside diameter of the real column. One-half the steel is assumed to be placed on one side of the equivalent column and one-half on the other. The distance between these two areas of steel is assumed to equal two-thirds of the diameter (D_s) of a circle passing through the center of the bars in the real column. These values are illustrated in Figure 10.9. Once the equivalent column is established, the calculations for P_n and M_n are made as for rectangular columns.

10.4   USE OF INTERACTION DIAGRAMS

We have seen that by statics the values of P_n and M_n for a given column with a certain set of strains can easily be determined. Preparing an interaction curve with a hand calculator for just one column, however, is quite tedious. Imagine the work involved in a design situation where various sizes, concrete strengths, and steel percentages need to be considered. Consequently, designers resort almost completely to computer programs, computer-generated interaction diagrams, or tables for their column calculations. The remainder of this chapter is concerned primarily with computer-generated interaction diagrams such as the one in Figure 10.10. As we have seen, such a diagram is drawn for a column as the load changes from one of a pure axial nature through varying combinations of axial loads and moments and on to a pure bending situation.

[2]Whitney, Charles S., 1942, "Plastic Theory of Reinforced Concrete Design," *Transactions ASCE*, 107, pp. 251–326.

Interaction diagrams are obviously useful for studying the strengths of columns with varying proportions of loads and moments. Any combination of loading that falls inside the curve is satisfactory, whereas any combination falling outside the curve represents failure.

If a column is loaded to failure with an axial load only, the failure will occur at point *A* on the diagram (Figure 10.10). Moving out from point *A* on the curve, the axial load capacity decreases as the proportion of bending moment increases. At the very bottom of the curve, point *C* represents the bending strength of the member if it is subjected to moment only with no axial load present. In between the extreme points *A* and *C*, the column fails due to a combination of axial load and bending. Point *B* is called the *balanced point* and represents the balanced loading case, where theoretically a compression failure and tensile yielding occur simultaneously.

Refer to point *D* on the curve. The horizontal and vertical dashed lines to this point indicate a particular combination of axial load and moment at which the column will fail. Should a radial line be drawn from point 0 to the interaction curve at any point (as to *D* in this case), it will represent a constant eccentricity of load, that is, a constant ratio of moment to axial load.

You may be somewhat puzzled by the shape of the lower part of the curve from *B* to *C*, where bending predominates. From *A* to *B* on the curve the moment capacity of a section increases as the axial load decreases, but just the opposite occurs from *B* to *C*. A little thought on this point, however, shows that the result is quite logical after all. The part of the curve from *B* to *C* represents

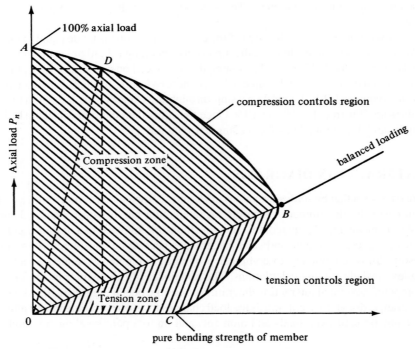

Figure 10.10 Column interaction diagram.

Massive reinforced concrete columns. (Courtesy of Bethlehem Steel Corporation.)

the range of tensile failures. Any axial compressive load in that range tends to reduce the stresses in the tensile bars, with the result that a larger moment can be resisted.

In Figure 10.11 an interaction curve is drawn for the $14''$ by $24''$ column with six #9 bars considered in Section 10.3. If eight #9 bars had been used in the same dimension column, another curve could be generated as shown in the figure; if ten #9 bars were used, still another curve would result. The shape of the new diagrams would be the same as for the six #9 curve, but the values of P_n and M_n would be larger.

10.5   CODE MODIFICATIONS OF COLUMN INTERACTION DIAGRAMS

If interaction curves for P_n and M_n values were prepared, they would be of the types shown in Figures 10.10 and 10.11. To use such curves to obtain design values, they would have to have three modifications made to them as specified in the Code. These modifications are as follows:

(a) The Code 9.3.2 specifies strength reduction or ϕ factors (0.65 for tied columns and 0.75 for spiral columns) that must be multiplied by P_n values. If a P_n curve for a particular column were multiplied by ϕ, the result would be a curve something like the ones shown in Figure 10.12.

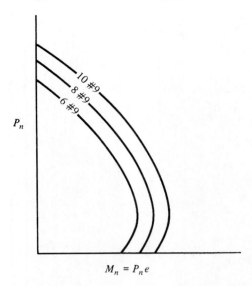

$$M_n = P_n e$$

Figure 10.11 Interaction curves for a rectangular column with different sets of reinforcing bars.

(b) The second modification also refers to ϕ factors. The Code specifies values of 0.65 and 0.75 for tied and spiral columns, respectively. Should a column have quite a large moment and a very small axial load so that it falls on the lower part of the curve between points B and C (see Figure 10.10), the use of these small ϕ values may be a little unreasonable. For instance, for a member in pure bending (point C on the same curve) the specified ϕ is 0.90, but if the same member has a very small axial load added, ϕ would immediately fall to 0.65 or 0.75. Therefore, the Code (9.3.2.2) states that when members subject to axial load and bending have net tensile strains (ϵ_t) between the limits for compression-controlled and tensile-controlled sections, they fall in the transition zone for ϕ. In this zone it is permissible to increase ϕ linearly from 0.65 or 0.75 to 0.90 as ϵ_t increases from the compression-controlled limit to 0.005. In this regard, the reader is again referred to Figure 3.5 in this text where the transition zone and the variation of ϕ values are clearly shown. This topic is continued in Section 10.9.

(c) As described in Chapter 9, maximum permissible column loads were specified for columns no matter how small their e values. As a result, the upper part of each design

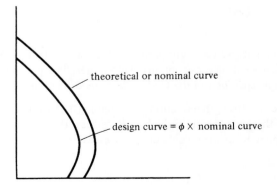

theoretical or nominal curve

design curve = $\phi \times$ nominal curve

Figure 10.12 Curves for P_n and ϕP_n for a single column.

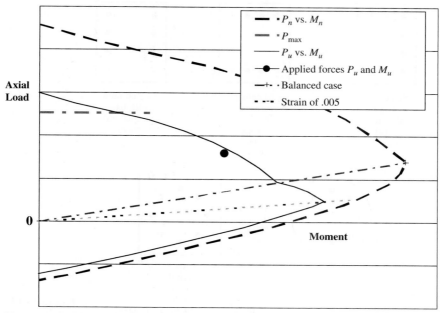

Figure 10.13 A column interaction curve adjusted for the three modifications described in this section (10.5).

interaction curve is shown as a horizontal line representing the appropriate value of

$$P_u = \phi P_{n\,max} \text{ for tied columns} = 0.80\phi[0.85f_c'(A_g - A_{st}) + f_y A_{st}] \quad \textbf{(ACI Equation 10-2)}$$

$$P_u = \phi P_{n\,max} \text{ for spiral columns} = 0.85\phi[0.85f_c'(A_g - A_{st}) + f_y A_{st}] \quad \textbf{(ACI Equation 10-1)}$$

These formulas were developed to be approximately equivalent to loads applied with eccentricities of $0.10h$ for tied columns and $0.05h$ for spiral columns.

Each of the three modifications described here is indicated on the design curve of Figure 10.13. In Figure 10.13, the solid curved line represents P_u and M_u, whereas the dashed curved line is P_n and M_n. The difference between the two curves is the ϕ factor. The two curves would have the same shape if the ϕ factor did not vary. Above the radial line labeled "balanced case," $\phi = 0.65$ (0.75 for spirals). Below the other radial line labeled "strain of 0.005," $\phi = 0.9$. It varies between the two values in between, and the P_u vs. M_u curve assumes a different shape.

10.6   DESIGN AND ANALYSIS OF ECCENTRICALLY LOADED COLUMNS USING INTERACTION DIAGRAMS

If individual column interaction diagrams were prepared as described in the preceding sections, it would be necessary to have a diagram for each different column cross section, for each different set of concrete and steel grades, and for each different bar arrangement. The result would be an astronomical number of diagrams. The number can be tremendously reduced, however, if the diagrams are plotted with ordinates of $K_n = P_n/f_c'A_g$ (instead of P_n) and with abscissas of $R_n = P_n e/f_c'A_g h$ (instead of M_n). The resulting normalized interaction diagrams can be used for

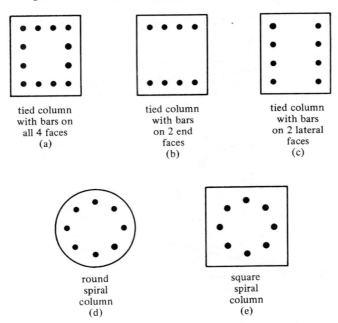

Figure 10.14

cross sections with widely varying dimensions. The ACI has prepared normalized interaction curves in this manner for the different cross section and bar arrangement situations shown in Figure 10.14 and for different grades of steel and concrete.[3]

Two of the ACI diagrams are given in Figures 10.15 and 10.16, while Appendix A (Graphs 2 to 13) presents several other ones for the situations given in parts (a), (b), and (d) of Figure 10.14. *Notice that these ACI diagrams do not include the three modifications described in the last section.*

The ACI column interaction diagrams are used in Examples 10.3 to 10.7 to design or analyze columns for different situations. In order to correctly use these diagrams, it is necessary to compute the value of γ (gamma), which is equal to the distance from the center of the bars on one side of the column to the center of the bars on the other side of the column divided by h, the depth of the column (both values being taken in the direction of bending). Usually the value of γ obtained falls in between a pair of curves, and interpolation of the curve readings will have to be made.

Caution

Be sure that the column picture at the upper right of the interaction curve being used agrees with the column being considered. In other words, are there bars on two faces of the column or on all four faces? If the wrong curves are selected, the answers may be quite incorrect.

Although several methods are available for selecting column sizes, a trial-and-error method is about as good as any. With this procedure the designer estimates what he or she thinks is a reasonable column size and then determines the steel percentage required for that column size

---

[3]American Concrete Institute, *Design Handbook*, 1997, Publication SP-17(97), Detroit, 482 pages.

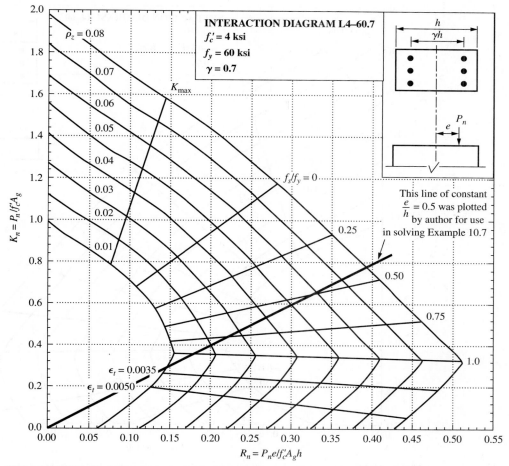

Figure 10.15 ACI rectangular column interaction diagrams when bars are placed on two faces only. (Permission of American Concrete Institute.)

from the interaction diagram. If it is felt that the ρ_g determined is unreasonably large or small, another column size can be selected and the new required ρ_g selected from the diagrams, and so on. In this regard, the selection of columns for which ρ_g is greater than 4 or 5% results in congestion of the steel, particularly at splices, and consequent difficulties in getting the concrete down into the forms.

A slightly different approach is used in Example 10.4 where the average compression stress at ultimate load across the column cross section is assumed to equal some value—say, 0.5 to $0.6 f_c'$. This value is divided into P_n to determine the column area required. Then cross-sectional dimensions are selected, and the value of ρ_g is determined from the interaction curves. Again, if the percentage obtained seems unreasonable, the column size can be revised and a new steel percentage obtained.

In Examples 10.3 to 10.5, reinforcing bars are selected for three columns. The values of $K_n = P_n/f_c'A_g$ and $R_n = P_n e/f_c'A_g h$ are computed. The position of those values is located on the appropriate graph, and ρ_g is determined and multiplied by the gross area of the column in question to determine the reinforcing area needed.

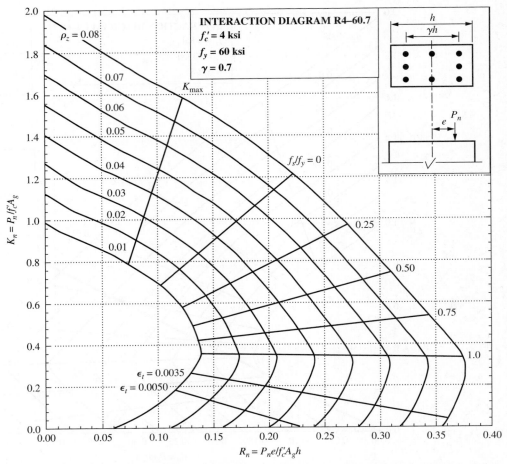

Figure 10.16 ACI rectangular column interaction diagram when bars are placed along all four faces. (Permission of American Concrete Institute.)

EXAMPLE 10.3

The short 14 × 20-in. tied column of Figure 10.17 is to be used to support the following loads and moments: $P_D = 125$ k, $P_L = 140$ k, $M_D = 75$ ft-k and $M_L = 90$ ft-k. If $f_c' = 4000$ psi and $f_y = 60,000$ psi, select reinforcing bars to be placed in its end faces only using appropriate ACI column interaction diagrams.

SOLUTION

$$P_u = (1.2)(125) + (1.6)(140) = 374 \text{ k}$$

$$P_n = \frac{374}{0.65} = 575.4 \text{ k}$$

$$M_u = (1.2)(75) + (1.6)(90) = 234 \text{ ft-k}$$

$$M_n = \frac{234}{0.65} = 360 \text{ ft-k}$$

$$e = \frac{(12)(360)}{575.4} = 7.51''$$

$$\gamma = \frac{15}{20} = 0.75$$

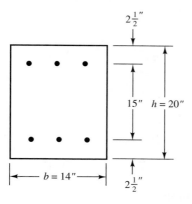

Figure 10.17

Compute values of K_n and R_n

$$K_n = \frac{P_n}{f'_c A_g} = \frac{575.4}{(4)(14 \times 20)} = 0.513$$

$$R_n = \frac{P_n e}{f'_c A_g h} = \frac{(575.4)(7.51)}{(4)(14 \times 20)(20)} = 0.193$$

The value of γ falls between γ values for Graphs 3 and 4 of Appendix A. Therefore interpolating between the two as follows:

γ	0.70	0.75	0.80
ρ_g	0.0220	0.0202	0.0185

$$A_s = \rho_g bh = (0.0202)(14)(20) = 5.66 \text{ in.}^2$$

Use 6 #9 bars = 6.00 in.^{2}

Notes

(a) Note that $\phi = 0.65$ as initially assumed since the graphs used show $\frac{f_s}{f_y}$ is < 1.0 and thus $\epsilon_t < 0.002$.

(b) Code requirements must be checked as in Example 9.1. (See Figure 10.25 to understand.)

EXAMPLE 10.4

Design a short square column for the following conditions: $P_u = 600$ k, $M_u = 80$ ft-k, $f'_c = 4000$ psi, and $f_y = 60,000$ psi. Place the bars uniformly around all four faces of the column.

SOLUTION   Assume the column will have an average compression stress = about $0.6 f'_c = 2400$ psi.

$$A_g \text{ required} = \frac{600}{2.400} = 250 \text{ in.}^2$$

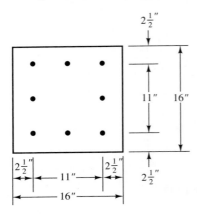

Figure 10.18

Try a 16 × 16-in. column ($A_g = 256$ in.^{2}) with the bar arrangement shown in Figure 10.18.

$$e = \frac{M_u}{P_u} = \frac{(12)(80)}{600} = 1.60''$$

$$P_n = \frac{P_u}{\phi} = \frac{600}{0.65} = 923.1 \text{ k}$$

$$K_n = \frac{P_n}{f'_c A_g} = \frac{923.1}{(4)(16 \times 16)} = 0.901$$

$$R_n = \frac{P_n e}{f'_c A_g h} = \frac{(923.1)(1.6)}{(4.0)(16 \times 16)(16)} = 0.0901$$

$$\gamma = \frac{11}{16} = 0.6875$$

Interpolating between values given in Graphs 6 and 7 of Appendix A.

γ	0.600	0.6875	0.700
ρ_g	0.025	0.023	0.022

$$A_s = (0.023)(16)(16) = 5.89 \text{ in.}^2$$

<u>Use eight #8 bars = 6.28 in.^{2}</u>

Notes

    (a) Note that $\phi = 0.65$ as initially assumed since the graphs used show $\frac{f_s}{f_y} < 1.0$ and thus $\epsilon_t < 0.002$.

    (b) Code requirements must be checked as in Example 9.1. (See Figure 10.25.)

EXAMPLE 10.5

    Using the ACI column interaction graphs, select reinforcing for the short round spiral column shown in Figure 10.19 if $f'_c = 4000$ psi, $f_y = 60,000$ psi, $P_u = 500$ k, and $M_u = 225$ ft-k.

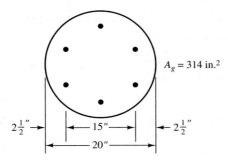

$A_g = 314 \text{ in.}^2$

$2\frac{1}{2}''$ $15''$ $2\frac{1}{2}''$

$20''$

Figure 10.19

SOLUTION

$$e = \frac{(12)(225)}{500} = 5.40 \text{ in.}$$

$$P_n = \frac{P_u}{\phi} = \frac{500}{0.75} = 666.7 \text{ k}$$

$$K_n = \frac{P_n}{f_c' A_g} = \frac{666.7}{(4)(314)} = 0.531$$

$$R_n = \frac{P_n e}{f_c' A_g h} = \frac{(666.7)(5.40)}{(4)(314)(20)} = 0.143$$

$$\gamma = \frac{15}{20} = 0.75$$

By interpolation between Graphs 11 and 12 of Appendix A, ρ_g is found to equal 0.0235 and $\frac{f_s}{f_y} < 1.0$.

$$\rho A_g = (0.0235)(314) = 7.38 \text{ in.}^2$$

<u>Use 8 #9 bars = 8.00 in.^{2}</u>

Same notes as for Examples 10.3 and 10.4.

In Example 10.6 it is desired to select a 14-in. wide column with approximately 2% steel. This is done by trying different column depths and then determining the steel percentage required in each case.

EXAMPLE 10.6

Design a 14-in. wide rectangular short tied column with bars only in the two end faces for $P_u = 500$ k, $M_u = 250$ ft-k, $f_c' = 4000$ psi, and $f_y = 60,000$ psi. Select a column with approximately 2% steel.

SOLUTION

$$e = \frac{M_u}{P_u} = \frac{(12)(250)}{500} = 6.00''$$

$$P_n = \frac{P_u}{\phi} = \frac{500}{0.65} = 769.2 \text{ k}$$

Trying several column sizes (14×20, 14×22, 14×24) and determining reinforcing.

Trial sizes	14×20	14×22	14×24
$K_n = \dfrac{P_n}{f'_c A_g}$	0.687	0.624	0.572
$R_n = \dfrac{P_n e}{f'_c A_g h}$	0.206	0.170	0.143
$\gamma = \dfrac{h - 2 \times 2.50}{h}$	0.750	0.773	0.792
ρ_g by interpolation	0.0315	0.020	0.011

Use 14×22 column

$$A_g = (0.020)(14 \times 22) = 6.16 \text{ in.}^2$$

Use 8 #8 bars $= 6.28$ in.^{2}

Same notes as for Examples 10.3 and 10.4.

One more illustration of the use of the ACI interaction diagrams is presented with Example 10.7. In this example, the nominal column load P_n at a given eccentricity which a column can support is determined.

With reference to the ACI interaction curves, the reader should carefully note that the value of R_n (which is $P_n e / f'_c A_g h$) for a particular column, equals e/h times the value of K_n ($P_n / f'_c A_g$) for that column. This fact needs to be understood when the user desires to determine the nominal load that a column can support at a given eccentricity.

In Example 10.7 the nominal load that the short column of Figure 10.20 can support at an eccentricity of 10 in. with respect to the x-axis is determined. If we plot on the interaction diagram the intersection point of K_n and R_n for a particular column and draw a straight line from that point to the lower left corner or origin of the figure, we will have a line with a constant e/h. For the column of Example 10.6 $e/h = 10/20 = 0.5$. Therefore a line is plotted from the origin through a set of assumed values for K_n and R_n in the proportion of $10/20$ to each other. In this case, K_n was set equal to 0.8 and $R_n = 0.5 \times 0.8 = 0.4$. Next a line was drawn from that intersection point to the origin of the diagram as shown in Figure 10.16. Finally, the intersection of this line with ρ_g (0.0316

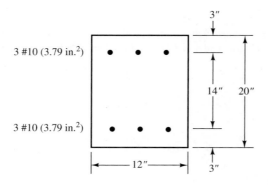

3 #10 (3.79 in.^{2})

3 #10 (3.79 in.^{2})

3″

14″   20″

12″

3″

Figure 10.20

in this example) was determined, and the value of K_n or R_n was read. This latter value enables us to compute P_n.

EXAMPLE 10.7

Using the appropriate interaction curves, determine the value of P_n for the short tied column shown in Figure 10.20 if $e_x = 10''$. Assume $f_c' = 4000$ psi and $f_y = 60,000$ psi.

SOLUTION

$$\frac{e}{h} = \frac{10}{20} = 0.500$$

$$\rho_g = \frac{(2)(3.79)}{(12)(20)} = 0.0316$$

$$\gamma = \frac{14}{20} = 0.700$$

Plotting a straight line through the origin and the intersection of assumed values of K_n and R_n (say 0.8 and 0.4, respectively).

For ρ_g of 0.0316 we read the value of R_n

$$R_n = \frac{P_n e}{f_c' A_g h} = 0.24$$

$$P_n = \frac{(0.24)(4)(12 \times 20)(20)}{10} = 460.8 \text{ k}$$

When the usual column is subjected to axial load and moment, it seems reasonable to assume initially that $\phi = 0.65$ for tied columns and 0.70 for spiral columns. It is to be remembered, however, that under certain conditions these ϕ values may be increased, as discussed in detail in Section 10.9.

10.7   SHEAR IN COLUMNS

The shearing forces in interior columns in braced structures are usually quite small and normally do not control the design. However, the shearing forces in exterior columns can be large, even in a braced structure, particularly in columns bent in double curvature. Section 11.3.1.2 of the ACI Code provides the following equations for determining the shearing force that can be carried by the concrete for a member subjected simultaneously to axial compression and shearing forces.

$$V_c = 2\left(1 + \frac{N_u}{2000A_g}\right)\lambda\sqrt{f_c'}b_w d \qquad \text{(ACI Equation 11-4)}$$

In SI units this equation is:

$$V_c = \left(1 + \frac{N_u}{14A_g}\right)\left(\frac{\lambda\sqrt{f_c'}}{6}\right)b_w d$$

In these equations, N_u is the factored axial force acting simultaneously with the factored shearing force, V_u, that is applied to the member. The value of N_u/A_g is the average factored axial stress in the column and is expressed in units of psi. Should V_u be greater than $\phi V_c/2$, it will be necessary to calculate required tie spacing using the stirrup spacing procedures described in Chapter 8. The results will be closer tie spacing than required by the usual column rules discussed earlier in Section 9.5.

Sections 11.2.3 and 11.4.7.3 of the ACI Code specify the method for calculating the contribution of the concrete to the total shear strength of circular columns and for calculating the contribution of shear reinforcement for cases where circular hoops, ties, or spirals are present. According to the Commentary of the Code in their Section 11.2.3, the entire cross section in circular columns is effective in resisting shearing forces. The shear area, $b_w d$, in ACI Equation 11-4 then would be equal to the gross area of the column. However, to provide for compatibility with other calculations requiring an effective depth, the ACI requires that the shear area in ACI Equation 11-4 when applied to circular columns be computed as an equivalent rectangular area in which

$$b_w = D$$

$$d = 0.8D$$

In these equations, D is the gross diameter of the column. If the constant modifying D in the effective depth equation were equal to $\pi/4$, which is equal to 0.7854, the effective rectangular area would be equal to the gross area of the circular column. As such, the area of the column is overestimated by a little less than 2% when using the equivalent area prescribed by the ACI.

10.8   BIAXIAL BENDING

Many columns are subjected to biaxial bending, that is, bending about both axes. Corner columns in buildings where beams and girders frame into the columns from both directions are the most common cases, but there are others, such as where columns are cast monolithically as part of frames in both directions or where columns are supporting heavy spandrel beams. Bridge piers are almost always subject to biaxial bending.

Circular columns have polar symmetry and thus the same ultimate capacity in all directions. The design process is the same, therefore, regardless of the directions of the moments. If there is bending about both the x and y axes, the biaxial moment can be computed by combining the two moments or their eccentricities as follows:

$$M_u = \sqrt{(M_{ux})^2 + (M_{uy})^2}$$

or

$$e = \sqrt{(e_x)^2 + (e_y)^2}$$

For shapes other than circular ones, it is necessary to consider the three-dimensional interaction effects. Whenever possible, it is desirable to make columns subject to biaxial bending circular in shape. Should it be necessary to use square or rectangular columns for such cases, the reinforcing should be placed uniformly around the perimeters.

You might quite logically think that you could determine P_n for a biaxially loaded column by using static equations, as was done in Example 10.2. Such a procedure will lead to the correct answer, but the mathematics involved is so complicated due to the shape of the compression side of the column that the method is not a practical one. Nevertheless, a few comments are made about this type of solution, and reference is made to Figure 10.21.

An assumed location is selected for the neutral axis, and the appropriate strain triangles are drawn as shown in the figure. The usual equations are written with $C_c = 0.85f_c'$ times the shaded area A_c and with each bar having a force equal to its cross-sectional area times its stress. The solution of the equation yields the load that would establish that neutral axis—but the

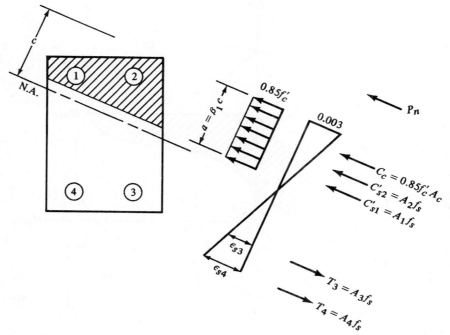

Figure 10.21

designer usually starts with certain loads and eccentricities and does not know the neutral axis location. Furthermore, the neutral axis is probably not even perpendicular to the resultant

$$e = \sqrt{(e_x)^2 + (e_y)^2}.$$

For column shapes other than circular ones, it is desirable to consider three-dimensional interaction curves such as the one shown in Figure 10.22. In this figure the curve labeled M_{nxo} represents the interaction curve if bending occurs about the x-axis only, while the one labeled M_{nyo} is the one if bending occurs about the y-axis only.

In this figure, for a constant P_n, the hatched plane shown represents the contour of M_n for bending about any axis.

Today, the analysis of columns subject to biaxial bending is primarily done with computers. One of the approximate methods that is useful in analysis and that can be handled with pocket calculators includes the use of the so-called reciprocal interaction equation, which was developed by Professor Boris Bresler of the University of California at Berkeley.[4] This equation, which is shown in Section R10.3.6 of the ACI Commentary, follows:

$$\frac{1}{P_{ni}} = \frac{1}{P_{nx}} + \frac{1}{P_{ny}} - \frac{1}{P_o}$$

where

$\quad P_{ni} =$ the nominal axial load capacity of the section when the load is placed at a given eccentricity along both axes.

---

[4]Bresler, B., 1960, "Design Criteria for Reinforced Concrete Columns under Axial Load and Biaxial Bending," *Journal ACI*, 57, p. 481.

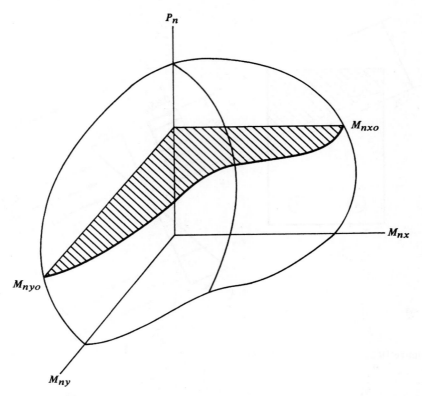

Figure 10.22

P_{nx} = the nominal axial load capacity of the section when the load is placed at an eccentricity e_x.

P_{ny} = the nominal axial load capacity of the section when the load is placed at an eccentricity e_y.

P_o = the nominal axial load capacity of the section when the load is placed with a zero eccentricity. It is usually taken as $0.85f_c'A_g + f_yA_s$.

The Bresler equation works rather well as long as P_{ni} is at least as large as $0.10\,P_o$. Should P_{ni} be less than $0.10\,P_o$, it is satisfactory to neglect the axial force completely and design the section as a member subject to biaxial bending only. This procedure is a little on the conservative side. For this lower part of the interaction curve, it will be remembered that a little axial load increases the moment capacity of the section. The Bresler equation does not apply to axial tension loads. Professor Bresler found that the ultimate loads predicted by his equation for the conditions described do not vary from test results by more than 10%.

Example 10.8 illustrates the use of the reciprocal theorem for the analysis of a column subjected to biaxial bending. The procedure for calculating P_{nx} and P_{ny} is the same as the one used for the prior examples of this chapter.

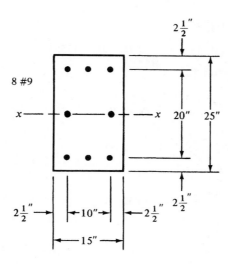

Figure 10.23

EXAMPLE 10.8

Determine the design capacity P_{ni} of the short tied column shown in Figure 10.23, which is subjected to biaxial bending. $f_c' = 4000$ psi, $f_y = 60,000$ psi, $e_x = 16$ in., and $e_y = 8$ in.

SOLUTION   **For Bending about x-Axis**

$$\gamma = \frac{20}{25} = 0.80$$

$$\rho_g = \frac{8.00}{(15)(25)} = 0.0213$$

$$\frac{e}{h} = \frac{16}{25} = 0.64$$

Drawing line of constant $\frac{e}{h} = 0.64$ in Graph 8 of Appendix A

$$R_n = \frac{P_n e}{f_c' A_g h} = 0.185$$

$$P_n = \frac{(4)(15 \times 25)(25)(0.185)}{16} = 434 \, \text{k}$$

For Bending about y-Axis

$$\gamma = \frac{10}{15} = 0.667$$

$$\rho_g = \frac{8.00}{(15)(25)} = 0.0213$$

$$\frac{e}{h} = \frac{8}{15} = 0.533$$

Drawing lines of constant $e/h(0.533)$ in Graphs 6 and 7 of Appendix A and interpolating between them for $\gamma = 0.667$.

$$R_n = \frac{P_{ny} e}{f_c' A_g h} = 0.163$$

$$P_{ny} = \frac{(4)(15 \times 25)(15)(0.163)}{8} = 458 \, \text{k}$$

Determining Axial Load Capacity of Section

$$P_o = (0.85)(4.0)(15 \times 25) + (8.00)(60) = 1755 \text{ k}$$

Using the Bresler Expression to Determine P_{ni}

$$\frac{1}{P_{ni}} = \frac{1}{P_{nx}} + \frac{1}{P_{ny}} - \frac{1}{P_o}$$

$$\frac{1}{P_{ni}} = \frac{1}{434} + \frac{1}{458} - \frac{1}{1755}$$

Multiplying through by 1755

$$\frac{1755}{P_{ni}} = 4.044 + 3.832 - 1$$

$$P_{ni} = \underline{\underline{255.3 \text{ k}}}$$

If the moments in the weak direction (y-axis here) are rather small compared to bending in the strong direction (x-axis), it is rather common to neglect the smaller moment. This practice is probably reasonable as long as e_y is less than about 20% of e_x, since the Bresler expression will show little reduction for P_{ni}. For the example just solved, an e_y equal to 50% of e_x caused the axial load capacity to be reduced by approximately 40%.

Example 10.9 illustrates the design of a column subject to biaxial bending. The Bresler expression, which is of little use in the proportioning of such members, is used to check the capacities of the sections selected by some other procedure. Exact theoretical designs of columns subject to biaxial bending are very complicated and, as a result, are seldom handled with pocket calculators. They are proportioned either by approximate methods or with computer programs.

10.9   DESIGN OF BIAXIALLY LOADED COLUMNS

During the past few decades, several approximate methods have been introduced for the design of columns with biaxial moments. For instance, quite a few design charts are available with which satisfactory designs may be made. The problems are reduced to very simple calculations in which coefficients are taken from the charts and used to magnify the moments about a single axis. Designs are then made with the regular uniaxial design charts.[5–7]

Another approximate procedure that works fairly well for design office calculations is used for Example 10.9. If this simple method is applied to square columns, the values of both M_{nx} and M_{ny} are assumed to act about both the x-axis and the y-axis (i.e., $M_x = M_y = M_{nx} + M_{ny}$). The steel is selected about one of the axes and is spread around the column, and the Bresler expression is used to check the ultimate load capacity of the eccentrically loaded column.

---

[5]Parme, A. L., Nieves, J. M., and Gouwens, A., 1966, "Capacity of Reinforced Rectangular Columns Subject to Biaxial Bending," *Journal ACI*, 63 (11), pp. 911–923.

[6]Weber, D. C., 1966, "Ultimate Strength Design Charts for Columns with Biaxial Bending," *Journal ACI*, 63 (11), pp. 1205–1230.

[7]Row, D. G., and Paulay, T., 1973, "Biaxial Flexure and Axial Load Interaction in Short Reinforced Concrete Columns," *Bulletin of New Zealand Society for Earthquake Engineering*, 6 (2), pp. 110–121.

Richmond Convention Center, Richmond, Virginia. Notice column size changes. (Courtesy of EFCO.)

Should a rectangular section be used where the y-axis is the weaker direction, it would seem logical to calculate $M_y = M_{nx} + M_{ny}$ and to use that moment to select the steel required about the y-axis and spread the computed steel area over the whole column cross section. Although such a procedure will produce safe designs, the resulting columns may be rather uneconomical because they will often be much too strong about the strong axis. A fairly satisfactory approximation is to calculate $M_y = M_{nx} + M_{ny}$ and multiply it by b/h, and with that moment design the column about the weaker axis.[8]

Example 10.9 illustrates the design of a short square column subject to biaxial bending. The approximate method described in the last two paragraphs is used, and the Bresler expression is used for checking the results. If this had been a long column, it would have been necessary to magnify the design moments for slenderness effects, regardless of the design method used.

---

[8]Fintel, M., ed., 1985, *Handbook of Concrete Engineering*, 2nd ed. (New York: Van Nostrand), pp. 37–39.

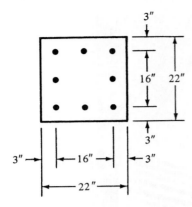

Figure 10.24

EXAMPLE 10.9

Select the reinforcing needed for the short square tied column shown in Figure 10.24 for the following: $P_D = 100$ k, $P_L = 200$ k, $M_{DX} = 50$ ft-k, $M_{LX} = 110$ ft-k, $M_{DY} = 40$ ft-k, $M_L = 90$ ft-k, $f'_c = 4000$ psi, and $f_y = 60,000$ psi.

SOLUTION   **Computing Design Values**

$$P_u = (1.2)(100) + (1.6)(200) = 440 \text{ k}$$

$$\frac{P_u}{f'_c A_g} = \frac{440}{(4)(484)} = 0.227$$

Assume, $\phi = 0.65$

$$P_n = \frac{440}{0.65} = 677 \text{ k}$$

$$M_{ux} = (1.2)(50) + (1.6)(110) = 236 \text{ ft-k}$$

$$M_{nx} = \frac{236}{0.65} = 363 \text{ ft-k}$$

$$M_{uy} = (1.2)(40) + (1.6)(90) = 192 \text{ ft-k}$$

$$M_{ny} = \frac{192}{0.65} = 295 \text{ ft-k}$$

As a result of biaxial bending, the design moment about the x- or y-axis is assumed to equal $M_{nx} + M_{ny} = 363 + 295 = 658$ ft-k.

Determining Steel Required

$$e_x = e_y = \frac{(12)(658)}{677} = 11.66''$$

$$\gamma = \frac{16}{22} = 0.727$$

By interpolation from interaction diagrams with bars on all four faces,

$$\rho_g = 0.0235$$
$$A_s = (0.0235)(22)(22) = 11.37 \text{ in.}^2$$

$$\underline{\text{Use 8 \#11} = 12.50 \text{ in.}^2}$$

A review of the column with the Bresler expression gives a $P_{ni} = 804$ k > 677 k, which is satisfactory. Should the reader go through the Bresler equation here, he or she must remember to calculate the correct e_x and e_y values for use with the interaction diagrams. For instance,

$$e_x = \frac{(12)(363)}{677} = 6.43 \text{ in.}$$

When a beam is subjected to biaxial bending, the following approximate interaction equation may be used for design purposes:

$$\frac{M_x}{M_{ux}} + \frac{M_y}{M_{uy}} \leq 1.0$$

In this expression M_x and M_y are the design moments, M_{ux} is the design moment capacity of the section if bending occurs about the x-axis only, and M_{uy} is the design moment capacity if bending occurs about the y-axis only. This same expression may be satisfactorily used for axially loaded members if the design axial load is about 15% or less of the axial load capacity of the section. For a detailed discussion of this subject, the reader is referred to the *Handbook of Concrete Engineering*.[9]

Numerous other methods are available for the design of biaxially loaded columns. One method that is particularly useful to the design profession is the PCA Load Contour Method, which is recommended in the ACI *Design Handbook*.[10]

10.10 CONTINUED DISCUSSION OF CAPACITY REDUCTION FACTOR, ϕ

As previously described, the value of ϕ can be larger than 0.65 for tied columns, or 0.75 for spiral columns, if ϵ_t is larger than f_y/E_s. The lower ϕ values are applicable to compression-controlled sections because of their smaller ductilities. Such sections are more sensitive to varying concrete strengths than are tensilely controlled sections. The Code (9.3.2.2) states that ϕ for a particular column may be increased linearly from 0.65 or 0.75 to 0.90 as the net tensile strain ϵ_t increases from the compression-controlled strain f_y/E_s to the tensilely controlled one of 0.005.

For this discussion, Figure 3.5 is repeated with slight modification as Figure 10.25. From this figure you can see the range of ϵ_t values for which ϕ may be increased.

The hand calculation of ϵ_t for a particular column is a long and tedious trial-and-error problem, and space is not taken here to present a numerical example. However, a description of the procedure is presented in the next few paragraphs. The average designer will not want to spend the time necessary to make these calculations and will either just use the smaller ϕ values or make use of a computer program such as the Excel spreadsheet provided for this chapter. This program uses a routine for computing ϵ_t and ϕ for columns.

The procedure described here can be used to make a long-hand determination of ϵ_t. As a beginning we assume $c/d_t = 0.60$ where $\epsilon_t = 0.002$ (assumed yield strain for Grade 60 reinforcement), as shown in Figure 10.25. With this value we can calculate c, a, ϵ_c, ϵ_t, f_s, and f'_c for our column. Then with reference to Figure 10.26, moments can be taken about the centerline of the column and the result solved for M_n and e determined.

$$M_n = T_s \left(\frac{d_t - d'}{2} \right) + C_s \left(\frac{d_t - d'}{2} \right) + C_c \left(\frac{h}{2} - \frac{a}{2} \right)$$

---

[9]Fintel, M., *Handbook of Concrete Engineering*, p. 38.

[10]American Concrete Institute, 1997, *Design Handbook in Accordance with the Strength Design Method*, Vol. 2, *Columns*, Publication SP-17(97), ACI, Farmington Hills, MI.

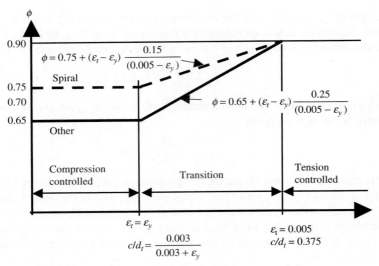

Figure 10.25 Variation of ϕ with net tensile ϵ_t and c/d_t.

As the next step, c/d_t can be assumed equal to 0.375 (where $\epsilon_t = 0.005$ as shown in Figure 10.25) and another value of ϵ_t determined. If the ϵ_t or our column falls between the two ϵ_t values we have just calculated, the column falls in the transition zone for ϕ. To determine its value we can try different c/d_t values between 0.600 and 0.375 until the calculated ϵ_t equals the actual ϵ_t of the column.

If you go through this process one time, you will probably have seen all you want to see of it and will no doubt welcome the fact that the Excel spreadsheet provided for this textbook may be used to determine the value of ϕ for a particular column.

When using the interaction diagrams in the Appendix, it is easy to see the region where the variable ϕ factor applies. In Figure 10.15 note that there are lines labeled f_s/f_y. If the coordinates of K_n and R_n are greater than the value of $f_s/f_y = 1$, the ϕ factor is 0.65 (0.75 for spiral columns). If the coordinates are below the line labeled $\epsilon_t = 0.005$, the ϕ factor is 0.90. Between these lines, the ϕ factor is variable, and you would have to resort to approximate methods or to the spreadsheet provided.

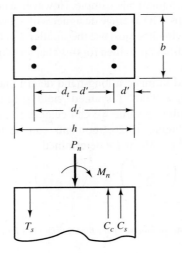

Figure 10.26

10.11   COMPUTER EXAMPLE

EXAMPLE 10.10

Using the Excel spreadsheet provided, plot the interaction diagram for the column obtained in Example 10.5

SOLUTION

Open the Excel spreadsheet called Chapter 9 and 10. Open the worksheet entitled Circular Column. In the cells in yellow highlight enter the values required. You do not have to input values for P_u and M_u, but it is helpful to see how the applied loads compare with the interaction diagram. Then open the worksheet called Interaction Diagram – Circular. The diagram shows that the applied load (single dot) is within the P_u vs. M_u diagram (smaller curved line), hence the column cross section is sufficient if it is a short column.

Circular Column Capacity

$P_u =$	500	k
$M_u =$	200	ft-k =
$h =$	20	in.
$\gamma =$	0.75	
$f'_c =$	4000	psi
$f_y =$	60000	psi
$A_{st} =$	6.28	in.^{2}
$A_g =$	314.2	in.^{2}
$\rho_t =$	0.0200	
$\beta_1 =$	0.85	
$\varepsilon_y =$	0.00207	
$E_s =$	29000	ksi
$c_{bal} =$	10.36	in.
$c_{.005} =$	6.5625	in.

(for M_u row: ft-k = 2400 k-in.)

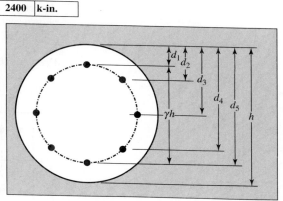

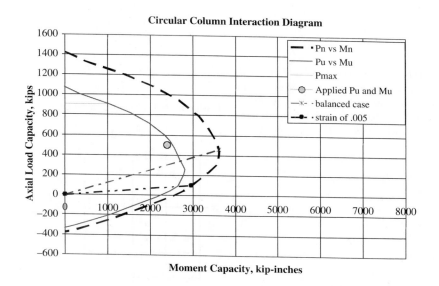

Circular Column Interaction Diagram

Legend:
- • Pn vs Mn
- Pu vs Mu
- Pmax
- ○ Applied Pu and Mu
- –×– balanced case
- ■ strain of .005

Axial Load Capacity, kips (vertical axis: −600 to 1600)

Moment Capacity, kip-inches (horizontal axis: 0 to 8000)

PROBLEMS

Location of Plastic Centroids
In Problems 10.1 and 10.2 locate the plastic centroids if $f'_c = 4000$ psi and $f_y = 60,000$ psi.

Problem 10.1 (*Ans.* 11.38 in. from left edge)

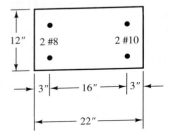

Problem 10.2

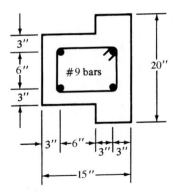

Analysis of Columns Subjected to Axial Load and Moment
10.3 Using statics equations, determine the values of P_n and M_n for the column shown, assuming it is strained to -0.00300 on the right-hand edge and to $+0.00200$ on its left-hand edge. $f'_c = 4000$ psi, $f_y = 60,000$ psi. (*Ans.* $P_n = 485.8$ k, $M_n = 335.9$ ft-k)

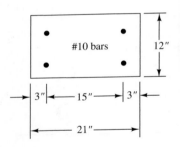

10.4 Repeat Problem 10.3 if the strain on the left edge is $+0.001$.

10.5 Repeat Problem 10.3 if the strain on the left edge is 0.000. (*Ans.* $P_n = 902.9$ k, $M_n = 165.4$ ft-k)

10.6 Repeat Problem 10.3 if the strain on the left edge is -0.001.

10.7 Repeat Problem 10.3 if the steel on the left side has a strain in tension of $\epsilon_y = f_y/E_s$ and the right edge is at 0.003. (*Ans.* $P_n = 360.8$ k, $M_n = 368.3$ ft-k)

Design of Columns for Axial Load and Moment
In Problems 10.8 to 10.10 use the interaction curves in Appendix A to select reinforcing for the short columns shown, with $f'_c = 4000$ psi and $f_y = 60,000$ psi.

Problem 10.8

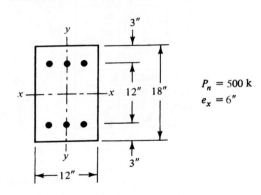

Problem 10.9 (*One ans.* 8 #9 bars)

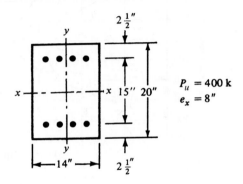

Problem 10.10

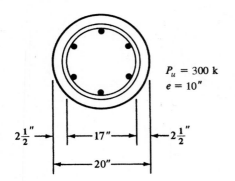

$P_u = 300$ k
$e = 10''$

$2\frac{1}{2}''$ $17''$ $2\frac{1}{2}''$

$20''$

Problem 10.14

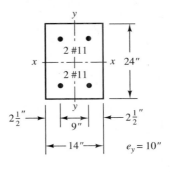

y
2 #11
x — — x 24''
2 #11
y

$2\frac{1}{2}''$ $9''$ $2\frac{1}{2}''$

$14''$ $e_y = 10''$

For Problems 10.11 to 10.16 use the interaction diagrams in Appendix A to determine ϕP_n values for the short columns shown which have bending about one axis. $f_y = 60{,}000$ psi and $f_c' = 4000$ psi.

Problem 10.11 (*Ans.* 328 k)

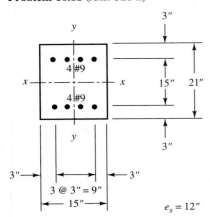

y
4 #9
x — — x 15''   21''
4 #9
y

$3''$

$3''$ $3''$

$3 @ 3'' = 9''$
$15''$ $e_x = 12''$

10.12 Repeat Problem 10.11 if $e_x = 9$ in.

Problem 10.13 (*Ans.* 199 k)

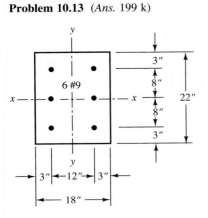

y
6 #9
x — — x 22'' $e_y = 10''$
$8''$
$8''$
$3''$
y

$3''$ $12''$ $3''$
$18''$

Problem 10.15 (*Ans.* 412 k)

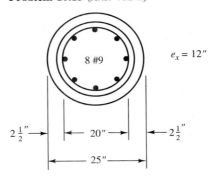

8 #9 $e_x = 12''$

$2\frac{1}{2}''$ $20''$ $2\frac{1}{2}''$

$25''$

Problem 10.16

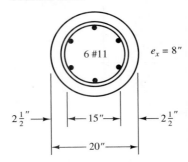

6 #11 $e_x = 8''$

$2\frac{1}{2}''$ $15''$ $2\frac{1}{2}''$

$20''$

In Problems 10.17 to 10.21, determine ϕP_n values for the short columns shown if $f_y = 60,000$ psi and $f'_c = 4000$ psi.

Problem 10.19 (*Ans.* 279 k)

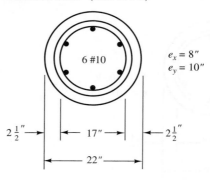

Problem 10.17 (*Ans.* 146 k)

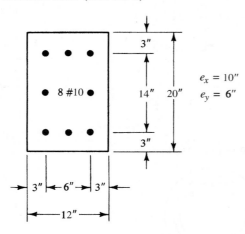

Problem 10.20

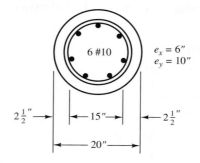

10.21 Repeat Problem 10.20 if $e_x = 8''$ and $e_y = 6''$ (*Ans.* 287 k)

In Problems 10.22 and 10.23 select reinforcing for the short columns shown if $f_y = 60,000$ psi and $f'_c = 4000$ psi. Check results with the Bresler equation.

Problem 10.18

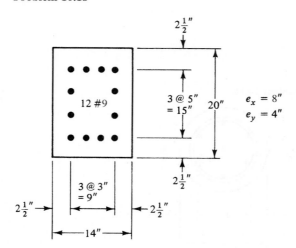

Problem 10.22

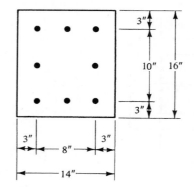

Problem 10.23 (*One ans.* 8 #7 bars)

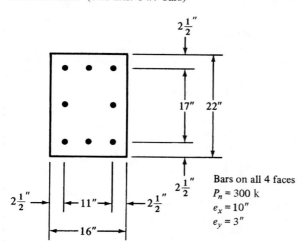

Bars on all 4 faces
$P_n = 300$ k
$e_x = 10''$
$e_y = 3''$

For Problems 10.24 to 10.27 use the Excel spreadsheet for Chapters 9 and 10.

10.24 If the column of Problem 10.8 is supporting a load $P_u = 250$ k, and $e_x = 0$, how large can M_{ux} be?

10.25 If the column of Problem 10.13 is supporting a load $P_u = 400$ k, and $e_x = 0$, how large can M_{uy} be if six #9 bars are used? (*Ans.* 264 ft-k)

10.26 If the column of Problem 10.15 is to support an axial load $P_u = 400$ k, and $e_x = 0$, how many #10 bars must be used to resist a design moment $M_{ux} = 300$ ft-k?

10.27 If the column in Problem 10.11 has a moment $M_{ux} = 300$ ft-k, what are the limits on P_u? (*Ans.* 420 k $\geq P_u \geq$ 20 k)

10.28 Prepare a flowchart for the preparation of an interaction curve for axial compression loads and bending for a short rectangular tied column.

Problems with SI Units

Column interaction curves are not provided in this text for the usual SI concrete strengths (21 MPa, 24 MPa, 28 MPa, etc.) or for the usual steel yield strength (420 MPa). Therefore the problems that follow are to be solved using the column curves for $f'_c = 4000$ psi and $f_y = 60,000$ psi. These diagrams may be applied for the corresponding SI units (27.6 MPa and 413.7 MPa), just as they are for U.S. customary units, but it is necessary to convert the results to SI values.

In Problems 10.29 to 10.31 use the column interaction diagrams in the Appendix to determine P_n values for the short columns shown if $f'_c = 28$ MPa and $f_y = 420$ MPa.

Problem 10.29 (*Ans.* 1855 kN)

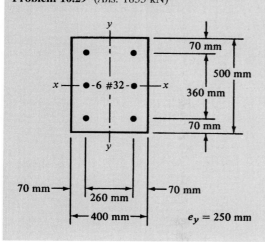

$e_y = 250$ mm

Problem 10.30

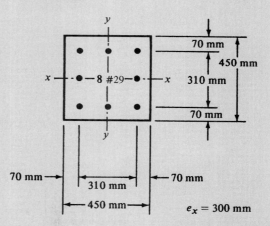

$e_x = 300$ mm

Problem 10.31   (*Ans.* $P_u = 1146$ kN)

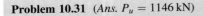

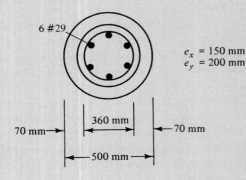

6 #29

$e_x = 150$ mm
$e_y = 200$ mm

70 mm→   360 mm   ←70 mm

←500 mm→

In Problems 10.32 to 10.34 select reinforcing for the short columns shown if $f_c' = 27.6$ MPa and $f_y = 413.7$ MPa. Remember to apply the conversion factor provided before Problem 10.28 when using the interaction curves.

Problem 10.33   (*One ans.* 6 #36)

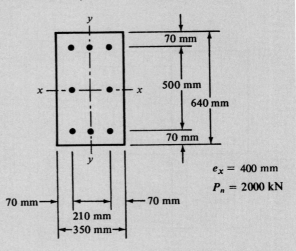

70 mm

500 mm

640 mm

70 mm

$e_x = 400$ mm
$P_n = 2000$ kN

70 mm→   ←70 mm

210 mm

←350 mm→

Problem 10.32

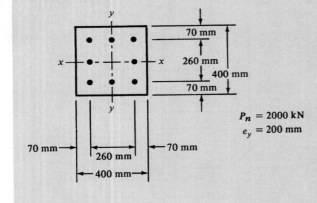

70 mm

260 mm

400 mm

70 mm

70 mm→   ←70 mm

260 mm

←400 mm→

$P_n = 2000$ kN
$e_y = 200$ mm

Problem 10.34

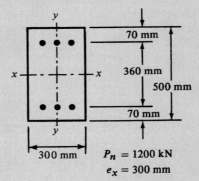

70 mm

360 mm

500 mm

70 mm

300 mm

$P_n = 1200$ kN
$e_x = 300$ mm

11

Slender Columns

11.1 INTRODUCTION

When a column bends or deflects laterally an amount Δ, its axial load will cause an increased column moment equal to $P\Delta$. This moment will be superimposed onto any moments already in the column. Should this $P\Delta$ moment be of such magnitude as to reduce the axial load capacity of the column significantly, the column will be referred to as a *slender column*.

Section 10.10.2 of the Code states that the design of a compression member should, desirably, be based on a theoretical analysis of the structure that takes into account the effects of axial loads, moments, deflections, duration of loads, varying member sizes, end conditions, and so on. If such a theoretical procedure is not used, the Code (10.10.5) provides an approximate method for determining slenderness effects. This method, which is based on the factors just mentioned for an "exact" analysis, results in a moment magnifier δ, which is to be multiplied by the larger moment at the end of the column denoted as M_2, and that value is used in design. If bending occurs about both axes, δ is to be computed separately for each direction and the values obtained multiplied by the respective moment values.

11.2 NONSWAY AND SWAY FRAMES

For this discussion it is necessary to distinguish between frames without sidesway and those with sidesway. In the ACI Code these are referred to respectively as *nonsway frames* and *sway frames*.

For the building story in question, the columns in nonsway frames must be designed according to Section 10.10.6 of the Code, while the columns of sway frames must be designed according to Section 10.10.7. As a result, it is first necessary to decide whether we have a nonsway frame or a sway frame. *You must realize that you will rarely find a frame that is completely braced against swaying or one that is completely unbraced against swaying. Therefore, you are going to have to decide which way to handle it.*

The question may possibly be resolved by examining the lateral stiffness of the bracing elements for the story in question. You may observe that a particular column is located in a story where there is such substantial lateral stiffness provided by bracing members, shear walls, shear trusses, and so on that any lateral deflections occurring will be too small to affect the strength of the column appreciably. You should realize while examining a particular structure that there may be some nonsway stories and some sway stories.

If we cannot tell by inspection whether we have a nonsway frame or a sway frame, the Code provides two ways of making a decision. First, in ACI Section 10.10.5.1, a story in a frame is said to be a nonsway one if the increase in column end moments due to second-order effects is 5% or less of the first-order end moments.

The second method presented by the Code for determining whether a particular frame is braced or unbraced is given in the Code (10.10.5.2). If the value of the so-called *stability index* which follows is ≤ 0.05, the Commentary states that the frame may be classified as a nonsway one. (Should V_u be equal to zero, this method will not apply.)

$$Q = \frac{\Sigma P_u \Delta_o}{V_u \ell_c} \qquad \text{(ACI Equation 10-10)}$$

Where

ΣP_u = total factored vertical load for all of the columns on the story in question

Δ_o = the elastically determined first-order lateral deflection due to V_u at the top of the story in question with respect to the bottom of that story

V_u = the total factored horizontal shear for the story in question

ℓ_c = the height of a compression member in a frame measured from center to center of the frame joints

Despite these suggestions from the ACI, the individual designer is going to have to make decisions as to what is adequate bracing and what is not, depending on the presence of structural walls and other bracing items. For the average size reinforced concrete building, load eccentricities and slenderness values will be small and frames will be considered to be braced. *Certainly, however, it is wise in questionable cases to err on the side of the unbraced.*

11.3   SLENDERNESS EFFECTS

The slenderness of columns is based on their geometry and on their lateral bracing. As their slenderness increases, their bending stresses increase, and thus buckling may occur. Reinforced concrete columns generally have small slenderness ratios. As a result, they can usually be designed as short columns without strength reductions due to slenderness. If slenderness effects are considered small, then columns can be considered "short," and can be designed according to Chapter 10. However, if they are "slender," the moment for which the column must be designed is increased or magnified. Once the moment is magnified, the column is then designed according to Chapter 10 using the increased moment.

Several items involved in the calculation of slenderness ratios are discussed in the next several paragraphs. These include unsupported column lengths, effective length factors, radii of gyration, and the ACI Code requirements. The ACI Code (10.10.2.1) limits second-order effects to not more than 40% of first-order effects.

Unsupported Lengths

The length used for calculating the slenderness ratio of a column, ℓ_u, is its unsupported length. This length is considered to be equal to the clear distance between slabs, beams, or other members that provide lateral support to the column. If haunches or capitals (see Figure 16.1) are present, the clear distance is measured from the bottoms of the capitals or haunches.

Effective Length Factors

To calculate the slenderness ratio of a particular column, it is necessary to estimate its effective length. This is the distance between points of zero moment in the column. For this initial

Round columns. (Digital Vision.)

discussion it is assumed that no sidesway or joint translation is possible. Sidesway or joint translation means that one or both ends of a column can move laterally with respect to each other.

If there were such a thing as a perfectly pinned end column, its effective length would be its unsupported length, as shown in Figure 11.1(a). The *effective length factor k* is the number that must be multiplied by the column's unsupported length to obtain its effective length. For a perfectly pinned end column, $k = 1.0$.

Columns with different end conditions have entirely different effective lengths. For instance, if there were such a thing as a perfectly fixed end column, its points of inflection (or points of zero moment) would occur at its one-fourth points, and its effective length would be $\ell_u/2$, as shown in Figure 11.1(b). As a result, its k value would equal 0.5.

Obviously, the smaller the effective length of a particular column, the smaller its danger of buckling and the greater its load-carrying capacity. Figure 11.1(c) shows a column with one end fixed and one end pinned. The k factor for this column is approximately 0.70.

The concept of effective lengths is simply a mathematical method of taking a column—whatever its end and bracing conditions—and replacing it with an equivalent pinned end braced column. A complex buckling analysis could be made for a frame to determine the critical stress in a particular column. The k factor is determined by finding the pinned end column with an equivalent length that provides the same critical stress. The k factor procedure is a method of making simple solutions for complicated frame-buckling problems.

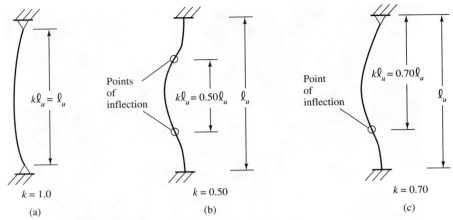

Figure 11.1 Effective lengths for columns in braced frames (sidesway prevented).

Reinforced concrete columns serve as parts of frames, and these frames are sometimes *braced* and sometimes *unbraced*. A braced frame is one for which sidesway or joint translation is prevented by means of bracing, shear walls, or lateral support from adjoining structures. An unbraced frame does not have any of these types of bracing supplied and must depend on the stiffness of its own members to prevent lateral buckling. For braced frames k values can never be greater than 1.0, but for unbraced frames the k values will always be greater than 1.0 because of sidesway.

An example of an unbraced column is shown in Figure 11.2(a). The base of this particular column is assumed to be fixed, whereas its upper end is assumed to be completely free to both rotate and translate. The elastic curve of such a column will take the shape of the elastic curve of a pinned end column of twice its length. Its effective length will therefore equal $2\ell_u$, as shown in the figure. In Figure 11.2(b) another unbraced column case is illustrated. The bottom of this column is connected to beams that provide resistance to rotation, but not enough to be considered a fixed end. In most buildings partial rotational restraint is common, not pinned or fixed ends. Section 11.4

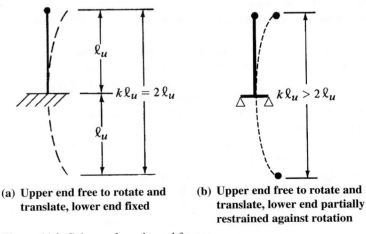

(a) Upper end free to rotate and translate, lower end fixed

(b) Upper end free to rotate and translate, lower end partially restrained against rotation

Figure 11.2 Columns for unbraced frames.

shows how to evaluate such partial restraint. For the case shown in Figure 11.2(b), if the beam at the bottom is flexible compared with the column, the k factor approaches infinity. If it is very stiff, k approaches 2.

The Code (10.10.6.3) states that the effective length factor is to be taken as 1.0 for compression members in frames braced against sidesway unless a theoretical analysis shows that a lesser value can be used. Should the member be in a frame not braced against sidesway, the value of k will be larger than 1.0 and must be determined with proper consideration given to the effects of cracking and reinforcing on the column stiffness. ACI-ASCE Committee 441 suggests that it is not realistic to assume that k will be less than 1.2 for such columns; therefore it seems logical to make preliminary designs with k equal to or larger than that value.

11.4   DETERMINING k FACTORS WITH ALIGNMENT CHARTS

The preliminary procedure used for estimating effective lengths involves the use of the alignment charts shown in Figure 11.3.[1,2] Before computerized analysis, use of such alignment charts was the traditional method for determining effective lengths of columns. The chart of part (a) of the figure is applicable to braced frames, whereas the one of part (b) is applicable to unbraced frames.

To use the alignment charts for a particular column, ψ factors are computed at each end of the column. The ψ factor at one end of the column equals the sum of the stiffness $[\Sigma(EI/\ell)]$ of the columns meeting at that joint, including the column in question, divided by the sum of all the stiffnesses of the beams meeting at the joint. Should one end of the column be pinned, ψ is theoretically equal to ∞, and if fixed, $\psi = 0$. Since a perfectly fixed end is practically impossible to have, ψ is usually taken as 1.0 instead of 0 for assumed fixed ends. When column ends are supported by, but not rigidly connected to a footing, ψ is theoretically infinity, but usually is taken as about 10 for practical design.

One of the two ψ values is called ψ_A and the other is called ψ_B. After these values are computed, the effective length factor k is obtained by placing a straightedge between ψ_A and ψ_B. The point where the straightedge crosses the middle nomograph is k.

It can be seen that the ψ factors used to enter the alignment charts and thus the resulting effective length factors are dependent on the relative stiffnesses of the compression and flexural members. If we have a very light flexible column and large stiff girders, the rotation and lateral movement of the column ends will be greatly minimized. The column ends will be close to a fixed condition, and thus the ψ values and the resulting k values will be small. Obviously if the reverse happens—that is, large stiff columns framing into light flexible girders—the column ends will rotate almost freely, approaching a pinned condition. Consequently, we will have large ψ and k values.

To calculate the ψ values it is necessary to use realistic moments of inertia. Usually, the girders will be appreciably cracked on their tensile sides, whereas the columns will probably have only a few cracks. If the I values for the girders are underestimated a little, the column k factors will be a little large and thus on the safe side.

---

[1]Structural Stability Research Council, *Guide to Stability Design Criteria for Metal Structures*, 4th ed., T. V. Galambos, ed. (New York: Wiley, 1988).

[2]Julian, O. G., and Lawrence, L. S., 1959, "Notes on J and L Nomograms for Determination of Effective Lengths," unpublished. These are also called the Jackson and Moreland Alignment Charts, after the firm with whom Julian and Lawrence were associated.

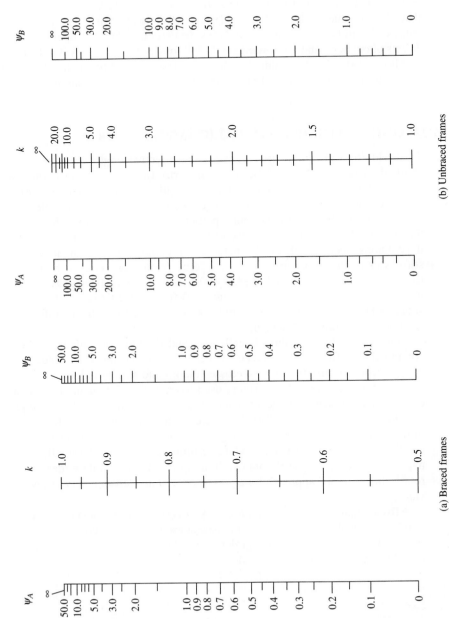

Figure 11.3 Effective length factors. ψ = ratio of $\Sigma(EI/\ell)$ of compression members to $\Sigma(EI/\ell)$ of flexural members in a plane at one end of a compression member. k = effective length factor.

Several approximate rules are in use for estimating beam and column rigidities. One common practice of the past for slenderness ratios of up to about 60 or 70 was to use gross moments of inertia for the columns and 50% of the gross moments of inertia for the beams.

In ACI Section 10.10.4.1, it is stated that for determining ψ values for use in evaluating k factors, the rigidity of the beams may be calculated on the basis of $0.35I_g$ to account for cracking and reinforcement, while $0.70I_g$ may be used for compression members. This practice is followed for the examples in this chapter. Other values for the estimated rigidity of walls and flat plates are provided in the same section.

11.5   DETERMINING k FACTORS WITH EQUATIONS

Instead of using the alignment charts for determining k values, an alternate method that involves the use of relatively simple equations. These equations, which were in the ACI 318-05 Code Commentary (R10.12.1) and taken from the British Standard Code of Practice,[3] are particularly useful with computer programs.

For braced compression members, an upper bound to the effective length factor may be taken as the smaller value determined from the two equations to follow in which ψ_A and ψ_B are the values just described for the alignment charts (commonly called the Jackson and Moreland alignment charts as described in footnote 2 of this chapter). ψ_{min} is the smaller of ψ_A and ψ_B.

$$k = 0.7 + 0.05(\psi_A + \psi_B) \leq 1.0$$

$$k = 0.85 + 0.05\psi_{min} \leq 1.0$$

The value of k for unbraced compression members restrained at both ends may be determined from the appropriate one of the following two equations, in which ψ_m is the average of ψ_A and ψ_B:

If $\psi_m < 2$

$$k = \frac{20 - \psi_m}{20}\sqrt{1 + \psi_m}$$

If $\psi_m \geq 2$

$$k = 0.9\sqrt{1 + \psi_m}$$

The value of the effective length factor of unbraced compression members that are hinged at one end may be determined from the following expression, in which ψ is the value at the restrained end:

$$k = 2.0 + 0.3\psi$$

As previously mentioned in Section 11.3 of this chapter the ACI Code in Section 10.10.6.3 states that k should be taken to be 1.0 for compression members in frames braced against sidesway unless a theoretical analysis shows that a lesser value can be used. *In the last paragraph of Section R10.10.6.3 of the Commentary, use of the alignment charts or the equations just presented is said to be satisfactory for justifying k values less than 1.0 for braced frames.*

---

[3]*Code of Practice for the Structural Use of Concrete* (CP110: Part 1), British Standards Institution, London, 1972, 154 pp.

Reinforced concrete columns. (Courtesy of RSM Advertising, representing Molded Fiber Glass Products Company.)

11.6 FIRST-ORDER ANALYSES USING SPECIAL MEMBER PROPERTIES

After this section, the remainder of this chapter is devoted to an approximate design procedure wherein the effect of slenderness is accounted for by computing moment magnifiers which are multiplied by the column moments. A magnifier for a particular column is a function of its factored axial load P_u and its critical buckling load P_c.

Before moment magnifiers can be computed for a particular structure, it is necessary to make a first-order analysis of the structure. The member section properties used for such an analysis should take into account the influence of axial loads, the presence of cracked regions in the members, and the effect of the duration of the loads. Instead of making such an analysis, ACI Code 10.10.4.1 permits use of the following properties for the members of the structure. These properties may be used for both nonsway and sway frames.

(a) *Modulus of elasticity* determined from the following expression given in Section 8.5.1 of the Code:
$E_c = w_c^{1.5} 33 \sqrt{f_c'}$ for values of w_c from 90 to 155 lb/ft^3 or $57,000 \sqrt{f_c'}$ for normal-weight concrete.

(b) *Moments of inertia* where I_g = moment of inertia of gross concrete section about centroidal axis neglecting reinforcing (ACI Section 10.10.4.1):

Beams	$0.35I_g$
Columns	$0.70I_g$
Walls—Uncracked	$0.70I_g$
—Cracked	$0.35I_g$
Flat plates and flat slabs	$0.25I_g$

(c) *Area* $1.0A_g$

As an alternative to the above approximate equations for columns and walls, the Code permits the following more complex value for moment of inertia:

$$I = \left(0.80 + 25\frac{A_{st}}{A_g}\right)\left(1 - \frac{M_u}{P_u h} - 0.5\frac{P_u}{P_0}\right)I_g \leq 0.875 I_g \qquad \text{(ACI Equation 10-8)}$$

P_u and M_u are to be from the load combination under consideration, or they can conservatively be taken as the values of P_u and M_u that result in the lowest value of I. In no case is the value of I for compression members required to be taken less than $0.35 I_g$. P_0 is the theoretical concentric axial load strength (see Chapter 9 of this textbook).

For flexural members (beams and flat plates and flat slabs) the following approximate equation is permitted:

$$I = (0.10 + 25\rho)\left(1.2 - 0.2\frac{b_w}{d}\right)I_g \leq 0.5 I_g \qquad \text{(ACI Equation 10-9)}$$

For continuous flexural members, it is permitted to use the average value of I from the positive and negative moment sections. In no case is the value of I for flexural members required to be taken less than $0.25 I_g$.

Often during the design process, the designer does not know the final values of member section dimensions or steel areas when making calculations such as those in ACI Equation 10-8. This leads to an iterative process where the last cycle of iteration assumes the same member properties as the final design. The Code (10.10.4.1) allows these values to be only within 10% of the final values in the final iteration.

11.7   SLENDER COLUMNS IN NONSWAY AND SWAY FRAMES

There is a major difference in the behavior of columns in nonsway or braced frames and those in sway or unbraced frames. In effect, each column in a braced frame acts by itself. In other words, its individual strength can be determined and compared to its computed factored loads and moments. In an unbraced or sway frame, a column will probably not buckle individually but will probably buckle simultaneously with all of the other columns on the same level. As a result, it is necessary in a sway frame to consider the buckling strength of all the columns on the level in question as a unit.

For a compression member in a nonsway frame, the effective slenderness ratio $k\ell_u/r$ is used to determine whether the member is short or slender. For this calculation, ℓ_u is the unbraced length of the member. The effective length factor k can be taken as 1.0 unless an analysis provides a lesser value. The radius of gyration r is equal to 0.25 times the diameter of a round column and 0.289 times the dimension of a rectangular column in the direction that stability is being considered. The ACI Code (10.10.1.2) permits the approximate value of 0.30 to be used in place of 0.289, and this is done herein. For other sections the value of r will have to be computed from the properties of the gross sections.

For nonsway frames, slenderness effects may be ignored if the following expression is satisfied:

$$\frac{k\ell_u}{r} \leq 34 - 12\left(\frac{M_1}{M_2}\right) \qquad \text{(ACI Equation 10-7)}$$

In this expression, M_1 is the smaller factored end moment in a compression member. It has a plus sign if the member is bent in single curvature (C-shaped) and a negative sign if the member is bent in double curvature (S-shaped). M_2 is the larger factored end moment in a compression member, and it always has a plus sign. In this equation the term $\left(34 - 12\frac{M_1}{M_2}\right)$ shall not be taken larger than 40, according to ACI Code 10.10.1.

For sway frames, slenderness effects may be ignored if

$$k\ell_u/r < 22 \qquad \text{(ACI Equation 10-6)}$$

Should $k\ell_u/r$ for a particular column be larger than the applicable ratio, we will have a slender column. For such a column the effect of slenderness must be considered. This may be done by using approximate methods or by using a theoretical second-order analysis that takes into account the effect of deflections. Second-order effects cannot exceed 40% of first-order effects (ACI 10.10.2.1).

A second-order analysis is one that takes into account the effect of deflections and also makes use of a reduced tangent modulus. The equations necessary for designing a column in this range are extremely complicated, and, practically, it is necessary to use column design charts or computer programs.

Avoiding Slender Columns

The design of slender columns is appreciably more complicated than the design of short columns. As a result, it may be wise to give some consideration to the use of certain minimum dimensions so that none of the columns will be slender. In this way they can be almost completely avoided in the average-size building.

If k is assumed equal to 1.0, slenderness can usually be neglected in braced frame columns if ℓ_u/h is kept to 10 or less on the first floor and 14 or less for the floors above the first one. To determine these values it was assumed that little moment resistance was provided at the footing–column connection and the first-floor columns were assumed to be bent in single curvature. Should the footing–column connection be designed to have appreciable moment resistance, the maximum ℓ_u/h value given above as 10 should be raised to about 14 or equal to the value used for the upper floors.[4]

Should we have an unbraced frame and assume $k = 1.2$, it is probably necessary to keep ℓ_u/h to 6 or less. So for a 10-ft clear floor height it is necessary to use a minimum h of about $10\,\text{ft}/6 = 1.67\,\text{ft} = 20\,\text{in.}$ in the direction of bending to avoid slender columns.

Example 11.1 illustrates the selection of the k factor and the determination of the slenderness ratio for a column in an unbraced frame. For calculating I/L values the author used 0.70 times the gross moments of inertia for the columns, 0.35 times the gross moments of inertia for the girders, and the full lengths of members center to center of supports.

EXAMPLE 11.1

(a) Using the alignment charts of Figure 11.3, calculate the effective length factor for column AB of the braced frame of Figure 11.4. Consider only bending in the plane of the frame.

(b) Compute the slenderness ratio of column AB. Is it a short or a slender column? *The maximum permissible slenderness ratio for a short unbraced column is 22, as will be described in Section 11.9 of this chapter.* End moments on the column are $M_1 = 45$ ft-k and $M_2 = 75$ ft-k, resulting in single curvature.

---

[4]Neville, G. B., ed., 1984, *Simplified Design Reinforced Concrete Buildings of Moderate Size and Height* (Skokie, IL: Portland Cement Association), pp. 5-10–5-12.

SOLUTION

(a) Effective Length Factor for Column AB

Using the Reduced Moments of Inertia from 11.6(b) and applying the method described in Section 11.4

$$\psi_A = \frac{\dfrac{0.7 \times 8000}{12 \times 10}}{\left[\dfrac{0.35 \times 5832}{12 \times 20} + \dfrac{0.35 \times 5832}{12 \times 24}\right]} = 2.99$$

$$\psi_B = \frac{\dfrac{0.7 \times 8000}{12 \times 10} + \dfrac{0.7 \times 8000}{12 \times 12}}{\left[\dfrac{0.35 \times 13{,}824}{12 \times 20} + \dfrac{0.35 \times 13{,}824}{12 \times 24}\right]} = 2.31$$

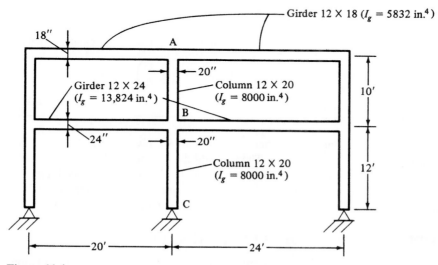

Girder 12 × 18 ($I_g = 5832$ in.⁴)

18″

A

20″

Girder 12 × 24
($I_g = 13{,}824$ in.⁴)

Column 12 × 20
($I_g = 8000$ in.⁴)

10′

B

24″

20″

Column 12 × 20
($I_g = 8000$ in.⁴)

12′

C

20′

24′

Figure 11.4

Adjustable steel column forming system. (Courtesy of Symons Corporation.)

From Figure 11.3(a)

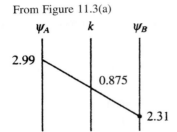

(b) Is it a Slender Column?

$$\ell_u = 10\ \text{ft} - \frac{9+12}{12} = 8.25\ \text{ft}$$

$$\frac{k\ell_u}{r} = \frac{(0.875) \times (12 \times 8.25)}{0.3 \times 20} = 14.44 < \text{Maximum}\ \frac{kl_u}{r}\ \text{for a short column in an braced}$$

$$\text{frame by ACI Equation 10-7} = 34 - 12\left(\frac{+45}{+75}\right) = 26.8$$

$$\therefore \text{It's not a slender column}$$

An experienced designer would first simply assume $k = 1$ and quickly see that $k\ell_u/r = \ell_u/r = 16.5 < 26.5$. There would then be no need to determine k.

If this column were in the same frame, but the frame unbraced, then k would be 1.78 and $k\ell_u/r = 29.37 > 22$. It would be a slender column. The only difference in determining k is the use of Figure 11.3(b) for sway columns instead of Figure 11.3(a) for nonsway columns.

11.8 ACI CODE TREATMENT OF SLENDERNESS EFFECTS

The ACI Code permits the determination of second-order effects by one of three methods. The first is by a *nonlinear second-order analysis* (ACI 10.10.3). Such an analysis must consider nonlinearity of materials, member curvature and lateral drift, load duration, volume changes in concrete due to creep and shrinkage, and foundation or support interaction. The analysis technique should predict the ultimate loads to within 15% or test results on statically indeterminate reinforced concrete structures. This technique would require sophisticated computer software that has been demonstrated to satisfy the 15% accuracy requirement mentioned previously.

The second method is by an *elastic second-order analysis* (ACI 10.10.4). This technique is simpler than the nonlinear method because it uses member stiffnesses immediately prior to failure. Values of E_c and moments of inertia and cross-sectional area for columns, beams, walls, flat plates, and flat slabs that are permitted to be used in the elastic second-order analysis are listed in Section 11.6. This method would also most likely require a computer analysis.

The third method is the *moment magnifier procedure* (ACI 10.10.5). Different procedures for this method are given for sway and nonsway structures. The next two sections describe the moment magnifier method for these two cases.

11.9 MAGNIFICATION OF COLUMN MOMENTS IN NONSWAY FRAMES

When a column is subjected to moment along its unbraced length, it will be displaced laterally in the plane of bending. The result will be an increased or secondary moment equal to the axial load times the lateral displacement or eccentricity. In Figure 11.5 the load P causes the column moment

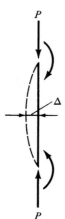

P

Figure 11.5 Moment magnification in a nonsway column.

to be increased by an amount $P\Delta$. This moment will cause δ to increase a little more, with the result that the $P\Delta$ moment will increase, which in turn will cause a further increase in Δ and so on until equilibrium is reached.

We could take the column moments, compute the lateral deflection, increase the moment by $P\Delta$, recalculate the lateral deflection and the increased moment, and so on. Although about two cycles would be sufficient, this would still be a tedious and impractical procedure.

It can be shown[5] that the increased moment can be estimated very well by multiplying the primary moment by $1/(1 - P/P_c)$, where P is the axial load and P_c is the Euler buckling load $\pi^2 EI/(k\ell_u)^2$.

In Example 11.2 this expression is used to estimate the magnified moment in a laterally loaded column. It will be noted that in this problem the primary moment of 75 ft-k is estimated to increase by 7.4 ft-k. If we computed the deflection due to the lateral load, we would get 0.445 in. For this value $P\Delta = (150)(0.445) = 66.75$ in.-k $= 5.6$ ft-k. This moment causes more deflection, which causes more moment, and so on.

EXAMPLE 11.2

(a) Compute the primary moment in the column shown in Figure 11.6 due to the lateral 20-k load.

(b) Determine the estimated total moment, including the secondary moment due to lateral deflection using the appropriate magnification factor just presented. $E = 3.16 \times 10^3$ ksi. Assume $k = 1.0$ and $\ell_u = 15$ ft.

SOLUTION   (a) Primary moment due to lateral load:

$$M_u = \frac{(20)(15)}{4} = \underline{\underline{75 \text{ ft-k}}}$$

(b) Total moment including secondary moment:

$$P_c = \text{Euler buckling load} = \frac{\pi^2 EI}{(k\ell_u)^2} \qquad \text{(ACI Equation 10-13)}$$

$$= \frac{(\pi^2)(3160)\left(\dfrac{1}{12} \times 12 \times 12^3\right)}{(1.0)(12 \times 15)^2} = 1663.4 \text{ k}$$

---

[5]Timoshenko, S. P., and Gere, J. M., 1961, *Theory of Elastic Stability*, 2nd ed. (New York: McGraw-Hill), pp. 319–356.

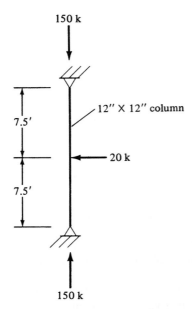

150 k

12″ X 12″ column

7.5′

20 k

7.5′

150 k   **Figure 11.6**

$$\text{Magnified moment} = 75\,\dfrac{1}{1-\dfrac{P}{P_c}}$$

$$= 75\,\dfrac{1}{1-\dfrac{150}{1663.4}} = \underline{\underline{82.4 \text{ ft-k}}}$$

As we have seen, it is possible to calculate approximately the increased moment due to lateral deflection by using the $(1 - P/P_c)$ expression. In ACI Code 10.10.16 the factored design moment for slender columns with no sway is increased by using the following expression, in which M_c is the magnified or increased moment and M_2 is the larger factored end moment on a compression member:

$$M_c = \delta M_2 \qquad \text{(ACI Equation 10-11)}$$

Should our calculations provide very small moments at both column ends, the Code provides an absolutely minimum value of M_2 to be used in design. In effect, it requires the computation of a moment based on a minimum eccentricity of $0.6 + 0.03h$, where h is the overall thickness of the member perpendicular to the axis of bending.

$$M_{2,\min} = P_u(0.6 + 0.03h) \qquad \text{(ACI Equation 10-17)}$$

Or in SI units

$$M_{2\,\min} = P_u(15 + 0.03h), \text{ where } h \text{ is in mm, as is the number 15.}$$

A moment magnifier δ is used to estimate the effect of member curvature between the ends of compression members. It involves a term C_m, which is defined later in this section.

$$\delta = \dfrac{C_m}{1 - \dfrac{P_u}{0.75P_c}} \geq 1.0 \qquad \text{(ACI Equation 10-12)}$$

The determination of the moment magnifier δ_{ns} involves the following calculations:

1. $E_c = 57,000\sqrt{f_c'}$ for normal-weight concrete (see Section 1.11 for other densities)
2. I_g = gross inertia of the column cross section about the centroidal axis being considered.
3. $E_s = 29 \times 10^6$ psi.
4. I_{se} = moment of inertia of the reinforcing about the centroidal axis of the section. (This value = the sum of each bar area times the square of its distance from the centroidal axis of the compression member.)
5. The term β_{dns} accounts for the reduction in stiffness caused by sustained axial loads and applies only to nonsway frames. It is defined as the ratio of the maximum factored sustained axial load divided by the total factored axial load associated with the same load combination. It is always assumed to have a plus sign and is never permitted to exceed 1.0.
6. Next it is necessary to compute *EI*. The two expressions given for *EI* in the Code were developed so as to account for creep, cracks, and so on. If the column and bar sizes have already been selected, or estimated, *EI* can be computed with the following expression, which is particularly satisfactory for columns with high steel percentages.

$$EI = \frac{(0.2E_cI_g + E_sI_{se})}{1 + \beta_{dns}} \qquad \text{(ACI Equation 10-14)}$$

The alternate expression for *EI* that follows is probably the better expression to use when steel percentages are low. Notice also that this expression will be the one used if the reinforcing has not been previously selected.

$$EI = \frac{0.4E_cI_g}{1 + \beta_{dns}} \qquad \text{(ACI Equation 10-15)}$$

7. The Euler buckling load is computed:

$$P_c = \frac{\pi^2 EI}{(k\ell_u)^2} \qquad \text{(ACI Equation 10-13)}$$

8. For some moment situations in columns, the amplification or moment magnifier expression provides moments that are too large. One such situation occurs when the moment at one end of the member is zero. For this situation the lateral deflection is actually about half of the deflection in effect provided by the amplification factor. Should we have approximately equal end moments that are causing reverse curvature bending, the deflection at mid-depth and the moment there are close to zero. As a result of these and other situations, the Code provides a modification factor (C_m) to be used in the moment expression that will result in more realistic moment magnification.

For braced frames without transverse loads between supports C_m can vary from 0.4 to 1.0 and is determined with the expression at the end of this paragraph. For all other cases it is to be taken as 1.0. (Remember the sign convention: M_1 is positive for single curvature and is negative for reverse curvature, and M_2 is always positive.)

$$C_m = 0.6 + 0.4\,\frac{M_1}{M_2} \geq 0.4 \qquad \text{(ACI Equation 10-16)}$$

Should $M_{2,\,\min}$ as computed with ACI Equation 10-17 be larger than M_2, the value of C_m above shall either be taken as equal to 1.0 or be based on the ratio of the computed end moments $\frac{M_1}{M_2}$ (ACI Section 10.10.6.4).

Example 11.3 illustrates the design of a column in a nonsway frame.

EXAMPLE 11.3

The tied column of Figure 11.7 has been approximately sized to the dimensions 12 in. × 15 in. It is to be used in a frame braced against sidesway. The column is bent in single curvature about its y-axis and has an ℓ_u of 16 ft. If $k = 0.83$, $f_y = 60,000$ psi, and $f_c' = 4000$ psi, determine the reinforcing required. Consider only bending in the plane of the frame. Note also that the unfactored dead axial load P_D is 30 k and concrete is normal weight.

SOLUTION

1. Is it a slender column?

$$\text{Max } \frac{k\ell_u}{r} \text{ for short columns} = 34 - 12\frac{M_1}{M_2} = 34 - 12\left(\frac{+82}{+86}\right)$$

$$= 22.56$$

$$\text{Actual } \frac{k\ell_u}{r} = \frac{(0.83)(12 \times 16)}{0.3 \times 15} = 35.41 > 22.56$$

∴ It's a slender column $k\ell_u/r$.

2. $E_c = 57,000\sqrt{f_c'} = 57,000\sqrt{4000} = 3,605,000 \text{ psi} = 3.605 \times 10^3 \text{ ksi}$

3. $I_g = \left(\frac{1}{12}\right)(12)(15)^3 = 3375 \text{ in.}^4$

4. $\beta_d = \dfrac{\text{factored axial dead load}}{\text{factored axial total load}} = \dfrac{(1.2)(30)}{110} = 0.327$

5. Because reinforcing has not been selected, we must use the second EI.

$$EI = \frac{0.4E_cI_g}{1+\beta_d} = \frac{(0.4)(3605)(3375)}{1+0.327} = 3.67 \times 10^6 \text{ k-in.}^2 \qquad \text{(ACI Equation 10-15)}$$

6. $P_c = \dfrac{\pi^2 EI}{(k\ell_u)^2} = \dfrac{(\pi^2)(3.67 \times 10^6)}{(0.83 \times 12 \times 16)^2} = 1426 \text{ k} \qquad \text{(ACI Equation 10-13)}$

7. $C_m = 0.6 + 0.4\dfrac{M_1}{M_2} = 0.6 + 0.4\left(\dfrac{+82}{+86}\right) = 0.981 \qquad \text{(ACI Equation 10-16)}$

8. $\delta = \dfrac{C_m}{1 - \dfrac{P_u}{0.75P_c}} = \dfrac{0.981}{1 - \dfrac{110}{(0.75)(1426)}} = 1.09 \qquad \text{(ACI Equation 10-12)}$

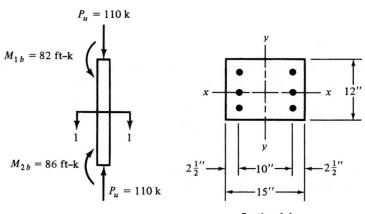

Section 1-1

Figure 11.7

9. $M_{2,\,min} = P_u(0.6 + 0.03h) = 110(0.6 + 0.03 \times 15)$

$\qquad = 115.5 \text{ in.-k} = 9.6 \text{ ft-k}$ $\qquad\qquad$ (ACI Equation 10-17)

10. $M_c = \delta M_2 = (1.09)(86) = 93.7 \text{ ft-k}$ $\qquad\qquad$ (ACI Equation 10-11)

11. Magnified $e = \delta e = \frac{(12)(93.7)}{110} = 10.22 \text{ in.}$

12. $\lambda = \frac{10}{15} = 0.667 \therefore \rho_g$ is determined by interpolation between values presented in Appendix A, Graphs 2 and 3.

$$P_n = \frac{P_u}{\phi} = \frac{110}{0.65} = 169.23 \text{ k}$$

$$K_n = \frac{P_n}{f_c' A_g} = \frac{169.23}{(4)(12 \times 15)} = 0.235$$

$$R_n = \frac{P_n}{f_c' A_g} \frac{\delta e}{h} = (0.235)\left(\frac{10.22}{15}\right) = 0.160$$

$$\rho_g = 0.0160$$

$$A_g = (0.0160)(12)(15) = 2.88 \text{ in.}^2 \qquad\qquad \text{Use 4 \#8 bars } (3.14 \text{ in.}^2)$$

Since K_n and R_n are between the radial lines labelled $f_s/f_y = 1.0$ and $\varepsilon_t = 0.005$ on the interaction diagrams, the ϕ factor is permitted to be increased from the 0.65 value used. If the Chapters 9 and 10 spreadsheet for rectangular columns is used, an area of reinforcing of only 1.80 in.^{2} is found to be sufficient. This significant reduction occurs because of the increased ϕ factor in this region.

Most columns are designed for multiple load combinations, and the designer must be certain that the column is able to resist all of them. Often there are some with high axial load and low moment, such as $1.2D + 1.6L$, and others with low axial load and high moment, such as $0.9D + 1.6E$. The first of these is likely to have a ϕ factor of 0.65. The second, however, is more likely to be eligible for the increase in the ϕ factor.

In this example the author assumed that the frame was braced, and yet we have said that frames are often in that gray area between being fully braced and fully unbraced. *Assuming a frame is fully braced clearly may be quite unconservative.*

11.10   MAGNIFICATION OF COLUMN MOMENTS IN SWAY FRAMES

Tests have shown that even though the lateral deflections in unbraced frames are rather small, their buckling loads are far less than they would be if the frames had been braced. As a result, the buckling strengths of the columns of an unbraced frame can be decidedly increased (perhaps by as much as two or three times) by providing bracing.

If a frame is unbraced against sidesway, it is first necessary to compute its slenderness ratio. If $k\ell_u/r$ is less than 22, slenderness may be neglected (ACI 10.10.1). For this discussion it is assumed that values > 22 are obtained.

When sway frames are involved, it is necessary to decide for each load combination which of the loads cause appreciable sidesway (probably the lateral loads) and which do not. The factored end moments that cause sidesway are referred to as M_{1s} and M_{2s}, and they must be magnified because of the $P\Delta$ effect. The other end moments, resulting from loads that do not cause appreciable sidesway, are M_{1ns} and M_{2ns}. They are determined by first-order analysis and will not have to be magnified.

The Code (10.10.7) states that the moment magnifier δ_s can be determined by one of the following two methods.

1. The moment magnifier may be calculated with the equation given at the end of this paragraph in which Q is the stability index previously presented in Section 11.2 of this chapter. Should the computed value of δ_s be > 1.5 it will be necessary to compute δ_s by ACI Section 10.10.7.4 or by a second-order analysis.

$$\delta_s = \frac{1}{1-Q} \geq 1 \qquad \text{(ACI Equation 10-20)}$$

2. With the second method and the one used in this chapter the magnified sway moments may be computed with the following expression:

$$\delta_s = \frac{1}{1 - \dfrac{\Sigma P_u}{0.75 \Sigma P_c}} \geq 1 \qquad \text{(ACI Equation 10-21)}$$

In this last equation ΣP_u is the summation of all the vertical loads in the story in question, and ΣP_c is the sum of all the Euler buckling loads $\left(P_c = \frac{\pi^2 EI}{(k\ell_u)^2}\right)$ for all of the sway resisting columns in the story with k values determined as described in ACI Section 10.10.7.2. This formula reflects the fact that the lateral deflections of all the columns in a particular story are equal and thus the columns are interactive.

Whichever of the preceding methods is used to determine the δ_s values, the design moments to be used must be calculated with the expressions that follow.

$$M_1 = M_{1ns} + \delta_s M_{1s} \qquad \text{(ACI Equation 10-18)}$$

$$M_2 = M_{2ns} + \delta_s M_{2s} \qquad \text{(ACI Equation 10-19)}$$

First Assembly Church in Wyoming, MI. (Courtesy of Veneklasen Concrete Construction Co.)

Sometimes the point of maximum moment in a slender column will fall between its ends. The ACI Commentary (R10.10.2.2) says the moment magnification for this case may be evaluated using the procedure described for nonsway frames (ACI 10.10.6).

Example 11.4 illustrates the design of a slender column subject to sway.

EXAMPLE 11.4

Select reinforcing bars using the moment magnification method for the 18 in. × 18 in. unbraced column shown in Figure 11.8 if $\ell_u = 17.5$ ft, $k = 1.3$, $f_y = 60$ ksi, and $f_c' = 4$ ksi. A first-order analysis has resulted in the following axial loads and moments:

$$P_D = 300 \text{ k} \qquad M_D = 48 \text{ k-ft}$$

$$P_L = 150 \text{ k} \qquad M_L = 25 \text{ k-ft}$$

$$P_W = 170 \text{ k} \qquad M_W = 29 \text{ k-ft}$$

The loading combination assumed to control for the case with no sidesway is ACI Equation 9.4 (Section 4.1 of this text).

$$P_U = 1.2P_D + 1.6P_L = 1.2(300) + 1.6(150) = 600 \text{ k}$$

$$M_U = 1.2M_D + 1.6M_L = 1.2(48) + 1.6(25) = 97.6 \text{ k-ft} = M_{2ns}$$

The loading combination assumed to control with sidesway is ACI Equation 9.6

$$P_U = 0.9P_D + 1.6P_W = 0.9(300) + 1.6(170) = 542 \text{ k}$$

$$M_U = 0.9M_D + 1.6M_W = 0.9(48) + 1.6(20) = 89.6 \text{ k-ft} = M_{2s}$$

Note that ACI Equation 9.3 may also control for sidesway, but in this case it is unlikely.

$$\Sigma P_u = 12,000 \text{ for all columns on floor}$$

$$\Sigma P_c = 60,000 \text{ for all columns on floor}$$

SOLUTION   *Is it a slender column? (ACI 10.13.2)*

$$\frac{k\ell_u}{r} = \frac{(1.3)(12)(17.5)}{(0.3)(18)} = 50.55 > 22 \qquad \underline{\underline{\text{Yes}}}$$

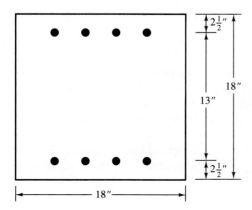

Figure 11.8

Calculating the Magnified Moment δ_s

$$\delta_s = \frac{1}{1 - \dfrac{\Sigma P_u}{0.75 \Sigma P_c}}$$

(ACI Equation 10-18)

$$= \frac{1}{1 - \dfrac{12,000}{(0.75)(60,000)}} = 1.364$$

Computing Magnified Moment M_2

$$M_2 = M_{2ns} + \delta_s M_{2s}$$

(ACI Equation 10-19)

$$= 97.6 + (1.364)(89.6) = 219.8 \text{ ft-k}$$

Is $M_{2ns} \geq$ minimum value permitted in ACI Section 10.10.6.5

$$M_{2\ \min} = P_u(0.6 + 0.03h)$$

$$= (544)(0.6 + 0.03 \times 18) = 620.2 \text{ in.-k}$$

(ACI Equation 10-17)

$$= 51.7 \text{ ft-k} \ < 97.6 \text{ ft-k}$$

Yes

Selecting Reinforcing

$$\gamma = \frac{13}{18} = 0.722 \text{ with reference to Figure 11.8}$$

$$P_n = \frac{P_u}{\phi} = \frac{542}{0.65} = 833.8$$

$$e = \frac{(12)(219.8)}{542} = 4.87$$

$$K_n = \frac{P_n}{f'_c A_g} = \frac{833.8}{(4)(18 \times 18)} = 0.643$$

$$R_n = \frac{P_n}{f'_c A_g}\frac{e}{h} = (0.643)\left(\frac{4.87}{18}\right) = 0.174$$

By interpolation between Appendix A Graphs 3 and 4, we find $\rho_z = 0.023$

$$A_z = \rho_z A_g = (0.023)(18 \times 18) = 7.45$$

Use 8 # 9 Bars (9.00)

11.11 ANALYSIS OF SWAY FRAMES

The frame of Figure 11.9 is assumed to be unbraced in the plane of the frame. It supports a uniform gravity load w_u and a short-term concentrated lateral load P_w. As a result, it is necessary to consider both the moments due to the loads that do not cause appreciable sidesway as well as the loads that do. It will therefore be necessary to compute both δ and δ_s values if the column proves to be slender.

The M_s values are obviously caused by the lateral load in this case. The reader should realize, however, that if the gravity loads and/or the frame are unsymmetrical, additional M_s or sidesway moments will occur.

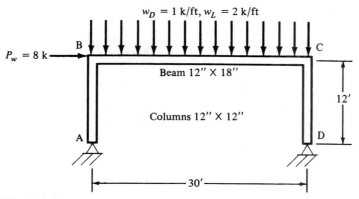

Figure 11.9

If we have an unbraced frame subjected to short-term lateral wind or earthquake loads, the columns will not have appreciable creep (which would increase lateral deflections and thus the $P\Delta$ moments). The effect of creep is accounted for in design by reducing the stiffness EI used to calculate P_c and thus δ_s by dividing EI by $1 + \beta_{dns}$ as specified in ACI Section 10.10.6.1. Both the concrete and steel terms in ACI Equation 10-14 are divided by this value. In the case of sustained lateral load, such as soil backfill or water pressure, ACI Section 10.10.4.2 requires that the moments of inertia for compression members in Section 11.6 be divided by $(1 + \beta_{ds})$. The term β_{ds} is the ratio of the maximum factored sustained shear within a story to the maximum factored shear in that story for the same load combination.

To illustrate the computation of the magnified moments needed for the design of a slender column in a sway frame, the authors have chosen the simple frame of Figure 11.9. We hope thereby that the student will not become lost in a forest of numbers as he or she might if a large frame were considered.

The beam and columns of the frame have been tentatively sized as shown in the figure. In Example 11.5 the frame is analyzed for each of the conditions specified in ACI Section 9.2 using $1.6W$.

In the example the magnification factors δ and δ_s are computed for each of the loading conditions and used to compute the magnified moments. Notice in the solution that different k values are used for determining δ and δ_s. The k for the δ calculation is determined from the alignment chart of Figure 11.3(a) for braced frames, whereas the k for the δ_s calculation is determined from the alignment chart of Figure 11.3(b) for unbraced frames.

EXAMPLE 11.5

Determine the moments and axial forces that must be used for the design of column CD of the unbraced frame of Figure 11.9. Consider only bending in the plane of the frame. The assumed member sizes shown in the figure are used for the analyses given in the problem. $f_y = 60{,}000$ psi and $f_c' = 4000$ psi. For this example, the authors considered the load factor cases of ACI Equations 9-1, 9-4, and 9-6. For other situations other appropriate ACI load factor equations will have to be considered.

SOLUTION

1. *Determine the effective length factor for the sway case using 0.35 I_g for the girder and 0.70 I_g for the columns.*

$$I_{\text{column}} = (0.70)\left(\frac{1}{12}\right)(12)(12)^3 = 1210 \text{ in.}^4$$

Note that if the lateral load were sustained, I_{column} would be divided by $(1 + \beta_{ds})$.

$$I_{\text{girder}} = (0.35)\left(\frac{1}{12}\right)(12)(18)^3 = 2041 \text{ in.}^4$$

$$\psi_B = \frac{\dfrac{1210}{12}}{\dfrac{2041}{30}} = 1.48$$

$$\psi_A = \infty \text{ for pinned ends} \qquad \text{(For practical purposes, use 10)}$$

$$k = 1.95 \text{ from Figure 11.3(b)}$$

2. *Is it a slender column?*

$$\ell_u = 12 - \frac{9}{12} = 11.25 \text{ ft}$$

$$\text{Max } \frac{k\ell_u}{4} \text{ to be a short column} = 22 \text{ for nonsway frames}$$

$$\frac{k\ell_u}{r} = \frac{(1.95)(12 \times 11.25)}{0.3 \times 12} = 73.12 > 22$$

$$\therefore \text{ It is a slender column}$$

3. *Consider the loading case U = 1.2D + 1.6L (See Figure 11.10).*

a. *Are column moments $\geq$ ACI minimum?*

$$e_{\min} = 0.6 + 0.03 \times 12 = 0.96 \text{ in.}$$

$$M_{2\,\text{Min}} = (66)(0.96) = 63.36 \text{ in.-k} = 5.28 \text{ ft-k} < 173.5 \text{ ft-k} \quad \text{(see Figure 11.10)} \qquad \underline{\underline{\text{OK}}}$$

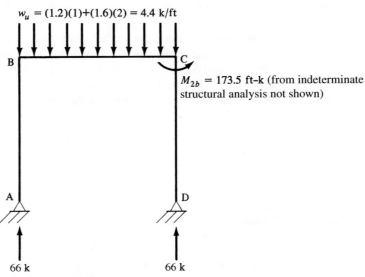

$w_u = (1.2)(1)+(1.6)(2) = 4.4$ k/ft

B   C

$M_{2b} = 173.5$ ft-k (from indeterminate structural analysis not shown)

A   D

66 k   66 k

Figure 11.10 Loading $1.2D + 1.6L$.

b. *Compute the magnification factor δ:*

$$E_c = 57,000\sqrt{4000} = 3,605,000 \text{ psi} = 3605 \text{ ksi}$$

$$\beta_d = \frac{(1.2)(1)}{(1.2)(1) + (1.6)(2)} = 0.273$$

$$EI = \frac{(0.4)(3605)(1728)}{1 + 0.273} = 1.96 \times 10^6 \text{ k-in.}^2$$

Assuming conservatively that $k = 1.0$ for computing P_c

$$P_c = \frac{(\pi^2)(1.96 \times 10^6)}{(1.0 \times 12 \times 11.25)^2} = 1061 \text{ k}$$

$$C_m = 0.6 + (0.4)\left(\frac{-0}{+173.5}\right) = 0.6$$

$$\delta_{ns} = \frac{0.6}{1 - \dfrac{66}{(0.75)(1061)}} = 0.65 < 1.0 \qquad\qquad \underline{\underline{\text{Use } 1.0}}$$

c. *Compute the magnification factor δ_s:*
   Using $k = 1.95$ as given for determining P_c

$$P_c = \frac{(\pi^2)(1.96 \times 10^6)}{(1.95 \times 12 \times 11.25)^2} = 279.1 \text{ k}$$

$$\delta_s = \frac{1}{1 - \dfrac{\Sigma P_u}{0.75\Sigma P_c}} = \frac{1}{1 - \dfrac{(2)(66)}{(0.75)(2 \times 279.1)}} = 1.46$$

d. *Compute the magnified moment:*

$$M_c = (1.0)(173.5) + (1.47)(0) = \underline{\underline{173.9 \text{ ft-k}}}$$

4. *Consider the loading case $U = (1.2D + 1.0L + 1.6W)$ as specified in ACI Code Section 9.2.1(b). Analysis results are shown in Figure 11.11.*

a. *Are column moments $\geq$ ACI minimum?*

$$e_{min} = 0.6 + 0.03 \times 12 = 0.96 \text{ in.}$$

$$M_{2\,min} = (48)(0.96) = 46.08 \text{ in.-k} = 3.84 \text{ ft-k} < 126.2 \text{ ft-k} \qquad\qquad \underline{\underline{\text{OK}}}$$

b. *Computing δ:*
β_{ns}, EI, and P_c are same as before

$$C_m = 0.6 + 0.4\left(\frac{-0}{+14}\right) = 0.6$$

$$\delta = \frac{0.6}{1 - \dfrac{48 + 5.12}{0.75 \times 1061}} = 0.64 \qquad\qquad \underline{\underline{\text{Use } 1.0}}$$

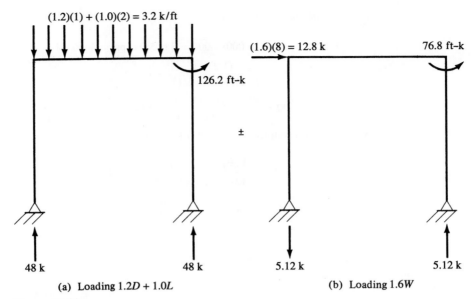

(a) Loading $1.2D + 1.0L$ (b) Loading $1.6W$

Figure 11.11

c. *Computing δ_s:*

$$\beta_{dns} = \frac{1.2D}{1.2D + 1.0L + 1.6W} = \frac{18}{18 + 30 + 5.12} = 0.339$$

$$EI = \frac{(0.4)(3605)(1728)}{1 + 0.339} = 1.86 \times 10^6 \text{ k-in.}^2$$

$$P_c = \frac{(\pi^2)(1.86 \times 10^6)}{(1.95 \times 12 \times 11.25)^2} = 264.9 \text{ k}$$

$$\delta_s = \frac{1}{1 - \dfrac{(2)(48) + 5.12 - 5.12}{0.75 \times 2 \times 264.9}} = 1.32$$

d. *Compute the magnified moment:*

$$M_c = (1.0)(126.2) + (1.32)(76.8) = \underline{227.6 \text{ ft-k}}$$

5. *Consider the loading case $0.9D + 1.6W$. Analysis results are shown in Figure 11.12.*

a. *Are column moments $\geq$ ACI minimum?*

$$e_{min} = 0.6 + (0.03)(12) = 0.96 \text{ in.}$$
$$M_{2\,min} = (13.5)(0.96) = 12.96 \text{ in.-k} = 1.08 \text{ ft-k} < 35.5 \text{ ft-k} \qquad \underline{\underline{OK}}$$

b. *Computing δ:*

$$\beta_{dns} = \frac{0.9D}{0.9D + 1.6W} = \frac{13.5}{13.5 + 5.12} = 0.725$$

$$EI = \frac{(0.4)(3605)(1728)}{1 + 0.725} = 1.44 \times 10^6 \text{ k-in.}^2$$

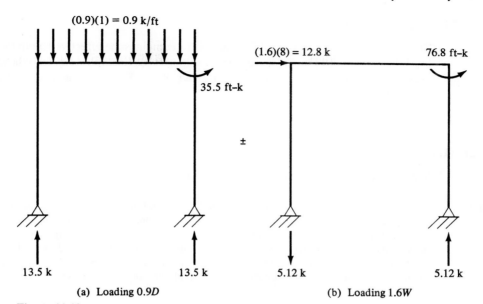

(a) Loading 0.9D   (b) Loading 1.6W

Figure 11.12

$$P_c = \frac{(\pi^2)(1.44 \times 10^6)}{(1.00 \times 12 \times 11.25)^2} = 780 \text{ k}$$

$$C_m = 0.6 + 0.4\left(\frac{-0}{35.5}\right) = 0.6$$

$$\delta = \frac{0.6}{1 - \dfrac{13.5 + 13.5}{0.75 \times 2 \times 780}} = 0.61 < 1.0 \qquad\qquad \underline{\underline{\text{Use 1.0}}}$$

c. *Computing δ_s:*

$$\beta_d = 0.725 \text{ from previous step}$$
$$EI = 1.44 \times 10^6 \text{ k-in.}^2$$
$$P_c = 780/1.95 = 400 \text{ k}$$

$$\delta_s = \frac{1}{1 - \dfrac{(2)(13.5) + 5.12 - 5.12}{0.75 \times 2 \times 400}} = 1.05$$

d. *Calculate moment*:

$$M_c = (1.0)(35.5) + (1.05)(76.8) = \underline{116.1 \text{ ft-k}}$$

6. *Summary of axial loads and moments to be used in design*:

Loading I: $P_u = 66 \text{ k}, M_c = 173.5 \text{ ft-k}$

Loading II: $P_u = 48 + 5.12 = 53.12 \text{ k}, \ M_c = 227.6 \text{ ft-k}$

Loading III: $P_u = 13.5 + 5.12 = 18.62 \text{ k}, \ M_c = 116.1 \text{ ft-k}$

Note: Should the reader now wish to determine the reinforcing needed for the above loads and moments, he or she will find that the steel percentage is much too high. As a result, a larger column will have to be used.

11.12 COMPUTER EXAMPLES

The Excel spreadsheet provided for Chapter 11 computes the effective length factor, k, for both braced and unbraced frames. It uses the same method as Example 11.1, with the exception that it uses the equations from Section 11.5 instead of the Jackson-Moreland Alignment Chart to determine k.

EXAMPLE 11.6

Repeat Example 11.1 using the Excel Spreadsheet for Chapter 11.

SOLUTION

Open the Chapter 11 spreadsheet and select the k factor tab Enter the values of the cells in yellow highlight. Note that a value of $b = 0$ is entered for member AB since it does not exist. The software determines a value of $\Psi_A = 2.99$ and $\Psi_B = 2.31$. These are the same as calculated by hand in Example 11.1. The value of k from the spreadsheet is 1.72 if the frame is unbraced compared with 1.74 from the Jackson-Moreland Alignment Chart. If the frame were braced, the software gives a value of $k = 0.96$. The Jackson-Moreland Alignment Chart gives a value of 0.87. The equations in Section 11.5 do not agree well with the chart in the case of braced frames.

This spreadsheet calculates the k factor for column BC.
Enter the values of cells in yellow highlight.
If you do not have all the members shown, enter $b = 0$ for the missing member.

Columns				Girders		
$f_c'=$	4000	psi		$f_c'=$	4000	
Col AB b	0	in.	**Beam EB** b	12	in.
h	12	in.		h	18	in.
col ht.	15	ft		span	20	
$0.70I_g$	0	in.^{4}		$0.35I_g$	2041	in.^{4}
Col BC b	12	in.^{2}	**Beam BF** b	12	
h	20			h	18	
col ht.	10	ft		span	24	in.
$0.70I_g$	5600			$0.35I_g$	2041	in.^{4}
Col CD b	12	in.	**Beam CG** b	12	in.
h	20	in.^{4}		h	24	
col ht.	12	ft		span	20	ft
$0.70I_g$	5600			$0.35I_g$	4838	in.^{4}
			Beam CH b	12	k
				h	24	k
				span	24	
				$0.35I_g$	4838	in.^{4}

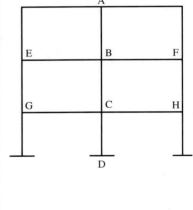

$$\Psi_B = \frac{E_c \times 0.70 I_g/l_{AB} + E_c \times 0.70 I_g/l_{BC}}{E_c \times 0.35 I_g/l_{BE} + E_c \times 0.35 I_g/l_{BF}} \quad 2.993$$

$$\Psi_C = \frac{E_c \times 0.70 I_g/l_{BC} + E_c \times 0.70 I_g/l_{CD}}{E_c \times 0.35 I_g/l_{CG} + E_c \times 0.35 I_g/l_{CH}} \quad 2.315$$

$$\Psi_{min} = 2.315$$

$$\Psi_m = 2.654$$

Braced

$k = 0.7 + 0.05(\Psi_A + \Psi_B) \leq 1.0 = 0.965$
$k = 0.85 + 0.05\Psi_{min} \leq 1.0 \quad = 0.966$

Braced

$k = 0.965$

Unbraced

$k = 1.72$

EXAMPLE 11.7

Repeat Example 11.3 using the Excel Spreadsheet for Chapter 11.

SOLUTION

Open the Chapter 11 Spreadsheet and select the "slender column braced rect." tab. Enter the values in the yellow colored cells (shaded below, yellow in the spreadsheet). The value of A_s entered is zero since the column is not yet designed. The software automatically uses ACI Equation 10-15 in this case. If a value for A_s is entered, the software compares the value of EI from ACI Equation 10-15 and 10-16 and uses the larger. The final value of δ is 1.093 which is in agreement with the solution obtained in Example 11.3. The final magnified moment, $\delta M_2 = 93.97$ ft-k. To complete the design, the Column Design Spreadsheets for Chapters 9 and 10 can be used.

Slender Column Braced - Rectangular			
	f'_c	4000	psi
	γ	145	pcf
Column	b	12	in.
	h	15	in.
	ℓ_u	16	ft.
	A_{st}	0	in.^{2}
	ρ_g	0	
Col loads	P_D	30	k
	P_L	46.25	k
	M_{2D}	10	ft-k
	M_{2L}	46.25	ft-k
	P_U	110	k
	M_2	86	ft-k
	M_{1D}	10	ft-k
	M_{1L}	43.75	ft-k
	M_1	82	ft-k
	C_m	0.9814	
	k	0.83	
	$k\ell_u/r$	35.41	
$34-12M_1/M_2$		22.56	
Is $k\ell_u/r < 34-12M_1/M_2$		consider slenderness	
	E_c	3644	psi
	n	7.96	
	I_g	3375	in.^{4}
	γ	0.67	
	β_{dns}	0.327	
	n	0.500	
	$.4E_cI_gn$	4920	k-in.^{2}
$EI = 0.4EIn/1+\beta_d$		3707	k-in.^{2}
	P_c	1440	k
	$M_{2\ min}$	9.625	ft-k
		-	
	δ	1.093	
	δM_2	93.97	ft-k

Use the Chapter 10 spreadsheet to design this column for

P_U	110	k
δM_2	93.97	ft-k

PROBLEMS

11.1 Using the alignment charts of Figure 11.3, determine the effective length factors for columns CD and DE for the braced frame shown. Assume that the beams are 12 in. × 20 in. and the columns are 12 in. × 16 in. Use 0.70 gross moments of inertia of columns and 0.35 gross moments of inertia of beams. Assume ψ_A and $\psi_C = 10$. (*Ans.* 0.94, 0.89)

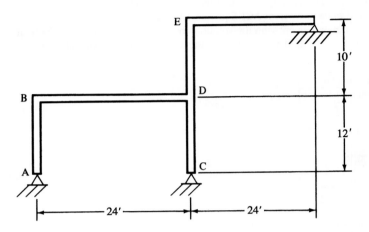

11.2 Repeat Problem 11.1 if the column bases are fixed.

11.3 Using the alignment charts of Figure 11.3, determine the effective length factors for columns AB and BC of the braced frame shown. Assume that all beams are 12 in. × 20 in. and all columns are 12 in. × 16 in. Use 0.70 gross moments of inertia of columns and 0.35 gross moments of inertia of beams. Assume far ends of beams are pinned and use $\psi_A = 10$. (*Ans.* 0.94, 0.92)

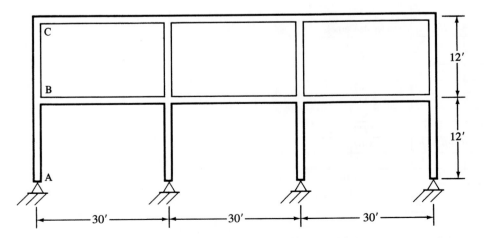

11.4 Repeat Problem 11.3 if the frame is unbraced.

11.5 The tied column shown is to be used in a braced frame. It is bent about its y-axis with the factored moments shown, and ℓ_u is 16 ft. If $k = 1.0$, $f_y = 60,000$ psi, and $f'_c = 4000$ psi, select the reinforcing required. Assume $P_D = 60$ k. (One *ans*. $\delta = 1.21$, 6 #8)

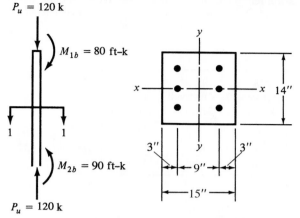

11.6 Repeat Problem 11.5 with *EI* based on the bar sizes given as the answer for that problem.

11.7 Repeat Problem 11.5 if the column is bent in reverse curvature and its length is 18 ft. Use ACI Equation 10-12 for EI. (One *ans*. $\delta = 1.0$, 6 #6)

For Problems **11.8** to **11.12** and the braced tied columns given, select reinforcing bars if the distance from the column edge to the c.g. of the bars is 2.5 in. $f_y = 60$ ksi and $f'_c = 4$ ksi for all problems. Place bars in two faces only. Rectangular columns are bent about their strong axis. Use ACI Equation 10-12 for *EI*.

Problem No.	Column size $b \times h$ (in.)	ℓ_u(ft)	k	P_u(k)	Not factored P_D(k)	Factored M_{1b}(ft-k)	Factored M_{2b}(ft-k)	Curvature	
11.8	14 × 14	12	1.0	400	100	75	85	Single	
11.9	15 × 15	16	1.0	500	120	100	140	Double	(*One ans*. $\delta = 1.0$, 8 #10)
11.10	16 × 18	15	0.85	250	40	100	120	Single	
11.11	12 × 18	16	0.80	500	130	90	120	Single	(*One ans*. $\delta = 1.20$, 8 #9)
11.12	14 × 18	18	0.90	600	150	100	130	Double	

11.13 Repeat Problem 11.9 using single curvature and ACI Equation 10-11 for *EI*. (One *ans*. $\delta = 1.28$, 8 #11)

For Problems **11.14** to **11.17** and the unbraced tied columns given, select reinforcing bars (placed in two faces) if the distance from the column edge to the c.g. of the bars is 2.5 in. $f_y = 60$ ksi and $f'_c = 4$ ksi. Use ACI Equation 10-12 for EI. Rectangular columns are bent about their strong axis. None of the wind load is considered sustained.

Problem No.	Column size $b \times h$ (in.)	l_u (ft)	k	P_u(k) for loads not consid. sway	P_u(k) due to wind	M_{1ns} (ft-k)	M_{2ns} (ft-k)	M_u (ft-k) due to wind	$\sum P_u$(k) for all columns on floor	$\sum P_c$(k) for all columns on floor	
11.14	16 × 20	15	1.3	500	110	90	80	100	12,000	34,000	
11.15	14 × 18	12	1.4	300	80	70	75	80	10,000	30,000	(*One ans*. 6 #10)
11.16	16 × 20	16	1.65	500	140	110	140	120	16,500	80,000	
11.17	12 × 18	12	1.5	480	140	90	120	110	14,000	36,200	(*One ans*. Larger column needed)

SI Problems

For Problems **11.18** to **11.20** and the braced tied columns given, select reinforcing bars (placed in two faces). Assume that distances from column edges to c.g. of bars are 75 mm each. To be able to use Appendix graphs, use $f_y = 413.7$ MPa and $f_c' = 27.6$ MPa. Also remember to apply the conversion factor.

Problem No.	Column size $b \times d$(mm)	ℓ_u(m)	k	P_u(kN)	P_D not factored (kN)	M_{1b} factored (kN·m)	M_{2b} factored (kN·m)	Curvature	
11.18	450 × 450	4	1.0	1800	400	80	100	Single	
11.19	300 × 400	5	0.92	2200	500	110	125	Double	(*One ans.* 6 #32)
11.20	300 × 500	6	0.88	2400	550	120	140	Single	

For Problems **11.21** to **11.22** and the unbraced tied columns given, select reinforcing bars (placed on two faces) if the distance from the column edge to the c.g. of the bars is 75 mm. $f_y = 420$ MPa and $f_c' = 28$ MPa. Remember to apply the conversion factor.

Problem No.	Column size $b \times d$(in.)	ℓ_u(m)	k	P_u(kN) not considering sway	P_u(kN) due to wind	M_{1ns} (kNm)	M_{2ns} (kNm)	M_u(kNm) dut to wind	$\sum P_u$(kN) for all columns on floor	$\sum P_c$(kN) for all columns on floor	
11.21	300 × 400	5	1.2	400	1200	40	50	60	40 000	110000	(*One ans.* 6 #32)
11.22	300 × 500	4	1.3	500	1800	50	60	70	44 000	125 000	
11.23	350 × 600	6	1.35	600	2500	65	90	110	50 000	156 000	(*One ans.* 6 #29)

12

Footings

12.1 INTRODUCTION

Footings are structural members used to support columns and walls and transmit their loads to the underlying soils. Reinforced concrete is a material admirably suited for footings and is used as such for both reinforced concrete and structural steel buildings, bridges, towers, and other structures.

The permissible pressure on a soil beneath a footing is normally a few tons per square foot. The compressive stresses in the walls and columns of an ordinary structure may run as high as a few hundred tons per square foot. It is therefore necessary to spread these loads over sufficient soil areas to permit the soil to support the loads safely.

Not only is it desired to transfer the superstructure loads to the soil beneath in a manner that will prevent excessive or uneven settlements and rotations, but it is also necessary to provide sufficient resistance to sliding and overturning.

To accomplish these objectives, it is necessary to transmit the supported loads to a soil of sufficient strength and then to spread them out over an area such that the unit pressure is within a reasonable range. If it is not possible to dig a short distance and find a satisfactory soil, it will be necessary to use piles or caissons to do the job. These latter subjects are not considered within the scope of this text.

The closer a foundation is to the ground surface, the more economical it will be to construct. There are two reasons, however, that may keep the designer from using very shallow foundations. First, it is necessary to locate the bottom of a footing below the ground freezing level to avoid vertical movement or heaving of the footing as the soil freezes and expands in volume. This depth varies from about 3 to 6 feet in the northern states and less in the southern states. Second, it is necessary to excavate a sufficient distance so that a satisfactory bearing material is reached, and this distance may on occasion be quite a few feet.

12.2 TYPES OF FOOTINGS

Among the several types of reinforced concrete footings in common use are the wall, isolated, combined, raft, and pile cap types. These are briefly introduced in this section; the remainder of the chapter is used to provide more detailed information about the simpler types of this group.

1. A *wall footing* [Figure 12.1(a)] is simply an enlargement of the bottom of a wall that will sufficiently distribute the load to the foundation soil. Wall footings are normally used around the perimeter of a building and perhaps for some of the interior walls.

2. An *isolated or single-column footing* [Figure 12.1(b)] is used to support the load of a single column. These are the most commonly used footings, particularly where the loads are relatively light and the columns are not closely spaced.

Placing concrete, Big Dig, Boston, MA. (Steve Dunwell/The Image Bank.)

3. *Combined footings* are used to support two or more column loads [Figure 12.1(c)]. A combined footing might be economical where two or more heavily loaded columns are so spaced that normally designed single-column footings would run into each other. Single-column footings are usually square or rectangular and, when used for columns located right at property lines, would extend across those lines. A footing for such a column combined with one for an interior column can be designed to fit within the property lines.

4. A *mat or raft or floating foundation* [Figure 12.1(d)] is a continuous reinforced concrete slab over a large area used to support many columns and walls. This kind of foundation is used where soil strength is low or where column loads are large but where piles or caissons are not used. For such cases, isolated footings would be so large that it is more economical to use a continuous raft or mat under the entire area. The cost of the formwork for a mat footing is far less than is the cost of the forms for a large number of isolated footings. If individual footings are designed for each column and if their combined area is greater than half of the area contained within the perimeter of the building, it is usually more economical to use one large footing or mat. The raft or mat foundation is particularly useful in reducing differential settlements between columns—the reduction being 50% or more. For these types of footings the excavations are often rather deep. The goal is to remove an amount of earth approximately equal to the building weight. If this is done, the net soil pressure after the building is constructed will theoretically equal what it was before the excavation was made. Thus the building will *float* on the raft foundation.

5. *Pile caps* [Figure 12.1(e)] are slabs of reinforced concrete used to distribute column loads to groups of piles.

12.3   ACTUAL SOIL PRESSURES

The soil pressure at the surface of contact between a footing and the soil is assumed to be uniformly distributed as long as the load above is applied at the center of gravity of the footing

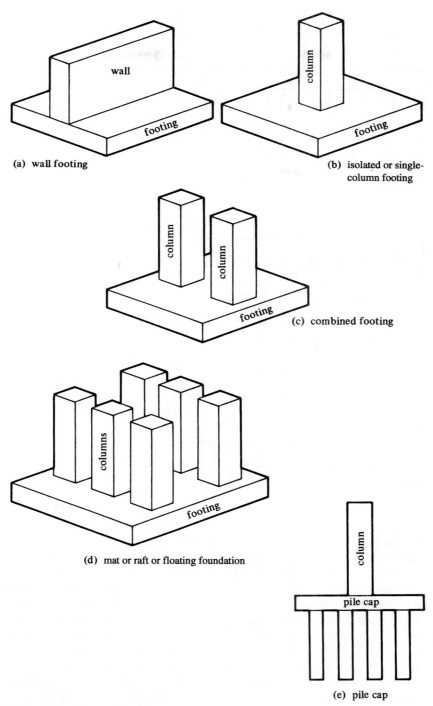

(a) wall footing

(b) isolated or single-
     column footing

(c) combined footing

(d) mat or raft or floating foundation

(e) pile cap

Figure 12.1 Types of footings.

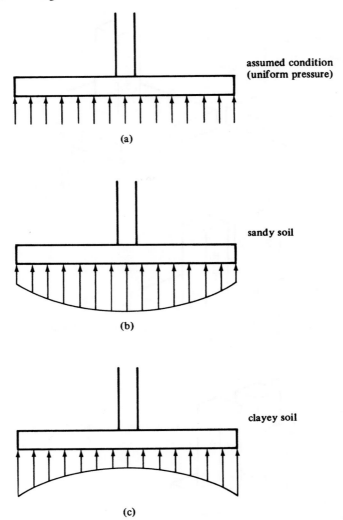

Figure 12.2 Soil conditions.

[Figure 12.2(a)]. This assumption is made even though many tests have shown that soil pressures are unevenly distributed due to variations in soil properties, footing rigidity, and other factors.

As an example of the variation of soil pressures, footings on sand and clay soils are considered. When footings are supported by sandy soils, the pressures are larger under the center of the footing and smaller near the edge [Figure 12.2(b)]. The sand at the edges of the footing does not have a great deal of lateral support and tends to move from underneath the footing edges, with the result that more of the load is carried near the center of the footing. Should the bottom of a footing be located at some distance from the ground surface, a sandy soil will provide fairly uniform support because it is restrained from lateral movement.

Just the opposite situation is true for footings supported by clayey soils. The clay under the edges of the footing sticks to or has cohesion with the surrounding clay soil. As a result, more of the load is carried at the edge of the footing than near the middle. [See Figure 12.2(c).]

The designer should clearly understand that the assumption of uniform soil pressure underneath footings is made for reasons of simplifying calculations and may very well have to be revised for some soil conditions.

Should the load be eccentrically applied to a footing with respect to the center of gravity of the footing, the soil pressure is assumed to vary uniformly in proportion to the moment, as illustrated in Section 12.12 and Figure 12.23.

12.4 ALLOWABLE SOIL PRESSURES

The allowable soil pressures to be used for designing the footings for a particular structure are desirably obtained by using the services of a geotechnical engineer. He or she will determine safe values from the principles of soil mechanics on the basis of test borings, load tests, and other experimental investigations. Other issues may enter into the determination of the allowable soil pressures, such as the sensitivity of the building frame to accommodate deflection of the footings. Also, cracking of the superstructure resulting from settlement of the footings would be much more important in a performing arts center than a warehouse.

Because such investigations often may not be feasible, most building codes provide certain approximate allowable bearing pressures that can be used for the types of soils and soil conditions occurring in that locality. Table 12.1 shows a set of allowable values that are typical of such building codes. It is thought that these values usually provide factors of safety of approximately 3 against severe settlements.

Section 15.2.2 of the ACI Code states that the required area of a footing is to be determined by dividing the anticipated total load, including the footing weight, by a permissible soil pressure or permissible pile capacity determined using the principles of soil mechanics. It will be noted that this total load is the *unfactored* load, and yet the design of footings described in this chapter is based on strength design, where the loads are multiplied by the appropriate load factors. It is obvious that an ultimate load cannot be divided by an allowable soil pressure to determine the bearing area required.

The designer can handle this problem in two ways. He or she can determine the bearing area required by summing up the actual or unfactored dead and live loads and dividing them by the allowable soil pressure. Once this area is determined and the dimensions are selected, an ultimate soil pressure can be computed by dividing the factored or ultimate load by the area provided. The

Table 12.1 Maximum Allowable Soil Pressure

Class of Material	Maximum Allowable Soil Pressure	
	U.S. Customary Units (kips/ft^2)	SI Units (kN/m^2)
Rock	20% of ultimate crushing strength	20% of ultimate crushing strength
Compact coarse sand, compact fine sand, hard clay, or sand clay	8	385
Medium stiff clay or sandy clay	6	290
Compact inorganic sand and silt mixtures	4	190
Loose sand	3	145
Soft sand clay or clay	2	95
Loose inorganic sand–silt mixtures	1	50
Loose organic sand–silt mixtures, muck, or bay mud	0	0

remainder of the footing can then be designed by the strength method using this ultimate soil pressure. This simple procedure is used for the footing examples here.

The 1971 ACI Commentary (15.2) provided an alternative method for determining the footing area required that will give exactly the same answers as the procedure just described. By this latter method the allowable soil pressure is increased to an ultimate value by multiplying it by a ratio equal to that used for increasing the magnitude of the service loads. For instance, the ratio for D and L loads would be

$$\text{Ratio} = \frac{1.2D + 1.6L}{D + L}$$

Or for $D + L + W$, and so on

$$\text{Ratio} = \frac{1.2D + 1.6W + 1.0L + 0.5(L_r \text{ or } S \text{ or } R)}{D + L + W + (L_r \text{ or } S \text{ or } R)}$$

The resulting ultimate soil pressure can be divided into the ultimate column load to determine the area required.

12.5   DESIGN OF WALL FOOTINGS

The theory used for designing beams is applicable to the design of footings with only a few modifications. The upward soil pressure under the wall footing of Figure 12.3 tends to bend the footing into the deformed shape shown. The footings will be designed as shallow beams for the moments and shears involved. In beams where loads are usually only a few hundred pounds per

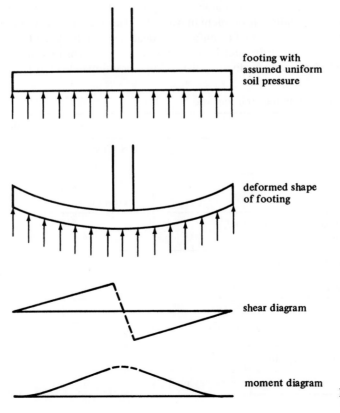

footing with assumed uniform soil pressure

deformed shape of footing

shear diagram

moment diagram

Figure 12.3

foot and spans are fairly large, sizes are almost always proportioned for moment. In footings, loads from the supporting soils may run several thousand pounds per foot and spans are relatively short. As a result, shears will almost always control depths.

It appears that the maximum moment in this footing occurs under the middle of the wall, but tests have shown that this is not correct because of the rigidity of such walls. If the walls are of reinforced concrete with their considerable rigidity, it is considered satisfactory to compute the moments at the faces of the walls (ACI Code 15.4.2). Should a footing be supporting a masonry wall with its greater flexibility, the Code states that the moment should be taken at a section halfway from the face of the wall to its center. (For a column with a steel base plate, the critical section for moment is to be located halfway from the face of the column to the edge of the plate.)

To compute the bending moments and shears in a footing, it is necessary to compute only the net upward pressure q_u caused by the factored wall loads above. In other words, the weight of the footing and soil on top of the footing can be neglected. These items cause an upward pressure equal to their downward weights, and they cancel each other for purposes of computing shears and moments. In a similar manner, it is obvious that there are no moments or shears existing in a book lying flat on a table.

Should a wall footing be loaded until it fails in shear, the failure will not occur on a vertical plane at the wall face but rather at an angle of approximately 45° with the wall, as shown in Figure 12.4. Apparently the diagonal tension, which one would expect to cause cracks in between the two diagonal lines, is opposed by the squeezing or compression caused by the downward wall load and the upward soil pressure. Outside this zone the compression effect is negligible in its effect on diagonal tension. Therefore, for nonprestressed sections shear may be calculated at a distance d from the face of the wall (ACI Code 11.1.3.1) due to the loads located outside the section.

The use of stirrups in footings is usually considered impractical and uneconomical. For this reason, the effective depth of wall footings is selected so that V_u is limited to the design shear strength ϕV_c that the concrete can carry without web reinforcing, that is, $\phi 2\lambda \sqrt{f_c'} b_w d$ (from ACI Section 11.3.1.1 and ACI Equation 11-3). Although the equation for V_c contains the term λ, it would be unusual to use lightweight concrete to construct a footing. The primary advantage for using lightweight concrete and its associated additional cost is to reduce the weight of the concrete superstructure. It would not be economical to use it in a footing. For this reason, the term λ will not be included in the example problems for this chapter. The following expression is used to select the depths of wall footings:

$$d = \frac{V_u}{(\phi)(2\lambda \sqrt{f_c'})(b_w)}$$

or for SI units

$$d = \frac{3V_u}{\phi \lambda \sqrt{f_c'} b_w}$$

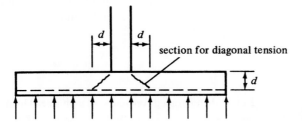

Figure 12.4

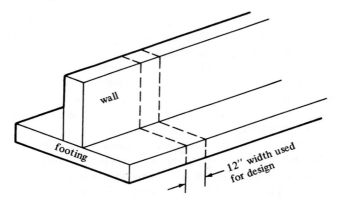

wall

footing

12″ width used
for design

Figure 12.5

The design of wall footings is conveniently handled by using 12-in. widths of the wall, as shown in Figure 12.5. Such a practice is followed for the design of a wall footing in Example 12.1. It should be noted in Section 15.7 of the Code that the depth of a footing above the bottom reinforcing bars may be no less than 6 in. for footings on soils and 12 in. for those on piles. Thus total minimum practical depths are at least 10 in. for the regular spread footings and 16 in. for pile caps.

In SI units is it convenient to design wall footings for 1-m-wide sections of the walls. The depth of such footings above the bottom reinforcing may not be less than 150 mm for footings on soils or 300 mm for those on piles. As a result, minimum footing depths are at least 250 mm for regular spread footings and 400 mm for pile caps.

The design of a wall footing is illustrated in Example 12.1. Although the example problems and homework problems of this chapter use various f_c' values, 3000 and 4000 psi concretes are commonly used for footings and are generally quite economical. Occasionally, when it is very important to minimize footing depths and weights, stronger concretes may be used. For most cases, however, the extra cost of higher strength concrete will appreciably exceed the money saved with the smaller concrete volume. The exposure category of the footing may control the concrete strength. ACI Section 4.2 requires that concrete exposed to sulfate have minimum f_c' values of 4000 or 4500 psi, depending on the sulphur concentration in the soil.

The determination of a footing depth is a trial-and-error problem. The designer assumes an effective depth d, computes the d required for shear, tries another d, computes the d required for shear, and so on, until the assumed value and the calculated value are within about 1 in. of each other.

You probably get upset when a footing size is assumed here. You say, "Where in the world did you get that value?" We think of what seems like a reasonable footing size and start there. We compute the d required for shear and probably find we've missed the assumed value quite a bit. We then try another value roughly two-thirds to three-fourths of the way from the trial value to the computed value (for wall footings) and compute d. (For column footings we probably go about half of the way from the trial value to the computed value.) Two trials are usually sufficient. We have the advantage over you in that often in preparing the sample problems for this textbook we made some scratch-paper trials before arriving at the assumed values used.

EXAMPLE 12.1

Design a wall footing to support a 12-in.-wide reinforced concrete wall with a dead load $D = 20$ k/ft and a live load $L = 15$ k/ft. The bottom of the footing is to be 4 ft below the final grade, the soil weighs 100 lb/ft^3, the allowable soil pressure q_a is 4 ksf, and there is no appreciable sulphur content in the soil. $f_y = 60$ ksi, and $f'_c = 3$ ksi, normal-weight concrete.

SOLUTION   Assume a 12-in.-thick footing $(d = 8.5$ in.$)$. The cover is determined by referring to the Code (7.7.1), which says that for concrete cast against and permanently exposed to the earth, a minimum of 3 in. clear distance outside any reinforcing is required. In severe exposure conditions, such as high sulfate concentration in the soil, the cover must be suitably increased (ACI Code Section 7.7.6).

The footing weight is $\left(\frac{12}{12}\right)(150) = 150$ psf, and the soil fill on top of the footing is $\left(\frac{36}{12}\right)(100) = 300$ psf. So 450 psf of the allowable soil pressure q_a is used to support the footing itself and the soil fill on top. The remaining soil pressure is available to support the wall loads. It is called q_e, the effective soil pressure

$$q_e = 4000 - \left(\frac{12}{12}\right)(150) - \left(\frac{36}{12}\right)(100) = 3550 \text{ psf}$$

$$\text{Width of footing required} = \frac{20 + 15}{3.55} = 9.86' \qquad \underline{\underline{\text{Say } 10'0''}}$$

Bearing Pressure for Strength Design

$$q_u = \frac{(1.2)(20) + (1.6)(15)}{10.00} = 4.80 \text{ ksf}$$

Depth Required for Shear at a Distance d from Face of Wall (Figure 12.6)

$$V_u = \left(\frac{10.00}{2} - \frac{6}{12} - \frac{8.5}{12}\right)(4.80) = 18.20 \text{ k}$$

$$d = \frac{18,200}{(0.75)(2\sqrt{3000})(12)} = 18.46'', h = 18.46 + 3.5'' = 21.96'' > 12'' \qquad \underline{\underline{\text{Try again}}}$$

Assume 20-in. Footing ($d = 16.5$ in.)

$$q_e = 4000 - \left(\frac{20}{12}\right)(150) - \left(\frac{28}{12}\right)(100) = 3517 \text{ psf}$$

$$\text{Width required} = \frac{35}{3.517} = 9.95 \text{ ft} \qquad \underline{\underline{\text{Say } 10'0''}}$$

Bearing Pressure for Strength Design

$$q_u = \frac{(1.2)(20) + (1.6)(15)}{10.00} = 4.80 \text{ ksf}$$

$q_u = 4.80$ ksf

$5' - \frac{8.5}{12} - \frac{6}{12}$ $6''$

$d = 8.5''$

$10'-0''$

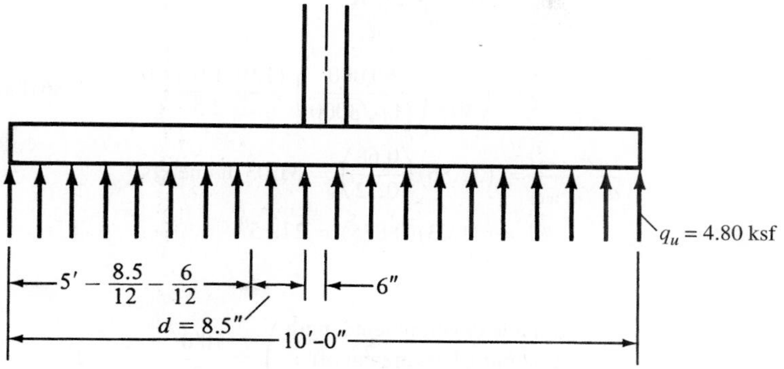

Figure 12.6

Depth Required for Shear

$$V_u = \left(\frac{10.00}{2} - \frac{6}{12} - \frac{16.50}{12}\right)(4.80) = 15.0\,\text{k}$$

$$d = \frac{15,000}{(0.75)(2\sqrt{3000})(12)} = 15.21'', \, h = 15.21 + 3.5'' = 18.71'' \qquad \underline{\text{Use 20'' total depth}}$$

(A subsequent check of a 19-in. footing shows it will not quite work as to depth.)

Steel Area (Using d = ~~15.5~~ in.)

16.5 in

Taking moments at face of wall,

$$\text{Cantilever length} = \frac{10.00}{2} - \frac{6}{12} = 4.50'$$

$$M_u = (4.50)(4.80)(2.25) = 48.6\,\text{ft-k}$$

$$\frac{M_u}{\phi bd^2} = \frac{(12)(48,600)}{(0.9)(12)(16.5)^2} = 198.3$$

From Appendix Table A.12, $\rho = 0.00345 < 0.0136$ from Appendix Table A.7. $\therefore$ Section is tension controlled and $\phi = 0.9$ as assumed.

$$A_s = (0.00345)(12)(16.5) = 0.68\,\text{in.}^2 \qquad \underline{\text{Use \#7 @ 10''(0.72 in.}^2 \text{ from Table A.6)}}$$

Development Length

From Table 7.1 in Chapter 7

$$\psi_t = \psi_e = \psi_s = \lambda = 1.0$$

$$c_b = \text{side cover} = 3.50\,\text{in.} \leftarrow$$

$$c_b = \text{one-half of center-to-center spacing of bars} = \frac{1}{2} \times 10 = 5''$$

Letting $K_{tr} = 0$

$$\frac{c_b + K_{tr}}{d_b} = \frac{3.5 + 0}{0.875} = 4.0 > 2.5 \qquad \underline{\therefore \text{Use 2.5}}$$

$$\frac{\ell_d}{d_b} = \frac{3}{40} \frac{f_y}{\lambda\sqrt{f_c'}} \frac{\psi_t \psi_e \psi_s}{\dfrac{c + K_{tr}}{d_b}}$$

$$= \left(\frac{3}{40}\right)\left(\frac{60,000}{(1)\sqrt{3000}}\right)\frac{(1.0)(1.0)(1.0)}{2.5} = 32.86\,\text{diameters}$$

$$\frac{\ell_d}{d_b}\frac{A_{s,\,\text{reqd}}}{A_{s,\,\text{furn}}} = (32.86)\left(\frac{0.68}{0.72}\right) = 31.03\,\text{diameters}$$

$$\ell_d = (31.03)(0.875) = 27.15'' \qquad \underline{\text{Say 28 in.}}$$

$$\left(\begin{array}{c}\text{Available development length}\\ \text{assuming bars are cut off}\\ 3'' \text{ from edge of footing}\end{array}\right) = \frac{10'0''}{2} - 6'' - 3''$$

$$= \begin{pmatrix} 4'3'' \text{ from face of} \\ \text{wall at section of} \\ \text{maximum moment} \end{pmatrix} > 28'' \qquad\qquad \underline{\underline{\text{OK}}}$$

(The bars should be extended to a point not less than 3 in. or more than 6 in. from the edge of the footing.)

Longitudinal Temperature and Shrinkage Steel

$$A_s = (0.0018)(12)(20) = 0.432 \text{ in.}^2 \qquad\qquad \underline{\underline{\text{Use \#5 at } 8''}}$$

12.6   DESIGN OF SQUARE ISOLATED FOOTINGS

Single-column footings usually provide the most economical column foundations. Such footings are generally square in plan, but they can just as well be rectangular or even circular or octagonal. Rectangular footings are used where such shapes are dictated by the available space or where the cross sections of the columns are very pronounced rectangles.

Most footings consist of slabs of constant thickness, such as the one shown in Figure 12.7(a), but if calculated thicknesses are greater than 3 or 4 ft, it may be economical to use stepped footings, as illustrated in Figure 12.7(b). The shears and moments in a footing are obviously larger near the column, with the result that greater depths are required in that area as compared to the outer parts of

Single column footing prior to placement of column reinforcing, Rhodes Annex – Clemson University, 2008.

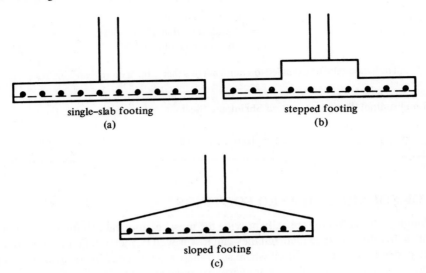

Figure 12.7 Shapes of isolated footings.

the footing. For very large footings, such as for bridge piers, stepped footings can give appreciable savings in concrete quantities.

Occasionally, sloped footings [Figure 12.7(c)] are used instead of the stepped ones, but labor costs can be a problem. Whether stepped or sloped, it is considered necessary to place the concrete for the entire footing in a single pour to ensure the construction of a monolithic structure, thus avoiding horizontal shearing weakness. If this procedure is not followed, it is desirable to use keys or shear friction reinforcing between the parts to ensure monolithic action. In addition, when sloped or stepped footings are used, it is necessary to check stresses at more than one section in the footing. For example, steel area and development length requirements should be checked at steps as well as at the faces of walls or columns.

Before a column footing can be designed, it is necessary to make a few comments regarding shears and moments. This is done in the paragraphs to follow, while a related subject, load transfer from columns to footings, is discussed in Section 12.8.

Shears

Two shear conditions must be considered in column footings, regardless of their shapes. The first of these is one-way or beam shear, which is the same as that considered in wall footings in the preceding section. For this discussion, reference is made to the footing of Figure 12.8. The total shear (V_{u1}) to be taken along section 1–1 equals the net soil pressure q_u times the hatched area outside the section. In the expression to follow, b_w is the whole width of the footing. The maximum value of V_{u1} if stirrups are not used equals ϕV_c, which is $\phi 2\sqrt{f_c'}b_w d$, and the maximum depth required is as follows:

$$d = \frac{V_{u1}}{\phi 2\sqrt{f_c'}b_w}$$

The second shear condition is two-way or punching shear, with reference being made to Figure 12.9. The compression load from the column tends to spread out into the footing, opposing diagonal tension in that area, with the result that a square column tends to punch out a piece of the slab, which has the shape of a truncated pyramid. The ACI Code (11.11.1.2) states that the critical section for two-way shear is located at a distance $d/2$ from the face of the column.

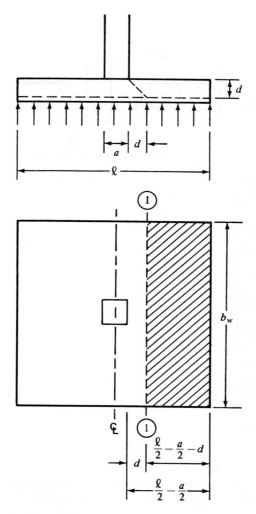

Figure 12.8 One-way or beam shear.

The shear force V_{u2} consists of all the net upward pressure q_u on the hatched area shown, that is, on the area outside the part tending to punch out. In the expressions to follow, b_o is the perimeter around the punching area, equal to $4(a+d)$ in Figure 12.9. The nominal two-way shear strength of the concrete V_c is specified as the smallest value obtained by substituting into the applicable equations that follow.

The first expression is the usual punching shear strength

$$V_c = 4\lambda\sqrt{f_c'}b_o d \qquad \text{(ACI Equation 11–35)}$$

Tests have shown that when rectangular footing slabs are subjected to bending in two directions and when the long side of the loaded area is more than two times the length of the short side, the shear strength $V_c = 4\lambda\sqrt{f_c'}b_o d$ may be much too high. In the expression to follow, β_c is the ratio of the long side of the column to the short side of the column, concentrated load, or reaction area.

$$V_c = \left(2 + \frac{4}{\beta_c}\right)\lambda\sqrt{f_c'}b_o d \qquad \text{(ACI Equation 11–33)}$$

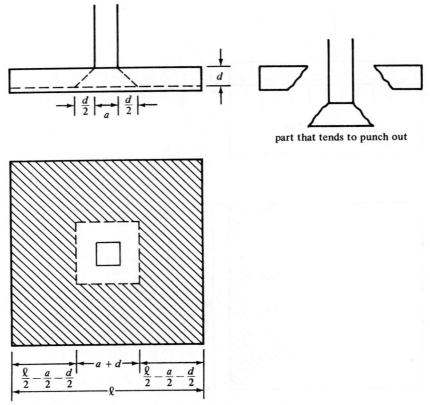

Figure 12.9 Two-way or punching shear.

The shear stress in a footing increases as the ratio b_o/d decreases. To account for this fact ACI Equation 11-34 was developed. The equation includes a term α_s which is used to account for variations in the ratio. In applying the equation α_s is to be used as 40 for interior columns (where the perimeter is four-sided), 30 for edge columns (where the perimeter is three-sided), and 20 for corner columns (where the perimeter is two-sided).

$$V_c = \left(\frac{\alpha_s d}{b_o} + 2\right)\lambda\sqrt{f_c'}b_o d \qquad \text{(ACI Equation 11–34)}$$

The d required for two-way shear is the largest value obtained from the following expressions:

$$d = \frac{V_{u2}}{\phi 4\lambda\sqrt{f_c'}b_o}$$

$$d = \frac{V_{u2}}{\phi\left(2 + \dfrac{4}{\beta_c}\right)\lambda\sqrt{f_c'}b_o} \quad \text{(not applicable unless } \beta_c \text{ is } > 2)$$

$$d = \frac{V_{u2}}{\phi\left(\dfrac{\alpha_s d}{b_o} + 2\right)\lambda\sqrt{f_c'}b_o}$$

Or in SI units, with f_c' in MPa and b_o and d in mm,

$$d = \frac{6V_{u2}}{\phi\lambda\sqrt{f_c'}b_o}$$

$$d = \frac{6V_{u2}}{\phi\left(1 + \dfrac{8}{\beta_c}\right)\lambda\sqrt{f_c'}b_o}$$

$$d = \frac{12V_{u2}}{\phi\left(\dfrac{\alpha_s d}{b_o} + 2\right)\lambda\sqrt{f_c'}b_o}$$

Moments

The bending moment in a square reinforced concrete footing with a square column is the same about both axes due to symmetry. If the column is not square, the moment will be larger in the direction of the shorter column dimension. It should be noted, however, that the effective depth of the footing cannot be the same in the two directions because the bars in one direction rest on top of the bars in the other direction. The effective depth used for calculations might be the average for the two directions or, more conservatively, the value for the bars on top. This lesser value is used for the examples in this text. Although the result is some excess of steel in one direction, it is felt that the steel in either direction must be sufficient to resist the moment in that direction. It should be clearly understood that having an excess of steel in one direction will not make up for a shortage in the other direction at a 90° angle.

The critical section for bending is taken at the face of a reinforced concrete column or halfway between the middle and edge of a masonry wall or at a distance halfway from the edge of the base plate and the face of the column if structural steel columns are used (Code 15.4.2).

The determination of footing depths by the procedure described here will often require several cycles of a trial-and-error procedure. There are, however, many tables and handbooks available with which footing depths can be accurately estimated. One of these is the previously mentioned *CRSI Design Handbook*. In addition, there are many rules of thumb used by designers for making initial thickness estimates, such as 20% of the footing width or the column diameter plus 3 in. Computer programs such as the spreadsheet provided for this chapter easily handle this problem.

The reinforcing steel area calculated for footings will often be appreciably less than the minimum values $(200b_w d/f_y)$ and $(3\sqrt{f_c'}b_w d/f_y)$ specified for flexural members in ACI Section 10.5.1. In Section 10.5.4, however, the Code states that in slabs of uniform thickness the minimum area and maximum spacing of reinforcing bars in the direction of bending need only be equal to those required for shrinkage and temperature reinforcement. The maximum spacing of this reinforcement may not exceed the lesser of three times the footing thickness, or 18 in. Many designers feel that the combination of high shears and low ρ values that often occurs in footings is not a good situation. Because of this, they specify steel areas at least as large as the flexural minimums of ACI Section 10.5.1. This is the practice we also follow herein.

Example 12.2 illustrates the design of an isolated column footing.

EXAMPLE 12.2

Design a square column footing for a 16-in. square tied interior column that supports a dead load $D = 200$ k and a live load $L = 160$ k. The column is reinforced with eight #8 bars, the base of the footing is 5 ft below grade, the soil weight is 100 lb/ft³, $f_y = 60,000$ psi, $f_c' = 3000$ psi, and $q_a = 5000$ psf.

SOLUTION   **After Two Previous Trials Assume 24-in. Footing ($d = 19.5$ in. estimated to c.g. of *top layer* of flexural steel)**

$$q_e = 5000 - \left(\frac{24}{12}\right)(150) - \left(\frac{36}{12}\right)(100) = 4400 \text{ psf}$$

$$A \text{ required} = \frac{200 + 160}{4.400} = 81.82 \text{ ft}^2$$

Use 9 ft 0 in. × 9 ft 0 in. footing = 81.0 ft²

Bearing Pressure for Strength Design

$$q_u = \frac{(1.2)(200) + (1.6)(160)}{81.0} = 6.12 \text{ ksf}$$

Depth Required for Two-way or Punching Shear (Figure 12.10)

$$b_o = (4)(35.5) = 142 \text{ in.}$$

$$V_{u2} = (81.0 - 2.96^2)(6.12) = 442.09 \text{ k}$$

$$d = \frac{442,090}{(0.75)(4\sqrt{3000})(142)} = 18.95 \text{ in.} < 19.5 \text{ in.} \qquad \underline{\underline{\text{OK}}}$$

$$d = \frac{442,090}{0.75\left(\dfrac{40 \times 19.5}{142} + 2\right)\sqrt{3000}(142)} = 10.12 \text{ in.} < 19.5 \text{ in.} \qquad \underline{\underline{\text{OK}}}$$

Since both values of d are less than the assumed value of 19.5, punching shear is OK.

Depth Required for One-way Shear (Figure 12.11)

$$V_{u1} = (9.00)(2.208)(6.12) = 121.62 \text{ k}$$

$$d = \frac{121,620}{(0.75)(2\sqrt{3000})(108)} = 13.71 \text{ in.} < 19.5 \text{ in.} \qquad \underline{\underline{\text{OK}}}$$

Use 24 in. total depth

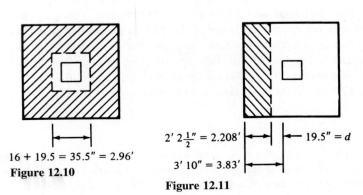

16 + 19.5 = 35.5" = 2.96'
Figure 12.10

2' 2$\frac{1}{2}$" = 2.208'   19.5" = d

3' 10" = 3.83'

Figure 12.11

$$M_u = (3.83)(9.00)(6.12)\left(\frac{3.83}{2}\right) = 404 \text{ ft-k}$$

$$\frac{M_u}{\phi b d^2} = \frac{(12)(404{,}000)}{(0.9)(108)(19.5)^2} = 131.2 \text{ psi, } \therefore \ \rho = 0.00225 < \rho_{\min} \text{ for flexure}$$

$$\text{Use } \rho = \text{larger of } \frac{200}{60{,}000} = 0.0033 \leftarrow$$

$$\text{or } \frac{3\sqrt{3000}}{60{,}000} = 0.00274$$

$$A_s = (0.0033)(108)(19.5) = 6.95 \text{ in.}^2$$

Use 9 #8 bars in both directions (7.07 in.^2)

Development Length

From Table 6.1 in Chapter 6

$$\psi_t = \psi_e = \psi_s = \lambda = 1.0$$

Assuming bars spaced $12''$ on center

leaving $6''$ on each side

$$c_b = \text{bottom cover} = 3.5''$$

$$c_b = \text{one-half of center-to-center spacing of bars} = \left(\tfrac{1}{2}\right)(12) = 6''$$

Letting $K_{tr} = 0$

$$\frac{c_b + K_{tr}}{d_b} = \frac{3.5 + 0}{1.0} = 3.5 > 2.5 \qquad\qquad \therefore \ \text{Use 2.5}$$

$$\frac{\ell_d}{d_b} = \frac{3}{40} \frac{f_y}{\lambda\sqrt{f_c'}} \frac{\psi_t \psi_e \psi_s}{\dfrac{c + K_{tr}}{d_b}}$$

$$= \frac{3}{40} \frac{60{,}000}{(1.0)\sqrt{3000}} \frac{(1.0)(1.0)(1.0)}{2.5}$$

$$= 32.86 \text{ diameters}$$

$$\frac{\ell_d}{d_b} \frac{A_{s\,\text{reqd}}}{A_{s\,\text{furn}}} = (32.86)\left(\frac{6.95}{7.07}\right) = 32.30 \text{ diameters}$$

$$\ell_d = (32.30)(1.00) = 32.30'' \qquad\qquad \text{Say } 33''$$

$$< \text{available } \ell_d = 4'6'' - \frac{16''}{2} - 3''$$

$$= 43'' \qquad\qquad \text{OK}$$

12.7 FOOTINGS SUPPORTING ROUND OR REGULAR POLYGON-SHAPED COLUMNS

Sometimes footings are designed to support round columns or regular polygon-shaped columns. If such is the case, Section 15.3 of the Code states that the column may be replaced with a square member having the same area as the round or polygonal one. Then the equivalent square is used for locating the critical sections for moment, shear, and development length.

12.8 LOAD TRANSFER FROM COLUMNS TO FOOTINGS

All forces acting at the base of a column must be satisfactorily transferred into the footing. Compressive forces can be transmitted directly by bearing, whereas uplift or tensile forces must be transferred to the supporting footing or pedestal by means of developed reinforcing bars or by mechanical connectors (which are often used in precast concrete).

A column transfers its load directly to the supporting footing over an area equal to the cross-sectional area of the column. The footing surrounding this contact area, however, supplies appreciable lateral support to the directly loaded part, with the result that the loaded concrete in the footing can support more load. Thus for the same grade of concrete, the footing can carry a larger bearing load than can the base of the column.

In checking the strength of the lower part of the column, only the concrete is counted. The column reinforcing bars at that point cannot be counted because they are not developed unless dowels are provided or unless the bars themselves are extended into the footing.

At the base of the column, the permitted bearing strength is $\phi(0.85f_c'A_1)$ (where ϕ is 0.65, but it may be multiplied by $\sqrt{A_2/A_1} \leq 2$ for bearing on the footing (ACI Code 10.14.1). In these

Pier bases for bridge from Prince Edward Island to mainland New Brunswick and Nova Scotia, Canada. (Courtesy of Economy Forms Corporation.)

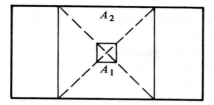

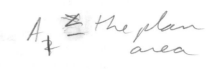

Figure 12.12

expressions, A_1 is the column area, and A_2 is the area of the portion of the supporting footing that is geometrically similar and concentric with the columns. (See Figure 12.12.)

If the computed bearing force is higher than the smaller of the two allowable value in the column or the footing, it will be necessary to carry the excess with dowels or with column bars extended into the footing. Instead of using dowels, it is also possible to increase the size of the column or increase f_c'. Should the computed bearing force be less than the allowable value, no dowels or extended reinforcing are theoretically needed, but the Code (15.8.2.1) states that there must be a minimum area of dowels furnished equal to no less than 0.005 times the gross cross-sectional area of the column or pedestal.

The development length of the bars must be sufficient to transfer the compression to the supporting member, as per the ACI Code (12.3). In no case may the area of the designed reinforcement or dowels be less than the area specified for the case where the allowable bearing force was not exceeded. As a practical matter in placing dowels, it should be noted that regardless of how small a distance they theoretically need to be extended down into the footing, they are usually bent at their ends and set on the main footing reinforcing, as shown in Figure 12.13. There the dowels can be tied firmly in place and not be in danger of being pushed through the footing during construction, as might easily happen otherwise. The bent part of the bar does not count as part of the compression development length (ACI Code 12.5.5). The reader should again note that the bar details shown in this figure are not satisfactory for seismic areas as the bars should be bent inward and not outward.

An alternative to the procedure described in the preceding paragraph is to place the footing concrete without dowels and then to push straight dowels down into the concrete while it is still in a plastic state. This practice is permitted by the Code in its Section 16.7.1 and is especially useful for plain concrete footings (to be discussed in Section 12.14 of this chapter). It is essential that the dowels be maintained in their correct position as long as the concrete is plastic. Before the licensed design professional approves the use of straight dowels as described here, he or she must be satisfied that the dowels will be properly placed and the concrete satisfactorily compacted around them.

The Code normally does not permit the use of lapped splices for #14 and #18 compression bars because tests have shown that welded splices or other types of connections are necessary. Nevertheless, based on years of successful use, the Code (15.8.2.3) states that #14 and #18 bars may be lap-spliced with dowels (no larger than #11) to provide for force transfer at the base of columns or walls or pedestals. These dowels must extend into the supported member a distance of

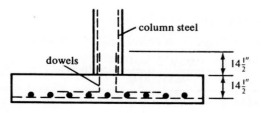

Figure 12.13

not less than ℓ_{dc} of #14 or #18 bars or the compression lap splice length of the dowels, whichever is greater; and into the footing a distance not less than ℓ_{dc} of the dowels.

If the computed development length of dowels is greater than the distance available from the top of the footing down to the top of the tensile steel, three possible solutions are available. One or more of the following alternatives may be selected:

1. A larger number of smaller dowels may be used. The smaller diameters will result in smaller development lengths.

2. A deeper footing may be used.

3. A cap or pedestal may be constructed on top of the footing to provide the extra development length needed.

Should bending moments or uplift forces have to be transferred to a footing such that the dowels would be in tension, the development lengths must satisfy the requirements for tension bars. For tension development length into the footing, a hook at the bottom of the dowel may be considered effective.

If there is moment or uplift, it will be necessary for the designer to conform to the splice requirements of Section 12.17 of the Code in determining the distance the dowels must be extended up into the wall or column.

Examples 12.3 and 12.4 provide brief examples of column-to-footing load transfer calculations for vertical forces only. Consideration is given to lateral forces and moments in Sections 12.12 and 12.13 of this chapter.

EXAMPLE 12.3

Design for load transfer from a 16-in. × 16-in. column to a 9-ft-0-in. × 9-ft-0-in. footing if $D = 200\,\text{k}$, $L = 160\,\text{k}$, $f_c' = 3000\,\text{psi}$ for the footing and 4000 psi for the column, and $f_y = 60{,}000\,\text{psi}$. The footing concrete is normal weight, but the column is constructed with sand-lightweight concrete.

SOLUTION

Bearing force at base of column $= (1.2)(200) + (1.6)(160) = 496\,\text{k}$

Allowable bearing force in concrete at base of column

$$= \phi(0.85 f_c' A_1) = (0.65)(0.85)(4.0)(16 \times 16)$$

Really don't need dowels

$$= 566\,\text{k} > 496\,\text{k}$$

0.85 f'c — *representative strength of concrete*

Allowable bearing force in footing concrete

$$= \phi(0.85 f_c' A_1)\sqrt{\frac{A_2}{A_1}} = (0.65)(0.85)(3.0)(16 \times 16)\sqrt{\frac{9 \times 9}{1.33 \times 1.33}}$$

$$= (0.65)(0.85)(3.0)(16 \times 16)(\text{Use } 2)[1] = 848.6\,\text{k} > 496\,\text{k} \qquad \underline{\text{OK}}$$

(.005)(area of column)

Minimum A_s for dowels $= (0.005)(16 \times 16) = 1.28\,\text{in.}^2$ $\qquad \underline{\text{Use 4 #6 bars}(1.77\,\text{in.}^2)}$

Development Lengths of Dowels (ACI 12.3)

For the column, using $\lambda = 0.85$ for sand-lightweight concrete and $f_c' = 4000\,\text{psi}$,

$$\ell_d = \frac{0.02 d_b f_y}{\lambda \sqrt{f_c'}} = \frac{0.02(0.75)(60{,}000)}{0.85\sqrt{4000}} = 16.74\,\text{in.}$$

---

[1] $\sqrt{A_2/A_1} < 2.0$ (see Section 12.8 and ACI Code Section 1.14.1)

For the footing, $\lambda = 1.0$, and $f_c' = 3000$ psi,

$$\ell_d = \frac{0.02 d_b f_y}{\lambda \sqrt{f_c'}} = \frac{0.02(0.75)(60,000)}{1.0\sqrt{3000}} = 16.43 \text{ in.}$$

In addition, the development must not be less than either

$$\ell_d = 0.0003 d_b f_y = 0.0003(0.75)(60,000) = 13.50 \text{ in.}$$
$$\ell_d = 8.00 \text{ in.}$$

In summary, the dowels must extend upward into the column at least $16.74''$, and down into the footing at least $16.43''$. *Use four #6 dowels extending $17''$ up into the column and $17''$ down into the footing and set on top of the reinforcing mat, as shown in Figure 12.13.*

EXAMPLE 12.4

Design for load transfer from a 14-in. × 14-in. column to a 13-ft-0-in. × 13-ft-0-in. footing with a P_u of 800 k, $f_c' = 3000$ psi in the footing and 5000 psi in the column, both normal weight, and $f_y = 60,000$ psi. The column has eight #8 bars.

SOLUTION

Bearing force at base of column $= P_u = 800 \text{ k}$

$$= \phi(\text{allowable bearing force in concrete}) + \phi(\text{strength of dowels})$$

Design bearing strength in concrete at base of column

$$= (0.65)(0.85)(5.0)(14 \times 14) = 541.5 \text{ k} < 800 \text{ k} \qquad \underline{\text{No good}}$$

Design bearing strength on footing concrete

$$= (0.65)(0.85)(3.0)(14 \times 14)(\text{Use } 2)^2 = 649.7 \text{ k} < 800 \text{ k} \qquad \underline{\text{No good}}$$

Therefore, the dowels must be designed for excess load.

$$\text{Excess load} = 800 - 541.5 = 258.5 \text{ k}$$

$$A_s \text{ of dowels} = \frac{258.5}{(0.9)(60)} = \underline{4.79 \text{ in.}^2} \text{ or } (0.005)(14)(14) = 0.98 \text{ in.}^2$$

$$\underline{\underline{\text{Use eight #7 bars}(4.80 \text{ in.}^2)}}$$

Development Length of #7 Dowels into Column

$$\ell_d = \frac{(0.02)(0.875)(60,000)}{(1.0)\sqrt{5000}} = 14.85''$$

$$\ell_d = (0.0003)(60,000)(0.875) = 15.75'' \leftarrow$$

$$\ell_d = 8''$$

Development Length of #7 Dowels into Footing (Different from Column Values because f_c' Values Are Different)

$$\ell_d = \frac{(0.02)(0.875)(60,000)}{(1.0)\sqrt{3000}} = 19.42'' \leftarrow$$

$$\ell_d = (0.0003)(0.875)(60,000) = 15.75''$$

Use eight #7 dowels extending 16 in. up into the column and 20 in. down into the footing.

[2] $\sqrt{A_2/A_1} < 2.0$ (see Section 12.8 and ACI Code Section 1.14.1)

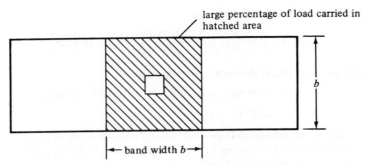

Figure 12.14

12.9   RECTANGULAR ISOLATED FOOTINGS

As previously mentioned, isolated footings may be rectangular in plan if the column has a very pronounced rectangular shape or if the space available for the footing forces the designer into using a rectangular shape. Should a square footing be feasible, it is normally more desirable than a rectangular one because it will require less material and will be simpler to construct.

The design procedure is almost identical with the one used for square footings. After the required area is calculated and the lateral dimensions are selected, the depths required for one-way and two-way shear are determined by the usual methods. One-way shear will very often control the depths for rectangular footings, whereas two-way shear normally controls the depths of square footings.

The next step is to select the reinforcing in the long direction. These longitudinal bars are spaced uniformly across the footing, but such is not the case for the short-span reinforcing. With reference to Figure 12.14, it can be seen that the support provided by the footing to the column will be concentrated near the middle of the footing, and thus the moment in the short direction will be concentrated somewhat in the same area near the column.

As a result of this concentration effect, it seems only logical to concentrate a large proportion of the short-span reinforcing in this area. The Code (15.4.4.2) states that a certain minimum percentage of the total short-span reinforcing should be placed in a band width equal to the length of the shorter direction of the footing. The amount of reinforcing in this band is to be determined with the following expression, in which β is the ratio of the length of the long side to the width of the short side of the footing:

$$\frac{\text{Reinforcing in band width}}{\text{Total reinforcing in short direction}} = \frac{2}{\beta + 1} = \gamma_s \qquad \text{(ACI Equation 15–1)}$$

The remaining reinforcing in the short direction should be uniformly spaced over the ends of the footing but the authors feel it should at least meet the shrinkage and temperature requirements of the ACI Code (7.12).

Example 12.5 presents the partial design of a rectangular footing in which the depths for one- and two-way shears are determined and the reinforcement selected.

EXAMPLE 12.5

Design a rectangular footing for an 18-in. square interior column with a dead load of 185 k and a live load of 150 k. Make the length of the long side equal to twice the width of the short side, $f_y = 60,000$ psi, $f'_c = 4000$ psi, normal weight and $q_a = 4000$ psf. Assume the base of the footing is 5 ft 0 in. below grade.

SOLUTION   **Assume 24-in. Footing (*d* = 19.5 in.)**

$$q_e = 4000 - \left(\frac{24}{12}\right)(150) - \left(\frac{36}{12}\right)(100) = 3400 \text{ psf}$$

$$A \text{ required} = \frac{185 + 150}{3.4} = 98.5 \text{ ft}^2 \qquad \underline{\text{Use } 7'0'' \times 14'0'' = 98.0 \text{ ft}^2}$$

$$q_u = \frac{(1.2)(185) + (1.6)(150)}{98.0} = 4.71 \text{ ksf}$$

Checking Depth for One-way Shear (Figure 12.15)

$$b = 7 \text{ ft}$$

$$V_{u1} = (7.0)(4.625)(4.71) = 152.49 \text{ k}$$

$$d = \frac{152,490}{(0.75)(1.0)(2\sqrt{4000})(84)} = 19.14'', h = d + 4.5'' = 23.64'' \qquad \underline{\text{Use } 24''}$$

Checking Depth for Two-way Shear (Figure 12.16)

$$b_o = (4)(37.5) = 150'' \qquad column\ with + \frac{d}{2}$$

$$V_{u2} = [98.0 - (3.125)^2](4.71) = 415.58 \text{ k}$$

$$d = \frac{415,580}{(0.75)(1.0)(4\sqrt{4000})(150)} = 14.60'' < 19.5 \qquad \underline{\text{OK}}$$

$$d = \frac{415,580}{0.75\left(\dfrac{40 \times 19.5}{150} + 2\right)(\sqrt{4000})(150)} = 11.23'' < 19.5 \qquad \underline{\text{OK}}$$

If either value of *d* in the last two equations had exceeded the assumed value of 19.5″, it would have been necessary to increase the trial value of *d* and start over.

Design of Longitudinal Steel

$$\text{Lever arm} = \frac{14}{2} - \frac{9}{12} = 6.25 \text{ ft}$$

$$M_u = (6.25)(7.0)(4.71)\left(\frac{6.25}{2}\right) = 643.9 \text{ ft-k}$$

$$\frac{M_u}{\phi b d^2} = \frac{(12)(643,900)}{(0.90)(84)(19.5)^2} = 268.8$$

$$\rho = 0.00467 \text{(from Appendix Table A.13)}$$

$$A_s = (0.00467)(84)(19.5) = 7.65 \text{ in.}^2 \qquad \underline{\text{Use 10 #8 bars}} \quad (7.85 \text{ in.}^2)$$

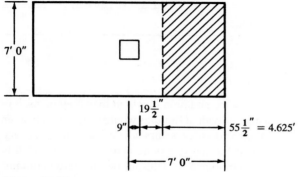

Figure 12.15

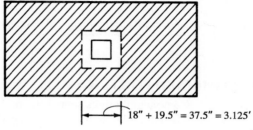

Figure 12.16

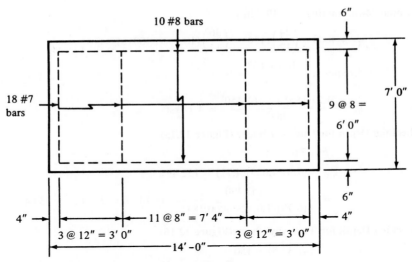

Figure 12.17 Two-way footing bar spacing diagram.

Design of Steel in Short Direction (Figure 12.17)

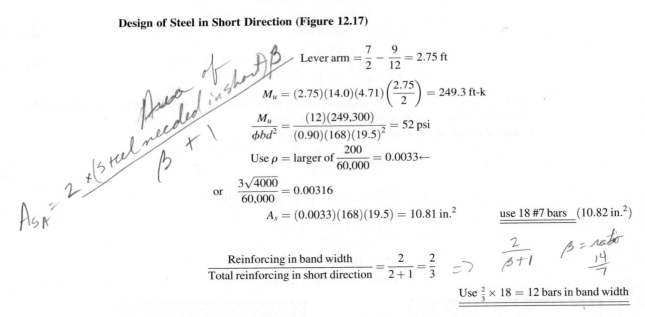

Lever arm $= \dfrac{7}{2} - \dfrac{9}{12} = 2.75$ ft

$$M_u = (2.75)(14.0)(4.71)\left(\dfrac{2.75}{2}\right) = 249.3 \text{ ft-k}$$

$$\dfrac{M_u}{\phi b d^2} = \dfrac{(12)(249,300)}{(0.90)(168)(19.5)^2} = 52 \text{ psi}$$

Use ρ = larger of $\dfrac{200}{60,000} = 0.0033 \leftarrow$

or $\dfrac{3\sqrt{4000}}{60,000} = 0.00316$

$A_s = (0.0033)(168)(19.5) = 10.81 \text{ in.}^2$ <u>use 18 #7 bars</u> (10.82 in.^2)

$$\dfrac{\text{Reinforcing in band width}}{\text{Total reinforcing in short direction}} = \dfrac{2}{2+1} = \dfrac{2}{3}$$

<u>Use $\frac{2}{3} \times 18 = 12$ bars in band width</u>

Subsequent check of required development lengths is OK.

12.10   COMBINED FOOTINGS

Combined footings support more than one column. One situation in which they may be used is when the columns are so close together that isolated individual footings would run into each other [Figure 12.18(a)]. Another frequent use of combined footings occurs where one column is very close to a property line, causing the usual isolated footing to extend across the line. For this situation the footing for the exterior column may be combined with the one for an interior column, as shown in Figure 12.18(b).

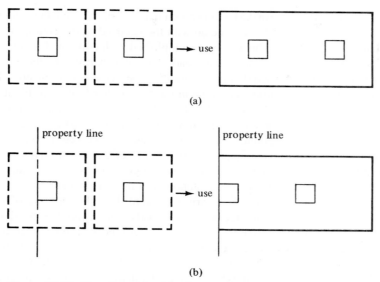

(a)

(b)

Figure 12.18   Use of combined footings.

On some occasions where a column is close to a property line and where it is desired to combine its footing with that of an interior column, the interior column will be so far away as to make the idea impractical economically. For such a case, counterweights or "deadmen" may be provided for the outside column to take care of the eccentric loading.

Combined footing (two column) after footing concrete is placed showing column reinforcing ready for splicing, Rhodes Annex – Clemson University 2008.

Because it is desirable to make bearing pressures uniform throughout the footing, the centroid of the footing should be made to coincide with the centroid of the column loads to attempt to prevent uneven settlements. This can be accomplished with combined footings that are rectangular in plan. Should the interior column load be greater than that of the exterior column, the footing may be so proportioned that its centroid will be in the correct position by extending the inward projection of the footing, as shown in the rectangular footing of Figure 12.18(b).

Other combined footing shapes that will enable the designer to make the centroids coincide are the trapezoid and strap or T footings shown in Figure 12.19(a) and 12.19(b). Footings with these shapes are usually economical when there are large differences between the magnitudes of the column loads or where the spaces available do not lend themselves to long rectangular footings. When trapezoidal footings are used, the longitudinal bars are usually arranged in a fan shape with alternate bars cut off some distance from the narrow end.

You probably realize that a problem arises in establishing the centroids of loads and footings when deciding whether to use service or factored loads. The required centroid of the footing will be slightly different for the two cases. The authors determine the footing areas and centroids with the service loads (ACI Code 15.2.2), but the factored loads could be used with reasonable results, too. The important item is to be consistent throughout the entire problem.

The design of combined footings has not been standardized as have the procedures used for the previous problems worked in this chapter. For this reason, practicing reinforced concrete designers use slightly varying approaches. One of these methods is described in the paragraphs to follow.

First, the required area of the footing is determined for the service loads, and the footing dimensions are selected so that the centroids coincide. Then the various loads are multiplied by the appropriate load factors, and the shear and moment diagrams are drawn along the long side of the footing for these loads. After the shear and moment diagrams are prepared, the depth for one- and two-way shear is determined and the reinforcing in the long direction is selected.

In the short direction it is assumed that each column load is spread over a width in the long direction equal to the column width plus $d/2$ on each side if that much footing is available. Then

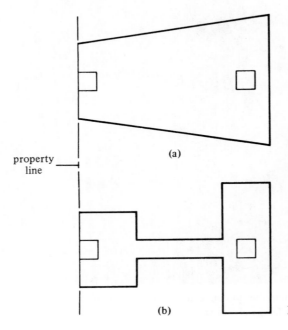

property line

(a)

(b)

Figure 12.19 Trapezoidal and strap or T footings.

the steel is designed, and a minimum amount of steel for temperature and shrinkage is provided in the remaining part of the footing.

The ACI Code does not specify an exact width for these transverse strips, and designers may make their own assumptions as to reasonable values. The width selected will probably have very little influence on the transverse bending capacity of the footing, but it can affect appreciably its punching or two-way shear resistance. If the flexural reinforcing is placed within the area considered for two-way shear, this lightly stressed reinforcing will reduce the width of the diagonal shear cracks and will also increase the aggregate interlock along the shear surfaces.

Space is not taken here to design completely a combined footing, but Example 12.6 is presented to show those parts of the design that are different from the previous examples of this chapter. A comment should be made about the moment diagram. If the length of the footing is not selected so that its centroid is located exactly at the centroid of the column loads, the moment diagrams will not close well at all since the numbers are very sensitive. Nevertheless, it is considered good practice to round off the footing lateral dimensions to the nearest 3 in. Another factor that keeps the moment diagram from closing is the fact that the average load factors of the various columns will be different if the column loads are different. We could improve the situation a little by taking the total column factored loads and dividing the result by the total working loads to get an average load factor. This value (which works out to be 1.375 in Example 12.6) could then be multiplied by the total working load at each column and used for drawing the shear and moment diagrams.

EXAMPLE 12.6

Design a rectangular combined footing for the two columns shown in Figure 12.20. $q_a = 5$ ksf, $f'_c = 3000$ psi, normal weight and $f_y = 60$ ksi. The bottom of the footing is to be 6 ft below grade.

SOLUTION   **Assume 27-in Footing ($d = 22.5$ in.)**

$$q_e = 5000 - \left(\frac{27}{12}\right)(150) - \left(\frac{45}{12}\right)(100) = 4287 \text{ psf}$$

$$A \text{ required} = \frac{570}{4.287} = 132.96 \text{ ft}^2$$

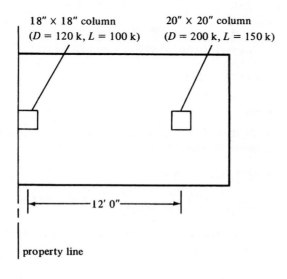

18″ × 18″ column
($D = 120$ k, $L = 100$ k)

20″ × 20″ column
($D = 200$ k, $L = 150$ k)

12′ 0″

property line

Figure 12.20

Locate Center of Gravity of Column Service Loads

$$X \text{ from c.g. of left column} = \frac{(350)(12)}{570} = 7.37'$$

$$\text{Distance from property line to c.g.} = 0.75 + 7.37 = 8.12'$$

$$\text{Length of footing} = (2 \times 8.12) = 16.24', \text{say } 16'3''$$

Use $16'3'' \times 8'3''$ footing $(A = 134 \text{ ft}^2)$.

$$q_u = \frac{(1.2)(320) + (1.6)(250)}{134} = 5.85 \text{ ksf}$$

Drawing Shear and Moment Diagrams (Figure 12.21) Depth Required for One-way Shear

$$V_{u1} \text{ a distance } d \text{ from interior face of right column}$$

$$= 271.1 - \left(\frac{22.5}{12}\right)(48.26) = 180.6 \text{ k}$$

$$d = \frac{(180,600)}{(0.75)(2\sqrt{3000})(99)} = 22.20 < 22.5' \qquad \underline{\underline{\text{OK}}}$$

Depth Required for Two-way Shear (ACI Equations 11.33 and 11.34 not shown as they do not control)

$$V_{u2} \text{ at right column} = 480 - \left(\frac{42.5}{12}\right)^2 (5.85) = 406.6 \text{ k}$$

$$d = \frac{406,600}{(0.75)\left((4)(1.0)\sqrt{3000}\right)(4 \times 42.5)}$$

$$= 14.56'' < 22.5'' \qquad \underline{\underline{\text{OK}}}$$

$$V_{u2} \text{ at left column} = 304 - \left(\frac{29.25 \times 40.5}{144}\right)(5.85) = 255.9 \text{ k}$$

$$d = \frac{255,900}{(0.75)\left((4)(1.0)\sqrt{3000}\right)(2 \times 29.25 + 40.5)} \qquad \underline{\underline{\text{OK}}}$$

$$= 15.73'' < 25.5$$

Depth Required for One-Way Shear

From the shear diagram in Figure 12.21, the largest shear force is 271.1 k at the left face of the right column. At a distance d to the left of this location, the value of shear is

$$V_{u1} = 271.1 - 48.26(22.5/12) = 180.61 \text{ k}$$

$$d = \frac{V_{u1}}{\phi 2\lambda \sqrt{f'_c} b} = \frac{180,610}{0.75(2)(1)\sqrt{3000}(8.25)(12)} = 22.2 \text{ in.} < 22.5 \text{ in.} \qquad \underline{\underline{\text{OK}}}$$

For this footing, one-way shear is more critical than two-way shear. This is not unusual for combined footings.

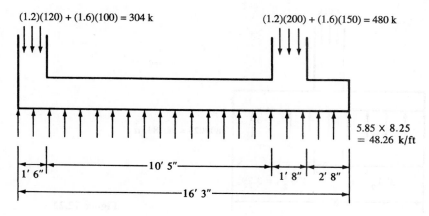

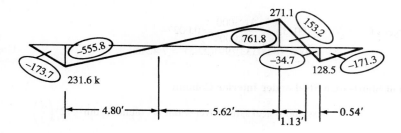

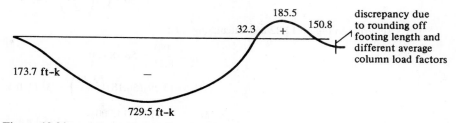

Figure 12.21

Design of Longitudinal Steel

$$-M_u = -729.5 \text{ ft-k}$$

$$\frac{M_u}{\phi bd^2} = \frac{(12)(729,500)}{(0.90)(99)(22.5)^2} = 194.1 \text{ psi}$$

ρ from Table A.12 by interpolation $= 0.00337$

$$-A_s = (0.00337)(99)(22.5) = 7.51 \text{ in.}^2 \qquad \underline{\text{Say 10 \#8}(7.85 \text{ in.}^2)}$$

$+M_u = +171.3$ ft-k (computed from right end of shear diagram, Figure 12.21)

$$\frac{M_u}{\phi bd^2} = \frac{(12)(173,700)}{(0.90)(99)(22.5)^2} = 46.2 \text{ psi}$$

$$\text{use } \rho = \rho_{\min}$$

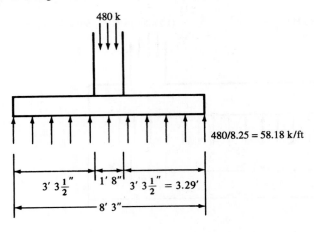

480/8.25 = 58.18 k/ft

3′ 3½″   1′ 8″   3′ 3½″ = 3.29′

8′ 3″

Figure 12.22

Use larger of $200/f_y = 0.00333$ or $\dfrac{3\sqrt{3000}}{60{,}000} = 0.00274$

$+A_s = (0.00333)(99)(22.5) = 7.42 \text{ in.}^2$ <u>Use 8 #9(8.00 in.²)</u>

Design of Short-Span Steel under Interior Column

Assuming steel spread over width $= \text{column width} + (2)\left(\dfrac{d}{2}\right)$

$$= 20 + (2)\left(\dfrac{22.5}{2}\right) = 42.5''$$

Referring to Figure 12.22 and calculating M_u:

$$q_u = \frac{480}{8.25} = 58.18 \text{ k/ft}$$

$$M_u = (3.29)(58.18)\left(\frac{3.29}{2}\right) = 314.9 \text{ ft-k}$$

$$\frac{M_u}{\phi b d^2} = \frac{(12)(314{,}900)}{(0.90)(42.5)(22.5)^2} = 195.1 \text{ psi}$$

$p = 0.00339$ from Appendix Table A.12

$A_s = (0.00339)(42.5)(22.5) = 3.24 \text{ in.}^2$ <u>Use 6 #7(3.61 in.²)</u>

Development lengths check out OK.

A similar procedure is used under the exterior column where the steel is spread over a width equal to 18 in. plus $d/2$, and not 18 in. plus $2(d/2)$, because sufficient room is not available on the property-line side of the column.

12.11   FOOTING DESIGN FOR EQUAL SETTLEMENTS

If three men are walking along a road carrying a log on their shoulders (a statically indeterminate situation) and one of them decides to lower his shoulder by 1 in., the result will be a drastic effect on the load supported by the other men. In the same way, if the footings of a building should settle by different amounts, the shears and moments throughout the structure will be greatly changed. In addition, there will be detrimental effects on the fitting of doors, windows, and partitions. Should all the footings settle by the same amount, however, these adverse effects will not occur. Thus equal settlement is the goal of the designer.

The footings considered in preceding sections have had their areas selected by taking the total dead plus live loads and dividing the sum by the allowable soil pressure. It would seem that if such a procedure were followed for all the footings of an entire structure, the result would be uniform settlements throughout—*but geotechnical engineers have clearly shown that this assumption may be very much in error.*

A better way to handle the problem is to attempt to design the footings so that the *usual loads* on each footing will cause approximately the same pressures. The usual loads consist of the dead loads plus the average percentage of live loads normally present. The usual percentage of live loads present varies from building to building. For a church it might be almost zero, perhaps 25% to 30% for an office building, and maybe 75% or more for some warehouses or libraries. Furthermore, the percentage in one part of a building may be entirely different from that in some other part (offices, storage, etc.).

One way to handle the problem is to design the footing that has the highest ratio of live to dead load, compute the usual soil pressure under that footing using dead load plus the estimated average percentage of live load, and then determine the areas required for the other footings so their usual soil pressures are all the same. It should be remembered that the dead load plus 100% of the live load must not cause a pressure greater than the allowable soil pressure under any of the footings.

A student of soil mechanics will realize that this method of determining usual pressures, though not a bad design procedure, will not ensure equal settlements. This approach at best will only lessen the amounts of differential settlements. He or she will remember first that large footings tend to settle more than small footings, even though their soil pressures are the same, because the large footings exert compression on a larger and deeper mass of soil. There are other items that can cause differential settlements. Different types of soils may be present at different parts of the building; part of the area may be in fill and part in cut: there may be mutual influence of one footing on another; and so forth.

Example 12.7 illustrates the usual load procedure for a group of five isolated footings.

EXAMPLE 12.7

Determine the footing areas required for the loads given in Table 12.2 so that the usual soil pressures will be equal. Assume that the usual live load percentage is 30% for all the footings, $q_e = 4$ ksf.

SOLUTION   The largest percentage of live load to dead load occurs for footing D.

$$\text{Area required for footing } D = \frac{100 + 150}{4} = 62.5 \text{ ft}^2$$

Table 12.2 Footings

Footing	Dead load (k)	Live load (k)
A	150	200
B	120	100
C	140	150
D	100	150
E	160	200

$$\text{Usual soil pressure under footing } D = \frac{100 + (0.30)(150)}{62.5} = 2.32 \text{ ksf}$$

Computing the areas required for the other footings and determining their soil pressures under total service loads, we show the results in Table 12.3. Note from the last column in Table 12.3 that footing D is the only footing that will be stressed to its allowable bearing stress.

Table 12.3 Areas and Soil Pressures

Footing	Usual load = $D + 0.30L$ (k)	Area required = usual load ÷ 2.32 (ft²)	Total soil pressure (ksf)
A	210	90.5	3.87
B	150	64.7	3.40
C	185	79.7	3.64
D	145	62.5	4.00
E	220	94.8	3.80

12.12 FOOTINGS SUBJECTED TO AXIAL LOADS AND MOMENTS

Walls or columns often transfer moments as well as vertical loads to their footings. These moments may be due to gravity loads or lateral loads. Such a situation is represented by the vertical load P and the bending moment M shown in Figure 12.23.

Moment transfer from columns to footings depends on how the column–footing connection is constructed. Many designers treat the connection between columns and footings as a pinned connection. Others treat it as fixed, and still others as somewhere in between. If it is truly pinned, no moment is transferred to the footing, and this section of the text is not applicable. If, on the other hand, it is treated as fixed or partially fixed, this section is applicable.

If a column–footing joint is to behave as a pin or hinge, it would have to be constructed accordingly. The reinforcing in the column might be terminated at the column base instead of continuing into the footing. Dowels would be provided, but these would not be adequate to provide a moment connection.

To provide continuity at the column–footing interface, the reinforcing steel would have to be continued into the footing. This is normally accomplished by embedding hooked bars into the footing and having them extend into the air where the columns will be located. The length they

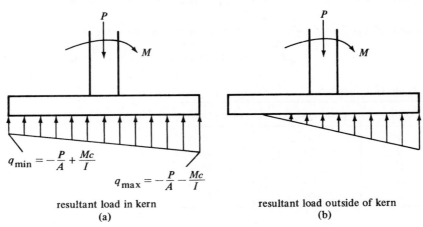

$$q_{min} = -\frac{P}{A} + \frac{Mc}{I}$$

$$q_{max} = -\frac{P}{A} - \frac{Mc}{I}$$

resultant load in kern
(a)

resultant load outside of kern
(b)

Figure 12.23

extend into the air must be at least the lap splice length; sometimes this can be a significant length. These bars are then lap spliced or mechanically spliced with the column bars, providing continuity of tension force in the reinforcing steel.

If there is a moment transfer from the column to the footing, the resultant force will not coincide with the centroid of the footing. Of course, if the moment is constant in magnitude and direction, it will be possible to place the center of the footing under the resultant load and avoid the eccentricity, but lateral forces such as wind and earthquake can come from any direction and symmetrical footings will be needed.

The effect of the moment is to produce a linearly varying soil pressure, which can be determined at any point with the expression

$$q = -\frac{P}{A} \pm \frac{Mc}{I}$$

In this discussion the term *kern* is used. If the resultant force strikes the footing base within the kern, the value of $-P/A$ is larger than $+Mc/I$ at every point and the entire footing base is in compression, as shown in Figure 12.23(a). If the resultant force strikes the footing base outside the kern, the value of $+Mc/I$ will at some points be larger than $-P/A$ and there will be uplift or tension. The soil–footing interface cannot resist tension, and the pressure variation will be as shown in Figure 12.23(b). The location of the kern can be determined by replacing Mc/I with Pec/I, equating it to P/A, and solving for e.

Should the eccentricity be larger than this value, the method described for calculating soil pressures $[(-P/A) \pm (Mc/I)]$ is not correct. To compute the pressure for such a situation it is necessary to realize that the centroid of the upward pressure must for equilibrium coincide with the centroid of the vertical component of the downward load. In Figure 12.24 it is assumed that the distance to this point from the right edge of the footing is a. Since the centroid of a triangle is located at one-third of its base, the soil pressure will be spread over the distance $3a$ as shown. For a rectangular footing with dimensions $\ell \times b$, the total upward soil pressure is equated to the downward load and the resulting expression solved for q_{max} as follows:

$$\left(\frac{1}{2}\right)(3ab)(q_{max}) = P$$

$$q_{max} = \frac{2P}{3ab}$$

Example 12.8 shows that the required area of a footing subjected to a vertical load and a lateral moment can be determined by trial and error. The procedure is to assume a size, calculate the maximum soil pressure, compare it with the allowable pressure, assume another size, and so on.

Once the area has been established, the remaining design will be handled as it was for other footings. Although the shears and moments are not uniform, the theory of design is unchanged. The factored loads are computed, the bearing pressures are determined, and the shears and

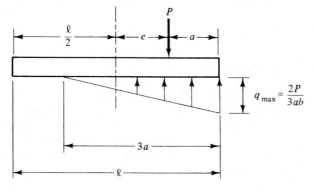

Figure 12.24

moments are calculated. For strength design, the footing must be proportioned for the effects of these loads as required in ACI Section 9.2.

EXAMPLE 12.8

Determine the width needed for a wall footing to support loads: $D = 18$ k/ft and $L = 12$ k/ft. In addition, a moment of 39 ft-k must be transferred from the column to the footing. Assume the footing is 18 in. thick, its base is 4 ft below the final grade, and $q_a = 4$ ksf.

SOLUTION   **First Trial**

Neglecting moment $q_e = 4000 - \left(\dfrac{18}{12}\right)(150) - \left(\dfrac{30}{12}\right)(100) = 3525$ psf

$$\text{Width required} = \frac{18 + 12}{3.525} = 8.51'$$ <u>Try 9'0"</u>

$$A = (9)(1) = 9 \text{ ft}^2$$

$$I = \left(\frac{1}{12}\right)(1)(9)^3 = 60.75 \text{ ft}^4$$

$$q_{max} = -\frac{P}{A} - \frac{Mc}{I} = -\frac{30}{9} - \frac{(39)(4.5)}{60.75}$$

$$= -6.22 \text{ ksf} > 3.525 \text{ ksf}$$ <u>No good</u>

$$q_{min} = -\frac{P}{A} + \frac{Mc}{I} = -\frac{30}{9} + \frac{(39)(4.5)}{60.75} = -0.44 \text{ ksf}$$

Second Trial

Assume 14-ft-wide footing (after a check of a 13-ft-wide footing proves it to be insufficient)

$$A = (14)(1) = 14 \text{ ft}^2$$

$$I = \left(\frac{1}{12}\right)(1)(14)^3 = 228.7 \text{ ft}^4$$

$$q_{max} = -\frac{30}{14} - \frac{(39)(7)}{228.7} = -3.34 \text{ ksf} < 3.525 \text{ ksf}$$

$$q_{min} = -\frac{30}{14} + \frac{(39)(7)}{228.7} = -0.95 \text{ ksf}$$ <u>Use 14'0" footing</u>

Note that in both trials, the sign for q_{min} is negative, meaning that the soil–footing interface is in compression. Had the value been positive, the equations used to calculate stress would not have been valid. Instead, the designer would have to use the equation

$$q_{max} = 2P/3ab$$

Footings must be designed to resist all applicable load combinations from ACI Section 9.2. The experienced designer can often guess which will be most critical, design the footing accordingly, and check to others to see that the footing can resist them. Experience and computer programs help immensely in this process.

12.13   TRANSFER OF HORIZONTAL FORCES

When it is necessary to transfer horizontal forces from walls or columns to footings, the shear friction method previously discussed in Section 8.12 of this text or other appropriate means should be used. (ACI Section 15.8.1.4) Sometimes shear keys (see Figure 13.1) are used between walls or columns and footings. This practice is of rather questionable value, however, because appreciable slipping has to occur to develop a shear key. A shear key may be thought of as providing an additional mechanical safety factor, but none of the lateral design force should be assigned to it.

The following example illustrates the consideration of lateral force transfer by the shear friction concept.

EXAMPLE 12.9

A 14-in. × 14-in. column is supported by a 13-ft-0-in. × 13-ft-0-in. footing ($f'_c = 3000$ psi (normal weight) and $f_y = 60,000$ psi for both). For vertical compression force transfer, six #6 dowels $(2.65$ in.$^2)$ were selected extending 18 in. up into the column and 18 in. down into the footing. Design for a horizontal factored force V_u of 65 k acting at the base of the column. Assume that the footing concrete has not been intentionally roughened before the column concrete is placed. Thus $\mu = 0.6\lambda = (0.6)(1.0)$ for normal-weight concrete $= 0.6$. (See ACI Code 11.6.4.)

SOLUTION   **Determine Minimum Shear Friction Reinforcement Required by ACI Section 11.7.4**

$$V_n = \frac{V_u}{\phi} = \frac{65}{0.75} = 86.7 \text{ k}$$

$$V_n = A_{vf}f_y\mu \qquad\qquad \text{(ACI Equation 11–25)}$$

$$A_{vf} = \frac{86.7}{(60)(0.6)} = 2.41 \text{ in.}^2 < 2.65 \text{ in.}^2 \qquad\qquad \text{OK}$$

The six #6 dowels $(2.65$ in.$^2)$ present may also be used as shear friction reinforcing. If their area had not been sufficient, it could have been increased and/or the value of μ could be increased significantly by intentionally roughening the concrete, as described in Section 11.6.4.3 of the Code.

Check Tensile Development Lengths of These Dowels

Shear friction reinforcing acts in tension and thus must have tensile anchorage on both sides of the shear plane. It must also engage the main reinforcing in the footing to prevent cracks from occurring between the shear reinforcing and the body of the concrete.

$$\text{Assuming} \frac{c + K_{tr}}{d_b} = 2.5$$

$$\frac{\ell_d}{d_b} = \left(\frac{3}{40}\right)\left(\frac{60,000}{(1.0)\sqrt{3000}}\right)\frac{(1.0)(1.0)(0.8)}{2.5}\left(\frac{2.41}{2.65}\right) = 23.91 \text{ diameters}$$

$$\ell_d = (23.91)(0.75) = 17.93 \text{ in.} \qquad\qquad \text{Say 18 in.}$$

Compute Maximum Shear Transfer Strength Permitted by the Code (11.6.5)

$$V_u \leq \phi0.2 f'_c A_c, \text{ but not} > \phi(800A_c)$$

$$\leq (0.75)(0.2)(3)(14 \times 14) = 88.2 \text{ k} > 65 \text{ k} \qquad\qquad \text{OK}$$

$$\text{but not} > (0.75)(800)(14 \times 14) = 117,600 \text{ lb} = 117.6 \text{ k}$$

12.14   PLAIN CONCRETE FOOTINGS

Occasionally, plain concrete footings are used to support light loads if the supporting soil is of good quality. Very often the widths and thicknesses of such footings are determined by rules of thumb such as "the depth of a plain footing must be equal to no less than the projection beyond the edges of the wall." In this section, however, a plain concrete footing is designed in accordance with the requirements of the ACI.

Chapter 22 of the ACI Code is devoted to the design of structural plain concrete. Structural plain concrete is defined as concrete that is completely unreinforced or that contains less than the minimum required amounts of reinforcing previously specified here for reinforced concrete members. The minimum compressive strength permitted for such concrete is 2500 psi,* as given in ACI Sections 22.2.3 and 1.1.1.

Structural plain concrete may only be used for (1) members continuously supported by soil or by other structural members that are capable of providing continuous support, (2) walls and pedestals, and (3) structural members with arch action where compression occurs for all loading cases (ACI 22.2.1).

The Code (22.7.3 and 22.7.4) states that when plain concrete footings are supported by soil, they cannot have an edge thickness less than 8 in. and they cannot be used on piles. The critical sections for shear and moment for plain concrete footings are the same as for reinforced concrete footings.

In ACI Code Section 22.5, nominal bending and shear strengths are specified for structural plain concrete. The proportions of plain concrete members will nearly always be controlled by tensile strengths rather than shear strengths.

In the equations that follow $\phi = 0.60$ (ACI 9.3.5) for all cases, S is the elastic section modulus of uncracked members, and β_c is the ratio of the long side to the short side of the column or loaded area. In computing the strengths, whether flexural or shear, for concrete cast against soil *the overall thickness h is to be taken as 2 in. less than the actual thickness* (ACI Section 22.4.7). This concrete is neglected to account for uneven excavation for the footing and for some loss of mixing water to the soil and other contamination.

Bending Strength:

$$M_n = 5\lambda\sqrt{f_c'}S \qquad \text{(ACI Equation 22–2)}$$

$$\phi M_n \geq M_u \qquad \text{(ACI Equation 22–1)}$$

Shear Strength for One-way or Beam Action:

$$V_n = \frac{4}{3}\lambda\sqrt{f_c'}bh \qquad \text{(ACI Equation 22–9)}$$

$$\phi V_n \geq V_u \qquad \text{(ACI Equation 22–8)}$$

Shear Strength for Two-way or Punching Action:

$$V_n = \left[\frac{4}{3} + \frac{8}{3\beta_c}\right]\lambda\sqrt{f_c'}b_oh \qquad \text{(ACI Equation 22–10)}$$

$$\leq 2.66\lambda\sqrt{f_c'}b_oh$$

In SI units

$$M_n = \frac{5}{12}\lambda\sqrt{f_c'}S \qquad \text{(ACI M Equation 22–2)}$$

$$V_n = \frac{1}{9}\lambda\sqrt{f_c'}bh \qquad \text{(ACI M Equation 22–9)}$$

$$V_n = \frac{1}{9}\left(1 + \frac{2}{\beta_c}\right)\lambda\sqrt{f_c'}b_oh \leq \frac{2}{9}\lambda\sqrt{f_c'}b_oh \qquad \text{(ACI M Equation 22–10)}$$

*In SI units it is 17 MPa.

A plain concrete footing will obviously require considerably more concrete than will a reinforced one. On the other hand, the cost of purchasing reinforcing and placing it will be eliminated. Furthermore, the use of plain concrete footings will enable us to save construction time in that we don't have to order the reinforcing and place it before the concrete can be poured. Therefore, plain concrete footings may be economical on more occasions than one might realize.

Even though plain footings are designed in accordance with the ACI requirements, they should at the very least be reinforced in the longitudinal direction to keep temperature and shrinkage cracks within reason and to enable the footing to bridge over soft spots in the underlying soil. Nevertheless, Example 12.10 presents the design of a plain concrete footing in accordance with the ACI Code.

EXAMPLE 12.10

Design a plain concrete footing for a 12-in. reinforced concrete wall that supports a dead load of 12 k/ft, including the wall weight, and a 6-k/ft live load. The base of the footing is to be 5 ft below the final grade, $f'_c = 3000$ psi, and $q_a = 4000$ psf.

SOLUTION

Assume 24-in. Footing (see Figure 12.25)

$$q_e = 4000 - \left(\frac{24}{12}\right)(145) - \left(\frac{36}{12}\right)(100) = 3410 \text{ psf}$$

$$\text{Width required} = \frac{18}{3.41} = 5.28 \text{ ft} \qquad\qquad \text{Say } 5'6''$$

Bearing Pressure for Strength Design

$$q_u = \frac{(1.2)(12) + (1.6)(6)}{5.5} = 4.36 \text{ ksf}$$

Checking Bending Strength, Neglecting Bottom 2 in. of Footing

M_u for 12 in. width of footing

$$= (4.36)(2.25)\left(\frac{2.25}{2}\right) = 11.04 \text{ ft-k}$$

$$S = \frac{bd^2}{6} = \frac{(12)(22)^2}{6} = 968 \text{ in.}^3$$

$$\phi M_n = \phi 5\sqrt{f'_c}S = (0.60)(5)(\sqrt{3000})(968)$$

$$= 159,058 \text{ in.-lb} = 13.25 \text{ ft-k}$$

$$> 11.04 \text{ ft-k} \qquad\qquad \text{OK}$$

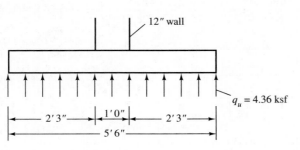

Figure 12.25

Checking Shearing Strength at a Distance 22 in. from Face of Wall

V_u for 12 in. width of footing

$$= \left(2.25 - \frac{22}{12}\right)(4.36) = 1.82\,\text{k}$$

$$V_u = \phi\left(\frac{4}{3}\right)(1.0)\sqrt{f'_c}bh = (0.60)\left(\frac{4}{3}\right)(1.0)(\sqrt{3000})(12)(22)$$

$$= 11{,}568\,\text{lb} = 11.57\,\text{k} > 1.82\,\text{k} \qquad \underline{\underline{\text{OK}}}$$

$$\underline{\underline{\text{Use } 24'' \text{ footing}}}$$

Note: A 23-in.-deep footing will work in this case, and a 22-in.-deep footing will almost work (within 1% moment capacity). The authors prefer to use the 24-in. depth for simplicity in this case.

12.15 SI EXAMPLE

EXAMPLE 12.11

Design a reinforced concrete wall footing to support a 300-mm-wide reinforced concrete wall with a dead load $D = 300\,\text{kN/m}^2$ and a live load $L = 200\,\text{kN/m}^2$. The bottom of the footing is to be 1 m below the final grade, the soil weight is $16\,\text{kN/m}^3$, the concrete weight is $24\,\text{kN/m}^3$, the allowable soil pressure q_u is $190\,\text{MPa/m}^2$, $f_y = 420\,\text{MPa}$, and $f'_c = 28\,\text{MPa}$.

SOLUTION **Assume 450-mm-Deep Footing ($d = 360$ mm)**

$$q_e = 190 - \left(\frac{450}{1000}\right)(24) - \left(\frac{550}{1000}\right)(16) = 170.4\,\text{kN/m}^2$$

$$\text{width required} = \frac{300 + 200}{170.4} = 2.93\,\text{m} \qquad \underline{\text{Say 3.00 m}}$$

Bearing Pressure for Strength Design

$$q_u = \frac{(1.2)(300) + (1.6)(200)}{3.00} = 226.7\,\text{kN/m}^2$$

Depth Required for Shear at a Distance d from Face of Wall (See Figure 12.26)

$$V_u = \left(\frac{3.00}{2} - \frac{150}{1000} - \frac{360}{1000}\right)(226.7) = 224.4\,\text{kN}$$

$$d = \frac{6V_u}{(0.75)\lambda(\sqrt{f'_c})(b_w)} = \frac{(6)(224.4)(10)^3}{(0.75)(1.0)(\sqrt{28})(1000)} = 339\,\text{mm} < 360\,\text{mm} \qquad \underline{\underline{\text{OK}}}$$

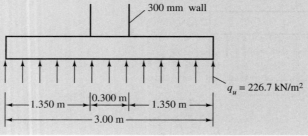

Figure 12.26

Steel Area (*d*=360 mm) Taking Moments at Face of Wall

$$\text{Cantilever length} = \frac{3.00}{2} - \frac{150}{1000} = 1.35 \text{ m}$$

$$M_u = (1.35)(226.7)\left(\frac{1.35}{2}\right) = 206.58 \text{ kN/m}$$

$$\frac{M_u}{\phi b d^2} = \frac{206.58 \times 10^6}{(0.9)(1000)(360)^2} = 1.771$$

$$\rho = 0.00439 \text{ from Appendix Table B.9}$$

$$A_s = (0.00439)(1000)(360) = 1580 \text{ mm}^2/\text{m}$$

Use 22 bars @ 225 mm o.c. (1720 mm^2)

Development Length

From Table 7.1 in Chapter 7

$$\psi_t = \psi_e = \psi_s = \lambda = 1.0$$

$$\text{Assum } c_b = \text{cover} = 90 \text{ mm} \leftarrow$$

$$c = \text{one-half } c. \text{ to } c. \text{ of bars} = 112.5 \text{ mm}$$

Letting $K_{tr} = 0$

$$\frac{c_b + K_{tr}}{d_b} = \frac{90 + 0}{22.2} = 4.05 > 2.50 \qquad\qquad \therefore \text{Use } 2.5$$

$$\frac{\ell_d}{d_b} = \frac{9 f_y \psi_1 \psi_2 \psi_3}{10 \lambda \sqrt{f_c'}\left(\dfrac{c_b + K_{tr}}{d_b}\right)}$$

$$= \frac{(9)(420)(1.0)(1.0)(1.0)}{\left(10(1.0)\sqrt{28}\right)(2.5)} = 28.57 \text{ diameters}$$

$$\frac{\ell_d}{d_b}\frac{A_{s,\,reqd}}{A_{s,\,furn}} = (28.57)\left(\frac{1580}{1720}\right) = 26.24 \text{ diameters}$$

$$\ell_d = (26.24)(22.2) = 583 \text{ mm} < 1350 \text{ mm} - 100 \text{ mm} = 1250 \text{ mm available} \qquad \text{OK}$$

Use 450 mm footing 3 m wide with 22 bars at 225 mm

12.16   COMPUTER EXAMPLES

EXAMPLE 12.12

Repeat Example 12.1 using the Excel spreadsheets provided for Chapter 12.

SOLUTION   Open the Chapter 12 spreadsheet and the Wall Footing worksheet. Enter values only for the cells in yellow highlight, beginning on the left side of the worksheet. When entering "*trial h*," estimate a reasonable value. In this case, $h = 12$ in. is the first try. Based on this assumption, d, q_e and ℓ_{min} are calculated. Look at the footing width ℓ_{min} (9.86 ft) and enter a value that is more practical and slightly larger for the actual width in the next cell for ℓ (10.0 ft). Now observe the values of d_{shear} and h_{shear} a few cells below. The correct theoretical answer for h_{shear} lies between the *trial h* and h_{shear}. So it is somewhere between *12 in* and *21.96 in*. Split the difference and go back to *trial h* with a value of *trial h* = 16 in. The footing width can remain 10 ft, and h_{shear} is now 20.34 in. Split the difference again, and enter *trial* $h = 19$ in. Now $h_{shear} = 19.12$ in. Trial and error can be avoided by using the Goal seek feature. In the first cycle, after trying $h = 12$ in., go to the cell called trial $h - h_{shear}$. Theoretically this should be zero. Highlight this cell, go to Tools on the menu bar and select Goal seek from the drop down menu. In "To value" enter 0, and in "By changing cell" enter C14. Click OK and the status window says there is a solution. Select OK and observe that cells C14 and C22 are now both 19.09 in. This may be faster than the trial-and-error method, but there is no need for the precision of the answer provided. Either method leads to the thickness of the footing of 20 in.

Now go to the right side of the worksheet and enter select $h = 20$ in. The required area of reinforcing steel is calculated and a list of possible choices of bar size and theoretical spacing is displayed. Pick the one you prefer, being sure to round the spacing down to a practical value. If #7 bars are selected, the theoretical spacing of 10.55 in. is rounded down to 10 in. (the same answer as Example 12.1).

A screen shot of the software for the first cycle with trial $h = 12$ in. is shown below.

Wall Footing Design		
$P_D =$	20	k/ft
$P_L =$	15	k/ft
$P_u =$	48	k/ft
$t =$	12	in.
cover =	3	in.
$f'_c =$	3000	psi
$\gamma_c =$	145	pcf
$\gamma_s =$	100	pcf
$f_y =$	60,000	psi
$\lambda =$	1.00	
$q_a =$	4.00	ksf
$d_{grade} =$	4.00	ft
trial $h =$	12.00	in.
$d =$	8.50	in.
$q_e =$	3.55	ksf
$\ell_{min} =$	9.86	ft
$\ell =$	10	ft
$q_u =$	4.8	ksf
$V_u =$	18.20	k/ft
$d_{shear} =$	18.46	in.
$h_{shear} =$	21.96	in.
trial $h - h_{shear} =$	9.96	in.

select $h =$	20	in.
$d =$	16.50	in.
$M_u =$	48.6	ft-k
$R_n =$	198.3471	psi
$\rho =$	0.00345	
$A_{s\ flexure} =$	0.682	in.^{2}/ft
$A_{s\ t\&s} =$	0.432	in.^{2}/ft
$A_{s\ min} =$	0.660	in.
$A_s =$	0.682	in.^{2}/ft

	s_{theor}	
#3	1.93	in.
#4	3.52	in.
#5	5.45	in.
#6	7.74	in.
#7	10.55	in.
#8	13.90	in.
#9	17.59	in.

Select a bar size and spacing from the table above.

EXAMPLE 12.13

Repeat Example 12.2 using the Excel spreadsheets provided for Chapter 12.

SOLUTION   Open the Chapter 12 spreadsheet and the Square Footing worksheet. Enter values only for the cells in yellow highlight, beginning on the left side of the worksheet. When entering "trial h," estimate a reasonable value. In Example 2 an initial value of $h = 24$ was used. This turned out to be the correct answer, which is not usually the case. Let's pick a value of $h = 18$ in. to illustrate the use of the spreadsheet's ability to converge on the correct answer. Note that ℓ_{min} is 9.02 ft and a value of $\ell = 9.00$ ft was selected. This slightly nonconservative choice is consistent with what was done in Example 12.2. Now look at the upper right side of the worksheet under Two-way shear. A value of $h_2 = 28.17$ in. means the trial h of 18 in. isn't enough. A quick look at the One-way shear results shows $h_1 = 21.32$ in. Since h_2 is larger, two-way shear is more critical. The correct value of h is somewhere between 18 in. and 28.17 in. Go back to trial h and split the difference by entering 23 in. A quick look at the required h_2 and h_1 shows this choice is acceptable for one-way shear, but not two way, which needs $h_2 = 24.14$ in. Now enter a trial h of 24 in., and both h_1 and h_2 are exceeded, indicating $h = 24$ in. is enough for shear. You can also do this by using Goal seek as described in Example 12.12.

Go to the lower part of the worksheet and enter 24 in. under select h. The required A_s in both directions is computed, and a table of possible choices is displayed. One of the choices is nine #8 bars, the same as selected in Example 12.2.

A screenshot of the software for the last cycle with trial $h = 24$ in. is shown below.

Square Single-Column Footing Design

$P_D =$	200	k	**Two-way Shear**		
$P_L =$	160	k	$V_{u2} =$	442.4	k
$P_u =$	496	k	$b_o =$	142	in.
$a =$	16	in.	$\alpha_s =$	40.00	
$b =$	16	in.	$\beta_c =$	1	
cover $=$	3	in.	$d_{2\,shear} =$	18.96	in.
$f'_c =$	3000	psi	$d_{2\,shear} =$	12.6405	in.
$\gamma_c =$	145	pcf	$d_{2\,shear} =$	10.12	in.
$\gamma_s =$	100	pcf	$d_2 =$	18.96	in.
$f_y =$	60,000	psi	$h_2 =$	23.46	in.
$\lambda =$	1.00		trial $h - h_{2\,shear} =$	0.54	
$q_a =$	5.00	ksf			
$d_{grade} =$	5.00	ft	**One-way Shear**		
trial $h =$	24.00	in.	$V_{u1} =$	121.70	in.^{2}
$d =$	19.50	in.	$d_{1\,shear} =$	13.72	in.
$q_e =$	4.40	ksf	$h_{1\,shear} =$	18.22	in.
$\ell_{min} =$	9.05	ksf	trial $h - h_{1\,shear} =$	5.78	in.
$\ell =$	9	ft			
$q_u =$	6.12	ksf			

select $h =$	24	in.
$d =$	19.50	in.

	Theoretical	No. of bars	Spacing, in.
#4	35.10	36	3.03
#5	22.65	23	4.67
#6	15.95	16	6.56
#7	11.70	12	8.85
#8	8.89	9	11.51
#9	7.02	8	14.36
#10	5.53	6	17.92

$M_u =$	404.914	ft-k
$R_n =$	131.464	psi
$\rho =$	0.00225	
$A_{S\,flexure} =$	4.74	in.^{2}
$A_{s\,t\&s} =$	4.67	in.^{2}
$A_{s\,min} =$	7.02	iin.^{2}
$A_s =$	7.02	in.^{2}

Select a bar size and number of bars from the table above, rounding up to the next integer.

PROBLEMS

For Problems 12.1 to 12.30, assume that reinforced concrete weighs 150 lb/ft^3, plain concrete 145 lb/ft^3, and soil 100 lb/ft^3.

Wall Footings

For Problems 12.1 to 12.5, design wall footings for the values given. The walls are to consist of reinforced concrete.

Problem	Wall thickness (in.)	D (k/ft)	L (k/ft)	f'_c (ksi)	f_y (ksi)	q_a (ksf)	Distance from bottom of footing to final grade (ft)
12.1	14	15	12	4	60	3	3
12.2	14	21	20	4	60	4	4
12.3	15	15	18	5	60	5	5
12.4	15	24	30	4	60	4	4
12.5	12	10	8	4	60	5	4

(Answer to Problem 12.1: 15 in. footing 10 ft 3 in. wide with #7 @ 9 in. main steel)
(Answer to Problem 12.3: 15 in. footing 7 ft 6 in. wide with #5 @ 6 in. main steel)
(Answer to Problem 12.5: 10 in. footing 4 ft 6 in. wide with #5 @ 11 in. main steel)

12.6 Repeat Problem 12.1 if a masonry wall is used.

Column Footings

In Problems 12.7 to 12.12, design square single-column footings for the values given. All columns are interior columns.

Problem	Column size (in.)	D (k)	L (k)	f'_c (ksi)	f_y (ksi)	q_a (ksf)	Distance from bottom of footing to final grade (ft)
12.7	12 × 12	100	160	4	60	4	4
12.8	12 × 12	100	80	3	60	5	5
12.9	15 × 15	160	150	4	60	4	4
12.10	15 × 15	150	120	3	60	4	4
12.11	16 × 16	110	100	3	60	5	5
12.12	Round, 18 dia.	240	140	4	60	5	6

(Answer to Problem 12.7: 21 in. footing, 9 ft 0 in. × 9 ft 0 in. with 8 #8 bars both directions)
(Answer to Problem 12.9: 22 in. footing, 9 ft 6 in. × 9 ft 6 in. with 12 #7 bars both directions)
(Answer to Problem 12.11: 18 in. footing, 7 ft 0 in. × 7 ft 0 in. with 7 #7 bars both directions)

12.13 Design for load transfer from an $16'' × 16''$ column with 6 #8 bars $(D = 200\,k, L = 350\,k)$ to an 8-ft-0-in. × 8-ft-0-in. footing. $f'_c = 4$ ksi for footing and 5 ksi for column. $f_y = 60$ ksi. (*Ans.* 4 #6, 13 in. into footing, 12.5 in. into column)

12.14 Repeat Problem 12.7 if a rectangular footing with one side of the footing limited to 7 ft. *due 12/1*

12.15 Design a footing with one side limited to 7 ft for the following: 12 in. × 12 in. edge column, $D = 130\,k$, $L = 155\,k, f'_c = 3000$ psi, $f_y = 60,000$ psi, $q_a = 4$ ksf, and a distance from top of backfill to bottom of footing bottom = 4 ft. (*Ans.* 7 ft 0 in. × 11 ft 8 in. footing, 24 in. deep, 7 #8 bars in long direction)

12.16 Design a footing limited to a maximum width of 7 ft 0 in. for the following: 15 in. × 15 in. interior column, $D = 180\,k$, $L = 160\,k, f'_c = 4000$ psi, $f_y = 60,000$ psi, $q_a = 4$ ksf, and a distance from top of backfill to bottom of footing = 5 ft.

12.17 Design a rectangular combined footing for the two columns shown in the accompanying illustration. The bottom of the footing is to be 5 ft below the final grade, $f'_c = 3.5$ ksi, $f_y = 50$ ksi, and $q_a = 5$ ksf. (*Ans.* 10 ft 6 in. × 12 ft 9 in., 26 in. deep, 11 #9 bars long direction)

12.18 Determine the footing areas required for the loads given in the accompanying table so that the usual soil pressures are equal. Assume $q_e = 5$ ksf and a usual live load percentage of 30% for all of the footings.

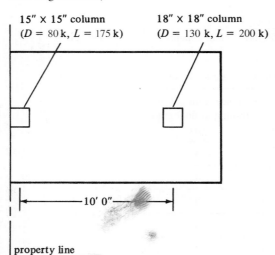

15" × 15" column
$(D = 80\,k, L = 175\,k)$

18" × 18" column
$(D = 130\,k, L = 200\,k)$

10' 0"

property line

Footing	Dead load	Live load
A	120 k	200 k
B	130 k	170 k
C	120 k	200 k
D	150 k	200 k
E	140 k	180 k
F	140 k	200 k

Footings with Moments

In Problems 12.19 and 12.20, determine the width required for the wall footings. Assume footings have total thicknesses of 24 in.

Problem	Reinforced concrete wall thickness	D	L	f'_c	f_y	q_a	Moment	Footing Thickness, h (in.)	Distance from bottom of footing to final grade
12.19	12″	14 k/ft	18 k/ft	4 ksi	60 ksi	4 ksf	32 ft-k	18	4′
12.20	14″	16 k/ft	24 k/ft	3 ksi	60 ksi	4 ksf	50 ft-k	24	5′

(*Answer to Problem 12.19:* 15 ft 0 in.)

12.21 Repeat Problem 12.13 if a lateral force $V_u = 100$ k acts at the base of the column. Use the shear friction concept. Assume that footing concrete has not been intentionally roughened before column concrete is placed and normal-weight concrete is used. ($\mu = 0.6\lambda$). (*Ans.* six #8 dowels, 27 in. into footing, 24 in. into column)

12.22 Repeat Problem 12.21 using intentionally roughened concrete ($\mu = 1.0\lambda$).

Plain Concrete Footings

In Problems 12.22 and 12.23, design plain concrete wall footings of uniform thickness.

Problem	Reinforced concrete wall thickness	D	L	f'_c	q_a	Distance from bottom of footing to final grade
12.23	12″	16 k/ft	14 k/ft	4 ksi	5 ksf	5′
12.24	14″	12 k/ft	10 k/ft	3 ksi	4 ksf	4′

(*Answer to Problem 12.23:* 32-in.-deep footing 7 ft 0 in. wide)

In Problems 12.25 and 12.26, design square plain concrete column footings of uniform thickness.

Problem	Reinforced concrete column size	D	L	f'_c	q_a	Distance from bottom of footing to final grade
12.25	12″ × 12″	70 k	60 k	3 ksi	4 ksf	5′
12.26	14″ × 14″	90 k	75 k	3.5 ksi	4 ksf	5′

(*Answer to Problem 12.25:* 6 ft 3 in. × 6 ft 3 in. footing 28 in. deep)

Computer Problems

For Problems 12.27 to 12.30, use the spreadsheet provided for Chapter 12.

12.27 Repeat Problem 12.2. (*Ans.* 11 ft 9 in. width, 21 in. depth with #7 at 8 in.)

12.28 Repeat Problem 12.8.

12.29 Repeat Problem 12.10. (*Ans.* 9 ft 0 in. × 9 ft 0 in., 21 in. depth with 10 #7 each way)

12.30 Repeat Problem 12.16.

Problems with SI Units

In Problems 12.31 and 12.32, design wall footings for the values given. The walls are to consist of reinforced concrete. Concrete weight = 24 kN/m³, soil weight = 16 kN/m³.

Problem	Wall thickness	D	L	f'_c	f_y	q_a	Distance from bottom of footing to final grade
12.31	300 mm	150 kN/m	200 kN/m	21 MPa	420 MPa	170 kN/m²	1.500 m
12.32	400 mm	180 kN/m	250 kN/m	28 MPa	420 MPa	210 kN/m²	1.200 m

(*Answer to Problem 12.31:* 370 mm footing, width = 2.5 m, and #19 @ 225 mm main steel)

In Problems 12.33 to 12.35, design square single-column footings for the values given. Concrete weight $= 24 \text{ kN/m}^3$, soil weight $= 16 \text{ kN/m}^3$, and all columns are interior ones.

Problem	Column size	D	L	f'_c	f_y	q_a	Distance from bottom of footing to final grade
12.33	350 mm × 350 mm	400 kN	500 kN	21 MPa	420 MPa	170 kN/m^2	1.200 m
12.34	400 mm × 400 mm	650 kN	800 kN	28 MPa	420 MPa	170 kN/m^2	1.200 m
12.35	450 mm × 450 mm	750 kN	1000 kN	28 MPa	420 MPa	210 kN/m^2	1.600 m

(*Answer to Problem 12.33:* 480 mm footing with 11 #19 bars both directions)
(*Answer to Problem 12.35:* 600 mm footing with 10 #25 bars both directions)

12.36 Design a plain concrete wall footing for a 300-mm-thick reinforced concrete wall that supports a 100-kN/m dead load (including its own weight) and a 120-kN/m live load. $f'_c = 21$ MPa, and $q_a = 170 \text{ kN/m}^2$. The base of the footing is to be 1.250 m below the final grade, Concrete weight $= 24 \text{ kN/m}^3$, and soil weight $= 16 \text{ kN/m}^3$.

12.37 Design a square plain concrete column footing to support a 300-mm × 300-mm reinforced concrete column that in turn is supporting a 130-kN dead load and a 200-kN live load. $f'_c = 28$ MPa, and $q_a = 210 \text{ kN/m}^2$. The base of the footing is to be 1.500 m below the final grade. Concrete weight $= 24 \text{ kN/m}^3$ and soil weight $= 16 \text{ kN/m}^3$. (*Ans.* 1400 mm wide and 520 mm thick)

13

Retaining Walls

13.1 INTRODUCTION

A retaining wall is a structure built for the purpose of holding back or retaining or providing one-sided lateral confinement of soil or other loose material. The loose material being retained pushes against the wall, tending to overturn and slide it. Retaining walls are used in many design situations where there are abrupt changes in the ground slope. Perhaps the most obvious examples to the reader occur along highway or railroad cuts and fills. Often, retaining walls are used in these locations to reduce the quantities of cut and fill as well as to reduce the right-of-way width required if the soils were allowed to assume their natural slopes. Retaining walls are used in many other locations as well, such as for bridge abutments, basement walls, and culverts.

Several different types of retaining walls are discussed in the next section, but whichever type is used, there will be three forces involved that must be brought into equilibrium: (1) the gravity loads of the concrete wall and any soil on top of the footing (the so-called *developed weight*), (2) the lateral pressure from the soil, and (3) the bearing resistance of the soil. In addition, the stresses within the structure have to be within permissible values, and the loads must be supported in a manner such that undue settlements do not occur. A retaining wall must be designed in such a way that the concrete elements that make up the wall comply with the code using, for the most part, principles already discussed in this text. In addition, the overall stability of the wall must be ensured. The wall may slide or tip over due to global instability without failure of the concrete elements.

13.2 TYPES OF RETAINING WALLS

Retaining walls are generally classed as being gravity or cantilever types, with several variations possible. These are described in the paragraphs to follow, with reference being made to Figure 13.1.

The *gravity retaining wall*, Figure 13.1(a), is used for walls of up to about 10 to 12 ft in height. It is usually constructed with plain concrete and depends completely on its own weight for stability against sliding and overturning. It is usually so massive that it is unreinforced. Tensile stresses calculated by the working-stress method are usually kept below $1.6\sqrt{f_c'}$. Gravity walls may also be constructed with stone or block masonry.

Semigravity retaining walls, Figure 13.1(b), fall between the gravity and cantilever types (to be discussed in the next paragraph). They depend on their own weights plus the weight of some soil behind the wall to provide stability. Semigravity walls are used for approximately the same range of heights as the gravity walls and usually have some light reinforcement.

The *cantilever retaining wall* or one of its variations is the most common type of retaining wall. Such walls are generally used for heights from about 10 to 25 ft. In discussing retaining walls, the vertical wall is referred to as the *stem*. The outside part of the footing that is pressed down into

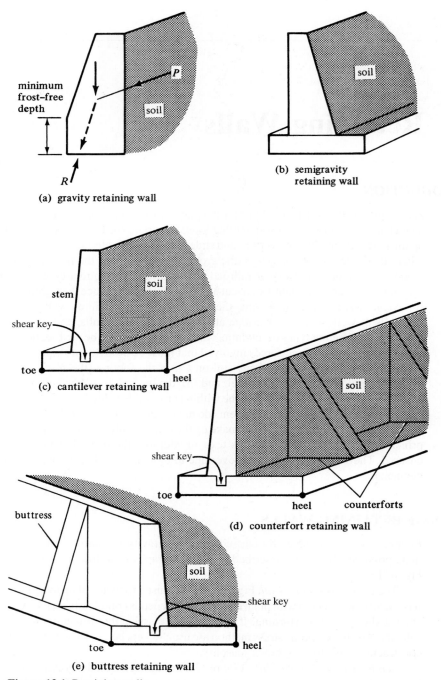

Figure 13.1 Retaining walls.

the soil is called the *toe*, while the part that tends to be lifted is called the *heel*. These parts are indicated for the cantilever retaining wall of Figure 13.1(c). The concrete and its reinforcing are so arranged that part of the material behind the wall is used along with the concrete weight to produce the necessary resisting moment against overturning. This resisting moment is generally referred to as the *righting moment.*

When it is necessary to construct retaining walls of greater heights than approximately 20 to 25 ft, the bending moments at the junction of the stem and footing become so large that the designer will, from economic necessity, have to consider other types of walls to handle the moments. This can be done by introducing vertical cross walls on the front or back of the stem. If the cross walls are behind the stem (that is, inside the soil) and not visible, the retaining walls are called *counterfort walls.* Should the cross walls be visible (that is, on the toe side), the walls are called *buttress walls.* These walls are illustrated in parts (d) and (e) of Figure 13.1. The stems for these walls are continuous members supported at intervals by the buttresses or counterforts. Counterforts or buttresses are usually spaced at distances approximately equal to one-half (or a little more) of the retaining wall heights.

The counterfort type is more commonly used because it is normally thought to be more attractive as the cross walls or counterforts are not visible. Not only are the buttresses visible on the toe side, but their protrusion on the outside or toe side of the wall will use up valuable space. Nevertheless, buttresses are somewhat more efficient than counterforts because they consist of concrete that is put in compression by the overturning moments, whereas the counterforts are concrete members used in a tension situation and they need to be tied to the wall with stirrups. Occasionally, high walls are designed with both buttresses and counterforts.

Figure 13.2 presents a few other retaining wall variations. When a retaining wall is placed at a property boundary or next to an existing building, it may be necessary to use a wall without a toe, as shown in part (a) of the figure, or without a heel, as shown in part (b). Another type of retaining wall very often encountered is the bridge abutment shown in part (c) of the figure. Abutments may very well have wing wall extensions on the sides to retain the soil in the approach area. The abutment, in addition to other loads, will have to support the end reactions from the bridge.

The use of precast retaining walls is becoming more common each year. The walls are built with some type of precast units, and the footings are probably poured in place. The results are very attractive, and the units are high-quality concrete members made under "plant controlled" conditions. Less site preparation is required and the erection of the walls is much quicker than cast-in-place ones. The precast units can later be disassembled and the units used again. Other types of precast retaining walls consist of walls or sheeting actually driven into the ground before excavation. Also showing promise are the *gabions* or wire baskets of stone used in conjunction with geotextile reinforced embankments.

13.3   DRAINAGE

One of the most important items in designing and constructing successful retaining walls is the prevention of water accumulation behind the walls. If water is allowed to build up there, the result can be great lateral water pressure against the wall and perhaps an even worse situation in cold climates due to frost action.

The best possible backfill for a retaining wall is a well-drained and cohesionless soil. Furthermore, this is the condition for which the designer normally plans and designs. In addition to a granular backfill material, weep holes of 4 in. or more in diameter (the large sizes are used for easy cleaning) are placed in the walls approximately 5 to 10 ft on center, horizontally and vertically, as shown in Figure 13.3(a). If the backfill consists of a coarse sand, it is desirable to put a few shovels of pea gravel around the weep holes to try to prevent the sand from stopping up the holes.

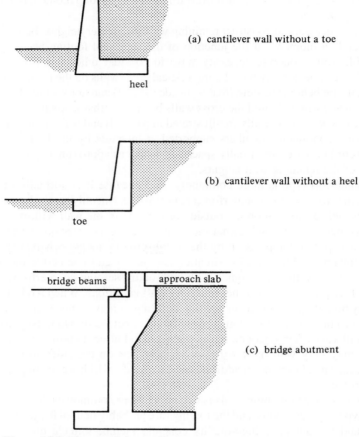

(a) cantilever wall without a toe

heel

(b) cantilever wall without a heel

toe

bridge beams   approach slab

(c) bridge abutment

Figure 13.2 More retaining walls.

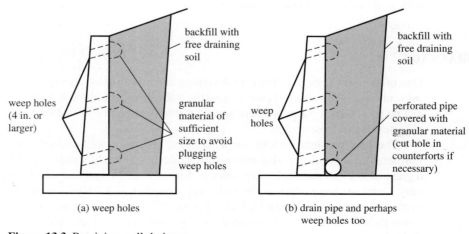

weep holes
(4 in. or
larger)

backfill with
free draining
soil

granular
material of
sufficient
size to avoid
plugging
weep holes

weep
holes

backfill with
free draining
soil

perforated pipe
covered with
granular material
(cut hole in
counterforts if
necessary)

(a) weep holes

(b) drain pipe and perhaps
weep holes too

Figure 13.3 Retaining wall drainage.

Weep holes have the disadvantages that the water draining through the wall is somewhat unsightly and also may cause a softening of the soil in the area of the highest soil pressure (under the footing toe). A better method includes the use of a 6- or 8-in. perforated pipe in a bed of gravel running along the base of the wall, as shown in Figure 13.3(b). Unfortunately, both weep holes and drainage pipes can become clogged, with the result that increased water pressure can occur. Manufactured drainage blankets or porous mats placed between the wall and the soil allow moisture to migrate freely to drainage systems, such as in Figure 13.3(b).

The drainage methods described in the preceding paragraphs are also quite effective for reducing frost action in colder areas. Frost action can cause very large movements of walls, not just in terms of inches but perhaps even in terms of a foot or two, and over a period of time can lead to failures. Frost action, however, can be greatly reduced if coarse, properly drained materials are placed behind the walls. The thickness of the fill material perpendicular to a wall should equal at least the depth of frost penetration in the ground in that area.

The best situation of all would be to keep the water out of the backfill altogether. Such a goal is normally impossible, but sometimes the surface of the backfill can be paved with asphalt or some other material, or perhaps a surface drain can be provided to remove the water, or it may be possible in some other manner to divert the water before it can get to the backfill.

Retaining wall for Long Island Railroad, Huntington, New York. Constructed with precast interlocking reinforced concrete modules. (Courtesy of Doublewal Corporation.)

13.4   FAILURES OF RETAINING WALLS

The number of failures or partial failures of retaining walls is rather alarming. The truth of the matter is that if large safety factors were not used, the situation would be even more severe. One reason for the large number of failures is the fact that designs are so often based on methods that are suitable only for certain special situations. For instance, if a wall that has a saturated clay behind it (never a good idea) is designed by a method that is suitable for a dry granular material, future trouble will be the result.

13.5   LATERAL PRESSURES ON RETAINING WALLS

The actual pressures that occur behind retaining walls are quite difficult to estimate because of the large number of variables present. These include the kinds of backfill materials and their compactions and moisture contents, the types of materials beneath the footings, the presence or absence of surcharge, and other items. As a result, the detailed estimation of the lateral forces applied to various retaining walls is clearly a problem in theoretical soil mechanics. For this reason the discussion to follow is limited to a rather narrow range of cases.

If a retaining wall is constructed against a solid rock face, there will be no pressure applied to the wall by the rock. But if the wall is built to retain a body of water, hydrostatic pressure will be applied to the wall. At any point the pressure (p) will equal wh, where w is the unit weight of the water and h is the vertical distance from the surface of the water to the point in question.

Gang-formed panels. (Courtesy of Burke Concrete Accessories, Inc.)

If a wall is built to retain a soil, the soil's behavior will generally be somewhere between that of rock and water (*but as we will learn, the pressure caused by some soils is much higher than that caused by water*). The pressure exerted against the wall will increase, as did the water pressure, with depth but usually not as rapidly. This pressure at any depth can be estimated with the following expression:

$$p = Cwh$$

In this equation, w is the unit weight of the soil, h is the distance from the surface to the point in question, and C is a constant that is dependent on the characteristics of the backfill. Unfortunately, the value of C can vary quite a bit, being perhaps as low as 0.3 or 0.4 for loose granular soils and perhaps as high as 0.9 or even 1.0 or more for some clay soils. Figure 13.4 presents charts that are sometimes used for estimating the vertical and horizontal pressures applied by soil backfills of up to 20-ft heights. Several different types of backfill materials are considered in the figure.

Unit weights of soils will vary roughly as follows: 90 to 100 lb/ft³ for soft clays, 100 to 120 lb/ft³ for stiff clays, 110 to 120 lb/ft³ for sands, and 120 to 130 lb/ft³ for sand and gravel mixes.

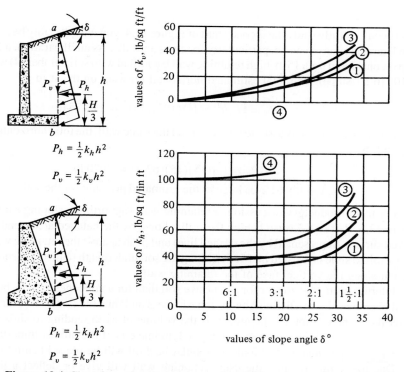

Figure 13.4 Chart for estimating pressure of backfill against retaining walls supporting backfills with plane surface. Use of this chart is limited to walls not over about 20 ft high. (1) Backfill of coarse-grained soil without admixture of fine particles, very permeable, as clean sand or gravel. (2) Backfill of coarse-grained soil of low permeability due to admixture of particles of silt size. (3) Backfill of fine silty sand, granular materials with conspicuous clay content, and residual soil with stones. (4) Backfill of soft or very soft clay, organic silt, or silty clay.[1]

---

[1]Peck, R. B., Hanson, W. E., and Thornburn, T. H., 1974, *Foundation Engineering*, 2nd ed. (New York: John Wiley & Sons), p. 425.

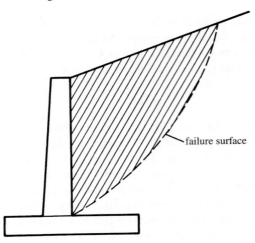

failure surface

Figure 13.5

If you carefully study the second chart of Figure 13.4, you will probably be amazed to see how high lateral pressures can be, particularly for clays and silts. As an illustration, a 1 ft-wide vertical strip is considered for a 15-ft-high retaining wall backfilled with soil number (4) with an assumed δ of $10°$ (6 : 1 slope). The total estimated horizontal pressure on the strip is

$$P_h = \tfrac{1}{2} k_h h^2 = \left(\tfrac{1}{2}\right)(102)(15)^2 = 11{,}475 \text{ lb}$$

If a 15-ft-deep lake is assumed to be behind the same wall, the total horizontal pressure on the strip will be

$$P_h = \left(\tfrac{1}{2}\right)(15)(15)(62.4) = 7020 \text{ lb}$$

(only 61% as large as the estimated pressure for the soil)

For this introductory discussion, a retaining wall supporting a sloping earth fill is shown in Figure 13.5. Part of the earth behind the wall (shown by the hatched area) tends to slide along a curved surface (represented by the dashed line) and push against the retaining wall. The tendency of this soil to slide is resisted by friction along the soil underneath (called *internal friction*) and by friction along the vertical face of the retaining wall.

Internal friction is greater for a cohesive soil than for a noncohesive one, but the wetter such a soil becomes, the smaller will be its cohesiveness and thus the flatter the plane of rupture. The flatter the plane of rupture, the greater is the volume of earth tending to slide and push against the wall. Once again it can be seen that good drainage is of the utmost importance. Usually the designer assumes that a cohesionless granular backfill will be placed behind the walls.

Due to lateral pressure, the usual retaining wall will give or deflect a little because it is constructed of elastic materials. Furthermore, unless the wall rests on a rock foundation, it will tilt or lean a small distance away from the soil due to the compressible nature of the supporting soils. For these reasons, retaining walls are frequently constructed with a slight batter, or inclination, toward the backfill so that the deformations described are not obvious to the passerby.

Under the lateral pressures described, the usual retaining wall will move a little distance and *active soil pressure* will develop, as shown in Figure 13.6. Among the many factors that affect the pressure applied to a particular wall are the kind of backfill material used, the drainage situation, the level of the water table, the seasonal conditions such as dry or wet or frozen, the presence of trucks or other equipment on the backfill, and so on.

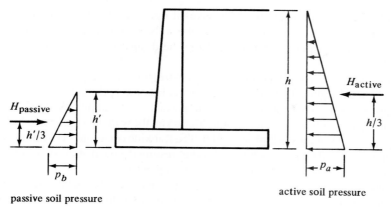

passive soil pressure

Figure 13.6

For design purposes it is usually satisfactory to assume that the active pressure varies linearly with the depth of the backfill. In other words, it is just as though (so far as lateral pressure is concerned) there is a liquid of some weight behind the wall that can vary from considerably less than the weight of water to considerably more. The chart of Figure 13.4 shows this large variation in possible lateral pressures. The assumed lateral pressures are often referred to as *equivalent fluid pressures. Values from 30 to 50 pcf are normally assumed but may be much too low for clay and silt materials.*

If the wall moves away from the backfill and against the soil at the toe, a passive soil pressure will be the result. Passive pressure, which is also assumed to vary linearly with depth, is illustrated in Figure 13.6. The inclusion or noninclusion of passive pressure in the design calculations is a matter of judgment on the designer's part. For effective passive pressure to be developed at the toe, the toe concrete must be placed against undisturbed earth without the use of vertical forms. Even if this procedure is followed, the designer will probably reduce the height of the undisturbed soil (h' in Figure 13.6) used in the calculations to account for some disturbance of the earth during construction operations.

As long as the backfills are granular, noncohesive, and dry, the assumption of an equivalent liquid pressure is fairly satisfactory. Formulas based on an assumption of dry sand or gravel backfills are not satisfactory for soft clays or saturated sands. Actually, clays should not be used for backfills because their shear characteristics change easily and they may tend to creep against the wall, increasing pressures as time goes by.

If a linear pressure variation is assumed, the active pressure at any depth can be determined as

$$p_a = k_a w h$$

or, for passive pressure,

$$p_p = k_p w h'$$

In these expressions, k_a and k_p are the approximate coefficients of active and passive pressures, respectively. These coefficients can be calculated by theoretical equations such as those of Rankine or Coulomb.[2] For a granular material, typical values of k_a and k_p are 0.3 and 3.3. The Rankine equation (published in 1857) neglects the friction of the soil on the wall, whereas the

[2]Terzaghi, K., and Peck, R. B., 1948, *Soil Mechanics in Engineering Practice* (New York: John Wiley & Sons), pp. 138–166.

Coulomb formula (published in 1776) takes it into consideration. These two equations were developed for cohesionless soils. For cohesive soils containing clays and/or silts it is necessary to use empirical values determined from field measurements (such as those given in Figure 13.4).

It has been estimated that the cost of constructing retaining walls varies directly with the square of their heights. Thus as retaining walls become higher, the accuracy of the computed lateral pressures becomes more and more important in providing economical designs. Since the Coulomb equation does take into account friction on the wall, it is thought to be the more accurate one and is often used for walls of over 20 ft. The Rankine equation is commonly used for ordinary retaining walls of 20 ft or less in height. It is interesting to note that the two methods give identical results if the friction of the soil on the wall is neglected.

The Rankine expressions for the active and passive pressure coefficients are given at the end of this paragraph, with reference being made to Figure 13.7. In these expressions δ is the angle the backfill makes with the horizontal, while ϕ is the angle of internal friction of the soil. For well-drained sand or gravel backfills, the angle of internal friction is often taken as the angle of repose of the slope. One common slope used is 1 vertically to $1\frac{1}{2}$ horizontally ($33°40'$).

$$k_a = \cos\delta \left(\frac{\cos\delta - \sqrt{\cos^2\delta - \cos^2\phi}}{\cos\delta + \sqrt{\cos^2\delta - \cos^2\phi}} \right)$$

$$k_p = \cos\delta \left(\frac{\cos\delta + \sqrt{\cos^2\delta - \cos^2\phi}}{\cos\delta - \sqrt{\cos^2\delta - \cos^2\phi}} \right)$$

Should the backfill be horizontal—that is, should δ be equal to zero—the expressions become

$$k_a = \frac{1 - \sin\phi}{1 + \sin\phi}$$

$$k_p = \frac{1 + \sin\phi}{1 - \sin\phi}$$

One trouble with using these expressions is in the determination of ϕ. It can be as small as $0°$ to $10°$ for soft clays and as high as $30°$ or $40°$ for some granular materials. As a result, the values of k_a can vary from perhaps 0.30 for some granular materials up to about 1.0 for some wet clays.

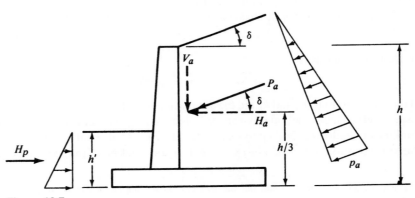

Figure 13.7

Once the values of k_a and k_p are determined, the total horizontal pressures, H_a and H_p, can be calculated as being equal to the areas of the respective triangular pressure diagrams. For instance, with reference made to Figure 13.7, the value of the active pressure is

$$H_a = \left(\tfrac{1}{2}\right)(p_a)(h) = \left(\tfrac{1}{2}\right)(k_a wh)(h)$$

$$H_a = \frac{k_a wh^2}{2}$$

and, similarly,

$$H_p = \frac{k_p wh'^2}{2}$$

In addition to these lateral pressures applied to the retaining wall, it is considered necessary in many parts of the country to add the effect of frost action at the top of the stem—perhaps as much as 600 or 700 lb per linear foot in areas experiencing extreme weather conditions.

13.6   FOOTING SOIL PRESSURES

Because of lateral forces, the resultant of the horizontal and vertical forces, R, intersects the soil underneath the footing as an eccentric load, causing greater pressure at the toe. This toe pressure should be less than the permissible value q_a of the particular soil. It is also desirable to keep the resultant force within the kern or the middle third of the footing base.

If the resultant force intersects the soil within the middle third of the footing, the soil pressure at any point can be calculated with the formula to follow exactly as the stresses are determined in an eccentrically loaded column.

$$q = -\frac{R_v}{A} \pm \frac{R_v ec}{I}$$

Retaining wall for United States Army Corps of Engineers. Colchester, Connecticut. Constructed with precast interlocking reinforced concrete modules. (Courtesy of Doublewal Corporation.)

In this expression, R_v is the vertical component of R or the total vertical load, e is the eccentricity of the load from the center of the footing, A is the area of a 1-ft-wide strip of soil of a length equal to the width of the footing base, and I is the moment of inertia of the same area about its centroid. This expression is correct only if R_v falls within the kern.

This expression can be reduced to the following expression, in which L is the width of the footing from heel to toe.

$$q = -\frac{R_v}{L} \pm \frac{R_v e(L/2)}{L^3/12} = -\frac{R_v}{L}\left(1 \pm \frac{6e}{L}\right)$$

If the resultant force falls outside of the middle third of the footing, the preceding expressions are not applicable because they indicate a tensile stress on one side of the footing—a stress the soil cannot supply. For such cases the soil pressures can be determined as previously described in Section 12.12 and Figure 12.24 of the preceding chapter. Such a situation should not be permitted in a retaining wall and is not considered further.

The soil pressures computed in this manner are only rough estimates of the real values and thus should not be valued too highly. The true pressures are appreciably affected by quite a few items other than the retaining wall weight. Included are drainage conditions, temperature, settlement, pore water, and so on.

13.7 DESIGN OF SEMIGRAVITY RETAINING WALLS

As previously mentioned, semigravity retaining walls are designed to resist earth pressure by means of their own weight plus some developed soil weight. Because they are normally constructed with plain concrete, stone, or perhaps some other type of masonry, their design is based on the assumption that only very little tension or none at all can be permitted in the structure. If the resultant of the earth pressure and the wall weight (including any developed soil weight) falls within the middle third of the wall base, tensile stresses will probably be negligible.

A wall size is assumed, safety factors against sliding and overturning are calculated, the point where the resultant force strikes the base is determined, and the soil pressures are calculated. It is normally felt that safety factors against sliding should be at least 1.5 for cohesionless backfills and 2.0 for cohesive ones. Safety factors of 2.0 for overturning are normally specified. A suitable wall is probably obtained after two or three trial sizes. Example 13.1 illustrates the calculations that need to be made for each trial.

Figure 13.8(a) shows a set of approximate dimensions that are often used for sizing semigravity walls. Dimensions may be assumed to be approximately equal to the values given and the safety factors against overturning and sliding computed. If the values are not suitable, the dimensions are adjusted and the safety factors are recalculated, and so on. Semigravity walls are normally trapezoidal in shape, as shown in Figure 13.8(a), but sometimes they may have broken backs, as illustrated in Figure 13.8(b).

EXAMPLE 13.1

A semigravity retaining wall consisting of plain concrete (weight = 145 lb/ft^3) is shown in Figure 13.9. The bank of supported earth is assumed to weigh 110 lb/ft^3, to have a ϕ of 30°, and to have a coefficient of friction against sliding on soil of 0.5. Determine the safety factors against overturning and sliding and determine the bearing pressure underneath the toe of the footing. Use the Rankine expression for calculating the horizontal pressures.

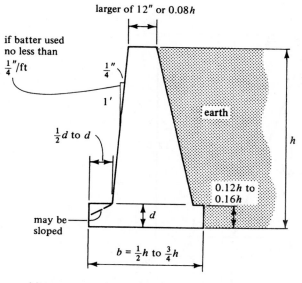

(a) some approximate dimensions for semigravity walls

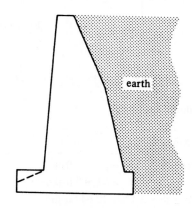

(b) broken-back semigravity wall

Figure 13.8

SOLUTION   **Computing the Soil Pressure Coefficients**

$$k_a = \frac{1 - \sin \phi}{1 + \sin \phi} = \frac{1 - 0.5}{1 + 0.5} = 0.333$$

$$k_p = \frac{1 + \sin \phi}{1 - \sin \phi} = \frac{1 + 0.5}{1 - 0.5} = 3.00$$

The Value of H_a

$$H_a = \frac{k_a w h^2}{2} = \frac{(0.333)(110)(12)^2}{2} = 2637 \text{ lb}$$

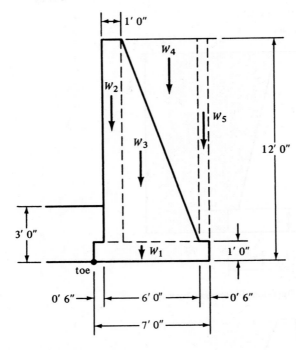

Figure 13.9

Overturning Moment

$$\text{O.T.M.} = (2637)\left(\tfrac{12}{3}\right) = 10{,}548 \text{ ft-lb}$$

Righting Moments (Taken about Toe)

Force		Moment arm	Moment
$W_1 = (7)(1)(145)$	$=$	$1015 \text{ lb} \times 3.5' =$	$3{,}552 \text{ ft-lb}$
$W_2 = (1)(11)(145)$	$=$	$1595 \text{ lb} \times 1.0' =$	$1{,}595 \text{ ft-lb}$
$W_3 = \left(\tfrac{1}{2}\right)(5)(11)(145)$	$=$	$3988 \text{ lb} \times 3.17' =$	$12{,}642 \text{ ft-lb}$
$W_4 = \left(\tfrac{1}{2}\right)(5)(11)(110)$	$=$	$3025 \text{ lb} \times 4.83' =$	$14{,}611 \text{ ft-lb}$
$W_5 = (0.5)(11)(110)$	$=$	$605 \text{ lb} \times 6.75' =$	$4{,}084 \text{ ft-lb}$
	$R_v =$	$10{,}228 \text{ lb}$	$M = 36{,}484 \text{ ft-lb}$

Safety Factor against Overturning (to be discussed at some length in Section 13.10)

$$\text{Safety factor} = \frac{36{,}484}{10{,}548} = 3.46 > 2.00 \qquad \underline{\underline{\text{OK}}}$$

Safety Factor against Sliding (also discussed at length in Section 13.10)

Assuming soil above the footing toe has eroded and thus the passive pressure is due only to soil of a depth equal to footing thickness,

$$H_p = \frac{k_p w h'^2}{2} = \frac{(3.0)(110)(1)^2}{2} = 165 \text{ lb}$$

$$\text{Safety factor against sliding} = \frac{(0.5)(10{,}228) + 165}{2637} = 2.00 > 1.50 \qquad \underline{\underline{\text{OK}}}$$

Distance of Resultant from Toe

$$\text{Distance} = \frac{36{,}484 - 10{,}548}{10{,}228} = 2.54' > 2.33' \qquad \therefore \text{Inside middle third}$$

Soil Pressure under Heel and Toe

$$A = (1)(7.0) = 7.0\,\text{ft}^2$$

$$I = \left(\frac{1}{12}\right)(1)(7)^3 = 28.58\,\text{ft}^4$$

$$f_\text{toe} = -\frac{R_v}{A} - \frac{R_v ec}{I} = -\frac{10{,}228}{7.0} - \frac{(10{,}228)(3.50 - 2.54)(3.50)}{28.58}$$

$$= -1461 - 1202 = -2663\,\text{psf}$$

$$f_\text{heel} = -\frac{R_v}{A} + \frac{R_v ec}{I} = -1461 + 1202 = -259\,\text{psf}$$

13.8   EFFECT OF SURCHARGE

Should there be earth or other loads on the surface of the backfill, as shown in Figure 13.10, the horizontal pressure applied to the wall will be increased. If the surcharge is uniform over the sliding area behind the wall, the resulting pressure is assumed to equal the pressure that would be caused by an increased backfill height having the same total weight as the surcharge. It is usually easy to handle this situation by adding a uniform pressure to the triangular soil pressure for a wall without surcharge, as shown in the figure.

If the surcharge does not cover the area entirely behind the wall, some rather complex soil theories are available to consider the resulting horizontal pressures developed. As a consequence, the designer usually uses a rule of thumb to cover the case, a procedure that works reasonably well.

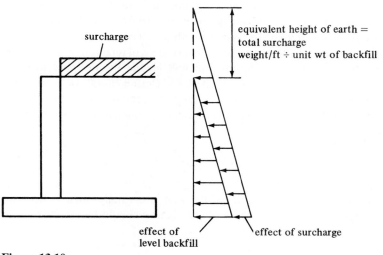

Figure 13.10

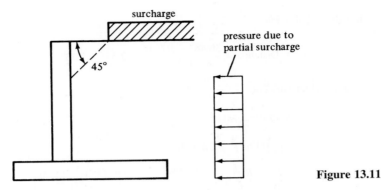

Figure 13.11

He or she may assume, as shown in Figure 13.11, that surcharge cannot affect the pressure above the intersection of a 45° line from the edge of the surcharge to the wall. The lateral pressure is increased, as by a full surcharge, below the intersection point. This is shown in the right side of the figure.

13.9 ESTIMATING THE SIZES OF CANTILEVER RETAINING WALLS

The statical analysis of retaining walls and consideration of their stability as to overturning and sliding are based on service-load conditions. In other words, the length of the footing and the position of the stem on the footing are based entirely on the actual soil backfill, estimated lateral pressure, coefficient of sliding friction of the soil, and so on.

On the other hand, the detailed designs of the stem and footing and their reinforcing are determined by the strength design method. To carry out these calculations, it is necessary to multiply the service loads and pressures by the appropriate load factors. From these factored loads the bearing pressures, moments, and shears are determined for use in the design.

Thus the initial part of the design consists of an approximate sizing of the retaining wall. Although this is actually a trial-and-error procedure, the values obtained are not too sensitive to slightly incorrect values, and usually one or two trials are sufficient.

Various rules of thumb are available with which excellent initial size estimates can be made. In addition, various handbooks present the final sizes of retaining walls that have been designed for certain specific cases. This information will enable the designer to estimate very well the proportions of a wall to be designed. The *CRSI Design Handbook* is one such useful reference.[3] In the next few paragraphs, suggested methods are presented for estimating sizes, without the use of a handbook. These approximate methods are very satisfactory as long as the conditions are not too much out of the ordinary.

Height of Wall

The necessary elevation at the top of the wall is normally obvious from the conditions of the problem. The elevation at the base of the footing should be selected so that it is below frost

---

[3]Concrete Reinforcing Steel Institute, 2002, *CRSI Design Handbook*, 9th ed. (Chicago), pp. 14-1–14-46.

penetration in the particular area—about 3 to 6 ft below ground level in the northern part of the United States. From these elevations the overall height of the wall can be determined.

Stem Thickness

Stems are theoretically thickest at their bases because the shears and moments are greatest there. They will ordinarily have total thicknesses somewhere in the range of 7% to 12% of the overall heights of the retaining walls. The shears and moments in the stem decrease from the bottom to the top; as a result, thicknesses and reinforcement can be reduced proportionately. Stems are normally tapered, as shown in Figure 13.12. The minimum thickness at the top of the stem is 8 in., with 12 in. preferable. As will be shown in Section 13.10, it is necessary to have a mat of reinforcing in the inside face of the stem and another mat in the outside face. To provide room for these two mats of reinforcing, for cover and spacing between the mats, a minimum total thickness of at least 8 in. is required.

The use of the minimum thickness possible for walls that are primarily reinforced in one direction (here it's the vertical bars) doesn't necessarily provide the best economy. The reason is that the reinforcing steel is a major part of the total cost. Making the walls as thin as possible will save some concrete but will substantially increase the amount of reinforcing needed. For fairly high and heavily loaded walls, greater thicknesses of concrete may be economical.

If ρ in the stem is limited to a maximum value of approximately $(0.18f'_c/f_y)$, the stem thickness required for moment will probably provide sufficient shear resistance without using stirrups. Furthermore, it will probably be sufficiently thick to limit lateral deflections to reasonable values.

For heights up to about 12 ft, the stems of cantilever retaining walls are normally made of constant thickness because the extra cost of setting the tapered formwork is usually not offset by the savings in concrete. Above 12-ft heights, concrete savings are usually sufficiently large to make tapering economical.

Actually, the sloping face of the wall can be either the front or the back, but if the outside face is tapered it will tend to counteract somewhat the deflection and tilting of the wall due to lateral pressures. A taper or batter of $\frac{1}{4}$ in. per foot of height is often recommended to offset deflection or the forward tilting of the wall.

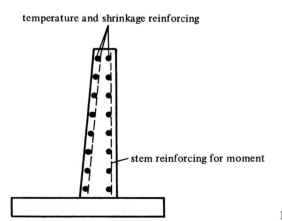

temperature and shrinkage reinforcing

stem reinforcing for moment

Figure 13.12

Base Thickness

The final thickness of the base will be determined on the basis of shears and moments. For estimating, however, its total thickness will probably fall somewhere between 7% and 10% of the overall wall height. Minimum thicknesses of at least 10 to 12 in. are used.

Base Length

For preliminary estimates, the base length can be taken to be about 40% to 60% of the overall wall height. A little better estimate, however, can be made by using the method described by the late Professor Ferguson in his reinforced concrete text.[4] For this discussion, reference is made to Figure 13.13. In this figure, W is assumed to equal the weight of all the material within area *abcd*. This area contains both concrete and soil, but the author assumes here that it is all soil. This means that a slightly larger safety factor will be developed against overturning than assumed. When surcharge is present, it will be included as an additional depth of soil, as shown in the figure.

If the sum of moments about point a due to W and the lateral forces H_1 and H_2 equal zero, the resultant force R will pass through point a. Such a moment equation can be written, equated to zero, and solved for x. Should the distance from the footing toe to point a be equal to one-half of the distance x in the figure and the resultant force R pass through point a, the footing pressure diagram will be triangular. In addition, if moments are taken about the toe of all the loads and forces for the conditions described, the safety factor against overturning will be approximately 2.

A summary of the preceding approximate first trial sizes for cantilever retaining walls is shown in Figure 13.14. These sizes are based on the dimensions of walls successfully constructed in the past. They often will be on the conservative side.

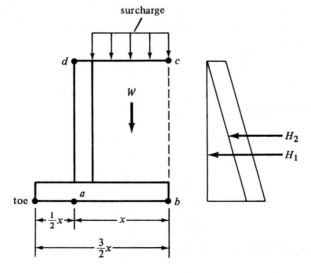

Figure 13.13

---

[4]Ferguson, P. M., 1979, *Reinforced Concrete Fundamentals*, 4th ed. (New York: John Wiley & Sons), p. 256.

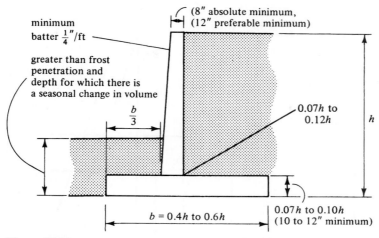

minimum batter $\frac{1}{4}$"/ft

(8" absolute minimum, (12" preferable minimum)

greater than frost penetration and depth for which there is a seasonal change in volume

$\frac{b}{3}$

0.07h to 0.12h

h

$b = 0.4h$ to $0.6h$

0.07h to 0.10h (10 to 12" minimum)

Figure 13.14

EXAMPLE 13.2

Using the approximate rules presented in this section, estimate the sizes of the parts of the retaining wall shown in Figure 13.15. The soil weighs 100 lb/ft³, and a surcharge of 300 psf is present. Assume $k_a = 0.32$. (For many practical soils such as clays or silts, k_a will be two or more times this large.)

SOLUTION   **Stem Thickness**

Assume 12-in. thickness at top.

$$\text{Assume bottom thickness} = 0.07h = (0.07)(21) = 1.47' \qquad \text{Say } 1'6''$$

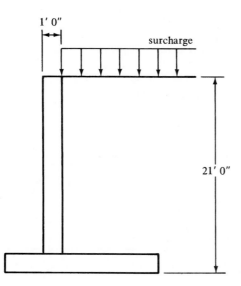

1' 0"

surcharge

21' 0"

Figure 13.15

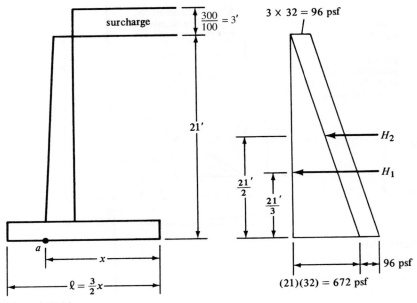

Figure 13.16

Base Thickness

Assume base $t = 7\%$ to 10% of overall wall height

$$t = (0.07)(21) = 1.47'$$

Say 1'6"

$$\text{Height of stem} = 21'0'' \text{ minus } 1'6'' = \underline{\underline{19'6''}}$$

Base Length and Position of Stem

Calculating horizontal forces without load factors, as shown in Figure 13.16.

$$\rho_a = k_a wh = (0.32)(100)(21) = 672 \text{ lb}$$

$$H_1 = \left(\frac{1}{2}\right)(21)(672) = 7056 \text{ lb}$$

$$H_2 = (21)(96) = 2016 \text{ lb}$$

$$W = (x)(24)(100) = 2400x$$

$\Sigma M_a = 0$

$$-(7056)(7.00) - (2016)(10.5) + (2400x)\left(\frac{x}{2}\right) = 0$$

$$x = 7.67'$$

$$b = \left(\frac{3}{2}\right)(7.67) = 11.505'$$

Say 11'6"

The final trial dimensions are shown in Figure 13.22.

13.10   DESIGN PROCEDURE FOR CANTILEVER RETAINING WALLS

This section is presented to describe in some detail the procedure used for designing a cantilever retaining wall. At the end of this section the complete design of such a wall is presented. Once the approximate size of the wall has been established, the stem, toe, and heel can be designed in detail. Each of these parts will be designed individually as a cantilever sticking out of a central mass, as shown in Figure 13.17.

Stem

The values of shear and moment at the base of the stem due to lateral earth pressures are computed and used to determine the stem thickness and necessary reinforcing. Because the lateral pressures are considered to be live load forces, a load factor of 1.6 is used.

It will be noted that the bending moment requires the use of vertical reinforcing bars on the soil side of the stem. In addition, temperature and shrinkage reinforcing must be provided. In Section 14.3 of the ACI Code, a minimum value of horizontal reinforcing equal to 0.0025 of the area of the wall bt is required as well as a minimum amount of vertical reinforcing (0.0015). These values may be reduced to 0.0020 and 0.0012 if the reinforcing is $\frac{5}{8}$ in. or less in diameter and if it consists of bars or welded wire fabric (not larger than $W31$ or $D31$), with f_y equal to or greater than 60,000 psi.

The major changes in temperature occur on the front or exposed face of the stem. For this reason most of the horizontal reinforcing (perhaps two-thirds) should be placed on that face with

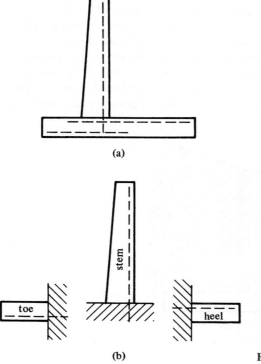

(a)

(b)   **Figure 13.17**

just enough vertical steel used to support the horizontal bars. The concrete for a retaining wall should be placed in fairly short lengths—not greater than 20- or 30-ft sections—to reduce shrinkage stresses.

Factor of Safety Against Overturning

Moments are taken about the toe of the unfactored overturning and righting forces. Traditionally, it has been felt that the safety factor against overturning should be at least equal to 2. In making these calculations, backfill on the toe is usually neglected because it may very well be eroded. Of course, there are cases where there is a slab (for instance, a highway pavement on top of the toe backfill) that holds the backfill in place over the toe. For such situations it may be reasonable to include the loads on the toe.

Factor of Safety Against Sliding

A consideration of sliding for retaining walls is a most important topic because a very large percentage of retaining wall failures occur due to sliding. To calculate the factor of safety against sliding, the estimated sliding resistance (equal to the coefficient of friction for concrete on soil times the resultant vertical force μR_v) is divided by the total horizontal force. The passive pressure against the wall is probably neglected, and the unfactored loads are used.

Typical design values of μ, the coefficient of friction between the footing concrete and the supporting soil, are as follows: 0.45 to 0.55 for coarse-grained soils, with the lower value applying if some silt is present, and 0.6 if the footing is supported on sound rock with a rough surface. Values of 0.3 to 0.35 are probably used if the supporting material is silt.

It is usually felt that the factor of safety against sliding should be at least equal to 1.5. When retaining walls are initially designed, the calculated factor of safety against sliding is very often considerably less than this value. To correct the situation, the most common practice is to widen the footing on the heel side. Another practice is to use a lug or key, as shown in Figure 13.18, with the front face cast directly against undisturbed soil. (Many designers feel that the construction of keys disturbs the soil so much that they are not worthwhile.) Keys are thought to be particularly

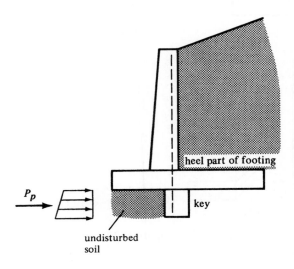

P_p

heel part of footing

key

undisturbed
soil

Figure 13.18

El Teniente Copper Mine, Rancagua, Chile. (Courtesy of EFCO.)

necessary for moist clayey soils. The purpose of a key is to cause the development of passive pressure in front of and below the base of the footing, as shown by P_p in the figure. The actual theory involved, and thus the design of keys, is still a question among geotechnical engineers. As a result, many designers select the sizes of keys by rules of thumb. One common practice is to give them a depth between two-thirds and the full depth of the footing. They are usually made approximately square in cross section and have no reinforcing provided other than perhaps the dowels mentioned in the next paragraph.

Keys are often located below the stem so that some dowels or extended vertical reinforcing may be extended into the key. If this procedure is used, the front face of the key needs to be at least 5 or 6 in. in front of the back face of the stem to allow room for the dowels. From a soil mechanics view, keys may be a little more effective if they are placed a little further toward the heel.

If the key can be extended down into a very firm soil or even rock, the result will be a greatly increased sliding resistance—that resistance being equal to the force necessary to shear the key off from the footing, that is, a shear friction calculated as described in Sections 8.12 and 12.13 of this text.

Heel Design

The lateral earth pressure tends to cause the retaining wall to rotate about its toe. This action tends to pick up the heel into the backfill. The backfill pushes down on the heel cantilever, causing tension in its top. The major force applied to the heel of a retaining wall is the downward weight of the backfill behind the wall. Although it is true that there is some upward soil pressure, many designers choose to neglect it because it is relatively small. The downward loads tend to push the heel of the footing down, and the necessary upward reaction to hold it attached to the stem is provided by the vertical tensile steel in the stem, which is extended down into the footing.

Because the reaction in the direction of the shear does not introduce compression into the heel part of the footing in the region of the stem, it is not permissible to determine V_u at a distance d from the face of the stem, as provided in Section 11.1.3.1 of the ACI Code. The value of V_u is determined instead at the face of the stem due to the downward loads. This shear is often of such magnitude as to control the thickness, but the moment at the face of the stem should be checked also. Because the load here consists of soil and concrete, a load factor of 1.2 is used for making the calculations.

It will be noted that the bars in the heel will be in the top of the footing. As a result, the required development length of these "top bars" may be rather large.

The percentage of flexural steel required for the heel will frequently be less than the ρ_{min} of $200/f_y$ and $3\sqrt{f_c'}/f_y$. Despite the fact that the ACI Code (10.5.4) exempts slabs of uniform thickness from these ρ_{min} values, the author recommends that these be used because the retaining wall is a major "beamlike" structure.

Toe Design

The toe is assumed to be a beam cantilevered from the front face of the stem. The loads it must support include the weight of the cantilever slab and the upward soil pressure beneath. Usually any earth fill on top of the toe is neglected (as though it has been eroded). Obviously, such a fill would increase the upward soil pressure beneath the footing, but because it acts downward and cancels out the upward pressure, it produces no appreciable changes in the shears and moments in the toe.

A study of Figure 13.19 shows that the upward soil pressure is the major force applied to the toe. Because this pressure is primarily caused by the lateral force H, a load factor of 1.6 is used for the calculations (Section 4.1 of this text shows that all load combinations including soil loads have

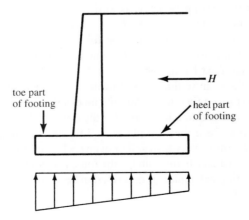

Figure 13.19

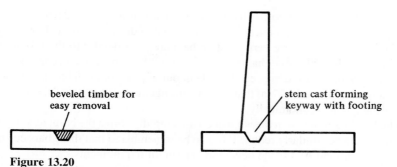

beveled timber for easy removal

stem cast forming keyway with footing

Figure 13.20

a load factor of 1.6 associated with H). The maximum moment for design is taken at the face of the stem, whereas the maximum shear for design is assumed to occur at a distance d from the face of the stem because the reaction in the direction of the shear does introduce compression into the toe of the footing. The average designer makes the thickness of the toe the same as the thickness of the heel, although such a practice is not essential.

It is a common practice in retaining wall construction to provide a shear keyway between the base of the stem and the footing. This practice, though definitely not detrimental, is of questionable value. The keyway is normally formed by pushing a beveled $2'' \times 4''$ or $2'' \times 6''$ into the top of the footing, as shown in Figure 13.20. After the concrete hardens, the wood member is removed, and when the stem is cast in place above, a keyway is formed. It is becoming more and more common simply to use a roughened surface on the top of the footing where the stem will be placed. This practice seems to be just as satisfactory as the use of a keyway.

In Example 13.3, #8 bars 6 in. on center are selected for the vertical steel at the base of the stem. Either these bars need to be embedded into the footing for development purposes or dowels equal to the stem steel need to be used for the transfer. This latter practice is quite common because it is rather difficult to hold the stem steel in position while the base concrete is placed.

The required development length of the #8 bars down into the footing or for #8 dowels is 33 in. when $f_y = 60,000$ psi and $f_c' = 3000$ psi. This length cannot be obtained vertically in the 1 ft 6 in. footing used unless the bars or dowels are either bent as shown in Figure 13.21(a) or extended

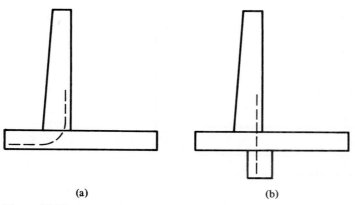

(a)   (b)

Figure 13.21

through the footing and into the base key as shown in Figure 13.21(b). Actually, the required development length can be reduced if more but smaller dowels are used. For #6 dowels, ℓ_d is 20 in.

If instead of dowels the vertical stem bars are embedded into the footing, they should not extend up into the wall more than 8 or 10 ft before they are spliced because they are difficult to handle in construction and may easily be bent out of place or even broken. Actually, you can see after examining Figure 13.21(a) that such an arrangement of stem steel can on some occasions be very advantageous economically.

The bending moment in the stem decreases rapidly above the base; as a result, the amount of reinforcing can be similarly reduced. It is to be remembered that these bars can be cut off only in accordance with the ACI Code development length requirements.

Example 13.3 illustrates the detailed design of a cantilever retaining wall. Several important descriptive remarks are presented in the solution, and these should be carefully read.

EXAMPLE 13.3

Complete the design of the cantilever retaining wall whose dimensions were estimated in Example 13.2 and are shown in Figure 13.22, if $f'_c = 3000$ psi, $f_y = 60,000$ psi, $q_a = 4000$ psf, and the coefficient of sliding friction equals 0.50 for concrete on soil. Use ρ approximately equal to $0.18 f'_c / f_y$ to maintain reasonable deflection control.

SOLUTION   The safety factors against overturning and sliding and the soil pressures under the heel and toe are computed using the actual unfactored loads.

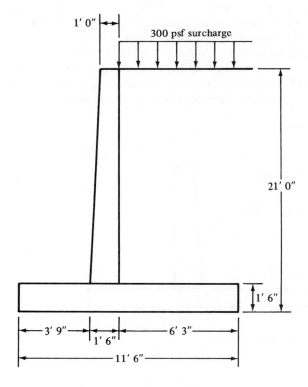

Figure 13.22

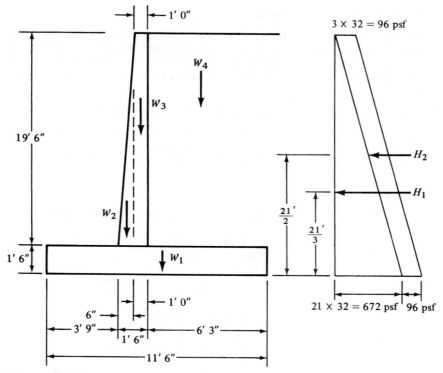

Figure 13.23

Safety Factor against Overturning (with Reference to Figure 13.23)

	Overturning moment	
Force	Moment arm	Moment
$H_1 = \left(\frac{1}{2}\right)(21)(672) = 7056 \text{ lb} \times 7.00'$		$= 49{,}392 \text{ ft-lb}$
$H_2 = (21)(96) = 2016 \text{ lb} \times 10.50'$		$= \underline{21{,}168 \text{ ft-lb}}$
Total		$70{,}560 \text{ ft-lb}$

	Righting moment	
Force	Moment arm	Moment
$W_1 = (1.5)(11.5)(150) = 2588 \text{ lb} \times 5.75'$		$= 14{,}881 \text{ ft-lb}$
$W_2 = \left(\frac{1}{2}\right)(19.5)\left(\frac{6}{12}\right)(150) = 731 \text{ lb} \times 4.08'$		$= 2982 \text{ ft-lb}$
$W_3 = (19.5)\left(\frac{12}{12}\right)(150) = 2925 \text{ lb} \times 4.75'$		$= 13{,}894 \text{ ft-lb}$
$W_4 = (22.5)(6.25)(100) = 14{,}062 \text{ lb} \times 8.37'$		$= 117{,}699 \text{ ft-lb}*$
$R_v = 20{,}306 \text{ lb}$	M	$= 149{,}456 \text{ ft-lb}$

*Includes surcharge.

$$\text{Safety factor against overturning} = \frac{149{,}456}{70{,}560} = 2.12 > 2.00 \qquad \underline{\underline{\text{OK}}}$$

Factor of Safety Against Sliding

Here the passive pressure against the wall is neglected. Normally, it is felt that the factor of safety should be at least 1.5. If it is not satisfactory, a little wider footing on the heel side will easily take care of the situation. In addition to or instead of this solution a key, perhaps 1 ft-6 in. × 1 ft-6 in. (size selected to provide sufficient development length for the dowels selected later in this design) can be used. Space is not taken here to improve this safety factor.

$$\text{Force causing sliding} = H_1 + H_2 = 9072 \text{ lb}$$

$$\text{Resisting force} = \mu R_v = (0.50)(20{,}306) = 10{,}153 \text{ lb}$$

$$\text{Safety factor} = \frac{10{,}153}{9072} = 1.12 < 1.50 \qquad \underline{\underline{\text{No good}}}$$

Footing Soil Pressures

$R_v = 20{,}306$ lb and is located a distance $\bar{x}$ from the toe of the footing

$$\bar{x} = \frac{149{,}456 - 70{,}560}{20{,}306} = \frac{78{,}896}{20{,}306} = 3.89' \qquad \underline{\underline{\text{Just inside middle third}}}$$

$$\text{Soil pressure} = -\frac{R_v}{A} \pm \frac{Mc}{I}$$

$$A = (1)(11.5) = 11.5 \text{ ft}^2$$

$$I = \left(\frac{1}{12}\right)(1)(11.5)^3 = 126.74 \text{ ft}^4$$

$$f_{\text{toe}} = -\frac{20{,}306}{11.5} - \frac{(20{,}306)(5.75 - 3.89)(5.75)}{126.74}$$

$$= -1766 - 1714 = -3480 \text{ psf}$$

$$f_{\text{heel}} = -1766 + 1714 = -52 \text{ psf}$$

Design of Stem

The lateral forces applied to the stem are calculated using a load factor of 1.6 as shown in Figure 13.24.

Design of Stem for Moment

$$M_u = (H_1)(6.50) + (H_2)(9.75) = (9734)(6.50) + (2995)(9.75)$$

$$M_u = 92{,}472 \text{ ft-lb}$$

Use

$$\rho = \text{approximately } \frac{0.18 f'_c}{f_y} = \frac{(0.18)(3000)}{60{,}000} = 0.009$$

$$\frac{M_u}{\phi b d^2} \text{ (from Table A.12)} = 482.6 \text{ psi}$$

$$bd^2 = \frac{(12)(92{,}472)}{(0.9)(482.6)} = 2555$$

$$d = \sqrt{\frac{2555}{12}} = 14.59''$$

$$h = 14.59 + 2'' + \frac{1''}{2} = 17.09'' \qquad \underline{\underline{\text{Say } 18'' (d = 15.50'')}}$$

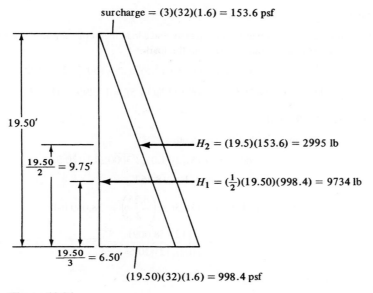

surcharge = (3)(32)(1.6) = 153.6 psf

19.50′

$\dfrac{19.50}{2} = 9.75'$

$H_2 = (19.5)(153.6) = 2995 \text{ lb}$

$H_1 = (\tfrac{1}{2})(19.50)(998.4) = 9734 \text{ lb}$

$\dfrac{19.50}{3} = 6.50'$

$(19.50)(32)(1.6) = 998.4 \text{ psf}$

Figure 13.24

$$\frac{M_u}{\phi bd^2} = \frac{(12)(92{,}472)}{(0.90)(12)(15.5)^2} = 427.7$$

$$\rho = 0.00786 (\text{from Appendix Table A.12})$$

$$A_s = (0.00786)(12)(15.5) = 1.46 \text{ in.}^2 \qquad \underline{\underline{\text{Use \#8 @ 6''}(1.57 \text{ in.}^2)}}$$

$$\text{Minimum vertical } \rho \text{ by ACI Section 14.3} = 0.0015 < \frac{1.57}{(12)(15.5)} = 0.0084 \qquad \underline{\underline{\text{OK}}}$$

$$\text{Minimum horizontal } A_s = (0.0025)(12)(\text{average stem } t)$$

$$= (0.0025)(12)\left(\frac{12 + 18}{2}\right) = 0.450 \text{ in.}^2$$

(say one-third inside face and two-thirds outside face)

$$\underline{\underline{\text{Use \#4 at } 7\tfrac{1}{2}'' \text{ outside face and \#4 at 15'' inside face}}}$$

Checking Shear Stress in Stem

Actually, V_u at a distance d from the top of the footing can be used, but for simplicity.

$$V_u = H_1 + H_2 = 9734 + 2995 = 12{,}729 \text{ lb}$$

$$\phi V_c = \phi 2\lambda\sqrt{f_c'}bd = (0.75)(2)(1.0)(\sqrt{3000})(12)(15.5)$$

$$= 15{,}281 \text{ lb} > 12{,}729 \text{ lb} \qquad \underline{\underline{\text{OK}}}$$

Design of Heel

The upward soil pressure is conservatively neglected, and a load factor of 1.2 is used for calculating the shear and moment because soil and concrete make up the load.

$$V_u = (22.5)(6.25)(100)(1.2) + (1.5)(6.25)(150)(1.2) = 18{,}563 \text{ lb}$$

$$\phi V_c = (0.75)(2)(1.0)(\sqrt{3000})(12)(14.5) = 14{,}295 < 18{,}563 \qquad \underline{\underline{\text{No good}}}$$

Try 24-in. Depth (d = 20.5 in.)

Neglecting slight change in V_u with different depth

$$\phi V_c = (0.75)(2)(1.0)(\sqrt{3000})(12)(20.5)$$

$$= 20{,}211 > 18{,}563 \qquad \underline{\underline{\text{OK}}}$$

$$M_u \text{ at face of stem} = (18{,}563)\left(\frac{6.25}{2}\right) = 58{,}009 \text{ ft-lb}$$

$$\frac{M_u}{\phi b d^2} = \frac{(12)(58{,}009)}{(0.9)(12)(20.5)^2} = 153$$

$$\rho = \rho_{\min} \qquad \text{Use } \rho = 0.00333$$

Using 0.00333,

$$A_x = (0.00333)(12)(20.5) = 0.82 \text{ in.}^2/\text{ft} \qquad \underline{\underline{\text{Use #8 @ 11''}}}$$

ℓ_d required calculated with ACI Equation 12-1 for #8 top bars with $c = 2.50$ in. and $K_{tr} = 0$ is 43 in. < 72 in. available. $\underline{\underline{\text{OK}}}$

Heel reinforcing is shown in Figure 13.25.

Note: Temperature and shrinkage steel is normally considered unnecessary in the heel and toe. However, the author has placed #4 bars at 18 in. on center in the long direction, as shown in Figures 13.25 and 13.27, to serve as spacers for the flexural steel and to form mats out of the reinforcing.

Design of Toe

For service loads, the soil pressures previously determined are multiplied by a load factor of 1.6 because they are primarily caused by the lateral forces, as shown in Figure 13.26.

$$V_u = 10{,}440 + 7086 = 17{,}526 \text{ lb}$$

(The shear can be calculated a distance d from the face of the stem because the reaction in the direction of the shear does introduce compression into the toe of the slab, but this advantage is neglected because 17,526 lb is already less than the 19,125 lb shear in the heel, which was satisfactory.)

$$M_u \text{ at face of stem} = (7086)\left(\frac{3.75}{3}\right) + (10{,}440)\left(\frac{2}{3} \times 3.75\right) = 34{,}958 \text{ ft-lb}$$

$$\frac{M_u}{\phi b d^2} = \frac{(12)(34{,}958)}{(0.9)(12)(20.5)^2} = 92$$

$$\rho = \text{less than } \rho_{\min}$$

Therefore, use

$$\frac{200}{60{,}000} = 0.00333$$

$$A_s = (0.00333)(12)(20.5) = 0.82 \text{ in.}^2/\text{ft} \qquad \underline{\underline{\text{Use #8 at 11''}}}$$

ℓ_d required calculated with ACI Equation 12-1 for #8 bottom bars with $c = 2.50$ in. and $K_{tr} = 0$ equals 33 in. < 42 in. available $\underline{\underline{\text{OK}}}$

Toe reinforcing is shown in Figure 13.27.

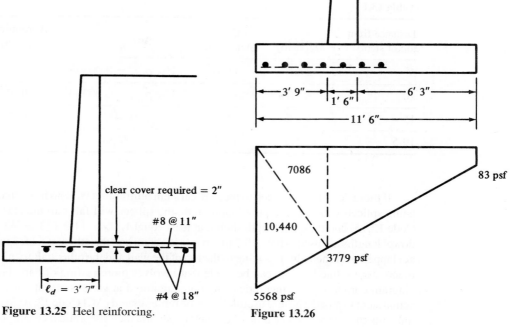

clear cover required = 2″

#8 @ 11″

#4 @ 18″

ℓ_d = 3′ 7″

Figure 13.25 Heel reinforcing.

7086

83 psf

10,440

3779 psf

5568 psf

Figure 13.26

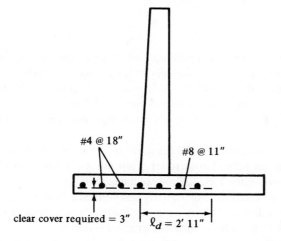

#4 @ 18″

#8 @ 11″

clear cover required = 3″

ℓ_d = 2′ 11″

Figure 13.27 Toe reinforcing.

Selection of Dowels and Lengths of Vertical Stem Reinforcing

The detailed selection of vertical bar lengths in the stem is omitted here to save space, and only a few general comments are presented. First, Table 13.1 shows the reduced bending moments up in the stem and also shows the corresponding reductions in reinforcing required.

After considering the possible arrangements of the steel in Figure 13.21 and the required areas of steel at different elevations in Table 13.1, the author decided to use dowels for load transfer at the stem base.

Use #8 dowels at 6 in. extending 33 in. down into footing and key

Table 13.1

Distance from top of stem	M_u(ft-lb)	Effective stem d (in.)	ρ	A_s required (in.^{2}/ft)	Bars needed
5′	2987	11.04	Use $\rho_{min} = 0.00333$	0.44	#8 @ 18″
10′	16,213	12.58	Use $\rho_{min} = 0.00333$	0.50	#8 @ 18″
15′	46,080	14.12	0.00452	0.77	#8 @ 12″
19.5′	92,472	15.50	0.00786	1.46	#8 @ 6″

If these dowels are spliced to the vertical stem reinforcing with no more than one-half the bars being spliced within the required lap length, the splices will fall into the class B category (ACI Code 12.15) and their lap length should at least equal $1.3\ell_d = (1.3)(33) = 43$ in. Therefore, two dowel lengths are used—half 3 ft 7 in. up into the stem and the other half 7 ft 2 in.—and the #7 bars are lapped over them, half running to the top of the wall and the other half to middepth. Actually, a much more refined design can be made that involves more cutting of bars. For such a design, a diagram comparing the theoretical steel area required at various elevations in the stem and the actual steel furnished is very useful. It is to be remembered (ACI Code 12.10.3) that the bars cut off must run at least a distance d or 12 diameters beyond their theoretical cutoff points and must also meet the necessary development length requirements.

13.11   CRACKS AND WALL JOINTS

Objectionable horizontal cracks are rare in retaining walls because the compression faces are the ones that are visible. When they do occur it is usually a sign of an unsatisfactory structural design and not shrinkage. In Chapter 6 of this book the ACI procedure (Section 10.6) for limiting crack sizes in tensile zones of one-way beams and slabs was presented. These provisions may be applied to vertical retaining wall steel. However, they are usually thought unnecessary because the vertical steel is on the earth side of the wall.

On the other hand, vertical cracks in walls are quite common unless sufficient construction joints are used. Vertical cracks are related to the relief of tension stresses due to shrinkage, with the resulting tensile forces exceeding the longitudinal steel capacity.

Construction joints may be used both horizontally and vertically between successive pours of concrete. The surface of the hardened concrete can be cleaned and roughened, or keys may be used as shown in Figure 13.28(a) to form horizontal construction joints.

If concrete is restrained from free movement when shrinking, as by being attached to more rigid parts of the structure, it will crack at points of weakness. Contraction joints are weakened places constructed so that shrinkage failures will occur at prepared locations. When the shrinkage tensile stresses become too large, they will pull these contraction joints apart and form neat cracks rather than the crooked unsightly ones that might otherwise occur. In addition to handling shrinkage problems, contraction joints are useful in handling differential settlements. They need to be spaced at intervals about 25 ft on center (the AASHTO says not greater than 30 ft). The joints are usually constructed with rubber strips that are left in place or with wood strips that are later removed and replaced with caulking.

Expansion joints are vertical joints that completely separate the different parts of a wall. They are placed approximately 50 to 100 ft on centers (the AASHTO says maximum spacing should not be greater than 90 ft). Reinforcing bars are generally run through all joints so that vertical and

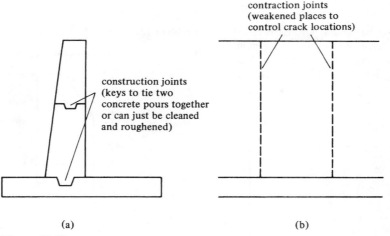

(a)   (b)

Figure 13.28

horizontal alignment is maintained. When the bars do run through a joint, one end of the bars on one side of the joint is either greased or sheathed so that the desired expansion can take place.

It is difficult to estimate the amount of shrinkage or expansion of a particular wall because the wall must slide on the soil beneath, and the resulting frictional resistance may be sufficient so that movement will be greatly reduced or even prevented. A rough value for the width of an expansion joint can be determined from the following expression, in which ΔL is the change in length, L is the distance between joints, ΔT is the estimated temperature change, and 0.000005 per unit length per degree Fahrenheit is the estimated coefficient of contraction of the wall.

$$\Delta L = (0.000005L)(\Delta T)$$

Box culvert. (Courtesy of Economy Forms Corporation.)

PROBLEMS

In Problems 13.1 to 13.4, use the Rankine equation to calculate the total horizontal active force and the overturning moment for the wall shown in the accompanying illustration. Assume that $\phi = 30°$ and the soil weighs 100 lb/ft^3. Neglect the fill on the toe for each wall.

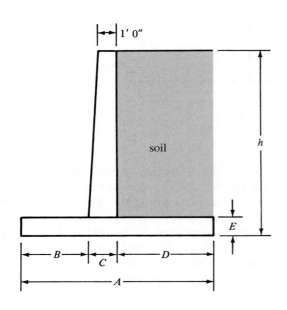

Problem	A	B	C	D	E	h
13.1	8'0"	2'0"	1'6"	4'6"	1'6"	14'0"
13.2	10'6"	2'6"	1'9"	6'3"	1'8"	18'0"
13.3	11'0"	3'6"	1'6"	6'0"	1'6"	20'0"
13.4	12'6"	4'0"	1'6"	7'0"	2'0"	22'0"

(Answer to Problem 13.1: 3266 lb; 15,242 ft-lb)
(Answer to Problem 13.3: 6666 lb; 44,440 ft-lb)

13.5 Repeat Problem 13.1 if δ is 20°. (*Ans.* 4059 lb; 18,943 ft-lb)

13.6 Repeat Problem 13.3 if δ is 23°40'.

In Problems 13.7 to 13.9, determine the safety factors against overturning and sliding for the gravity and semigravity walls shown if $\phi = 30°$ and the coefficient of friction (concrete on soil) is 0.5. Compute also the soil pressure under the toe and heel of each footing. The soil weighs 100 lb/ft^3, and the plain concrete used in the footing weighs 145 lb/ft^3. Determine horizontal pressures using the Rankine equation.

Problem 13.7   (*Ans.* 5.69, 2.67, 2193 psf, − 1015 psf)

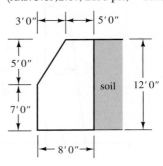

Problem 13.8

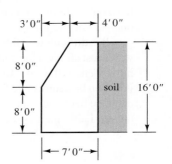

Problem 13.9   (*Ans.* 1.73, 1.32, − 4739 psf, − 0 psf)

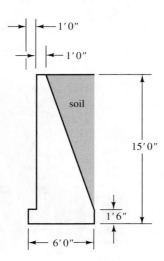

In Problems 13.10 to 13.13, if Rankine's coefficient k_a is 0.75, the soil weight 110 lb/ft³, the concrete weight 150 lb/ft³, and the coefficient of friction (concrete on soil) is 0.55, determine the safety factors against overturning and sliding for the wall shown in the accompanying illustration.

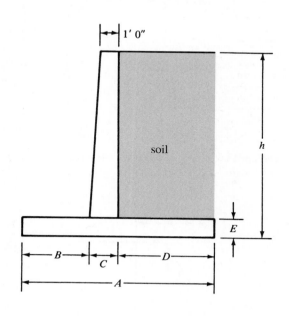

Problem	A	B	C	D	E	h
13.10	8'0"	2'0"	1'0"	5'0"	1'3"	14'0"
13.11	11'0"	2'6"	1'6"	7'0"	1'6"	15'0"
13.12	13'6"	4'0"	1'6"	8'0"	1'6"	18'0"
13.13	14'0"	3'6"	1'6"	9'0"	1'9"	20'0"

(Answer to Problem 13.11: 2.70, 0.93)
(Answer to Problem 13.13: 2.41, 0.86)

13.14 Repeat Problem 13.4 assuming a surcharge of 200 psf. Calculate overturning moment.

13.15 Repeat Problem 13.9 assuming a surcharge of 200 psf. *(Ans. 1.36, 1.12, − 8089 psf, 0 psf)*

13.16 Repeat Problem 13.12 assuming a surcharge of 330 psf. Also determine toe and heel soil pressures.

In Problems 13.17 to 13.20, determine approximate dimensions of retaining walls, check safety factors against overturning and sliding, and calculate soil pressures for the wall shown. Also determine the required stem thickness at their bases and select vertical reinforcing there, using $f_y = 60,000$ psi, $f'_c = 3000$ psi,

$q_a = 5000$ psf, $\rho =$ approximately $0.18f'_c/f_y$, angle of internal friction = 33°40′, and coefficient of sliding friction (concrete on soil) = 0.45. Soil weight = 100 lb/ft³. Concrete weight = 150 lb/ft³.

Problem	h	Surcharge
13.17	12'0"	None
13.18	16'0"	None
13.19	18'0"	None
13.20	15'0"	200 psf

(Answer to Problem 13.17: 6 ft wide, O.T. safety factor = 2.62)
(Answer to Problem 13.19: 8 ft 6 in. wide, O.T. safety factor = 2.30)

In Problems 13.21 to 13.23, determine the same information required for Problems 13.17 to 13.20 with same data, but design heels instead of stems.

Problem	h	Surcharge
13.21	14'0"	None
13.22	18'0"	300 psf
13.23	20'0"	300 psf starting 4'0" from inside face of wall

(Answer to Problem 13.21: 6 ft 6 in. wide, O.T. safety factor = 2.24)
(Answer to Problem 13.23: 10 ft 3 in. wide, O.T. safety factor = 2.16)

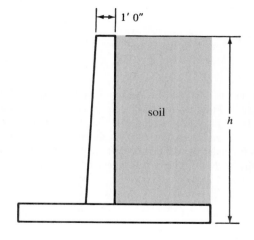

Problems with SI Units

In Problems 13.24 to 13.26, use the Rankine equation to calculate the total horizontal force and the overturning moment for the wall shown. Assume $\sin \phi = 0.5$ and the soil weighs 16 kN/m^3.

Problem	A	B	C	D	E	h
13.24	2.400 m	600 mm	500 mm	1.300 m	450 mm	4 m
13.25	2.700 m	700 mm	500 mm	1.500 m	500 mm	6 m
13.26	3.150 m	800 mm	550 mm	1.800 m	500 mm	8 m

(Answer to Problem 13.25: 95.904 kN, 191.908 kN-m)

Problem 13.27 *(Ans. 3.06, 1.68, -141.87 kN/m^2, -17.03 kN/m^2)*

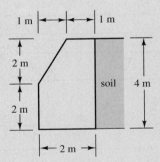

Problem 13.28

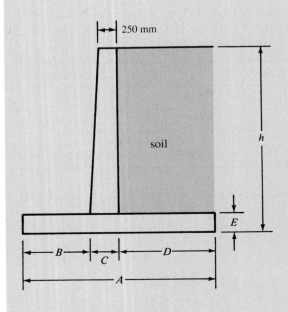

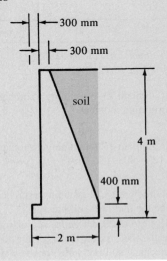

In Problems 13.27 and 13.28, determine the safety factors against overturning and sliding for the gravity and semi-gravity walls shown if $\phi = 30°$ and the coefficient of sliding (concrete on soil) is 0.45. Compute also the soil pressure under the toe and heel of each footing. The soil weighs 16 kN/m^3, and the plain concrete used in the footing weighs 22.7 kN/m^3.

In Problems 13.29 and 13.30, if Rankine's coefficient is 0.35, the soil weight 16 kN/m³, the concrete weight 23.5 kN/m³, and the coefficient of friction (concrete on soil) is 0.50, determine the safety factors against overturning and sliding for the wall shown.

Problem	A	B	C	D	E	h
13.29	4 m	1.5 m	300 mm	2.2 m	700 mm	5 m
13.30	5 m	1.5 m	500 mm	3.0 m	800 mm	7 m

(Answer to Problem 13.29: 5.32, 1.77)

In Problems 13.31 to 13.33, select approximate dimensions for the cantilever retaining wall shown and determine reinforcing required at base of stem using those dimensions and the following data: $f'_c = 21$ MPa, $f_y = 420$ MPa, $\rho = $ approximately $\frac{3}{8}\rho_{bal}$, angle of internal friction $33°40'$. Soil weight $= 16$ kN/m³ and reinforced concrete weight $= 23.5$ kN/m³.

Problem	h	Surcharge
13.31	4 m	None
13.32	6 m	None
13.33	7 m	4 kN/m

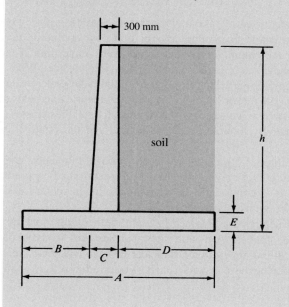

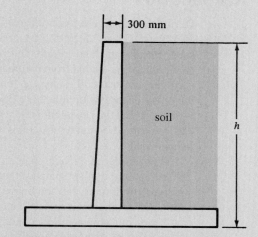

(Answer to Problem 13.31: Use 320 mm stem at base with $d = 250$ mm and #16 bars @ 225 mm vertical steel)*(Answer to Problem 13.33:* Use 560 mm stem at base with $d = 490$ mm and #25 bars @ 225 mm vertical steel)

14

Continuous Reinforced Concrete Structures

14.1 INTRODUCTION

During the construction of reinforced concrete structures, as much concrete as possible is placed in each pour. For instance, the concrete for a whole floor or for a large part of it, including the supporting beams and girders and parts of the columns, may be placed at the same time. The reinforcing bars extend from member to member, as from one span of a beam into the next. When there are construction joints, the reinforcing bars are left protruding from the older concrete so they may be lapped or spliced to the bars in the newer concrete. In addition, the old concrete is cleaned so that the newer concrete will bond to it as well as possible. The result of all these facts is that reinforced concrete structures are generally monolithic or continuous and thus statically indeterminate.

A load placed in one span of a continuous structure will cause shears, moments, and deflections in the other spans of that structure. Not only are the beams of a reinforced concrete structure continuous, but the entire structure is continuous. In other words, loads applied to a column affect the beams, slabs, and other columns, and vice versa.

The result is that more economical structures are obtained because the bending moments are smaller and thus member sizes are smaller. Although the analyses and designs of continuous structures are more complicated than they are for statically determinate structures, this fact has become less important because of the constantly increasing availability of good software.

14.2 GENERAL DISCUSSION OF ANALYSIS METHODS

In reinforced concrete design today, we use elastic methods to analyze structures loaded with factored or ultimate loads. Such a procedure probably doesn't seem quite correct to the reader, but it does yield satisfactory results. The reader might very well ask, "Why don't we use ultimate or inelastic analyses for reinforced concrete structures?" The answer is that our theory and tests are just not sufficiently advanced.

It is true that under certain circumstances some modifications of moments are permitted to recognize ultimate or inelastic behavior as described in Section 14.5 of this chapter. In general, however, we will discuss elastic analyses for reinforced concrete structures. Actually no method of analysis, elastic or inelastic, will give exact results because of the unknown effects of creep, settlement, shrinkage, workmanship, and so on.

Confinazas Financial Center, Caracas, Venezuela. (Courtesy of Economy Forms Corporation.)

14.3   QUALITATIVE INFLUENCE LINES

Many methods might be used to analyze continuous structures. The most common hand calculation method is moment distribution, but other methods are frequently used, such as matrix methods, computer solutions, and others. Whichever method is used, you should understand that to determine maximum shears and moments at different sections in the structure, it is necessary to consider different positions of the live loads. As a background for this material, a brief review of *qualitative influence lines* is presented.

Qualitative influence lines are based on a principle introduced by the German professor Heinrich Müller-Breslau. This principle is as follows: *The deflected shape of a structure represents to some scale the influence line for a function such as reaction, shear, or moment if the function in question is allowed to act through a small distance.* In other words, the structure draws its own influence line when the proper displacement is made.

The shape of the usual influence line needed for continuous structures is so simple to obtain with the Müller-Breslau principle that in many situations it is unnecessary to compute the numerical values of the coordinates. It is possible to sketch the diagram roughly with sufficient accuracy to locate the critical positions for live loads for various functions of the structure. These diagrams are referred to as *qualitative* influence lines, whereas those with numerical values are referred to as *quantitative* influence lines.[1]

---

[1]McCormac, J. C., 2007, *Structural Analysis: Using Classical and Matrix Methods*, 4th ed. (Hoboken, NJ: John Wiley & Sons), pp. 189–194.

If the influence line is desired for the left reaction of the continuous beam of Figure 14.1(a), its general shape can be determined by letting the reaction act upward through a unit distance, as shown in Figure 14.1(b). If the left end of the beam is pushed up, the beam will take the shape shown. This distorted shape can be sketched easily by remembering that the other supports are considered to be unyielding. The influence line for V_c, drawn in a similar manner, is shown in Figure 14.1(c).

Figure 14.1(d) shows the influence line for positive moment at point x near the center of the left-hand span. The beam is assumed to have a pin or hinge inserted at x and a couple is applied adjacent to each side of the pin, which will cause compression in the top fibers. Bending the beam on each side of the pin causes the left span to take the shape indicated, and the deflected shape of the remainder of the beam may be roughly sketched. A similar procedure is used to draw the influence line for negative moment at point y in the third span, except that a moment couple is applied at the assumed pins, which will tend to cause compression in the bottom beam fibers, corresponding with negative moment.

Finally, qualitative influence lines are drawn for positive shear at points x and y. At point x the beam is assumed to be cut, and the two vertical forces of the nature required to give positive shear are applied to the beam on the sides of the cut section. The beam will take the shape shown in Figure 14.1(f). The same procedure is used in Figure 14.1(g) to draw a diagram for positive shear at point y. (Theoretically, for qualitative shear influence lines, it is necessary to have a moment on each side of the cut section sufficient to maintain equal slopes. Such moments are indicated in parts (f) and (g) of the figure by the letter M.)

From these diagrams, considerable information is available concerning critical live loading conditions. If a maximum positive value of V_A were desired for a uniform live load, the load would

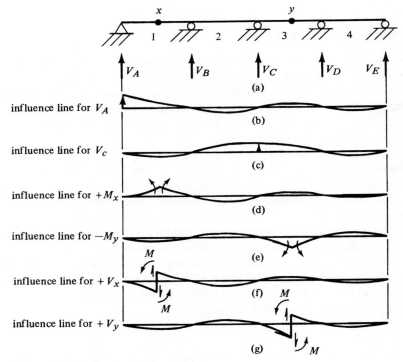

Figure 14.1 Qualitative influence lines.

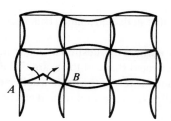

Figure 14.2

be placed in spans 1 and 3, where the diagram has positive ordinates, if maximum negative moment were required at point *y*, spans 2 and 4 would be loaded, and so on.

Qualitative influence lines are particularly valuable for determining critical load positions for buildings, as illustrated by the moment influence line for the building of Figure 14.2. In drawing diagrams for an entire frame, the joints are assumed to be free to rotate, but the members at each joint are assumed to be rigidly connected to each other so that the angles between them do not change during rotation. The influence line shown in the figure is for positive moment at the center of beam *AB*.

The spans that should be loaded to cause maximum positive moment are obvious from the diagram. It should be realized that loads on a member more than approximately three spans away have little effect on the function under consideration.

In the last few paragraphs, influence lines have been used to determine the critical positions for placing live loads to cause maximum moments. The same results can be obtained (and perhaps more easily) by considering the deflected shape or curvature of a member under load. If the live loads are placed so that they cause the greatest curvature at a particular point, they will have bent the structure the greatest amount at that point, which means that the greatest moment will have been obtained.

For the continuous beam of Figure 14.3(a), it is desired to cause the maximum negative moment at support *B* by the proper placement of a uniform live load. In part (b) of the figure, the deflected shape of the beam is sketched as it would be when a negative moment occurs at *B*, and the rest of the beam's deflected shape is drawn as shown by the dashed line. Then the live uniform load

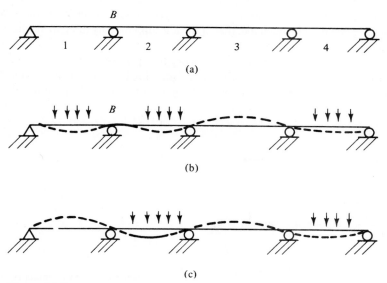

Figure 14.3

is placed in the locations that would exaggerate that deflected shape. This is done by placing the load in spans 1, 2, and 4.

A similar situation is shown in Figure 14.3(c), where it is desired to obtain maximum positive moment at the middle of the second span. The deflected shape of the beam is sketched as it would be when a positive moment occurs in that span, and the rest of the beam's deflected shape is drawn in. To exaggerate this positive or downward bending in the second span, it can be seen that the live load should be placed in spans 2 and 4.

14.4   LIMIT DESIGN

It can be clearly shown that a statically indeterminate beam or frame normally will not collapse when its ultimate moment capacity is reached at just one section. Instead, there is a redistribution of the moments in the structure. Its behavior is rather similar to the case where three men are walking along with a log on their shoulders and one of the men gets tired and lowers his shoulder just a little. The result is a redistribution of loads to the other men and thus changes in the shears and moments throughout the log.

It might be well at this point to attempt to distinguish between the terms *plastic design* as used in structural steel and *limit design* as used in reinforced concrete. In structural steel, plastic design involves both (a) the increased resisting moment of a member after the extreme fiber of the member is stressed to its yield point and (b) the redistribution or change in the moment pattern in the member. (Load and resistance factor design [LRFD] is a steel design method that incorporates much of the theory associated with plastic design.) In reinforced concrete the increase in resisting moment of a section after part of the section has been stressed to its yield point has already been accounted for in the strength design procedure. Therefore, limit design for reinforced concrete structures is concerned only with the change in the moment pattern after the steel reinforcing at some cross section is stressed to its yield point.

The basic assumption used for limit design of reinforced concrete structures and for plastic design of steel structures is the ability of these materials to resist a so-called yield moment while an appreciable increase in local curvature occurs. In effect, if one section of a statically indeterminate member reaches this moment, it begins to yield but does not fail. Rather, it acts like a hinge (called a *plastic hinge*) and throws the excess load off to sections of the members that have lesser stresses. The resulting behavior is much like that of the log supported by three men when one man lowered his shoulder.

To apply the limit design or plastic theory to a particular structure, it is necessary for that structure to behave plastically. For this initial discussion it is assumed that an ideal plastic material, such as a ductile structural steel, is involved. Figure 14.4 shows the relationship of moment to the

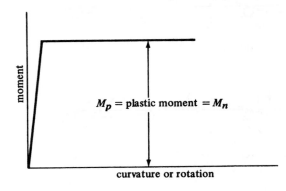

M_p = plastic moment = M_n

curvature or rotation

Figure 14.4 Moment–curvature relationship for an ideal plastic material.

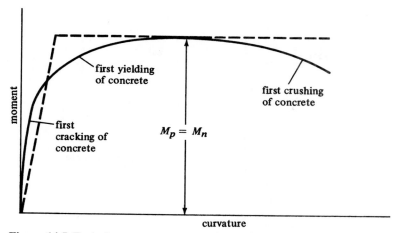

Figure 14.5 Typical moment–curvature relationship for a reinforced concrete member.

resulting curvature of a short length of a ductile steel member. The theoretical ultimate or nominal resisting moment of a section is referred to in this text as M_n (it's the same as the plastic moment M_p).

Although the moment-to-curvature relationship for reinforced concrete is quite different from the ideal one pictured in Figure 14.4, the actual curve can be approximated reasonably well by the ideal one, as shown in Figure 14.5. The dashed line in the figure represents the ideal curve, while the solid line is a typical one for reinforced concrete. Tests have shown that the lower the reinforcing percentage in the concrete ρ or $\rho - \rho'$ (where ρ' is the percentage of compressive reinforcing), the closer will the concrete curve approach the ideal curve. This is particularly true when very ductile reinforcing steels such as grade 40 are used. Should a large percentage of steel be present in a reinforced concrete member, the yielding that actually occurs before failure will be so limited that the ultimate or limit behavior of the member will not be greatly affected by yielding.

The Collapse Mechanism

To understand moment redistribution in steel or reinforced concrete structures, it is necessary first to consider the location and number of plastic hinges required to cause a structure to collapse. A statically determinate beam will fail if one plastic hinge develops. To illustrate this fact, the simple beam of constant cross section loaded with a concentrated load at midspan shown in Figure 14.6(a) is considered. Should the load be increased until a plastic hinge is developed at the point of maximum moment (underneath the load in this case), an unstable structure will have been created, as shown in Figure 14.6(b). Any further increase in load will cause collapse.

The plastic theory is of little advantage for statically determinate beams and frames, but it may be of decided advantage for statically indeterminate beams and frames. For a statically indeterminate structure to fail, it is necessary for more than one plastic hinge to form. The number of plastic hinges required for failure of statically indeterminate structures will be shown to vary from structure to structure, but may never be less than two. The fixed-end beam of Figure 14.7 cannot fail unless the three plastic hinges shown in the figure are developed.

Although a plastic hinge may be formed in a statically indeterminate structure, the load can still be increased without causing failure if the geometry of the structure permits. The plastic hinge will act like a real hinge insofar as increased loading is concerned. As the load is increased, there is a redistribution of moment because the plastic hinge can resist no more moment. As more plastic

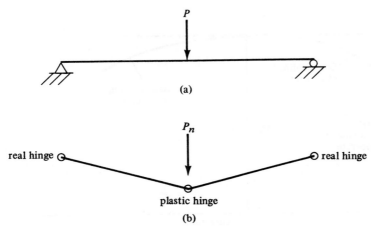

Figure 14.6

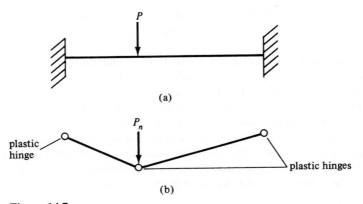

Figure 14.7

hinges are formed in the structure, there will eventually be a sufficient number of them to cause collapse.

The propped beam of Figure 14.8 is an example of a structure that will fail after two plastic hinges develop. Three hinges are required for collapse, but there is a real hinge at the right end. In this beam the largest elastic moment caused by the design-concentrated load is at the fixed end. As the magnitude of the load is increased, a plastic hinge will form at that point.

The load may be further increased until the moment at some other point (here it will be at the concentrated load) reaches the plastic moment. Additional load will cause the beam to collapse. The arrangement of plastic hinges, and perhaps real hinges that permit collapse in a structure, is called the *mechanism*. Parts (b) of Figures 14.6, 14.7, and 14.8 show mechanisms for various beams.

Plastic Analysis by the Equilibrium Method

To analyze a structure plastically, it is necessary to compute the plastic or ultimate moments of the sections, to consider the moment redistribution after the ultimate moments develop, and finally to

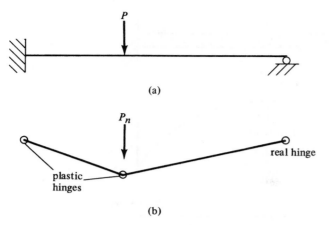

(a)

(b)   **Figure 14.8**

determine the ultimate loads that exist when the collapse mechanism is created. The method of plastic analysis known as the *equilibrium method* will be illustrated in this section.

As the first illustration, the fixed-end beam of Figure 14.9 is considered. It is desired to determine the value of w_n, the theoretical ultimate load the beam can support. The maximum moments in a uniformly loaded fixed-end beam in the elastic range occur at the fixed ends, as shown in the figure.

If the magnitude of the uniform load is increased, the moments in the beam will be increased proportionately until a plastic moment is eventually developed at some point. Due to symmetry, plastic moments will be developed at the beam ends, as shown in Figure 14.10(b). Should the loads be further increased, the beam will be unable to resist moments larger than M_n at its ends. Those points will rotate through large angles, and thus the beam will be permitted to deflect more and permit the moments to increase out in the span. Although the plastic moment has been reached at the ends and plastic hinges are formed, the beam cannot fail because it has, in effect, become a simple end-supported beam for further load increases as shown in Figure 14.10(c).

The load can now be increased on this "simple" beam, and the moments at the ends will remain constant; however, the moment out in the span will increase as it would in a uniformly loaded simple beam. This increase is shown by the dashed line in Figure 14.11(b). The load may be increased until the moment at some other point (here the beam centerline) reaches the plastic moment. When this happens, a third plastic hinge will have developed and a mechanism will have been created, permitting collapse.

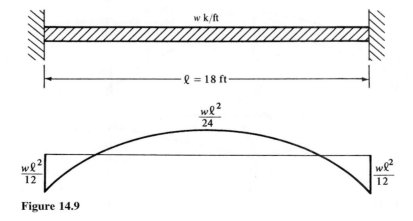

Figure 14.9

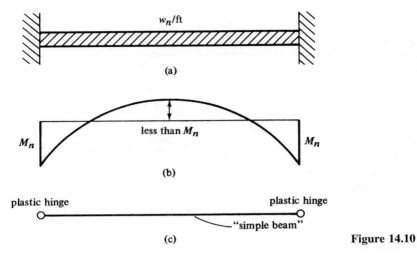

(a)

(b)

(c)

Figure 14.10

One method of determining the value of w_n is to take moments at the centerline of the beam (knowing the moment there is M_n at collapse). Reference is made here to Figure 14.11(a) for the beam reactions.

$$M_n = -M_n + \left(w_n \frac{\ell}{2}\right)\left(\frac{\ell}{2} - \frac{\ell}{4}\right) = -M_n + \frac{w_n\ell^2}{8}$$

$$w_n = \frac{16M_n}{\ell^2}$$

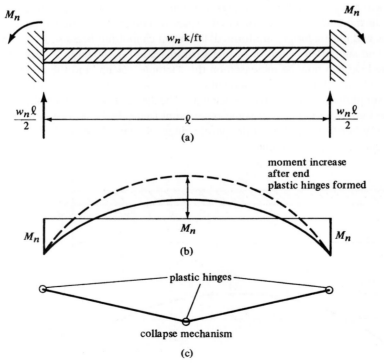

(a)

(b)

(c)

Figure 14.11

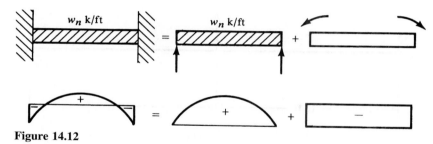

Figure 14.12

The same value could be obtained by considering the diagrams shown in Figure 14.12. You will remember that a fixed-end beam can be replaced with a simply supported beam plus a beam with end moments. Thus the final moment diagram for the fixed-end beam equals the moment diagram if the beam had been simply supported plus the end moment diagram.

For the beam under consideration, the value of M_n can be calculated as follows (see Figure 14.13):

$$2M_n = \frac{w_n \ell^2}{8}$$

$$M_n = \frac{w_n \ell^2}{16}$$

The propped beam of Figure 14.14, which supports a concentrated load, is presented as a second illustration of plastic analysis. It is desired to determine the value of P_n, the theoretical ultimate load the beam can support before collapse. The maximum moment in this beam in the elastic range occurs at the fixed end, as shown in the figure. If the magnitude of the concentrated load is increased, the moments in the beam will increase proportionately until a plastic moment is eventually developed at some point. This point will be at the fixed end, where the elastic moment diagram has its largest ordinate.

After this plastic hinge is formed, the beam will act as though it is simply supported insofar as load increases are concerned, because it will have a plastic hinge at the left end and a real hinge at the right end. An increase in the magnitude of the load P will not increase the moment at the left end but will increase the moment out in the beam, as it would in a simple beam. The increasing simple beam moment is indicated by the dashed line in Figure 14.14(c). Eventually, the moment at the concentrated load will reach M_n and a mechanism will form, consisting of two plastic hinges and one real hinge, as shown in Figure 14.14(d).

The value of the theoretical maximum concentrated load P_n that the beam can support can be determined by taking moments to the right or left of the load. Figure 14.14(e) shows the beam reactions for the conditions existing just before collapse. Moments are taken to the right of the load as follows:

$$M_n = \left(\frac{P_n}{2} - \frac{M_n}{20}\right) 10$$

$$P_n = 0.3 M_n$$

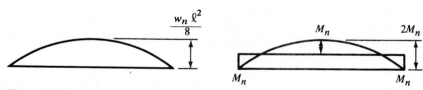

Figure 14.13

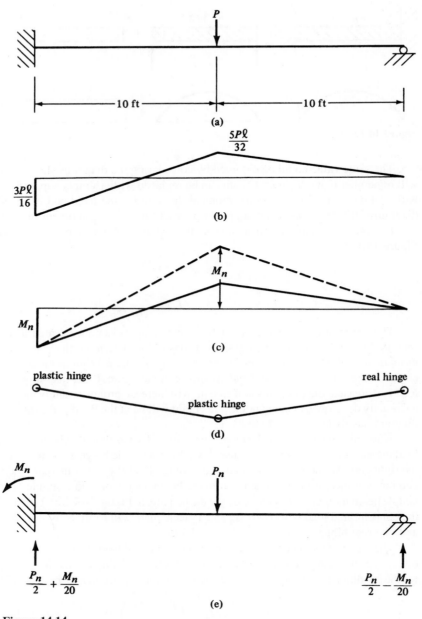

Figure 14.14

The subject of plastic analysis can be continued for different types of structures and loadings, as described in several textbooks on structural analysis or steel design.[2] The method has been proved to be satisfactory for ductile structural steels by many tests. Concrete, however, is a relatively brittle material, and the limit design theory has not been fully accepted by the ACI Code.

---

[2]McCormac, J. C., 2008, *Structural Steel Design: LRFD Method*, 4th ed. (Hoboken, NJ: Pearson Prentice Hall), pp. 239–251.

The Code does recognize that there is some redistribution of moments and permits partial redistribution based on a rule of thumb that is presented in the next section of this chapter.

14.5   LIMIT DESIGN UNDER THE ACI CODE

Tests of reinforced concrete frames have shown that under certain conditions there is definitely a redistribution of moments before collapse occurs. Recognizing this fact, the ACI Code (8.4.1) permits factored moments calculated by elastic theory (*not by an approximate analysis*) to be decreased at locations of maximum negative or positive moment in any span of a continuous structure for any loading arrangement. The amount by which moments can be decreased cannot exceed $1000\epsilon_t$ percent with a maximum of 20%. Redistribution of moments as described herein is not permissible unless ϵ_t is equal to or greater than 0.0075 at the section where moment is reduced (ACI Section 8.4.2). Appendix Tables A.7 and B.7 of this book provide percentages of steel for which ϵ_t will be equal to 0.0075. If ρ is greater than these values, redistribution is not permitted as ϵ_t will be less than 0.0075. The values given in the table are provided for rectangular sections with tensile reinforcing.

According to ACI Section 18.10.4.1 negative or positive moments may be decreased for prestressed sections using the same rules if bonded reinforcing (as described in Chapter 19 herein) is used.

The ACI Code's percentage of moment redistribution has purposely been limited to a very conservative value to be sure that excessively large concrete cracks do not occur at high steel stresses and to ensure adequate ductility for moment redistribution at the plastic hinges. The ACI Code likely will expand its presently conservative redistribution method after the behavior of plastic hinges is better understood, particularly as regards shears, deflections, and development of reinforcing. It is assumed here that the sections are satisfactorily reinforced for shears so that the ultimate moments can be reached without shear failure occurring. The adjustments are applied to the moments resulting from each of the different loading conditions. The member in question will then be proportioned on the basis of the resulting moment envelope. Figures 14.15 through 14.18 illustrate the application of the moment redistribution permitted by the Code to a three-span continuous beam. It will be noted in these figures that factored loads and elastic analyses are used for all the calculations.

Three different live-load conditions are considered in these figures. To determine the maximum positive moment in span 1, live load is placed in spans 1 and 3 (Figure 14.15). Similarly, to produce maximum positive moment in span 2, the live load is placed in that span only (Figure 14.16). Finally, maximum negative moment at the first interior support from the left end is caused by placing the live load in spans 1 and 2 (Figure 14.17).

For this particular beam it is assumed that the Code permits a 10% decrease in the negative or positive moments. This will require that $1000\epsilon_t$ exceed 10 at the sections where the moment is reduced to provide the needed ductility. The result will be smaller design moments at the critical sections. Initially, the loading for maximum positive moment in span 1 is considered as shown in Figure 14.15. If the maximum calculated positive moments of 425 ft-k near midspan of the end spans are each decreased by 10% to 383 ft-k, the negative moments at both interior supports will be increased to 406 ft-k. Even though the negative moment has increased significantly, from 308 to 406 ft-k, this higher value still will not control the required moment capacity at this location. Hence, the positive design moment is decreased, but the negative moment is not increased.

In the same fashion, in Figure 14.16, where the beam is loaded to produce maximum positive moment in span 2, a 10% decrease in positive moment from 261 to 235 ft-k will increase the negative moment at both interior supports from 339 to 365 ft-k.

Finally, in Figure 14.17, the live-load placement causes a maximum negative moment at the first interior support of 504 ft-k. If this value is reduced by 10%, the maximum moment there will

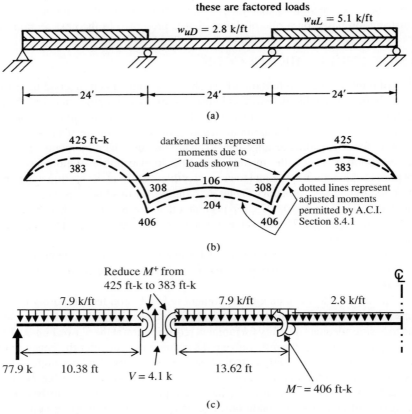

Figure 14.15 Maximum positive moment in end spans. (a) Loading patterns for maximum positive moment in both end spans. (b) Moment diagram before and after reducing positive moment in end spans. (c) Shears and moments in left span after M^+ is reduced 10% in end spans.

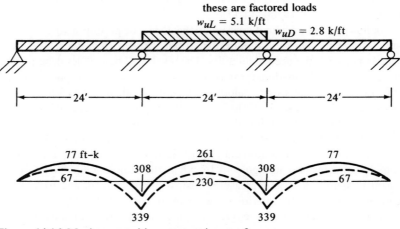

Figure 14.16 Maximum positive moment in span 2.

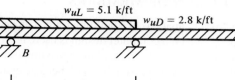

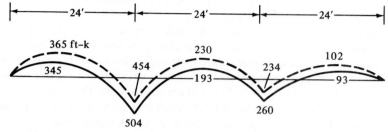

Figure 14.17 Maximum negative moment at support *B*.

be −454 ft-k. In this figure the authors have reduced the negative moment at the other interior support by 10% also.

It will be noticed that the net result of all of the various decreases in the positive or negative moments is a net reduction in both the maximum positive and the maximum negative values. The result of these various redistributions is actually an envelope of the extreme values of the moments at the critical sections. The envelope for the three-span beam considered in this section is presented in Figure 14.18. You can see at a glance the parts of the beams that need positive reinforcement, negative reinforcement, or both.

The reductions in bending moments due to moment redistribution as described here do not mean that the safety factors for continuous members will be less than those for simple spans. Rather, the excess strength that such members have due to this continuity is reduced so that the overall factors of safety are nearer but not less than those of simple spans.

Various studies have shown that cracking and deflection of members selected by the limit design process are no more severe than those for the same members designed without taking advantage of the permissible redistributions.[3,4]

(Appendix B of the ACI Code presents quite a few variations which may be used in design for flexure and axial loads. There are changes in the moment redistribution percentages permitted for continuous members, in the reinforcing limits and in the strength reduction or ϕ factors. These latter changes are dependent on the strain conditions, including whether the sections are compression or tension controlled.)

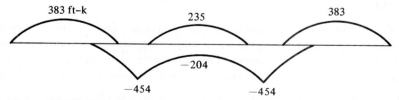

Figure 14.18 Moment envelope.

[3]Cohn, M. Z., 1964, "Rotational Compatibility in the Limit Design of Reinforced Concrete Continuous Beams," *Proceedings of the International Symposium on the Flexural Mechanics of Reinforced Concrete*, ASCE-ACI (Miami), pp. 359–382.

[4]Mattock, A. H., 1959, "Redistribution of Design Bending Moments in Reinforced Concrete Continuous Beams," *Proceedings of the Institution of Civil Engineers*, 113, pp. 35–46.

14.6   PRELIMINARY DESIGN OF MEMBERS

Before an "exact" analysis of a building frame can be made, it is necessary to estimate the sizes of the members. Even if a computer design is used, it is often economically advisable to make some preliminary estimates as to sizes. If an approximate analysis of the structure has been made, it will be possible to make very reasonable member size estimates. The result will be appreciable saving of both computer time and money.

An experienced designer can usually make very satisfactory preliminary size estimates based upon his or her previous experience. In the absence of such experience, however, the designer can still make quite reasonable size estimates based on his or her knowledge of structural analysis. For instance, to approximately size columns, a designer can neglect moments and assume an average axial stress or P_u/A_g value of about 0.4 to $0.6f'_c$. This rough value can be divided into the estimated column load to obtain its estimated area. If moments are large, lower values of average stress (0.4 to $0.5f'_c$) may be used; if moments are small, higher values (0.55 to $0.6f'_c$) may be used.

Preliminary beam sizes can be obtained by considering their approximate moments. A uniformly loaded simple beam will have a maximum bending moment equal to $w_u\ell^2/8$, whereas a uniformly loaded fixed-end beam will have a maximum moment of $w_u\ell^2/12$. For a continuous uniformly loaded beam, the designer might very well estimate a maximum moment somewhere between the values given, perhaps $w_u\ell^2/10$, and use that value to estimate the beam size.

For many structures it is necessary to conduct at least two different analyses. One analysis is made to consider the effect of gravity loads as described in Section 14.7 of this chapter, while another might be made to consider the effect of lateral loads as discussed in Section 14.8. For the gravity loads only, U usually equals $1.2D + 1.6L$.

Because the gravity loads affect only the floor to which they are applied, each floor can probably be analyzed independently of the others. Such is not the case for lateral loads because lateral loads applied anywhere on the frame affect the lateral displacements throughout the frame and thus affect the forces in the frame below. For this situation the load factor equations involving lateral forces (ACI Equations 9-3, 9-4, etc.) must be applied.

Sometimes a third analysis should be made—one that involves the possibility of force reversals on the windward side or even overturning of the structure. If overturning is being considered, the dead and live gravity loads should be reduced to their smallest possible values (that is, zero live load and $0.9D$, in case the dead loads have been overestimated a little) while the lateral loads are acting. For this case ACI load factor equations 9-6 and 9-7 must be considered.

14.7   APPROXIMATE ANALYSIS OF CONTINUOUS FRAMES FOR VERTICAL LOADS

Statically indeterminate structures may be analyzed "exactly" or "approximately." Some approximate methods involving the use of simplifying assumptions are presented in this section. Despite the increased use of computers for making "exact" analyses, approximate methods are used about as much or more than ever, for several reasons. These include the following:

1. The structure may be so complicated that no one who has the knowledge to make an "exact" analysis is available or no suitable computer software is available.

2. For some structures, either method may be subject to so many errors and imperfections that approximate methods may yield values as accurate as those obtained with an "exact" analysis. A specific example is the analysis of a building frame for wind loads where the walls, partitions, and floors contribute an indeterminate amount to wind resistance. Wind forces calculated in the frame by either method are not accurate.

3. To design the members of a statically indeterminate structure, it is sometimes necessary to make an estimate of their sizes before structural analysis can begin by an exact method. Approximate analysis of the structure will yield forces from which reasonably good initial estimates can be made as to member sizes.

4. Approximate analyses are quite useful in rough-checking exact solutions.

From the discussion of influence lines in Section 14.3 you can see that unless a computer is used (a very practical alternative today), an exact analysis involving several different placements of the live loads would be a long and tedious affair. For this reason it is common when a computer is not readily available to use some approximate methods of analysis, such as the ACI moment and shear coefficients, the equivalent rigid-frame method, the assumed point-of-inflection-location method, and others discussed in the pages to follow.

ACI Coefficients for Continuous Beams and Slabs

A very common method used for the design of continuous reinforced concrete structures involves the use of the ACI coefficients given in Section 8.3.3 of the Code. These coefficients, which are reproduced in Table 14.1, provide estimated maximum shears and moments for buildings of normal proportions. The values calculated in this manner will usually be somewhat larger than those that would be obtained with an "exact" analysis. As a result, appreciable economy can normally be obtained by taking the time or effort to make such an analysis. In this regard, it should be realized that these coefficients are considered best applied to continuous frames having more than three or four continuous spans.

In developing the coefficients, the negative-moment values were reduced to take into account the usual support widths and also some moment redistribution, as described in Section 14.5 of this chapter. In addition, the positive-moment values have been increased somewhat to account for the moment redistribution. It will also be noted that the coefficients account for the fact that in

Table 14.1 ACI Coefficients

Positive moment	
End spans	
If discontinuous end is unrestrained	$\frac{1}{11} w_u \ell_n^2$
If discontinuous end is integral with the support	$\frac{1}{14} w_u \ell_n^2$
Interior spans	$\frac{1}{16} w_u \ell_n^2$
Negative moment at exterior face of first interior support	
Two spans	$\frac{1}{9} w_u \ell_n^2$
More than two spans	$\frac{1}{10} w_u \ell_n^2$
Negative moment at other faces of interior supports	$\frac{1}{11} w_u \ell_n^2$
Negative moment at face of all supports for (a) slabs with spans not exceeding 10 ft and (b) beams and girders where ratio of sum of column stiffnesses to beam stiffness exceeds eight at each end of the span	$\frac{1}{12} w_u \ell_n^2$
Negative moment at interior faces of exterior supports for members built integrally with their supports	
Where the support is a spandrel beam or girder	$\frac{1}{24} w_u \ell_n^2$
Where the support is a column	$\frac{1}{16} w_u \ell_n^2$
Shear in end members at face of first interior support	$1.15(w_n \ell_n / 2)$
Shear at face of all other supports	$w_n \ell_n / 2$

One Shell Plaza, Houston, Texas, which has 52 stories and is 714 ft high.
(Courtesy of Master Builders.)

monolithic construction the supports are not simple and moments are present at end supports, such as where those supports are beams or columns.

In applying the coefficients, w_u is the design load while ℓ_n *is the clear span for calculating positive moments and the average of the adjacent clear spans for calculating negative moments.*

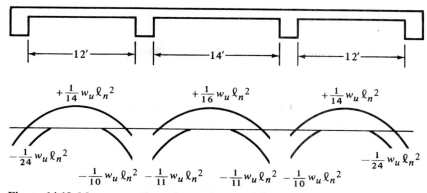

$$+\tfrac{1}{14} w_u \ell_n{}^2 \qquad\qquad +\tfrac{1}{16} w_u \ell_n{}^2 \qquad\qquad +\tfrac{1}{14} w_u \ell_n{}^2$$

$$-\tfrac{1}{24} w_u \ell_n{}^2 \qquad\qquad\qquad\qquad\qquad\qquad\qquad -\tfrac{1}{24} w_u \ell_n{}^2$$

$$-\tfrac{1}{10} w_u \ell_n{}^2 \;\; -\tfrac{1}{11} w_u \ell_n{}^2 \quad -\tfrac{1}{11} w_u \ell_n{}^2 \;\; -\tfrac{1}{10} w_u \ell_n{}^2$$

Figure 14.19 Moment envelopes for continuous slab constructed integrally with beams.

These values can be applied only to members with approximately equal spans (the larger of two adjacent spans not exceeding the smaller by more than 20%) and for cases where the ratio of the uniform service live load to the uniform service dead load is not greater than three. In addition, the values are not applicable to prestressed concrete members. Should these limitations not be met, a more precise method of analysis must be used.

For the design of a continuous beam or slab, the moment coefficients provide in effect two sets of moment diagrams for each span of the structure. One diagram is the result of placing the live loads so that they will cause maximum positive moment out in the span, while the other is the result of placing the live loads so as to cause maximum negative moments at the supports. To be truthful, however, it is not possible to produce maximum negative moments at both ends of a span simultaneously. It takes one placement of the live loads to produce maximum negative moment at one end of the span and another placement to produce maximum negative moment at the other end. The assumption of both maximums occurring at the same time is on the safe side, however, because the resulting diagram will have greater critical values than are produced by either one of the two separate loading conditions.

The ACI coefficients give maximum points for a moment envelope for each span of a continuous frame. Typical envelopes are shown in Figure 14.19 for a continuous slab, which is assumed to be constructed integrally with its exterior supports, which are spandrel girders.

Example 14.1 presents the design of the slab of Figure 14.20 using the moment coefficients of the ACI Code. The calculations for this problem can be conveniently set up in some type of table such as the one shown in Figure 14.21. For this particular slab the authors used an arrangement of reinforcement that included bent bars. It is quite common, however, in slabs—particularly those 5 in. or less in thickness—to use straight bars only in the top and bottom of the slab.

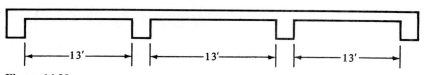

Figure 14.20

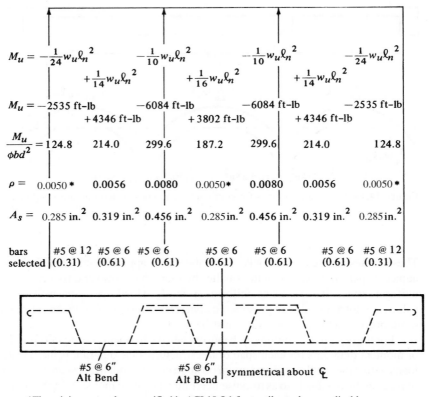

$$M_u = -\frac{1}{24}w_u\ell_n^2 \qquad -\frac{1}{10}w_u\ell_n^2 \qquad -\frac{1}{10}w_u\ell_n^2 \qquad -\frac{1}{24}w_u\ell_n^2$$
$$+\frac{1}{14}w_u\ell_n^2 \qquad +\frac{1}{16}w_u\ell_n^2 \qquad +\frac{1}{14}w_u\ell_n^2$$

| $M_u =$ | −2535 ft-lb | | −6084 ft-lb | | −6084 ft-lb | | −2535 ft-lb |
| | | +4346 ft-lb | | +3802 ft-lb | | +4346 ft-lb | |

$\dfrac{M_u}{\phi b d^2} =$	124.8	214.0	299.6	187.2	299.6	214.0	124.8
$\rho =$	0.0050 *	0.0056	0.0080	0.0050*	0.0080	0.0056	0.0050 *
$A_s =$	0.285 in.^{2}	0.319 in.^{2}	0.456 in.^{2}	0.285 in.^{2}	0.456 in.^{2}	0.319 in.^{2}	0.285 in.^{2}
bars selected	#5 @ 12 (0.31)	#5 @ 6 (0.61)	#5 @ 6 (0.61)	#5 @ 6 (0.61)	#5 @ 6 (0.61)	#5 @ 6 (0.61)	#5 @ 12 (0.31)

#5 @ 6"
Alt Bend

#5 @ 6"
Alt Bend

symmetrical about ℄

*The minimum ρ values specified in ACI 10.5.1 for tensile steel are applicable
to both positive and negative moment regions (ACI Commentary R10.5).

Figure 14.21

EXAMPLE 14.1

Design the continuous slab of Figure 14.20 for moments calculated with the ACI coefficients. The slab is to support a service live load of 165 psf, and a superimposed dead load of 5 psf in addition to its own dead weight. $f'_c = 3000$ psi and $f_y = 40,000$ psi. The slab is to be constructed integrally with its spandrel girder supports, and the spandrel supports are 12 in. wide.

SOLUTION

Minimum t for Deflection by ACI Code (9.5.2.3) Note that this table uses ℓ, not ℓ_n as the span.

$$\text{Deflection multiplier for } 40,000-\text{psi steel} = 0.4 + \frac{40,000}{100,000} = 0.80$$

$$\text{Minimum } t \text{ for end span} = 0.8\frac{\ell}{24} = \frac{(0.8)(12)(13+1)}{24} = 5.6''$$

$$\text{Minimum } t \text{ for interior span} = 0.8\frac{\ell}{28} = \frac{(0.8)(12)(13+1)}{28} = 4.80''$$

Loads and Maximum Moment Assuming 6-in. Slab ($d = 4\frac{3}{4}$ in.)

$$w_D = \text{slab weight} = \left(\frac{6}{12}\right)(150) = 75 \text{ psf}$$

$$w_L = 165 \text{ psf}$$

$$w_u = (1.2)(75+5) + (1.6)(165) = 360 \text{ psf}$$

$$\text{Max } M_u = \left(\frac{1}{10}\right)w_u\ell_n^2 = \left(\frac{1}{10}\right)(360)(13)^2 = 6084 \text{ ft-lb}$$

Computing moments, ρ values, A_s requirements, and selecting bars at each section, as shown in Figure 14.21.

For floor slabs we are not concerned with the design of web reinforcing. For continuous beams, however, web reinforcing must be carefully designed. Such designs are based on the maximum shears occurring at various sections along the span.

From previous discussions you will remember that to determine the maximum shear occurring at Section 1-1 in the beam of Figure 14.22, the uniform dead load would extend all across the span while the uniform live load would be placed from the section to the most distant support.

If the live load is placed so as to cause maximum shears at various points along the span and the shear is calculated for each point, a maximum shear curve can be drawn. Practically speaking, however, it is unnecessary to go through such a lengthy process for buildings of normal proportions because the values that would be obtained do not vary significantly from the values given by the ACI Code, which are shown in Figure 14.23.

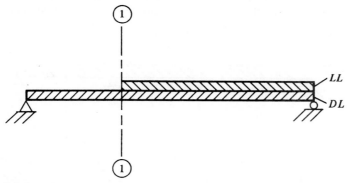

Figure 14.22

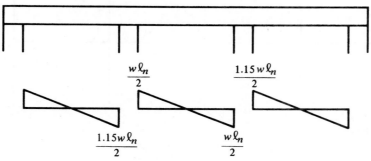

Figure 14.23

Equivalent Rigid-Frame Method

When continuous beams frame into and are supported by girders, the normal assumption made is that the girders provide only vertical support. Thus they are analyzed purely as continuous beams, as shown in Figure 14.24. The girders do provide some torsional stiffness, and if the calculated torsional moments exceed $\phi\lambda\sqrt{f_c'}(A_{cp}^2/\rho_{cp})$ as specified in ACI Section 11.6.1 they must be considered, as will be described in Chapter 15.

Where continuous beams frame into columns, the bending stiffnesses of the columns together with the torsional stiffnesses of the girders are of such magnitude that they must be considered. An approximate method frequently used for analyzing such reinforced concrete members is the *equivalent rigid-frame method*. In this method, which is applicable only to gravity loads, the loads are assumed to be applied only to the floor or roof under consideration and the far ends of the columns are assumed to be fixed, as shown in Figure 14.25. The sizes of the members are estimated, and an analysis is frequently made with moment distribution.

For this type of analysis it is necessary to estimate the sizes of the members and compute their relative stiffness or I/ℓ, values. From these values, distribution factors can be computed and the method of moment distribution applied. The moments of inertia of both columns and beams are normally calculated on the basis of gross concrete sections, with no allowance made for reinforcing.

There is a problem involved in determining the moment of inertia to be used for continuous T beams. The moment of inertia of a T beam is much greater where there is positive moment with the flanges in compression than where there is negative moment with the flanges cracked due to tension. Because the moment of inertia varies along the span, it is necessary to use an equivalent value. A practice often used is to assume that the equivalent moment of inertia equals twice the moment of inertia of the web, assuming that the web depth equals the full effective depth of the beam.[5] Some designers use other equivalent values, such as assuming an equivalent T section

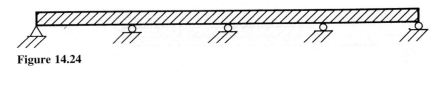

Figure 14.24

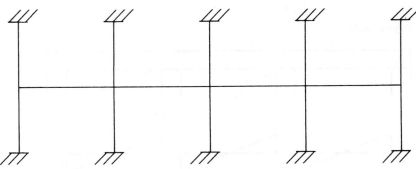

Figure 14.25

---

[5]Portland Cement Association, 1959, *Continuity in Concrete Building Frames*, 4th ed. (Chicago), pp. 17–20.

with flanges of effective widths equal to so many (say, 2 to 6) times the web width. These equivalent sections can be varied over a rather wide range without appreciably affecting the final moments.

The ACI Code (8.9.2) states that for such an approximate analysis, only two live-load combinations need to be considered. These are (1) live load placed on two adjacent spans and (2) live load placed on alternate spans. Example 14.2 illustrates the application of the equivalent rigid-frame method to a continuous T beam.

Computer results appear to indicate that the model shown in Figure 14.25 (as permitted by the ACI Code) may not be trustworthy for unsymmetrical loading. Differential column shortening can completely redistribute the moments obtained from the model (i.e., positive moments can become negative moments). As a result, designers should take into account possible axial deformations in their designs.

EXAMPLE 14.2

Using the equivalent rigid-frame method, draw the shear and moment diagrams for the continuous T beam of Figure 14.26. The beam is assumed to be framed into 16-in. × 16-in. columns and is to support a service dead load of 2.33 k/ft (including beam weight) and a service live load of 3.19 k/ft. Assume that the live load is applied in the center span only. The girders are assumed to have a depth of 24 in. and a web width of 12 in. Assume that the I of the T beam equals 2 times the I of its web.

SOLUTION   **Computing Fixed-End Moments**

$$w_u \text{ in first and third spans} = (1.2)(2.33) = 2.8 \text{ k/ft}$$

$$M_u = \frac{(2.8)(24)^2}{12} = 134.4 \text{ ft-k}$$

$$w_u \text{ in center span} = (1.2)(2.33) + (1.6)(3.19) = 7.9 \text{ k/ft}$$

$$M_u = \frac{(7.9)(24)^2}{12} = 379.2 \text{ ft-k}$$

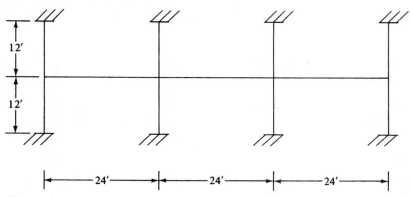

Figure 14.26

Computing Stiffness Factors

$$I \text{ of columns} = \left(\frac{1}{12}\right)(16)(16)^3 = 5461 \text{ in.}^4$$

$$k \text{ of columns} = \frac{I}{\ell} = \frac{5461}{12} = 455$$

$$\text{Equivalent } I \text{ of T beam} = (2)\left(\frac{1}{12}b_w h^3\right) = (2)\left(\frac{1}{12}\right)(12)(24)^3 = 27{,}648 \text{ in.}^4$$

$$k \text{ of T beam} = \frac{27{,}648}{24} = 1152$$

Record stiffness factors on the frame and compute distribution factors, as shown in Figure 14.27.

Balance fixed-end moments and draw shear and moment diagrams, as shown in Figure 14.28.

Assumed Points of Inflection

Another approximate method of analyzing statically indeterminate building frames is to assume the locations of the points of inflection in the members. Such assumptions have the effect of creating simple beams between the points of inflection in each span, and the positive moments in each span can be determined by statics. Negative moments occur in the girders between their ends and the points of inflection. They may be computed by considering the portion of each beam out to the point of inflection (P.I.) to be a cantilever. The shear at the end of each of the girders contributes to the axial forces in the columns. Similarly, the negative moments at the ends of the girders are transferred to the columns.

In Figure 14.29, beam *AB* of the building frame shown is analyzed by assuming points of inflection at the one-fifth points and assuming fixed supports at the beam ends.

14.8 APPROXIMATE ANALYSIS OF CONTINUOUS FRAMES FOR LATERAL LOADS

Building frames are subjected to lateral loads as well as to vertical loads. The necessity for careful attention to these forces increases as buildings become taller. Buildings must not only have sufficient lateral resistance to prevent failure, but also must have sufficient resistance to deflections to prevent injuries to their various parts. Rigid-frame buildings are highly statically indeterminate;

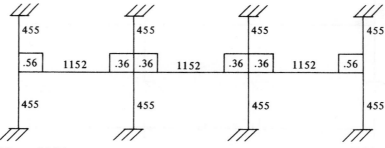

Figure 14.27

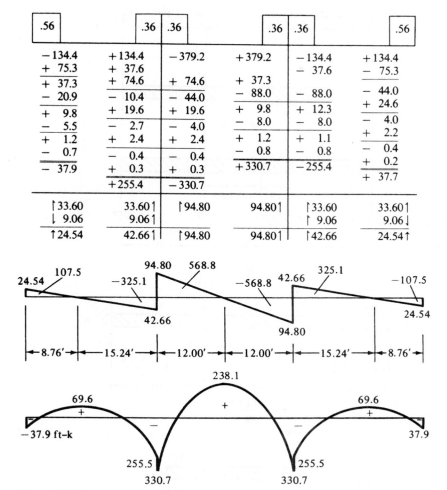

Figure 14.28

their analysis by "exact" methods (unless computers are used) is so lengthy as to make the approximate methods very popular.

The approximate method presented here is called the *portal method*. Because of its simplicity, it has probably been used more than any other approximate method for determining wind forces in building frames. This method, which was presented by Albert Smith in the *Journal of the Western Society of Engineers* in April 1915, is said to be satisfactory for most buildings up to 25 stories in height. Another method very similar to the portal method is the cantilever method. It is thought to give slightly better results for high narrow buildings and can be used for buildings not in excess of 25 to 35 stories.[6]

The portal method is merely a variation of the method described in Section 14.7 for analyzing beams in which the location of the points of inflection was assumed. For the portal method, the loads are assumed to be applied at the joints only. If this loading condition is correct, the moments

[6]"Wind Bracing in Steel Buildings," 1940, *Transactions of the American Society of Civil Engineers*, 105, pp. 1723–1727.

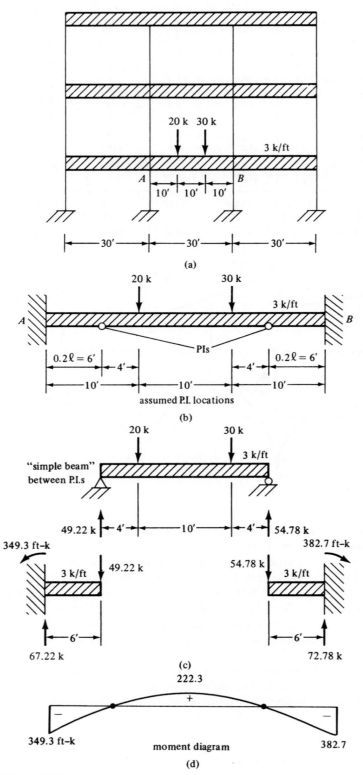

Figure 14.29

The 16-story Apoquindo Tower, Santiago, Chile. (Courtesy of Economy Forms Corporation.)

will vary linearly in the members, and points of inflection will be located fairly close to member midpoints.

No consideration is given in the portal method to the elastic properties of the members. These omissions can be very serious in unsymmetrical frames and in very tall buildings. To illustrate the seriousness of the matter, the changes in member sizes are considered in a very tall building. In such a building, there will probably not be a great deal of change in beam sizes from the top floor to the bottom floor. For the same loadings and spans, the changed sizes would be due to the large wind moments in the lower floors. The change, however, in column sizes from top to bottom would be tremendous. The result is that the relative sizes of columns and beams on the top floors are entirely different from the relative sizes on the lower floors. When this fact is not considered, it causes large errors in the analysis.

In the portal method, the entire wind loads are assumed to be resisted by the building frames, with no stiffening assistance from the floors, walls, and partitions. Changes in the lengths of girders and columns are assumed to be negligible. They are not negligible, however, in tall slender buildings, the height of which is five or more times the least horizontal dimension.

If the height of a building is roughly five or more times its least lateral dimension, it is generally felt that a more precise method of analysis should be used. There are several approximate methods that make use of the elastic properties of structures and give values closely approaching the results of the "exact" methods. These include the factor method,[7] the Witmer method of K percentages, and the Spurr method.[8]

The building frame shown in Figure 14.30 is analyzed by the portal method, as described in the following paragraphs.

---

[7]Norris, C. H., Wilbur, J. B., and Utku, S., 1976, *Elementary Structural Analysis*, 3rd ed. (New York: McGraw-Hill), pp. 207–212.

[8]"Wind Bracing in Steel Buildings," pp. 1723–1727.

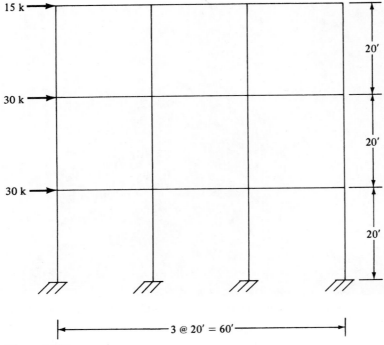

Figure 14.30

At least three assumptions must be made for each individual portal or for each girder. In the portal method, the frame is theoretically divided into independent portals (Figure 14.31), and the following three assumptions are made:

1. The columns bend in such a manner that there is a point of inflection at middepth.

2. The girders bend in such a manner that there is a point of inflection at their centerlines.

3. The horizontal shears on each level are arbitrarily distributed between the columns. One commonly used distribution (and the one illustrated here) is to assume that the shear divides among the columns in the ratio of one part to exterior columns and two parts to interior columns.

The reason for the ratio in assumption 3 can be seen in Figure 14.31. Each of the interior columns is serving two bents, whereas the exterior columns are serving only one. Another common distribution is to assume that the shear V taken by each column is in proportion to the floor area it supports. The shear distribution by the two procedures would be the same for a building with equal bays, but for one with unequal bays the results would differ, with the floor area method probably giving more realistic results.

Figure 14.31 One level of frame of Figure 14.30.

Frame Analysis by Portal Method

The frame of Figure 14.30 is analyzed in Figure 14.32 on the basis of the preceding assumptions. The arrows shown on the figure give the direction of the girder shears and the column axial forces. You can visualize the stress condition of the frame if you assume that the wind is tending to push it over from left to right, stretching the left exterior columns and compressing the right exterior columns. Briefly, the calculations were made as follows:

1. *Column shears:* The shears in each column on the various levels were first obtained. The total shear on the top level is 15 k. Because there are two exterior and two interior columns, the following expression may be written:

$$x + 2x + 2x + x = 15 \text{ k}$$

$$x = 2.5 \text{ k}$$

$$2x = 5.0 \text{ k}$$

The shear in column *CD* is 2.5 k; in *GH* it is 5.0 k; and so on. Similarly, the member shears were determined for the columns on the first and second levels, where the total shears are 75 k and 45 k, respectively.

2. *Column moments:* The columns are assumed to have points of inflection at their middepths; therefore, their moments, top and bottom, equal the column shears times half the column heights.

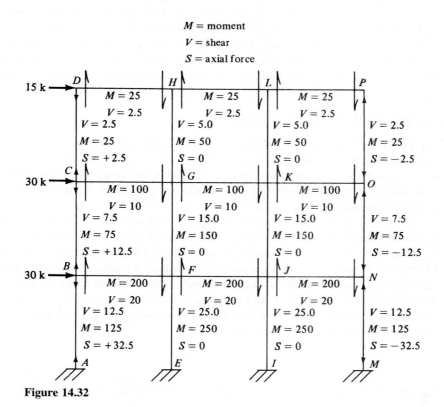

Figure 14.32

3. *Girder moments and shears:* At any joint in the frame, the sum of the moments in the girders equals the sum of the moments in the columns. The column moments have been previously determined. By beginning at the upper left-hand corner of the frame and working across from left to right, adding or subtracting the moments as the case may be, the girder moments were found in this order: *DH, HL, LP, CG, GK,* and so on. It follows that with points of inflection at girder centerlines, the girder shears equal the girder moments divided by half-girder lengths.

4. *Column axial forces:* The axial forces in the columns may be directly obtained from the girder shears. Starting at the upper left-hand corner, the column axial force in *CD* is equal to the shear in girder *DH*. The axial force in column *GH* is equal to the difference between the two girder shears *DH* and *HL*, which equals zero in this case. (If the width of each of the portals is the same, the shears in the girder on one level will be equal and the interior columns will have no axial force, since only lateral loads are considered.)

14.9   COMPUTER ANALYSIS OF BUILDING FRAMES

All structures are three-dimensional, but theoretical analyses of such structures by hand calculation methods are so lengthy as to be impractical. As a result, such systems are normally assumed to consist of two-dimensional or planar systems, and they are analyzed independently of each other. The methods of analysis presented in this chapter were handled in this manner.

Today's electronic computers have greatly changed the picture, and it is now possible to analyze complete three-dimensional structures. As a result, more realistic analyses are available and the necessity for high safety factors is reduced. The application of computers is not restricted merely to analysis; they are used in almost every phase of concrete work from analysis to design, to detailing, to specification writing, to material takeoffs, to cost estimating, and so on.

Another major advantage of computer analysis for building frames is that the designer is able to consider many different loading patterns quickly. The results are sometimes rather surprising.

14.10   LATERAL BRACING FOR BUILDINGS

For the usual building, the designer will select relatively small columns. Although such a procedure results in more floor space, it also results in buildings with small lateral stiffnesses or resistance to wind and earthquake loads. Such buildings may have detrimental lateral deflections and vibrations during windstorms unless definite lateral stiffness or bracing is otherwise provided in the structure.

To provide lateral stiffness it will be necessary for the roof and floor slabs to be attached to rigid walls, stairwells, or elevator shafts. Sometimes structural walls, called *shear walls* are added to a structure to provide the necessary lateral resistance. (The design of shear walls is considered in Chapter 18 of this text.) If it is not possible to provide such walls, stairwells or elevator shafts may be designed as large, box-shaped beams to transmit lateral loads to the supporting foundations. These members will behave as large cantilever beams. In designing such members, the designer should try to keep resistance symmetrical so as to prevent uneven lateral twisting or torsion in the structure when lateral loads are applied.[9]

---

[9]Leet, K., 1991, *Reinforced Concrete Design*, 2nd ed. (New York: McGraw-Hill), pp. 453–454.

14.11   DEVELOPMENT LENGTH REQUIREMENTS FOR CONTINUOUS MEMBERS

In Chapter 7 a general introduction to the subject of development lengths for simple and cantilever beams was presented. This section examines the ACI development length requirements for continuous members for both positive and negative reinforcing. After studying this information, the reader probably will be pleased to find at the end of this section a discussion of simplified design office practices for determining bar lengths.

Positive-Moment Reinforcement

Section 12.11 of the Code provides several detailed requirements for the lengths of positive-moment reinforcement. These are briefly summarized in the following paragraphs.

1. At least one-third of the positive steel in simple beams and one-fourth of the positive steel in continuous members must extend uninterrupted along the same face of the member at least 6 in. in the support (12.11.1). The purpose of this requirement is to make sure that the moment resistance of a beam will not be reduced excessively in parts of beams where the moments may be changed due to settlements, lateral loads, and so on.

2. The positive reinforcement required in the preceding paragraph must, *if the member is part of a primary lateral load resisting system,* be extended into the support a sufficient distance to develop the yield stress in tension of the bars at the face of the support. This requirement is included by the Code (12.11.2) to assure a ductile response to severe overstress as might occur with moment reversal during an earthquake or explosion. As a result of this requirement, it is necessary to have bottom bars lapped at interior supports and to use additional embedment lengths and hooks at exterior supports.

3. Section 12.11.3 of the Code says that at simple supports and at points of inflection, the positive-moment tension bars must have their diameters limited to certain maximum sizes. The purpose of the limitation is to keep bond stresses within reason at these points of low moments and large shears. (It is to be remembered that when bar diameters are smaller, the bars have greater surface area in proportion to their cross-sectional areas. Thus for bonding to concrete, the larger the bars' diameters the larger must be their development lengths. This fact is reflected in the expressions for ℓ_d.) It has not been shown that long anchorage lengths are fully effective in developing bars in a short distance between a P.I. and a point of maximum bar stress, a condition that might occur in heavily loaded short beams with large bottom bars. It is specified that ℓ_d as computed by the requirements presented in Chapter 7 may not exceed the following:

$$\ell_d \leq \frac{M_n}{V_u} + \ell_a$$

(ACI Equation 12-5)

In this expression, M_n is the computed theoretical flexural strength of the member if all reinforcing in that part of the beam is assumed stressed to f_y and V_u is the factored shear at the section. At a support, ℓ_a is equal to the sum of the embedment length beyond the ℄ of the support and the equivalent embedment length of any furnished hooks or mechanical anchorage. Headed and mechanically anchored deformed bars, which are permitted in Section 12.6 of the ACI Code, consist of bars or plates or angles or other pieces welded or otherwise attached transversely to the flexural bars in locations where sufficient anchorage length is not available. See Section 1.15 and 7.15 of this text for details on headed anchorage. At a point of inflection, ℓ_a is equal to the larger of the effective depth of the member or $12d$ (ACI 12.11.3). When the ends of the reinforcement are

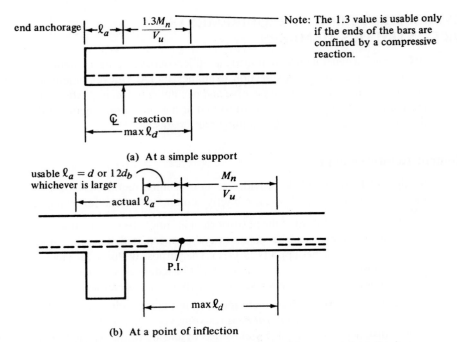

Note: The 1.3 value is usable only if the ends of the bars are confined by a compressive reaction.

(a) At a simple support

(b) At a point of inflection

Figure 14.33 Development length requirements for positive-moment reinforcing.

confined by a compression reaction, such as when there is a column below but not when a beam frames into a girder, an increase of 30% in the value M_n/V_u to $1.3M_n/V_u$ is allowed. The values described here are summarized in Figure 14.33, which is similar to Figure R12.11.3 of the ACI Code. A brief numerical illustration for a simple end-supported beam is provided in Example 14.3.

EXAMPLE 14.3

At the simple support shown in Figure 14.34, two uncoated #9 bars have been extended from the maximum moment area and into the support. Are the bar sizes satisfactory if $f_y = 60$ psi, $f'_c = 3$ ksi, $b = 12$ in., $d = 24$ in., and $V_u = 65$ k, if normal sand-gravel concrete is used, and if the reaction is compressive? Assume cover for bars $= 2''$ clear.

SOLUTION

$$a = \frac{A_s f_y}{0.85 f'_c b} = \frac{(2.00)(60)}{(0.85)(3)(12)} = 3.92''$$

$$M_n = A_s f_y \left(d - \frac{a}{2}\right) = (2.00)(60)\left(24 - \frac{3.92}{2}\right) = 2644.8 \text{ in.-k}$$

$$c = \text{side cover} = 2 + \frac{1.128}{2} = 2.564'' \leftarrow$$

$$c = \text{one-half to c. to c. spacing of bars} = \left(\frac{1}{2}\right)(12 - 2 \times 2 - 1.128) = 3.436''$$

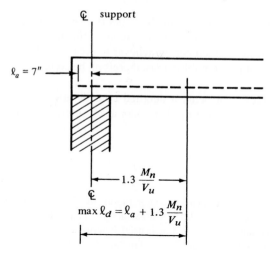

Figure 14.34

Assume $K_{tr} = 0$

$$\frac{\ell_d}{d_b} = \frac{3}{40} \frac{f_y}{\lambda \sqrt{f_c'}} \frac{\Psi_t \Psi_e \Psi_s}{\dfrac{c_b + K_{tr}}{d_b}}$$

$$= \left(\frac{3}{40}\right)\left(\frac{60{,}000}{(1.0)\sqrt{3000}}\right)\frac{(1.0)(1.0)(1.0)}{\dfrac{2.564 + 0}{1.128}} = 36.14 \text{ diameters}$$

$$\ell_d = (36.14)(1.128) = 41''$$

Maximum permissible $\ell_d = \dfrac{M_n}{V_u} + \ell_a$

$$= (1.3)\left(\frac{2644.8}{65}\right) + 7 = 59.90'' > 41'' \qquad \underline{\underline{\text{OK}}}$$

Note: If this condition has not been satisfied, the permissible value of ℓ_d could have been increased by using smaller bars or by increasing the end anchorage ℓ_a, as by the use of hooks.

Negative-Moment Reinforcement

Section 12.12 of the Code says that the tension reinforcement required by negative moments must be properly anchored into or through the supporting member. Section 12.12.1 of the Code says that the tension in these negative bars shall be developed on each side of the section in question by embedment length, hooks, or mechanical anchorage. Hooks may be used because these are tension bars.

Section 12.10.3 of the Code says that the reinforcement must extend beyond the point where it is no longer required for moment by a distance equal to the effective depth d of the member or 12 bar diameters, whichever is greater, except at the supports of simple beams or at the free ends of cantilevers. Section 12.12.3 of the Code says that at least one-third of the total reinforcement provided for negative moment at the support must have an embedment length beyond the point of inflection no less than the effective depth of the member, 12 bar diameters, or one-sixteenth of the clear span of the member, whichever is greatest. The location of the point of inflection can vary for different load combinations, therefore, be certain to use the most critical case.

Example 14.4, which follows, presents a detailed consideration of the lengths required by the Code for a set of straight negative-moment bars at an interior support of a continuous beam.

EXAMPLE 14.4

At the first interior support of the continuous beam shown in Figure 14.35, six #8 straight uncoated top bars (4.71 in.^{2}) are used to resist a moment M_u of 390 ft-k for which the calculated A_s required is 4.24 in.^{2} If $f_y = 60$ ksi, $f'_c = 3$ ksi, and normal sand-gravel concrete is used, determine the length of the bars as required by the ACI Code. The dimensions of the beam are $b_w = 14$ in. and $d = 24$ in. Assume bar cover $= 2d_b$ and clear spacing $= 3d_b$.

Note: The six negative-tension bars are actually placed in one layer in the top of the beam, but they are shown for illustration purposes in Figures 14.35 and 14.36 as though they are arranged in three layers of two bars each. In the solution that follows, the "bottom" two bars are cut off first, at the point where the moment plus the required development length is furnished; the "middle" two are cut when the moment plus the needed development length permits; and the "top" two are cut off at the required distance beyond the P.I.

SOLUTION   1. Required development length

$$c_b = \text{side cover} = (2)(1.0) + \frac{1.0}{2} = 2.5''$$

$$c_b = \text{one-half of c. to c. spacing of bars} = \left(\frac{3}{2}\right)(1.0) + \left(\frac{1.0}{2}\right) = 2.0'' \leftarrow$$

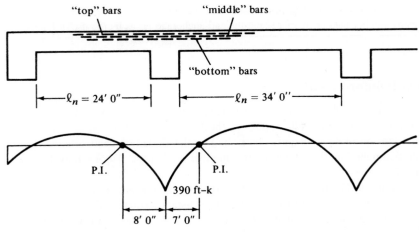

Figure 14.35

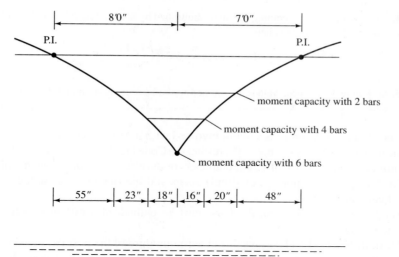

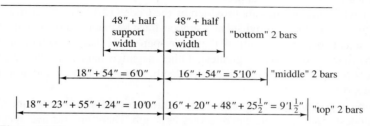

Figure 14.36

Assume $K_{tr} = 0$

$$\frac{\ell_d}{d_b} = \left(\frac{3}{40}\right)\left(\frac{60{,}000}{(1.0)\sqrt{3000}}\right)\frac{(1.3)(1.0)(1.0)}{\dfrac{2.0+0}{1.0}} = 53.4 \text{ diameters}$$

$$\ell_d = (53.41)(1.0) = 53.4'' \quad \text{Say } 54'' \qquad\qquad \underline{\underline{\text{OK}}}$$

2. Section 12.12.3 of the Code requires one-third of the bars to extend beyond P.I. $(1/3 \times 6 = 2 \text{ bars})$ for a distance equal to the largest of the following values (applies to "top" bars in figure).

  (a) d of web $= 24''$
  (b) $12d_b = (12)(1) = 12''$
  (c) $\left(\frac{1}{16}\right)(24 \times 12)$ for $24''$ span $= 18''$
  $\quad\ \left(\frac{1}{16}\right)(34 \times 12)$ for $34''$ span $= 25\frac{1}{2}''$

$$\underline{\underline{\text{Use } 24'' \text{ for short span and } 25\tfrac{1}{2}'' \text{ for long span}}}$$

3. Section 12.10.3 of the Code requires that the bars shall extend beyond the point where they are no longer required by moment for a distance equal to the greater of

  (a) $d = 24''$
  (b) $12d_b = (12)(1) = 12''$

4. Section 12.3.3 says required ℓ_d can be multiplied by $A_{s\,reqd}/A_{s\,furn}$. This only applies to the first bars cut off ("bottom" bars here).

$$\text{Reduced } \ell_d = (53.4)\left(\frac{4.24}{4.71}\right) = 48''$$

5. In Figure 14.36 these values are applied to the bars in question and the results are given.

The structural designer is seldom involved with a fixed-moment diagram—the loads move and the moment diagram changes. Therefore, the Code (12.10.3) says that reinforcing bars should be continued for a distance of 12 bar diameters or the effective depth of the member, whichever is greater (except at the supports of simple spans and the free ends of cantilevers), beyond their theoretical cutoff points.

As previously mentioned, the bars must be embedded a distance ℓ_d from their point of maximum stress.

Next, the Code (12.11.1) says that at least one-third of the positive steel in simple spans and one-fourth of the positive steel in continuous spans must be continued along the same face of the beam at least 6 in. into the support.

Somewhat similar rules are provided by the Code (12.12.3) for negative steel. At least one-third of the negative steel provided at a support must be extended beyond its point of inflection a distance equal to one-sixteenth of the clear span or 12 bar diameters or the effective depth of the member, whichever is greatest. Other negative bars must be extended beyond their theoretical point of cutoff by the effective depth, 12 bar diameters and at least ℓ_d from the face of the support.

Trying to go through these various calculations for cutoff or bend points for all of the bars in even a modest-sized structure can be a very large job. Therefore, the average designer or perhaps the structural draftsperson will cut off or bend bars by certain rules of thumb, which have been developed to meet the Code rules described here. In Figure 14.37 a sample set of such rules is given for continuous beams. In the *CRSI Handbook*,[10] such rules are provided for several different types of structural members, such as solid one-way slabs, one-way concrete joists, two-way slabs, and so forth.

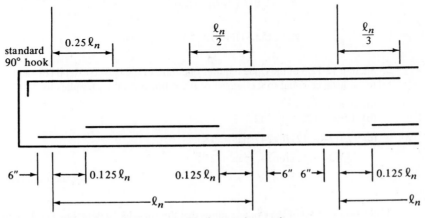

Figure 14.37 Recommended bar details for continuous beams.

---

[10]Concrete Reinforcing Steel Institute, 2002 *CRSI Handbook*, 9th ed., (Chicago), p. 12-1.

PROBLEMS

Qualitative Influence Lines

In Problems 14.1 to 14.3, draw qualitative influence lines for the functions indicated in the structures shown.

14.1 Reactions at A and B, positive moment and positive shear at X.

14.2 Reaction at B and D and negative moment and negative shear at X.

14.3 Positive moment at X, positive shear at X, negative moment just to the left of Y.

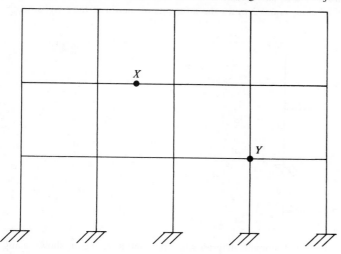

Moment Envelopes

14.4 For the continuous beam shown and for a service dead load of 2 k/ft and a service live load of 4 k/ft, draw the moment envelope using factored loads, assuming a permissible 10% up or down redistribution of the maximum negative moment. Do not use ACI coefficients.

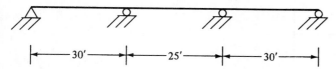

Slab Designs with ACI Coefficients

In Problems 14.5 and 14.6, design the continuous slabs shown using the ACI moment coefficients assuming that a service live load of 200 psf is to be supported in addition to the weight of the slabs. The slabs are to be built integrally with the end supports, which are spandrel beams. $f_y = 60,000$ psi, $f'_c = 4000$ psi normal-weight concrete. Clear spans are shown in the figure.

Problem 14.5 (*One ans.* $7\frac{1}{2}''$ slab, $d = 6\frac{1}{4}''$, #4 bars @ 6-1/2″ positive A_s, all 3 spans)

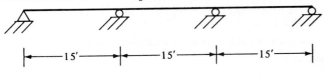

Problem 14.6

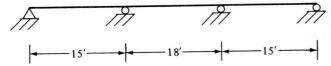

Equivalent Frame Analysis

14.7 With the equivalent rigid-frame method, draw the shear and moment diagrams for the continuous beam shown using factored loads. Service dead load including beam weight is 1.5 k/ft, and service live load is 3 k/ft. Place the live load in the center span only. Assume the moments of inertia of the T beams equal 2 times the moments of inertia of their webs. (*Ans.* max $-M = 184.8$ ft-k, max $+M = 145.2$ ft-k)

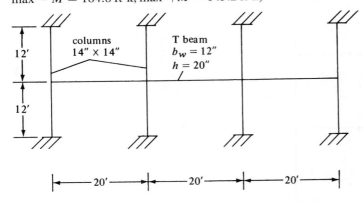

14.8 With the equivalent rigid-frame method, draw the shear and moment diagrams for the continuous beam shown using factored loads. Service dead load including beam weight is 2.4 k/ft, and service live load is 3.2 k/ft. Place the live load in spans 1 and 2 only. Assume the moments of inertia of the T beams equal 2 times the moments of inertia of their webs.

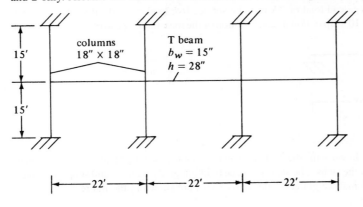

Analyses by Assuming P.I. Locations

14.9 Draw the shear and moment diagrams for member AB of the frame shown in the accompanying illustration if points of inflection are assumed to be located at $0.15L$ from each end of the span. (*Ans.* max. $M = 304$ ft-k @ ₵)

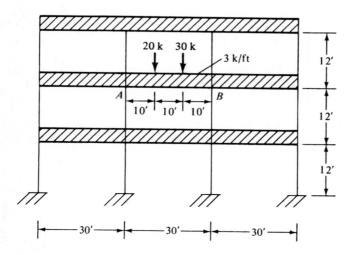

Portal Method of Analysis

In Problems 14.10 to 14.12, compute moments, shears, and axial forces for all the members of the frames shown, using the portal method.

Problem 14.10

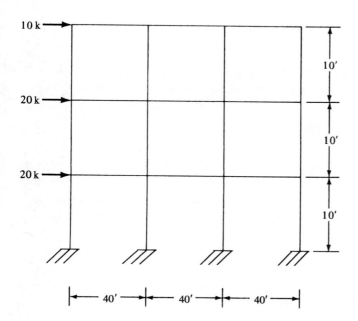

Problem 14.11   (*Ans.* for lower left column $V = 12.5$ k, $M = 75$ ft-k, and $S = 6.6$ k)

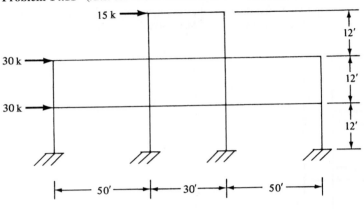

Problem 14.12

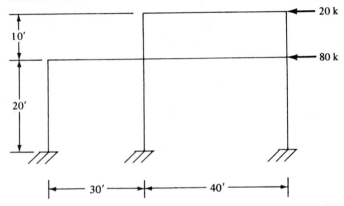

Problems in SI Units

14.13 For the continuous beam shown and for a service dead load of 26 kN/m and a service live load of 32 kN/m, draw the moment envelope, assuming a permissible 10% up or down redistribution of the maximum negative moment. (*Ans.* max $-M = 1110.7$ N.m, max $+M = 748.8$ N · m)

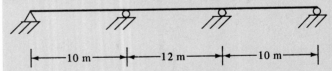

14.14 Design the continuous slab shown using the ACI moment coefficients, assuming the concrete weighs 23.5 kN/m³ and assuming that a service live load of 9 kN/m is to be supported. The slabs are built integrally with the end supports, which are spandrel beams, $f_y = 420$ MPa and $f'_c = 21$ MPa. Clear spans are given.

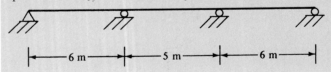

14.15 Using the equivalent rigid-frame method, draw the shear and moment diagrams for the continuous beams shown, using factored loads. Service dead load including beam weight is 10 kN/m, and service live load is 14 kN/m. Place the live load in spans 1 and 3 only. Assume the I of T beams equals 1.5 times the I of their webs. (*Ans.* max $-M = 171.7$ N $\cdot$ m, max $+M = 127.4$ N $\cdot$ m)

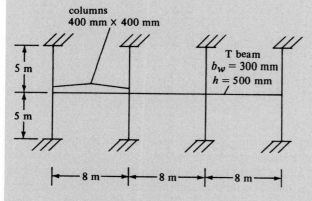

14.16 Compute moments, shears, and axial forces for all the members of the frame shown, using the portal method.

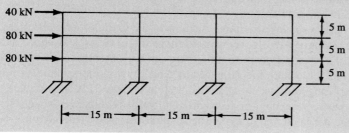

15

Torsion

15.1 INTRODUCTION

The average designer probably does not worry about torsion very much. He or she thinks almost exclusively of axial forces, shears, and bending moments, and yet most reinforced concrete structures are subject to some degree of torsion. Until recent years the safety factors required by codes for the design of reinforced concrete members for shear, moment, and so forth were so large that the effects of torsion could be safely neglected in all but the most extreme cases. Today, however, overall safety factors are less than they used to be and members are smaller, with the result that torsion is a more common problem.

Appreciable torsion does occur in many structures, such as in the main girders of bridges, which are twisted by transverse beams or slabs. It occurs in buildings where the edge of a floor slab and its beams are supported by a spandrel beam running between the exterior columns. This situation is illustrated in Figure 15.1, where the floor beams tend to twist the spandrel beam laterally. Earthquakes can cause dangerous torsional forces in all buildings. This is particularly true in asymmetrical structures where the centers of mass and rigidity do not coincide. Other cases where torsion may be significant are in curved bridge girders, spiral stairways, balcony girders, and whenever large loads are applied to any beam "off center." An off-center case where torsional stress can be very large is illustrated in Figure 15.2. It should be realized that if the supporting member is able to rotate, the resulting torsional stresses will be fairly small. If, however, the member is restrained, the torsional stresses can be quite large.

Should a plain concrete member be subjected to pure torsion, it will crack and fail along 45° spiral lines due to the diagonal tension corresponding to the torsional stresses. For a very effective demonstration of this type of failure, you can take a piece of chalk in your hands and twist it until it breaks. Although the diagonal tension stresses produced by twisting are very similar to those caused by shear, they will occur on all faces of a member. As a result, they add to the stresses caused by shear on one side and subtract from them on the other.[1]

In recent years there have been more reports of structural failures attributed to torsion. As a result, a rather large amount of research has been devoted to the subject, and thus there is a much improved understanding of the behavior of structural members subjected to torsion. On the basis of this rather extensive experimental work, the ACI Code includes very specific requirements for the design of reinforced concrete members subjected to torsion or to torsion combined with shear and bending. It should be realized that maximum shears and torsional forces may occur in areas where bending moments are small. For such cases, the interaction of shear and torsion can be particularly important as it relates to design.

---

[1]White, R. N., Gergely, P., and Sexsmith, R. G., 1974, *Structural Engineering*, vol. 3 (New York: John Wiley & Sons), pp. 423–424.

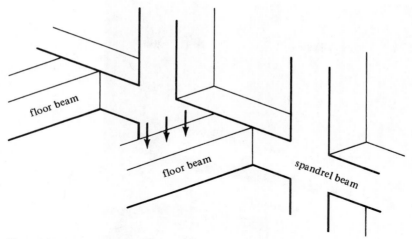

Figure 15.1 Torsion in spandrel beams.

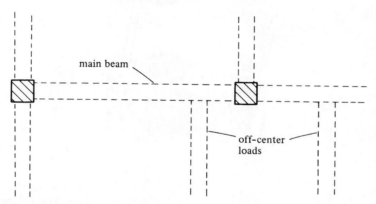

Figure 15.2 Off-center loads causing torsion in main beam.

15.2   TORSIONAL REINFORCING

Reinforced concrete members subjected to large torsional forces may fail quite suddenly if they are not specially provided with torsional reinforcing. The addition of torsional reinforcing does not change the magnitude of the torsion that will cause diagonal tension cracks, but it does prevent the members from tearing apart. As a result, they will be able to resist substantial torsional moments without failure. Tests have shown that both longitudinal bars and closed stirrups (or spirals) are necessary to intercept the numerous diagonal tension cracks that occur on all surfaces of members subject to appreciable torsional forces.

The normal ⊔-shaped stirrups are not satisfactory. They must be closed either by welding their ends together to form a continuous loop, as illustrated in Figure 15.3(a), or by bending their ends around a longitudinal bar, as shown in part (b) of the same figure. If one-piece stirrups such as these are used, the entire beam cage may have to be prefabricated and placed as a unit (and that may not be feasible if the longitudinal bars have to be passed between column bars) or the longitudinal bars will have to be threaded one by one through the closed stirrups and perhaps the column bars. It is easy to see that some other arrangement is usually desirable.

Elevated highway bridge, Wichita
Falls, Texas. (Courtesy of EFCO.)

In the recent past it was rather common to use two overlapping ⊔ stirrups arranged as shown in
Figure 15.4. Although this arrangement simplifies the placement of longitudinal bars, it has not
proved very satisfactory. As described in the ACI Commentary (R11.5.4.1), members primarily
subjected to torsion lose their concrete side cover by spalling at high torques. Should this happen,
the lapped spliced ⊔ stirrups of Figure 15.4 will prove ineffective, and premature torsion failure
may occur.

A much better type of torsion reinforcement consists of ⊔ stirrups, each with a properly
anchored top bar, such as the ones shown in Figure 15.5. It has been proved by testing that the use

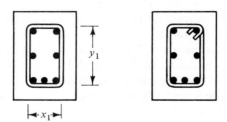

(a) (b)

Figure 15.3 Closed stirrups (these types frequently
impractical).

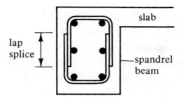

Figure 15.4 Overlaping ⊔ stirrups used as torsion reinforcing *but not desirable.*

of torsional stirrups with 90° hooks results in spalling of the concrete outside the hooks. The use of 135° hooks for both the ⊔ stirrups and the top bars is very helpful in reducing this spalling.

Should lateral confinement of the stirrups be provided as shown in parts (b) and (c) of Figure 15.5, the 90° hooks will be satisfactory for the top bars.

Should beams with wide webs (say, 2 ft or more) be used, it is common to use multiple-leg stirrups. For such situations the outer stirrup legs will be proportioned for both shear and torsion, while the interior legs will be designed to take only vertical shear.

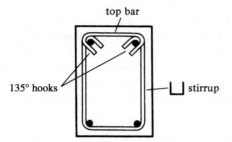

(a) No confining concrete either side—thus 135° hooks needed for both ends of top bar.

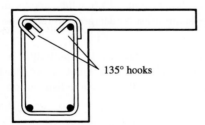

(b) Lateral confinement provided by slab on right-hand side— thus 90° hook permissible for top bar on that side.

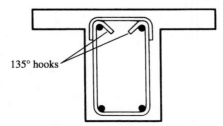

(c) Lateral confinement provided on both sides by concrete slab—thus 90° hooks permissible at both ends of top bar.

Figure 15.5 Recommended torsion reinforcement.

The strength of closed stirrups cannot be developed unless additional longitudinal reinforcing is supplied. Longitudinal bars should be spaced uniformly around the insides of the stirrups, not more than 12 in. apart. There must be at least one bar in each corner of the stirrups to provide anchorage for the stirrup legs (Code 11.6.6.2); otherwise, if the concrete inside the corners were to crush, the stirrups would slip and the result would be even larger torsional cracks. These longitudinal bars must have diameters at least equal to 0.042 times the stirrup spacing. Their size may not be less than #3.

15.3 THE TORSIONAL MOMENTS THAT HAVE TO BE CONSIDERED IN DESIGN

The reader is well aware from his or her studies of structural analysis that if one part of a statically indeterminate structure "gives" when a particular force is applied to that part, the amount of force that the part will have to resist will be appreciably reduced. For instance, if three men are walking along with a log on their shoulders (a statically indeterminate situation) and one of them lowers his shoulder a little under the load, there will be a major redistribution of the internal forces in the "structure" and a great deal less load for him to support. On the other hand, if two men are walking along with a log on their shoulders (a statically determinate situation) and one of them lowers his shoulder slightly, there will be little change in force distribution in the structure. These are similar to the situations that occur in statically determinate and indeterminate structures subject to torsional moments. They are referred to respectively as *equilibrium torsion* and *compatibility torsion*.

1. *Equilibrium torsion.* For a statically determinate structure, there is only one path along which a torsional moment can be transmitted to the supports. This type of torsional moment, which is referred to as *equilibrium torsion* or *statically determinate torsion*, cannot be reduced by a redistribution of internal forces or by a rotation of the member. Equilibrium torsion is illustrated in Figure 15.6, which shows an edge beam supporting a concrete canopy. The edge beam must be designed to resist the full calculated torsional moment.

2. *Compatibility torsion.* The torsional moment in a particular part of a statically indeterminate structure may be substantially reduced if that part of the structure cracks under the torsion and "gives" or rotates. The result will be a redistribution of forces in the structure.

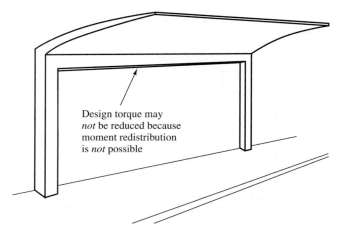

Design torque may *not* be reduced because moment redistribution is *not* possible

Figure 15.6 Equilibrium torsion.

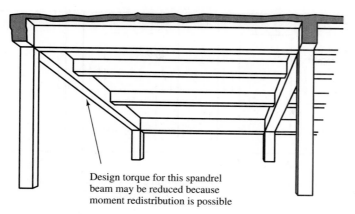

Design torque for this spandrel
beam may be reduced because
moment redistribution is possible

Figure 15.7 Compatibility
torsion.

This type of torsion, which is illustrated in Figure 15.7, is referred to as *statically indeterminate torsion* or *compatibility torsion,* in the sense that the part of the structure in question twists in order to keep the deformations of the structure compatible.

Where a reduction or redistribution of torsion is possible in a statically indeterminate structure, the maximum factored moment T_u can be reduced as follows for nonprestressed members according to ACI Section 11.6.2.2. In the expression to follow, A_{cp} is the area enclosed by the outside perimeter of the concrete cross section and p_{cp} is the outside perimeter of that cross section. It is assumed that torsional cracking will occur when the principal tension stress reaches the tensile strength of the concrete in biaxial tension-compression. This cracking value is taken to be $4\lambda\sqrt{f_c'}$ and thus the torque at cracking, T_{cr}, is

$$T_{cr} = \phi 4\lambda\sqrt{f_c'}\left(\frac{A_{cp}^2}{p_{cp}}\right)$$

When reinforced concrete members are subjected to axial tensile or compressive forces, T_{cr} is to be computed with the expression to follow in which N_u is the factored axial force taken as positive if the force is compressive and negative if it is tensile.

$$T_{cr} = \phi 4\lambda\sqrt{f_c'}\left(\frac{A_{cp}^2}{p_{cp}}\right)\lambda\sqrt{1 + \frac{N_u}{4\sqrt{f_c'}}}$$

After cracking occurs, the torsional moments in the spandrel beam shown in Figure 15.7 are reduced as a result of the redistribution of the internal forces. Consequently, the torsional moment used for design in the spandrel beam can be reduced.

15.4   TORSIONAL STRESSES

As previously mentioned, the torsional stresses add to the shear stresses on one side of a member and subtract from them on the other. This situation is illustrated for a hollow beam in Figure 15.8.

Torsional stresses are quite low near the center of a solid beam. Because of this, hollow beams (assuming the wall thicknesses meet certain ACI requirements) are assumed to have almost exactly the same torsional strengths as solid beams with the same outside dimensions.

In solid sections the shear stresses due to the torsion T_u are concentrated in an outside "tube" of the member, a shown in Figure 15.9(a), while the shear stresses due to V_u are spread across the

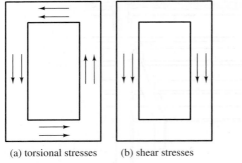

(a) torsional stresses   (b) shear stresses

Figure 15.8 Torsion and shear stresses in a hollow beam.

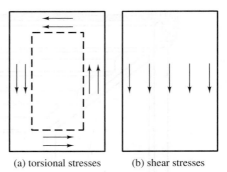

(a) torsional stresses   (b) shear stresses

Figure 15.9 Torsion and shear stresses in a solid beam.

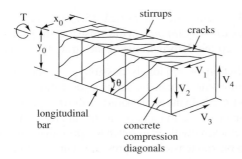

Figure 15.10 Imaginary space truss.

width of the solid section, as shown in part (b) of the figure. As a result, the two types of shear stresses (due to shear and torsion) are combined, using a square root expression shown in the next section of this chapter.

After cracking, the resistance of concrete to torsion is assumed to be negligible. The torsion cracks tend to spiral around members (hollow or solid) located at approximately 45° angles with the longitudinal edges of those members. Torsion is assumed to be resisted by an imaginary space truss located in the outer "tube" of concrete of the member. Such a truss is shown in Figure 15.10. The longitudinal steel in the corners of the member and the closed transverse stirrups act as tension members in the "truss," while the diagonal concrete between the stirrups acts as struts of compression members. The cracked concrete is still capable of taking compression stresses.

15.5   WHEN TORSIONAL REINFORCING IS REQUIRED BY THE ACI

The design of reinforced concrete members for torsion is based on a thin-walled tube space truss analogy in which the inside or core concrete of the members is neglected. After torsion has caused a member to crack, its resistance to torsion is provided almost entirely by the closed stirrups and the longitudinal reinforcing located near the member surface. Once cracking occurs, the concrete is assumed to have negligible torsional strength left. (This is not the case in shear design, where the concrete is assumed to carry the same amount of shear as it did before cracking.)

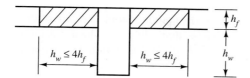

Figure 15.11 Portions of monolithic *T* beam that may be used for torsion calculations.

If torsional stresses are less than about one-fourth of the cracking torque T_{cr} of a member, they will not appreciably reduce either its shear or flexural strengths. Torsional cracking is assumed to occur when the principal tensile stress reaches $4\lambda\sqrt{f_c'}$. In ACI Section 11.6.1 it is stated that torsion effects may be neglected for nonprestressed members if

$$T_u < \phi\lambda\sqrt{f_c'}\left(\frac{A_{cp}^2}{p_{cp}}\right) = \frac{1}{4}T_{cr}$$

For statically indeterminate structures where reductions in torsional moments can occur due to redistribution of internal forces, Section 11.6.2.2 of the ACI Code permits the maximum factored torsional moment to be reduced to the following value:

$$\phi 4\lambda\sqrt{f_c'}\left(\frac{A_{cp}^2}{p_{cp}}\right)$$

In other words, the applied torque may be limited to a member's calculated cracking moment. (If the computed torque for a particular member is larger than the above value, the above value may be used in design.) Should the torsional moments be reduced as described here, it will be necessary to redistribute these moments to adjoining members. The ACI Commentary (R11.5.2.1 and R11.5.2.2) does say that when the layout of structures is such as to impose significant torsional rotations within a short length of a member (as where a large torque is located near a stiff column), a more exact analysis should be used.

For isolated members with or without flanges, A_{cp} equals the area of the entire cross sections (including the area of any voids in hollow members) and p_{cp} represents the perimeters of the entire cross sections. Should a beam be cast monolithically with a slab, the values of A_{cp} and p_{cp} may be assumed to include part of the adjacent slabs of the resulting T- or L-shaped sections. The widths of the slabs that may be included as parts of the beams are described in ACI Section 13.2.4 and illustrated in Figure 15.11. Those widths or extensions may not exceed the projections of the beams above or below the slab or four times the slab thickness, whichever is smaller.

When appreciable torsion is present, it may be more economical to select a larger beam than would normally be selected so that torsion reinforcing does not have to be used. Such a beam may very well be more economical than a smaller one with the closed stirrups and additional longitudinal steel required for torsion design. On other occasions such a practice may not be economical, and sometimes architectural considerations may dictate the use of smaller sections.

15.6   TORSIONAL MOMENT STRENGTH

The sizes of members subject to shear and torsion are limited by the ACI Code so that unsightly cracking is reduced and crushing of the surface concrete caused by inclined compression stresses is prevented. This objective is accomplished with the equations that follow, in which the left-hand portions represent the shear stresses due to shear and torsion. The sum of these two stresses in a particular member may not exceed the stress that will cause shear cracking ($8\sqrt{f_c'}$ as per ACI R11.6.3). In these expressions, $V_c = 2\lambda\sqrt{f_c'}b_w d$ (ACI Equation 11-3). For solid sections

$$\sqrt{\left(\frac{V_u}{b_w d}\right)^2 + \left(\frac{T_u p_h}{1.7 A_{oh}^2}\right)^2} \le \phi\left(\frac{V_c}{b_w d} + 8\sqrt{f_c'}\right) \qquad \text{(ACI Equation 11-18)}$$

For hollow sections

$$\left(\frac{V_u}{b_w d}\right) + \left(\frac{T_u p_h}{1.7 A_{oh}^2}\right) \le \phi\left(\frac{V_c}{b_w d} + 8\sqrt{f_c'}\right) \qquad \text{(ACI Equation 11-19)}$$

Should the wall thickness of a hollow section be less than A_{oh}/p_h, the second term in ACI Equation 11-19 is to be taken not as $T_u p_h/1.7 A_{oh}^2$ but as $T_u/1.7 A_{oh}$, where t is the thickness of the wall where stresses are being checked (ACI 11.5.3.3).

Another requirement given in ACI Section 11.5.4.4 for hollow sections is that the distance from the centerline of the transverse torsion reinforcing to the inside face of the wall must not be less than $0.5 A_{oh}/p_h$. In this expression, p_h is the perimeter of the centerline of the outermost closed torsional reinforcing, while A_{oh} is the cross-sectional area of the member that is enclosed within this centerline. The letters *oh* stand for *outside hoop* (of stirrups).

15.7 DESIGN OF TORSIONAL REINFORCING

The torsional strength of reinforced concrete beams can be greatly increased by adding torsional reinforcing consisting of closed stirrups and longitudinal bars. If the factored torsional moment for a particular member is larger than the value given in ACI Section 11.5.1 [$\phi\lambda\sqrt{f_c'}(A_{cp}^2/P_h)$], the Code provides an expression to compute the absolute minimum area of transverse closed stirrups that may be used.

$$(A_v + 2A_t) = 0.75\sqrt{f_c'}\,\frac{b_w s}{f_{yt}} \ge \frac{50 b_w s}{f_{yt}} \qquad \text{(ACI Equation 11-23)}$$

In this expression, A_v is the area of reinforcing required for shear in a distance s (which represents the stirrup spacing). You will remember from shear design that the area A_v obtained is for both legs of a two-legged stirrup (or for all legs of a four-legged stirrup, etc.). The value A_t, which represents the area of the stirrups needed for torsion, is for only one leg of the stirrup. Therefore the value $A_v + 2A_t$ is the total area of both legs of the stirrup (for two-legged stirrups) needed for shear plus torsion. It is considered desirable to use equal volumes of steel in the stirrups and the added longitudinal steel so that they will participate equally in resisting torsional moments. This theory was followed in preparing the ACI equations used for selecting torsional reinforcing. The ACI Code requires that the area of stirrups A_t used for resisting torsion be computed with the equation that follows:

$$T_n = \frac{2A_o A_t f_{yt}}{s}\cot\theta \qquad \text{(ACI Equation 11-21)}$$

This equation is usually written in the following form:

$$\frac{A_t}{s} = \frac{T_n}{2A_o f_{yt}\cot\theta}$$

The transverse reinforcing is based on the torsional moment strength T_n, which equals T_u/ϕ. The term A_o represents the gross area enclosed by the shear flow path around the perimeter of the

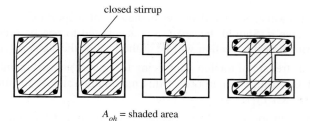

A_{oh} = shaded area

Figure 15.12 Values of A_{oh}.

tube. This area is defined in terms of A_{oh}, which is the area enclosed by the outermost closed hoops. Figure 15.12 illustrates this definition of A_{oh} for several beam cross sections.

The value of A_o may be determined by analysis, or it may be taken as $0.85A_{oh}$. The term θ represents the angle of the concrete "compression diagonals" in the analogous space truss. It may not be smaller than 30° or larger than 60°, and it may be taken equal to 45°, according to ACI Section 11.5.3.6. In ACI Section 11.5.3.6, the value of θ may be 45° for nonprestressed members, and that practice is followed herein. Suggested θ values for prestressed concrete are given in this same ACI section.

As given in the ACI Commentary (R11.5.3.8), the required stirrup areas for shear and torsion are added together as follows for a two-legged stirrup:

$$\text{Total}\left(\frac{A_{v+t}}{s}\right) = \frac{A_v}{s} + \frac{2A_t}{s}$$

The spacing of transverse torsional reinforcing may not be larger than $p_h/8$ or 12 in., where p_h is the perimeter of the centerline of the outermost closed transverse reinforcing (ACI 11.5.6.1). Remember also the maximum spacings of stirrups for shear $\frac{d}{2}$ and $\frac{d}{4}$ given in ACI Sections 11.4.5.1 and 11.4.5.3.

It has been found that reinforced concrete specimens with less than about 1% torsional reinforcing by volume which are loaded in pure torsion fail as soon as torsional cracking occurs. The percentage is smaller for members subject to both torsion and shear. The equation to follow, which provides a minimum total area of longitudinal torsional reinforcing, is based on using about 0.5% torsional reinforcing by volume. In this expression, A_{cp} is the area enclosed by the outside concrete cross section. The value A_t/s may not be taken as less than $25b_w/f_{yv}$, according to ACI Section 11.5.5.3.

$$A_{\ell\,\min} = \frac{5\sqrt{f'_c}A_{cp}}{f_y} - \left(\frac{A_t}{s}\right)p_h\frac{f_{yt}}{f_y} \qquad \text{(ACI Equation 11-24)}$$

The longitudinal torsion reinforcement must be developed at both ends, states ACI Section 11.5.4.3.

Maximum torsion generally acts at the ends of beams, and as a result the longitudinal torsion bars should be anchored for their yield strength at the face of the supports. To do this it may be necessary to use hooks or horizontal ⊔-shaped bars lap spliced with the longitudinal torsion reinforcing. *A rather common practice is to extend the bottom reinforcing in spandrel beams subjected to torsion 6 in. into the supports. Usually this is insufficient.*

15.8 ADDITIONAL ACI REQUIREMENTS

Before we present numerical examples for torsion design, it is necessary to list several other ACI requirements. These are:

1. Sections located at a distance less than d from the face of support may be designed for the torque at a distance d. Should, however, a concentrated torque be present within this distance, the critical design section will be at the face of the support (ACI 11.5.2.4).

2. The design yield strength of torsion reinforcing for nonprestressed members may not be greater than 60,000 psi. The purpose of this maximum value is to limit the width of diagonal cracks (ACI 11.5.3.4).

3. The longitudinal tension created by torsion moments is partly offset in the flexural compression zones of members. In these zones the computed area of longitudinal torsional reinforcing may be reduced by an amount equal to $M_u/0.9df_y$, according to ACI Section 11.5.3.9. In this expression, M_u is the factored moment acting at the section in combination with T_u. The reinforcing provided, however, may not be less than the minimum values required in ACI Sections 11.5.5.3 and 11.5.6.2.

4. The longitudinal reinforcing must be distributed around the inside perimeter of the closed stirrups and must be spaced no farther apart than 12 in. At least one bar must be placed in each corner of the stirrups to provide anchorage for the stirrup legs. These bars have to be #3 or larger in size and they must have diameters no less than 0.042 times the stirrup spacings (ACI 11.5.6.2).

5. Torsional reinforcing must be provided for a distance no less than $b_t + d$ beyond the point where it is theoretically no longer required. The term b_t represents the width of that part of the member cross section which contains the closed torsional stirrups (ACI 11.5.6.3).

15.9 EXAMPLE PROBLEMS USING U.S. CUSTOMARY UNITS

In this section the design of torsional reinforcing for a beam is presented using U.S. customary units; an example using SI units is presented in the next section.

EXAMPLE 15.1

Design the torsional reinforcing needed for the beam shown in Figure 15.13 if $f'_c = 4000$ psi, $f_y = 60,000$ psi, $T_u = 30$ ft-k, and $V_u = 60$ k. Assume 1.5 in. clear cover, #4 stirrups, and a required A_s for M_u of 3.52 in.^{2} Select #8 bars for flexural reinforcing. Normal-weight concrete is specified.

SOLUTION

1. Is torsion reinforcing needed?

$$A_{cp} = \text{area enclosed by outside perimeter of}$$
$$\text{concrete cross section} = (16)(26) = 416 \text{ in.}^2$$
$$\text{(the letter } c \text{ stands for concrete, and the}$$
$$\text{latter } p \text{ stands for perimeter of cross section)}$$

$$p_{cp} = \text{outside perimeter of the cross section}$$
$$= (2)(16+26) = 84 \text{ in.}$$

Torsion T_u can be neglected if less than

$$\phi\lambda\sqrt{f'_c}\frac{A_{cp}^2}{p_{cp}} = (0.75)(1.0)(\sqrt{4000})\left(\frac{416^2}{84}\right)$$
$$= 97,723 \text{ in.-lb} = 97.72 \text{ in.-k} < 30 \times 12 = 360 \text{ in.-k}$$

$$\therefore \text{ Torsion reinforcing is required}$$

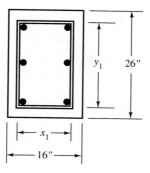

Note x_1 and y_1 run from c.g. of stirrup on one edge to c.g. of stirrup on other edge

Figure 15.13

2. Compute sectional properties

$$A_{oh} = \text{area enclosed by centerline of the outermost closed stirrups}$$

Noting 1.5 in. clear cover and #4 stirrups

$$\left.\begin{aligned}x_1 &= 16 - (2)(1.5 + 0.25) = 12.5 \text{ in.}\\ y_1 &= 26 - (2)(1.5 + 0.25) = 22.5 \text{ in.}\end{aligned}\right\} \text{See Figure 15.13}$$

$$A_{oh} = (12.5)(22.5) = 281.25 \text{ in.}^2$$

$$A_o = \text{gross area enclosed by shear flow path}$$

$$= 0.85 A_{oh} \text{ from ACI Section 11.5.3.6}$$

$$= (0.85)(281.25) = 239.06 \text{ in.}^2$$

$$d = \text{effective depth of beam}$$

$$= 26 - 1.50 - 0.50 - \frac{1.00}{2} = 23.50 \text{ in.}$$

$$p_h = \text{perimeter of centerline of outermost closed torsional reinforcing}$$

$$= (2)(x_1 + y_1)$$

$$= (2)(12.5 + 22.5) = 70 \text{ in.}$$

3. Is the concrete section sufficiently large to support the torsion?

$$V_c = \text{nominal shear strength of concrete section}$$

$$= 2\lambda\sqrt{f_c'}b_w d = (2)(1.0)(\sqrt{4000})(16)(23.50) = 47{,}561 \text{ lb}$$

Applying ACI Equation 11-18

$$\sqrt{\left(\frac{V_u}{b_w d}\right)^2 + \left(\frac{T_u p_h}{1.7 A_{oh}^2}\right)^2} \leq \phi\left(\frac{V_c}{b_w d} + 8\sqrt{f_c'}\right)$$

$$\sqrt{\left(\frac{60{,}000}{16 \times 23.50}\right)^2 + \left(\frac{360{,}000 \times 70}{1.7 \times 281.25^2}\right)^2} \leq 0.75\left(\frac{47{,}561}{16 \times 23.50} + 8\sqrt{4000}\right)$$

$$246 \text{ psi} < 474 \text{ psi}$$

$$\therefore \text{Section is sufficiently large}$$

4. Determine the transverse torsional reinforcing required

$$T_n = \frac{T_u}{\phi} = \frac{30}{0.75} = 40 \text{ ft-k} = 480{,}000 \text{ in.-lb}$$

Assuming $\theta = 45°$ as per ACI Section 11.5.3.6(a).

$$\frac{A_t}{s} = \frac{T_n}{2A_o f_{yt} \cot \theta} = \frac{480,000}{(2)(239.06)(60,000)(\cot 45°)}$$

$$= 0.0167 \text{ in.}^2/\text{in. for one leg of stirrups}$$

5. **Calculate the area of shear reinforcing required**

$$V_u = 60,000 \text{ lbs} > \frac{1}{2}\phi V_c = \left(\frac{1}{2}\right)(0.75)(47,561) = 17,835 \text{ lb}$$

∴ Shear reinforcing is required

$$V_s = \frac{V_u - \phi V_c}{\phi} = \frac{60,000 - (0.75)(1.0)(47,561)}{0.75} = 32,439 \text{ lb}$$

Applying ACI Equation 11-15

$$\frac{A_v}{s} = \frac{V_s}{f_{yt} d} = \frac{32,439}{(60,000)(23.50)}$$

$$= 0.0230 \text{ in.}^2/\text{in. for 2 legs of stirrup}$$

6. **Select stirrups**

Total web reinforcing required for 2 legs $= \dfrac{A_{v+t}}{s}$

$$= \frac{2A_t}{s} + \frac{A_v}{s} = (2)(0.0167) + 0.0230$$

$$= 0.0564 \text{ in.}^2/\text{in. for 2 legs}$$

Using #4 stirrups

$$s = \frac{(2)(0.20)}{0.0564} = 7.09 \text{ in. o.c.}$$

Say 7 in.

Maximum allowable spacing of stirrups

$$\text{from ACI Section 11.5.6.1} = \frac{p_h}{8} = \frac{70}{8} = \text{Say 8.75 in. or 12 in.}$$

Minimum area of stirrups A_v by ACI Equation 11-23

$$A_v + 2A_t = 0.75\sqrt{f_c'} \frac{b_w s}{f_{yt}} < \frac{50 b_w s}{f_y} = \frac{(50)(16)(7)}{60,000} = 0.0933 \text{ in}$$

$$= 0.75\sqrt{4000} \frac{(16)(7)}{60,000} = 0.089 \text{ in.}^2$$

$$< (2)(0.20) = 0.40 \text{ in.}$$

OK

Use #4 stirrups @ 7 in.

7. **Selection of longitudinal torsion reinforcing**

Additional longitudinal reinforcing required for torsion

$$A_\ell = \frac{A_t}{s} p_h \left(\frac{f_{yt}}{f_y}\right) \cot^2 \theta \qquad \text{(ACI Equation 11-22)}$$

$$= (0.0167)(70)\left(\frac{60,000}{60,000}\right)(1.00)^2 = 1.17 \text{ in.}^2$$

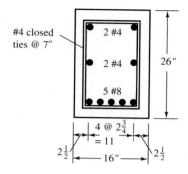

#4 closed
ties @ 7"

2 #4

2 #4

5 #8

26"

4 @ $2\frac{3}{4}$
= 11

$2\frac{1}{2}$

$2\frac{1}{2}$

16"

Figure 15.14 Reinforcing selected for beam of Example 15.1.

$$\text{Min. } A_\ell = \frac{5\sqrt{f_c'}A_{cp}}{f_y} - \left(\frac{A_t}{s}\right)p_h\frac{f_{yt}}{f_y}$$

(ACI Equation 11-24)

$$\left[\text{with min } \frac{A_t}{s} = \frac{25b_w}{f_{yv}} = \frac{(25)(16)}{60,000} = 0.00667\right]$$

$$= \frac{(5\sqrt{4000})(416)}{60,000} - (0.0167)(70)\left(\frac{60,000}{60,000}\right)$$

$$= 1.02 \text{ in.}^2 < 1.17 \text{ in.}^2$$

OK

This additional longitudinal steel is spread out into three layers to the four inside corners of the stirrups and vertically in between, with a spacing not in excess of 12 in.

Assume one-third in top = $1.17/3 = 0.39$ in.^{2}, $0.39 + 3.52 = 3.91$ in.^{2} in bottom, and remainder 0.39 in.^{2} in between.

Use two #4 bars (0.40 in.^{2}) in top corner, two #4 bars in the middle, and five #8 bars (3.93 in.^{2}) at the bottom, as shown in Figure 15.14.

15.10   SI EQUATIONS AND EXAMPLE PROBLEM

The SI equations necessary for the design of reinforced concrete members which are different from those needed for similar designs using U.S. customary units are listed here, and their use is illustrated in Example 15.2.

Torsional moments can be neglected if

$$T_u \le \frac{\phi\lambda\sqrt{f_c'}A_{cp}^2}{12}\frac{}{p_{cp}}$$

(ACI Section 11.6.1[a])

$$V_c = \frac{\lambda\sqrt{f_c'}}{6}b_wd$$

(ACI Equation 11-3)

Maximum torsional moment strength for solid sections

$$\sqrt{\left(\frac{V_u}{b_wd}\right)^2 + \left(\frac{T_up_h}{1.7A_{oh}^2}\right)^2} \le \phi\left(\frac{V_c}{b_wd} + \frac{2\sqrt{f_c'}}{3}\right)$$

(ACI Equation 11-18)

Maximum torsional moment strength for hollow sections

$$\frac{V_u}{b_wd} + \frac{T_up_h}{1.7A_{oh}^2} \le \phi\left(\frac{V_c}{b_wd} + \frac{2\sqrt{f_c'}}{3}\right)$$

(ACI Equation 11-19)

The expression used to calculate the required area of torsional reinforcing needed for 1 leg of stirrups

$$T_n = \frac{2A_o A_t f_{yt}}{s} \cot\theta \qquad \text{(ACI Equation 11-21)}$$

from which

$$\frac{A_t}{s} = \frac{T_n}{2A_o f_{yv} \cot\theta}$$

Minimum area of transverse reinforcing required

$$A_v + 2A_t = \frac{1}{16}\sqrt{f_c'}\frac{b_w s}{f_{yt}}$$

$$\geq \frac{0.33 b_w s}{f_{yt}} \qquad \text{(ACI Equation 11-23)}$$

Additional longitudinal reinforcing required for torsion

$$A_\ell = \frac{A_t}{s} p_h \left(\frac{f_{yt}}{f_y}\right)\cot^2\theta \qquad \text{(ACI Equation 11-22)}$$

Minimum total area of additional longitudinal reinforcing required

$$A_{\ell,\,min} = \frac{5\sqrt{f_c'}A_{cp}}{12 f_y} - \frac{A_t}{s}p_h\frac{f_{yt}}{f_y} \qquad \text{(ACI Equation 11-24)}$$

In this last expression, the value of (A_t/s) may not be taken as less than $(b_w/6f_{yv})$. Other SI requirements for design of torsional reinforcing are:

1. Max spacing permitted for transverse torsional reinforcing $= \frac{1}{8}p_h$ or 300 mm or $\frac{d}{2}$ or $\frac{d}{4}$ as required for shear design (ACI 11.5.6.1 and 11.4.4).
2. The diameter of stirrups may not be less than 0.042 times their spacing, and stirrups smaller than #10 may not be used. (ACI Section 11.5.6.2.)
3. Maximum yield stresses f_{yv} or $f_{y\ell}$=420 Mpa $\qquad$ (ACI Section 11.5.3.4)

Example 15.2 illustrates the design of the torsional reinforcing for a member using SI units.

EXAMPLE 15.2

Design the torsional reinforcing for the beam shown in Figure 15.15, for which $f_c' = 28$ MPa, $f_y = 420$ MPa, $V_u = 190$ kN, $T_u = 30$ kN-m, and A_s required for M_u is 2050 mm². Assume #13 stirrups and a clear cover equal to 40 mm.

SOLUTION
1. **Is torsion reinforcing necessary?**

$$A_{cp} = (350)(650) = 227\,500 \text{ mm}^2$$

$$p_{cp} = (2)(350 + 650) = 2000 \text{ mm}$$

$$\frac{\phi\sqrt{f_c'}}{12}\frac{A_{cp}^2}{p_{cp}} = \frac{(0.75)\sqrt{28}}{12}\frac{(227\,500)^2}{2000} \qquad \text{[ACI Equation 11.5.1(a)]}$$

$$= 8.558 \times 10^6 \text{ N} \cdot \text{mm} = 8.56 \text{ kN} \cdot \text{m} < 30 \text{ kN} \cdot \text{m}$$

∴ Torsion reinforcing is required

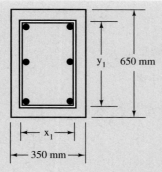

Figure 15.15

2. **Compute sectional properties**

   With 40-mm clear cover and #13 stirrups (diameter = 12.7 mm)

   $$x_1 = 350 - (2)\left(40 + \frac{12.7}{2}\right) = 257.3 \text{ mm}$$

   $$y_1 = 650 - (2)\left(40 + \frac{12.7}{2}\right) = 557.3 \text{ mm}$$

   $$A_{oh} = (257.3)(557.3) = 143\,393 \text{ mm}^2$$

   $$A_o = 0.85 A_{oh} = (0.85)(143\,393) = 121\,884 \text{ mm}^2$$

   Assuming bottom reinforcing consists of #25 bars (diameter = 25.4 mm)

   $$d = 650 - 40 - 12.7 - \frac{25.4}{2} = 584.6 \text{ mm}$$

   $$p_h = 2(x_1 + y_1) = (2)(257.3 + 557.3) = 1629 \text{ mm}$$

3. **Is the concrete section sufficiently large to support T_u?**

   $$V_c = \frac{\lambda \sqrt{f_c'}}{6} b_w d = (1.0)\frac{\sqrt{28}}{6}(350)(584.6) = 180\,449 \text{ N} \qquad \text{(ACI Equation 11-3)}$$

   $$= 180.45 \text{ kN}$$

   $$\sqrt{\left(\frac{V_u}{b_w d}\right)^2 + \left(\frac{T_u p_h}{1.7 A_{oh}^2}\right)^2} \le \phi\left(\frac{V_c}{b_w d} + \frac{2\sqrt{f_c'}}{3}\right) \qquad \text{(ACI Equation 11-18)}$$

   $$\sqrt{\left(\frac{190 \times 10^3}{350 \times 584.6}\right)^2 + \left(\frac{30 \times 10^6 \times 1629}{1.7(143\,393)^2}\right)^2} < 0.75\left(\frac{180.45 \times 10^3}{350 \times 584.6} + \frac{2\sqrt{28}}{3}\right)$$

   $$1.678 \text{ N/mm}^2 < 3.307 \text{ N/mm}^2$$

   $$\therefore \text{Section is sufficiently large}$$

4. **Determine the transverse torsional reinforcing required**

   $$T_n = \frac{T_u}{\phi} = \frac{30}{0.75} = 40 \text{ kN} \cdot \text{m}$$

   Assuming $\theta = 45°$ (as per ACI 11.5.3.6(a))

$$\frac{A_t}{s} = \frac{T_n}{2A_o f_{yt} \cot \theta} = \frac{40 \times 10^6}{(2)(121\,884)(420)(1.0)}$$ (ACI Equation 11-21)

$$= 0.391 \text{ mm}^2/\text{mm for 1 leg of stirrup}$$

5. Calculate the area of shear reinforcing required

$$V_u = 190 \text{ kN} > \frac{1}{2} V_c = \left(\frac{1}{2}\right)(180.45) = 90.22 \text{ kN}$$

∴ Shear reinforcing is required

$$V_s = \frac{V_u - \phi V_c}{\phi} = \frac{190 - (0.75)(180.45)}{0.75} = 72.88 \text{ kN}$$

$$\frac{A_v}{s} = \frac{V_s}{f_y d} = \frac{72.88 \times 10^3}{(420)(584.6)} = 0.297 \text{ mm}^2/\text{mm for 2 legs of stirrup}$$

6. Select stirrups

$$\frac{2A_t}{s} + \frac{A_v}{s} = (2)(0.391) + 0.297 = 1.079 \text{ mm}^2/\text{mm for 2 legs of stirrup}$$

Using #13 stirrup ($A_s = 129 \text{ mm}^2$)

$$s = \frac{(2)(129)}{1.079} = 239 \text{ mm}$$

Maximum allowable spacing of stirrup

$$= \frac{p_h}{8} = \frac{1629}{8} = 204 \text{ mm}$$

Use 200 mm

Minimum area of stirrup

$$A_v + 2A_t = \frac{1}{16}\sqrt{28}\frac{(350)(200)}{420} = 55.11 \text{ mm}^2$$

$$\leq \frac{1}{3}\frac{b_w s}{f_y} = \left(\frac{1}{3}\right)\frac{(350)(200)}{420}$$ (ACI Equation 11-23)

$$= 55.56 \text{ mm}^2 < (2)(129) = 258 \text{ mm}^2$$

OK

7. Selection of longitudinal torsion reinforcing

$$A_\ell = \frac{A_t}{s} p_h \frac{f_{yt}}{f_y} \cot^2\theta = (0.391)(1629)\left(\frac{420}{420}\right)(1.0)^2$$ (ACI Equation 11-22)

$$= 637 \text{ mm}^2$$

$$\text{Min } A_\ell = \frac{5\sqrt{f_c'}A_{cp}}{12f_y} - \frac{A_t}{s} p_h \frac{f_{yt}}{f_y}$$ (ACI Equation 11-24)

$$= \frac{(5\sqrt{28})(227\,500)}{(12)(420)} - (0.391)(1629)\left(\frac{420}{420}\right)$$

$$= 557 \text{ mm}^2$$

OK

Additional longitudinal steel is spread out to the four inside corners of the stirrups and vertically in between. Assume one-third in top $= \frac{557}{3} = 186 \text{ mm}^2$, $186 + 2050 = 2236 \text{ mm}^2$ in bottom, and remainder 186 mm^2 in between.

Use two #13 bars (258 mm^2) in top corners and at middepth, and five #25 (2550 mm^2) in bottom.

15.11   COMPUTER EXAMPLE

EXAMPLE 15.3

Repeat Example 15.1 using the Excel spreadsheet provided for Chapter 15

SOLUTION   Open the Excel Spreadsheet for Chapter 15 and the worksheet entitled Torsion. Observe that this sheet has U.S. customary units. Enter values for all cells in yellow highlight (shaded below, but yellow in the spreadsheet). Note that the results are the same as those obtained in Example 15.1.

Shear and Torsion Design — Rectangular Beams
U. S. Customary Units

$V_u =$	60	k				
$T_u =$	30	ft-k				
$N_u =$	0	k				
$f'_c =$	4000	psi				
$\lambda =$	1					
$b =$	16	in.				
$h =$	26	in.				
cover =	1.5	in.				
stirrup dia. =	0.5	in.				
$A_v =$	0.40	in.^{2}				
$f_{yt} =$	60,000	psi				
$f_y =$	60,000	psi				
$\phi =$	0.75					
$A_{cp} =$	416.00	in.^{2}				
$p_{cp} =$	84.00	in.				
Neglect torsion if $T_u <$	97723	in.-lb=	8.144	ft-k	$< T_u$ consider torsion	
$x_1 =$	12.50	in.				
$y_1 =$	22.50	in.				
$A_{oh} =$	281.25	in.^{2}				
$A_o =$	239.06	in.^{2}				
$d =$	23.5	in.				
$p_h =$	70.00	in.				
$V_c =$	47561	lb				
Eq. 11-18 left	246					
Eq. 11-18 right	474	section is large enough				
$A_t/s =$	0.0167	in.^{2}/in. for 2 legs of stirrup				
$V_u > \phi V_c/2$?	yes, shear reinforcing needed					
$V_s =$	32439	lb				
$A_v/s =$	0.0230	in.^{2}/in. for 2 legs of stirrup				
$A_{v+t}/s =$	0.0565	in.^{2}/in. for 2 legs of stirrup				
$s =$	7.08	in.				
$s \le p_h/8 =$	8.75	in.				
$s \le 12 =$	12.00	in.				
$s =$	7.08	in.				
Use $s =$	7.00	in.				
Eq.11-23 $0.75S_{qrt}(f'_c)bs/f_{yt} =$	0.089	$< A_v$ – OK				
$A_\ell =$	1.17	in.^{2}	Select additional longitudinal bars			
Min $A_\ell =$	1.02	OK				

PROBLEMS

For Problems 15.1 to 15.3, determine the equilibrium torsional capacity of the sections if no torsional reinforcing is used. $f'_c = 4000$ psi and $f_y = 60,000$ psi., normal-weight concrete.

Problem 15.1 (*Ans.* 5.62 ft-k)

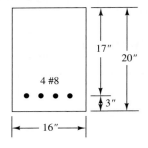

Problem 15.2

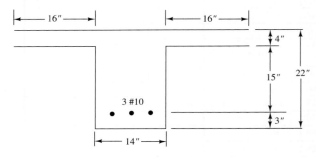

Problem 15.3 (*Ans.* 5.21 ft-k)

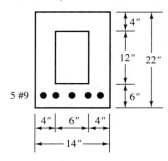

15.4 Repeat Problem 15.1 if the width is changed from 16 in. to 12 in. and the depth from 20 in. to 28 in.

15.5 Repeat Problem 15.3 if $f'_c = 3000$ psi sand-light-weight concrete. (*Ans.* 3.83 ft-k)

15.6 What minimum total theoretical depth is needed for the beam shown if no torsional reinforcing is to be used? The cross-section is not shown, but it is rectangular, with b = 20 in., and the depth to be determined. The concentrated load is located at the end of the cantilever 8 in. to one side of the beam ℄, $f_y = 60,000$ psi, $f'_c = 4000$ psi., all lightweight concrete.

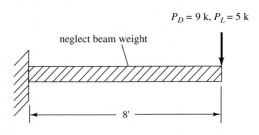

15.7 If the reinforced concrete spandrel beam shown in the accompanying illustration has $f'_c = 3000$ psi sand-lightweight concrete and $f_y = 60,000$ psi, determine the theoretical spacing required for #3 stirrups at a distance d from the face of the support where $V_u = 32$ k and $T_u = 12$ ft-k. Clear cover = $1\frac{1}{2}$ in. (*Ans.* 55.87 ft-k > 10 ft-k ∴ Torsion reinf. reqd @ 7.74 in. on center)

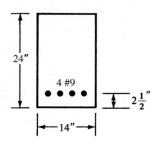

15.8 Design the torsion reinforcement for the beam shown in the accompanying figure at a section a distance d from the face of the support for a torsional moment of 24 ft-k. $V_u = 90$ k, $f_c' = 3000$ psi, and $f_y = 60,000$ psi. Clear cover $= 1.50$ in. Use #4 stirrups.

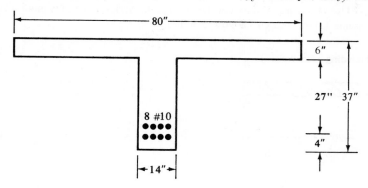

15.9 Determine the theoretical spacing of #4 closed □ stirrups at a distance d from the face of the support for the edge beam shown in the accompanying illustration if $f_c' = 4000$ psi and $f_y = 60,000$ psi. T_u equals 36 ft-k at the face of the support and is assumed to vary along the beam in proportion to the shear. Clear cover $= 1\frac{1}{2}$ in., normal-weight concrete. (*Ans.* 6.82 in.)

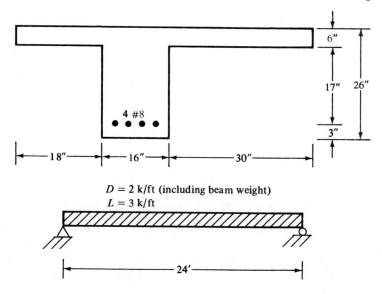

15.10 A 12-in. × 22-in. spandrel beam ($d = 19.5$ in.) with a 20-ft simple span has a 4-in. slab 16 in. wide on one side acting as a flange. It must carry a maximum V_u of 60 k and a maximum T_u of 20 ft-k at the face of the support. Assuming these values are zero at the beam ₵, select #4 stirrups if $f_c' = 3000$ psi normal weight concrete, $f_y = 60,000$ psi, and clear cover $= 1.5$ in.

15.11 Determine the theoretical spacing of #4 closed □ stirrups at a distance d from the face of the support for the beam shown if the load is acting 6 in. off center of the beam. Assume that the torsion equals the uniform load times the 3 in., $f'_c = 3000$ psi, all lightweights aggregate concrete and $f_y = 60,000$ psi. Use #4 stirrups. Assume that the torsion value varies from a maximum at the support to 0 at the beam Ç, as does the shear. Assume 1.5 in. clear cover. (*Ans.* 8.50 in.)

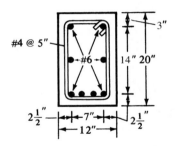
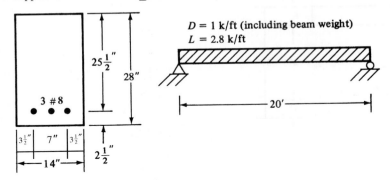

15.12 Is the beam shown satisfactory to resist a T_u of 15 ft-k and a V_u of 60 k if $f'_c = 4000$ psi and $f_y = 60,000$ psi? The bars shown are used in addition to those provided for bending moment.

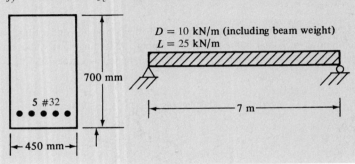

15.13 Using the Excel Spreadsheet for Chapter 15 determine the required spacing at a distance d from the support for Problem 15.7. Use the same materials, but change to concrete to all-lightweight aggregate. (*Ans.* 7.12 in)

15.14 Repeat Problem 15.11 using #3 closed stirrups $f'_c = 4000$ psi semi-lightweight concrete and the Chapter 15 spreadsheet.

Problems in SI Units

15.15 Determine the required spacing of #13 closed □ stirrups at a distance d from the face of the support for the beam shown in the accompanying figure, assuming that the torsion decreases uniformly from the beam end to the beam Ç. The member is subjected to a 34−kN · m service dead load torsion and a 40 kN · m service line load torsion at the face of the support, $f_y = 420$ MPa and $f'_c = 24$ MPa, clear cover = 40 mm. (*Ans.* 175 mm)

15.16 Determine stirrup spacing at distance d from support for the beam shown if the load is acting 100 mm off center of the beam. Assume the torsion at the support equals the uniform load times 100 mm, $f'_c = 28$ MPa, and $f_y = 420$ MPa. Use #13 stirrups and assume that the torsion and shear vary from a maximum at the support to 0 at the beam $\mathbb{C}$. Clear cover = 40 mm.

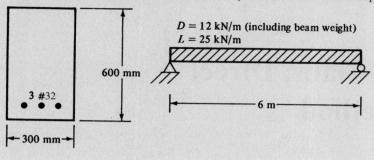

600 mm

3 #32

300 mm

$D = 12$ kN/m (including beam weight)
$L = 25$ kN/m

6 m

16

Two-Way Slabs, Direct Design Method

16.1 INTRODUCTION

In general, slabs are classified as being one-way or two-way. Slabs that primarily deflect in one direction are referred to as *one-way slabs* [see Figure 16.1(a)]. Simple-span, one-way slabs have previously been discussed in Section 4.7 of this text, while the design of continuous one-way slabs was considered in Section 14.7. When slabs are supported by columns arranged generally in rows so that the slabs can deflect in two directions, they are usually referred to as *two-way slabs*.

Two-way slabs may be strengthened by the addition of beams between the columns, by thickening the slabs around the columns (*drop panels*), and by flaring the columns under the slabs (*column capitals*). These situations are shown in Figure 16.1 and discussed in the next several paragraphs.

Flat plates [Figure 16.1(b)] are solid concrete slabs of uniform depths that transfer loads directly to the supporting columns without the aid of beams or capitals or drop panels. Flat plates can be constructed quickly due to their simple formwork and reinforcing bar arrangements. They need the smallest overall story heights to provide specified headroom requirements, and they give the most flexibility in the arrangement of columns and partitions. They also provide little obstruction to light and have high fire resistance because there are few sharp corners where spalling of the concrete might occur. Flat plates are probably the most commonly used slab system today for multistory reinforced concrete hotels, motels, apartment houses, hospitals, and dormitories.

Flat plates present a possible problem in transferring the shear at the perimeter of the columns. In other words, there is a danger that the columns may punch through the slabs. As a result, it is frequently necessary to increase column sizes or slab thicknesses or to use *shearheads*. Shearheads consist of steel I or channel shapes placed in the slab over the columns, as discussed in Section 16.5 of this chapter. Although such procedures may seem expensive, it is noted that the simple formwork required for flat plates will usually result in such economical construction that the extra costs required for shearheads are more than canceled. For heavy industrial loads or long spans, however, some other type of floor system may be required.

Flat slabs [Figure 16.1(c)] include two-way reinforced concrete slabs with capitals, drop panels, or both. These slabs are very satisfactory for heavy loads and long spans. Although the formwork is more expensive than for flat plates, flat slabs will require less concrete and reinforcing than would be required for flat plates with the same loads and spans. They are particularly economical for warehouses, parking and industrial buildings, and similar structures where exposed drop panels or capitals are acceptable.

In Figure 16.1(d) a *two-way slab with beams* is shown. This type of floor system is obviously used where its cost is less than the costs of flat plates or flat slabs. In other words, when the loads or spans or both become quite large, the slab thickness and column sizes required for flat plates or flat

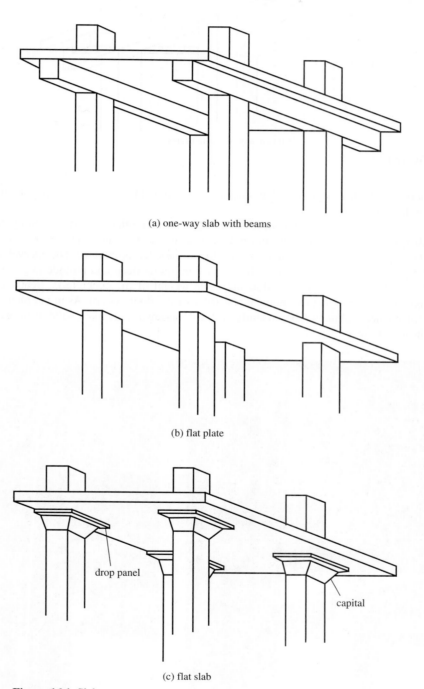

(a) one-way slab with beams

(b) flat plate

drop panel

capital

(c) flat slab

Figure 16.1 Slabs.

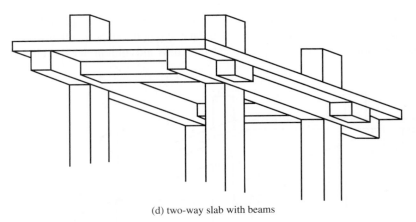

(d) two-way slab with beams

Figure 16.1 (*continued*)

slabs are of such magnitude that it is more economical to use two-way slabs with beams, despite the higher formwork costs.

Another type of floor system is the *waffle slab*. The floor is constructed by arranging square fiberglass or metal pans with tapered sides with spaces. When the concrete is placed over and between the pans and the forms removed, the waffle shape is obtained. The intervals or gaps between the pans form the beam webs. These webs are rather deep and provide large moment arms for the reinforcing bars. With waffle slabs, the weight of the concrete is greatly reduced without significantly changing the moment resistance of the floor system. As in flat plates, shear can be a problem near columns. Consequently, waffle floors are usually made solid in those areas to increase shear resistance.

Flat slab with drop panels and no column capitals. (Courtesy of Portland Cement Association.)

16.2   ANALYSIS OF TWO-WAY SLABS

Two-way slabs bend under load into dish-shaped surfaces, so there is bending in both principal directions. As a result, they must be reinforced in both directions by layers of bars that are perpendicular to each other. A theoretical elastic analysis for such slabs is a very complex problem due to their highly indeterminate nature. Numerical techniques such as finite difference and finite elements are required, but such methods require sophisticated software to be practical in design. The methods described in this chapter can be done by hand or with simple spreadsheets, and are sufficiently accurate for most design problems.

Actually, the fact that a great deal of stress redistribution can occur in such slabs at high loads makes it unnecessary to make designs based on theoretical analyses. Therefore the design of two-way slabs is generally based on empirical moment coefficients, which, though they might not accurately predict stress variations, result in slabs with satisfactory overall safety factors. In other words, if too much reinforcing is placed in one part of a slab and too little somewhere else, the resulting slab behavior will probably still be satisfactory. *The total amount of reinforcement in a slab seems more important than its exact placement.*

You should clearly understand that though this chapter and the next are devoted to two-way slab design based on approximate methods of analysis, there is no intent to prevent the designer from using more exact methods. Designers may design slabs on the basis of numerical solutions, yield-line analysis, or other theoretical methods, provided that it can be clearly demonstrated that they have met all the necessary safety and service ability criteria required by the ACI Code.

Although it has been the practice of designers for many years to use approximate analyses for design and to use average moments rather than maximum ones, two-way slabs so designed have proved to be very satisfactory under service loads. Furthermore, they have been proved to have appreciable overload capacity.

16.3   DESIGN OF TWO-WAY SLABS BY THE ACI CODE

The ACI Code (13.5.1.1) specifies two methods for designing two-way slabs for gravity loads. These are the direct design method and the equivalent frame method.

Direct Design Method

The Code (13.6) provides a procedure with which a set of moment coefficients can be determined. The method, in effect, involves a single-cycle moment distribution analysis of the structure based on (a) the estimated flexural stiffnesses of the slabs, beams (if any), and columns and (b) the torsional stiffnesses of the slabs and beams (if any) transverse to the direction in which flexural moments are being determined. Some types of moment coefficients have been used satisfactorily for many years for slab design. They do not, however, give very satisfactory results for slabs with unsymmetrical dimensions and loading patterns.

Equivalent Frame Method

In this method a portion of a structure is taken out by itself, as shown in Figure 16.2, and analyzed much as a portion of a building frame was handled in Example 14.2. The same stiffness values used for the direct design method are used for the equivalent frame method. This latter method, which is very satisfactory for symmetrical frames as well as for those with unusual dimensions or loadings, is discussed in Chapter 17 of this text.

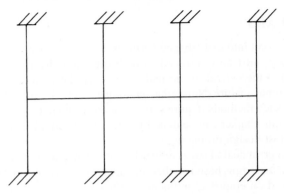

Figure 16.2 Equivalent frame method.

Design for Lateral Loads

The ACI Code permits considerable freedom for the designer to model two-way slab systems for lateral loads. The method must satisfy equilibrium and geometric compatibility and be in reasonable agreement with test data. The effects of cracking and such parameters as the slab aspect ratio and ratio of column-to-slab span dimensions should be considered[1] (ACI Section R13.5.1.2).

16.4   COLUMN AND MIDDLE STRIPS

After the design moments have been determined by either the direct design method or the equivalent frame method, they are distributed across each panel. The panels are divided into column and middle strips, as shown in Figure 16.3, and positive and negative moments are estimated in each strip. The *column strip* is a slab with a width on each side of the column

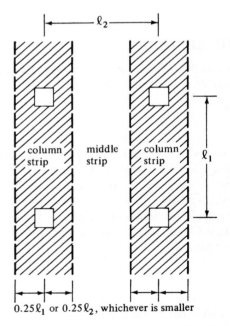

$0.25\ell_1$ or $0.25\ell_2$, whichever is smaller

Figure 16.3 Column and middle strips.

[1]Vanderbilt, M. D., and Corley, W. G., 1983 "Frame Analysis of Concrete Buildings," *Concrete International: Design and Construction*, V.5, No. 12, Dec. 1983, pp. 33–43.

Flat Plate Construction – Pharr Road Condominiums, Atlanta, GA. (Courtesy Prof. Larry Kahn, Georgia Inst. of Technology.)

centerline equal to one-fourth the smaller of the panel dimensions ℓ_1 or ℓ_2. It includes beams if they are present. The middle strip is the part of the slab between the two column strips.

The part of the moments assigned to the column and middle strips may be assumed to be uniformly spread over the strips. As will be described later in this chapter, the percentage of the moment assigned to a column strip depends on the effective stiffness of that strip and on its *aspect ratio* ℓ_2/ℓ_1 (where ℓ_1 is the length of span, center to center, of supports in the direction in which moments are being determined and ℓ_2 is the span length, center to center, of supports in the direction transverse to ℓ_1). Note that Figure 16.3 shows column and middle strips in only one direction. A similar analysis must be performed in the perpendicular direction. The resulting analysis will result in moments in both directions.

16.5   SHEAR RESISTANCE OF SLABS

For two-way slabs supported by beams or walls, shears are calculated at a distance d from the faces of the walls or beams. The value of ϕV_c is, as for beams, $\phi 2\lambda \sqrt{f_c'} b_w d$. Shear is not usually a problem for these types of slabs.

For flat slabs and flat plates supported directly by columns, shear may be the critical factor in design. In almost all tests of such structures, failures have been due to shear or perhaps shear and torsion. These conditions are particularly serious around exterior columns.

Two kinds of shear must be considered in the design of flat slabs and flat plates. These are the same two that were considered in column footings—one-way and two-way shears (that is, beam shear and punching shear). For beam shear analysis, the slab is considered to act as a wide beam running between the supports. The critical sections are taken at a distance d from the face of the column or capital. For punching shear, the critical section is taken at a distance $d/2$ from the face of the column, capital, or drop panel and the shear strength, as usually used in footings, is $\phi 4\lambda \sqrt{f_c'} b_w d$.

If shear stresses are too large around interior columns, it is possible to increase the shearing strength of the slabs by as much as 75% by using shearheads. A shearhead, as defined in Section 11.11.4 of the Code, consists of four steel I or channel shapes fabricated into cross arms and placed in the slabs, as shown in Figure 16.4(a). The Code states that shearhead designs of this type do not apply at exterior columns. Thus special designs are required, and the Code does not provide specific requirements. Shearheads increase the effective b_o for two-way shear, and they also increase the negative moment resistance of the slab, as described in the Code (11.11.4.9). The negative moment reinforcing bars in the slab are usually run over the top of the steel shapes, while the positive reinforcing is normally stopped short of the shapes.

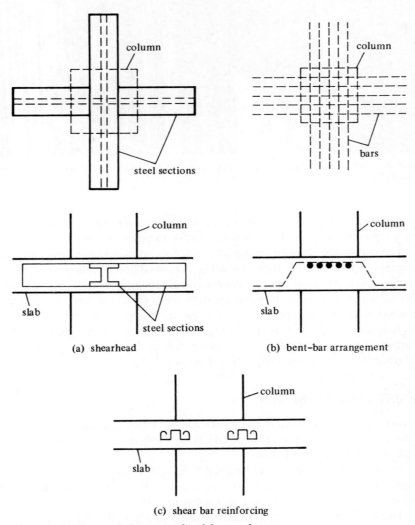

(a) shearhead   (b) bent–bar arrangement

(c) shear bar reinforcing

Figure 16.4 Shear reinforcement for slabs at columns.

Another type of shear reinforcement permitted in slabs by the Code (11.11.3) involves the use of groups of bent bars or wires. One possible arrangement of such bars is shown in Figure 16.4(b). The bars are bent across the potential diagonal tension cracks at 45° angles, and they are run along the bottoms of the slabs for the distances needed to fully develop the bar strengths. Another type of bar arrangement that might be used is shown in Figure 16.4(c). When bars (or wires) are used as shear reinforcement, the Code (11.11.3.2) states that the nominal two-way shear strength permitted on the critical section at a distance $d/2$ from the face of the column may be increased from $4\sqrt{f_c'}b_od$ to $6\sqrt{f_c'}b_od$.

The main advantage of shearheads is that they push the critical sections for shear farther out from the columns, thus giving a larger perimeter to resist the shear, as illustrated in Figure 16.5. In this figure, ℓ_v is the length of the shearhead arm from the centroid of the concentrated load or reaction, and c_1 is the dimension of the rectangular or equivalent rectangular column or capital or bracket, measured in the direction in which moments are being calculated. The Code (11.11.4.7) states that the critical section for shear shall cross the shearhead arm at a distance equal to $\frac{3}{4}[\ell_v - (c_1/2)]$ from the column face, as shown in Figure 16.5(b). Although this critical section is to be located so that its perimeter is a minimum, it does not have to be located closer to the column face or edges of capitals or drop panels than $d/2$ at any point. When shearhead reinforcing is

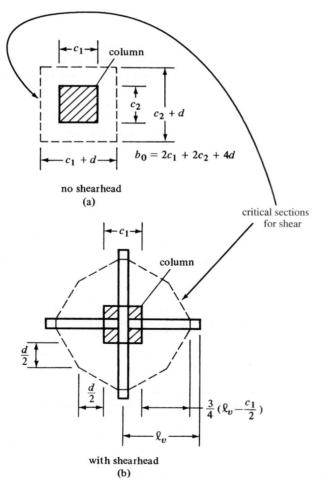

no shearhead
(a)

with shearhead
(b)

Figure 16.5 Critical sections for shear.

provided with reinforcing bars or steel I or channel shapes, the maximum shear strength can be increased to $7\sqrt{f_c'}b_o d$ at a distance $d/2$ from the column. According to the Code (11.11.4.8), this is only permissible if the maximum computed shear does not exceed $4\sqrt{f_c'}b_o d$ along the dashed critical section for shear shown in Figure 16.5(b).

In Section 16.12 the subject of shear stresses is continued with a consideration of the transfer of moments and shears between slabs and columns. The maximum load that a two-way slab can support is often controlled by this transfer strength.

16.6 DEPTH LIMITATIONS AND STIFFNESS REQUIREMENTS

It is obviously very important to keep the various panels of a two-way slab relatively level (that is, with reasonably small deflections). Thin reinforced two-way slabs have quite a bit of moment resistance, but deflections are often large. As a consequence, their depths are very carefully controlled by the ACI Code so as to limit these deflections. This is accomplished by requiring the designer to either (a) compute deflections and make sure they are within certain limitations or (b) use certain minimum thicknesses as specified in Section 9.5.3 of the Code. Deflection computations for two-way slabs are rather complicated, so the average designer usually uses the minimum ACI thickness values, presented in the next few paragraphs of this chapter.

Slabs without Interior Beams

For a slab without interior beams spanning between its supports and with a ratio of its long span to short span not greater than 2.0, the minimum thickness can be taken from Table 16.1 of this chapter [Table 9.5(c) in the Code]. The values selected from the table, however, must not be less than the following values (ACI 9.5.3.2):

1. Slabs without drop panels 5 in.
2. Thickness of those slabs with drop panels outside the panels 4 in.

Table 16.1 Minimum Thickness of Slabs without Interior Beams

Yield strength, f_y, psi[*]	Without drop panels[†]			With drop panels[†]		
	Exterior panels		Interior panels	Exterior panels		Interior panels
	Without edge beams	With edge beams[‡]		Without edge beams	With edge beams[‡]	
40,000	$\dfrac{\ell_n{}^{\S}}{33}$	$\dfrac{\ell_n}{36}$	$\dfrac{\ell_n}{36}$	$\dfrac{\ell_n}{36}$	$\dfrac{\ell_n}{40}$	$\dfrac{\ell_n}{40}$
60,000	$\dfrac{\ell_n}{30}$	$\dfrac{\ell_n}{33}$	$\dfrac{\ell_n}{33}$	$\dfrac{\ell_n}{33}$	$\dfrac{\ell_n}{36}$	$\dfrac{\ell_n}{36}$
75,000	$\dfrac{\ell_n}{28}$	$\dfrac{\ell_n}{31}$	$\dfrac{\ell_n}{31}$	$\dfrac{\ell_n}{31}$	$\dfrac{\ell_n}{34}$	$\dfrac{\ell_n}{34}$

[*]For values of reinforcement yield strength between the values given in the table, minimum thickness shall be determined by linear interpolation.

[†]Drop panel is defined in ACI Sections 13.3.7 and 13.2.5.

[‡]Slabs with beams between columns along exterior edges. The value of α_f for the edge beam shall not be less than 0.8.

[§]For two-way construction ℓ_n is the length of the clear span in the long direction, measured face to face of the supports in slabs without beams and face to face of beams or other supports in other cases.

In Table 16.1 some of the values are given for slabs with drop panels. To be classified as a drop panel, according to Sections 13.3.7 and 13.2.5 of the Code, a panel must (a) extend horizontally in each direction from the centerline of the support no less than one-sixth the distance, center to center, of supports in that direction and (b) project vertically below the slab a distance no less than one-fourth the thickness of the slab away from the drop panel. In this table, ℓ_n is the length of the clear span in the long direction of two-way construction, measured face to face of the supports in slabs without beams and face to face of beams or other supports in other cases.

Very often slabs are built without interior beams between the columns but with edge beams running around the perimeter of the building. These beams are very helpful in stiffening the slabs and reducing the deflections in the exterior slab panels. The stiffness of slabs with edge beams is expressed as a function of α_f, which follows.

Throughout this chapter the letter α_f is used to represent the ratio of the flexural stiffness $(E_{cb}I_b)$ of a beam section to the flexural stiffness of the slab $(E_{cs}I_s)$ whose width equals the distance between the centerlines of the panels on each side of the beam. If no beams are used, as for the flat plate, α will equal 0. For slabs with beams between columns along exterior edges, α for the edge beams may not be < 0.8 as specified in a footnote to Table 16.1.

$$\alpha_f = \frac{E_{cb}I_b}{E_{cs}I_s} \qquad\qquad \text{(ACI Equation 13-3)}$$

Where

E_{cb} = the modulus of elasticity of the beam concrete

E_{cs} = the modulus of elasticity of the column concrete

I_b = the gross moment of inertia about the centroidal axis of a section made up of the beam and the slab on each side of the beam extending a distance equal to the projection of the beam above or below the slab (whichever is greater) but not exceeding four times the slab thickness (ACI 13.2.4)

I_s = the moment of inertia of the gross section of the slab taken about the centroidal axis and equal to $h^3/12$ times the slab width, where the width is the same as for α

EXAMPLE 16.1

Using the ACI Code, determine the minimum permissible total thicknesses required for the slabs in panels ③ and ② for the floor system shown in Figure 16.6. Edge beams are used around the building perimeter, and they are 12 in. wide and extend vertically for 8 in. below the slab, as shown in Figure 16.7. They also extend 8 in. out into the slab as required by ACI Section 13.2.4. No drop panels are used, and the concrete in the slab is the same as that used in the edge beams. $f_y = 60{,}000$ psi.

SOLUTION   **For Interior Panel ③**

$$\alpha_f = 0 \text{ (since the interior panels have no perimeter beams)}$$

$$\ell_n = 20 - \frac{16}{12} = 18.67 \text{ ft (clear distance between columns)}$$

$$\text{Min } h \text{ from Table 16.1} = \frac{\ell_n}{33}$$

$$= \frac{18.67}{33} = 0.566 \text{ ft} = 6.79 \text{ in.}$$

May not be less than 5 in., according to Section 9.5.3.2   Try 7 in.

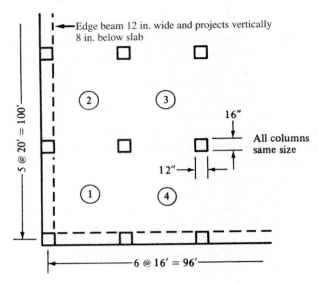

Figure 16.6 A flat-plate floor slab.

For Exterior Panel ②

Assume $h = 7$ in. and compute α_f with reference made to Figure 16.7(a). Centroid of cross-hatched beam section located by statics 6.55 in. from top.

$$I_b = \left(\frac{1}{3}\right)(20)(6.55)^3 + \left(\frac{1}{3}\right)(12)(8.45)^3 + \left(\frac{1}{3}\right)(8)(0.45)^3$$

$$= 4287 \text{ in.}^4$$

$$I_s = \left(\frac{1}{12}\right)(102)(7)^3 = 2915.5 \quad \text{See Figure 16.7(b)}$$

$$\alpha = \frac{EI_b}{EI_s} = \frac{(E)(4287)}{(E)(2915.5)} = 1.47 > 0.8$$

$\therefore$ This is an edge beam as defined in footnote in Table 16.1.

$$\text{Min } h = \frac{\ell_n}{33} = \frac{20 - \dfrac{16}{12}}{33} = 0.566 \text{ ft} = 6.79 \text{ in.}$$

<u>Try 7 in.</u>

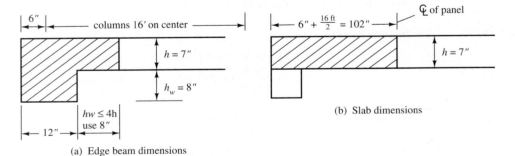

(a) Edge beam dimensions

(b) Slab dimensions

Figure 16.7

Slabs with Interior Beams

To determine the minimum thickness of slabs with beams spanning between their supports on all sides, Section 9.5.3.3 of the Code must be followed. Involved in the expressions presented there are span lengths, panel shapes, flexural stiffness of beams if they are used, steel yield stresses, and so on. In these equations the following terms are used:

$\ell_n =$ the clear span in the long direction, measured face to face, of (a) columns for slabs without beams and (b) beams for slab with beams

$\beta =$ the ratio of the long to the short clear span

$\alpha_{fm} =$ the average value of the ratios of beam-to-slab stiffness on all sides of a panel

The minimum thickness of slabs or other two-way construction may be obtained by substituting into the equations to follow, which are given in Section 9.5.3.3 of the Code. In the equations, the quantity β is used to take into account the effect of the shape of the panel on its deflection, while the effect of beams (if any) is represented by α_{fm}. If there are no beams present (as is the case for flat slabs), α_{fm} will equal 0.

1. For $\alpha_{fm} \leq 0.2$, the minimum thicknesses are obtained as they were for slabs without interior beams spanning between their supports.

2. For $0.2 \leq \alpha_{fm} \leq 2.0$, the thickness may not be less than 5 in. or

$$ h = \frac{\ell_n \left(0.8 + \dfrac{f_y}{200{,}000} \right)}{36 + 5\beta(\alpha_{fm} - 0.2)} \qquad \text{(ACI Equation 9-12)} $$

3. For $\alpha_{fm} > 2.0$, the thickness may not be less than 3.5 in. or

$$ h = \frac{\ell_n \left(0.8 + \dfrac{f_y}{200{,}000} \right)}{36 + 9\beta} \qquad \text{(ACI Equation 9-13)} $$

where ℓ_n and f_y are in inches and psi, respectively.

For panels with discontinuous edges, the Code (9.5.3.3d) requires that edge beams be used, which have a minimum stiffness ratio α_f equal to 0.8, or else that the minimum slab thicknesses, as determined by ACI Equations 9-12 and 9-13, must be increased by 10%.

The designer may use slabs of lesser thicknesses than those required by the ACI Code as described in the preceding paragraphs if deflections are computed and found to be equal to or less than the limiting values given in Table 9.5(b) of the ACI Code (Table 6.1 in this text).

Should the various rules for minimum thickness be followed but the resulting slab be insufficient to provide the shear capacity required for the particular column size, column capitals will probably be required. Beams running between the columns may be used for some slabs where partitions or heavy equipment loads are placed near column lines. A very common case of this type occurs where exterior beams are used when the exterior walls are supported directly by the slab. Another situation where beams may be used occurs where there is concern about the magnitude of slab vibrations. Example 16.2 illustrates the application of the minimum slab thickness rules for a two-way slab with beams.

EXAMPLE 16.2

The two-way slab shown in Figure 16.8 has been assumed to have a thickness of 7 in. Section A-A in the figure shows the beam cross section. Check the ACI equations to determine if the slab thickness is satisfactory for an interior panel. $f'_c = 3000$ psi, $f_y = 60,000$ psi, normal-weight concrete.

SOLUTION

(Using the same concrete for beams and slabs)

Computing α_1 for Long (Horizontal) Span for Interior Beams

$$I_s = \text{gross moment of inertia of slab 20 ft wide}$$

$$= \left(\frac{1}{12}\right)(12 \times 20)(7)^3 = 6860 \text{ in.}^4$$

$$I_b = \text{gross } I \text{ of } T \text{ beam cross section shown in Figure 16.8}$$

$$\text{about centroidal axis} = 18,060 \text{ in.}^4$$

$$\alpha_1 = \frac{EI_b}{EI_s} = \frac{(E)(18,060)}{(E)(6860)} = 2.63$$

Computing α_2 for Long Interior Beams

$$I_s \text{ for 24-ft-wide slab} = \left(\frac{1}{12}\right)(12 \times 24)(7)^3 = 8232 \text{ in.}^4$$

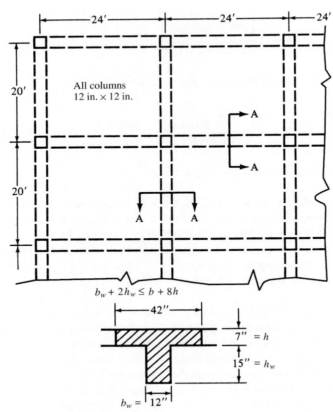

Section A-A

Figure 16.8 A two-way slab.

$$I_b = 18{,}060 \text{ in.}^4$$

$$\alpha_2 = \frac{(E)(18{,}060)}{(E)(8232)} = 2.19$$

$$\alpha_{fm} = \frac{\alpha_1 + \alpha_2}{2} = \frac{2.63 + 2.19}{2} = 2.41$$

Determining Slab Thickness as per ACI Section 9.5.3.3

$$\alpha_{fm} = 2.41 > 2 \qquad\qquad \therefore \text{Use ACI Equation 9-13}$$

$$h = \frac{\ell_n\left(0.8 + \dfrac{f_y}{200{,}000}\right)}{36 + 9\beta}$$

$$\ell_{n \text{ long}} = 24 - 1 = 23 \text{ ft}$$

$$\ell_{n \text{ short}} = 20 - 1 = 19 \text{ ft}$$

$$\beta = \frac{23}{19} = 1.21$$

$$h = \frac{(23)\left(0.8 + \dfrac{60{,}000}{200{,}000}\right)}{36 + (9)(1.21)} = 0.540 \text{ ft} = 6.47 \text{ in.}$$

Use 7-in. slab

Note that the interior panel will generally not control the required slab thickness. Usually, it will be an edge or corner panel. The interior panel was chosen here to illustrate the calculations and to avoid excess complexity. Had a corner panel been selected, each edge of the panel would have had a different α_f.

16.7   LIMITATIONS OF DIRECT DESIGN METHOD

For the moment coefficients determined by the direct design method to be applicable, the Code (13.6.1) says that the following limitations must be met, unless a theoretical analysis shows that the strength furnished after the appropriate capacity reduction or ϕ factors are applied is sufficient to support the anticipated loads and provided that all serviceability conditions such as deflection limitations are met:

1. There must be at least three continuous spans in each direction.
2. The panels must be rectangular, with the length of the longer side of any panel not being more than 2.0 times the length of its shorter side lengths being measured c to c of supports.
3. Span lengths of successive spans in each direction may not differ in length by more than one-third of the longer span.
4. Columns may not be offset by more than 10% of the span length in the direction of the offset from either axis between center lines of successive columns.
5. The unfactored live load must not be more than two times the unfactored dead load. All loads must be due to gravity and must be uniformly distributed over an entire panel.

6. If a panel is supported on all sides by beams, the relative stiffness of those beams in the two perpendicular directions, as measured by the following expression, shall not be less than 0.2 or greater than 5.0.

$$\frac{\alpha_{f1}\ell_2^2}{\alpha_{f2}\ell_1^2}$$

The terms ℓ_1 and ℓ_2 were shown in Figure 16.3.

16.8   DISTRIBUTION OF MOMENTS IN SLABS

The total moment M_o that is resisted by a slab equals the sum of the maximum positive and negative moments in the span. It is the same as the total moment that occurs in a simply supported beam. For a uniform load per unit area, q_u, it is as follows:

$$M_o = \frac{(q_u\ell_2)(\ell_1)^2}{8}$$

In this expression ℓ_1 is the span length, center to center, of supports in the direction in which moments are being taken and ℓ_2 is the length of the span transverse to ℓ_1, measured center to center of the supports.

The moment that actually occurs in such a slab has been shown by experience and tests to be somewhat less than the value determined by the above M_o expression. For this reason ℓ_1 is replaced with ℓ_n, the clear span measured face to face of the supports in the direction in which moments are taken. The Code (13.6.2.5) states that ℓ_n may not be taken to be less than 65% of the span ℓ_1 measured center to center of supports. If ℓ_1 is replaced with ℓ_n, the expression for M_o, which is called the *static moment*, becomes

$$M_o = \frac{(q_u\ell_2)(\ell_n)^2}{8} \qquad\qquad \text{(ACI Equation 13-4)}$$

When the static moment is being calculated in the long direction, it is convenient to write it as $M_{o\ell}$, and in the short direction as M_{os}.

It is next necessary to know what proportions of these total moments are positive and what proportions are negative. If a slab was completely fixed at the end of each panel, the division would be as it is in a fixed-end beam, two-thirds negative and one-third positive, as shown in Figure 16.9.

This division is reasonably accurate for interior panels where the slab is continuous for several spans in each direction with equal span lengths and loads. In effect, the rotation of the interior

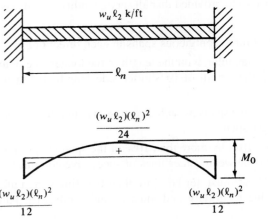

Figure 16.9

columns is assumed to be small, and moment values of $0.65M_o$ for negative moment and $0.35M_o$ for positive moment are specified by the Code (13.6.3.2). For cases where the span lengths and loadings are different, the proportion of positive and negative moments may vary appreciably, and the use of a more detailed method of analysis is desirable. The equivalent frame method (Chapter 17) will provide rather good approximations for such situations.

The relative stiffnesses of the columns and slabs of exterior panels are of far greater significance in their effect on the moments than is the case for interior panels. The magnitudes of the moments are very sensitive to the amount of torsional restraint supplied at the discontinuous edges. This restraint is provided both by the flexural stiffness of the slab and by the flexural stiffness of the exterior column.

Should the stiffness of an exterior column be quite small, the end negative moment will be very close to zero. If the stiffness of the exterior column is very large, the positive and negative moments will still not be the same as those in an interior panel unless an edge beam with a very large torsional stiffness is provided that will substantially prevent rotation of the discontinuous edge of the slab.

If a 2-ft-wide beam were to be framed into a 2-ft-wide column of infinite flexural stiffness in the plane of the beam, the joint would behave as would a perfectly fixed end, and the negative beam moment would equal the fixed-end moment.

If a two-way slab 24 ft wide were to be framed into this same 2-ft-wide column of infinite stiffness, the situation of no rotation would occur along the part of the slab at the column. For the remaining 11-ft widths of slab on each side of the column, there would be rotation varying from zero at the side face of the column to maximums 11 ft on each side of the column. As a result of this rotation, the negative moment at the face of the column would be less than the fixed-end moment. Thus the stiffness of the exterior column is reduced by the rotation of the attached transverse slab.

To take into account the fact that the rotation of the edge of the slab is different at different distances from the column, the exterior columns and slab edge beam are replaced with an equivalent column that has the same estimated flexibility as the column plus the edge beam. It can be seen that this is quite an involved process; therefore, instead of requiring a complicated analysis, the Code (13.6.3.3) provides a set of percentages for dividing the total factored static moment into its positive and negative parts in an end span. These divisions, which are shown in Table 16.2, include values for unrestrained edges (where the slab is simply supported on a masonry or concrete wall) and for restrained edges (where the slab is constructed integrally with a very stiff reinforced concrete wall so that the little rotation occurs at the slab-to-wall connection).

Table 16.2 Distribution of Total Span Moment in an End Span (ACI Code 13.6.3.3)

	(1)	(2)	(3)	(4)	(5)
			Slab without beams between interior supports		
	Exterior edge unrestrained	Slab with beams between all supports	Without edge beam	With edge beam	Exterior edge fully restrained
Interior negative factored moment	0.75	0.70	0.70	0.70	0.65
Positive factored moment	0.63	0.57	0.52	0.50	0.35
Exterior negative factored moment	0	0.16	0.26	0.30	0.65

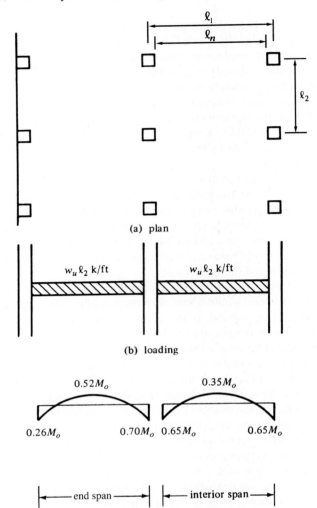

Figure 16.10 Sample moments for a flat plate with no edge beams.

In Figure 16.10 the distribution of the total factored moment for the interior and exterior spans of a flat-plate structure is shown. The plate is assumed to be constructed without beams between interior supports and without edge beams.

The next problem is to estimate what proportion of these moments is taken by the column strips and what proportion is taken by the middle strips. For this discussion a flat-plate structure is assumed, and the moment resisted by the column strip is estimated by considering the tributary areas shown in Figure 16.11.

To simplify the mathematics, the load to be supported is assumed to fall within the dashed lines shown in either part (a) or (b) of Figure 16.12. The corresponding load is placed on the simple span, and its centerline moment is determined as an estimate of the portion of the static moment taken by the column strip.[2]

---

[2]White, R. N., Gergely, P., and Sexsmith, R. G., 1974, *Structural Engineering*, vol. 3: *Behavior of Members and Systems* (New York: John Wiley & Sons), pp. 456–461.

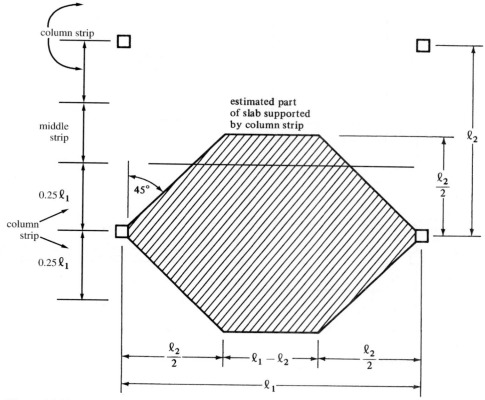

Figure 16.11

In Figure 16.12(a), the load is spread uniformly over a length near the midspan of the beam, thus causing the moment to be overestimated a little, while in Figure 16.12(b) the load is spread uniformly from end to end, causing the moment to be underestimated. Based on these approximations, the estimated moments in the column strips for square panels will vary from $0.5M_o$ to $0.75M_o$, where M_o = the absolute sum of the positive and average negative factored moments in each direction = $q_u \ell_2 \ell_n^2 / 8$. As ℓ_1 becomes larger than ℓ_2, the column strip takes a larger proportion of the moment. For such cases, about 60% to 70% of M_o will be resisted by the column strip.

If you sketch in the approximate deflected shape of a panel, you will see that a larger portion of the positive moment is carried by the middle strip than by the column strip, and vice versa for the negative moments. As a result, about 60% of the positive M_o and about 70% of the negative M_o are expected to be resisted by the column strip.[3]

Section 13.6.4.1 of the Code states that the column strip shall be proportioned to resist the percentages of the total interior negative design moment given in Table 16.3.

In the table, α_1 is again the ratio of the stiffness of a beam section to the stiffness of a width of slab bounded laterally by the ₵ of the adjacent panel, if any, on each side of the beam and equals $E_{cb}I_b / E_{cs}I_s$.

Section 13.6.4.2 of the Code states that the column strip is to be assumed to resist percentages of the exterior negative design moment, as given in Table 16.4. In this table β_t is the ratio of

---

[3]White, Gergely, and Sexsmith, *Structural Engineering*, vol. 3: *Behavior of Members and Systems*.

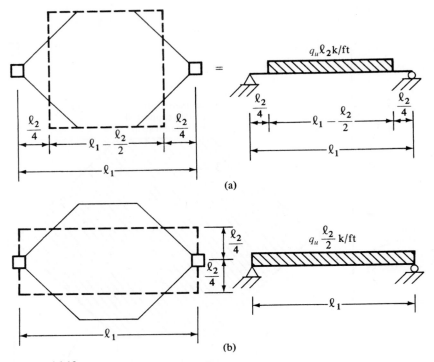

Figure 16.12

Table 16.3 Percentages of Interior Negative Design Moments to Be Resisted by Column Strip

$\dfrac{\ell_2}{\ell_1}$	0.5	1.0	2.0
$\dfrac{\alpha_{f1}\ell_2}{\ell_1} = 0$	75	75	75
$\dfrac{\alpha_{f1}\ell_2}{\ell_1} \geq 1.0$	90	75	45

the torsional stiffness of an edge beam section to the flexural stiffness of a width of slab equal to the span length of the beam center to center of supports ($\beta_t = E_{cb}C/2E_{cs}I_s$). The computation of the cross-sectional constant C is described in Section 16.11 of this chapter.

The column strip (Section 13.6.4.4 of the Code) is to be proportioned to resist the portion of the positive moments given in Table 16.5.

Equations can be used instead of the two-way interpolation sometimes required by Tables 16.3, 16.4, and 16.5. Instead of Table 16.3, the percentage of interior negative moment to be resisted by the column strip ($\%_{\text{int col}}^{-}$) can be determined by

$$\%_{\text{int col}}^{-} = 75 + 30\left(\frac{\alpha_{f1}\ell_2}{\ell_1}\right)\left(1 - \frac{\ell_2}{\ell_1}\right)$$

Table 16.4 Percentages of Exterior Negative Design Moment to Be Resisted by Column Strip

$\dfrac{\ell_2}{\ell_1}$		0.5	1.0	2.0
$\dfrac{\alpha_{f1}\ell_2}{\ell_1} = 0$	$\beta_t = 0$	100	100	100
	$\beta_t \geq 2.5$	75	75	75
$\dfrac{\alpha_{f1}\ell_2}{\ell_1} \geq 1.0$	$\beta_t = 0$	100	100	100
	$\beta_t \geq 2.5$	90	75	45

Table 16.5 Percentages of Positive Design Moment to Be Resisted by Column Strip

$\dfrac{\ell_2}{\ell_1}$	0.5	1.0	2.0
$\dfrac{\alpha_{f1}\ell_2}{\ell_1} = 0$	60	60	60
$\dfrac{\alpha_{f1}\ell_2}{\ell_1} \geq 1.0$	90	75	45

The percentage of exterior negative design moment resisted by the column ($\%^-_{\text{ext col}}$) strip given in Table 16.4 can be found by

$$\%^-_{\text{ext col}} = 100 - 10\beta_t + 12\left(\frac{\alpha_{f1}\ell_2}{\ell_1}\right)\left(1 - \frac{\ell_2}{\ell_1}\right)$$

And finally, for positive design moment in either an interior or exterior span (Table 16.5), the percentage resisted by the column strip ($\%^+$) is

$$\%^+ = 60 + 30\left(\frac{\alpha_{f1}\ell_2}{\ell_1}\right)\left(1.5 - \frac{\ell_2}{\ell_1}\right)$$

In the above three equations, if $\beta_t > 2.5$, use 2.5 and if $\alpha_{f1}\ell_2/\ell_1 > 1$, use 1.

In Section 13.6.5, the Code requires that the beam be allotted 85% of the column strip moment if $\alpha_{f1}(\ell_2/\ell_1) \geq 1.0$. Should $\alpha_{f1}(\ell_2/\ell_1)$ be between 1.0 and zero, the moment allotted to the beam is determined by linear interpolation from 85% to 0%. The part of the moment not given to the beam is allotted to the slab in the column strip.

Finally, the Code (13.6.6) requires that the portion of the design moments not resisted by the column strips as previously described is to be allotted to the corresponding half middle strip. The middle strip will be designed to resist the total of the moments assigned to its two half middle strips.

16.9   DESIGN OF AN INTERIOR FLAT PLATE

In this section an interior flat plate is designed by the direct design method The procedure specified in Chapter 13 of the ACI Code is applicable not only to flat plates but also to flat slabs, waffle slabs, and two-way slabs with beams. The steps necessary to perform the designs are briefly summarized at the end of this paragraph. The order of the steps may have to be varied somewhat for different types of slab designs. Either the direct design method or the equivalent frame method may be used to determine the design moments. The design steps are as follows:

1. Estimate the slab thickness to meet the Code requirements.
2. Determine the depth required for shear.
3. Calculate the total static moments to be resisted in the two directions.
4. Estimate the percentages of the static moments that are positive and negative, and proportion the resulting values between the column and middle strips.
5. Select the reinforcing.

Example 16.3 illustrates this method of design applied to a flat plate.

EXAMPLE 16.3

Design an interior flat plate for the structure considered in Example 16.1. This plate is shown in Figure 16.13. Assume a service live load equal to 80 psf, a service dead load equal to 110 psf (including slab weight), $f_y = 60,000$ psi, $f'_c = 3000$ psi, normal-weight concrete, and column heights of 12 ft.

SOLUTION
Estimate Slab Thickness
When shear is checked, the 7-in. slab estimated in Example 16.1 is not quite sufficient. One alternative is to increase f'_c from 3000 psi, which is a fairly low strength. However, we will increase the slab thickness. The calculations for the 7-in. thick slab are the same as those that follow for the 7.5-in. slab thickness with the exception of the slab thickness change.

$$\therefore \text{Try } 7\tfrac{1}{2} \text{ in. slab}$$

Determine Depth Required for Shear
Using d for shear equal to the estimated average of the d values in the two directions, we obtain

$$d = 7.50'' - \frac{3''}{4} \text{ cover} - 0.50'' = 6.25''$$

$$q_u = (1.2)(110) + (1.6)(80) = 260 \text{ psf}$$

Checking One-Way or Beam Shear (Seldom Controls in Two-way Floor Systems)
Using the dimensions shown in Figure 16.14, we obtain

$$V_{u1} = (8.81)(260) = 2291 \text{ lb for a } 12'' \text{ width}$$

$$\phi V_c = \phi 2\lambda \sqrt{f'_c} bd$$

$$= (0.75)(1.0)(2\sqrt{3000})(12)(6.25)$$

$$= 6162 \text{ lb} > 2291 \text{ lb} \qquad \underline{\underline{\text{OK}}}$$

Checking Two-Way or Punching Shear around the Column

$$b_o = (2)(16 + 6.25) + (2)(12 + 6.25) = 81''$$

$$V_{u2} = \left[(20)(16) - \left(\frac{16 + 6.25}{12} \right) \left(\frac{12 + 6.25}{12} \right) \right] (0.260)$$

$$= 82.47 \text{ k} = 82,470 \text{ lb}$$

$$\phi V_c = (0.75)(1.0)(4\sqrt{3000})(81)(6.25)$$

$$= 83,185 \text{ lb} > 82,470 \text{ lb} \qquad \underline{\underline{\text{OK}}}$$

$$\text{Use } h = 7\tfrac{1}{2}''$$

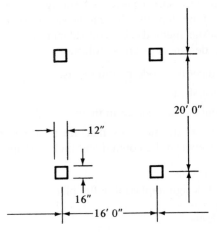

Figure 16.13 Interior panel of flat-plate structure of Example 16.1.

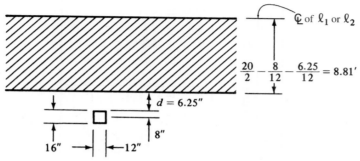

Figure 16.14

Calculate Static Moments in the Long and Short Directions

$$M_{o\ell} = \frac{q_u \ell_2 \ell_n^2}{8} = \frac{(0.260)(16)\left(20 - \frac{16}{12}\right)^2}{8} = 181.2 \text{ ft-k}$$

$$M_{os} = \frac{q_u \ell_1 \ell_n^2}{8} = \frac{(0.260)(20)\left(16 - \frac{12}{12}\right)^2}{8} = 146.2 \text{ ft-k}$$

Proportion the Static Moments to the Column and Middle Strips and Select the Reinforcing
The static moment is divided into positive and negative moments in accordance with ACI 13.6.2. Since this is an interior span, 65% of the total static moment is negative, and 35% positive. See also the right-hand side of Figure 16.10(c). If this example were an end span, then the total static moment would be divided into positive and negative values in accordance with Table 16.2.

The next step is to divide the moments determined in the previous paragraph into column and middle strips. Again, since this is an interior span, Table 16.3 applies. Since $\alpha_{f1} = 0$ (no interior beams), 75% of the moment goes to the column strip. This value is independent of ℓ_2/ℓ_1 for the case where there are no interior beams. The remaining 25% of the moment is assigned to the middle strip, half on each side of the column strip. Table 16.5 is used to determine how much of the total positive moment is assigned to the column strip. In this case, since $\alpha_{f1} = 0$, 60% goes to the column strip, and the remaining 40% is assigned to the middle strip, half on each side of the column strip.

These calculations can be conveniently arranged, as in Table 16.6. This table is very similar to the one used for the design of the continuous one-way slab in Chapter 14. To assist in the interpretation of Table 16.6, the numbers in the first column will be discussed. The first is the determination of M_u. This calculation uses the 0.65 factor from ACI 13.6.2 and the 0.75 factor from Table 16.3, both applied to the total static moment of 181.2 ft-k. Dividing this value of $M_u = -88.4$ ft-k by $\phi b d^2$ ($\phi = 0.9$, $b = 8$ ft. $= 96$ in., $d = 6.5$ in.) results in $M_u/\phi b d^2 = 290.6$ psi. From Table A12, $\rho = 0.00516$ (by interpolation). *The area of reinforcing steel in the column strip is $A_s = \rho b d = 0.00516(96)(6.5) = 3.22$ in.^{2}.* A bar selection of 17 #4 bars is chosen, having a total $A_s = 3.34$ in.^{2}. The remaining entries in Table 16.6 follow a similar procedure.

As the different percentages of moments are selected from the tables for the column and middle strips of this slab, it will be noted that $\alpha_f = 0$.

In the solution to Example 16.3 it will be noted that the $M_u/\phi b d^2$ values are sometimes quite small, and thus most of the ρ values do not fall within Table A.12 (see Appendix). For such cases the authors use the temperature and shrinkage minimum $0.0018bh$.

Actually, the temperature reinforcing includes bars in the top and bottom of the slab. In the negative moment region, some of the positive steel bars have been extended into the support region and are also available for temperature and shrinkage steel. If desirable, these positive bars can be lapped instead of being stopped in the support.

The selection of the reinforcing bars is the final step taken in the design of this flat plate. The Code Figure 13.3.8 (given as Figure 16.15 here) shows the minimum lengths of slab reinforcing bars for flat plates and for flat slabs with drop panels. This figure shows that some of the positive reinforcing must be run into the support area.

Table 16.6

	Long span (estimate $d = 6.50''$)				Short span (estimate $d = 6.00''$)			
	Column strip (8′)		Middle strip (8′)		Column strip (8′)		Middle strip (12′)	
	−	+	−	+	−	+	−	+
M_u	(0.65)(0.75) (181.2) $= -88.4$ ft-k	(0.35)(0.60) (181.2) $= +38.1$	(0.65)(181.2) -88.4 $= -29.4$ ft-k	(0.35)(181.2) -38.1 $= +25.3$ ft-k	(0.65)(0.75) (146.2) $= -71.3$ ft-k	(0.35)(0.60) (146.2) $= +30.7$ ft-k	(0.65)(146.2) -71.3 $= -23.8$ ft-k	(0.35)(146.2) -30.7 $= +20.5$ ft-k
$\dfrac{M_u}{\phi bd^2}$	290.6 psi	125.2 psi	97.6 psi	83.2 psi	275.1 psi	118.4 psi	61.2	52.7 psi
ρ^*	$0.00516bd$	$0.00214bd$	$0.0018bh$	$0.0018bh$	$0.00486bd$	$0.00202bd$	$0.0018bh$	$0.0018bh$
A_s	3.22 in.²	1.34 in.²	1.30 in.²	1.30 in.²	2.80 in.²	1.16 in.²	1.94 in.²	1.94 in.²
Bars selected	17 #4	7 #4	7 #4	7 #4	15 #4	6 #4	10 #4	10 #4

*Values may not be less than the temperature and shrinkage minimum $0.0018bh$.

506

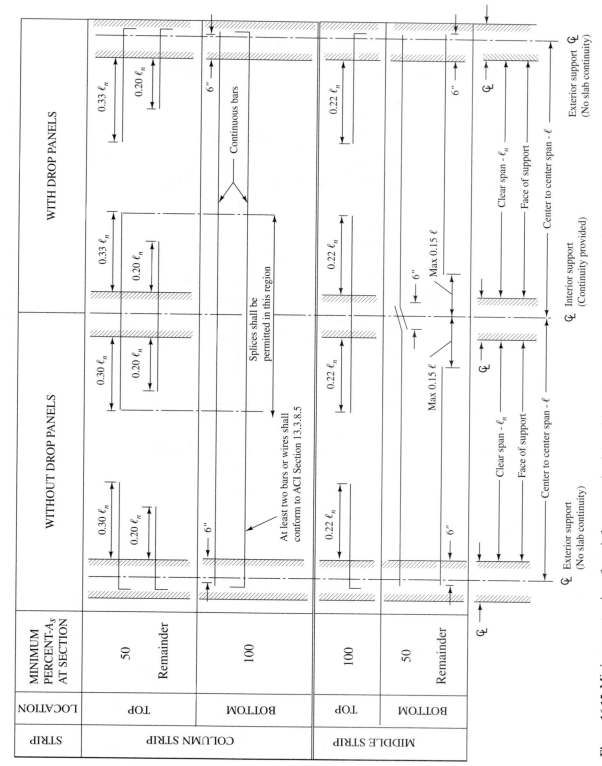

Figure 16.15 Minimum extensions for reinforcement in slabs without beams (see ACI Section 12.11.1 for reinforcement extension into supports).

507

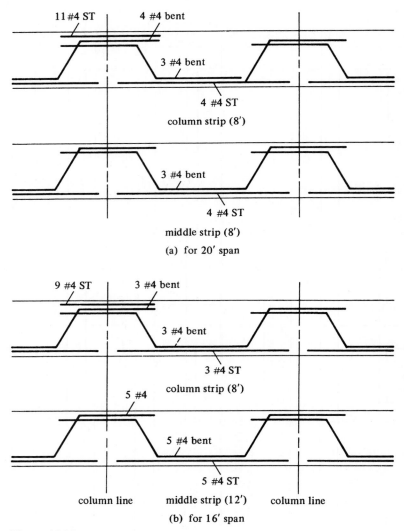

Figure 16.16

The bars selected for this flat plate are shown in Figure 16.16. Bent bars are used in this example, but straight bars could have been used just as well. There seems to be a trend among designers in the direction of using more straight bars in slabs and fewer bent bars.

16.10 PLACING OF LIVE LOADS

The moments in a continuous floor slab are appreciably affected by different positions or patterns of the live loads. The usual procedure, however, is to calculate the total static moments, assuming that all panels are subjected to full live load. When different loading patterns are used, the moments can be changed so much that overstressing may occur in the slab.

Section 13.7.6.2 of the Code states that if a variable unfactored live load does not exceed three-fourths of the unfactored dead load, or if it is of a type such that all panels will be loaded simultaneously, it is permissible to assume that full live load placed over the entire area will cause maximum moment values throughout the entire slab system.

For other loading conditions it may be assumed (according to ACI Section 13.7.6.3) that maximum positive moment at the midspan of a panel will occur when three-fourths of the full factored live load is placed on the panel in question and on alternate spans. It may be further assumed that the maximum negative moment at a support will occur when three-fourths of the full factored live load is placed only on the adjacent spans.

The Code permits the use of the three-fourths factor because the absolute maximum positive and negative moments cannot occur simultaneously under a single loading condition and also because some redistribution of moments is possible before failure will occur. Although some local overstress may be the result of this procedure, it is felt that the ultimate capacity of the system after redistribution will be sufficient to resist the full factored dead and live loads in every panel.

The moment determined as described in the last paragraph may not be less than moments obtained when full factored live loads are placed in every panel (ACI 13.7.6.4).

16.11   ANALYSIS OF TWO-WAY SLABS WITH BEAMS

In this section the moments are determined by the direct design method for an exterior panel of a two-way slab with beams. The example problem presented in this section is about as complex as any that may arise in flat plates, flat slabs, or two-way slabs with beams using the direct design method.

The requirements of the Code are so lengthy and complex that in Example 16.4, which follows, the steps and appropriate Code sections are spelled out in detail. The practicing designer should obtain a copy of the *CRSI Design Handbook*, because the tables therein will be of tremendous help in slab design.

EXAMPLE 16.4

Determine the negative and positive moments required for the design of the exterior panel of the two-way slab with beam structure shown in Figure 16.17. The slab is to support a live load of 120 psf and a dead load of 100 psf including the slab weight. The columns are 15 in. × 15 in. and 12 ft long. The slab is supported by beams along the column line with the cross section shown. Determine the slab thickness and check the shear stress if $f'_c = 3000$ psi and $f_y = 60,000$ psi.

SOLUTION

1. Check ACI Code limitations (13.6.1). These conditions, which are discussed in Section 16.7 of this text, are met. The first five of these criteria are easily satisfied by inspection. The sixth requires calculations that follow.

2. Minimum thickness as required by Code (9.5.3)

   (a) Assume $h = 6$ in.

   (b) Effective flange projection of column line beam as specified by the Code (13.2.4)

   $$= 4h_f = (4)(6) = 24'' \text{ or } h - h_f = 20 - 6 = \underline{\underline{14''}}$$

   (c) Gross moments of inertia of T beams. The following values are the gross moments of inertia of the edge and interior beams computed, respectively, about their centroidal axes. Many designers use approximate values for these moments of inertia, I_s, with almost identical results for slab thicknesses. One common practice is to use 2 times the gross moment of inertia of the stem (using

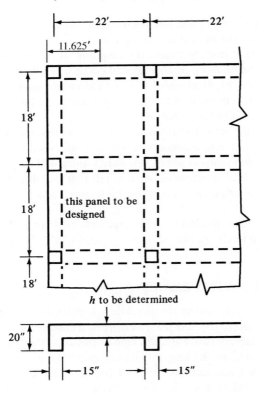

Figure 16.17

a depth of stem running from top of slab to bottom of stem) for interior beams and 1.5 times the stem gross moment of inertia for edge beams.

$$I \text{ for edge beams} = 13{,}468 \text{ in.}^4$$

$$I \text{ for interior beams} = 15{,}781 \text{ in.}^4$$

(d) Calculating α values (where α is the ratio of the stiffness of the beam section to the stiffness of a width of slab bounded laterally by the centerline of the adjacent panel, if any, on each side of the beam).

For edge beam $\left(\text{width} = \dfrac{1}{2} \times 22 + \dfrac{7.5}{12} = 11.625 \text{ ft} \right)$

$$I_s = \left(\frac{1}{12} \right)(12 \times 11.625)(6)^3 = 2511 \text{ in.}^4$$

$$\alpha_f = \frac{13{,}468}{2511} = 5.36$$

For 18-ft interior beam (with 22-ft slab width)

$$I_s = \left(\frac{1}{12} \right)(12 \times 22)(6)^3 = 4752 \text{ in.}^4$$

$$\alpha_f = \frac{15{,}781}{4752} = 3.32$$

For 22-ft interior beam (with 18-ft slab width)

$$I_s = \left(\frac{1}{12}\right)(12 \times 18)(6)^3 = 3888 \text{ in.}^4$$

$$\alpha_f = \frac{15{,}781}{3888} = 4.06$$

$$\text{Avg. } \alpha_f = \alpha_{fm} = \frac{5.36 + 3.32 + (2)(4.06)}{4} = 4.20$$

$$\beta = \text{ratio of long to short clear span} = \frac{22 - \dfrac{15}{12}}{18 - \dfrac{15}{12}} = 1.24$$

(e) Now that we have determined the values of α_f in the two perpendicular directions, the sixth and final limitation for use of the direct design method (ACI 13.6.1.6) can now be checked.

$$\frac{\alpha_{f1}\ell_2^2}{\alpha_{f2}\ell_1^2} = \frac{4.06(18)^2}{3.32(22)^2} = 0.818$$

Since this value is between 0.2 and 5.0, this condition is satisfied. Note that the directions that are designated as ℓ_1 and ℓ_2 are arbitrary. Had the short direction been used as ℓ_1, the calculation above would simply be inverted, and the ratio would have been 1.22 instead of 0.818. This value would also have been between the limits of 0.2 and 5.0.

(f) Thickness limits by Code Section 9.5.3

$$\ell_n = 22.0 - \frac{15}{12} = 20.75 \text{ ft}$$

As $\alpha_{fm} > 2.0$, use ACI Equation 9-13.

$$h = \frac{\ell_n\left(0.8 + \dfrac{f_y}{200{,}000}\right)}{36 + 9\beta}$$

$$= \frac{(12)(20.75)\left(0.8 + \dfrac{60{,}000}{200{,}000}\right)}{36 + (9)(1.24)} = 5.81 \text{ in.} \leftarrow$$

h not less than 3.5 in. (as per ACI Section 9.5.3.3(c))

Try $h = 6$ in. (shear checked later)

3. Moments for the short-span direction centered on interior column line

$$q_u = (1.2)(100) + (1.6)(120) = 312 \text{ psf}$$

$$M_o = \frac{(q_u\ell_2)(\ell_n)^2}{8} = \frac{(0.312)(22)(16.75)^2}{8} = 241 \text{ ft-k}$$

(a) Dividing this static design moment into negative and positive portions, as per Section 13.6.3.2 of the Code

$$\text{negative design moment} = (0.65)(241) = -157 \text{ ft-k}$$

$$\text{positive design moment} = (0.35)(241) = +84 \text{ ft-k}$$

(b) Allotting these interior moments to beam and column strips, as per Section 13.6.4 of the Code

$$\frac{\ell_2}{\ell_1} = \frac{22}{18} = 1.22$$

$$\alpha_{f1} = \alpha_f \text{ in direction of short span} = 3.32$$

$$\alpha_{f1}\frac{\ell_2}{\ell_1} = (3.32)(1.22) = 4.05$$

The portion of the interior negative moment to be resisted by the column strip, as per Table 16.3 of this chapter, by interpolation is $(0.68)(-157) = -107$ ft-k. This result can also be obtained from the equation

$$\%_{\text{int col}}^- = 75 + 30\left(\frac{\alpha_{f1}\ell_2}{\ell_1}\right)\left(1 - \frac{\ell_2}{\ell_1}\right) = 75 + 30(1)\left(1 - \frac{22}{18}\right) = 68.3\%$$

$$M_{\text{int col}}^- = 0.683(157) = 107 \text{ ft-k}$$

Note that since $\alpha_{f1}\ell_2/\ell_1 = 4.05 > 1$, a value of 1 was used in the above equation.

This -107 ft-k is allotted 85% to the beam (Code 13.6.5), or -91 ft-k, and 15% to the slab, or -16 ft-k. The remaining negative moment, $157 - 107 = 50$ ft-k, is allotted to the middle strip.

The portion of the interior positive moment to be resisted by the column strip, as per Table 16.5 of this chapter, by interpolation is $(0.68)(+84) = +57$ ft-k. The 68% value can also be obtained from the equation for Table 16.5. This 57 ft-k is allotted 85% to the beam (Code 13.6.5), or $+48$ ft-k, and 15% to the slab, or $+9$ ft-k. The remaining positive moment, $84 - 57 = 27$ ft-k, goes to the middle strip.

4. Moments for the short-span direction centered on the edge beam

$$M_o = \frac{(q_u\ell_2)(\ell_n)^2}{8} = \frac{(0.312)(11.625)(16.75)^2}{8} = 127 \text{ ft-k}$$

(a) Dividing this static design moment into negative and positive portions, as per Section 13.6.3.2 of the Code

$$\text{Negative design moment} = (0.65)(127) = -83 \text{ ft-k}$$

$$\text{Positive design moment} = (0.35)(127) = +44 \text{ ft-k}$$

(b) Allotting these exterior moments to beam and column strips, as per Section 13.6.4 of the Code

$$\frac{\ell_2}{\ell_1} = \frac{22}{18} = 1.22$$

$$\alpha_{f1} = \alpha_{f1} \text{ for edge beam} = 5.36$$

$$\alpha_{f1}\frac{\ell_2}{\ell_1} = (5.36)(1.22) = 6.54$$

5. The portion of the exterior negative moment going to the column strip, from Table 16.4 of this chapter, by interpolation is $(0.68)(-83) = -56$ ft-k. This -56 ft-k is allotted 85% to the beam (Code 13.6.5), or -48 ft-k, and 15% to the slab, or -8 ft-k. The remaining negative moment, $83 - 56 = -27$ ft-k, is allotted to the middle strip.

The portion of the exterior positive moment to be resisted by the column strip, as per Table 16.5 or the equation for Table 16.5, is $(0.68)(+44) = +30$ ft-k. This 30 ft-k is allotted 85% to the beam, or $+26$ ft-k, and 15% to the slab, or $+5$ ft-k. The remaining positive moment, $44 - 30 = +14$ ft-k, goes to the middle strip.

A summary of the short-span moments is presented in Table 16.7.

6. Moments for the long-span direction

$$M_o = \frac{(q_u\ell_2)(\ell_n)^2}{8} = \frac{(0.312)(18)(20.75)^2}{8} = 302.3 \text{ ft-k}$$

Table 16.7 Short-Span Moments (ft-k)

	Column strip moments		Middle strip slab moments
	Beam	Slab	
Interior slab-beam strip			
Negative	− 91	−16	− 50
Positive	+48	+9	+30
Exterior slab-beam strip			
Negative	− 48	−9	− 27
Positive	+26	+5	+ 16

(a) From Table 16.2 of this chapter (Code 13.6.3.3) for an end span with beams between all interior supports:

$$\text{Interior negative factored moment} = 0.70M_o = -(0.70)(302.3) = -212 \text{ ft-k}$$
$$\text{Positive factored moment} = 0.57M_o = +(0.57)(302.3) = 172 \text{ ft-k}$$
$$\text{Exterior negative factored moment} = 0.16M_o = -(0.16)(302.3) = -48 \text{ ft-k}$$

These factored moments may be modified by 10%, according to Section 13.6.7 of the Code, but this reduction is neglected here.

(b) Allotting these moments to beam and column strips

$$\frac{\ell_2}{\ell_1} = \frac{18}{22} = 0.818$$

$$\alpha_{f1} = \alpha_{f1} \text{ for the 22' beam} = 4.06$$

$$\alpha_{f1} \frac{\ell_2}{\ell_1} = (4.08)(0.818) = 3.32$$

Next an expression is given for β_t. It is the ratio of the torsional stiffness of an edge beam section to the flexural stiffness of a width of slab equal to the span length of the beam measured center to center of supports.

$$\beta_t = \frac{E_{cb}C}{2E_{cs}I_s}$$

Involved in the equation is a term C, which is a property of the cross-sectional area of the torsion arm estimating the resistance to twist.

$$C = \Sigma\left(1 - 0.63\frac{x}{y}\right)\frac{x^3 y}{3}$$

where x is the length of the short side of each rectangle and y is the length of the long side of each rectangle. The exterior beam considered here is described in ACI Section 13.2.4 and is shown in Figure 16.18, together with the calculation of C. The beam cross section could be divided into rectangles in other ways, but the configuration shown results in the greatest value for C.

$$\beta_t = \frac{E_{cb}C}{2E_{cs}I_s} = \frac{(E_c)(12,605)}{(2)(E_c)(3888)} = 1.62$$

The portion of the interior negative design moment allotted to the column strip, from Table 16.3, by interpolation or by equation is $(0.80)(-212) = -170$ ft-k. This −170 ft-k is allotted 85% to the beam (Code 13.6.5), or −145 ft-k, and 15% to the slab, or −26 ft-k. The remaining negative moment, $-212 + 170 = -42$ ft-k, is allotted to the middle strip.

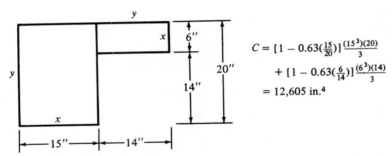

$$C = [1 - 0.63(\tfrac{15}{20})]\tfrac{(15^3)(20)}{3}$$
$$+ [1 - 0.63(\tfrac{6}{14})]\tfrac{(6^3)(14)}{3}$$
$$= 12{,}605 \text{ in.}^4$$

Figure 16.18 Evaluation of C.

The portion of the positive design moment to be resisted by the column strip, as per Table 16.5, is $(0.80)(172) = +138$ ft-k. This 138 ft-k is allotted 85% to the beam, or $(0.85)(138) = 117$ ft-k, and 15% to the slab, or $+21$ ft-k. The remaining positive moment, $172 - 138 = 34$ ft-k, goes to the middle strip.

The portion of the exterior negative moment allotted to the column strip is obtained by double interpolation from Table 16.4, and is $(0.86)(-48) = -42$ ft-k. This -42 ft-k is allotted 85% to the beam, or -36 ft-k, and 15% to the slab, or -6 ft-k. The remaining negative moment, -6 ft-k, is allotted to the middle strip.

A summary of the long-span moments is presented in Table 16.8.

7. Check shear strength in the slab at a distance d from the face of the beam. Shear is assumed to be produced by the load on the tributary area shown in Figure 16.19, working with a 12-in.-wide strip as shown.

$$\text{average } d = h - \text{cover} - \text{one bar diam.} = 6.00 - \frac{3}{4} - \frac{1}{2} = 4.75 \text{ in.}$$

Table 16.8 Long-Span Moments (ft-k)

	Column strip moments		Middle strip
	Beam	Slab	slab moments
Interior negative	−145	−26	−42
Positive	+117	+21	+34
Exterior negative	−36	−6	−6

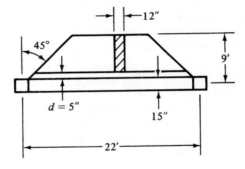

Figure 16.19

$$V_u = (0.312)\left(9 - \frac{7.5}{12} - \frac{4.75}{12}\right) = 2.49 \text{ k} = 2490 \text{ lb}$$

$$\phi V_c = (0.75)(1.0)(2\sqrt{3000})(12)(4.75) = 4684 \text{ lb} > 2490 \text{ lb} \qquad \underline{\underline{\text{OK}}}$$

16.12   TRANSFER OF MOMENTS AND SHEARS BETWEEN SLABS AND COLUMNS

On many occasions the maximum load that a two-way slab can support is dependent upon the strength of the joint between the column and the slab. Not only is the load transferred by shear from the slab to the column along an area around the column, but also there may be moments that have to be transferred as well. The moment situation is usually most critical at the exterior columns.

If there are moments to be transferred, they will cause shear stresses of their own in the slabs, as will be described in this section. Furthermore, shear forces resulting from moment transfer must be considered in the design of the lateral column reinforcement (that is, ties and spirals), as stated in Section 11.10.1 of the Code.

When columns are supporting slabs without beams (that is, flat plates or flat slabs), the load transfer situation between the slabs and columns is extremely *critical*. Perhaps if we don't have the exact areas and positions of the flexural reinforcing designed just right throughout the slab, inelastic redistribution of moments (ACI 13.6.7) may still allow the system to perform adequately; however, if we handle the shear strength situation incorrectly, the results may very well be disastrous.

The serious nature of this problem is shown in Figure 16.20, where it can be seen that if there is no spandrel beam, all of the total exterior slab moment has to be transferred to the column. The transfer is made by both flexure and eccentric shear, the latter being located at a distance of about $d/2$ from the column face.

Section 13.6.3.6 of the Code states that for moment transfer between the slab and edge column, the gravity load moment to be transferred shall be $0.3M_o$ (where M_o is the factored statical moment).

When gravity loads, wind or earthquake loads, or other lateral forces cause a transfer of an unbalanced moment between a slab and a column, a part of the moment equal to $\gamma_f M_u$ shall be transferred by flexure, according to ACI Section 13.5.3.2. Based on both tests and experience, this

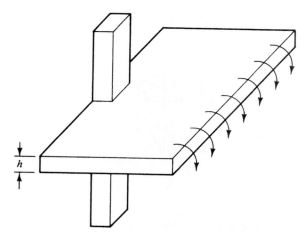

Figure 16.20

transfer is to be considered to be made within an effective slab width between lines that are located 1.5 times the slab or drop panel thickness outside opposite faces of the column or capital. The value γ_f is to be taken as

$$\gamma_f = \cfrac{1}{1 + \cfrac{2}{3}\sqrt{\cfrac{b_1}{b_2}}} \qquad \text{(ACI Equation 13-1)}$$

With reference made to Figure 16.21, b_1 is the length of the shear perimeter, which is perpendicular to the axis of bending $(c_1 + d)$, and b_2 is the length of the shear perimeter parallel to the axis of bending $(c_2 + d)$. Also, c_1 is the width of the column perpendicular to the axis of bending, while c_2 is the column width parallel to the axis of bending.

The remainder of the unbalanced moment referred to as $\gamma_v M_u$ by the Code is to be transferred by eccentricity of shear about the centroid of the critical section.

$$\gamma_v = 1 - \gamma_f \qquad \text{(ACI Equation 11-39)}$$

From this information, the shear stresses due to moment transfer by eccentricity of shear are assumed to vary linearly about the centroid of the critical section described in the last paragraph and are to be added to the usual factored shear forces. (In other words, there is the usual punching shear situation plus a twisting due to the moment transfer that increases the shear.) The resulting shear stresses may not exceed $\phi V_n = \frac{\phi V_c}{b_o d}$ for members without shear reinforcement, and $\phi V_n = \frac{\phi(V_c + V_s)}{b_o d}$ for members with shear reinforcement other than shear heads. V_c in the two previous equations is the lesser of ACI Equations 11-33, 11-34, or 11-35 (Section 12.6 of this text).

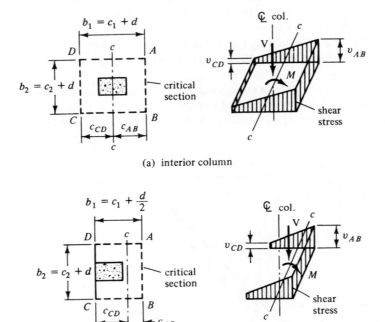

(a) interior column

(b) edge column

Figure 16.21 Assumed distribution of shear stress. (ACI Figure R11.11.7.2.)

The combined stresses are calculated by the expressions to follow, with reference being made to Figure 16.21 and ACI Commentary R11.11.7.2:

$$v_u \text{ along } AB = \frac{V_u}{A_c} + \frac{\gamma_v M_u c_{AB}}{J_c}$$

$$v_u \text{ along } CD = \frac{V_u}{A_c} - \frac{\gamma_v M_u c_{CD}}{J_c}$$

In these expressions, A_c is the area of the concrete along the assumed critical section. For instance, for the interior column of Figure 16.21(a) it would be equal to $(2a + 2b)d$, and for the edge column of Figure 16.21(b) it would equal $(2a + b)d$.

J_c is a property analogous to the polar moment of inertia about the z–z axis of the shear areas located around the periphery of the critical section. First, the centroid of the shear area A_f is located by taking moments. The centroid is shown with the distances c_{AB} and c_{CD} in both parts (a) and (b) of Figure 16.21. Then the value of J_c is computed for the shear areas. For the interior column of part (c), it is

$$J_c = d\left[\frac{a^3}{6} + \frac{ba^2}{2}\right] + \frac{ad^3}{6}$$

and for the edge column of part (b), it is

$$J_c = d\left[\frac{2a^3}{3} - (2a + b)(c_{AB})^2\right] + \frac{ad^3}{6}$$

Example 16.5 shows the calculations involved for shear and moment transfer for an exterior column.

The Commentary (R11.11.7.3) states that the critical section described for two-way action for slabs located at $d/2$ from the perimeter of the column is appropriate for the calculation of shear stresses caused by moment transfer even when shearheads are used. Thus the critical sections for direct shear and shear due to moment transfer will be different from each other.

The total reinforcing that will be provided in the column strip must include additional reinforcing concentrated over the column to resist the part of the bending moment transferred by flexure $= \gamma_f M_u$.

EXAMPLE 16.5

For the flat slab of Figure 16.22, compute the negative steel required in the column strip for the exterior edge indicated. Also check the slab for moment and shear transfer at the exterior column; $f_c' = 3000$ psi, $f_y = 60{,}000$ psi, and $LL = 100$ psf. An 8-in. slab has already been selected with $d = 6.75$ in.

SOLUTION

1. Compute w_u and $M_{o\ell}$.

$$w_d = \left(\frac{8}{12}\right)(150) = 100 \text{ psf}$$

$$w_\ell = 100 \text{ psf}$$

$$w_u = (1.2)(100) + (1.6)(100) = 280 \text{ psf}$$

$$M_{o\ell} = \frac{q_u \ell_2 \ell_n^2}{8} = \frac{(0.280)(18)(18.75)^2}{8} = 221.5 \text{ ft-k}$$

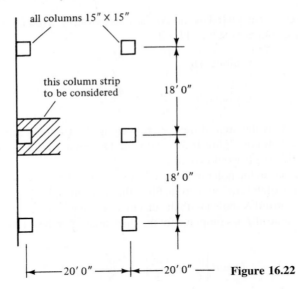

Figure 16.22

2. Determine the exterior negative moment as per Section 13.6.3.3.

$$-0.26 M_{o\ell} = -(0.26)(221.5) = 57.6 \text{ ft-k}$$

$$\text{Width of column strip} = (0.50)(18) = 9' - 0'' = 108''$$

3. ACI Section 13.6.4.2 shows that 100% of the exterior negative moment is to be resisted by the column strip.

4. Design of steel in column strip.

$$\frac{M_u}{\phi b d^2} = \frac{(12)(57{,}600)}{(0.9)(108)(6.75)^2} = 156.1$$

$$\rho = 0.0027 \text{ from Appendix Table A.12}$$

$$A_s = (0.0027)(108)(6.75) = 1.97 \text{ in.}^2 \qquad \underline{\underline{\text{Use 10 \#4}}}$$

5. Moment transfer design.

   (a) The Code (13.5.3.2) states that additional bars must be added over the column in a width = column width + (2)(1.5h) = 15 + (2)(1.5 × 8) = 39''.

   (b) The additional reinforcing needed over the columns is to be designed for a moment = $\gamma_f M_u$. In ACI Equation 13-1 below, b_1 and b_2 are the side dimensions of the perimeter b_o in ACI Section 11.11.1.2 (see Figure 16.21). In this case $b_1 = c_1 + d/2 = 15 + 6.75/2 = 18.375$ in. In the perpendicular direction, $b_2 = c_2 + d = 15 + 6.75 = 21.75$ in.

$$\gamma_f = \frac{1}{1 + \dfrac{2}{3}\sqrt{\dfrac{b_1}{b_2}}} = \frac{1}{1 + \dfrac{2}{3}\sqrt{\dfrac{18.375}{21.75}}} = 0.62 \qquad \textbf{(ACI Equation 13-1)}$$

$$\gamma_f M_u = (0.62)(57.6) = 35.7 \text{ ft-k}$$

   (c) Add four #4 bars in the 39'' width and check to see whether the moment transfer situation is satisfactory. To resist the 35.7 ft-k, we now have the four #4 bars just added plus four #4. This number of bars is obtained by taking the ratio of $39''/108''$ times 10 bars to get 3.6 bars and rounding to four. The total number of bars put in for the column strip design is eight #4 bars (1.60 in.^2).

$$a = \frac{A_s f_y}{0.85 f_c' b} = \frac{(1.57)(60)}{(0.85)(3)(39)} = 0.947''$$

$$\phi M_n = M_u = \phi A_s f_y \left(d - \frac{a}{2} \right)$$

$$= \frac{(0.9)(1.57)(60)\left(6.75 - \dfrac{0.947}{2} \right)}{12}$$

$$= 44.3 \text{ ft-k} > 35.7 \text{ ft-k} \qquad\qquad \text{OK}$$

6. Compute combined shear stress at exterior column due to shear and moment transfer.

(a) ACI Section 13.6.3.5 requires that a moment of $(0.3M_o)(\gamma_v)$ be transferred from the slab to the column by eccentricity of shear. The total moment to be transferred is $0.3(M_o) = 0.3(221.5) = 66.45$ ft-k.

(b) Fraction of unbalanced moment carried by eccentricity of shear $= \gamma_v(0.3M_o)$.

$$\gamma_v = 1 - \frac{1}{1 + \dfrac{2}{3}\sqrt{\dfrac{18.375}{21.75}}} = 0.38$$

$$\gamma_v M_n = (0.38)(66.45) = 25.25 \text{ ft-k}$$

(c) Compute properties of critical section for shear (Figure 16.23).

$$A_c = (2a + b)d = (2 \times 18.375 + 21.75)(6.75) = 394.875 \text{ in.}^2$$

$$c_{AB} = \frac{(2)(18.375)(6.75)\left(\dfrac{18.375}{2}\right)}{394.875} = 5.77 \text{ in.}$$

$$J_c = d\left[\frac{2a^3}{3} - (2a + b)(c_{AB})^2\right] + \frac{ad^3}{6}$$

$$= 6.75\left[\frac{(2)(18.375)^3}{3} - (2 \times 18.375 + 21.75)(5.77)^2\right] + \frac{(18.375)(6.75)^3}{6}$$

$$= 15{,}714 \text{ in.}^4$$

$a = 15 + \dfrac{6.75}{2} = 18.375''$

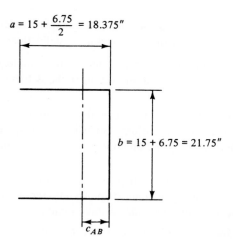

$b = 15 + 6.75 = 21.75''$

c_{AB}

Figure 16.23

(d) Compute gravity load shear to be transferred at the exterior column.

$$V_u = \frac{q_u \ell_1 \ell_2}{2} = \frac{(0.280)(18)(20)}{2} = 50.4 \text{ k}$$

(e) Combined stresses.

$$
\begin{aligned}
v_u &= \frac{V_u}{A_c} + \frac{\gamma_v M_n c_{AB}}{J_c} \\
&= \frac{50,400}{394.875} + \frac{(12 \times 25,250)(5.77)}{15,714} \\
&= 128 + 111 = 239 \text{ psi} \\
&> 4\sqrt{3000} = 219 \text{ psi} \qquad\qquad \underline{\underline{\text{No good}}}
\end{aligned}
$$

$\therefore$ It is necessary to do one or more of the following: increase depth of slab, use higher-strength concrete, use drop panel, or install shearhead reinforcing.

Factored Moments in Columns and Walls

If there is an unbalanced loading of two adjoining spans, the result will be an additional moment at the connection of walls and columns to slabs. The Code (13.6.9.2) provides the approximate equation listed at the end of this paragraph to consider the effects of such situations. This particular equation was derived for two adjoining spans, one longer than the other. It was assumed that the longer span was loaded with dead load plus one-half live load and that only dead load was applied to the shorter span.

$$M_u = 0.07[(q_{du} + 0.5q_{\ell u})\ell_2 \ell_n^2 - q_{Du'} \ell_2' (\ell_n')^2] \qquad \text{(ACI Equation 13-7)}$$

In this expression q'_{du}, ℓ_2', and ℓ_n' are for the shorter spans. The resulting approximate value should be used for unbalanced moment transfer by gravity loading at interior columns unless a more theoretical analysis is used.

16.13   OPENINGS IN SLAB SYSTEMS

According to the Code (13.4), openings can be used in slab systems if adequate strength is provided and if all serviceability conditions of the ACI, including deflections, are met.

1. If openings are located in the area common to intersecting middle strips, it will be necessary to provide the same total amount of reinforcing in the slab that would have been there without the opening.

2. For openings in intersecting column strips, the width of the openings may not be more than one-eighth the width of the column strip in either span. An amount of reinforcing equal to that interrupted by the opening must be placed on the sides of the opening.

3. Openings in an area common to one column strip and one middle strip may not interrupt more than one-fourth of the reinforcing in either strip. An amount of reinforcing equal to that interrupted shall be placed around the sides of the opening.

4. The shear requirements of Section 11.11.6 of the Code must be met.

16.14   COMPUTER EXAMPLES

EXAMPLE 16.6

Use the Chapter 16 Excel Spreadsheet to solve Example 16.4.

SOLUTION   Open the Excel spreadsheet provided for Chapter 16 and open the worksheet Two-Way Slabs. Enter values only for cells in yellow highlight. Note that values for β_t and α_{f1} must be entered. To obtain β_t, open the worksheet C Torsional Constant. Enter values for cells in yellow highlight.

torsional constant C and β_1

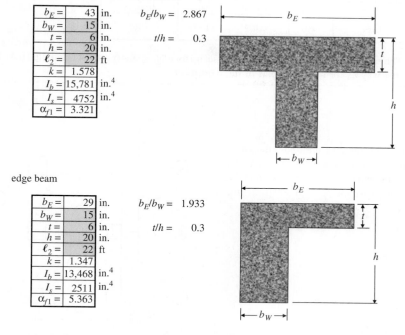

x	y
15	20
11,868.75	
6	14
735.84	
0.0	0.0
0.00	

$C = (1 - 0.63\, x/y)(x^3 y/3) =$

$12{,}605 = C$ in.^{4} y

ℓ_2 (ft)	t (in.)
18	6

$I_s = \ell_2 t^3/12 =$ 3888   in.^{4}

$\beta_t = C/2I_s =$ 1.621

If there are flanges on both sides of the web, a third value of x and y can be entered. In this case, their values are zero.

The value of *af* is obtained from the worksheet Alpha T Beam. Results of the α_f calculations in step 2(d) of Example 16.4 are shown below.

interior beam (flange on both sides)

$b_E =$	43	in.
$b_W =$	15	in.
$t =$	6	in.
$h =$	20	in.
$\ell_2 =$	22	ft
$k =$	1.578	
$I_b =$	15,781	in.^{4}
$I_s =$	4752	in.^{4}
$\alpha_{f1} =$	3.321	

$b_E/b_W = 2.867$

$t/h = 0.3$

edge beam

$b_E =$	29	in.
$b_W =$	15	in.
$t =$	6	in.
$h =$	20	in.
$\ell_2 =$	22	ft
$k =$	1.347	
$I_b =$	13,468	in.^{4}
$I_s =$	2511	in.^{4}
$\alpha_{f1} =$	5.363	

$b_E/b_W = 1.933$

$t/h = 0.3$

The other α_f values are obtained simply by changing ℓ_2 from 22 to 18 ft. Return to the Example 16.6 worksheet and enter $\beta_t = 1.62$ and $\alpha_{f1} = 3.32$. Now determine from Table 16.2 which case (1 through 5) applies to your example. In this situation, case (2) applies since there are beams between all supports. Enter 0.16, 0.57, and 0.70 in the highlighted cells in row 11. In addition to the information determined in Example 16.4, the spreadsheet also determines the required area of reinforcing steel throughout the slab.

Two-way slab

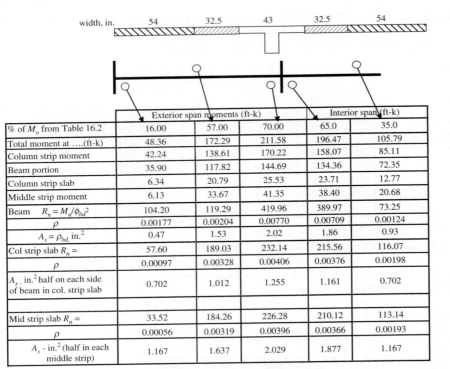

	Exterior span moments (ft-k)			Interior span (ft-k)	
% of M_o from Table 16.2	16.00	57.00	70.00	65.0	35.0
Total moment at ....(ft-k)	48.36	172.29	211.58	196.47	105.79
Column strip moment	42.24	138.61	170.22	158.07	85.11
Beam portion	35.90	117.82	144.69	134.36	72.35
Column strip slab	6.34	20.79	25.53	23.71	12.77
Middle strip moment	6.13	33.67	41.35	38.40	20.68
Beam $R_n = M_u/\phi bd^2$	104.20	119.29	419.96	389.97	73.25
ρ	0.00177	0.00204	0.00770	0.00709	0.00124
$A_s = \rho bd$, in.^{2}	0.47	1.53	2.02	1.86	0.93
Col strip slab $R_n =$	57.60	189.03	232.14	215.56	116.07
ρ	0.00097	0.00328	0.00406	0.00376	0.00198
A_s in.^{2} half on each side of beam in col. strip slab	0.702	1.012	1.255	1.161	0.702
Mid strip slab $R_n =$	33.52	184.26	226.28	210.12	113.14
ρ	0.00056	0.00319	0.00396	0.00366	0.00193
A_s - in.^{2} (half in each middle strip)	1.167	1.637	2.029	1.877	1.167

$\ell_1 =$ 22 ft. (centerline span)
$\ell_2 =$ 18 ft. (transverse span)
$c_1 =$ 15 in. column
$c_2 =$ 15 in. dimension
$t =$ 6 in. (slab thk.)
$d_s =$ 4.75 in. (slab "d")
$\ell_n =$ 20.75 ft. (clear span)
$b_w =$ 15 in. (beam web width)
$h =$ 20 in. (beam depth)
$b_{eff} =$ 43 in.
$y =$ 7.487 in.
$I_g =$ 15781 in.^{4}
$I_s =$ 3888 in.^{4}
$d_b =$ 17.5 in. (beam "d")
$f'_c =$ 3 ksi
$f_y =$ 60 ksi
$m =$ 23.53
$w_D =$ 100 psf
$w_L =$ 120 psf
$w_u =$ 312 psf

$M_{ol} =$ 302.3 ft-k

$\beta_t =$ 1.62 not more than 2.5
$\alpha_{f1} =$ 4.06

$\alpha_{f12}\ell/\ell_1 =$ 1.00
T & S % 0.0018

The edge span part of the spreadsheet is set up for edge spans (part 4 of Example 16.4). Enter information the same way as for the upper part of the worksheet.

PROBLEMS

16.1 Using the ACI Code, determine the minimum thickness required for panels ① and ③ of the flat plate floor shown in the accompanying illustration. Edge beams are not used along the exterior floor edges. $f'_c = 3000$ psi and $f_y = 60,000$ psi. (*Ans.* 9.07″ and 8.12″)

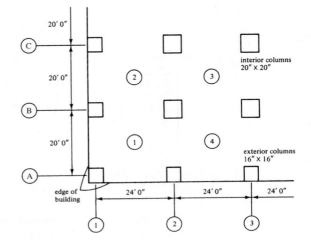

16.2 Assume that the floor system of Problem 16.1 is to support a service live load of 80 psf and a service dead load of 60 psf in addition to its own weight. If the columns are 10 ft long, determine depth required for an interior flat plate for panel ③.

16.3 Repeat Problem 16.2 for panel ① in the structure of Problem 16.1. Include depth required for one-way and two-way shear. (*Ans.* Use 10-in. slab.)

16.4 Determine the required reinforcing in the column strip and middle strips for column line ⑧ in Problem 16.1. Use a slab thickness of 10 in. and #6 bars. Determine theoretical spacings and practical spacings in the exterior span (from column line ① to ②) and in the interior span (from column line ② to ③).

16.5 Use the Flat Plate Worksheet of the Chapter 16 spreadsheet to work Problem 16.4. (*Ans.* $A_s = 6.53$ in.^{2} in column strip at column line ②.∴. Use #6 @ 8 in. in top of slab)

17

Two-Way Slabs, Equivalent Frame Method

17.1 MOMENT DISTRIBUTION FOR NONPRISMATIC MEMBERS

Most of the moment distribution problems the student has previously faced have dealt with prismatic members for which carryover factors of $\frac{1}{2}$, fixed-end moments for uniform loads of $w_u \ell^2/8$, stiffness factors of I/ℓ, and so on were used. Should nonprismatic members be encountered, such as the continuous beam of Figure 17.1, none of the preceding values apply.

Carryover factors, fixed-end moments, and so forth can be laboriously obtained by various methods such as the moment-area and column-analogy methods.[1] There are various tables available however, from which many of these values can be obtained. The tables numbered A.16 through A.20 of Appendix A cover most situations encountered with the equivalent frame method.

Before the equivalent frame method is discussed, an assumed set of fixed-end moments are balanced in Example 17.1 for the nonprismatic beam of Figure 17.1. The authors' purpose in presenting this example is to show the reader how moment distribution can be applied to the analysis of structures consisting of nonprismatic members.

EXAMPLE 17.1

The carryover factors (C.O.), stiffness factors (K), and fixed-end moments (FEM) shown in Figure 17.2 have been assumed for the continuous nonprismatic member of Figure 17.1. Balance these moments by moment distribution. It will be shown later in Example 17.2 how to determine these factors for two-way slab systems. The purpose of this example is to demonstrate the moment distribution method when nonprismatic members are involved.

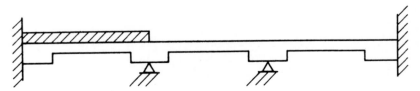

Figure 17.1

---

[1]McCormac, J. C., 1984, *Structural Analysis*, 4th ed. (New York: Harper & Row), pp. 333–334, 567–582.

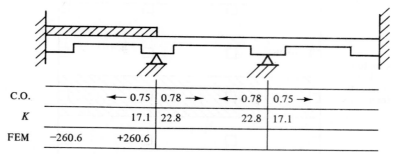

C.O.		← 0.75	0.78 →	← 0.78	0.75 →
K			17.1	22.8	22.8	17.1
FEM	−260.6	+260.6			

Figure 17.2

SOLUTION

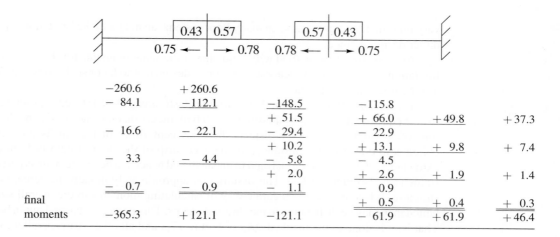

	−260.6	+ 260.6				
	− 84.1	−112.1	−148.5	−115.8		
		+ 51.5	+ 66.0	+49.8	+ 37.3
	− 16.6	− 22.1	− 29.4	− 22.9		
		+ 10.2	+ 13.1	+ 9.8	+ 7.4
	− 3.3	− 4.4	− 5.8	− 4.5		
		+ 2.0	+ 2.6	+ 1.9	+ 1.4
	− 0.7	− 0.9	− 1.1	− 0.9		
final			+ 0.5	+ 0.4	+ 0.3
moments	−365.3	+ 121.1	−121.1	− 61.9	+61.9	+ 46.4

17.2   INTRODUCTION TO THE EQUIVALENT FRAME METHOD

With the preceding example, the authors hoped to provide the reader with a general idea of the kinds of calculations he or she will face with the equivalent frame method. A part of a two-way slab building will be taken out by itself and analyzed by moment distribution. The slab-beam members of this part of the structure will be nonprismatic because of the columns, beams, drop panels, and so on of which they consist. As a result, it will be quite similar to the beam of Figure 17.1, and we will analyze it in the same manner.

The only difference between the direct design method and the equivalent frame method is in the determination of the longitudinal moments in the spans of the equivalent rigid frame. Whereas the direct design method involves a one-cycle moment distribution, the equivalent frame method involves a normal moment distribution of several cycles. The design moments obtained by either method are distributed to the column and middle strips in the same fashion.

It will be remembered that the range in which the direct design method can be applied is limited by a maximum 2-to-1 ratio of live-to-dead load and a maximum ratio of the longitudinal span length to the transverse span length of 2 to 1. In addition, the columns may not be offset by more than 10% of the span length in the direction of the offset from either axis between centerlines of successive columns. There are no such limitations on the equivalent frame method. This is a

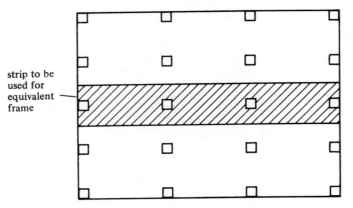

Figure 17.3 Slab beam.

strip to be
used for
equivalent —
frame

very important matter because so many floor systems do not meet the limitations specified for the direct design method.

Analysis by either method will yield almost the same moments for those slabs that meet the limitations required for application of the direct design method. For such cases, it is then simpler to use the direct design method.

The equivalent frame method involves the elastic analysis of a structural frame consisting of a row of equivalent columns and horizontal slab members that are each one panel long and have a transverse width equal to the distance between centerlines of the panels on each side of the columns in question. For instance, the hatched strip of the floor system in Figure 17.3 can be extracted, and the combined slab and beam analyzed to act as a beam element as part of a structural frame (see also Figure 17.4). This assumption approximately models the actual behavior of the structure. It is a reasonably accurate way of calculating moments in the overall structural frame which then may be distributed to the slab and beams. This process is carried out in both directions. That is, it is done to the hatched strip in Figure 17.3 and all other strips parallel to it. Then, it is carried out on strips that are perpendicular to these strips, hence the term two-way floor system.

For vertical loads, each floor, together with the columns above and below, is analyzed separately. For such an analysis, the far ends of the columns are considered fixed. Figure 17.4 shows a typical equivalent slab beam as described in this chapter.

Should there be a large number of panels, the moment at a particular joint in a slab beam can be satisfactorily obtained by assuming that the member is fixed two panels away. This simplification

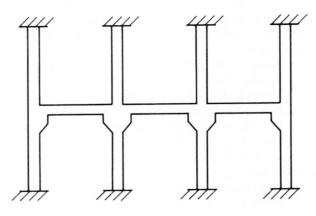

Figure 17.4 Equivalent frame for the crosshatched strip of Figure 17.3 for vertical loading.

is permissible because vertical loads in one panel only appreciably affect the forces in that panel and in the one adjacent to it on each side.

For lateral loads it is necessary to consider an equivalent frame that extends for the entire height of the building, because the forces in a particular member are affected by the lateral forces on all the stories above the floor being considered. When lateral loads are considered, it will be necessary to analyze the frame for them and combine the results with analyses made for gravity loads (ACI Code 13.5.1.3).

The equivalent frame is made up of the horizontal slab, any beams spanning in the direction of the frame being considered, the columns or other members that provide vertical support above and below the slab, and any parts of the structure that provide moment transfer between the horizontal and vertical members. You can see there will be quite a difference in moment transfer from the case where a column provides this transfer and where there is a monolithic reinforced concrete wall extending over the full length of the frame. For cases in between, the stiffnesses of the torsional members such as edge beams will be estimated.

The same minimum ACI slab thicknesses must be met as in the direct design method. The depths should be checked for shear at columns and other supports, as specified in Section 11.12 of the Code. Once the moments have been computed, it will also be necessary to check for moment shear transfer at the supports.

The analysis of the frame is made for the full design live load applied to all spans, unless the actual unfactored live load exceeds 0.75 times the unfactored dead load (ACI Code 13.7.6). When the live load is greater than 0.75 times the dead load, a pattern loading with three-fourths times the live load is used for calculating moments and shears.

The maximum positive moment in the middle of a span is assumed to occur when three-fourths of the full design load is applied in that panel and in alternate spans. The maximum negative moment in the slab at a support is assumed to occur when three-fourths of the full design live load is applied only to the adjacent panels. The values so obtained may not be less than those calculated, assuming full live loads in all spans.

17.3   PROPERTIES OF SLAB BEAMS

The parts of the frame are the slabs, beams, drop panels, columns, and so on. Our first objective is to compute the properties of the slab beams and the columns (that is, the stiffness factors, distribution factors, carryover factors, and fixed-end moments). To simplify this work, the properties of the members of the frame are permitted by the ACI Code (13.7.3.1) to be based on their gross moments of inertia rather than their transformed or cracked sections. Despite the use of the gross dimensions of members, the calculations involved in determining the properties of nonprismatic members still represent a lengthy task, and we will find the use of available tables very helpful.

Figures 17.5 and 17.6 present sketches of two-way slab structures, together with the equivalent frames that will be used for their analysis. A flat slab with columns is shown in Figure 17.5(a). Cross sections of the structure are shown through the slab in part (b) of the figure and through the column in part (c). Then in part (d) the equivalent frame that will be used for the actual numerical calculations is shown. In this figure, E_{cs} is the modulus of elasticity of the concrete slab. With section 2-2, a fictitious section is shown that will have a stiffness approximately equivalent to that of the actual slab and column. An expression for I for the equivalent section is also given. In this expression, c_2 is the width of the column in a direction perpendicular to the direction of the span, and ℓ_2 is the width of the slab beam. The gross moment of inertia at the face of the support is calculated and is divided by $(1 - c_2/\ell_2)^2$. This approximates the effect of the large increase in depth provided by the column for the distance in which the slab and column are in contact.

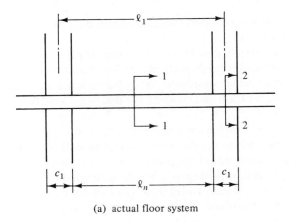

(a) actual floor system

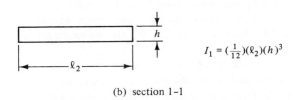

$$I_1 = (\tfrac{1}{12})(\ell_2)(h)^3$$

(b) section 1–1

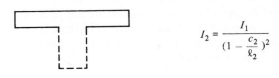

$$I_2 = \frac{I_1}{(1 - \dfrac{c_2}{\ell_2})^2}$$

(c) equivalent section 2–2

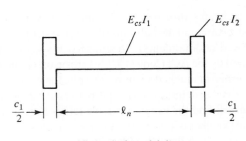

(d) equivalent slab beam
stiffness diagram

Figure 17.5 Slab system without beams.

In Figure 17.6, similar sketches and I values are shown for a slab with drop panels. Figure 13.7.3 of the 1983 ACI Commentary showed such information for slab systems with beams and for slab systems with column capitals.

With the equivalent slab beam stiffness diagram it is possible, using the conjugate beam method, column analogy, or some other method, to compute stiffness factors, distribution factors,

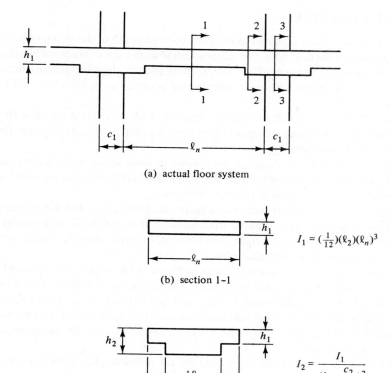

(a) actual floor system

$$I_1 = (\tfrac{1}{12})(\ell_2)(\ell_n)^3$$

(b) section 1-1

$$I_2 = \frac{I_1}{(1 - \dfrac{c_2}{\ell_2})^2}$$

(c) section 2-2

$$I_3 = \frac{I_2}{(1 - \dfrac{c_2}{\ell_2})^2}$$

(d) equivalent section 3-3

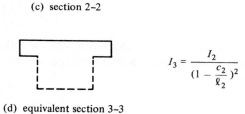

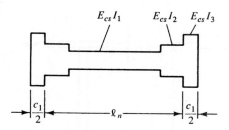

(e) equivalent slab beam stiffness diagram

Figure 17.6 Slab systems with drop panels.

carryover factors, and fixed-end moments for use in moment distribution. Tables A.16 through A.20 of Appendix A of this text provide tabulated values of these properties for various slab systems. The numerical examples of this chapter make use of this information.

17.4 PROPERTIES OF COLUMNS

The length of a column is assumed to run from the mid-depth of the slab on one floor to the mid-depth of the slab on the next floor. For stiffness calculations, the moments of inertia of columns are based on their gross dimensions. Thus if capitals are present, the effect of their dimensions must be used for those parts of the columns. Columns are assumed to be infinitely stiff for the depth of the slabs.

Figure 17.7 shows a sample column, together with its column stiffness diagram. Similar diagrams are shown for other columns (where there are drop panels, capitals, etc.) in Figure 13.7.4 of the 1983 ACI Commentary.

With a column stiffness diagram, the column flexural stiffness K_c can be determined by the conjugate beam procedure or other methods. Tabulated values of K_c are given in Table A.20 of the Appendix of this text for typical column situations.

In applying moment distribution to a particular frame, we need the stiffnesses of the slab beam, the torsional members, and the equivalent column so that the distribution factors can be calculated. For this purpose the equivalent column, the equivalent slab beam, and the torsional members are needed at a particular joint.

For this discussion, reference is made to Figure 17.8, where it is assumed that there is a column above and below the joint in question. Thus the column stiffness (K_c) here is assumed to include the stiffness of the column above (K_{ct}) and the one below (K_{cb}). Thus $\Sigma K_c = K_{ct} + K_{cb}$. In a similar fashion, the total torsional stiffness is assumed to equal that of the torsional members on both sides of the joint ($\Sigma K_t = K_{t1} + K_{t2}$). For an exterior frame, the torsional member will be located on one side only.

The following approximate expression for the stiffness (K_t) of the torsional member was determined using a three-dimensional analysis for various slab configurations (ACI R13.7.5).

$$K_t = \Sigma \frac{9E_{cs}C}{\ell_2\left(1 - \dfrac{c_2}{\ell_2}\right)^3}$$

In this formula C is to be determined with the following expression by dividing the cross section of the torsional member into rectangular parts and summing up the C values for the different parts.

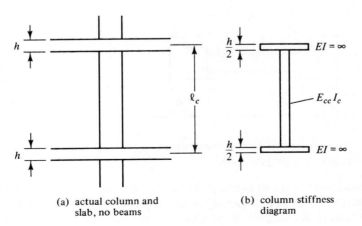

(a) actual column and slab, no beams

(b) column stiffness diagram

Figure 17.7

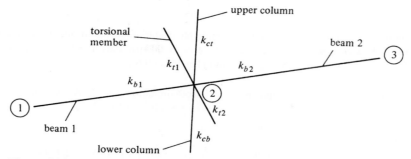

Figure 17.8

$$C = \Sigma\left(1 - 0.63\frac{x}{y}\right)\frac{x^3 y}{3} \qquad \text{(ACI Equation 13-6)}$$

If there is no beam framing into the column in question, a part of the slab equal to the width of the column or capital shall be used as the effective beam. If a beam frames into the column, a T beam or L beam will be assumed, with flanges of widths equal to the projection of the beam above or below the slab but not more than four times the slab thickness.

The flexibility of the equivalent column is equal to the reciprocal of its stiffness, as follows:

$$\frac{1}{K_{ec}} = \frac{1}{\Sigma K_c} + \frac{1}{\Sigma K_t}$$

$$\frac{1}{K_{ec}} = \frac{1}{K_{ct} + K_{cb}} + \frac{1}{K_c + K_t}$$

Solving this expression for the equivalent column stiffness and multiplying through by K_c

$$K_{ec} = \frac{(K_{ct} + k_{cb})(k_t + k_t)}{(k_{ct} + k_{cb}) + (k_t + k_t)}$$

An examination of this brief derivation shows that the torsional flexibility of the slab column joint reduces the joint's ability to transfer moment.

After the value of K_{ec} is obtained, the distribution factors can be computed as follows, with reference again made to Figure 17.8:

$$\text{DF for beam 2-1} = \frac{K_{b1}}{K_{b1} + K_{b2} + K_{ec}}$$

$$\text{DF for beam 2-3} = \frac{K_{b2}}{K_{b1} + K_{b2} + K_{ec}}$$

$$\text{DF for column above} = \frac{K_{ec}/2}{K_{b1} + K_{b2} + K_{ec}}$$

17.5   EXAMPLE PROBLEM

Example 17.2 illustrates the determination of the moments in a flat-plate structure by the equivalent frame method.

EXAMPLE 17.2

Using the equivalent frame method, determine the design moments for the hatched strip of the flat-plate structure shown in Figure 17.9 if $f'_c = 4000$ psi, $f_y = 60{,}000$ psi, and (unfactored dead load) $q_D = 120$ psf and (unfactored live load) $q_L = 82.5$ psf. Column lengths $= 9'6''$.

SOLUTION

1. Determine the depth required for ACI depth limitations (9.5.3). Assume that this has been done and that a preliminary slab $h = 8''$ ($d = 6.75''$) has been selected.

2. Check beam shear for exterior column

$$q_u = 1.2q_D + 1.6q_L = 1.2(120) + 1.6(82.5) = 276 \text{ psf}$$

$$V_u \text{ for } 12'' \text{ width} = (0.276)\left(11.0 - \frac{7.5}{12} - \frac{6.75}{12}\right) = 2.708 \text{ k/ft}$$

$$\phi V_c = \frac{(0.75)(2)(1.0)(\sqrt{4000})(12)(6.75)}{1000} = 7.684 \text{ k/ft} > 2.708 \text{ k/ft} \qquad \underline{\underline{\text{OK}}}$$

3. Check two-way shear around interior columns

$$V_u = \left[(18)(22) - \left(\frac{15+6.75}{12}\right)^2\right](0.276) = 108.39 \text{ k}$$

$$\phi V_c = \frac{(0.75)(4)(1.0)(\sqrt{4000})(4)(15+6.75)(6.75)}{1000} = 111.42 \text{ k} > 108.39 \text{ k} \qquad \underline{\underline{\text{OK}}}$$

4. Using tables in the Appendix, determine stiffness factor and fixed-end moments for the 22-foot spans

$$I_s = \frac{\ell_2 h^3}{12} = \frac{(12 \times 18)(8)^3}{12} = 9216 \text{ in.}^4$$

$$E_{cs} = 3.64 \times 10^6 \text{ psi from Table A.1 of Appendix}$$

With reference to Appendix Table A.16, notice that C values are column dimensions as shown in the figures accompanying Appendix Tables A.16 to A.19. The tables are rather difficult to read.

$$C_{1A} = C_{2A} = C_{1B} = C_{2B} = 15'' = 1.25'$$

$$\frac{C_{1A}}{\ell_1} = \frac{1.25}{22.0} = 0.057$$

$$\frac{C_{1B}}{\ell_1} = \frac{1.25}{22.0} = 0.057$$

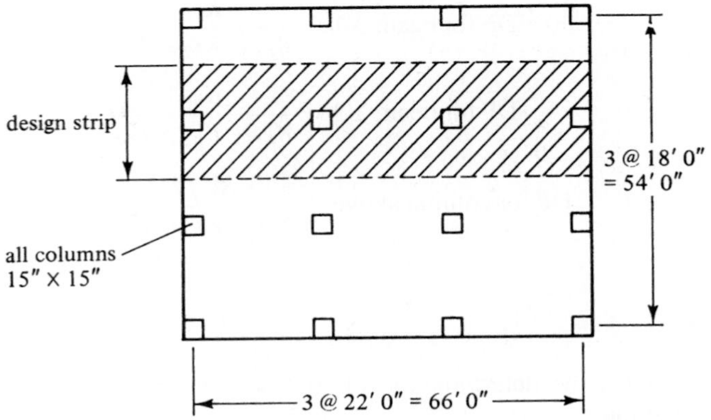

Figure 17.9

By interpolation in the table (noting that A is for near end and B is for far end). The values from the table are very rough.

$$k_{AB} = 4.17$$

$$k_{AB} = \frac{4.17E_{cs}I_s}{\ell_1} = \frac{(4.17)(3.64 \times 10^6)(9216)}{(12)(22)}$$

$$= 529.9 \times 10^6 \text{ in.-lb.}$$

$$\text{FEM}_{AB} = \text{FEM}_{BA} = 0.084q_u\ell_2\ell_1^2$$

$$= (0.084)(0.276)(18)(22)^2 = 202 \text{ ft-k}$$

$$C_{AB} = C_{BA} = \text{carryover factor}$$

$$= 0.503$$

5. Determine column stiffness

$$I_c = \left(\frac{1}{12}\right)(15)(15)^3 = 4219 \text{ in.}^4$$

$$E_{cc} = 3.64 \times 10^6 \text{ psi}$$

Using Table A.20 of Appendix,

$$\ell_n = 9'6'' = 9.50 \text{ ft}$$

$$\ell_c = 9.50 - \frac{8}{12} = 8.833 \text{ ft}$$

$$\frac{\ell_n}{\ell_c} = \frac{8.833}{9.50} = 0.930 = \frac{\ell_u}{\ell_c}$$

With reference to the figure given with Table A.20,

$$\frac{a}{b} = \frac{4}{4} = 1.00$$

$$k_{AB} = 4.81 \text{ by interpolation}$$

$$K_c = \frac{4.81E_{cc}I_c}{H} = \frac{(4.81)(3.64 \times 10^6)(4219)}{(9.5)(12)}$$

$$= 648 \times 10^6 \text{ in.-lb}$$

$$C_{AB} = 0.55 \text{ by interpolation}$$

6. Determine the torsional stiffness of the slab section (see Figure 17.10)

$$C = \Sigma\left(1 - 0.63\frac{x}{y}\right)\left(\frac{x^3y}{3}\right)$$

$$= \left(1 - \frac{0.63 \times 8}{15}\right)\left(\frac{8^3 \times 15}{3}\right)$$

$$= 1700 \text{ in.}^4$$

$$K_t = \frac{9E_{cs}C}{\left[\ell_2\left(1 - \frac{c_2}{\ell_2}\right)^3\right]} = \frac{(9)(3.64 \times 10^6)(1700)}{\left[(12)(18)\left(1 - \frac{15}{(12)(22)}\right)^3\right]} = 307.3 \times 10^6 \text{ psi}$$

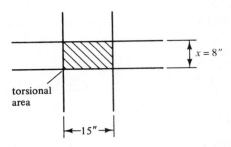

Figure 17.10

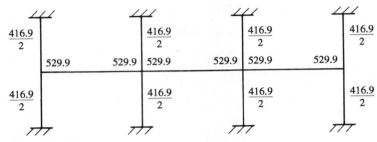

Figure 17.11

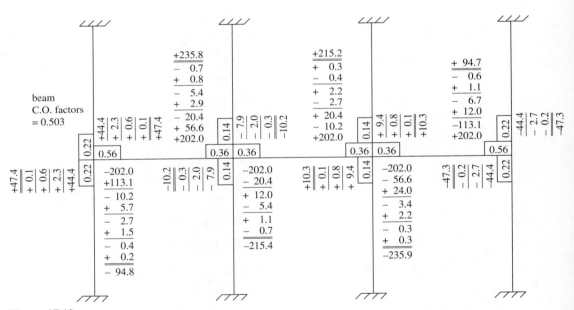

Figure 17.12

7. Compute K_{ec}, the stiffness of the equivalent column

$$K_{ec} = \frac{\Sigma K_c \Sigma K_t}{\Sigma K_c + \Sigma K_t} = \frac{(2 \times 648.0)(2 \times 307.3)}{2 \times 648.0 + 2 \times 307.3}$$

$$= 416.9 \times 10^6 \text{ in-lb}$$

A summary of the stiffness values is shown in Figure 17.11.

8. Computing distribution factors and balancing moments (see Figure 17.12): The author does not show moments at tops and bottoms of columns, but this could easily be done by multiplying the balanced column moments at the joints with the slabs by the carryover factor for the columns, which is 0.55.

A summary of the moment values for Example 17.2 is given in Figure 17.13. The positive moments shown in each span are assumed to equal the simple beam centerline moments plus the average of the end negative moments. This is correct if the end moments in a particular span are equal and is approximately correct if the end moments are unequal. For span 1,

$$^+M = \frac{(0.276)(18)(22)^2}{8} - \left(\frac{94.8 + 235.8}{2}\right) = 135.3 \text{ ft-k}$$

The negative moments shown in Figures 17.12 and 17.13 were calculated at the centerlines of the supports. At these supports, the cross section of the slab beam is very large due to the presence of the column. At the face of the column, however, the cross section is far smaller, and the Code (13.7.7) specifies that negative reinforcing be designed for the moment there. (If the column is not rectangular, it is replaced with a square column of the same total area and the moment is computed at the face of that fictitious column.) Because the ratio of unfactored dead to live load is less than 0.75, ACI Section 13.7.6.2 permits a single analysis with live load for all spans. No pattern load analysis is required.

The design moments shown in Figure 17.14 were determined by drawing the shear diagram and computing the area of that diagram between the centerline of each support and the face of the column.

For interior columns, the critical section (for both column and middle strips) is to be taken at the face of the supports, but not at a distance greater than $0.175\ell_1$ from the center of the column. At exterior supports with brackets or capitals, the moment used in the span perpendicular to the edge shall be computed at a distance from the face of the support element not greater than one-half of the projection of the bracket or capital beyond the face of the supporting element.

Sometimes the total of the design moments (that is, the positive moment plus the average of the negative end moments) obtained by the equivalent frame method for a particular span may be greater than $M_o = q_u \ell_2 \ell_n^2 / 8$, as used in Chapter 16. Should this happen, the Code (13.7.7.4) permits a reduction in those moments proportionately, so their sum does equal M_o.

17.6   COMPUTER ANALYSIS

The *equivalent frame method* was developed with the intention that the moment distribution method was to be used for the structural analysis. Truthfully, the method is so involved that it is not satisfactory for hand calculations. It is possible, however, to use computers and plane frame

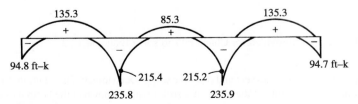

Figure 17.13

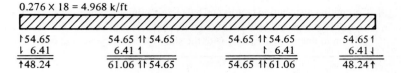

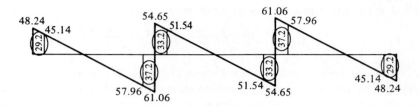

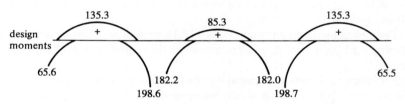

Figure 17.14

analysis programs if the structure is especially modeled. (In other words, we must establish various nodal points in the structure so as to account for the changing moments of inertia along the member axes.) There are also some computer programs on the market especially written for these frames. One of the best known is called ADOSS (analysis and design of concrete floor systems) and was prepared by the Portland Cement Association. The moments shown in Figure 17.14 can be entered in the Chapter 16 Excel spreadsheet, Two Way Slab, in cells 12C to 12G for interior spans and 45C to 45G for edge spans.

17.7 COMPUTER EXAMPLES

EXAMPLE 17.3

Use the Excel spreadsheet provided for Chapter 17 to solve Example 17.2.

SOLUTION   Open the Chapter 17 Excel spreadsheet and open the worksheet Moment Distribution. Enter values only for cells in yellow highlight. Results of the moment distribution are shown at the bottom of each column. Note that they are in agreement with those in Example 17.2

Moment Distribution for User Input Distribution and Carryover Factors and Fixed-end Moments

Joint	A	B		C		D
Member	AB	BA	BC	CB	CD	DC
Distribution factor	0.56	0.36	0.36	0.36	0.36	0.56
Fixed-end moment	−202	202	−202	202	−202	202
Carryover	0.503	0.503	0.503	0.503	0.503	0.503
Balance	113.12	0	0	0	0	−113.12
Carryover	0.000	56.899	0.000	0.000	−56.899	0.000
Balance	0.000	−20.484	−20.484	20.484	20.484	0.000
Carryover	−10.303	0.000	10.303	−10.303	0.000	10.303
Balance	5.770	−3.709	−3.709	3.709	3.709	−5.770
Carryover	−1.866	2.902	1.866	−1.866	−2.902	1.866
Balance	1.045	−1.716	−1.716	1.716	1.716	−1.045
Carryover	−0.863	0.526	0.863	−0.863	−0.526	0.863
Balance	0.483	−0.500	−0.500	0.500	0.500	−0.483
Carryover	−0.252	0.243	0.252	−0.252	−0.243	0.252
Balance	0.141	−0.178	−0.178	0.178	0.178	−0.141
Carryover	−0.090	0.071	0.090	−0.090	−0.071	0.090
Balance	0.050	−0.058	−0.058	0.058	0.058	−0.050
Carryover	−0.029	0.025	0.029	−0.029	−0.025	0.029
Balance	0.016	−0.020	−0.020	0.020	0.020	−0.016
	−94.78	236.00	−215.26	215.26	−236.00	94.78

PROBLEMS

17.1 Determine the end moments for the beams and columns of the frame shown for which the fixed-end moments, carryover factors, and distribution factors (circled) have been computed. Use the moment distribution method. (*Ans.* 44.9 ft-k and 21.0 ft-k at column bases)

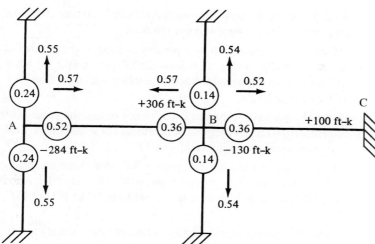

17.2 Repeat Problem 16.4 using the equivalent frame method instead of the direct design method.

17.3 Use the Chapter 17 spreadsheet to determine the end moments in member *AB* and *BC* of Problem 17.1. (*Ans.* $M_{AB} = -163.6$ ft-k, $M_{BA} = 307.2$ ft-k, $M_{BC} = -229.7$ ft-k, $M_{CB} = 48.1$ ft-k)

18

Walls

18.1 INTRODUCTION

Before the advent of frame construction during the nineteenth century, most walls were of a load-bearing type. Since the late 1800s, however, the non–load-bearing wall has become quite common because other members of the structural frame can be used to provide stability. As a result, today we have walls that serve all sorts of purposes, such as retaining walls, basement walls, partition walls, fire walls, and so on. These walls may or may not be of a load-bearing type.

In this chapter the following kinds of concrete walls will be considered: non–load-bearing, load-bearing, and shear walls (the latter being either load-bearing or non–load-bearing).

18.2 NON–LOAD-BEARING WALLS

Non–load-bearing walls are those that support only their own weights and perhaps some lateral loads. Falling into this class are retaining walls, facade-type walls, and some basement walls. For non–load-bearing walls the ACI Code provides several specific limitations, which are listed at the end of this paragraph. The values given for minimum reinforcing quantities and wall thicknesses do not have to be met if lesser values can be proved satisfactory by structural analysis (14.2.7). The numbers given in parentheses are ACI section numbers.

1. The thickness of a non–load-bearing wall cannot be less than 4 in. or $1/30$ times the least distance between members that provide lateral support (14.6.1).

2. The minimum amount of vertical reinforcement as a percent of gross concrete area is 0.0012 for deformed bars #5 or smaller with $f_y =$ at least 60,000 psi, 0.0015 for other deformed bars, and 0.0012 for plain or deformed welded wire fabric not larger than $W31$ or $D31$—that is, $\frac{5}{8}$ in. in diameter (14.3.2).

3. The vertical reinforcement does not have to be enclosed by ties unless the percent of vertical reinforcing is greater than 0.01 times the gross concrete area or where the vertical reinforcing is not required as compression reinforcing (14.3.6).

4. The minimum amount of horizontal reinforcing as a percent of gross concrete area is 0.0020 for deformed bars #5 or smaller with $f_y \geq 60,000$ psi, 0.0025 for other deformed bars, and 0.0020 for plain or deformed welded wire fabric not larger than $W31$ or $D31$ (14.3.3).

5. The spacing of vertical and horizontal reinforcement may not exceed three times the wall thickness, or 18 in. (14.3.5).

6. Reinforcing for walls more than 10 in. thick (not including basement walls) must be placed in two layers as follows: one layer containing from one-half to two-thirds of the

Retaining Wall with stepped wall thickness, Clemson University 2008.

total reinforcing placed in the exterior surface not less than 2 in. nor more than one-third times the wall thickness from the exterior surface; the other layer placed not less than $\frac{3}{4}$ in. nor more than one-third times the wall thickness from the interior surface (14.3.4).

7. For walls less than 10 in. thick the Code does not specify two layers of steel, but to control shrinkage it is probably a good practice to put one layer on the face of walls exposed to view and one on the nonstressed side of foundation walls 10 ft or more in height.

8. In addition to the reinforcing specified in the preceding paragraphs, at least two #5 bars in walls having two layers of reinforcement in both directions, and one #5 bar in walls having a single layer of reinforcement in both directions, must be provided around all window, door and similar-sized openings. These bars must be anchored to develop f_y in tension at the corners of the openings (14.3.7).

9. For cast-in-place walls, the area of reinforcing across the interface between a wall and a footing must be no less than the minimum vertical wall reinforcing given in 14.3.2 (15.8.2.2).

10. For precast, nonprestressed walls, the reinforcement must be designed in accordance with the preceding requirements on this list as well as the requirements of Chapters 10 or 14 of the Code, except that the area of the horizontal and vertical reinforcing must not be less than 0.001 times the gross cross-sectional area of the wall. In addition, the spacing of the reinforcing may not be greater than 5 times the wall thickness or 30 in. for interior walls, or 18 in. for exterior ones (16.4.2).

18.3 LOAD-BEARING CONCRETE WALLS—EMPIRICAL DESIGN METHOD

Most of the concrete walls in buildings are load-bearing walls that support not only vertical loads, but also some lateral moments. As a result of their considerable in-plane stiffnesses, they are quite important in resisting wind and earthquake forces.

Load-bearing walls with solid rectangular cross sections may be designed as were columns subject to axial load and bending, or they may be designed by an empirical method given in Section 14.5 of the Code. The empirical method may be used only if the resultant of all the factored loads falls within the middle third of the wall (that is, the eccentricity must be equal to or less than one-sixth the thickness of the wall). *Whichever of the two methods is used, the design must meet the minimum requirements given in the preceding section of this chapter for non–load-bearing walls.*

This section is devoted to the empirical design method, which is applicable to relatively short vertical walls with approximately concentric loads. The Code (14.5.2) provides an empirical formula for calculating the design axial load strength of solid rectangular cross-sectional walls with *e* less than one-sixth of wall thicknesses. Should walls have nonrectangular cross sections (such as ribbed wall panels) and/or should *e* be greater than one-sixth of wall thicknesses, the rational design procedure for columns subject to axial load and bending (Code 14.4) must be followed.

The practical use of the empirical wall formula, which is given at the end of this paragraph, is for relatively short walls with small moments. When lateral loads are involved, *e* will quickly exceed one-sixth of wall thicknesses. The number 0.55 in the equation is an eccentricity factor that causes the equation to yield a strength approximately equal to that which would be

Thirty-two-foot-tall foundation walls for the MCI Mid-Continent Data Center in Omaha, Nebraska. (Courtesy of Economy Forms Corporation.)

obtained by the axial load and bending procedure of Chapter 10 of the Code if the eccentricity is $h/6$.

$$\phi P_{nw} = 0.55\phi f'_c A_g \left[1 - \left(\frac{k\ell_c}{32h} \right)^2 \right]$$ (ACI Equation 14-1)

Where

$\phi = 0.65$

A_g = gross area of the wall section (in.^{2})

ℓ_c = vertical distance between supports (in.)

h = overall thickness of member (in.)

k = effective length factor determined in accordance with the values given in Table 18.1

Table 18.1 Effective Length Factors for Load-Bearing Walls (14.5.2)

1. Walls braced top and bottom against lateral translation and	
(a) Restrained against rotation at one or both ends (top and/or bottom)	0.80
(b) Not restrained against rotation at either end	1.0
2. For walls not braced against lateral translation	2.0

Other ACI requirements for load-bearing concrete walls designed by the empirical formula follow.

1. The thickness of the walls may not be less than 1/25 the supported height or length, whichever is smaller, or less than 4 in. (14.5.3.1).

2. The thickness of exterior basement walls and foundation walls may not be less than $7\frac{1}{2}$ in. (14.5.3.2).

3. The horizontal length of a wall that can be considered effective for each concentrated load may not exceed the smaller of the center-to-center distance between loads or the bearing width plus four times the wall thickness. This provision may be waived if a larger value can be proved satisfactory by a detailed analysis (14.2.4).

4. Load-bearing walls must be anchored to intersecting elements such as floors or roofs, or they should be anchored to columns, pilasters, footings, buttresses, and intersecting walls (14.2.6).

The empirical method is quite easy to apply because only one calculation has to be made to determine the design axial strength of a wall. Example 18.1, which follows, illustrates the design of a bearing wall with a small moment.[1]

EXAMPLE 18.1

Design a concrete-bearing wall using the ACI empirical equation 14-1 to support a set of precast concrete roof beams 7'0" on center, as shown in Figure 18.1. The bearing width of each beam is 10". The wall is considered to be laterally restrained top and bottom and is further assumed to be restrained against rotation at the footing; thus $k = 0.8$. Neglect wall weight. Other data: $f'_c = 3000$ psi, $f_y = 60,000$ psi, beam reaction, $D = 30$ k, $L = 18$ k.

---

[1]For the example problems presented in this chapter, the authors have followed the general procedures used in *Notes on ACI 318-05 Building Code Requirements for Structural Concrete* 2005. B. G. Rabbat, ed. (Skokie, IL: Portland Cement Association), pp. 21-19 through 21-20.

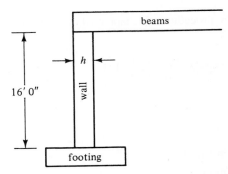

Figure 18.1

SOLUTION

1. Determine minimum wall thickness (14.5.3.1)

    (a) $h = \left(\frac{1}{25}\right)(12 \times 16) = 7.68'' \leftarrow$ Try $8''$

    (b) $h = 4''$

    Compute factored beam reactions

    $$P_u = (1.2)(30) + (1.6)(18) = 64.8 \text{ k}$$

2. Is the bearing strength of wall concrete satisfactory under beam reactions (10.17.1)?

    $$\phi(0.85 f'_c A_1) = (0.65)(0.85)(3)(8 \times 10)$$
    $$= 132.6 \text{ k} > 64.8 \text{ k} \qquad \underline{\underline{\text{OK}}}$$

3. Horizontal length of wall to be considered as effective in supporting each concentrated load (14.2.4)

    (a) Center-to-center spacing of beams $7'0'' = 84''$

    (b) Width of bearing $+ 4h = 10 + (4)(8) = 42'' \leftarrow$

4. Design strength of wall

    $$\phi P_{nw} = 0.55 \phi f'_c A_g \left[1 - \left(\frac{k\ell_c}{32h}\right)^2\right] \qquad \text{(ACI Equation 14-1)}$$

    $$= (0.55)(0.65)(3)(8 \times 42)\left[1 - \left(\frac{0.80 \times 12 \times 16}{32 \times 8}\right)^2\right]$$

    $$= 230.6 \text{ k} > 64.8 \text{ k} \qquad \underline{\underline{\text{OK}}}$$

5. Select reinforcing (14.3.5, 14.3.2, and 14.3.3)

    $$\text{Maximum spacing} = (3)(8) = 24'' \text{ or } \underline{\underline{18''}}$$

    $$\text{Vertical } A_s = (0.0012)(12)(8) = 0.115 \text{ in.}^2/\text{ft} \qquad \underline{\underline{\#4 @ 18''}}$$

    $$\text{Horizontal } A_s = (0.0020)(12)(8) = 0.192 \text{ in.}^2/\text{ft} \qquad \underline{\underline{\#4 @ 12''}}$$

Although the Code is not specific on this issue, it would be prudent to provide continuity of the vertical wall reinforcement into the footing. This is usually accomplished by using a hooked bar embedded in the wall footing that is lap-spliced with the vertical wall bars. If a value of $k = 1$ had been used in this example, there would have been no assumption of continuity at the base of the wall. It would still be necessary to ensure that adequate shear capacity be provided at the base of the wall.

18.4   LOAD-BEARING CONCRETE WALLS—RATIONAL DESIGN

Reinforced concrete-bearing walls may be designed as columns by the rational method of Chapter 10 of the Code whether the eccentricity is smaller or larger than $h/6$ (they must be designed rationally if $e > h/6$). The minimum vertical and horizontal reinforcing requirements of Section 14.3 of the Code must be met.

The design of walls as columns is difficult unless design aids are available. Various wall design aids are available from the Portland Cement Association, but the designer can prepare his or her own aids such as axial load and bending interaction diagrams. The designs may be complicated by the fact that walls often will be classified as "long columns" with the result that the slenderness requirements of Section 10.10 of the Code will have to be met. An alternative procedure for slender walls is presented in ACI Section 14.8.

Very slender walls are rather common, particularly in tilt-up wall construction. The Portland Cement Association has available a design aid that is particularly useful for such cases.[2]

The interaction diagrams discussed in Section 10.6 of this text can be used to design walls with steel in two layers and subject to out-of-plane bending combined with axial loads. However, the reinforcement ratio is limited to 0.01, unless the compression reinforcement is laterally tied (ACI Section 14.3.6), which is impractical in many cases. Graphs 2 through 5 in Appendix A are applicable to walls having two layers of steel. The values of γ for these graphs may be too large, however, especially for thinner walls. Example 18.2 illustrates how to use these design aids when designing walls.

EXAMPLE 18.2

Design a reinforced concrete foundation wall having the following conditions: $\ell = 15$ ft between lateral supports, backfill height is also 15 ft, $P_D = 520$ plf at $e = 2$ in., $P_L = 250$ plf at $e = 2$ in. Assume the base is pinned.

$f_c' = 4000$ psi, normal weight aggregate concrete, Grade 60 reinforcing steel

Soil properties, $k_a = 0.40$, $\gamma = 100$ pcf

SOLUTION

Assuming that the wall is simply supported at both the top and bottom, the soil pressure exerted on the wall increases linearly from zero at the top of bakfill to $k_a\ell$ at the base. The maximum bending moment[3] is

$$M_H = 0.064\,k_a\gamma\ell^3 = (0.064)(0.4)(100)(15)^3$$

$$= 8640\,\text{ft-lb/ft} = 103,680\,\text{in.-lb/ft}$$

Maximum moment[3] occurs at $0.577\ell = 8.66$ ft from top of wall. Note that the eccentric axial loads cause a reduction in the moment. In the case of dead axial load, the moment at the top of the wall is $P \times e = 520(2) = 1040$ in.-lb/ft. This moment varies linearly to zero at the base of the wall, so at the location of maximum soil moment, its value is $1040(8.66)/15 = 600$ in.-lb/ft. The same analysis applied to the live load results in a moment of 288 in.-lb/ft. In this case the reduction is small (less than 3%) and could be ignored. At the location of maximum bending moment, the axial loads are

$P_D = 520 + 8.66(150)(8)(12)/144 = 1386$ plf (assuming an 8-in. wall thickness). $P_L = 250$ plf.

Three load combinations applicable to this situation are shown in the table below along with K_n and R_n values used in conjuction with Graph 2 in Appendix A

---

[2]Portland Cement Association, 1980. Tilt-Up Load Bearing Walls—A Design Aid., EBO74D (Skokie, IL), 28pp.

[3]McCormac, J.C., Structural Analysis: Using Classical and Matrix Methods, 4th ed. (Hoboken, NJ: John Wiley & Sons), pp. 104–106.

	P_D	P_L	P_H	M_D	M_L	M_H	P_U	M_U	$K_n = P_U/\phi f_c' A_g$	$R_n = M_U/\phi f_c' A_g h$	ρ_t
Unfactored	1386	250	0	600	288	103,680					
$U = 1.4D$	1.4			1.4			1940	840	0.0056	0.0003	< 0.01
$U = 1.2D + 1.6L + 1.6H$	1.2	1.6	1.6	1.2	1.6	1.6	2063	167,069	0.0060	0.0604	0.01
$U = 0.9D + 1.6H$	0.9		1.6	0.9		1.6	1247	166,428	0.0036	0.0602	0.01

There is so little difference between the second and third load case that the difference in ρ_t is indistinguishable when reading the graph. Also, a value of $\phi = 0.9$ was used because when reading the graph it is obvious that the controlling points are below the line for $\varepsilon_t = 0.005$. Using the resulting value of ρ_t, $A_{st} = \rho_t bh = 0.010(12)(8) = 0.96$ in.^{2}/ft, half in each face (0.48 in.^{2}/ft). From Table A.6, select #5 @ $7\frac{1}{2}$-in. o.c. vertically in both faces. The #5 bar size was picked because the required cover increases for larger bars from $1\frac{1}{2}$ to 2 in. Horizontal reinforcing must be provided in accordance with ACI 14.3.3(b), $A_s \geq 0.0025bh = 0.0025(12)(8) = 0.24$ in.^{2}/ft. Choose #4 @ 18 in. o.c. horizontally in both faces. Since this wall is less than 10 in. thick, ACI 14.3.4 permits the reinforcement to be placed in a single layer. However, Graph 2 is based on having steel in two layers separated by a distance of $\gamma h = 0.6(8) = 4.8$ in. Actually, a value of $\gamma = 0.5$ would have been a better choice in this problem in order to provide sufficient cover, but it is not available.

Because the axial load is so small in this case, the wall could have been designed as a beam with compression steel. The beam width would be 12 in., $h = 8$ in, $d = 6.4$ in. and the moment taken as 167 in-k. The authors carried out this design approach using the Chapter 5 spreadsheet and calculated a required area of reinforcing steel of 0.98 in.^{2}/ft (compared with 0.96 in.^{2}/ft from Graph 2).

The wall exerts reactions at the bottom against the footing and at the top against the floor. The floor system must be designed to resist this horizontal force which is called a diaphragm force. At the bottom, the interface of the wall and the footing must be designed for shear transfer between these elements. Shear friction (Section 8.12 of this text) is the best way to accommodate this transfer.

Note that this wall has a slenderness ratio of $kl_u/r = 1.0(15)12/(0.3)(8) = 75 > 34 - 12(0/M_2) = 34$, so slenderness must be considered. However, since the axial loads are so small, the moment magnifier is not greater than 1.0 and hence the moment is not magnified.

If this wall were not laterally supported at the top by the floor shown in Figure 18.2, it would be a retaining wall. The bending moment due to soil pressure on the retaining wall would be $M_H = k_a \gamma \ell^3/6$ which is 260% of the moment in Example 18.2. Be sure to use the correct moment for the boundary conditions that apply to your type of construction.

If the reinforcement ratio had exceeded 0.01, this method of solution would not be valid. It would be necessary to use mechanics to determine the wall capacity without using steel in compression. The design aids used in this example use the compression force in the reinforcing steel, hence the steel has to be laterally tied for this solution to be valid.

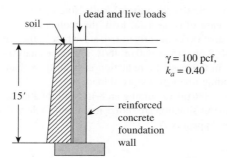

Figure 18.2 Foundation wall.

18.5 SHEAR WALLS

For tall buildings it is necessary to provide adequate stiffness to resist the lateral forces caused by wind and earthquake. When such buildings are not properly designed for these forces there may be very high stresses, vibrations, and sidesway when the forces occur. The results may include not only severe damages to the buildings but also considerable discomfort for their occupants.

When reinforced concrete walls with their very large in-plane stiffnesses are placed at certain convenient and strategic locations, they can often be economically used to provide the needed resistance to horizontal loads. Such walls, called *shear walls*, are in effect deep vertical cantilever beams that provide lateral stability to structures by resisting the in-plane shears and bending moments caused by the lateral forces.

As the strength of shear walls is almost always controlled by their flexural resistance, their name is something of a misnomer. It is true, however, that on some occasions they may require some shear reinforcing to prevent diagonal tension failures. Indeed, one of the basic requirements of shear walls designed for high seismic forces is to ensure flexure rather than shear-controlled design.

The usual practice is to assume that the lateral forces act at the floor levels. The stiffnesses of the floor slabs horizontally are quite large as compared to the stiffnesses of the walls and columns. Thus it is assumed that each floor is displaced in its horizontal plane as a rigid body.

Figure 18.3 shows the plan of a building that is subjected to horizontal forces. The lateral forces, usually from wind or earthquake loads, are applied to the floor and roof slabs of the building, and those slabs, acting as large beams lying on their sides or diaphragms, transfer the loads primarily to the shear walls A and B. Should the lateral forces be coming from the other (perpendicular) direction, they would be resisted primarily by the shear walls C and D.

Shear Wall with integral end columns – Rhodes Annex, Clemson University, 2008.

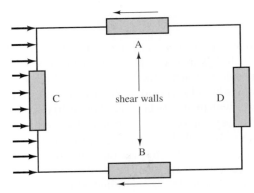

Figure 18.3 Plan view of a floor supported by shear walls.

The walls must be sufficiently stiff so as to limit deflections to reasonable values.

Shear walls are commonly used for buildings with flat-plate floor slabs. In fact, this combination of slabs and walls is the most common type of construction used today for tall apartment buildings and other residential buildings.

Shear walls span the entire vertical distances between floors. If the walls are carefully and symmetrically placed in plan, they will efficiently resist both vertical and lateral loads and do so without interfering substantially with the architectural requirements. Reinforced concrete buildings of up to 70 stories have been constructed with shear walls as their primary source of lateral

Buttress Shear Wall, New York Hilton, New York, NY (Courtesy Prof. Larry Kahn, Georgia Institute of Technology).

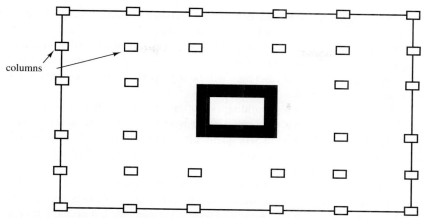

Figure 18.4 Shear walls around elevators and stairwells.

stiffness. In the horizontal direction full shear walls may be used—that is, they will run for the full panel or bay lengths. When forces are smaller, they need only run for partial bay lengths.

Shear walls may be used to resist lateral forces only, or they may be used in addition as bearing walls. Furthermore they may be used to enclose elevators, stairwells, and perhaps restrooms, as shown in Figure 18.4. These box-type structures shown are very satisfactory for resisting horizontal forces.

Another possible arrangement of shear walls is shown in Figure 18.5. Although shear walls may be also be needed in the long direction of this building, they are not included in this figure.

On most occasions it is not possible to use shear walls without some openings in them for doors, windows, and penetrations for mechanical services. Usually it is possible, however, with careful planning to place these openings so they do not seriously affect stiffnesses or stresses in the walls. When the openings are small, their overall effect is minor, but this is not the case when large openings are present.

Usually the openings (windows, doors, etc.) are placed in vertical and symmetrical rows in the walls throughout the height of the structure. The wall sections on the sides of these openings are tied together either by beams enclosed in the walls, by the floor slabs, or by a combination of both. As you can see, the structural analysis for such a situation is extremely complicated. Though shear wall designs are generally handled with empirical equations, they can be appreciably affected by the designer's previous experience.

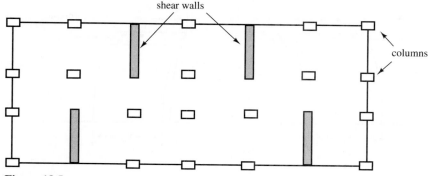

Figure 18.5

When earthquake-resistant construction is being considered, note that the relatively stiff parts of a structure will attract much larger forces than will the more flexible parts. A structure with reinforced concrete shear walls is going to be quite stiff and thus will attract large seismic forces. If the shear walls are brittle and fail, the rest of the structure may not be able to take the shock. But if the shear walls are ductile (and they will be if properly reinforced), they will be very effective in resisting seismic forces.

Tall reinforced concrete buildings are often designed with shear walls to resist seismic forces, and such buildings have performed quite well in recent earthquakes. During an earthquake, properly designed shear walls will decidedly limit the amount of damage to the structural frame. They will also minimize damages to the nonstructural parts of a building, such as the windows, doors, ceilings and partitions.

Figure 18.6 shows a shear wall subjected to a lateral force V_u. The wall is in actuality a cantilever beam of width h and overall depth ℓ_w. In part (a) of the figure the wall is being bent from left to right by V_u, with the result that tensile bars are needed on the left or tensile side. If V_u is applied from the right side as shown in part (b) of the figure, tensile bars will be needed on the right-hand end of the wall. Thus it can be seen that a shear wall needs tensile reinforcing on both sides because V_u can come from either direction. For horizontal shear calculations the depth of the beam from the compression end of the wall to the center of gravity of the tensile bars is estimated to be about 0.8 times the wall length ℓ_w as per ACI Section 11.10.4. (If a larger value of d is obtained by a proper strain compatibility analysis, it may be used.)

The shear wall acts as a vertical cantilever beam and in providing lateral support is subjected to both bending and shear forces. For such a wall the maximum shear V_u and the maximum moment M_u can be calculated at the base. If flexural stresses are calculated, their magnitude will be affected by the design axial load N_u, and thus its effect should be included in the analysis.

Shear is more important in walls with small height-to-length ratios. Moments will be more important for higher walls, particularly those with uniformly distributed reinforcing.

It is necessary to provide both horizontal and vertical shear reinforcing for shear walls. The Commentary (R11.9.9) says that in low walls the horizontal shear reinforcing is less effective and the vertical shear reinforcing is more effective. For high walls the situation is reversed. This situation is reflected in ACI Equation 11-32, which is presented in the next section. The vertical shear reinforcing contributes to the shear strength of a wall by shear friction.

Reinforcing bars are placed around all openings, whether or not structural analysis indicates a need for them. Such a practice is deemed necessary to prevent diagonal tension cracks, which tend to develop radiating from the corners of openings.

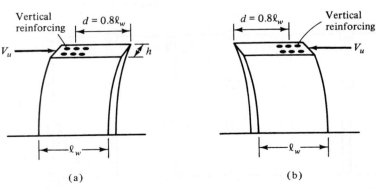

Figure 18.6 Shear wall.

18.6   ACI PROVISIONS FOR SHEAR WALLS

1. The factored beam shear must be equal to or less than the design shear strength of the wall.

$$V_u \leq \phi V_n$$

2. The design shear strength of a wall is equal to the design shear strength of the concrete plus that of the shear reinforcing.

$$V_u \leq \phi V_c + \phi V_s$$

3. The nominal shear strength V_n at any horizontal section in the plane of the wall may not be taken greater than $10\sqrt{f'_c}hd$ (11.9.3).

4. In designing for the horizontal shear forces in the plane of a wall, d is to be taken as equal to $0.8\ell_w$, where ℓ_w is the horizontal wall length between faces of the supports, unless it can be proved to be larger by a strain compatibility analysis (11.9.4).

5. ACI Section 11.10.5 states that unless a more detailed calculation is made (as described in the next paragraph), the value of the nominal shear strength V_c used may not be larger than $2\lambda\sqrt{f'_c}hd$ for walls subject to a factored axial compressive load N_u. Should a wall be subject to a tensile load N_u, the value of V_c may not be larger than the value obtained with the following equation:

$$V_c = 2\left(1 + \frac{N_u}{500A_g}\right)\lambda\sqrt{f'_c}b_w d \geq 0 \qquad \text{(ACI Equation 11-8)}$$

6. Using a more detailed analysis, the value of V_c is to be taken as the smaller value obtained by substituting into the two equations that follow, in which N_u is the factored axial load normal to the cross section occurring simultaneously with V_u. N_u is to be considered positive for compression and negative for tension (11.10.6).

$$V_c = 3.3\lambda\sqrt{f'_c}hd + \frac{N_u d}{4\ell_w} \qquad \text{(ACI Equation 11-29)}$$

or

$$V_c = \left[0.6\lambda\sqrt{f'_c} + \frac{\ell_w(1.25\lambda\sqrt{f'_c} + 0.2N_u/\ell_w h)}{\dfrac{M_u}{V_u} - \dfrac{\ell_w}{2}}\right]hd \qquad \text{(ACI Equation 11-30)}$$

The first of these equations was developed to predict the inclined cracking strength at any section through a shear wall corresponding to a principal tensile stress of about $4\lambda\sqrt{f'_c}$ at the centroid of the wall cross section. The second equation was developed to correspond to an occurrence of a flexural tensile stress of $6\lambda\sqrt{f'_c}$ at a section $\ell_w/2$ above the section being investigated. Should $M_u/V_u - \ell_w/2$ be negative, the second equation will have no significance and will not be used.

In SI units, these last three equations are as follows:

$$V_c = \left(1 + \frac{0.3N_u}{A_g}\right)\frac{\lambda\sqrt{f'_c}}{6}b_w d \geq 0 \qquad \text{(ACI Equation 11-8)}$$

$$V_c = \frac{1}{4}\lambda\sqrt{f_c'}hd + \frac{N_u d}{4\ell_w} \qquad \text{(ACI Equation 11-29)}$$

$$V_c = \left[\frac{1}{2}\lambda\sqrt{f_c'} + \frac{\ell_w\left(\lambda\sqrt{f_c'} + 2\frac{N_u}{\ell_w h}\right)}{\dfrac{M_u}{V_u} - \dfrac{\ell_w}{2}}\right]\frac{hd}{10} \qquad \text{(ACI Equation 11-30)}$$

7. The values of V_c computed by the two preceding equations at a distance from the base equal to $\ell_w/2$ or $h_w/2$ (whichever is less) are applicable for all sections between this section and one at the wall base (11.9.7).

8. Should the factored shear V_u be less than $\phi V_c/2$ computed as described in the preceding two paragraphs, it will not be necessary to provide a minimum amount of both horizontal and vertical reinforcing, as described in Section 11.9.9 or Chapter 14 of the Code.

9. Should V_u be greater than ϕV_c, shear wall reinforcing must be designed as described in Section 11.9.9 of the Code.

10. If the factored shear force V_u exceeds the shear strength ϕV_c, the value of V_s is to be determined from the following expression, in which A_v is the area of the horizontal shear reinforcement and s is the spacing of the shear or torsional reinforcing in a direction perpendicular to the horizontal reinforcing (11.9.9.1).

$$V_s = \frac{A_v f_y d}{s} \qquad \text{(ACI Equation 11-31)}$$

11. The amount of horizontal shear reinforcing ρ_t (as a percent of the gross vertical concrete area) shall not be less than 0.0025 (11.9.9.2).

12. The maximum spacing of horizontal shear reinforcing s_2 shall not be greater than $\ell_w/5$, $3h$, or 18 in. (11.9.9.3).

13. The amount of vertical shear reinforcing ρ_n (as a percent of the gross horizontal concrete area) shall not be less than the value given by the following equation, in which h_w is the total height of the wall (11.9.9.4).

$$\rho_\ell = 0.0025 + 0.5\left(2.5 - \frac{h_w}{\ell_w}\right)(\rho_h - 0.0025) \qquad \text{(ACI Section 11-32)}$$

It shall not be less than 0.0025 but need not be greater than the required horizontal shear reinforcing ρ_t.

For high walls, the vertical reinforcing is much less effective than it is in low walls. This fact is reflected in the preceding equation, where for walls with a height/length ratio less than 0.5, the amount of vertical reinforcing required equals the horizontal reinforcing required. If the ratio is larger than 2.5, only a minimum amount of vertical reinforcing is required (that is $0.0025sh$).

14. The maximum spacing of vertical shear reinforcing shall not be greater than $\ell_w/3$, $3h$, or 18 in. (11.9.9.5).

EXAMPLE 18.3

Design the reinforced concrete wall shown in Figure 18.7 if $f'_c = 3000$ psi and $f_y = 60,000$ psi.

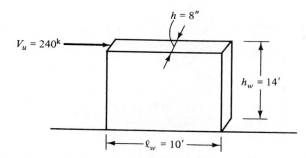

$V_u = 240^k$

$h = 8''$

$h_w = 14'$

$\ell_w = 10'$

Figure 18.7

SOLUTION

1. Is the wall thickness satisfactory?

$$V_u = \phi 10\sqrt{f'_c}hd \qquad \text{(ACI Section 11.9.3)}$$

$$d = 0.8\ell_w = (0.8)(12 \times 10) = 96'' \qquad \text{(ACI Section 11.9.4)}$$

$$V_u = (0.75)(10)(\sqrt{3000})(8)(96)$$

$$V_u = 315{,}488 \text{ lb} = 315.5 \text{ k} > 240 \text{ k} \qquad \underline{\underline{\text{OK}}}$$

2. Compute V_c for wall (lesser of two values)

(a) $V_c = 3.3\lambda\sqrt{f'_c}hd + \dfrac{N_u d}{4\ell_w} = (3.3)(1.0)(\sqrt{3000})(8)(96) + 0$

$= 138{,}815 \text{ lb} = 138.8 \text{ k} \leftarrow \qquad \text{(ACI Equation 11-29)}$

(b) $V_c = \left[0.6\lambda\sqrt{f'_c} + \dfrac{\ell_w\left(1.25\lambda\sqrt{f'_c} + 0.2N_u/\ell_w h\right)}{\dfrac{M_u}{V_u} - \dfrac{\ell_w}{2}}\right] hd \qquad \text{(ACI Equation 11-30)}$

Computing V_u and M_u at the lesser of $\ell_w/2 = 10/2 = 5'$ or $h_w/2 = 14/2 = 7'$ from base (ACI 11.9.7):

$$V_u = 240 \text{ k}$$

$$M_u = 240(14 - 5) = 2160 \text{ ft-k} = 25{,}920 \text{ in.-k}$$

$$V_c = \left[(0.6)(1.0)(\sqrt{3000}) + \dfrac{(12 \times 10)(1.25)(1.0)(\sqrt{3000}) + 0}{\dfrac{25{,}920}{240} - \dfrac{(12)(10)}{2}}\right](8)(96)$$

$$= 156{,}692 \text{ lb} = 156.7 \text{ k}$$

3. Is shear reinforcing needed?

$$\frac{\phi V_c}{2} = \frac{(0.75)(1.0)(138.8)}{2} = 52.05 \text{ k} > 240 \text{ k} \qquad\qquad \underline{\underline{\text{Yes}}}$$

4. Select horizontal shear reinforcing

$$V_u = \phi V_c + \phi V_s$$

$$V_u = \phi V_c + \phi \frac{A_v f_y d}{s}$$

$$\frac{A_v}{s} = \frac{V_u - \phi V_c}{\phi f_y d} = \frac{240 - (0.75)(138.8)}{(0.75)(60)(96)} = 0.0315$$

Try different-sized horizontal bars with A_v = two bar cross sectional areas. Two layers of horizontal bars will be placed at the calculated spacing, hence A_v = twice the bar area. Compute s_2 = vertical spacing of horizontal stirrups.

$$\text{Try \#3 bars}: \; s = \frac{(2)(0.11)}{0.0315} = 6.98''$$

$$\text{Try \#4 bars}: \; s = \frac{(2)(0.20)}{0.0315} = 12.70''$$

Maximum vertical spacing of horizontal stirrups

$$\frac{\ell_w}{5} = \frac{(12)(10)}{5} = 24''$$

$$3h = (3)(8) = 24''$$

$$18'' = 18'' \leftarrow \qquad\qquad \underline{\underline{\text{Try \#4 @ 12}''}}$$

$$\rho_t = \frac{A_v}{A_s}$$

where A_g = wall thickness times the vertical spacing of the horizontal stirrups

$$\rho_t = \frac{(2)(0.20)}{(8)(12)} = 0.00417$$

which is greater than the minimum ρ_n of 0.0025 required by Code.

$\underline{\underline{\text{Use \#4 horizontal stirrups 12}'' \text{ o.c. vertically}}}$

5. Design vertical shear reinforcing

$$\text{min. } \rho_\ell = 0.0025 + 0.5\left(2.5 - \frac{h_w}{\ell_w}\right)(\rho_h - 0.0025) \qquad\qquad \text{(ACI Equation 11-32)}$$

$$= 0.0025 + 0.5\left(2.5 - \frac{12 \times 14}{12 \times 10}\right)(0.00417 - 0.0025)$$

$$= 0.00342$$

Assume #4 closed vertical bars with A_v = two bar cross-sectional areas and with s = horizontal spacing of vertical stirrups.

$$s = \frac{(2)(0.20)}{(8)(0.00342)} = 14.62''$$

Maximum horizontal spacing of vertical stirrups

$$\frac{\ell_w}{3} = \frac{(12)(10)}{3} = 40''$$

$$3h = (3)(8) = 24''$$

$$18'' = 18'' \leftarrow$$

Use #4 vertical stirrups 14" o.c. horizontally

6. Design vertical flexural reinforcing

$$M_u = (240)(14) = 3360 \text{ ft-k @ base of wall}$$

$$\frac{M_u}{\phi bd^2} = \frac{(12)(3360)(1000)}{(0.9)(8)(96)^2} = 607.6$$

$$\rho = 0.0118 \text{ from Appendix Table A.12}$$

$$A_s = \rho bd$$

where b is wall thickness and d is approximated by $0.80\ell_w = (0.8)(12 \times 10) = 96''$

$$A_s = (0.0118)(8)(96) = 9.06 \text{ in.}^2$$

Use 10 #9 bars each end(assuming V_u could come from either direction)

7. A sketch of the wall cross section is given in Figure 18.8. If this same wall had been subjected to significant axial load, the method used to calculate A_s for flexure would have to be revised to include its effect. Spreadsheets to calculate the coordinates of the interaction diagram using the assumptions in Chapter 10 can be developed for this purpose.

The centroid of the bar group at either end of the wall is approximately 7 in. from the wall end. Assuming all of the tension bars are yielding, the resultant tension force is also located at 7 in. from the wall end. The assumed value of $d = 0.8\ell_w$ was overly conservative. It can be taken as $120 - 7 = 113$ in. Revising the calculation for A_s using this value of d results in a new $A_s = 7.32$ in.^{2} As a result the bar size can be reduced to #8 with the same number of bars (10 #8 bars at each end).

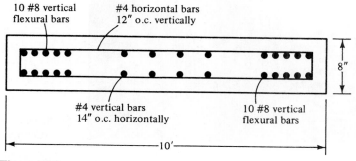

10 #8 vertical flexural bars

#4 horizontal bars 12" o.c. vertically

8"

#4 vertical bars 14" o.c. horizontally

10 #8 vertical flexural bars

10'

Figure 18.8

18.7 ECONOMY IN WALL CONSTRUCTION

To achieve economical reinforced concrete walls, it is necessary to consider such items as wall thicknesses, openings, footing elevations, and so on.

The thicknesses of walls should be sufficient to permit the proper placement and vibration of the concrete. All of the walls in a building should have the same thickness if practical. Such a practice will permit the reuse of forms, ties, and other items. Furthermore, it will reduce the possibilities of field mistakes.

As few openings as possible should be placed in concrete walls. Where openings are necessary it is desirable to repeat the sizes and positions of openings in different walls rather than using different sizes and positions. Furthermore, a few large openings are more economical than a larger number of smaller ones.

Much money can be saved if a footing elevation can be kept constant for any given wall. Such a practice will appreciably simplify the use of wall forms. If steps are required in a footing, their number should be kept to the minimum possible.[4]

South point water facility, Durham, NC. (Courtesy of EFCO.)

---

[4]Neville, G. B., ed., 1984, *Simplified Design Reinforced Concrete Buildings of Moderate Size* (Skokie, IL: Portland Cement Association), pp. 9-12–9-13.

18.8   COMPUTER EXAMPLES

Concrete walls can be designed using the Excel spreadsheet provided for Chapters 9 and 10 with appropriate input values and interpretation of results. A value of $b = 12$ in. is used, thus the loads per foot of wall length are simply the input loads. Only a value of A_{s1} is input as it is uncommon to have more than two layers of steel in a wall. If only one layer is used, then input a value of $\gamma = 0$. In order to avoid having to laterally tie the compression reinforcement, the total vertical reinforcement area is limited to 0.01 times the gross concrete area. The spreadsheet could easily be modified to neglect the contribution of the compression steel, and if this were done, the 0.01 limit would not have to be applied.

EXAMPLE 18.4

Work Example 18.2 using the spreadsheets for Chapters 9 and 10.

SOLUTION   Open the Excel Spreadsheet for Chapter 9 and 10, and the Rect Col Worksheet. Enter one load case at a time. Refer to the table of load combinations in Example 18.2 and look under the heading P_u and M_u. Only the values of P_u and M_u for loading combination $U = 0.9D + 1.6H$ are shown in the screenshot below. It is not possible in this example to distinguish between the value of A_{s1} required for load cases $U = 0.9D + 1.6H$ and $U = 1.2D + 1.6L + 1.6H$, so only the former is shown. Once values are entered for P_u, M_u, b (always 12 in. for walls), h, f'_c, and f_y, enter trial values for A_{s1} (A_{s2} is always zero for walls). Then look at the interaction diagram and see if the loading "dot" is within the contour of the interaction diagram. Use the smaller diagram that has been reduced by the ϕ factor. If the dot falls well outside the contour, you may need to increase the wall thickness. In working this problem with the interaction diagrams in the Appendix, it was necessary to use $\gamma = 0.6$. For walls with two steel layers, $\gamma = 0.5$ is more realistic, and this example could be easily worked using this value. The value of A_{s1} obtained by trial and error is the total area of steel per foot of wall length. Half goes in each layer, so enter Appendix Table A.6 seeking a value close to and exceeding 0.48 in.^{2}/ft. Select #5 @ $7\frac{1}{2}$ in. If the steel is in only one layer, enter a value of $\gamma = 0$. Steel placed in this fashion is less efficient, and often a thicker wall is needed. Refer to Example 18.2 for horizontal steel requirements.

Rectangular Column Capacity

$P_u =$	1.247	k
$M_u =$	13.8667	ft-k =
$b =$	12	in.
$h =$	8	in.
$\gamma =$	0.6	
$f'_c =$	4000	psi
$f_y =$	60,000	psi
$A_{s1} =$	0.96	in.^{2}
$A_{s2} =$	0.00	in.^{2}
$A_{st} =$	0.96	in.^{2}
$A_g =$	96.0	in.^{2}
$\rho_t =$	0.0100	
$\beta_1 =$	0.85	
$\varepsilon_y =$	0.00207	
$E_s =$	29,000	ksi
$c_{bal} =$	3.79	in.
$c_{.005} =$	2.4	in.

166.4 k-in.

γ in cell C6 may be too large to meet cover requirements

Interaction Diagram—Rectangular Column

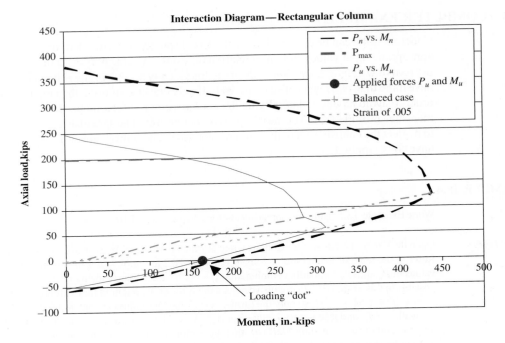

Loading "dot"

PROBLEMS

18.1 Design a reinforced concrete bearing wall using the ACI empirical formula to support a set of precast concrete roof beams $6'0''$ on center, as shown in the accompanying illustration. The bearing width of each beam is $8''$. The wall is considered to be laterally restrained top and bottom and is further assumed to be restrained against rotation at the footing. Neglect wall weight. Other data: $f'_c = 3000$ psi, $f_y = 60,000$ psi, beam reaction, $D = 35$ k, $L = 25$ k. (*Ans.* $7\frac{1}{2}$ in. wall with #3 bars @ 12 in. vertical steel).

18.2 Repeat Problem 18.1 if the wall is not restrained against rotation at top or bottom, is $8'0''$ in height, and has an $f'_c = 4000$ psi.

18.3 Design the reinforced concrete wall shown if $f'_c = 4000$ psi and $f_y = 60,000$ psi. (*One ans.* 10 in. wall with 8 #9 flexural bars each end).

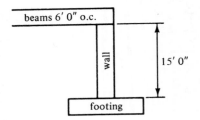

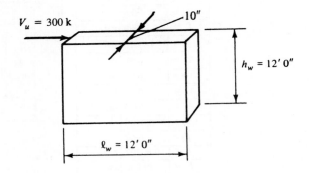

18.4 Repeat Problem 18.3 if $h_w = 15'0''$ and if $f_c' = 3000$ psi.

18.5 Design the wall in Problem 18.1 using the column interaction diagrams (See Example 18-2). Change the 15-ft wall height to 20 ft. Replace the beams with a solid, one-way slab. The slab exerts a dead load of 6 k/ft and a live load of 4 k/ft, both at a 3-in. eccentricity measured from the center of the wall toward the left. Soil back fill is placed to a depth of 20 ft on the right-hand side of the wall ($\gamma_s = 100$ pcf, $k_a = 0.33$). (*Ans* 8-in. wall with #6 bars @ 5 in., 2 layers)

18.6 Repeat Problem 18.5 using the Excel spreadsheet provided for Chapters 8 and 9.

18.7 Repeat Problem 18.6 using steel in one layer. (*Ans* 12-in. wall with #9 @ 4 in.)

Problem in SI Units

18.8 Design a reinforced concrete bearing wall using the ACI empirical formula to support a set of precast roof beams 2 m on center, as shown in the accompanying illustration. The bearing width of each beam is 200 mm. The wall is considered to be laterally restrained top and bottom and is further assumed to be restrained against rotation at the footing. Neglect wall weight. Other data: $f_c' = 21$ MPa, $f_y = 420$ MPa, beam reaction $D = 120$ kN, $L = 100$ kN. (*Ans.* 160 mm thick wall with #10 bars @ 200 mm o.c. horizontal reinforcing)

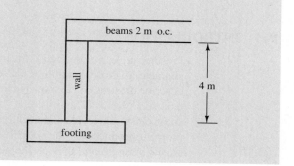

19

Prestressed Concrete

19.1 INTRODUCTION

Prestressing can be defined as the imposition of internal stresses into a structure that are of opposite character to those that will be caused by the service or working loads. A common method used to describe prestressing is shown in Figure 19.1, where a row of books has been squeezed together by a person's hands. The resulting "beam" can carry a downward load as long as the compressive stress due to squeezing at the bottom of the "beam" is greater than the tensile stress there due to the moment produced by the weight of the books and the superimposed loads. Such a beam has no tensile strength and thus no moment resistance until it is squeezed together or prestressed. You might very logically now expand your thoughts to a beam consisting of a row of concrete blocks squeezed together and then to a plain concrete beam with its negligible tensile strength similarly prestressed.

The theory of prestressing is quite simple and has been used for many years in various kinds of structures. For instance, wooden barrels have long been made by putting tightened metal bands around them, thus compressing the staves together and making a tight container with resistance to the outward pressures of the enclosed liquids. Prestressing is primarily used for concrete beams to counteract tension stresses caused by the weight of the members and the superimposed loads. Should these loads cause a positive moment in a beam, it is possible by prestressing to introduce a negative moment that can counteract part or all of the positive moment. An ordinary beam has to have sufficient strength to support itself as well as the other loads, but it is possible with prestressing to produce a negative loading that will eliminate the effect of the beam's weight, thus producing a "weightless beam."

From the preceding discussion it is easy to see why prestressing has captured the imagination of so many persons and why it has all sorts of possibilities now and in the future.

In the earlier chapters of this book, only a portion of the concrete cross sections of members in bending could be considered effective in resisting loads because a large part of those cross sections were in tension and thus the concrete cracked. If, however, concrete flexural members can be prestressed so that their entire cross sections are kept in compression, then the properties of the entire sections are available to resist the applied forces.

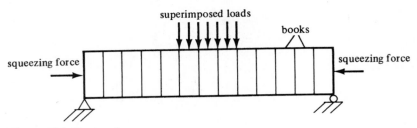

Figure 19.1 Prestressing.

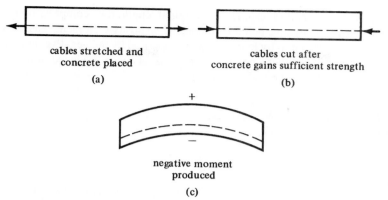

cables stretched and
concrete placed

(a)

cables cut after
concrete gains sufficient strength

(b)

+

negative moment
produced

(c)

Figure 19.2

For a more detailed illustration of prestressing, reference is made to Figure 19.2. It is assumed that the following steps have been taken with regard to this beam:

1. Steel strands (represented by the dashed lines) were placed in the lower part of the beam form.

2. The strands were tensioned to a very high stress.

3. The concrete was placed in the form and allowed to gain sufficient strength for the prestressed strands to be cut.

4. The strands were cut.

Prestressed concrete channels, John A. Denies Son Company Warehouse #4, Memphis, Tennessee. (Courtesy of Master Builders.)

The cut strands tend to resume their original length, thus compressing the lower part of the beam and causing a negative bending moment. The positive moment caused by the beam weight and any superimposed gravity loads is directly opposed by the negative moment. Another way of explaining this is to say that a compression stress has been produced in the bottom of the beam opposite in character to the tensile stress that is caused there by the working loads.

19.2 ADVANTAGES AND DISADVANTAGES OF PRESTRESSED CONCRETE

Advantages

As described in Section 19.1, it is possible with prestressing to utilize the entire cross sections of members to resist loads. Thus, smaller members can be used to support the same loads, or the same-size members can be used for longer spans. This is a particularly important advantage because member weights make up a substantial part of the total design loads of concrete structures.

Prestressed members are crack-free under working loads and, as a result, look better and are more watertight, providing better corrosion protection for the steel. Furthermore, crack-free prestressed members require less maintenance and last longer than cracked reinforced concrete members. Therefore, for a large number of structures, prestressed concrete provides the lowest first-cost solution, and when its reduced maintenance is considered, prestressed concrete provides the lowest overall cost for many additional cases.

The negative moments caused by prestressing produce camber in the members, with the result that total deflections are reduced. Other advantages of prestressed concrete include the following: reduction in diagonal tension stresses, sections with greater stiffnesses under working loads, and increased fatigue and impact resistance as compared to ordinary reinforced concrete.

Disadvantages

Prestressed concrete requires the use of higher-strength concretes and steels and the use of more complicated formwork, with resulting higher labor costs. Other disadvantages include the following:

1. Closer quality control required in manufacture.
2. Losses in the initial prestressing forces. When the compressive forces due to prestressing are applied to the concrete, it will shorten somewhat, partially relaxing the cables. The result is some reduction in cable tension with a resulting loss in prestressing forces. Shrinkage and creep of the concrete add to this effect.
3. Additional stress conditions must be checked in design, such as the stresses occurring when prestress forces are first applied and then after prestress losses have taken place, as well as the stresses occurring for different loading conditions.
4. Cost of end anchorage devices and end-beam plates that may be required.

19.3 PRETENSIONING AND POSTTENSIONING

The two general methods of prestressing are pretensioning and posttensioning. *Pretensioning* was illustrated in Section 19.1, where the prestress tendons were tensioned before the concrete was placed. After the concrete had hardened sufficiently, the tendons were cut and the prestress force

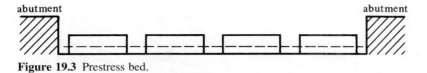

Figure 19.3 Prestress bed.

Figure 19.4 Posttensioned beam.

was transmitted to the concrete by bond. This method is particularly well suited for mass production because the casting beds can be constructed several hundred feet long. The tendons can be run for the entire bed lengths and used for casting several beams in a line at the same time, as shown in Figure 19.3.

In *posttensioned* construction (see Figure 19.4), the tendons are stressed after the concrete is placed and has gained the desired strength. Plastic or metal tubes, conduits, sleeves, or similar devices with unstressed tendons inside (or later inserted) are located in the form and the concrete is placed. After the concrete has sufficiently hardened, the tendons are stretched and mechanically attached to end anchorage devices to keep the tendons in their stretched positions. Thus by posttensioning, the prestress forces are transferred to the concrete not by bond, but by end bearing.

It is actually possible in posttensioning to have either bonded or unbonded tendons. If bonded, the conduits are often made of aluminum, steel, or other metal sheathing. In addition, it is possible to use steel tubing or rods or rubber cores that are cast in the concrete and removed a few hours after the concrete is placed. After the steel is tensioned, cement grout is injected into the duct for bonding. The grout is also useful in protecting the steel from corrosion. If the tendons are to be unbonded, they should be greased to facilitate tensioning and to protect them from corrosion.[1]

19.4   MATERIALS USED FOR PRESTRESSED CONCRETE

The materials ordinarily used for prestressed concrete are concrete and high-strength steels. The concrete used is normally of a higher strength than that used for reinforced concrete members, for several reasons, including the following:

1. The modulus of elasticity of such concretes is higher, with the result that the elastic strains in the concrete are smaller when the tendons are cut. Thus the relaxations or losses in the tendon stresses are smaller.

2. In prestressed concrete, the entire members are kept in compression, and thus all the concrete is effective in resisting forces. Hence, it is reasonable to pay for a more expensive but stronger concrete if all of it is going to be used. (In ordinary reinforced concrete beams, more than half of the cross sections are in tension and thus assumed to be cracked. As a result, more than half of a higher-strength concrete used there would be wasted.)

---

[1]Nawy, E. G., 2005, *Prestressed Concrete*, 5th ed. (Upper Saddle River, NJ: Prentice Hall), pp. 62–69.

Prestressed concrete segmental bridge over the River Trent near Scunthorpe, Lincolnshire, England. (Courtesy of Cement and Concrete Association.)

3. Most prestressed work in the United States is of the precast, pretensioned type done at the prestress yard where the work can be carefully controlled; consequently, dependable higher-strength concrete can readily be obtained.

4. For pretensioned work, the higher-strength concretes permit the use of higher bond stresses between the cables and the concrete.

High-strength steels are necessary to produce and keep satisfactory prestress forces in members. The strains that occur in these steels during stressing are much greater than those that can be obtained with ordinary reinforcing steels. As a result, when the concrete elastically shortens in compression and also shortens due to creep and shrinkage, the losses in strain in the steel (and thus stress) represent a smaller percentage of the total stress. Another reason for using high-strength steels is that a large prestress force can be developed in a small area.

Early work with prestressed concrete using ordinary-strength bars to induce the prestressing forces in the concrete resulted in failure because the low stresses that could be put into the bars were completely lost due to the concrete's shrinkage and creep. Should a prestress of 20,000 psi be put into such rods, the resulting strains would be equal to $20,000/(29 \times 10^6) = 0.00069$. This value is less than the long-term creep and shrinkage strain normally occurring in concrete, roughly 0.0008, which would completely relieve the stress in the steel. Should a high-strength steel be stressed to about 150,000 psi and have the same creep and shrinkage, the stress reduction will be of the order of $(0.0008)(29 \times 10^6) = 23,000$ psi, leaving $150,000 - 23,200 = 126,800$ psi in the steel (a loss of only 15.47% of the steel stress).[2]

Three forms of prestressing steel are used: single wires, wire strands, and bars. The greater the diameter of the wires, the smaller become their strengths and bond to the concrete. As a result, wires are manufactured with diameters from 0.192 in. up to a maximum of 0.276 in. (about $\frac{9}{32}$ in.). In posttensioning work, large numbers of wires are grouped in parallel into tendons. Strands that are made by twisting wires together are used for most pretensioned work. They are of the seven-wire type, where a center wire is tightly surrounded by twisting the other six wires helically around it. Strands are manufactured with diameters from $\frac{1}{4}$ to $\frac{1}{2}$ in. Sometimes large-size, high-strength, heat-treated alloy steel bars are used for posttensioned sections. they are available with diameters running from $\frac{3}{4}$ to $1\frac{3}{8}$ in.

High-strength prestressing steels do not have distinct yield points (see Figure 19.10) as do the structural carbon reinforcing steels. The practice of considering yield points, however, is so firmly

[2]Winter, G., and Nilson, A. H., 1991, *Design of Concrete Structures*, 11th ed. (New York: McGraw-Hill), pp. 759–760.

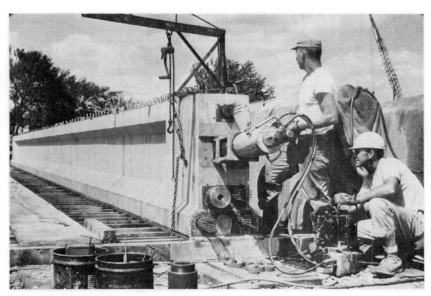

Prestressing girders for bridge in Butler, Pennsylvania. (Courtesy of Portland Cement Association.)

embedded in the average designer's mind that high-strength steels are normally given an arbitrary yield point anyway. The yield stress for wires and strands is usually assumed to be the stress that causes a total elongation of 1% to occur in the steel. For high-strength bars the yield stress is assumed to occur when a 0.2% permanent strain occurs.

19.5   STRESS CALCULATIONS

For a consideration of stresses in a prestressed rectangular beam, reference is made to Figure 19.5. For this example the prestress tendons are assumed to be straight, although it will later be shown that a curved shape is more practical for most beams. The tendons are assumed to be located an eccentric distance e below the centroidal axis of the beam. As a result, the beam is subjected to a combination of direct compression and a moment due to the eccentricity of the prestress. In addition, there will be a moment due to the external load, including the beam's own weight. The resulting stress at any point in the beam caused by these three factors can be written as follows where P is the prestressing force:

$$f = -\frac{P}{A} \pm \frac{Pec}{I} \pm \frac{Mc}{I}$$

In the above expression, P is the prestress force, e is the eccentricity of the prestress force with respect to the centroid of the cross section, c is the distance from the centroidal axis to the extreme fiber (top or bottom depending on where the stresses are being determined), M is the applied moment due to unfactored loads at the stage at which stresses are being calculated, A is the uncracked concrete cross-sectional area, and I is the moment of inertia of the gross concrete cross section. In Figure 19.5, a stress diagram is drawn for each of these three items, and all three are combined to give the final stress diagram.

The usual practice is to base the stress calculations in the elastic range on the properties of the gross concrete section. The gross section consists of the concrete external dimensions with no additions made for the transformed area of the steel tendons nor subtractions made for the duct

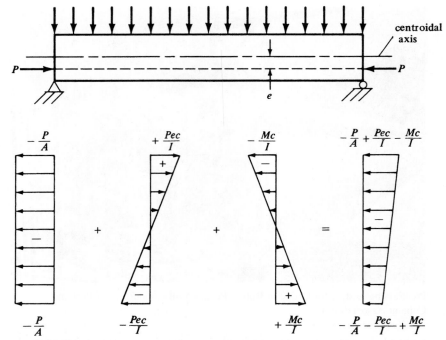

Figure 19.5

areas in posttensioning. This method is considered to give satisfactory results because the changes in stresses obtained if net or transformed properties are used are usually not significant.

Example 19.1 illustrates the calculations needed to determine the stresses at various points in a simple-span prestressed rectangular beam. It will be noted that, as there are no moments at the ends of a simple beam due to the external loads or to the beams own weight, the Mc/I part of the stress equation is zero there and the equation reduces to

$$f = -\frac{P}{A} \pm \frac{Pec}{I}$$

EXAMPLE 19.1

Calculate the stresses in the top and bottom fibers at the centerline and ends of the beam shown in Figure 19.6.

SOLUTION   **Section Properties**

$$I = \left(\frac{1}{12}\right)(12)(24)^3 = 13{,}824 \text{ in.}^4$$

$$A = (12)(24) = 288 \text{ in.}^2$$

$$M = \frac{(3)(20)^2}{8} = 150 \text{ ft-k}$$

Stresses at Beam Centerline

$$f_{\text{top}} = -\frac{P}{A} + \frac{Pec}{I} - \frac{Mc}{I} = -\frac{250}{288} + \frac{(250)(9)(12)}{13{,}824} - \frac{(12)(150)(12)}{13{,}824}$$

$$= -0.868 + 1.953 - 1.562 = -0.477 \text{ ksi}$$

$$f_{\text{bottom}} = -\frac{P}{A} - \frac{Pec}{I} + \frac{Mc}{I} = -0.868 - 1.953 + 1.562 = -1.259 \text{ ksi}$$

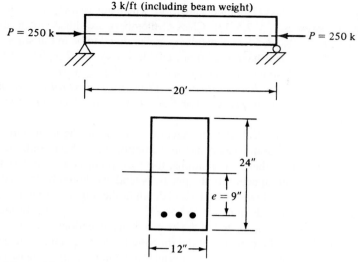

Figure 19.6

Stresses at Beam Ends

$$f_{\text{top}} = -\frac{P}{A} + \frac{Pec}{I} = -0.868 + 1.953 = +1.085 \text{ ksi}$$

$$f_{\text{bottom}} = -\frac{P}{A} - \frac{Pec}{I} = -0.868 - 1.953 = -2.821 \text{ ksi}$$

In Example 19.1 it was shown that when the prestress tendons are straight, the tensile stress at the top of the beam at the ends will be quite high. If, however, the tendons are draped, as shown in Figure 19.7, it is possible to reduce or even eliminate the tensile stresses. Out in the span, the centroid of the strands may be below the lower kern point (see Example 19.2 for determination of the kern point for this section), but if at the ends of the beam, where there is no stress due to dead-load moment, it is below the kern point, tensile stresses in the top will be the result. If the tendons are draped so that at the ends they are located at or above this point, tension will not occur in the top of the beam.

In posttensioning, the sleeve or conduit is placed in the forms in the curved position desired. The tendons in pretensioned members can be placed at or above the lower kern points and then can be pushed down to the desired depth at the centerline or at other points. In Figure 19.7 the tendons are shown held down at the one-third points. Two alternatives to draped tendons that have been used are to use straight tendons, located below the lower kern point but which are encased in tubes at their ends, or have their ends greased. Both methods are used to prevent the development of negative moments at the beam ends.

In ACI Section 18.3.3 bonded and unbonded prestressed members are designated as being Class U, T or C members. These classifications are based on computed tensile stresses in members

Figure 19.7 Draped tendons.

subject to service loads. Class U members are those which are assumed to be uncracked and have maximum tensile stresses $f_t \leq 7.5\sqrt{f_c'}$. Class C members are those which are assumed to be cracked and have $f_t > 12\sqrt{f_c'}$. Class T members are assumed to be in transition between cracked and uncracked members and have maximum tensile stresses $> 7.5\sqrt{f_c'} \leq 12\sqrt{f_c'}$. Prestressed two-way slabs must be designed as Class U sections with $f_t \leq 6\sqrt{f_c'}$.

ACI Section 18.3.4 states that for Class U and T members flexural stresses may be computed using the uncracked section properties. For Class C sections, however, it is necessary to use cracked section properties.

Example 19.2 shows the calculations necessary to locate the kern point for the beam of Example 19.1. In addition, the stresses at the top and bottom of the beam ends are computed. It will be noted that, according to these calculations, the kern point is 4 in. below the mid-depth of the beam, and it would thus appear that the prestress tendons should be located at the kern point at the beam ends and pushed down to the desired depth farther out in the beam. Actually, however, the tendons at the beam ends do not have to be as high as the kern points because the ACI Code (18.4.1) permits some tension in the top of the beam when the tendons are cut. This value is $3\sqrt{f_{ci}'}$, where f_{ci}' is the strength of the concrete at the time the tendons are cut, as determined by testing concrete cylinders. The subscript i denotes "initial," meaning at initial release of the prestressing tendon, before the concrete gains its full 28-day strength. This permissible value equals about 40% of the cracking strength or modulus of rupture of the concrete $(7.5\sqrt{f_{ci}'})$ at that time. The stress at the bottom of the beam, which is compressive, is permitted to go as high as $0.60f_{ci}'$.

The Code actually permits tensile stresses at the ends of simple beams to go as high as $6\sqrt{f_{ci}'}$. These allowable tensile values are applicable to the stresses that occur immediately after the transfer of the prestressing forces and after the losses occur due to elastic shortening of the concrete and relaxation of the tendons and anchorage seats. It is further assumed that the time-dependent losses of creep and shrinkage have not occurred. A discussion of these various losses is presented in Section 19.7 of this chapter. If the calculated tensile stresses are greater than the permissible values, it is necessary to use some additional bonded reinforcing (prestressed or unprestressed) to resist the *total* tensile force in the concrete computed on the basis of an uncracked section.

Section 18.4.2 of the Code provides allowable stresses at service loads for Class U and Class T members after all prestress losses have occurred. An extreme fiber compression stress equal to $0.60f_c'$ is permitted for prestress plus sustained loads. The allowable compression stress for prestress plus total loads is $0.70f_c'$. In effect, the ACI here provides a one-third increase in allowable compression stress when a large percentage of the service loads are deemed to be transient or of short duration.

The allowable tensile stress at ends of simply supported members immediately after prestress transfer is $6\sqrt{f_{ci}'}$. Section 18.4.3 of the Code permits higher permissible stresses than those presented here under certain conditions. The Commentary on this section of the Code states that it is the intent of the Code writers to permit higher stress values when justified by the development of newer and better products, materials, and prestress techniques. Approval of such increases must be in accordance with the procedures of Section 1.4 of the Code.

Only compressive stresses should be allowed in prestressed sections that are to be used in severe corrosive conditions. If tension cracks occur, the result may very well be increased cable corrosion.

EXAMPLE 19.2

Determine the location of the lower kern point at the ends of the beam of Example 19.1. Calculate the stresses at the top and bottom of the beam ends, assuming the tendons are placed at the kern point.

SOLUTION   **Locating the Kern Point**

$$f_{top} = -\frac{P}{A} + \frac{Pec}{I} = 0$$

$$-\frac{250}{288} + \frac{(250)(e)(12)}{13,824} = 0$$

$$-0.868 + 0.217e = 0$$

$$e = 4''$$

Computing Stresses

$$f_{top} = -\frac{P}{A} - \frac{Pec}{I} = -\frac{250}{288} + \frac{(250)(4)(12)}{13,824}$$

$$= -0.868 + 0.868 = 0$$

$$f_{bottom} = -\frac{P}{A} - \frac{Pec}{I} = -0.868 - 0.868 = -1.736 \text{ ksi}$$

19.6   SHAPES OF PRESTRESSED SECTIONS

For simplicity in introducing prestressing theory, rectangular sections are used for most of the examples of this chapter. From the viewpoint of formwork alone, rectangular sections are the most economical, but more complicated shapes, such as I's and T's, will require smaller quantities of concrete and prestressing steel to carry the same loads and, as a result, they frequently have the lowest overall costs.

If a member is to be made only one time, a cross section requiring simple formwork (thus often rectangular) will probably be used. For instance, simple formwork is essential for most cast-in-place work. Should, however, the forms be used a large number of times to make many identical members, more complicated cross sections, such as I's and T's, channels, or boxes, will be used. For such sections the cost of the formwork as a percentage of each member's total cost will be greatly reduced. Several types of commonly used prestressed sections are shown in Figure 19.8. The same general theory used for the determination of stresses and flexural strengths applies to shapes such as these, as it does to rectangular sections.

The usefulness of a particular section depends on the simplicity and reusability of the formwork, the appearance of the sections, the degree of difficulty of placing the concrete, and the theoretical properties of the cross section. The greater the amount of concrete located near the extreme fibers of a beam, the greater will be the lever arm between the *C* and *T* forces and thus the greater the resisting moment. Of course, there are some limitations on the widths and thicknesses of the flanges. In addition, the webs must be sufficiently large to resist shear and to allow the proper placement of the concrete and at the same time be sufficiently thick to avoid buckling.

A prestressed *T* such as the one shown in Figure 19.8(a) is often a very economical section because a large proportion of the concrete is placed in the compression flange, where it is quite effective in resisting compressive forces. The double T shown in Figure 19.8(b) is used for schools, office buildings, stores, and so on and is probably the most used prestressed section in the United States today. The total width of the flange provided by a double T is in the range of about 5 to 8 ft, and spans of 30 to 50 ft are common. You can see that a floor or roof system can be erected easily and quickly by placing a series of precast double T's side by side ⊓⊓ ⊓⊓ ⊓⊓. The sections serve as both the beams and slabs for the floor or roof system. Single T's are normally used for heavier loads and longer spans up to as high as 100 or 120 ft. Double T's for such spans would be very heavy and difficult to handle. The single T is not used as much today as it was in the recent past due to stability difficulties in both shipping and erection.

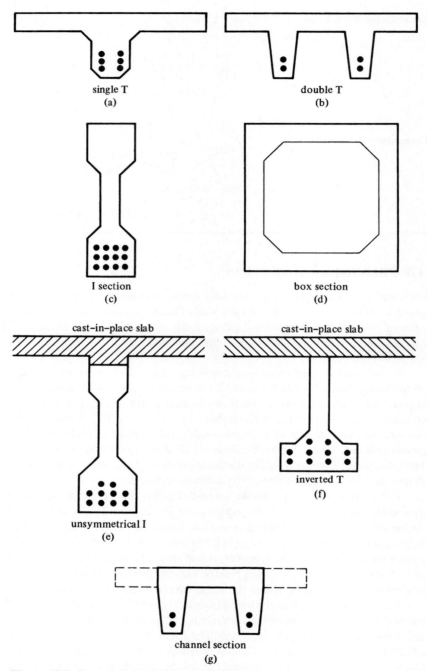

Figure 19.8 Commonly used prestressed sections.

The I and box sections, shown in parts (c) and (d) of Figure 19.8, have a larger proportion of their concrete placed in their flanges, with the result that larger moments of inertia are possible (as compared to rectangular sections with the same amounts of concrete and prestressing tendons). The formwork, however, is complicated, and the placing of concrete is difficult. Box girders are

Prestressed Bridge Girders – the right hand girder illustrates tendons with both upward and downward ecentricity. (Courtesy Portland Cement Assoc.)

Posttensioned segmental precast concrete for East Moors Viaduct, Lanbury Way, Cardiff, South Wales. (Courtesy of Cement and Concrete Association.)

frequently used for bridge spans. Their properties are the same as for I sections. Unsymmetrical I's [Figure 19.8(e)], with large bottom flanges to contain the tendons and small top flanges, may be economical for certain composite sections where they are used together with a slab poured in place to provide the compression flange. A similar situation is shown in Figure 19.8(f), where an inverted T is used with a cast-in-place slab.

Many variations of these sections are used, such as the channel section shown in Figure 19.8(g). Such a section might be made by blocking out the flanges of a double-T form as shown, and the resulting members might be used for stadium seats or similar applications.

19.7 PRESTRESS LOSSES

The flexural stresses calculated for the beams of Examples 19.1 and 19.2 were based on initial stresses in the prestress tendons. These stresses, however, become smaller with time (over a period of roughly five years) due to several factors. These factors, which are discussed in the paragraphs to follow, include:

1. Elastic shortening of the concrete
2. Shrinkage and creep of the concrete
3. Relaxation or creep in the tendons
4. Slippage in posttensioning end anchorage systems
5. Friction along the ducts used in posttensioning

Although it is possible to calculate prestress losses individually for each of the factors listed above, it is usually more practical and often just as satisfactory to use single lump-sum estimates for all the items together. There are just too many interrelated factors affecting the estimates to achieve accuracy.

Such lump-sum estimates of total prestress losses are applicable only to average prestress members made with normal concrete, construction procedures, and quality control. Should conditions be decidedly different from these and/or if the project is extremely significant, it would be well to consider making detailed loss estimates such as those introduced in the next few paragraphs.

The ultimate strength of a prestressed member is almost completely controlled by the tensile strength and cross-sectional area of the cables. Consequently, losses in prestress will have very little effect on its ultimate flexural strength. However, losses in prestress will cause more cracking to occur under working loads, with the result that deflections will be larger. Furthermore, the member's shear and fatigue strength will be somewhat reduced.

Elastic Shortening of the Concrete

When the tendons are cut for a pretensioned member, the prestress force is transferred to the concrete, with the result that the concrete is put in compression and shortens, thus permitting some relaxation or shortening of the tendons. The stress in the concrete adjacent to the tendons can be computed as described in the preceding examples. The strain in the concrete, ϵ_c, which equals f_c/E_c, is assumed due to bond to equal the steel strain ϵ_s. Thus the loss in prestress can be computed as $\epsilon_s E_s$. An average value of prestress loss in pretensioned members due to elastic shortening is about 3% of the initial value.

An expression for the loss of prestress due to elastic shortening of the concrete can be derived as shown in the paragraphs to follow.

It can be seen that the compressive strain in the concrete due to prestress must equal the lessening of the steel strain

$$\epsilon_c = \Delta\epsilon_s$$

These values can be written in terms of stresses as follows:

$$\frac{f_c}{E_c} = \frac{\Delta f_s}{E_s}$$

Thus we can write

$$\Delta f_s = \frac{E_s f_c}{E_c} = n f_c$$

where f_c is the stress in the concrete at the level of the tendon centroid after transfer of stresses from the cables.

If we express Δf_s as being the initial tendon stress f_{si} minus the tendon stress after transfer, we can write

$$f_{si} - f_s = n f_c$$

Then letting P_0 be the initial total cable stress and P_f the stress afterward, we obtain

$$P_0 - P_f = n \frac{P_f}{A_c} A_{ps}$$

$$P_0 = n \frac{P_f}{A_c} A_{ps} + P_f$$

$$P_0 = P_f \left(\frac{n A_{ps}}{A_c} + 1 \right) = \frac{P_f}{A_c}(n A_{ps} + A_c)$$

Then

$$f_c = \frac{P_0}{A_c + A_{ps}} = \text{approximately } \frac{P_0}{A_g}$$

and finally

$$\Delta f_s = n f_c = \frac{n P_0}{A_g}$$

a value that can easily be calculated.

For posttensioned members, the situation is a little more involved because it is rather common to stress a few of the strands at a time and connect them to the end plates. As a result, the losses vary, with the greatest losses occurring in the first strands stressed and the least losses occurring in the last strands stressed. For this reason, an average loss may be calculated for the different strands. Losses due to elastic shortening average about .5% for posttensioned members. It is, by the way, often possible to calculate the expected losses in each set of tendons and overstress them by that amount so the net losses will be close to zero.

Shrinkage and Creep of the Concrete

The losses in prestressing due to the shrinkage and creep in the concrete are quite variable. For one thing, the amount of shrinkage that occurs in concrete varies from almost zero up to about 0.0005 in./in. (depending on dampness and on the age of the concrete when it is loaded), with an average value of about 0.0003 in./in. being the usual approximation.

The loss in prestress due to shrinkage can be said to equal $\epsilon_{sh} E_s$, where ϵ_{sh} is the shrinkage strain of the concrete. A recommended value of ϵ_{sh} is given in Zia et al. (1979) which is to be determined by taking the basic shrinkage strain times a correction factor based on the volume (V)-to-surface (S) ratio times a relative humidity correction (H).[3]

$$\epsilon_{sh} = (0.00055)\left(1 - 0.06\,\frac{V}{S}\right)(1.5 - 0.15H)$$

Should the member be posttensioned, an additional multiplier is provided in Zia et al. to take into account the time between the end of the moist curing until the prestressing forces are applied.

The amount of creep in the concrete depends on several factors, which have been previously discussed in this text and can vary from 1 to 5 times the instantaneous elastic shortening. Prestress forces are usually applied to pretensioned members much earlier in the age of the concrete than for posttensioned members. Pretensioned members are normally cast in a bed at the prestress yard, where the speed of production of members is an important economic matter. The owner wants to tension the steel, place the concrete, and take the members out of the prestress bed as quickly as the concrete gains sufficient strength so that work can start on the next set of members. As a result, creep and shrinkage are larger, as are the resulting losses. Average losses are about 6% for pretensioned members and about 5% for posttensioned members.

The losses in cable stresses due to concrete creep strain can be determined by multiplying an experimentally determined creep coefficient C_t by nf_c.

$$\Delta f_s = C_t n f_c$$

In Zia et al. (1979) a value of $C_t = 2.0$ is recommended for pretensioned sections, while 1.6 is recommended for posttensioned ones. These values should be reduced by 20% if lightweight concrete is used. The value f_c is defined as the stress in the concrete adjacent to the centroid of the tendons due to the initial prestress ($-P/A$) and due to the permanent dead loads that are applied to the member after prestressing ($-Pec/I$), where e is measured from the centroid of the section to the centroid of the tendons.

Relaxation or Creep in the Tendons

The plastic flow or relaxation of steel tendons is quite small when the stresses are low, but the percentage of relaxation increases as stresses become higher. In general, the estimated losses run from about 2% to 3% of the initial stresses. The amount of these losses actually varies quite a bit for different steels and should be determined from test data available from the steel manufacturer in question. A formula is available with which this loss can be computed.

Slippage in Posttensioning End Anchorage Systems

When the jacks are released and the prestress forces are transferred to the end anchorage system, a little slippage of the tendons occurs. The amount of the slippage depends on the system used and tends to vary from about 0.10 in. to 0.20 in. Such deformations are quite important if the members and thus the tendons are short, but if they are long, the percentage is much less important.

---

[3]Zia, P., Preston, H. K., Scott, N. L., and Workman, E. B., 1979, "Estimating Prestress Losses," *Concrete International: Design & Construction*, vol. 1, no. 6 (Detroit: American Concrete Institute), pp. 32–38.

Friction along the Ducts Used in Posttensioning

There are losses in posttensioning due to friction between the tendons and the surrounding ducts. In other words, the stress in the tendons gradually falls off as the distance from the tension points increases due to friction between the tendons and the surrounding material. These losses are due to the so-called length and curvature effects.

The *length effect* is the friction that would have existed if the cable had been straight and not curved. Actually, it is impossible to have a perfectly straight duct in posttensioned construction, and the result is friction, called the length effect or sometimes the *wobble effect*. The magnitude of this friction is dependent on the stress in the tendons, their length, the workmanship for the particular member in question, and the coefficient of friction between the materials.

The *curvature effect* is the amount of friction that occurs in addition to the unplanned wobble effect. The resulting loss is due to the coefficient of friction between the materials caused by the pressure on the concrete from the tendons, which is dependent on the stress and the angle change in the curved tendons.

It is possible to reduce frictional losses substantially in prestressing by several methods. These include jacking from both ends, overstressing the tendons initially, and lubricating unbonded cables.

The ACI Code (18.6.2.2) requires that frictional losses for posttensioned members be computed with wobble and curvature coefficients experimentally obtained and verified during the prestressing operation. Furthermore, the Code provides Equations 18-1 and 18-2 (in Section 18.6.2.1) for making the calculations. The ACI Commentary (R18.6.2) provides values of the friction coefficients for use in the equations.

19.8   ULTIMATE STRENGTH OF PRESTRESSED SECTIONS

Considerable emphasis is given to the ultimate strength of prestressed sections, the objective being to obtain a satisfactory factor of safety against collapse. You might wonder why it is necessary in prestress work to consider *both* working-stress and ultimate-strength situations. The answer lies in the tremendous change that occurs in a prestressed member's behavior after tensile cracks occur. Before the cracks begin to form, the entire cross section of a prestressed member is effective in resisting forces, but after the tensile cracks begin to develop, the cracked part is not effective in resisting tensile forces. Cracking is usually assumed to occur when calculated tensile stresses equal the modulus of rupture of the concrete (about $7.5\sqrt{f_c'}$).

Another question that might enter your mind at this time is this: "What effect do the prestress forces have on the ultimate strength of a section?" The answer to the question is quite simple. An ultimate-strength analysis is based on the assumption that the prestressing strands are stressed above their yield point. If the strands have yielded, the tensile side of the section has cracked and the theoretical ultimate resisting moment is the same as for a nonprestressed beam constructed with the same concrete and reinforcing.

The theoretical calculation of ultimate capacities for prestressed sections is not such a routine thing as it is for ordinary reinforced concrete members. The high-strength steels from which prestress tendons are manufactured do not have distinct yield points. Despite this fact, the strength method for determining the ultimate moment capacities of sections checks rather well with load tests as long as the steel percentage is sufficiently small as to ensure a tensile failure and as long as bonded strands are being considered.

In the expressions used here, f_{ps} is the average stress in the prestressing steel at the design load. This stress is used in the calculations because the prestressing steels usually used in prestressed beams do not have well-defined yield points (that is, the flat portions that are common to stress–strain curves for ordinary structural steels). Unless the yield points of these steels are determined from detailed studies, their values are normally specified. For instance, the ACI Code (18.7.2) states that the following approximate expression may be used for calculating f_{ps}. In this expression f_{pu} is the ultimate strength of the prestressing steel, ρ_p is the percentage of prestress reinforcing A_{ps}/bd_p, and f_{se} is the effective stress in the prestressing steel after losses. If more accurate stress values are available, they may be used instead of the specified values. In no case may the resulting values be taken as more than the specified yield strength f_{py}, or $f_{se} + 60,000$. For bonded members,

$$f_{ps} = f_{pu}\left(1 - \frac{\gamma_p}{\beta_1}\left[\rho_p\frac{f_{pu}}{f_c'} + \frac{d}{d_p}(\omega - \omega')\right]\right) \qquad \text{if } f_{se} \geq 0.5f_{pu} \qquad \text{(ACI Equation 18-3)}$$

where γ_p is a factor for the type of prestress tendon whose values are specified in ACI Section 18.0 ($\gamma_p = 0.55$ for f_{py}/f_{pu} not less than 0.80, 0.40 for f_{py}/f_{pu} not less than 0.85, and 0.28 for f_{py}/f_{pu} not less than 0.90), $d_p =$ distance from the extreme compression fiber to the centroid of the prestress reinforcement, $\omega = \rho f_y/f_c'$, and $\omega' = \rho' f_y/f_c'$.

If any compression reinforcing is considered in calculating f_{ps}, the terms in brackets may not be taken less than 0.17 (see Commentary R18.7.2). Should compression reinforcing be taken into account and if the term in brackets is small, the depth to the neutral axis will be small and thus the compression reinforcing will not reach its yield stress. For this situation the results obtained with ACI Equation 18-3 are not conservative, thus explaining why the ACI provides the 0.17 limit.

Should the compression reinforcing be neglected in using the equation, ω' will equal zero and the term in brackets may be less than 0.17. Should d' be large, the strain in the compression steel may be considerably less than the yield strain, and as a result the compression steel will not influence f_{ps} as favorably as implied by the equation. As a result, ACI Equation 18-3 may only be used for beams in which $d' \leq 0.15\, d_p$.

For unbonded members with span to depth ≤ 35,

$$f_{ps} = f_{se} + 10,000 + \frac{f_c'}{100\rho_p} \quad \text{but not greater than } f_{py} \text{ nor } (f_{se} + 60,000)$$

$$\text{(ACI Equation 18-4)}$$

For unbonded members with span to depth > 35,

$$f_{ps} = f_{se} + 10,000 + \frac{f_c'}{300\rho_p} \qquad \text{(ACI Equation 18-5)}$$

However, f_{ps} may not exceed f_{py}, or $f_{se} + 30,000$.

As in reinforced concrete members, the amount of steel in prestressed sections is limited to ensure tensile failures. The limitation rarely presents a problem except in members with very small amounts of prestressing or in members that have not only prestress strands, but also some regular reinforcing bars.

Example 19.3 illustrates the calculations involved in determining the permissible ultimate capacity of a rectangular prestressed beam. Some important comments about the solution and about ultimate-moment calculations in general are made at the end of the example.

EXAMPLE 19.3

Determine the permissible ultimate moment capacity of the prestressed bonded beam of Figure 19.9 if $f_{py} = 240{,}000$ psi, f_{pu} is $275{,}000$ psi, and f_c' is 5000 psi.

SOLUTION   **Approximate Value of f_{ps} from ACI Code**

$$\rho_p = \frac{A_{ps}}{bd_p} = \frac{1.40}{(12)(21.5)} = 0.00543$$

$$\frac{f_{py}}{f_{pu}} = \frac{240{,}000}{275{,}000} = 0.873$$

$\therefore \gamma_p = 0.40$, as given immediately after the presentation of ACI Equation 18-3 earlier in this section. $f_{ps} =$ estimated stress in prestressed reinforcement at nominal strength. Note that $\beta_1 = 0.80$ for 5000 psi concrete and d, the distance from the extreme compression fiber of the beam to the centroid of any nonprestressed tension reinforcement is zero since there is no such reinforcement in this beam.

$$f_{ps} = f_{pu}\left\{1 - \frac{\gamma_p}{\beta_1}\left[\rho_p \frac{f_{pu}}{f_c'} + \frac{d}{d_p}(\omega - \omega')\right]\right\} \qquad \text{(ACI Equation 18-3)}$$

$$= 275\left\{1 - \frac{0.40}{0.80}\left[0.00543\left(\frac{275}{5}\right) + 0\right]\right\} = 233.9 \text{ ksi}$$

Moment Capacity

$$a = \frac{A_{ps}f_{ps}}{0.85f_c'b} = \frac{(1.40)(233.9)}{(0.85)(5)(12)} = \underline{\underline{6.42''}}$$

$$c = \frac{a}{\beta_1} = \frac{6.42}{0.80} = 8.03''$$

$$\epsilon_t = \frac{d_p - c}{c}\,0.003 = \frac{21.5 - 8.03}{8.03}(0.003)$$

$$= 0.0050 \qquad\qquad\qquad\qquad \underline{\therefore \text{ The member is}}$$
$$\underline{\text{tension controlled and } \phi = 0.9.}$$

$$\phi M_n = \phi A_{ps}f_{ps}\left(d - \frac{a}{2}\right) = (0.9)(1.40)(233.9)\left(21.5 - \frac{6.42}{2}\right)$$

$$= 5390 \text{ in.-k} = \underline{\underline{449.2 \text{ ft-k}}}$$

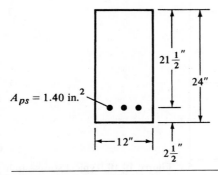

$A_{ps} = 1.40 \text{ in.}^2$

$21\frac{1}{2}''$

$24''$

$12''$

$2\frac{1}{2}''$

Figure 19.9

Discussion

The approximate value of f_{ps} obtained by the ACI formula is very satisfactory for all practical purposes. Actually, a slightly more accurate value of f_{ps} and thus of the moment capacity of the section can be obtained by calculating the strain in the prestress strands due to the prestress and adding to it the strain due to the ultimate moment. This latter strain can be determined from the values of a and the strain diagram, as frequently used in earlier chapters for checking to see if tensile failures control in reinforced concrete beams. With the total strain, a more accurate cable stress can be obtained by referring to the stress–strain curve for the prestressing steel being used. Such a curve is shown in Figure 19.10.

The analysis described herein is satisfactory for pretensioned beams or for bonded posttensioned beams, but is not so good for unbonded posttensioned members. In these latter beams the steel can slip with respect to the concrete; as a result, the steel stress is almost constant throughout the member. The calculations for M_u for such members are less accurate than for bonded members. Unless some ordinary reinforcing bars are added to these members, large cracks may form which are not attractive and which can lead to some corrosion of the prestress strands.

If a prestressed beam is satisfactorily designed with service loads, then checked by strength methods and found to have insufficient strength to resist the factored loads ($M_u = 1.2M_D + 1.6M_L$), nonprestressed reinforcement may be added to increase the factor of safety. The increase in T due to these bars is assumed to equal $A_s f_y$ (Code 18.7.3). The Code (18.8.2) further states that the total amount of prestressed and nonprestressed reinforcement shall be sufficient to develop an ultimate moment equal to at least 1.2 times the cracking moment of the section. This cracking moment is calculated with the modulus of rupture of the concrete, except for flexural members with a shear and flexural strength equal to at least twice that required to support the factored loads and for two-way, unbonded posttensioned slabs. This additional steel also will serve to reduce cracks. (The 1.2 requirement may be waived for two-way unbonded posttensioned slabs and for flexural members with shear and flexural strength at least equal to twice that required by ACI Section 9.2.)

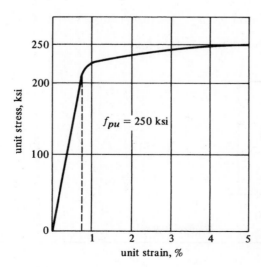

Figure 19.10 Typical stress–strain curve for high-tensile steel wire.

19.9   DEFLECTIONS

The deflections of prestressed concrete beams must be calculated very carefully. Some members that are completely satisfactory in all other respects are not satisfactory for practical use because of the magnitudes of their deflections.

In previous chapters, one method used for limiting deflections was to specify minimum depths for various types of members (as in Table 4.1 of this textbook). These minimum depths, however, are applicable only to nonprestressed sections. The actual deflection calculations are made as they are for members made of other materials, such as structural steel, reinforced concrete, and so on. However, the same problem exists for reinforced concrete members, and that is the difficulty of determining the modulus of elasticity to be used in the calculations. The modulus varies with age, with different stress levels, and with other factors. Usually the gross moments of inertia are used for immediate deflection calculations for members whose calculated extreme fiber stresses at service loads in the precompressed tensile zone are $\leq 7.5\sqrt{f_c'}$ (ACI 18.3.3). Transformed I values may be used for other situations as described in ACI Sections 18.3.3, 18.3.4, and 18.3.5.

The deflection due to the force in a set of straight tendons is considered first in this section, with reference being made to Figure 19.11(a). The prestress forces cause a negative moment equal to Pe and thus an upward deflection or camber of the beam. This $\mathcal{C}$ deflection can be calculated by taking moments at the point desired when the conjugate beam is loaded with the M/EI diagram. At the $\mathcal{C}$ the deflection equals

$$-\left(\frac{Pe\ell}{2EI}\right)\left(\frac{\ell}{2}-\frac{\ell}{4}\right) = -\frac{Pe\ell^2}{8EI}\uparrow$$

Should the cables not be straight, the deflection will be different due to the different negative moment diagram produced by the cable force. If the cables are bent down or curved, as shown in parts (b) and (c) of Figure 19.11, the conjugate beam can again be applied to compute the deflections. The resulting values are shown in the figure.

Bridge construction. (PhotoDisc, Inc./Getty Images.)

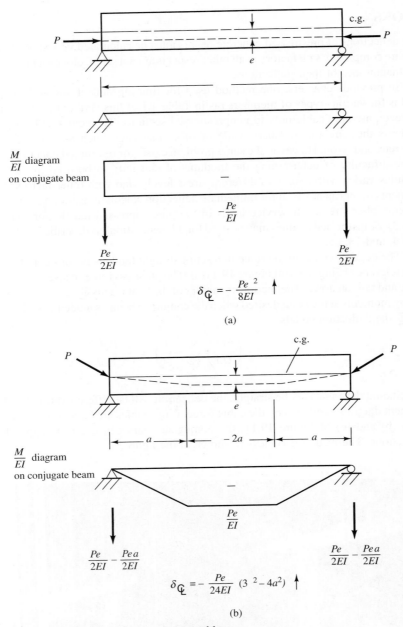

Figure 19.11 Deflections in prestressed beams.

The deflections due to the tendon stresses will change with time. First of all, the losses in stress in the prestress tendons will reduce the negative moments they produce and thus the upward deflections. On the other hand, the long-term compressive stresses in the bottom of the beam due to the prestress negative moments will cause creep and therefore increase the upward deflections.

In addition to the deflections caused by the tendon stresses, there are deflections due to the beam's own weight and due to the additional dead and live loads subsequently applied to the beam.

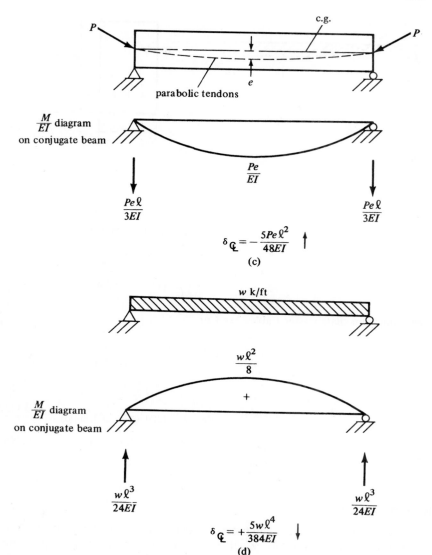

$$\delta_{\mathbb{C}} = -\frac{5Pe\ell^2}{48EI} \uparrow$$

(c)

$$\delta_{\mathbb{C}} = +\frac{5w\ell^4}{384EI} \downarrow$$

(d)

Figure 19.11 (*continued*)

These deflections can be computed and superimposed on the ones caused by the tendons. Figure 19.11(d) shows the $\mathbb{C}$ deflection of a uniformly loaded simple beam obtained by taking moments at the $\mathbb{C}$ when the conjugate beam is loaded with the M/EI diagram.

Example 19.4 shows the initial and long-term deflection calculations for a rectangular pretensioned beam.

EXAMPLE 19.4

The pretensioned rectangular beam shown in Figure 19.12 has straight cables with initial stresses of 175 ksi and final stresses after losses of 140 ksi. Determine the deflection at the beam $\mathbb{C}$ immediately after the cables are cut. $E = 4 \times 10^6$ psi. Assume concrete is uncracked.

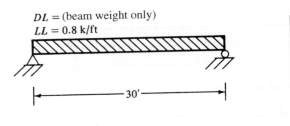

Figure 19.12

SOLUTION

$$I_g = \left(\frac{1}{12}\right)(12)(20)^3 = 8000 \text{ in.}^4$$

$$e = 6''$$

$$Bm \text{ weight} = \frac{(12)(20)}{144}(150) = 250 \text{ lb/ft}$$

Deflection Immediately after Cables Are Cut

$$\delta \text{ due to cable} = -\frac{Pe\ell^2}{8EI} = -\frac{(1.2 \times 175{,}000)(6)(12 \times 30)^2}{(8)(4 \times 10^6)(8000)} = -0.638''\uparrow$$

$$\delta \text{ due to beam weight} = +\frac{5w\ell^4}{384EI} = \frac{(5)\left(\dfrac{250}{12}\right)(12 \times 30)^4}{(384)(4 \times 10^6)(8000)} = \underline{+0.142''\downarrow}$$

$$\text{Total deflection} = -0.496''\uparrow$$

Additional Deflection Comments

Long-term deflections can be computed as previously described in Chapter 6. From the preceding example it can be seen that, not counting external loads, the beam is initially cambered upward by 0.496 in.; as time goes by, this camber increases due to creep in the concrete. Such a camber is often advantageous in offsetting deflections caused by the superimposed loads. In some members, however, the camber can be quite large, particularly for long spans and where lightweight aggregates are used. If this camber is too large, the results can be quite detrimental to the structure (warping of floors, damage to roofing, cracking and warping of partitions, and so on).

To illustrate one problem that can occur, it is assumed that the roof of a school is being constructed by placing 50-ft double T's made with a lightweight aggregate side by side over a classroom. The resulting cambers may be rather large, and, worse, they may not be equal in the different sections. It then becomes necessary to force the different sections to the same deflection and tie them together in some fashion so that a smooth surface is provided for roofing. Once the surface is even, the members may be connected by welding together metal inserts, such as angles that were cast in the edges of the different sections for this purpose.

Both reinforced concrete members and prestressed members with overhanging or cantilevered ends will often have rather large deflections. The total deflections at the free end of these members are due to the sum of the normal deflections plus the effect of support rotations. This latter effect may frequently be the larger of the two, and, as a result, the sum of the two deflections may be so large as to affect the appearance of the structure detrimentally. For this reason, many designers try to avoid cantilevered members in prestressed construction.

19.10   SHEAR IN PRESTRESSED SECTIONS

Web reinforcement for prestressed sections is handled in a manner similar to that used for a conventional reinforced concrete beam. In the expressions that follow, b_w is the web width or the diameter of a circular section, and d_p is the distance from the extreme fiber in compression to the centroid of the tensile reinforcement. Should the reaction introduce compression into the end region of a prestressed member, sections of the beam located at distances less than $h/2$ from the face of the support may be designed for the shear computed at $h/2$ where h is the overall thickness of the member.

$$v_u = \frac{V_u}{\phi b_w d_p}$$

The Code (11.3) provides two methods for estimating the shear strength that the concrete of a prestressed section can resist. There is an approximate method, which can be used only when the effective prestress force is equal to at least 40% of the tensile strength of the flexural reinforcement f_{pu}, and a more detailed analysis, which can be used regardless of the magnitude of the effective prestress force. These methods are discussed in the paragraphs to follow.

Approximate Method

With this method, the nominal shear capacity of a prestressed section can be taken as

$$V_c = \left(0.6\lambda \sqrt{f_c'} + \frac{700 V_u d_p}{M_u} \right) b_w d \qquad \text{(ACI Equation 11-9)}$$

The Code (11.3.2) states that regardless of the value given by this equation, V_c need not be taken as less than $2\lambda \sqrt{f_c'} b_w d$, nor may it be larger than $5\lambda \sqrt{f_c'} b_w d$. In this expression, V_u is the maximum design shear at the section being considered, M_u is the design moment at the same section occurring simultaneously with V_u, and d is the distance from the extreme compression fiber to the centroid of the prestressed tendons. The value of $V_u d_p / M_u$ is limited to a maximum value of 1.0.

More Detailed Analysis

If a more detailed analysis is desired (it will have to be used if the effective prestressing force is less than 40% of the tensile strength of the flexural reinforcement), the nominal shear force carried by the concrete is considered to equal the smaller of V_{ci} or V_{cw}, to be defined here. The term V_{ci} represents the nominal shear strength of a member provided by the concrete when diagonal cracking results from combined shear and moment. The term V_{cw} represents the nominal shear strength of the member provided by the concrete when diagonal cracking results from excessive principal tensile stress in the concrete. In both expressions to follow, d is the distance from the extreme compression fiber to the centroid of the prestressed tendons or is $0.8h$, whichever is greater (Code 11.3.3.2).

The estimated shear capacity V_{ci} can be computed by the following expression, given by the ACI Code (11.3.3.1):

$$V_{ci} = 0.6\lambda\sqrt{f_c'}b_w d_p + V_d + \frac{V_i M_{cre}}{M_{max}} \quad \text{but need not be taken as less than } 1.7\lambda\sqrt{f_c'}b_w d$$

(ACI Equation 11-10)

In this expression, V_d is the shear at the section in question due to service dead load, M_{max} is the factored maximum bending moment at the section due to externally applied design loads, V_i is the shear that occurs simultaneously with M_{max}, and M_{cr} is the cracking moment, which is to be determined as follows:

$$M_{cr} = \left(\frac{I}{y_t}\right)(6\lambda\sqrt{f_c'} + f_{pe} - f_d) \qquad \text{(ACI Equation 11-11)}$$

where

I = the moment of inertia of the section that resists the externally applied loads

y_t = the distance from the centroidal axis of the gross section (neglecting the reinforcing) to the extreme fiber in tension

f_{pe} = the compressive stress in the concrete due to prestress after all losses at the extreme fiber of the section where the applied loads cause tension

f_d = the stress due to unfactored dead load at the extreme fiber where the applied loads cause tension

From a somewhat simplified principal tension theory, the shear capacity of a beam is equal to the value given by the following expression but need not be less than $1.7\lambda\sqrt{f_c'}b_w d$.

$$V_{cw} = (3.5\lambda\sqrt{f_c'} + 0.3f_{pc})b_w d_p + V_p \qquad \text{(ACI Equation 11-12)}$$

In this expression, f_{pc} is the calculated compressive stress (in pounds per square inch) in the concrete at the centroid of the section resisting the applied loads due to the effective prestress after all losses have occurred. (Should the centroid be in the flange, f_{pc} is to be computed at the junction of the web and flange.) V_p is the vertical component of the effective prestress at the section under consideration. Alternately, the Code (11.3.3.2) states that V_{cw} may be taken as the shear force that corresponds to a multiple of dead load plus live load, which results in a calculated principal tensile stress equal to $4\lambda\sqrt{f_c'}$ at the centroid of the member or at the intersection of the flange and web if the centroid falls in the web.

A further comment should be made here about the computation of f_{pc} for pretensioned members, since it is affected by the transfer length. The Code (11.3.4) states that the transfer length can be taken as 50 diameters for strand tendons and 100 diameters for wire tendons. The prestress force may be assumed to vary linearly from zero at the end of the tendon to a maximum at the aforesaid transfer distance. If the value of $h/2$ is less than the transfer length, it is necessary to consider the reduced prestress when V_{cw} is calculated (ACI 11.3.4).

19.11   DESIGN OF SHEAR REINFORCEMENT

Should the computed value of V_u exceed ϕV_c, the area of vertical stirrups (the Code not permitting inclined stirrups or bent-up bars in prestressed members) must not be less than A_v as determined by the following expression from the Code (11.4.7.2):

Prestressed concrete Runnymede Bridge over Thames River near Egham, Surrey, England. (Courtesy of Cement and Concrete Association.)

$$V_s = \frac{A_v f_y d}{s}$$ (ACI Equation 11-15)

As in conventional reinforced concrete design, a minimum area of shear reinforcing is required at all points where V_u is greater than $\frac{1}{2}\phi V_c$. This minimum area is to be determined from the expression to follow if the effective prestress is less than 40% of the tensile strength of the flexural reinforcement (ACI Code 11.4.6.4):

$$A_{v\,min} = 0.75\sqrt{f_c'}\,\frac{b_w s}{f_{yt}} \text{ but shall not be less than } \frac{50 b_w s}{f_{yt}}$$ (ACI Equation 11-13)

where b_w and s are in inches.

If the effective prestress is equal to or greater than 40% of the tensile strength of the flexural reinforcement, the following expression, in which A_{ps} is the area of prestressed reinforcement in the tensile zone, is to be used to calculate A_v:

$$A_{v,m} = \left(\frac{A_{ps}}{80}\right)\left(\frac{f_{pu}}{f_{yt}}\right)\left(\frac{s}{d}\right)\sqrt{\left(\frac{d}{b_w}\right)}$$ (ACI Equation 11-14)

Section 11.4.5.1 of the ACI Code states that in no case may the maximum spacing exceed $0.75h$ or 24 in. Examples 19.5 and 19.6 illustrate the calculations necessary for determining the shear strength and for selecting the stirrups for a prestressed beam.

EXAMPLE 19.5

Calculate the shearing strength of the section shown in Figure 19.13 at 4 ft from the supports, using both the approximate method and the more detailed method allowed by the ACI Code. Assume that the area of the prestressing steel is 1.0 in.^{2}, the effective prestress force is 250 k, and $f'_c = 4000$ psi.

SOLUTION **Approximate Method**

$$w_u = (1.2)(1.2) + (1.6)(2.1) = 4.8 \text{ k/ft}$$
$$V_u = (10)(4.8) - (4)(4.8) = 28.8 \text{ k}$$
$$M_u = (10)(4.8)(4) - (4)(4.8)(2) = 153.6 \text{ ft-k}$$
$$\frac{V_u d}{M_u} = \frac{(28.8)(24-3-3)}{(12)(153.6)} = 0.281 < 1.0 \qquad \underline{\underline{\text{OK}}}$$

$$V_c = \left(0.6\lambda\sqrt{f'_c} + 700\frac{V_u d_p}{M_u}\right)b_w d \qquad \text{(ACI Equation 11-9)}$$

$$= [(0.6)(1.0)(\sqrt{4000}) + (700)(0.281)](12)(18) = 50{,}684 \text{ lb}$$

$$\text{Minimum } V_c = (2)(1.0)(\sqrt{4000})(12)(18) = 27{,}322 \text{ lb} < 50{,}684 \text{ lb}$$
$$\text{Maximum } V_c = (5)(1.0)(\sqrt{4000})(12)(18) = 68{,}305 \text{ lb} > 50{,}684 \text{ lb}$$
$$V_c = 50{,}684 \text{ lb}$$

More Detailed Method

$$I = \left(\frac{1}{12}\right)(12)(24)^3 = 13{,}824 \text{ in.}^4$$

$$y_t = 12''$$

$f_{pe} = $ compressive stress in concrete due to prestress after all losses

$$= \frac{P}{A} + \frac{Pec}{I}$$

$$f_{pe} = \frac{250{,}000}{(12)(24)} + \frac{(250{,}000)(6)(12)}{13{,}824} = 2170 \text{ psi}$$

$$M_d = \text{dead load moment at } 4' \text{ point} = (10)(1.2)(4) - (4)(1.2)(2)$$
$$= 38.4 \text{ ft-k}$$

$$f_d = \text{stress due to the dead load moment} = \frac{(12)(38{,}400)(12)}{13{,}824}$$
$$= 400 \text{ psi}$$

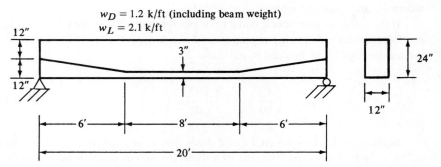

Figure 19.13

$$M_{cr} = \text{cracking moment} = \left(\frac{I}{y_t}\right)\left(6\lambda\sqrt{f_c'} + f_{pe} - f_d\right) \qquad \text{(ACI Equation 11-11)}$$

$$= \left(\frac{13,824}{12}\right)(6)(1.0)(\sqrt{4000} + 2170 - 400) = 2,476,193 \text{ in.-lb}$$

$$= 206,349 \text{ ft-lb}$$

$$\text{Beam weight} = \frac{(12)(24)}{144}(150) = 300 \text{ lb/ft}$$

$$w_u \text{ not counting beam weight} = (1.2)(1.2 - 0.3) + (1.6)(2.1) = 4.44 \text{ k/ft}$$

$$M_{\max} = (10)(4.44)(4) - (4)(4.44)(2) = 142.08 \text{ ft-k} = 142,080 \text{ ft-lb}$$

$$V_i \text{ due to } w_u \text{ occurring same time as } M_{\max} = (10)(4.44) - (4)(4.44)$$

$$= 26.64 \text{ k} = 26,640 \text{ lb}$$

$$V_d = \text{dead load shear} = (10)(1.2) - (4)(1.2) = 7.2 \text{ k} = 7200 \text{ lb}$$

$$d = 24 - 3 - 3 = 18'' \text{ or } (0.8)(24) = \underline{\underline{19.2''}}$$

$$V_{ci} = 0.6\lambda\sqrt{f_c'}b_w d_p + V_d + \frac{V_i M_{cr}}{M_{\max}} \qquad \text{(ACI Equation 11-10)}$$

$$= (0.6)(1.0)(\sqrt{4000})(12)(19.2) + 7200 + \frac{(26,640)(206,349)}{142,080} = 54,634 \text{ lb}$$

$$\text{but need not be less than}(1.7)(1.0)(\sqrt{4000})(12)(19.2) = 24,772 \text{ lb}$$

Computing V_{cw}

f_{pc} = calculated compressive stress in psi at the centroid of the concrete due to the effective prestress

$$= \frac{250,000}{(12)(24)} = 868 \text{ psi}$$

V_p = vertical component of effective prestress at section $= \dfrac{9}{\sqrt{9^2 + 72^2}}(250,000)$

$$= \left(\frac{9}{72.56}\right)(250,000) = 31,009 \text{ lb}$$

$$V_{cw} = (3.5\lambda\sqrt{f_c'} + 0.3f_{pc})b_w d + V_p \qquad \text{(ACI Equation 11-12)}$$

$$= (3.5)(1.0)(\sqrt{4000} + 0.3 \times 868)(12)(19.2) + 31,009 = 142,006 \text{ lb}$$

Using Lesser of V_{ci} or V_{cw}

$$V_c = 54,634 \text{ lb}$$

EXAMPLE 19.6

Determine the spacing of #3 ⊔ stirrups required for the beam of Example 19.5 at 4 ft from the end support if f_{pu} is 250 ksi for the prestressing steel and f_y for the stirrups is 40 ksi. Use the value of V_c obtained by the approximate method, 54,634 lb.

SOLUTION

$$w_u = (1.2)(1.2) + (1.6)(2.1) = 4.8 \text{ k/ft}$$

$$V_u = (10)(4.8) - (4)(4.8) = 28.8 \text{ k}$$

$$\phi V_c = (0.75)(54,634) = 40,976 \text{ lb}$$

$$> V_u = 28,800 \text{ lb}$$

$$V_u > \frac{\phi V_c}{2} = 20,488 \text{ lb} < \phi V_c$$

A minimum amount of reinforcement is needed.
Since effective prestress is greater than 40% of tensile strength of reinforcing,

$$A_v = \left(\frac{A_{ps}}{80} \right) \left(\frac{f_{pu}}{f_{yt}} \right) \left(\frac{s}{d} \right) \sqrt{\left(\frac{d}{b_w} \right)} \qquad \text{(ACI Equation 11-14)}$$

$$(2)(0.11) = \left(\frac{1.0}{80} \right) \left(\frac{250,000}{40,000} \right) \left(\frac{s}{18} \right) \sqrt{\frac{18}{12}}$$

$$s = 41.38'', \text{ but maximum } s = \left(\frac{3}{4} \right)(24) = 18'' \qquad \underline{\underline{\text{Use } 18''}}$$

19.12   ADDITIONAL TOPICS

This chapter has presented a brief discussion of prestressed concrete. A number of other important topics have been omitted from this introductory material. Several of these items are briefly mentioned in the paragraphs that follow.

Stresses in End Blocks

The part of a prestressed member around the end anchorages of the steel tendons is called the *end block*. In this region the prestress forces are transferred from very concentrated areas out into the whole beam cross section. It has been found that the length of transfer for posttensioned members is less than the height of the beam and in fact is probably much less.

For posttensioned members, there is direct-bearing compression at the end anchorage; therefore, solid end blocks are usually used there to spread out the concentrated prestress forces. To prevent bursting of the block, either wire mesh or a grid of vertical and horizontal reinforcing bars is placed near the end face of the beam. In addition, both vertical and horizontal reinforcing is placed throughout the block.

For pretensioned members where the prestress is transferred to the concrete by bond over a distance approximately equal to the beam depth, a solid end block is probably not necessary, but spaced stirrups are needed. A great deal of information on the subject of end block stresses for posttensioned and pretensioned members is available.[4]

Composite Construction

Precast prestressed sections are frequently used in buildings and bridges in combination with cast-in-place concrete. Should such members be properly designed for shear transfer so the two parts will act together as a unit, they are called *composite sections*. Examples of such members were previously shown in parts (e) and (f) of Figure 19.8. In composite construction, the parts that are difficult to form and that contain most of the reinforcing are precast, whereas the slabs and perhaps the top of the beams, which are relatively easy to form, are cast in place.

The precast sections are normally designed to support their own weights plus the green cast-in-place concrete in the slabs plus any other loads applied during construction. The dead and live loads applied after the slab hardens are supported by the composite section. The combination of the two parts will yield a composite section that has a very large moment of inertia and thus a very large resisting moment. It is usually quite economical to use (a) a precast prestressed beam made with a high-strength concrete and (b) a slab made with an ordinary grade of concrete. If this practice is followed, it will be necessary to account for the different moduli of elasticity of the two materials in calculating the composite properties (thus it becomes a transformed area problem).

Continuous Members

Continuous prestressed sections may be cast in place completely with their tendons running continuously from one end to the other. It should be realized for such members that where the service loads tend to cause positive moments, the tendons should produce negative moments and vice versa. This means that the tendons should be below the member's center of gravity in normally positive moment regions and above the center of gravity in normally negative moment regions. To produce the desired stress distributions, it is possible to use curved tendons and members of constant cross section or straight tendons with members of variable cross section. In Figure 19.14 several continuous beams of these types are shown.

Another type of continuous section that has been used very successfully in the United States, particularly for bridge construction, involves the use of precast prestressed members made into continuous sections with cast-in-place concrete and regular reinforcing steel. Figure 19.14(d) shows such a case. For such construction the precast section resists a portion of the dead load, while the live load and the dead load that is applied after the cast-in-place concrete hardens are resisted by the continuous member.

Partial Prestressing

During the early days of prestressed concrete, the objective of the designer was to proportion members that could never be subject to tension when service loads were applied. Such members are said to be *fully prestressed*. Subsequent investigations of fully prestressed members

---

[4]Nawy, *Prestressed Concrete*, p. 173.

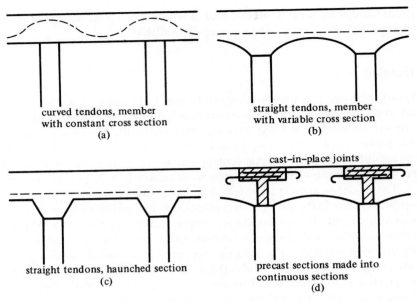

Figure 19.14 Continuous beams.

have shown that they often have an appreciable amount of extra strength. As a result, many designers now believe that certain amounts of tensile stresses can be permitted under service loads. Members that are permitted to have some tensile stresses are said to be *partially prestressed*.

A major advantage of a partially prestressed beam is a decrease in camber. This is particularly important when the beam load or the dead load is quite low compared to the total design load.

To provide additional safety for partially prestressed beams, it is common practice to add some conventional reinforcement. This reinforcement will increase the ultimate flexural strength of the members as well as help to carry the tensile stresses in the beam.[5]

19.13 COMPUTER EXAMPLES

EXAMPLE 19.7

Use the Excel spreadsheet provided for Chapter 19 to solve Example 19.2.

SOLUTION Open the Chapter 19 Excel spreadsheet and open the worksheet Stress Calculations. Enter values only for cells in yellow highlight. Results are shown below. This spreadsheet does many more calculations. The ones to be compared to Example 19.2 are encircled.

---

[5]Lin, T. Y., and Burns, N. H., 1981, *Design of Prestressed Concrete Structures*, 3rd ed. (New York: John Wiley & Sons), pp. 325–344.

Stresses in prestressed concrete beams

$P_i =$	**294.118**	kips
$R =$	**0.85**	
$e_{midspan} =$	**9**	in.
$e_{support} =$	**9**	in.
tendon dia. =	**0.5**	in.
$S_1 =$	**1152**	in.³
$S_2 =$	**1152**	in.³
$A =$	**288**	in.²
$f'_{ci} =$	**3500**	psi
$f'_c =$	**5000**	psi
$\gamma_c =$	**145**	pcf
$\ell =$	**20**	ft
$w_D =$	**1700**	plf
$w_L =$	**1000**	plf
% of w_L sustained	**50.00%**	
$e_{0.4\ell} =$	**9**	inches
$e_{50\ diam} =$	**9.00**	inches
$w_o =$	**300.00**	plf
$M_0 =$	**180**	in.-kips
$M_s =$	**1500**	in.-kips
$M_T =$	**1800**	in.-kips

$f_{ti} =$ **177** psi
$f_{ts} =$ **849** psi
$\beta_1 =$ **0.8**

				stress					
Stress calculation at release	$f_1 =$ $-P_i/A$	$+P_i e/S_1$	$-M_o/S_1$	at midspan	at $x=0.4$	at 50 dia.	at support	allowable	
	−1021	2298	−156.25	1120	1127	1277	1277	177	$=f_{ti}$ midspan
								355	$=f_{ti}$ ends
	$f_2 =$ $-P_i/A$	$-P_i e/S_2$	$+M_o/S_2$						
	−1021	−2298	156	−3163	−3169	−3319	−3319	−2100	$=f_{ci}$
Loaded(M_t)	$f_1 =$ $-P_e/A$	$+P_e e/S_1$	$-M_T/S_1$						
	−868	1953	−1562.5	−477	−415	1085	1085	−3000	$=f_{cs}$(for M_t)
Loaded(M_s)	$f_1 =$ $-P_e/A$	$+P_e e/S_1$	$-M_s/S_1$						
	−868	1953	−1302.1	−217	−165	1085	1085	−2250	$=f_{cs}$(for M_s)
	$f_2 =$ $-P_e/A$	$-P_e e/S_2$	$+M_T/S_1$						
	−868	−1953	1563	−1259	−1321	−2821	−2821	849	

PROBLEMS

For all problems, unless otherwise stated, assume concrete is normal weight.

19.1 The beam shown in the accompanying illustration has an effective total prestress of 240 k. Calculate the fiber stresses in the top and bottom of the beam at the ends and ℄. The tendons are assumed to be straight. (*Ans.* $f_{top} = -1.291$ ksi, $f_{bott} = +0.148$ ksi at the ℄.)

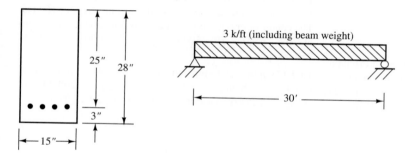

3 k/ft (including beam weight)

25" 28" 3" 15"

30′

19.2 Compute the stresses in the top and bottom of the beam shown at the ends and centerline immediately after the cables are cut. Assume straight cable and 10% losses. Initial prestress in 170 ksi.

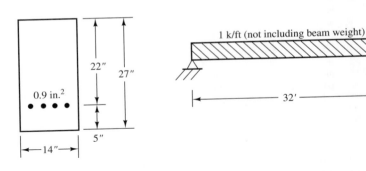

19.3 The beam shown has a 30-ft simple span; $f_c' = 5000$ psi, $f_{pu} = 250,000$ psi, and the initial prestress is $160,000$ psi, $f_{py} = 0.85 f_{pu}$, and 10% losses.

(a) Calculate the concrete stresses in the top and bottom of the beam at midspan immediately after the tendons are cut. (*Ans.* $f_{top} = +0.183$ ksi, $f_{bott} = -1.041$ ksi)

(b) Recalculate the stresses at midspan after assumed prestress losses in the tendons of 20%. (*Ans.* $f_{top} = +0.076$ ksi, $f_{bott} = -0.762$ ksi)

(c) What maximum service live load can this beam support in addition to its own weight if allowable stresses of $0.6 f_c'$ in compression and $12\sqrt{f_c'}$ in tension are permitted? (*Ans.* 1.605 k/ft.)

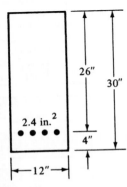

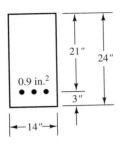

19.4 Using the same allowable stresses permitted in Problem 19.3, what total uniform load can the beam shown in the accompanying illustration support for a 50-ft simple span in addition to its own weight? Assume 20% prestress loss.

19.5 Compute the cracking moment and the permissible ultimate moment capacity of the beam of Problem 19.3 if $f_c' = 5000$ psi and $f_{py} = 200,000$ psi. (*Ans.* $M_{cr} = 144.74$ ft-k, $\phi M_n = 300$ ft-k)

19.6 Compute the stresses in the top and bottom at the ℄ of the beam of Problem 19.1 if it is picked up at the ℄. Assume 100% impact. Concrete weighs 150 lb-ft³.

19.7 Determine the stresses at the one-third points of the beam of Problem 19.6 if the beam is picked up at those points. (*Ans.* $f_{top} = +1.009$ ksi, $f_{bott} = -2.151$ ksi)

19.8 Calculate the design moment capacity of a 12-in. × 20-in. pretensioned beam that is prestressed with 1.2 in.² of steel tendons stressed to an initial stress of 160 ksi. The center of gravity of the tendons is 3 in. above the bottom of the beam; $f_c' = 5000$ psi, $f_{py} = 200,000$ psi, and $f_{pu} = 250,000$ psi.

19.9 Compute the design moment capacity of the pretensioned beam of Problem 19.2 if $f_c' = 5000$ psi, $f_{py} = 200,000$ psi, and $f_{pu} = 225$ ksi. Assume $\frac{f_{py}}{f_{pu}} > 0.85$. (*Ans.* 293.1 ft-k)

19.10 Compute the design moment capacity of the bonded T beam shown in the accompanying illustration if $f'_c = 5000$ psi, $f_{pu} = 250,000$ psi, and the initial stress in the cables is 160,000 psi. Also $f_{py} = 200,000$ psi.

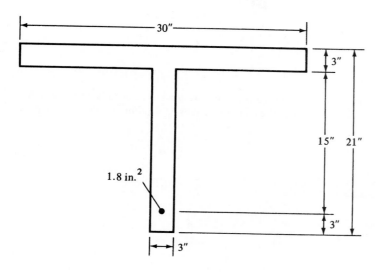

19.11 Calculate the deflection at the ℄ of the beam in Problem 19.3 immediately after the cables are cut, assuming cable stress $= 160,000$ psi. (*Ans.* -0.225 in.↑)

19.12 Calculate the deflection at the ℄ of the beam in Problem 19.1 immediately after the cables are cut, assuming P initial $= 240$ k and P after losses $= 190$ k. Repeat the calcu-

lation after losses if 20-k concentrated live loads are located at the one-third points of the beam and $f'_c = 5000$ psi. Assume that cables are straight and that no loads are present other than the beam weight and the two 20-k loads. Use I_g for all calculations.

19.13 Determine the shearing strength of the beam shown in the accompanying illustration 3 ft from the supports using the approximate method allowed by the ACI Code. Determine the required spacing of #3 ⊔ stirrups at the same section, if $f'_c = 5000$ psi, $f_{pu} = 250,000$ psi, f_y for stirrups $= 50,000$ psi, and $f_{se} = 200,000$ psi. (*Ans. s* $= 21.73''$)

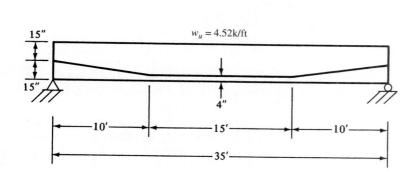
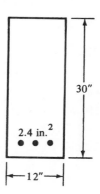

19.14 Repeat Problem 19.1 using the Chapter 19 spreadsheet.

19.15 Repeat Problem 19.3 using the Chapter 19 spreadsheet (*Ans.* Same as Problem 19.3)

Problems in SI Units

19.16 Immediately after cutting the cables in the beam shown, they have an effective prestress of 1.260 GPa. Determine the stresses at the top and bottom of the beam at the ends and centerline. The concrete weighs 23.5 kN/m³. Cables are straight. $E = 27{,}800$ MPa. $f_{py} = 0.8 f_{pu}$.

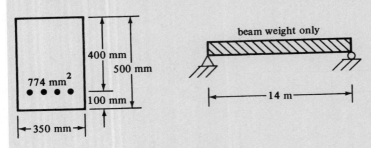

19.17 The beam shown has a 12-m simple span: $f_c' = 35$ MPa, $f_{pu} = 1.725$ GPa, and the initial prestress is 1.10 GPa.

(a) Calculate the concrete stresses in the top and bottom of the beam at midspan immediately after the tendons are cut. (*Ans.* $f_{top} = +1.766$ MPa, $f_{bott} = -12{,}086$ (MPa)

(b) Recalculate the stresses after assumed losses in the tendons of 18%. (*Ans.* $f_{top} = +0.618$ MPa, $f_{bott} = -9.08$ MPa)

(c) What maximum service uniform live load can the beam support in addition to its own weight if allowable stresses of $0.45 f_c'$ in compression and $0.5\sqrt{f_c'}$ in tension are permitted? (*Ans.* 10.095 kN/m)

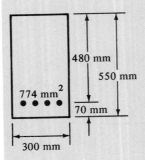

19.18 Using the same allowable stresses permitted and cable stresses as in Problem 19.17, what total uniform load, including beam weight, can the beam shown in the accompanying illustration support for a 15-m simple span?

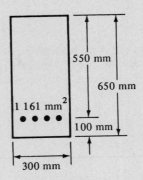

19.19 Compute the cracking moment and the design moment capacity of the bonded beam of Problem 19.17 if $f_{py} = 0.8 f_{pu}$ (*Ans.* 181.68 kN · m, 438.07 kN · m)

19.20 Compute the stresses in the top and bottom of the beam of Problem 19.16 if it is picked up at its one-third points. Assume an impact of 100%.

19.21 Compute the design moment capacity of the bonded T beam shown in the accompanying illustration $f'_c = 35$ MPa, $f_{pu} = 1.725$ GPa, $f_{py} = 0.8f_{pu}$ and the initial stress in the cables is 1.100 GPa. (*Ans*. 843.6 kN · m)

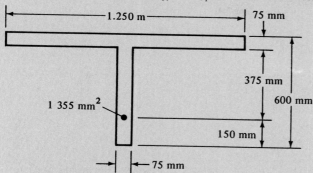

19.22 Calculate the deflection at the ℄ of the beam of Problem 19.16 immediately after the cables are cut. Use I_g.

20

Formwork

20.1 INTRODUCTION

Concrete forms are molds into which semiliquid concrete is placed. The molds need to be sufficiently strong to hold the concrete in the desired size and shape until the concrete hardens. Since forms are structures, they should be carefully and economically designed to support the imposed loads using the methods required for the design of other engineering structures.

Safety is a major concern in formwork because a rather large percentage of the accidents that occur during the construction of concrete structures is due to formwork failures. Normally, formwork failures are not caused by the application of excessive gravity loading. Although such failures do occasionally occur, the usual failures are due to lateral forces that cause the supporting members to be displaced. These lateral forces may be caused by wind, by moving equipment on the forms, by vibration from passing traffic, or by the lateral pressure of freshly placed and vibrated concrete. Sadly, most of these failures could have been prevented if only a little additional lateral bracing had been used. There are, of course, other causes of failure, such as stripping the forms too early and improper control of the placement rate of the concrete.

Although you might think that shape, finish, and safety are the most important items in concrete formwork, you should realize that economy is also a major consideration. The cost of the formwork, which can range from one-third to almost two-thirds of the total cost of a concrete structure, is often more than the cost of both the concrete and the reinforcing steel. For the average concrete structure, the formwork is considered to represent about 50% of the total cost. *From this discussion it is obvious that any efforts made to improve the economy of concrete structures should be primarily concentrated on reducing formwork costs.* Formwork must be treated as an integral part of the overall job plan, and the lowest bidder is likely to be the contractor who has planned the most economical forming job.

When designers are considering costs, they tend to think only of quantities of materials. As a result, they will sometimes carefully design a structure with the lightest possible members and end up with some complicated and expensive formwork.

20.2 RESPONSIBILITY FOR FORMWORK DESIGN

Designers of temporary structures such as concrete formwork must meet the requirements of the Occupational Safety and Health Act (OSHA) of the U.S. government. In addition, there may be state and local safety codes to which designers must adhere.

Normally, the structural designer is responsible for the design of reinforced concrete structures, while the design of the formwork and its construction is the responsibility of the contractor. For some structures, however, where the formwork is very complex, such as for shells,

Wastewater treatment plant, Santa Rosa, California. (Courtesy of Simpson Timber Company.)

folded plates, and arches, it is desirable for the designer to take responsibility for the formwork design. There does seem to be an increasing trend, particularly on larger jobs, for the contractor to employ structural engineers to design and detail the formwork.

The contract documents should clearly specify the responsibility of each party for the work. It is wise to allow the contractor as much room as possible to plan his or her own formwork and construction details, because more economical bids will often be the result.

It is to be remembered in contract law that if the contract documents describe exactly how something is to be done and at the same time spell out the results to be accomplished, it is not generally enforceable. In other words, if the exact form of the construction details is specified in the contract and followed by the contractor, and the formwork fails, the contractor cannot be held responsible.

20.3   MATERIALS USED FOR FORMWORK

Several decades ago, lumber was the universal material used for constructing concrete forms. The forms were constructed, used one time, and torn down, and there was little salvage other than perhaps a board or timber here or there. Such a practice still exists in some parts of the world where labor costs are very low. In a high labor-cost area, such as the United States, the trend for several decades has been toward reusable forms made from many different materials. Wood is still the most widely used material, however, and no matter what types of materials are used, some wood will still probably be required, whether lumber or plywood.

The lumber used for formwork usually comes from the softwood species such as southern pine, ponderosa pine, Douglas fir, and spruce. These woods are relatively light and are commonly available. Southern pine and Douglas fir are two of the strongest of these woods, are usually available, and thus are most often used for formwork. Form lumber should be only partially seasoned because if it is too green, it shrinks and warps in hot dry weather. If it is too dry, it swells a great deal when it becomes wet. The lumber should be planed if it is going to be in contact with the concrete, but it is possible to use rough lumber for braces and shoring.

Lumber is graded according to the rules established by various agencies and published in the American Softwood Standard. It can be graded visually in accordance with ASTM-D245, "Methods for Establishing Standard Grades and Related Allowable Properties for Visually Graded Lumber." It may also be graded by a machine with nondestructive testing.

The introduction of plywood made great advances in formwork possible. Large sheets of plywood save a great deal of labor in the construction of the forms and at the same time result in large areas of joint-free concrete surfaces with corresponding reductions in costs of finishing and rubbing of the exposed concrete. Plywood also has considerable resistance to changes in shape when it becomes wet, can withstand rather rough usage, and can be bent to a certain degree, making it possible to construct curved surfaces.

Steel forms are extremely important for today's concrete construction. Not only are steel panel systems important for buildings, but steel bracing and framing are important where wood or plywood forms may be used. If steel forms are properly maintained, they may be reused many times. In addition, steel forms with their great strength may be used in places where other materials are not feasible, as in forming long spans. Forms for some other types of structures are frequently made with steel simply as a matter of expediency. Falling into this class are the forms for round columns, tunnels, and so on.

The Sonotube paper-fiber forms are another type of form often used for round columns. These patented forms are lightweight one-piece units that can be sawed to fit beams, utility outlets, and other items. They have built-in moisture barriers that aid in curing. Because they are disposable, there are no cleaning, reshipping, or inventory costs. They can be quickly stripped with electric saws or hand tools.

Two other materials often used for concrete forms are glass-fiber–reinforced plastic and insulation boards. The glass-fiber–reinforced plastic can be sprayed over wooden base forms or can be used to fabricate special forms such as the pans for waffle-type floors. Quite a few types of insulation board used as form liners are on the market. They are fastened to the forms, and when the forms are removed, the boards are left in place either bonded to the concrete or held in place by some type of clips.

20.4 FURNISHING OF FORMWORK

There are companies that specialize in formwork and from whom forms, shoring, and needed accessories can be rented or purchased. The contractor may very well take bids from these types of companies for forms for columns, floors, and walls. If structures are designed with standard dimensions, the use of rented forms may provide the most economical solution. Such a process makes the contractor's job a little easier. He or she can regulate and smooth out the workload of his or her company by assigning the responsibility and risk involved in the formwork to a company that specializes in that particular area. Renting formwork permits the contractor to reduce investment in the many stock items, such as spacers, fasteners, and so forth, that would be necessary if the contractor did all of the formwork.

If forms of the desired sizes are not available for rent or purchase, it is possible for the contractor to make forms or to have them made by one of the companies that specialize in form manufacture. If this latter course is chosen, the result is usually very high-quality formwork.

Finally, the forms can be constructed in place. Although the quality may not be as high as for the last two methods mentioned, it may be the only option available for complicated structures for which the forms cannot be reused. Of course, for many jobs a combination of rented or built-in-place or shop-built forms may be used for best economy.

20.5   ECONOMY IN FORMWORK

The first opportunities for form economy occur during the design of the structure. Concrete designers can often save the owner a great deal of money if they take into account factors that permit formwork economy in the design of the structure. Among the items that can be done in this regard are the following:

1. Attempt to coordinate the design with the architectural design, such as varying room sizes a little to accommodate standard forms.

2. Keep story heights as well as the sizes of beams, slabs, and columns the same for as many floors as possible to permit the reuse of forms without change from floor to floor. Without a doubt the most important item affecting the total cost of formwork is the number of times the forms can be used.

3. Make beams as wide or wider than columns so as to make connections simpler. When beams are narrower than columns, formwork is complicated and expensive.

4. Remember in the design of columns for multistory buildings that their sizes may be kept constant for a good many floors by changing the percentage of steel used from floor to floor, starting at the top with approximately 1% steel and increasing it floor by floor until the maximum percentage is reached. In addition, the strength of the concrete can be increased as we move down in the building.

5. Keep columns the same size; the more this is done, the smaller costs will be. Should it be necessary to change the size of a particular column as we move vertically in a building, it is desirable to do it in 2 in. increments—2 in. on one side of a column on one floor and then 2 in. on the other side on another floor.

6. Design structures that permit the use of commercial forms, such as metal pans or corrugated metal deck sheets for floor or roof slabs.

7. It is desirable to use the same thickness for all of the reinforced concrete walls in a particular project. Wall openings should be kept to a minimum and should if possible be placed in the same locations in each wall. It is also desirable to keep footing elevations constant under a particular wall. Where this is not possible, as few footing steps as possible should be used.

8. Use beam and column shapes that do not have complicated haunches, offsets, and cutouts.

9. Usually wide flat beams are more economical than deep narrow beams because it may be difficult to place the concrete and the reinforcement in the latter types.

10. For reinforced concrete frames, the floor slabs and their required formwork are usually the most expensive items. From the standpoint of formwork, one-way flat slabs and two-way flat plates are the most economical floor systems. On the other extreme, the two-way and one-way beam and slab floors require the most expensive formwork. In between these two extremes fall the one-way and two-way pan floors.

11. Allow the contractor to use his or her own methods in constructing the forms, holding the contractor responsible only for their adequacy.

20.6   FORM MAINTENANCE

To be able to reuse forms many times, it is obvious that they should be properly removed, maintained, and stored. One thing the contractor can do to lengthen form lives appreciably is to require the same crews to do both erection and stripping. If while stripping a set of forms the workers know that they are going to have to erect those same forms for the next job, they will be much more careful in their handling and maintenance. (Using the same crews for both jobs may not be possible in some areas because of labor contracts.)

Metal bars or pries should not be used for stripping plywood forms because they may easily cause splitting. Wooden wedges, which are gradually tapped to separate the forms from the hardened concrete, are preferable. After the forms are removed, they should be thoroughly cleaned and oiled. Any concrete that has adhered to metal parts of the forms should be thoroughly scraped off. The holes in wooden form faces caused by nails, ties (to be described in Section 20.7), and other items should be carefully filled with metal plates, cork, or plastic materials. After the cleaning and repairs are completed, the forms should be coated with oil or other preservative. Steel forms should be coated both front and back to keep them from rusting and to prevent spilled concrete from sticking to them.

Plywood surfaces should be lightly oiled, and the forms should be stored in a flat position with the oiled faces together. The stored panels should be kept out of the sun and rain or should be lightly covered so as to permit air circulation and prevent heat buildup. In addition, wood strips

Formwork for Water Tower Place, Chicago, Illinois. (Courtesy of Symons Corporation.)

may be placed between forms to increase this ventilation. Although some plywood forms have reportedly been used successfully for 200 or more times, it is normal with proper maintenance to be able to use them approximately 30 or 40 times. After plywood panels are deemed no longer suitable for formwork, they can often be used by the wise contractor for subflooring or wall or roof sheathing.

Before forms are used, their surfaces should be wetted and oiled or coated with some type of material that will not stain or soften the concrete. The coating is primarily used to keep the concrete from sticking to the forms. A large number of compounds are on the market that will reduce sticking and at the same time serve as sealers or protective coatings to reduce substantially the absorption of water from the concrete by the forms.

Various types of oil or oil compounds have been the most commonly used form-release agents in the past for both wood and metal surfaces. Many form-release and sealer agents are available, but before trying a new one on a large job, the contractor would be wise to experiment with a small application. No material should be used that will form a coating on the concrete that will, after stripping, interfere with the wetting of the concrete for curing purposes. In the same manner, if the concrete is later to be plastered or painted, the coating should not be one that will leave a waxy or oily surface that will interfere with the sticking of paint or plaster.

Various kinds of products are used to coat the plywood during its manufacture. Some manufacturers of these products say that the oiling of forms coated with their products is not necessary in the field. Glass-fiber–reinforced plastic forms, which are used a great deal for precast concrete and for architectural concrete because of the very excellent surfaces produced, need to be oiled only very lightly before use. In fact, they have been used satisfactorily by some contractors with no oiling whatsoever.

20.7   DEFINITIONS

In this section several terms relating to formwork, with which you need to be familiar, are introduced. They are not listed alphabetically but rather are given in the order in which the author thinks they will help the reader understand the material in subsequent sections of this chapter.

Double-headed nails: A tremendous number of devices have been developed to simplify the erection and stripping of forms. For instance, there are the very simple double-headed nails. These nails can be easily removed, permitting the quick removal of braces and dismantling of forms.

Double-headed nail.

Sheathing: The material that forms the contact face of the forms to the concrete is called sheathing or lagging or sheeting. (See Figure 20.1.)

Joists: The usually horizontal beams that support the sheathing for floor and roof slabs are called joists. (See Figure 20.1.)

Stringers: The usually horizontal beams that support the joists and that rest on the vertical supports are called stringers. (See Figure 20.1.)

Shores: The temporary members (probably wood or metal) that are used for vertical support for formwork holding up fresh concrete are called shores. (See Figure 20.1.) Other names

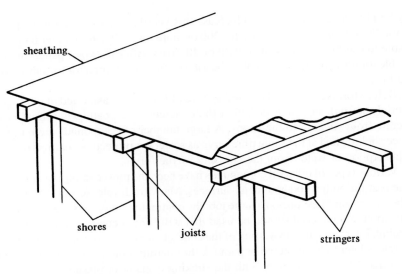

Figure 20.1 Wood formwork for a floor slab.

for these members are struts, posts, or props. The whole system of vertical supports for a particular structure is called the shoring.

Ties: Devices used to support both sides of wall forms against the lateral pressure of the concrete are called ties. They pass through the concrete and are fastened on each side. (See Figure 20.2.) The lateral pressure caused by the fresh concrete, which tends to push out against the forms, is resisted in tension by the ties. They can, when used correctly, eliminate

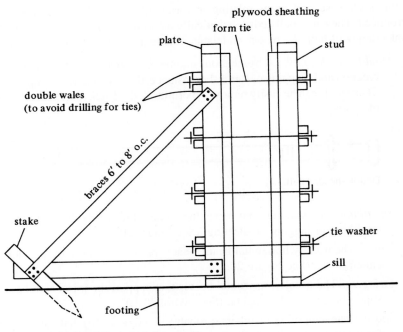

Figure 20.2 Wood formwork for a wall.

most of the external braces otherwise required for wall forms. They may or may not have a provision for holding the forms the desired distance apart.

Spreaders: Braces, usually made of wood, that are placed in a form to keep its two sides from being drawn together are called spreaders. They should be removed when the concrete is placed so they are not cast in the concrete.

Snap ties: Rather than using spreaders separate from the ties, it is more common to use a type of spreader called a snap tie, which is combined with the spreader. The ends of these patented devices can be twisted or snapped off a little distance from the face of the concrete. After the forms are removed, the small holes resulting can easily be repaired with cement mortar.

Column clamps: These devices, which are placed around column forms, withstand the outward pressure of the fresh concrete.

Studs: The vertical members that support sheathing for wall forms are called studs. (See Figure 20.2.)

Wales: The long horizontal members that support the studs are called wales. (See Figure 20.2.) Notice that each wale usually consists of two timbers with the ties passing between so that holes do not have to be drilled.

20.8   FORCES APPLIED TO CONCRETE FORMS

Formwork must be designed to resist all vertical and lateral loads applied to it until the concrete gains sufficient strength to do the job itself. The vertical loads applied to the forms include the weights of the concrete, the reinforcing steel, and the forms themselves, as well as the construction live loads. The lateral loads include the liquid pressure of freshly placed concrete, wind forces, and the lateral forces caused by the movement of the construction equipment. These forces are discussed in the paragraphs to follow.

Vertical Loads

The vertical dead loads that must be supported by the formwork in addition to its own weight include the weight of the concrete and the reinforcing bars. The weight of ordinary concrete, including the reinforcing, is taken as 150 lb/ft^3. The weight of the formwork, which is frequently neglected in the design calculations, may vary from 3 or 4 psf up to 12 to 15 psf.

The vertical live load to be supported includes the weight of the workers, the construction equipment, and the storage of materials on freshly hardened slabs. A minimum load of 50 psf of horizontal projection is recommended by ACI Committee 347.[1] This figure includes the weight of the workers, equipment, runways, and impact. When powered concrete buggies are used, this value should be increased to a minimum of at least 75 psf. Furthermore, to make allowance for the storage of materials on freshly hardened slabs (a very likely circumstance), it is common in practice to design slab forms for vertical live loads as high as 150 psf.

Lateral Loads

For walls and columns, the loads are different from those of slabs. The semiliquid concrete is placed in the form and exerts lateral pressures against the form, as does any liquid. The amount of

---

[1]ACI, Committee 347, *Guide to Formwork for Concrete* (ACI 347 R-94) (Detroit: American Concrete Institute), 33 pp.

Pulp mill kiln near Castlegar, British Columbia. (Courtesy of Economy Forms Corporation.)

this pressure is dependent on the concrete weight per cubic foot, the rate of placing, the temperature of the concrete, and the method of placing.

The exact pressures applied to forms by freshly placed concrete are difficult to estimate due to several factors. These factors include the following:

1. The rate of placing the concrete. Obviously, the faster the concrete is placed, the higher the pressure will be.

2. The temperature of the concrete. The liquid pressure on the forms varies quite a bit with different temperatures. For instance, the pressure when the concrete is at 50°F is appreciably higher than when the concrete is at 70°F. The reason for this is that concrete placed at 50°F hardens at a slower rate than does concrete deposited at 70°F. As a result, the colder concrete remains in a semiliquid state for a longer time and thus to a greater depth, and pressures may be as much as 25% or more higher. Because of this fact, forms for winter-placed concrete should be designed for much higher lateral pressures than those designed for summer-placed concrete.

3. The method of placing the concrete. If high-frequency vibration is used, the concrete is kept in a liquid state to a fairly high depth, and it acts very much like a liquid with a weight per cubic foot equal to that of the concrete. Vibration may increase lateral pressures by as much as 20% over pressures caused by spading.

4. Size and shape of the forms and the consistency and proportions of the concrete. These items affect lateral pressures to some degree, but for the usual building are felt to be negligible. The use of self-consolidating concrete will produce lateral pressures that are higher than those of other concrete mix proportions. Navcik[2] recommends using full liquid head pressure. When pumping from the bottom, increase full liquid head pressure by 25% to account for pump pressures.

The pressures that occur in formwork due to the semiliquid concrete are often of such magnitude that other people do not believe it when the designer says the lateral pressure is 1500 psf or 1800 psf or more. ACI Committee 347 has published recommended formulas for calculating lateral concrete pressures for different temperatures and rates of placing the concrete. In these expressions, which follow, p is the maximum equivalent liquid pressure (in pounds per square foot) at any elevation in the form, R is the vertical rate of concrete placement (in feet per hour), T is the temperature (Fahrenheit) of the concrete in the forms, and h is the maximum height of fresh concrete in the forms, in ft, above the point being considered. These expressions are to be used for walls for which concrete with a slump no greater than 4 in. is placed at a rate not exceeding 7 ft/hr. Vibration depth is limited to 4 ft below the concrete surface.[3]

In Walls with R Not Exceeding 7 ft/hr

$$p = 150 + \frac{9000R}{T} \quad \text{(maximum 2000 psf or 150}h\text{, whichever is less)}$$

In Walls with R Greater than 7 ft/hr but Not Greater than 10 ft/hr

$$p = 150 + \frac{43{,}400}{T} + \frac{2800R}{T} \quad \text{(maximum 2000 psf or 150}h\text{, whichever is less)}$$

Column forms are often filled very quickly. They may in fact be completely filled in less time than is required for the bottom concrete to set. Furthermore, vibration will frequently extend for the full depth of the form. As a result, greater lateral pressures are produced than would be expected for the wall conditions just considered. ACI Committee 347 presents the following equation for estimating the maximum pressures for column form design.

Columns with Maximum Horizontal Dimension of 6 ft or Less

$$p = 150 + \frac{9000R}{T} \quad \text{(maximum 3000 psf or 150}h\text{, whichever is less)}$$

In Figure 20.3 a comparison of the pressures calculated with the preceding expressions is presented for temperatures ranging from 30°F to 100°F.

In addition to the lateral pressures on the forms caused by the fresh concrete, it is necessary for the braces and shoring to be designed to resist all other possible lateral forces, such as wind, dumping of concrete, movement of equipment, bumping of equipment, guy cable tension, uneven placing of concrete, and so on.

---

[2]Formwork for self-consolidating concrete, Nasvik, Joe Concrete Construction, October 2004.
[3]ACI, Committee 347, *Guide to Formwork for Concrete* (ACI 347 R-94).

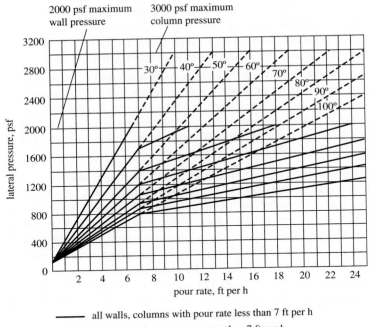

2000 psf maximum
wall pressure

3000 psf maximum
column pressure

——— all walls, columns with pour rate less than 7 ft per h

- - - - columns with pour rate greater than 7 ft per h

Figure 20.3 Lateral pressure of fresh concrete in columns and walls. (Courtesy of American Plywood Association.)

20.9 ANALYSIS OF FORMWORK FOR FLOOR AND ROOF SLABS

This section is devoted to the calculations needed for analyzing formwork for concrete floor and roof slabs; their design is presented in the following section. Because wood formwork is the most common type in practice, the discussion and examples of this chapter are concerned primarily with wood.

Wood members have a very useful property with which you should be familiar, and that is their ability to support excessive loads for short periods of time. As a result of this characteristic, it is a common practice to allow them a 25% increase over their normal allowable stresses if the applied loads are of short duration.

Formwork is usually considered to be a temporary type of structure because it remains in place only a short time. Furthermore, the loads supported by the forms reach a peak during pouring activity and then rapidly fall off as the concrete hardens around the reinforcing and begins to support the load. As a result, the formwork designer will want to use increased allowable stresses where possible. The ability of wood to take overloads is based on the total time of application of the loads. In other words, if forms are to be used many times, the amount of overload they can resist is not as large as if they are to be used only one time.

The paragraphs that follow discuss flexure, shear, and deflections in formwork.

Flexure

Normally, sheathing, joists, and stringers are continuous over several spans. For such cases it seems reasonable to assume that the maximum moment is equal to the maximum moment that would occur in a uniformly loaded span continuous over three or more spans. This value is approximately equal to

$$M = \frac{w\ell^2}{10}$$

Shear

The horizontal and vertical shear stresses at any one point in a beam are equal. For materials for which strengths are the same in every direction, no distinction is made between shear values acting in different directions. Materials such as wood, however, have entirely different shear strengths in the different directions. Wood tends to split or shear between its fibers (usually parallel to the beam axis). Since horizontal shear is rather critical for wood members, it is common in talking about wood formwork to use the term *horizontal shear*.

The horizontal shearing stress in a rectangular wooden member can be calculated by the usual formulas as follows:

$$f_v = \frac{VQ}{Ib} = \frac{(V)[b \times (h/2) \times (h/4)]}{\left(\frac{1}{12}bh^3\right)(b)} = \frac{3V}{2bh} = \frac{3V}{2A}$$

In this expression, V is equal to $w\ell/2$ for uniformly loaded simple spans. As in earlier chapters relating to reinforced concrete, it is permissible to calculate the shearing stress at a distance h from the face of the support. If h is given in inches and if w is the uniform load per foot, the external shear at a distance h from the support can be calculated as follows:

$$V = 0.5w\ell - \frac{h}{12}w$$

$$V = 0.5w\left(\ell - \frac{2h}{12}\right)$$

The sheathing, joists, stringers, wales, and so on for formwork are normally continuous. For uniformly loaded beams continuous over three or more spans, V is approximately equal to $0.6w\ell$. At a distance h from the support, it is *assumed* to equal the following value:

$$V = 0.6w\left(\ell - \frac{2h}{12}\right)$$

The allowable shear stresses for construction grades of lumber are quite low, and the designer is not really out of line if he or she increases the given allowable shear stresses by something more than 25% if the forms in question are to be used only once. Such increases are included in the allowable stresses given later in this chapter. It is not usually necessary to check the shearing stresses in sheathing where they are normally very low.

Deflections

Formwork must be designed to limit deflections to certain maximum values. If deflections are not controlled, the end result will be unsightly concrete, with bulges and perhaps cracks marring its appearance. The amounts of deflection permitted are dependent on the desired finish of the concrete as well as its location. For instance, where a rough finish is used, small deflections might not be very obvious, but where smooth surfaces are desired, small deflections are easily seen and are quite objectionable.

For some concrete forms, the deflection limitations of $\ell/240$ or $\ell/270$ are considered satisfactory, while for others, particularly for horizontal surfaces, a limitation of $\ell/360$ may be required. For some cases, particularly for architectural exposure, even tighter limitations may be required, perhaps $\ell/480$.

It should be realized that perfectly straight beams and girders seem to the eye of a person below to sag downwards; consequently, just a little downward deflection seems to be very large and quite detrimental to the appearance of the structure. For this reason it is considered good practice to camber the forms of such members so they do not appear to be deflecting. A camber frequently used is about $\frac{1}{4}$ in. for each 10 ft of length.

For uniformly loaded equal-span beams continuous over three or more spans, the maximum centerline deflections can be approximately determined with the following expression:

$$\delta = \frac{w\ell^4}{128EI}$$

In using deflection expressions, it is to be noted that the deflections that occur in the formwork are going to be permanent in the completed structure, and thus the modulus of elasticity E should not be increased even if the forms are to be used just one time. As a matter of fact, it may be necessary to reduce E rather than increase it for some parts of the formwork. When wood becomes wet, it becomes more flexible, with the result that larger deflections occur. You can see that wet formwork, particularly the sheathing, is the normal situation. As a result, it is considered wise to use a smaller E than normally permitted, and a reduction factor of $\frac{10}{11}$ is commonly specified.

Three tables useful for both the analysis and design of formwork are presented in this section. The first of these, Table 20.1, gives the properties of several sizes of lumber that are commonly used for formwork. These properties include the dressed dimensions, cross-sectional areas,

Table 20.1 Properties of American Standard Board, Plank, Dimension, and Timber Sizes Commonly Used for Form Construction (Courtesy of the American Concrete Institute)

Nominal size in inches $b \times h$	American Standard size in inches $b \times h$ S4S* 19% maximum moisture	Area of section, $A = bh$ (sq. in.)		Moment of inertia (in.^{4}) $I = \dfrac{bh^3}{12}$		Section modulus (in.^{3}) $S = \dfrac{bh^2}{6}$		Board feet per linear foot of piece
		Rough	S4S	Rough	S4S	Rough	S4S	
4×1	$3\frac{1}{2} \times \frac{3}{4}$	3.17	2.62	0.20	0.12	0.46	0.33	$\frac{1}{3}$
6×1	$5\frac{1}{2} \times \frac{3}{4}$	4.92	4.12	0.31	0.19	0.72	0.52	$\frac{1}{2}$
8×1	$7\frac{1}{4} \times \frac{3}{4}$	6.45	5.44	0.41	0.25	0.94	0.68	$\frac{2}{3}$
10×1	$9\frac{1}{4} \times \frac{3}{4}$	8.20	6.94	0.52	0.32	1.20	0.87	$\frac{5}{6}$
12×1	$11\frac{1}{4} \times \frac{3}{4}$	9.95	8.44	0.63	0.39	1.45	1.05	1
$4 \times 1\frac{1}{4}$	$3\frac{1}{2} \times 1$	4.08	3.50	0.43	0.29	0.76	0.58	$\frac{5}{12}$
$6 \times 1\frac{1}{4}$	$5\frac{1}{2} \times 1$	6.33	5.50	0.68	0.46	1.19	0.92	$\frac{5}{8}$
$8 \times 1\frac{1}{4}$	$7\frac{1}{4} \times 1$	8.30	7.25	0.87	0.60	1.56	1.21	$\frac{5}{6}$
$10 \times 1\frac{1}{4}$	$9\frac{1}{4} \times 1$	10.55	9.25	1.11	0.77	1.98	1.54	$1\frac{1}{24}$
$12 \times 1\frac{1}{4}$	$11\frac{1}{4} \times 1$	12.80	11.25	1.35	0.94	2.40	1.87	$1\frac{1}{4}$
$4 \times 1\frac{1}{2}$	$3\frac{1}{2} \times 1\frac{1}{4}$	4.98	4.37	0.78	0.57	1.14	0.91	$\frac{1}{2}$
$6 \times 1\frac{1}{2}$	$5\frac{1}{2} \times 1\frac{1}{4}$	7.73	6.87	1.22	0.89	1.77	1.43	$\frac{3}{4}$
$8 \times 1\frac{1}{2}$	$7\frac{1}{4} \times 1\frac{1}{4}$	10.14	9.06	1.60	1.18	2.32	1.89	1

Table 20.1 *(Continued)*

Nominal size in inches $b \times h$	American Standard size in inches $b \times h$ S4S[*] 19% maximum moisture	Area of section, $A = bh$ (sq. in.)		Moment of inertia (in.^{4}) $I = \dfrac{bh^3}{12}$		Section modulus (in.^{3}) $S = \dfrac{bh^2}{6}$		Board feet per linear foot of piece
		Rough	S4S	Rough	S4S	Rough	S4S	
$10 \times 1\frac{1}{2}$	$9\frac{1}{4} \times 1\frac{1}{4}$	12.89	11.56	2.03	1.50	2.95	2.41	$1\frac{1}{4}$
$12 \times 1\frac{1}{2}$	$11\frac{1}{4} \times 1\frac{1}{4}$	15.64	14.06	2.46	1.83	3.58	2.93	$1\frac{1}{2}$
4×2	$3\frac{1}{2} \times 1\frac{1}{2}$	5.89	5.25	1.30	0.98	1.60	1.31	$\frac{2}{3}$
6×2	$5\frac{1}{2} \times 1\frac{1}{2}$	9.14	8.25	2.01	1.55	2.48	2.06	1
8×2	$7\frac{1}{4} \times 1\frac{1}{2}$	11.98	10.87	2.64	2.04	3.25	2.72	$1\frac{1}{3}$
10×2	$9\frac{1}{4} \times 1\frac{1}{2}$	15.23	13.87	3.35	2.60	4.13	3.47	$1\frac{2}{3}$
12×2	$11\frac{1}{4} \times 1\frac{1}{2}$	18.48	16.87	4.07	3.16	5.01	4.21	2
2×4	$1\frac{1}{2} \times 3\frac{1}{2}$	5.89	5.25	6.45	5.36	3.56	3.06	$\frac{2}{3}$
2×6	$1\frac{1}{2} \times 5\frac{1}{2}$	9.14	8.25	24.10	20.80	8.57	7.56	1
2×8	$1\frac{1}{2} \times 7\frac{1}{4}$	11.98	10.87	54.32	47.63	14.73	13.14	$1\frac{1}{3}$
2×10	$1\frac{1}{2} \times 9\frac{1}{4}$	15.23	13.87	111.58	98.93	23.80	21.39	$1\frac{2}{3}$
2×12	$1\frac{1}{2} \times 11\frac{1}{4}$	18.48	16.87	199.31	177.97	35.04	31.64	2
3×4	$2\frac{1}{2} \times 3\frac{1}{2}$	9.52	8.75	10.42	8.93	5.75	5.10	1
3×6	$2\frac{1}{2} \times 5\frac{1}{2}$	14.77	13.75	38.93	34.66	13.84	12.60	$1\frac{1}{2}$
3×8	$2\frac{1}{2} \times 7\frac{1}{4}$	19.36	18.12	87.74	79.39	23.80	21.90	2
3×10	$2\frac{1}{2} \times 9\frac{1}{4}$	24.61	23.12	180.24	164.89	38.45	35.65	$2\frac{1}{2}$
3×12	$2\frac{1}{2} \times 11\frac{1}{4}$	29.86	28.12	321.96	296.63	56.61	52.73	3
4×4	$3\frac{1}{2} \times 3\frac{1}{2}$	13.14	12.25	14.39	12.50	7.94	7.15	$1\frac{1}{3}$
4×6	$3\frac{1}{2} \times 5\frac{1}{2}$	20.39	19.25	53.76	48.53	19.12	17.65	2
4×8	$3\frac{1}{2} \times 7\frac{1}{4}$	26.73	25.38	121.17	111.15	32.86	30.66	$2\frac{2}{3}$
4×10	$3\frac{1}{2} \times 9\frac{1}{4}$	33.98	32.38	248.91	230.84	53.10	49.91	$3\frac{1}{3}$
6×3	$5\frac{1}{2} \times 2\frac{1}{2}$	14.77	13.75	8.48	7.16	6.46	5.73	$1\frac{1}{2}$
6×4	$5\frac{1}{2} \times 3\frac{1}{2}$	20.39	19.25	22.33	19.65	12.32	11.23	2
6×6	$5\frac{1}{2} \times 5\frac{1}{2}$	31.64	30.25	83.43	76.26	29.66	27.73	3
6×8	$5\frac{1}{2} \times 7\frac{1}{2}$	42.89	41.25	207.81	193.36	54.51	51.56	4
8×8	$7\frac{1}{2} \times 7\frac{1}{2}$	58.14	56.25	281.69	263.67	73.89	70.31	$5\frac{1}{3}$

[*]Rough dry sizes are $\frac{1}{8}$-in. larger, both dimensions.

moments of inertia, and section moduli. Most lumber has been planed on all four sides so that the surfaces and dimensions are uniform. Such lumber is referred to as S4S (surfaced on four sides).

Plywood is the standard material used in practice for floor and wall sheathing for formwork, and it is so used for the examples presented in this chapter. Almost all exterior plywoods manufactured with waterproof glue can be satisfactorily used, but the industry produces a particular type called *plyform* intended especially for formwork. It is this product that is referred to in the remainder of this chapter.

Plyform is manufactured by gluing together an odd number of sheets of wood with the grain of each sheet being perpendicular to the grain of the sheet adjoining it. The alternating of the grains of the wood sheets, which is called *cross-banding*, greatly reduces the shrinkage and warping of the resulting panels. It also results in a material whose adjacent layers do not have the same properties (at least in a common direction). Consequently, the section properties of plywood (such as moments of inertia and section moduli) are calculated using a transformed-area approach. Plyform can be obtained in thicknesses from $\frac{1}{8}$ to $1\frac{1}{8}$ in. Actually, however, many suppliers only carry the $\frac{5}{8}$ and $\frac{3}{4}$-in. sizes, and the others may have to be specially ordered.

The various types of wood used for manufacturing plywood (fir, spruce, larch, redwood, cedar, etc.) have different strengths and stiffnesses. Those types of woods having similar properties are assigned to a particular species group. To simplify design calculations, the plywood industry has accounted for the different species and the effects of gluing the wood sheets together with alternating grains in developing the properties given in Table 20.2. As a result, the designer needs only to consider the allowable stresses for the plyform and the plyform properties given in the table.

It will be noted that section properties are given in Table 20.2 for stresses applied both parallel to the face grain and perpendicular to it. Should the face grain be placed parallel to the direction of bending, the section is stronger; it is weaker if placed in the opposite direction. This situation is shown in Figure 20.4.

In plyform the shearing stresses between the plies or along the glue lines are calculated with the usual expression $H = f_v = VQ/Ib$ and are referred to as the *rolling shear stresses*. For convenience in making the calculations, the values of Ib/Q are presented in the table for the

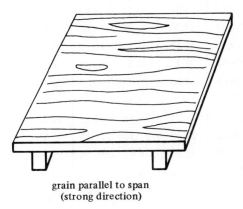

grain parallel to span
(strong direction)

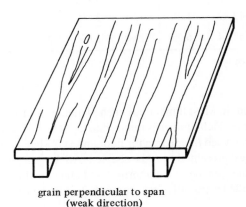

grain perpendicular to span
(weak direction)

Figure 20.4

Table 20.2 Section Properties and Allowable Stresses for Certain Sizes of Plyform (Courtesy of the American Plywood Association)

Thickness (inches)	Approximate weight (psf)	Properties for stress applied parallel with face grain			Properties for stress applied perpendicular to face grain		
		Moment of inertia, I (in.^{4}/ft)	Effective section modulus, K_s (in.^{3}/ft)	Rolling shear constant, lb/Q (in.^{2}/ft)	Moment of inertia, I (in.^{4}/ft)	Effective section modulus, K_s (in.^{3}/ft)	Rolling shear constant, lb/Q (in.^{2}/ft)
Class I							
$\frac{1}{2}$	1.5	0.077	0.268	5.127	0.035	0.167	2.919
$\frac{5}{8}$	1.8	0.130	0.358	6.427	0.064	0.250	3.692
$\frac{3}{4}$	2.2	0.199	0.455	7.854	0.136	0.415	4.565
$\frac{7}{8}$	2.6	0.280	0.553	8.204	0.230	0.581	5.418
1	3.0	0.427	0.737	8.871	0.373	0.798	7.242
$1\frac{1}{8}$	3.3	0.554	0.894	9.872	0.530	0.986	8.566
Class II							
$\frac{1}{2}$	1.5	0.075	0.267	4.868	0.029	0.182	2.671
$\frac{5}{8}$	1.8	0.130	0.357	6.463	0.053	0.320	3.890
$\frac{3}{4}$	2.2	0.198	0.455	7.892	0.111	0.530	4.814
$\frac{7}{8}$	2.6	0.280	0.553	8.031	0.186	0.742	5.716
1	3.0	0.421	0.754	8.614	0.301	1.020	7.645
$1\frac{1}{8}$	3.3	0.566	0.869	9.571	0.429	1.260	9.032
Structural I							
$\frac{1}{2}$	1.5	0.078	0.271	5.166	0.042	0.229	3.076
$\frac{5}{8}$	1.8	0.131	0.361	6.526	0.077	0.343	3.887
$\frac{3}{4}$	2.2	0.202	0.464	7.926	0.162	0.570	4.812
$\frac{7}{8}$	2.6	0.288	0.569	7.539	0.275	0.798	6.242
1	3.0	0.479	0.827	7.978	0.445	1.098	7.639
$1\frac{1}{8}$	3.3	0.623	0.955	8.841	0.634	1.356	9.031

Note: All properties are adjusted to account for reduced effectiveness of plies with grain perpendicular to applied stress.

different plyform sizes, and *H* can be simply determined for a particular case by dividing *V* by the appropriate *Ib/Q* value from the table.

The normal framing applications of plyform are for wet conditions, and adjustments in allowable stresses should be made for those conditions as well as for duration of load and other experience factors. As a result of these items, the following allowable stresses are normally given for plyform:

	Class I plyform	Class II plyform	Structural I plyform
Modulus of elasticity (psi)	1,650,000	1,430,000	1,650,000
Bending stress (psi)	1930	1330	1930
Rolling shear stress (psi)	80	72	102

Table 20.3 shows allowable bending stresses, allowable compression stresses perpendicular and parallel to the grain, allowable shearing stresses, and moduli of elasticity for several types of lumber often used in formwork.

Example 20.1 illustrates the calculations necessary to determine the bending and shear stresses and the maximum deflections that occur in the sheathing, joists, and stringers used as the formwork for a certain concrete slab and live loading. In an actual design problem, it may be necessary to consider several different arrangements of the joists, stringers, and so on and the different grades of lumber available with their different allowable stresses before a final design is selected.

EXAMPLE 20.1

The formwork for a 6-in.-thick reinforced concrete floor slab of normal weight consists of $\frac{3}{4}$-in. Class II plyform sheathing supported by 2- × 6-in. joists spaced 2 ft 0 in. on center, which in turn are supported by 2- × 8-in. stringers spaced 4 ft 0 in. on centers. The stringers are themselves supported by shores spaced 4 ft 0 in. on centers.

The joists and stringers are constructed with Douglas fir, coastal construction, and are to be used many times. The allowable flexural stress in the fir members is 1875 psi, the allowable horizontal shear stress is 180 psi, and the modulus of elasticity is 1,760,000 psi. For the plyform the corresponding values are 1330 psi, 72 psi, and 1,430,000 psi, with the stress applied parallel with the face grain.

Using a live load of 150 psf, check the bending and shear stresses in the sheathing, joists, and stringers. Check deflections assuming that the maximum permissible deflection of each of these elements is $\ell/360$ for all loads.

SOLUTION

Sheathing

Properties of 12-in.-wide piece of $\frac{3}{4}$ plyform from Table 20.2

$$S = 0.455 \text{ in.}^3, \quad I = 0.198 \text{ in.}^4, \quad \frac{Ib}{Q} = 7.892 \text{ in.}^2$$

Loads (neglecting weight of forms)

$$\text{Concrete} = \left(\frac{6}{12}\right)(150) = 75 \text{ psf}$$

$$LL = 150$$

$$\text{Total load} = \overline{225} \text{ psf}$$

Table 20.3 Suggested Working Stresses for Design of Wood Formwork

Species and grade of form framing lumber	Class I formwork, single use or light construction					Class II formwork, multiple use or heavy construction				
	Allowable unit stress, psi					Allowable unit stress, psi				
	Extreme fiber bending	Compression ⊥ to grain	Compression ∥ to grain	Horizontal shear	Modulus of elasticity, psi	Extreme fiber bending	Compression ⊥ to grain	Compression ∥ to grain	Horizontal shear	Modulus of elasticity, psi
Douglas fir, coastal construction	1875	490	1500	180	1,760,000	1500	390	1200	145	1,760,000
Douglas fir, inland common structural	1815	475	1565	180	1,760,000	1450	380	1250	145	1,760,000
Hemlock, west coast construction	1875	455	1380	150	1,540,000	1500	365	1100	120	1,540,000
Larch, common structural	1815	490	1660	180	1,650,000	1450	390	1325	145	1,650,000
Pine, southern No. 1 SR	1875	490	1625–1875	220	1,760,000	1500	390	1300–1500	175	1,760,000
Redwood, heart structural	1625	400	1380	135	1,760,000	1500	390	1100	175	1,760,000
Spruce, eastern, 1450 f structural	1815	375	1310	160	1,320,000	1300	320	1100	110	1,320,000
Plywood Douglas fir, concrete form B-B	2000	(Bearing on face) 435	1500	*	1,600,000	1450	300	1050	130	1,320,000
						1500	(Bearing on face) 325	1100	*	1,600,000
Board sheathing all species	Use 100% of value for species, shown above	Use 67% of value for species, shown above	Use 90% of value for species, shown above	Use 100% of value for species, shown above	Use $\frac{10}{11}$ of value for species, shown above	Use 100% of value for species, shown above	Use 67% of value for species, shown above	Use 90% of value for species, shown above	Use 100% of value for species, shown above	Use $\frac{10}{11}$ of value for species, shown above

*Shear is not a governing design consideration for plywood form panels except in the case of very short, heavily loaded spans. If 1-in. or $1\frac{1}{8}$-in. panels are being used, check for rolling shear according to plywood manufacturer's suggestions.

Source: Derived from "National Design Specifications for Stress-Grade Lumber and Its Fastenings," 1960 edition, as amended, and from "Recommendations of the Douglas Fir Plywood Association."

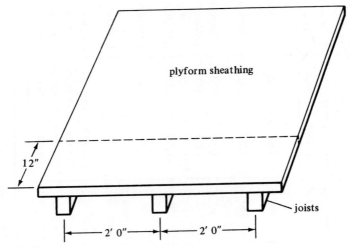

Figure 20.5

Bending stresses (with reference made to Figure 20.5)

$$M = \frac{w\ell^2}{10} = \frac{(225)(2)^2}{10} = 90 \text{ ft-lb}$$

$$f_b = \frac{(12)(90)}{0.455} = 2374 \text{ psi} > 1330 \text{ psi} \qquad \underline{\underline{\text{No good}}}$$

Shear stresses (not usually checked in sheathing)

$$V = 0.6w\left(\ell - \frac{2h}{12}\right) = (0.6)(225)\left(2 - \frac{2 \times \frac{3}{4}}{12}\right) = 253 \text{ lb}$$

$$f_v = \frac{VQ}{Ib} = \frac{V}{Ib/Q} = \frac{253}{7.892} = 32 \text{ psi} < 72 \text{ psi} \qquad \underline{\underline{\text{OK}}}$$

Deflection

$$\text{Permissible deflection} = \left(\frac{1}{360}\right)(12 \times 2) = 0.067''$$

$$\text{Actual } \delta = \frac{w\ell^4}{128EI} = \frac{\left(\frac{225}{12}\right)(12 \times 2)^4}{(128)(1.43 \times 10^6)(0.198)} = 0.172'' > 0.067'' \qquad \underline{\underline{\text{No good}}}$$

Joists

Properties of 2×6 joist with finished dimensions $1\frac{1}{2} \times 5\frac{1}{2}$ (Table 20.1)

$$A = 8.250 \text{ in.}^2, \quad I = 20.80 \text{ in.}^4, \quad S = 7.56 \text{ in.}^3$$

$$\text{Load per foot} = (2)(225) = 450 \text{ lb/ft}$$

Bending and shear stresses

$$M = \frac{w\ell^2}{10} = \frac{(450)(4)^2}{10} = 720 \text{ ft-lb}$$

$$f_b = \frac{(12)(720)}{7.56} = 1143 \text{ psi} < 1875 \text{ psi} \qquad \underline{\underline{\text{OK}}}$$

$$V = 0.6w\left(\ell - \frac{2h}{12}\right) = (0.6)(450)\left(4 - \frac{2 \times 5.50}{12}\right) = 832.5 \text{ lb}$$

$$f_v = \frac{3V}{2A} = \frac{(3)(832.5)}{(2)(8.25)} = 151 \text{ psi} < 180 \text{ psi} \qquad \underline{\underline{\text{OK}}}$$

Deflection

$$\text{Permissible deflection} = \left(\frac{1}{360}\right)(12 \times 4) = 0.133''$$

$$\text{Actual } \delta = \frac{w\ell^4}{128EI} = \frac{\left(\frac{450}{12}\right)(12 \times 4)^4}{(128)(1.76 \times 10^6)(20.80)} = 0.0425'' < 0.133'' \qquad \underline{\underline{\text{OK}}}$$

Stringers

Properties of 2×8 stringer with finished dimensions $1\frac{1}{2} \times 7\frac{1}{4}$ (Table 20.1)

$$A = 10.87 \text{ in.}^2, \quad I = 47.63 \text{ in.}^4, \quad S = 13.14 \text{ in.}^3$$

$$\text{load/ft} = (4)(225) = 900 \text{ lb/ft}$$

Bending and shear stresses

$$M = \frac{w\ell^2}{10} = \frac{(900)(4)^2}{10} = 1440 \text{ ft-lb}$$

$$f_b = \frac{(12)(1440)}{13.14} = 1315 \text{ psi} < 1875 \text{ psi} \qquad \underline{\underline{\text{OK}}}$$

$$V = 0.6w\left(\ell - \frac{2h}{12}\right) = (0.6)(900)\left(4 - \frac{2 \times 7.25}{12}\right) = 1507 \text{ lb}$$

$$f_v = \frac{(3)(1507)}{(2)(10.87)} = 208 \text{ psi} > 180 \text{ psi} \qquad \underline{\underline{\text{No good}}}$$

Deflection

$$\text{Permissible deflection} = \left(\frac{1}{360}\right)(12 \times 4) = 0.133''$$

$$\text{Actual } \delta = \frac{\left(\frac{900}{12}\right)(12 \times 4)^4}{(128)(1.76 \times 10^6)(47.63)} = 0.0371'' < 0.133'' \qquad \underline{\underline{\text{OK}}}$$

20.10   DESIGN OF FORMWORK FOR FLOOR AND ROOF SLABS

From the same principles used for calculating the flexural and shear stresses and the deflections in the preceding section, it is possible to calculate maximum permissible spans for certain sizes of sheathing, joists, or stringers. As an illustration, it is assumed that a rectangular section is used to support a total uniform load of w lb/ft over three or more continuous equal spans. The term ℓ used

The truss table system for deck forming. (Courtesy of Symons Corporation.)

in the equations to follow is the span, center to center, of supports in inches, while f is the allowable flexural stress.

Moment

The bending moment can be calculated in inch-pounds and equated to the resisting moment, also in inch-pounds. The resulting expression can be solved for ℓ, the maximum permissible span, in inches.

$$M = \frac{w\ell^2}{10} \text{ in ft-lb} = \frac{w\ell^2}{120} \text{ in in.-lb}$$

$$M_{\text{res}} = fS \text{ in in.-lb}$$

Equating and solving for ℓ

$$\frac{w\ell^2}{120} = fS$$

$$\ell = 10.95\sqrt{\frac{fS}{w}}$$

Shear

In the same fashion, an expression can be written for the maximum permissible span from the standpoint of horizontal shear. In the expression to follow, H is the allowable shearing stress in pounds per square inch, while V is the maximum external shear applied to the member at a distance h from the support previously assumed to equal $0.6w[\ell - (2h/12)]$.

$$H = \frac{3V}{2A} = \frac{(3)(0.6w)[\ell - (2h/12)]}{2bh}$$

The value of ℓ used in the preceding expression was given in feet; therefore, when the expression is solved for ℓ, it must be multiplied by 12 to give an answer in inches to coincide with the units of the moment and deflection expressions.

$$\ell = 12\left(\frac{Hbh}{0.9w} + \frac{2h}{12}\right)$$

Deflection

If the maximum centerline deflection is equated to the maximum permissible deflection, the resulting equation can be solved for ℓ. For this case the maximum permissible deflection is assumed to equal $\ell/360$.

$$\delta = \frac{w\ell^4}{128EI} = \frac{(w/12)(\ell^4)}{128EI} = \frac{\ell}{360}$$

$$\ell = 1.62\sqrt[3]{\frac{EI}{w}}$$

EXAMPLE 20.2

It is desired to support a total uniform load of 200 psf for a floor slab with $\frac{3}{4}$-in. Class II plyform. If the forms are to be reused many times and the maximum permissible deflection is $\ell/360$, determine the maximum permissible span of the sheathing center to center, of the joists if $f = 1330$ psi and $E = 1,430,000$ psi. Neglect shear in the sheathing. Make similar calculations but consider shear values for 2- × 4-in. joists spaced 1 ft 6 in. on center if $f = 1500$ psi, $H = 180$ psi, and $E = 1,700,000$ psi.

SOLUTION

Sheathing

Assuming a 12-in.-wide piece of plywood (Table 20.2)

$$S = 0.455 \text{ in.}^3, \quad I = 0.198 \text{ in.}^4$$

Moment

$$\ell = 10.95\sqrt{\frac{fS}{w}} = 10.95\sqrt{\frac{(1330)(0.455)}{200}} = 19.05''$$

Deflection

$$\ell = 1.62\sqrt[3]{\frac{EI}{w}} = 1.62\sqrt[3]{\frac{(1,430,000)(0.198)}{200}} = \underline{\underline{18.19''}}$$

Joists

Properties of 2 × 4 joists with finished dimensions $1\frac{1}{2} \times 3\frac{1}{2}$ (Table 20.2)

$$A = 5.250 \text{ in.}^2, \quad I = 5.36 \text{ in.}^4, \quad S = 3.063 \text{ in.}^3$$

Load per foot if joists spaced $1'6''$ on center $= (1.50)(200) = 300$ lb/ft

Moment

$$\ell = 10.95\sqrt{\frac{(1500)(3.063)}{300}} = \underline{\underline{42.85''}}$$

Shear

$$\ell = (12)\left(\frac{Hbh}{0.9w} + \frac{2h}{12}\right) = (12)\left(\frac{180 \times 1.50 \times 3.50}{0.9 \times 300} + \frac{2 \times 3.50}{12}\right) = 49.00''$$

Deflection

$$\ell = 1.62\sqrt[3]{\frac{1,700,000 \times 5.36}{300}} = 50.55'' \quad \text{ANS: } \ell = 18 \text{ in.}$$

For practical design work, many tables are available to simplify and expedite the design calculations. Table 20.4 which shows the maximum permissible spacings, center to center, of supports for joists, stringers, and other beams continuous over four or more supports for one particular grade of lumber, is one such example.

20.11 DESIGN OF SHORING

Wood shores are usually designed as simply supported columns using a modified form of the Euler equation. If the Euler equation is divided by a factor of safety of 3, and if r is replaced with 0.3 times d (the least lateral dimension of square or rectangular shores), as follows:

$$\frac{P}{A} = \frac{\pi^2 E}{3(\ell/0.3d)^2}$$

the so-called National Forest Products Association formula results

$$\frac{P}{A} = \frac{0.3E}{(\ell/d)^2}$$

The maximum ℓ/d value normally specified is 50. If a round shore is being used, it may be replaced for calculation purposes with a square shore having the same cross-sectional area. Should a shore be braced at different points laterally so that it has different unsupported lengths along its different faces, it will be necessary to calculate the ℓ/d ratios in each direction and use the largest one to determine the allowable stress.

The allowable stress used may not be greater than the value obtained from this equation, nor greater than the allowable unit stress in compression parallel to the grain for the grade and type of lumber in question. If the formwork is to be used only once, the allowable stress in the column can reasonably be increased by 25%. Example 20.3 shows the calculation of the permissible column load for a particular shore. Tables are readily available for making these calculations for shores, as they are for sheathing and beam members. For example, Table 20.5 gives the allowable column loads as determined here for a set of simple shores.

EXAMPLE 20.3

Are 4- × 6-in. shores (S4S) 10 ft long and 4 ft on centers satisfactory for supporting the floor system of Example 20.1? Assume the allowable compression stress parallel to the grain is 1000 psi and $E = 1.70 \times 10^6$ psi.

Table 20.4 Safe Spacing (in Inches) of Supports for Joists, Studs (or Other Beam Components of Formwork), Continuous Over Three or More Spans (Courtesy of the American Concrete Institute)

$f = 1875$ psi $E = 1,700,000$ psi $H = 225$ psi

Uniform load, lb per lineal ft (equals uniform load on forms times spacing between joists or studs, ft)	2 × 4	2 × 6	2 × 8	2 × 10	2 × 12	3 × 4	3 × 6	3 × 8	3 × 10	4 × 2	4 × 4	4 × 6	4 × 8	6 × 2	6 × 4	6 × 6	6 × 8	8 × 2	8 × 8	10 × 2
								Normal size of S4S lumber												
100	76	111	137	164	190	90	126	156	187	43	98	138	169	50	110	154	195	55	210	60
200	59	92	115	138	160	72	106	131	157	34	80	116	142	40	92	130	164	44	177	47
300	48	75	99	125	145	62	96	118	142	30	70	105	129	35	81	117	148	38	160	41
400	41	65	86	110	133	54	84	110	132	27	63	97	120	32	74	109	138	35	149	38
500	37	58	77	98	119	48	75	99	125	24	57	89	113	29	69	103	130	32	141	35
600	33	52	69	88	107	44	69	91	116	22	52	81	107	28	65	98	124	30	134	33
700	29	46	61	78	95	40	64	84	107	20	48	75	99	26	60	94	120	29	129	31
800	27	42	55	70	86	38	59	78	100	19	45	70	93	24	56	88	116	28	125	30
900	24	38	51	65	79	36	56	74	94	18	42	66	87	23	53	83	112	26	121	29
1000	23	36	47	60	73	33	52	69	88	17	40	63	83	21	50	79	109	25	118	28
1100	21	33	44	56	68	31	48	64	82	16	38	60	79	20	48	75	103	24	115	27
1200	20	32	42	53	65	29	45	60	76	16	37	57	76	20	46	72	99	23	113	25
1300	19	30	40	50	61	27	43	56	72	15	35	55	73	19	44	69	94	22	110	24
1400	18	29	38	48	59	26	40	53	68	14	33	52	69	18	42	67	91	21	106	24
1500	17	27	36	46	56	24	38	51	65	13	31	49	65	18	41	64	88	20	103	23
1600	17	26	35	44	54	23	37	48	62	13	30	47	62	17	40	62	85	19	100	22
1700	16	26	34	43	52	22	35	46	59	12	29	45	59	16	38	61	83	19	97	21
1800	16	25	33	42	51	21	34	45	57	12	27	43	57	16	37	59	80	18	94	21
1900	15	24	32	40	49	20	33	43	55	11	26	41	56	16	36	57	78	18	91	20
2000	15	23	31	39	48	19	32	42	53	11	25	40	53	15	35	56	76	17	89	20
2100	14	23	30	38	47	19	31	40	51	10	24	38	51	15	34	54	74	17	87	19
2200	14	22	29	37	45	19	30	39	50	10	24	37	49	14	33	52	71	17	85	19

(continued)

Table 20.4 (Continued)

Uniform load, lb per lineal ft (equals uniform load on forms times spacing between joists or studs, ft)	Normal size of S4S lumber																			
	2 × 4	2 × 6	2 × 8	2 × 10	2 × 12	3 × 4	3 × 6	3 × 8	3 × 10	4 × 2	4 × 4	4 × 6	4 × 8	6 × 2	6 × 4	6 × 6	6 × 8	8 × 2	8 × 8	10 × 2
2300	14	22	29	37	44	18	29	38	49	10	23	36	48	14	32	50	69	16	83	18
2400	14	21	28	36	44	18	28	37	47	10	22	35	46	13	31	49	66	16	81	18
2500	13	21	27	35	43	17	27	36	46	9	22	34	45	13	30	47	63	16	80	18
2600	13	20	27	34	42	17	27	35	45	9	21	33	44	12	29	46	61	15	78	17
2700	13	20	27	34	41	17	26	35	44	9	20	32	43	12	28	45	60	15	77	17
2800	13	20	26	33	41	16	26	34	43	9	20	32	42	12	28	43	59	15	75	17
2900	12	19	26	32	40	16	25	33	42	8	20	31	41	11	27	42	58	14	73	16
3000	12	19	25	32	39	16	25	32	41	8	19	30	40	11	26	41	56	14	71	16
3200	12	19	25	31	38	15	24	31	40	8	18	29	38	11	25	39	54	13	68	16
3400	12	18	24	31	37	15	23	30	39	8	18	28	37	10	24	38	51	13	65	15
3600	11	18	24	30	37	14	22	30	38	7	17	27	36	10	23	36	49	12	62	15
3800	11	17	23	30	36	14	22	29	37	7	17	26	34	9	22	35	48	12	59	14
4000	11	17	23	29	35	14	21	28	36	7	16	25	33	9	21	34	46	11	57	13
4500	10	16	22	28	34	13	20	27	34	6	15	24	31	8	20	31	43	10	53	12
5000	10	16	21	27	33	12	19	25	32	6	14	22	30	8	18	29	40	9	49	11

$f = 1875$ psi $\qquad E = 1,700,000$ psi $\qquad H = 225$ psi

Note: Span values above the solid line are governed by deflection. Values within dashed line box are spans governed by shear. $\Delta_{max} = \ell/360$, but not to exceed $\frac{1}{4}$ in.

618

Table 20.5 Allowable Load in Pounds on Simple Wood Shores* For Lumber of the Indicated Strength, Based on Unsupported Length (Courtesy of the American Concrete Institute)

$c \parallel$ to grain = 1150 psi $E = 1,400,000$ psi $\dfrac{\ell}{d_{max}} = 50$ $\dfrac{P}{A_{max}} = \dfrac{0.30E}{(\ell/d)^2}$

Nominal lumber size, in.	2 × 4		3 × 4		4 × 4		4 × 2		4 × 3		4 × 6		6 × 6	
Unsupported length, ft	R**	S4S**	R	S4S	R	S4S	R	S4S	R	S4S	R	S4S	R	S4S
							Bracing needed†							
4	2,800	2,200	10,900	10,000	15,100	15,000	6,800	6,000	10,900	10,100	23,400	22,100	36,400	34,800
5	1,800	1,400	7,700	6,400	15,100	15,000	6,800	6,000	10,900	10,100	23,400	22,100	36,400	34,800
6	1,300	1,000	5,300	4,400	14,400	12,200	6,300	5,200	10,100	8,700	21,700	19,100	36,400	34,800
7			3,900	3,300	10,300	8,900	4,600	3,800	7,400	6,400	15,900	14,000	36,400	34,800
8			3,000	2,500	7,870	6,800	3,500	2,900	5,700	4,900	12,200	10,700	36,400	34,800
9			2,400	2,000	6,200	5,400	2,800	2,300	4,500	3,900	9,700	8,500	36,200	32,900
10			1,900	1,600	5,000	4,400	2,300	1,900	3,600	3,100	7,800	6,900	29,200	26,700
11			—	—	4,200	3,600	1,900	1,500	3,000	2,600	6,500	5,700	24,100	22,100
12					3,500	3,000	1,600	1,300	2,500	2,200	5,400	4,800	20,300	18,500
13					3,000	2,600	1,300	1,100	2,200	1,800	4,600	4,100	17,300	15,800
14					2,600	2,200	1,200	1,000	1,900	1,600	4,000	3,500	14,900	13,600
15					2,200	—	1,000	—	1,600	—	3,500	—	13,000	11,900
16					—	—	—	—	—	—	—	—	11,400	10,400
17					—		—		—				10,100	9,200
18													9,200	8,200
19													8,100	7,100
20													7,300	6,700

*Calculated to nearest 100 lb.

**R indicates rough lumber; S4S indicates lumber finished on all four sides.

†The dimension used in determining ℓ/d is that shown first in the size column. Where this is the larger dimension, the column must be braced in the other direction so that ℓ/d is equal to or less than that used in arriving at the loads shown. For 4 × 2's, bracing in the plane of the 2-in. dimension must be at intervals not greater than 0.4 times the unsupported length. For 4 × 3's, bracing in the plane of the 3-in. dimension must be at intervals not more than 0.7 times the unsupported length.

619

Table 20.5 (Continued)

	c‖to grain = 1100 psi				E = 1,600,000 psi								$\dfrac{P}{A_{max}} = \dfrac{0.30E}{(\ell/d)^2}$	
	2 × 4		3 × 4		4 × 4		4 × 2		4 × 3		4 × 6		6 × 6	
							$\dfrac{\ell}{d_{max}} = 50$							
Nominal lumber size, in.	R**	S4S**	R	S4S	R	S4S	R	S4S	R	S4S	R	S4S	R	S4S
Unsupported length, ft							Bracing needed†							
4	3,200	2,500	9,500	8,700	13,100	12,200	5,900	5,200	9,500	8,700	20,400	19,200	31,600	30,200
5	2,100	1,600	8,700	7,300	13,100	12,200	5,900	5,200	9,500	8,700	20,400	19,200	31,600	30,200
6	1,400	1,100	6,100	5,100	13,100	12,200	5,900	5,200	9,500	8,700	20,400	19,200	31,600	30,200
7	—	—	4,500	3,700	11,700	10,200	5,300	4,400	8,500	7,300	18,200	16,000	31,600	30,200
8			3,400	2,800	9,000	7,800	4,000	3,300	6,500	5,600	13,900	12,300	31,600	30,200
9			2,700	2,200	7,100	6,200	3,200	2,600	5,100	4,400	11,000	9,700	31,600	30,200
10			2,200	1,800	5,800	5,000	2,600	2,100	4,200	3,600	8,900	7,900	31,600	30,200
11			—	—	4,800	4,100	2,100	1,800	3,400	2,900	7,400	6,500	27,600	25,200
12					4,000	3,500	1,800	1,500	2,900	2,500	6,200	5,500	23,200	21,200
13					3,400	3,000	1,500	1,300	2,500	2,100	5,300	4,600	19,700	18,000
14					2,900	2,500	1,300	1,100	2,100	1,800	4,600	4,000	17,000	15,600
15					2,600	—	1,100	—	1,800	—	4,000	—	14,800	13,600
16					—		—		—		—		13,000	11,900
17													11,500	10,500
18													10,300	9,400
19													9,200	8,400
20													8,300	7,600

*Calculated to nearest 100 lb.

**R indicates rough lumber; S4S indicates lumber finished on all four sides.

†The dimension used in determining ℓ/d is that shown first in the size column. Where this is the larger dimension, the column must be braced in the other direction so that ℓ/d is equal to or less than that used in arriving at the loads shown. For 4 × 2's, bracing in the plane of the 2-in. dimension must be at intervals not greater than 0.4 times the unsupported length. For 4 × 3's, bracing in the plane of the 3-in. dimension must be at intervals not more than 0.7 times the unsupported length.

Table 20.5 (Continued)

	c∥ to grain = 1100 psi				E = 1,700,000 psi								$\dfrac{\ell}{d_{max}} = 50$ $\quad$ $\dfrac{P}{A_{max}} = \dfrac{0.30E}{(\ell/d)^2}$	
	2 × 4		3 × 4		4 × 4		4 × 2		4 × 3		4 × 6		6 × 6	
Nominal lumber size, in.	R**	S4S**	R	S4S	R	S4S	R	S4S	R	S4S	R	S4S	R	S4S
Unsupported length, ft							Bracing needed†							
4	3,400	2,600	9,500	8,800	13,100	12,300	5,900	5,300	9,500	8,800	20,400	19,200	31,600	30,200
5	2,200	1,700	9,300	7,700	13,100	12,300	5,900	5,300	9,500	8,800	20,400	19,200	31,600	30,200
6	1,500	1,200	6,500	5,400	13,100	12,300	5,900	5,300	9,500	8,800	20,400	19,200	31,600	30,200
7	—	—	4,700	4,000	12,500	10,800	5,600	4,600	9,000	7,700	19,400	17,000	31,600	30,200
8			3,600	3,000	9,600	8,300	4,300	3,600	6,900	5,900	14,800	13,100	31,600	30,200
9			2,900	2,400	7,600	6,600	3,400	2,800	5,500	4,700	11,700	10,300	31,600	30,200
10			2,300	1,900	6,100	5,300	2,700	2,300	4,400	3,800	9,500	8,400	31,600	30,200
11			—	—	5,100	4,400	2,300	1,900	3,700	3,100	7,900	6,900	29,300	28,000
12					4,200	3,700	1,900	1,600	3,100	2,600	6,600	5,800	24,600	22,500
13					3,600	3,100	1,600	1,300	2,600	2,200	5,600	4,900	21,000	19,200
14					3,100	2,700	1,400	1,200	2,300	1,900	4,800	4,300	18,100	16,500
15					2,700	—	1,200	—	2,000	—	4,200	—	15,800	14,400
16					—		—		—		—		13,800	12,600
17													12,300	11,200
18													10,900	10,000
19													9,800	9,000
20													8,900	8,100

*Calculated to nearest 100 lb.

**R indicates rough lumber; S4S indicates lumber finished on all four sides.

†The dimension used in determining ℓ/d is that shown first in the size column. Where this is the larger dimension, the column must be braced in the other direction so that ℓ/d is equal to or less than that used in arriving at the loads shown. For 4 × 2's, bracing in the plane of the 2-in. dimension must be at intervals not greater than 0.4 times the unsupported length. For 4 × 3's, bracing in the plane of the 3-in. dimension must be at intervals not more than 0.7 times the unsupported length.

SOLUTION **Using 4 × 6 Shores (Dressed Dimensions $3\frac{1}{2} \times 5\frac{1}{2}$, $A = 19.250$ in.^{2})**

$$\text{Load applied to each shore} = (4)(4)(225) = 3600 \text{ lb}$$

$$\frac{\ell}{d} = \frac{(12)(10)}{3.5} = 34.29 < 50 \qquad \underline{\underline{\text{OK}}}$$

$$\text{Allowable } \frac{P}{A} = \frac{(0.3)(1.70 \times 10^6)}{(34.29)^2} = 434 \text{ psi} < 1000 \text{ psi} \qquad \underline{\underline{\text{OK}}}$$

$$\text{Allowable } P = (434)(19.250) = 8354 > 3600 \text{ lb} \qquad \underline{\underline{\text{OK}}}$$

Wood shores may consist of different pieces of lumber that are nailed or bolted together to form larger built-up columns or shores. Such members are usually given lower allowable stresses than those available for solid-sawn or round columns. These built-up shores may be spaced columns consisting of two pieces of lumber with spacer blocks in between.

Spaced shores are very important for formwork because of economic factors. Logs can be efficiently sawed into 2 × material (2 × 4's, 2 × 6's, etc.). The cost of larger pieces, such as 4 ×, 6 ×, and so on, does not make such efficient use of the logs, and prices are appreciably higher. As a result, built-up or spaced shores are frequently used. The allowable stresses, maximum ℓ/d ratios, and other factors are somewhat different from single shores.

A number of adjustable shoring systems are available. The simplest type is made by overlapping two wood members. The workers use a portable jacking tool to make vertical adjustments. Various hardware is available for joining the shores to the stringers with a minimum amount of nailing. Another type of adjustable shore consists of dimension lumber combined with a steel column section and a jacking device. Several manufacturers have available all-metal shores, called *jack shores*, which are adjustable for heights from 4 to 16 ft.[4]

20.12 BEARING STRESSES

The bearing stresses produced when one member rests upon another may be critical in the design of formwork. These stresses need to be carefully checked where joists rest on stringers, where stringers rest on shores, and where studs rest on wales, as well as where form ties bear on the wales through brackets or washers. In each of the cases mentioned, the bearing forces applied to the horizontal timber members cause compression stresses perpendicular to the grain. The allowable stresses for compression perpendicular to the grain are much less than they are for compression parallel to the grain and, in fact, will often be less than the allowable compression stresses in shores as determined by the National Forest Products Association formula.

As a first illustration, the bearing stresses in stringers resting on shores are considered. One particular point that should be noted is that the bearing area may well be less than the full cross-sectional area of the shore. For instance, in Examples 20.1 and 20.3, the 2- × 8-in. stringers are supported by 4- × 6-in. shores, as shown in Figure 20.6. Obviously, the bearing area does not equal the full cross-sectional area of the 4 × 6 ($3.50 \times 5.50 = 19.25$ in.^{2}), but rather equals the smaller crosshatched area ($1.50 \times 5.50 = 8.25$ in.^{2}).

---

[4]Hurd, M. K., 1989, *Formwork for Concrete*, 5th ed. (Detroit: American Concrete Institute), pp. 68–69.

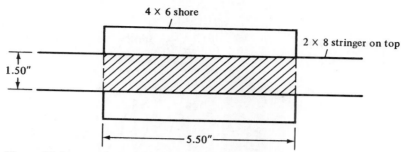

Figure 20.6

The bearing stress can be calculated by dividing the total load applied to the shore (3600 lb from Example 20.3) by the hatched area shown in Figure 20.6.

$$\text{Bearing stress} = \frac{3600}{8.25} = 436\,\text{psi}$$

From Table 20.3, the allowable bearing stress perpendicular to the grain in the stringer is 490 psi for the Douglas fir, coastal construction lumber used in the examples in this chapter. The values given in this table for compression perpendicular to the grain are actually applicable to bearing spread over any length at the end of a member and for bearing at 6 in. or more in length at interior locations.

Should the bearing be located more than 3 in. from the end of a beam and be less than 6 in. long, the allowable compressive stresses perpendicular to the grain may be safely increased by multiplying them by the following factor, in which ℓ is the length of bearing measured along the grain. Should the bearing area be circular, as for a washer, the length of bearing is considered to equal the diameter of the circle.

$$\text{Multiplier} = \frac{\ell + \dfrac{3}{8}}{\ell}$$

If the shore is bearing at an interior point of the stringer in Figure 20.6, the allowable compression stress perpendicular to the grain equals

$$\left(\frac{\ell + \dfrac{3}{8}}{\ell}\right)(490) = \left(\frac{5.50 + 0.375}{5.50}\right)(490) = 523\,\text{psi} > 436\,\text{psi} \qquad \underline{\underline{\text{OK}}}$$

If the calculated bearing stress exceeds the allowable stress, it is necessary to spread the load out, as by using larger members or by using a bearing plate or a hardwood cap on top of the shore.[5]

Another bearing stress situation is presented in Example 20.4, where the bearing stress caused in the wales of a wall form by the form ties is considered.

---

[5]Hurd, *Formwork for Concrete*, pp. 88–93.

Pier forms, Steamboat Rock, Iowa.
(Courtesy of EFCO.)

EXAMPLE 20.4

Form ties are subjected to an estimated force of 5000 lb from the fresh concrete in a wall form. The load is transferred to double 2- × 4-in. wales through $3\frac{1}{2}$-in. brackets or washers, as shown in Figure 20.7. Are the calculated bearing stresses within the permissible values if the wales are Douglas fir, coastal construction and if the bearing locations are at interior points in the wales?

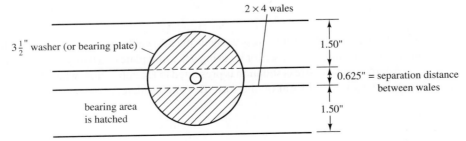

Figure 20.7

SOLUTION

$$\text{Bearing area} = \frac{(\pi)(3.50)^2}{4} - (0.625)(3.50)$$

$$= 7.43 \text{ in.}^2$$

$$\text{Bearing stress} = \frac{5000}{7.43} = 673 \text{ psi}$$

$$\text{Allowable compression} \perp \text{to grain} = \left(\frac{\ell + \dfrac{3}{8}}{\ell}\right)(490)$$

$$= \left(\frac{3.50 + 0.375}{3.50}\right)(490)$$

$$= 543 \text{ psi} < 673 \text{ psi} \qquad\qquad \underline{\underline{\text{No good}}}$$

Use larger washers or smaller spacing of ties because allowable bearing is not satisfactory.

20.13   DESIGN OF FORMWORK FOR WALLS

In Section 20.8 the lateral pressures on wall and column forms were discussed. These pressures increase from a minimum at the top to a maximum at the bottom of the semiliquid concrete. Occasionally in design this varying pressure is recognized, but most of the time it is practical to consider that the maximum pressure exists for the entire height. For convenience in construction, the sheathing, studs, wales, and ties are usually kept at the same size and spacings throughout the entire height of a wall, as are the wales and ties. In effect, therefore, uniform maximum pressures are assumed.

The sheathing and studs for a wall are designed as they are for the sheathing and joists for roof and floor slabs. In addition, the wales are designed as they are for the stringers for slabs, the spans being the center-to-center spacing of the ties.

If the wales are framed on both sides, the ties will extend through the formwork and be attached to the wales on each side, as was shown in Figure 20.2. The tensile force applied to each tie is then dependent on the pressure from the liquid concrete and on the horizontal and vertical spacing of the ties. If the spacing of ties for a certain wall is 2 ft vertically and 3 ft horizontally, and the calculated wall pressure is 800 psf, the total tensile force on each tie equals $2 \times 3 \times 800 = 4800$ lb, assuming a uniform pressure.

Should concrete be placed at 50°F at a rate of 4 ft/hr in a 12-ft high wall, the lateral pressure for the semiliquid concrete could be calculated from the ACI expression as follows:

$$p = 150 + \frac{9000R}{T} = 150 + \frac{(9000)(4)}{50} = 870 \text{ psf}$$

The resulting pressure diagram for the wall would be as shown in Figure 20.8.

Assuming a uniform pressure of 870 psf, the spacing of the studs could be selected as was the spacing for the joists in the floor slab formwork. Example 20.5 shows the design procedure for the sheathing, studs, wales, and ties for a wall form.

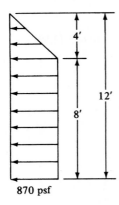

870 psf   **Figure 20.8**

EXAMPLE 20.5

Design the forms for a 12-ft-high concrete wall for which the concrete is to be placed at a rate of 4 ft/hr at a temperature of 50°F. (Refer to Figure 20.8.) Use the following data:

1. Sheathing is to be $\frac{3}{4}$-in. Class I plyform, $f = 1930$ psi, and $E = 1,650,000$ psi.

2. Studs and wales are to consist of Douglas fir, coastal construction, $f = 1875$ psi, $H = 180$ psi, and $E = 1,760,000$ psi. Allowable compression perpendicular to the grain is 490 psi.

3. Ties can carry 5000 lb each, and the tie washers are the same size as the one shown in Figure 20.7.

4. Double wales are used to avoid drilling for ties.

5. Maximum deflection in any form component is $\frac{1}{360}$ of the span.

SOLUTION   **Design of Sheathing**

Properties of 12-in.-wide piece of $\frac{3}{4}$ plyform from Table 20.2

$$S = 0.455 \text{ in.}^3, \quad I = 0.199 \text{ in.}^4$$

$$\text{Load} = 870 \text{ lb/ft}$$

Moment

$$\ell = 10.95\sqrt{\frac{fS}{w}} = 10.95\sqrt{\frac{(1930)(0.455)}{870}} = \underline{\underline{11.00''}}$$

Deflection

$$\ell = 1.62\sqrt[3]{\frac{EI}{w}} = 1.62\sqrt[3]{\frac{(1,650,000)(0.199)}{870}} = 11.71'' \qquad \underline{\text{Use studs at } 0'11'' \text{ on center}}$$

Design of Studs (Spaced 0 ft 11 in. on Center)

Assuming 2×4 studs $\left(1\frac{1}{2} \times 3\frac{1}{2}\right)$, from Table 20.1

$$I = 5.36 \text{ in.}^4, \quad S = 3.06 \text{ in.}^3$$

$$\text{Load} = \left(\frac{11}{12}\right)(870) = 797.5 \text{ lb/ft}$$

Moment

$$\ell = 10.95\sqrt{\frac{(1875)(3.06)}{797.5}} = 29.37''$$

Deflection

$$\ell = 1.62\sqrt[3]{\frac{(1,760,000)(5.36)}{797.5}} = 36.91''$$

Shear

$$\ell = 12\left(\frac{Hbh}{0.9w} + \frac{2h}{12}\right) = (12)\left(\frac{180 \times 1.50 \times 3.50}{0.9 \times 797.5} + \frac{2 \times 3.50}{12}\right)$$

$$= \underline{\underline{22.80''}}$$ Use 1′10″ spacing of wales on center

$$\text{Load on wales} = \left(\frac{22}{12}\right)(870) = 1595 \text{ lb/ft}$$

$$\text{Maximum tie spacing} = \frac{5000}{1595} = 3.13'$$

Design of Wales

Assuming two 2 × 4 wales ($I = 2 \times 5.36 \times 10.72$ in.^{4}, $S = 2 \times 3.06 = 6.12$ in.^{3})
Moment

$$\ell = 10.95\sqrt{\frac{(1875)(6.12)}{1595}} = 29.37''$$

Deflection

$$\ell = 1.62\sqrt[3]{\frac{(1,760,000)(6.12)}{1595}} = 30.62''$$

Shear

$$\ell = (12)\left(\frac{180 \times 3.00 \times 3.50}{0.9 \times 1595} + \frac{2 \times 3.50}{12}\right) = 22.80'' \leftarrow$$

Use ties at every other stud on 1′10″ center

Check Bearing for Studs to Wales

$$\text{Maximum load transferred} = \frac{(22)(11)}{144}(870) = 1462 \text{ lb}$$

$$\text{Bearing area (see Figure 20.9)} = (2)(1.50)(1.50) = 4.50 \text{ in.}^2$$

$$\text{Bearing stress} = \frac{1462}{4.50} = 325 \text{ psi} < 490 \text{ psi}$$ $\underline{\underline{OK}}$

Check Bearing from Ties to Wedges

Assume same dimensions as in Figure 20.7

$$\text{Bearing load} = \frac{(22)(22)}{144}(870) = 2924 \text{ lb}$$

$$\text{Bearing stress} = \frac{2924}{7.43} = 394 \text{ psi} < 543 \text{ psi}$$

as determined in Example 20.4

OK

Selected references for formwork:

1. Chen, W. F., and Mosallam, K. H., 1991, *Concrete Buildings, Analysis for Safe Construction* (Boca Raton, FL: CRC Press).
2. *Concreter Forming*, 1988 (Tacoma, WA: American Plywood Association).
3. Hurd, M. K., 1989, *Formwork to Concrete*, 5th ed. (Detroit: American Concrete Institute).
4. Moore, C. E., 1977, *Concrete Form Construction* (New York: Van Nostrand Reinhold).
5. *Plywood Design Specification* (Tacoma, WA: American Plywood Association).

PROBLEMS

20.1 Repeat Example 20.1 if the joists are spaced 1 ft 6 in. on centers, the stringers at 6 ft 0 in. on centers as are the shores, and the live load is 100 psf. (*Ans.* sheathing and joists OK but stringers unsatisfactory in bending, shear, and deflection)

20.2 Repeat Example 20.1 if $1\frac{1}{8}$-in. Class II plyform, 2×8 joists spaced 2 ft 0 in. on centers, and 2×8 stringers spaced 6 ft 0 in. on centers are used.

20.3 It is desired to support a total uniform load of 150 psf for a floor slab with $\frac{5}{8}$-in. Class I plyform. If the forms are to be used many times and the maximum permissible deflection is $\ell/360$, determine the maximum permissible span of the sheathing, center to center of the joists, if $f = 1930$ psi and $E = 1,650,000$ psi. Neglect shear in the sheathing. Make similar calculations but consider shear values for 2×4 joists (S4S) spaced 1 ft 6 in. on centers if $f = 1815$ psi, $H = 180$ psi, and $E = 1,760,000$ psi. (*Ans.* sheathing $18.25''$, joists $54.40''$)

20.4 Repeat Problem 20.3 if $\frac{3}{4}$-in. Structural I plyform is used.

20.5 Repeat Problem 20.3 if 1-in. Class II plyform is used with $f = 1330$ psi. Assume joists are spaced 2 ft 0 in. on center. (*Ans.* sheathing $27.00''$, joists $47.11''$)

20.6 Is a 4- $\times$ 4-in. shore (S4S) 10 ft long satisfactory to support an axial compression load of 4000 lb if the allowable compression stress parallel to the grain is 1150 psi and $E = 1,400,000$ psi?

20.7 Repeat Problem 20.6 if 6- $\times$ 6-in. shores made from rough lumber 14 ft long are used, each to support an axial compression load of 16,000 lb. (*Ans.* $P = 14{,}893$ lb $< 16{,}000$ lb, No good)

20.8 Design the forms for a 15-ft-high concrete wall for which the concrete is to be placed at a rate of 5 ft/hr at a temperature of 70°F. Use the following data:

1. Sheathing is to be $\frac{3}{4}$-in. Class II plyform, $f = 1330$ psi, and $E = 1,430,000$ psi.
2. Studs and wales are to consist of Douglas fir, coastal construction, $f = 2000$ psi, $H = 180$ psi, and $E = 1,600,000$ psi. Allowable compression perpendicular to the grain is 435 psi.
3. Ties can carry 5000 lb each and the tie washers are the same size as those shown in Figure 20.7.
4. Double wales are used to avoid drilling for ties.
5. Maximum deflection in any form component is $\frac{1}{360}$ of the span.

21

Seismic Design of Reinforced Concrete Structures

21.1 INTRODUCTION

Seismic design of reinforced concrete structures is a subject that could easily fill an entire textbook. Many organizations are dedicated to studying the earthquake response and design of structures. Each earthquake teaches us new lessons, and we continually refine our code requirements based on such lessons.

Earthquakes produce horizontal and vertical ground motions which shake the base of a structure. Because the movement of the rest of the structure is resisted by the structure's mass (inertia), ground shaking creates deformations in the structure, and these deformations produce forces in the structure. Earthquake motions produce seismic loads on structures, even those that are not part of the lateral load-resisting system. These forces can be both horizontal and vertical, and can subject structural elements to axial forces, moments, and shears whose magnitudes depend on many of the properties of the structure, such as its mass, its stiffness, and its ductility. Also important is the structure's period of vibration (the time that the structure takes to vibrate back and forth laterally). In this chapter, the seismic design of reinforced concrete structures is approached from the viewpoint of code application. Calculation of seismic design forces is discussed; element design and detailing for those forces is explained; and examples are provided.

The seismic design of reinforced concrete structures is addressed by the general design provisions of ACI 318, and also by the special seismic-design provisions of Chapter 21 of ACI 318. Reinforced concrete structures designed and detailed according to ACI 318 are intended to resist earthquakes without structural collapse. In general terms, the strength of an earthquake depends on the accelerations, velocities, and displacements of the ground motion that it produces. Seismic design loads are prescribed by *Minimum Design Loads for Buildings and Other Structures* (ASCE/SEI 7–05).[1] In that document, the severity of the design earthquake motion for a concrete structure is described in terms of the structure's *seismic design category* (SDC), which depends on the structure's geographic location and also the soil on which it is built. Structures assigned to the lowest seismic design category, SDC A, must meet only the general design provisions of ACI 318, and do not have to meet the special requirements of Chapter 21. Structures assigned to higher SDCs (B, C, D, E, or F) have increasing seismic demands, however, and must meet the requirements of Chapter 21, which increase in severity with higher SDC. For those higher seismic design categories, the requirements of Chapter 21 of ACI 318 are based on the assumption that a reinforced concrete structure responds inelastically. Inelastic behavior is characterized by

---

[1]American Society of Civil Engineers, *Minimum Design Loads for Buildings and Other Structures*, Chapters 11–23, Reston, Virginia.

yielding of the reinforcing steel as described in Section 3.6 of this textbook. Structural members whose reinforcing steel yields can dissipate some of the energy imparted to the structure by an earthquake, and the forces that develop in such members during an earthquake are less than they would be if the structure responded elastically. Seismic design categories are discussed in more detail in Section 21.5 of this textbook. For now, let's continue with a discussion of the fundamental steps of earthquake design according to the load provisions of ASCE7-05 and the element design and detailing provisions of Chapter 21 of ACI 318–08.

21.2   MAXIMUM CONSIDERED EARTHQUAKE

Areas with high risk of significant ground motion, such as the west coast of the United States, have the highest seismic hazard level. Most areas of the United States have at least some level of seismic risk, however. A large part of ASCE 7–05 is dedicated to determining seismic design forces. These forces are based on the "maximum considered earthquake" (MCE), which is an extreme earthquake, considered to occur only once every 2,500 years. The severity of MCE-level ground shaking is described in terms of the spectral response acceleration parameters S_S and S_1, whose values are given in contour maps provided within ASCE 7 and also available from the United States Geological Service (USGS) Web site (www.usgs.gov). The parameter S_S, a measure of how strongly the MCE affects structures with a short period (0.2 sec). The parameter S_1 is a measure of how strongly the MCE affects structures with a longer period (1 sec). These are called "spectral response parameters," and their values are provided in Figures 22-1 through 22–14 of ASCE/SEI 7–05.[2] If S_1 is less than or equal to 0.04 and S_S is less than or equal to 0.15, the structure is assigned to SDC A. Higher values of S_1 and S_S correspond to successively higher seismic design categories. S_S and S_1 are proportions or ratios of gravity. For example, in parts of southern California, the value of S_S may be 1.0 (100% of the acceleration of gravity), whereas in parts of the Midwest it may be only a few percent.

21.3   SOIL SITE CLASS

The spectral response parameters determined above are modified based on the structure's *soil site class*. The soil at the site is classified into soil site class A through F in accordance with Table 20.3-1 and Section 20.3 of ASCE/SEI 7, using only the upper 100 ft of the site profile. The lowest soil site class, site class A (hard rock), gives a relatively low seismic design force. Higher soil site classes give higher seismic design forces. If such site-specific data are not available, ASCE/SEI 7 permits the registered design professional preparing the soil investigation report to estimate soil properties from known geologic conditions. If the soil properties are not sufficiently known, site class D is used unless the authority having jurisdiction or geotechnical data determines that site class E or F is appropriate. Once the soil site class is assigned, the corresponding site coefficients for short and long periods, F_a an F_v, respectively, are determined using Table 21.1 and the values of S_S and S_1 as described above.

MCE Spectral Response Accelerations and Design Response Accelerations

The MCE spectral response accelerations (related to design forces) for short periods (S_{MS}) and for longer $(1-s)$ periods (S_{M1}) are obtained by multiplying each spectral response acceleration

---

[2]ASCE/SEI 7-05, pp. 210–227.

Table 21.1 Maximum Considered Earthquake Spectral Response Acceleration Parameters

Site Class	Mapped Maximum Considered Earthquake Spectral Response Acceleration Parameter at Short Period				
	$S_S \leq 0.25$	$S_S = 0.5$	$S_S = 0.75$	$S_S = 1.0$	$S_S \geq 1.25$
A	0.8	0.8	0.8	0.8	0.8
B	1.0	1.0	1.0	1.0	1.0
C	1.2	1.2	1.1	1.0	1.0
D	1.6	1.4	1.2	1.1	1.0
E	2.5	1.7	1.2	0.9	0.9
F	A site response analysis must be performed (See Section 11.4.7, ASCI/ACI 7-05).				

Note: Use straight-line interpolation for intermediate values of S_S.

(a) Site Coefficient, F_a, based on Site Class and Mapped Maximum Considered Earthquake Spectral Response Acceleration Parameter at Short Period (from ASCE/SEI 7-05,[1] American Society of Civil Engineers/Structural Engineers Institute, Minimum Design Loads for Buildings and Other Structures. ASCE 7-05 (Reston, VA: American Society of Civil Engineers) Table 11.4-1, Site Coefficient, F_a).

Site Class	Mapped Maximum Considered Earthquake Spectral Response Acceleration Parameter at 1-Second Period				
	$S_1 \leq 0.1$	$S_1 = 0.2$	$S_1 = 0.3$	$S_1 = 0.4$	$S_1 \geq 0.5$
A	0.8	0.8	0.8	0.8	0.8
B	1.0	1.0	1.0	1.0	1.0
C	1.7	1.6	1.5	1.4	1.3
D	2.4	2.0	1.8	1.6	1.5
E	3.5	3.2	2.8	2.4	2.4
F	A site response analysis must be performed (See Section 11.4.7 ASCI/ACI 7-05).				

Note: Use straight-line interpolation for intermediate values of S_1.

(b) Mapped Maximum Considered Earthquake Spectral Response Acceleration Parameter at 1-second Period, (from ASCE/SEI 7-05,[1] American Society of Civil Engineers/Structural Engineers Institute, Minimum Design Loads for Buildings and Other Structures. ASCE 7-05 (Reston, VA: American Society of Civil Engineers) Table 11.4-2, Site Coefficient, F_v).

parameter (S_S and S_1) by its corresponding site coefficient:

$$S_{MS} = F_a S_s \qquad \text{(ASCE/SEI Equation 11.4-1)}$$

$$S_{M1} = F_v S_1 \qquad \text{(ASCE/SEI Equation 11.4-2)}$$

The site coefficients can be as high as 2.5 for S_S, and as high as 3.5 for S_1 (site class E). If the designer uses the default site class D instead of a lower site class to avoid the expense of a soil report, the required seismic design forces may be significantly increased.

Design forces are based on a design earthquake (less severe than the maximum considered earthquake, considered to occur only once every 500 years). The design spectral acceleration parameters, S_{DS} and S_{D1}, are obtained by multiplying the values of S_{MS} and S_{M1} by 2/3:

$$S_{DS} = \frac{2}{3} S_{MS} \qquad \text{(ASCE/SEI Equation 11.4-3)}$$

$$S_{D1} = \frac{2}{3} S_{M1} \qquad \text{(ASCE/SEI Equation 11.4-4)}$$

21.4   OCCUPANCY AND IMPORTANCE FACTORS

The occupancy of a building is an important consideration in determining its SDC. A lean-to shed on a farm is obviously less important than a hospital, fire station, or police station. Chapter 1 of ASCE/SEI 7 lists four occupancy categories in Table 1.1. These occupancy categories are correlated to importance factors that range from 1.0 to 1.5 (ASCE/SEI 7-05, Table 11.5-1). Occupancy categories and importance factors are combined into a single table (Table 21.2) below.

21.5   SEISMIC DESIGN CATEGORIES

Seismic design categories are assigned using Table 21.3 of ASCE/SEI 7–05, and depend on the *seismic hazard level*, *soil type*, *occupancy*, and *use*. The seismic hazard level depends on the geographic location of the structure. Where S_1 is less than 0.75, the seismic design category can be determined from Table 21.3 (a) of this textbook alone where certain conditions apply.[3] When Table 21.3 (a) and (b) give different results for the same structure, the more severe SDC is used. Table 21.3 does not contain SDC E or SDC F. Structures with occupancy category I, II, or III that are located where the mapped spectral response acceleration parameter at $1-s$ period, S_1, is greater than or equal to 0.75 are assigned to SDC E. Structures with occupancy category IV that are located where $S_1 \geq 0.75$ are assigned to SDC F.

21.6   SEISMIC DESIGN LOADS

Vertical Forces

Vertical seismic loads, E_v, are based on the value of S_{DS} (the design spectral response acceleration parameter) and the dead load, D.

$$E_v = 0.2 S_{DS} D \qquad \text{(ASCE/SEI 7-05 Equation 12.4-4)}$$

The vertical seismic load must be considered to act either upward or downward, whichever is more critical for design. The critical design load combination for most reinforced concrete columns usually occurs below their balanced point. In this region columns generally have less moment capacity if axial compression is decreased (Figure 10.8 and 10.8 of this textbook). Hence, an upward seismic load would result in reduced moment capacity.

Lateral Forces

Structures assigned to seismic design category A are designed for the effects of static lateral forces applied independently in each of two orthogonal plan directions. In each direction, the design lateral forces are applied simultaneously at all levels. The design lateral force at each level is determined as follows:

$$F_x = 0.01 \, w_x \qquad \text{(ASCE/SEI 7-05 Equation 11.7-1)}$$

Where

$F_x = $ the design lateral force applied at story x

$w_x = $ the portion of the total dead load of the structure, D, located or assigned to level x

---

[3]ASCE/SEI 7-05, Section 11.4.

Table 21.2 Type of Occupancy and Importance Factors (from ASCE/SEI 7-05, Tables 1.1 and 11.5-1)

Type of Occupancy	Occupancy Category	Importance Factor, I
Buildings and other structures that represent a low hazard to human life in the event of failure, including, but not limited to: agricultural facilities, certain temporary facilities, minor storage facilities.	I	1.0
All buildings and other structures except those listed in Occupancy Categories I, III, and IV.	II	1.0
Buildings and other structures that represent a substantial hazard to human life in the event of failure, including, but not limited to:	III	1.25
• Buildings and other structures where more than 300 people congregate in one area		
• Buildings and other structures with daycare facilities with a capacity for more than 150		
• Buildings and other structures with elementary school or secondary school facilities with a capacity for more than 250		
• Buildings and other structures with a capacity for more than 500 for colleges or adult education facilities		
• Healthcare facilities with a capacity of 50 or more resident patients, but not having surgery or emergency treatment facilities		
• Jails and detention facilities		
Buildings and other structures, not included in Occupancy Category IV, with potential to cause a substantial economic impact and/or mass disruption of day-to-day civilian life in the event of failure, including, but not limited to: power generating stations, water treatment facilities, sewage treatment facilities, telecommunication centers.		
Buildings and other structures not included in Occupancy Category IV (including, but not limited to, facilities that manufacture, process, handle, store, use, or dispose of such substances as hazardous fuels, hazardous chemicals, hazardous waste, or explosives) containing sufficient quantities of toxic or explosive substances to be dangerous to the public if released.		
Buildings and other structures containing toxic or explosive substances shall be eligible for classification as Occupancy Category II structures if it can be demonstrated to the satisfaction of the authority having jurisdiction by a hazard assessment as described in Section 1.5.2 that a release of the toxic or explosive substances does not pose a threat to the public.		

(Continued)

Table 21.2 (Continued)

Type of Occupancy	Occupancy Category	Importance Factor, I
Buildings and other structures designated as essential facilities, including, but not limited to:	IV	1.5
• Hospitals and other healthcare facilities having surgery or emergency treatment facilities		
• Fire, rescue, ambulance, and police stations, and emergency vehicle garages		
• Designated earthquake, hurricane, or other emergency shelters		
• Designated emergency preparedness, communication, and operation centers, and other facilities required for emergency response		
• Power generating stations and other public utility facilities required in an emergency		
• Ancillary structures (including, but not limited to, communication towers, fuel storage tanks, cooling towers, electrical substation structures, fire water storage tanks or other structures housing or supporting water, or other fire-suppression material or equipment) required for operation of Occupancy Category IV structures during an emergency		
• Aviation control towers, air traffic control centers, and emergency aircraft hangars		
• Water storage facilities and pump structures required to maintain water pressure for fire suppression		
• Buildings and other structures having critical national defense functions		
Buildings and other structures (including, but not limited to, facilities that manufacture, process, handle, store, use, or dispose of such substances as hazardous fuels, hazardous chemicals, or hazardous waste) containing highly toxic substances where the quantity of the material exceeds a threshold quantity established by the authority having jurisdiction.		
Buildings and other structures containing highly toxic substances shall be eligible for classification as Occupancy Category II structures if it can be demonstrated to the satisfaction of the authority having jurisdiction by a hazard assessment as described in Section 1.5.2 that a release of the highly toxic substances does not pose a threat to the public. This reduced classification shall not be permitted if the buildings or other structures also function as essential facilities.		

*Cogeneration power plants that do not supply power on the national grid shall be designated Occupancy Category II.

634

Table 21.3 Seismic Design Category (SDC) Based on Occupancy Category and Response Acceleration Parameter

Value of S_{DS}	Occupancy Category		
	I or II	III	IV
$S_{DS} < 0.167$	A	A	A
$0.167 \leq S_{DS} < 0.33$	B	B	C
$0.33 \leq S_{DS} < 0.50$	C	C	D
$0.50 = S_{DS}$	D	D	D

(a) Based on Short-Period Response Acceleration Parameter (from ASCE/SEI 7-05, Table 11.6-1).

Value of S_{D1}	Occupancy Category		
	I or II	III	IV
$S_{D1} < 0.067$	A	A	A
$0.067 \leq S_{D1} < 0.133$	B	B	C
$0.133 \leq S_{D1} < 0.20$	C	C	D
$0.20 \leq S_{D1}$	D	D	D

(b) Based on 1-second Period Response Acceleration Parameter (from ASCE/SEI 7-05, Table 11.6-2).

Quite simply, a structure assigned to SDC A is designed for a lateral seismic load equal to 1% of it's design dead load. Structures assigned to SDC A must also meet requirements for load path connections, connection to supports, and anchorage of concrete or masonry walls.[4]

Structures assigned to SDC B through SDC F must be designed using a more detailed method. One such method is the *equivalent lateral force procedure*, in which the design seismic base shear, V, in each principal plan direction is determined as:

$$V = C_s W \qquad \text{(ASCE/SEI 7-05 Equation 12.8-1)}$$

Where

$C_s = $ the seismic response coefficient determined in accordance with ASCE/SEI Section 12.8.1.1

$W = $ the effective seismic weight (ASCE/SEI Section 12.7.2)

It includes the total dead load and other loads that are likely to be present during an earthquake. For example, at least 25 % of the floor live load in storage areas must be included. Where partitions are present, the larger of the actual partition weight or 10 psf (0.48 kN/m^2) must be included. The total operating weight of permanent equipment must be included. Where the flat roof snow load, P_f, exceeds 30 psf (1.44 kN/m^2), 20% of the uniform design snow load, regardless of actual roof slope is included.

---

[4]ASCE/SEI 7, Section 11.7

The seismic response coefficient, C_s, is determined by

$$C_S = \frac{S_{DS}}{R/I}$$ (ASCE/SEI 7-05 Equation 12.8-2)

and need not exceed

$$C_S = \frac{S_{DI}}{T\left(\dfrac{R}{I}\right)} \quad \text{for } T \leq T_L$$ (ASCE/SEI 7-05 Equation 12.8-3)

or

$$C_S = \frac{S_{D1}T_L}{T^2\left(\dfrac{R}{I}\right)} \quad \text{for } T > T_L$$ (ASCE/SEI 7-05 Equation 12.8-4)

In no case is C_S permitted to be less than 0.01. When $S_1 \geq 0.6$ g,

$$C_S = \frac{0.5S_1}{R/I}$$ (ASCE/SEI 7-05 Equation 12.8-6)

Supplement 2 to ASCE 7-05 changed the lower limit on C_S back to the original value of $0.044IS_{DS}$.

The fundamental period of the structure, T, in the direction under consideration is established using the structural properties of the resisting elements in a properly substantiated analysis. The fundamental period, T, must not exceed the product of the coefficient for upper limit on calculated period (C_u) from Table 12.8-1 and the approximate fundamental period, T_a, determined from Eq. 12.8-7. As an alternative to performing an analysis to determine the fundamental period, T, it is permitted to use the approximate building period, T_a, calculated in accordance with Section 12.8.2.1, directly.

The approximate fundamental period (T_a), in s, can be determined from the following equation:

$$T_a = C_t h_n^x$$ (ASCE/SEI 7-05 Equation 12.8-7)

where h_n is the height in feet above the base to the highest level of the structure. For concrete moment resisting frames, the coefficient C_t is 0.016 (0.0466 in SI units) and x is 0.9.

As an alternative, the approximate fundamental period (T_a), in seconds, can be found from the following equation for structures not exceeding 12 stories in height in which the seismic force-resisting system consists entirely of concrete moment resisting frames and the story height is at least 10 ft (3 m):

$$T_a = 0.1 N$$ (ASCE/SEI 7-05 Equation 12.8-8)

where N = number of stories.

whereas T_a for concrete shear wall structures can be determined by

$$T_a = \frac{0.0019}{\sqrt{C_w}} h_n$$ (ASCE/SEI 7-05 Equation 12.8-9)

where h_n is as defined previously and C_w is calculated as follows:

$$C_w = \frac{100}{A_B} \sum_{i=1}^{x} \left(\frac{h_n}{h_i}\right)^2 \frac{A_i}{\left[1 + 0.83\left(\dfrac{h_i}{D_i}\right)^2\right]}$$ (ASCE/SEI 7-05 Equation 12.8-10)

Where

A_B = area of base of structure, ft^2

A_i = web area of shear wall "i" in ft^2

D_i = length of shear wall "i" in ft

h_i = height of shear wall "i" in ft

x = number of shear walls in the building effective in resisting lateral forces in the direction under consideration

The total design seismic base shear, V, is distributed to each building level in accordance with the following expression which is obtained by combining ASCE/SEI 7-05 Equations 12.8-11 and 12.8-12.

$$F_x = \frac{w_x h_x^k}{\sum_{i=1}^{n} w_i h_i^k} V$$

Where

w_x or w_i = the portion of the total effective weight of the structure, W, assigned to level x or i, respectively

k = an exponent related to the structure period as follows:

for structures having a period of 0.5 sec or less, $k = 1$
for structures having a period of 2.5 sec or more, $k = 2$
for structures having a period between 0.5 and 2.5 sec, k shall be 2 or shall be determined by linear interpolation between 1 and 2

Structures that respond elastically to earthquakes generally incur large seismic forces. If a structure is designed and detailed to be capable of nonlinear inelastic response, it will be subjected to lower seismic forces, however, even for the same earthquake at the same site. The *response modification coefficient*, R, reduces the design seismic force for structures capable of responding inelastically. As shown in Table 21.4, this coefficient is 3.0 for ordinary concrete moment frames, 5.0 for intermediate concrete moment frames, and 8.0 for special concrete moment frames. In this table, the terms "ordinary," "intermediate," and "special" refer to increasingly severe levels of

Table 21.4 Response Modification Coefficients for Different Seismic Force-Resisting Systems (from ASCE/SEI 7-05, Table 12.2-1, abridged)

	Seismic Force-Resisting System	R[*]
Bearing Wall System	Special Reinforced Concrete Shearwall	5
	Ordinary Reinforced Concrete Shearwall	4
	Detailed Plain Concrete Shearwall	2
	Ordinary Plain Concrete Shearwall	1.5
Building Frame System	Special Reinforced Concrete Shearwall	6
	Ordinary Reinforced Concrete Shearwall	5
	Detailed Plain Concrete Shearwall	2
	Ordinary Plain Concrete Shearwall	1.5
Moment Resistant Frames	Special Reinforced Concrete Moment Frames	8
	Intermediate Reinforced Concrete Moment Frames	5
	Ordinary Reinforced Concrete Moment Frames	3

[*]Response Modification Coefficient, R.

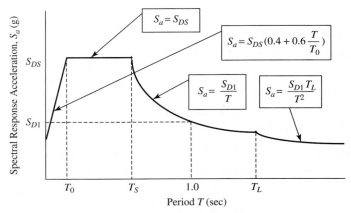

Figure 21.1 Design response spectrum (from ASCE/SEI 7-05, Figure 11.4-1).

seismic detailing, and are discussed later in this chapter. Higher values of R correspond to lower seismic design forces, since R appears in the denominator of the equation for seismic design base shear. A special concrete moment frame must be designed for only 3/8 the seismic base shear of a geometrically identical ordinary concrete moment frame.

Structures assigned to SDC A need not comply with the requirements of Chapter 21 of ACI 318. Structures assigned to SDC B and higher must comply with successively more severe requirements within that chapter. For example, structures assigned to SDC B must satisfy ACI 318 Section 21.1.2; structures assigned to SDC C must satisfy ACI 318 Sections 21.1.2 and 21.1.8; and structures assigned to SDC D through F must satisfy ACI 318 Sections 21.1.2, 21.1.8, and Sections 21.11 through 21.13.

More complex structures must be designed using the *general response spectra method* or site-specific, ground-motion procedures. In the general response spectra method, the design response acceleration, S_a, depends on the fundamental building period, T, as shown in Figure 21.1. That figure has four distinct regions, each with its own equation relating S_a to S_{DS} or S_{D1} and to T.

Where

S_{DS} = the design spectral response acceleration parameter at short periods

S_{D1} = the design spectral response acceleration parameter at $1-s$ period

T = the fundamental period of the structure, seconds

$T_0 = 0.2S_{D1}/S_{DS}$

$T_S = S_{D1}/S_{DS}$

T_L = long-period transition period(s) shown in Chapter 22 of ASCE/SEI 7-05

21.7   DETAILING REQUIREMENTS FOR DIFFERENT CLASSES OF REINFORCED CONCRETE MOMENT FRAMES

Ordinary moment frames that are part of the seismic force-resisting system are permitted only in SDC B,[5] and must meet seismic design and detailing requirements for beams and columns as prescribed in ACI 318, Chapter 21. Beams must have at least two of their longitudinal bars

---

[5]ASCE/SEI 7-05, Table 12.2-1.

continuous along both the top and bottom faces, and these bars must be developed at the faces of supports. Such bars provide the frame with seismic load-resisting capability that may not have been required by analysis. Columns with clear height less than or equal to 5 times the dimension c_1 must be designed for shear in accordance with ACI Section 21.3.3. The term c_1 is the dimension of a rectangular (or equivalent rectangular column, capital, or bracket) in the direction of the span for which moments are being calculated. The design shear is determined as the summation of the moment capacity at the faces of the joints at each end of the column, divided by the distance between those faces. This approach, referred to as "capacity design for shear," is intended to ensure that the columns do not fail in shear during an earthquake. If the moment capacity of a column is larger than that required based on analysis (due, for example, to reinforcing bars with cross-sectional areas greater than the theoretically required areas), then the shear capacity of the column must be correspondingly increased. This increased design shear need not exceed the shear corresponding to a value of earthquake load E twice that required by the applicable code, however.

Intermediate moment frames that are part of the seismic force-resisting system are permitted in SDC B and C[5] only, and must satisfy the more stringent requirements of ACI 318 Section 21.3. These include the shear requirements of ACI 21.3.3 described previously for certain columns in ordinary moment frames. In addition, beams (members with axial compressive loads, P_u, less than $A_g f_c'/10$) must be designed for shear using capacity design as required in ACI 318 Section 21.3.4 and illustrated in Figure 21.2. Beams must also be detailed for ductility, using closed spirals, closed hoops, or closed rectangular ties to confine the concrete so that it will be stronger and more ductile. Members with larger values of P_u must meet the requirements for columns in ACI 318 Section 21.3.5. These also include more stringent requirements for concrete confinement.

Special moment frames are permitted in any seismic design category (ASCE 7–05, Table 12.2-1), and must satisfy ACI 318 Sections 21.5 through 21.8. ACI 318 Section 21.5 applies only to flexural members in special moment frames. As with intermediate moment frames, a flexural member in a special moment frame is defined as one having a factored axial compressive force on the member, P_u, that does not exceed $A_g f_c'/10$. Such a flexural member must have a clear span, l_n, not less than 4 times its effective depth. Its width, b_w, cannot be less than the smaller of $0.3h$ and 10 in. Additionally its width, b_w, must not exceed the width of the supporting member, c_2, plus a distance on each side of supporting member equal to the smaller of (a) the width of supporting member, c_2, and (b) 0.75 times the overall dimension of supporting member, c_1. These geometric limits are intended to provide greater ductility. A limit of 0.025 is imposed on the longitudinal reinforcement ratio, to enhance flexural ductility and avoid congestion. A minimum of two bars must be provided continuously at both top and bottom. The positive moment strength at any joint face must be at least one-half the negative moment strength of the flexural member. The negative and the positive moment strength at any section along the member length must be at least one-fourth of the maximum moment strength provided at the face of either joint.

Lap splices of flexural reinforcement are permitted only if confinement reinforcement (hoops or spiral reinforcement) is provided over the entire lap length. The spacing of such transverse reinforcement cannot exceed the smaller of $d/4$ and 4 in. Lap splices are not permitted in regions where flexural yielding is expected, including

 a. Within the joints
 b. Within a distance of 2 times the member depth from the face of the joint
 c. Where analysis shows flexural yielding caused by inelastic lateral displacements of the frame

Requirements for transverse confinement are similar to but more stringent than those for intermediate concrete moment frames. They are intended to confinement of the concrete within the hoop and to provide lateral support to resist buckling of yielded longitudinal reinforcement under reversed cyclic loading. Hoops are required in regions expected to experience hinging. Where hoops are not required, stirrups having seismic hooks at both ends must be provided, spaced at a distance not more than $d/2$ throughout the length of the member.

Members of special moment frame must be designed for shear using the capacity design procedures explained above. When members of special moment frames are subjected to combined flexure and factored axial compressive forces exceeding $A_g f'_c / 10$, additional requirements must be met. Geometric requirements include the following:

1. The smallest cross-sectional dimension, measured on a line passing through the geometric centroid, must be at least 12 in.

2. The ratio of the shortest cross-sectional dimension to the perpendicular dimension must be at least 0.4 in.

ACI 318 Section 21.6.2 requires that columns of special moment frames be designed so that their nominal flexural strengths are 20% stronger than those of the beams framing into a beam-column joint. This requirement is intended to ensure that if hinges should form at a beam-column joint, they would occur in beams rather than columns. If hinges form in columns, the result may be collapse of the frame. This requirement is waived if the columns' lateral strength and stiffness are ignored in determining the structure's strength and stiffness, such as in a braced frame.

The longitudinal reinforcement in columns of special moment frames must be between 1% and 6% of the gross cross-sectional area. In addition, lap splices must meet the requirements in ACI 318 Section 21.6.3.2.

Transverse reinforcement requirements (ACI 318 Section 21.6.4) for special moment frames are more stringent than those of ordinary or intermediate moment frames, and are intended to provided even higher ductility.

Shear requirements for intermediate moment frames are increased for special moment frames by changing M_{nl} and M_{nr} in Figure 21.2 to M_{pr1} and M_{pr2}. M_{pr} is the probable flexural strength at the face of the joint considering axial load, if any, using a reinforcing steel stress of $1.25f_y$ and a ϕ

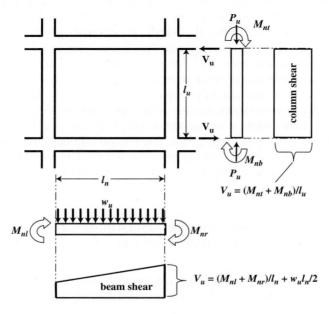

$V_u = (M_{nt} + M_{nb})/l_u$

$V_u = (M_{nl} + M_{nr})/l_n + w_u l_n/2$

Figure 21.2 Column and beam design shear for intermediate moment frames.

factor of 1.0. The subscripts 1 and 2 on M_{pr} denote the left and right ends of the flexural member, respectively. Similarly, the moments M_{nt} and M_{nb} in Figure 21.2 are changed to M_{pr3} and M_{pr4}, where the subscripts 3 and 4 denote the top and bottom of the column, respectively. While Chapter 21 of ACI 318 contains additional requirements for shear strength and development length in tension in special moment frames, these are more complex than necessary for this introductory text in reinforced concrete design. Additional information on these and other seismic design provisions for reinforced concrete structures is provided in Chapter 29 of PCA's Notes on ACI 318–08, Chapter 29.

EXAMPLE 21.1

Determine the design lateral forces due to earthquake on a 6-story concrete frame hospital using the equivalent lateral force procedure. The structure is selected as a hospital to illustrate the use of importance factors in calculating seismic design loads using the procedures of ASCE/SEI 7-05. Some states have additional requirements for hospitals, which are not addressed by this example. The structure is located in Memphis, Tennessee, for which MCE values are determined from USGS maps to be $S_S = 2.0$ and $S_1 = 0.9$. The structure is located on soil determined to be site class C. Each floor is 12 ft in height. The value of W for each floor is determined to be $450\ k$, and for the roof, $200\ k$.

1. Determine F_a and F_v

   From Table 21.1(a), using $S_s \geq 1.25$ and site class C, $F_a = 1.0$.
   From Table 21.1(b), using $S_1 \geq 0.5$ and site class C, $F_v = 1.3$.

2. Determine S_{MS} and S_{M1}

$$S_{MS} = F_a S_S = (1.0)(2.0) = 2.0 \qquad \text{(ASCE/ACI Equation 11.4-1)}$$

$$S_{M1} = F_v S_1 = (1.3)(0.9) = 1.17 \qquad \text{(ASCE/ACI Equation 11.4-2)}$$

3. Determine S_{DS} and S_{D1} (page 631)

$$S_{DS} = 2S_{MS}/3 = 1.33 \qquad \text{(ASCE/ACI Equation 11.4-3)}$$

$$S_{D1} = 2S_{M1}/3 = 0.78 \qquad \text{(ASCE/ACI Equation 11.4-4)}$$

4. Occupancy and importance factors—Table 21.2 lists the occupancy category for hospitals as IV. This corresponds to an importance factor of 1.5, also from the same table. This is a critical facility that requires the highest level of consideration, hence the highest importance factor. Imagine the consequences if our fire stations, police stations, and hospitals could not function after a serious earthquake.

5. Determine the seismic design category—Table 21.3(a) requires SDC D for $S_{DS} \geq 0.50$ and Occupancy category IV. Table 21.3(b) likewise requires SDC D for $S_{D1} \geq 0.20$ and Occupancy category IV.

6. Determine the response modification coefficient, R—Since a special moment frame is required for SDC D, $R = 8$ from Table 21.4. Note that the frame must be detailed in accordance with the requirements for special moment frames.

7. Determine the fundamental period of the structure—The approximate value of T is

$$T_a = 0.1N = 0.1(6) = 0.6\ \text{sec} \qquad \text{(ASCE/ACI Equation 12.8-8)}$$

(for frames with floor to floor heights exceeding 10 ft. and with fewer than 12 stories)

8. Determine T_S and T_L

$$T_S = S_{D1}/S_{DS} = 0.78/1.33 = 0.59$$

$$T_L = 12 \text{ (from ASCE/SEI 7-05, Figure 22-15)}$$

9. Determine the total design lateral seismic force on the structure

$$C_S = \frac{S_{DS}}{R/I} = \frac{1.33}{8/1.5} = 0.25 \qquad \text{(ASCE/ACI Equation 12.8-2)}$$

and since $T_a < T_L$, C_S need not exceed

$$C_S = \frac{S_{D1}}{T\left(\frac{R}{I}\right)} = \frac{0.78}{0.6\left(\frac{8}{1.5}\right)} = 0.244 \qquad \text{(ASCE/ACI Equation 12.8-3)}$$

the controlling value is $C_S = 0.244$.

$$V = C_S W = (0.244)(450)(5) + (0.244)(200)(1) = 598 \text{ kips} \qquad \text{(ASCE/ACI Equation 12.8-1)}$$

The force at the top floor (roof level) is determined using Equation 12.8-12 of ASCE/SEI 7-05.

$$F_R = \frac{w_r h_R^{1.05}}{\sum_{i=1}^{n} w_i h_i^{1.05}} V = \frac{(200)(72)^{1.05}}{(450)(12)^{1.05} + (450)(24)^{1.05} + \dots\dots + (200)(72)^{1.05}} (598)$$

$$= \frac{17,833}{115,333}(598) = 92.4\,k$$

The coefficient k is determined to be *1.05* by interpolation, using a value of $T = 0.6$ sec.

At the fifth floor level

$$F_6 = \frac{w_6 h_6^{1.05}}{\sum_{i=1}^{n} w_i h_i^{1.05}} V = \frac{(450)(60)^{1.05}}{(450)(12)^{1.05} + (450)(24)^{1.05} + \dots\dots + (200)(72)^{1.05}} (598)$$

$$= \frac{33,134}{115,333}(598) = 171.8\,k$$

The remaining forces at other floor levels are calculated using the technique shown above for the roof and sixth levels, and the results are shown below.

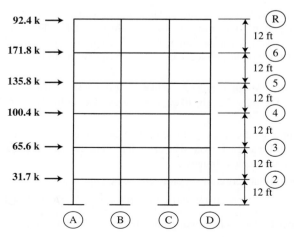

Figure 21.3 Design lateral seismic forces for Example 21.1.

EXAMPLE 21.2

Determine the design column shear for the column shown below if it is part of an intermediate concrete moment frame.

Determine the moment capacity of the column.

$$K_n = \frac{P_n}{f_c' A_g} = \frac{P_u}{\phi f_c' A_g} = \frac{120}{0.65(4)16^2} = 0.18$$

$$\rho_z = \frac{6}{16^2} = 0.0234$$

Using the Column Interaction diagrams in Appendix A, Graph 3, $R_n = 0.18$. However, the location of the coordinates of K_n and R_n appears to be on the radial line corresponding to $\varepsilon_t = 0.005$. The ϕ factor for this value of K_n and R_n is 0.9, not 0.65 as assumed above. Repeating the calculation of K_n using $\phi = 0.9$ results in $K_n = 0.130$. From Graph 3, $R_n = 0.17$.

$$M_n = R_n f_c' A_g h = 0.17(4)16^2 16 = 2785 \, k{-}in = 232.1 \, k{-}ft$$

Since the moment capacities at the top and bottom of the column are the same, $M_{nt} = M_{nb} = 232.1$ k-ft

$$V_u = \frac{M_{nt} + M_{nb}}{l_u} = \frac{232.1 + 232.1}{12} = 38.68 \, k$$

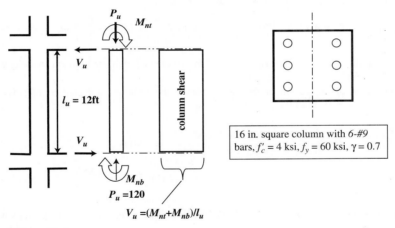

16 in. square column with 6-#9 bars, $f_c' = 4$ ksi, $f_y = 60$ ksi, $\gamma = 0.7$

Figure 21.4

EXAMPLE 21.3

The beam-column joint shown below is part of a special moment frame. Determine if the joint is in compliance with ACI 318-08, Section 21.6.2.2. If not, redesign the columns to comply with this provision.

SOLUTION   ACI 318-08, Section 21.6.2.2 requires that the sum of the column moments at a joint $(M_{nt} + M_{nb})$ be not less than 120% of the sum of the beam moments framing into the same joint $(M_{nl} + M_{nr})$. The axial force is included in determining the column's flexural capacity.

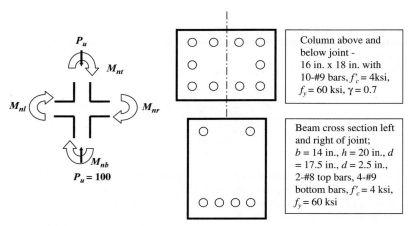

Figure 21.5

Beam capacity: The beam cross section at the left side of the joint is subjected to positive moment. Hence, $A_s = 4$ in.^{2} and $A'_s = 1.57$ in.^{2}.

$$a = \frac{A_s f_y}{0.85 f'_c b} = \frac{4.00 \text{ in.}^2 \times 60 \text{ ksi}}{0.85 \times 4 \text{ ksi} \times 14 \text{ in.}} = 5.04 \text{ in.}$$

$$M_n = A_s f_y (d - a/2) = 4.00 \text{ in.}^2 \times 60 \text{ ksi} \times \left(17.5 - \frac{5.04}{2}\right) \text{ in.} = 3595 \text{ k} - \text{in.}$$

Note: If the compression steel ($A'_s = 1.57$ in.^{2}) were included, the moment capacity would be 3710 in.-k (only 3% more).

The beam cross section at the right side of the joint is subjected to negative moment. Hence, $A_s = 1.57$ in.^{2} and $A'_s = 4.00$ in.^{2}.

$$a = \frac{A_s f_y}{0.85 f'_c b} = \frac{1.57 \text{ in.}^2 \times 60 \text{ ksi}}{0.85 \times 4 \text{ ksi} \times 14 \text{ in.}} = 1.98 \text{ in}$$

$$M_n = A_s f_y (d - a/2) = 1.57 \text{ in.}^2 \times 60 \text{ ksi} \times \left(17.5 - \frac{1.98}{2}\right) \text{ in.} = 1555 \text{ k} - \text{in.}$$

Note: If the compression steel ($A'_s = 4.00$ in.^{2}) were included, the moment capacity would be 1558 in.-k (only 0.2% more).

Column capacity: Using the interaction diagrams in Appendix A, the moment capacity corresponding to an axial compressive load of 300 kips is determined as follows:

Assume $\phi = 0.65$. $\rho_z = A_{st}/bh = 10/(16 \cdot 18) = 0.035$.

$$K_n = P_u/\phi f'_c bh = 300/(0.65 \cdot 4 \cdot 16 \cdot 18) = 0.40$$

$$\gamma = (h-5)/h = 13/18 = 0.72 \text{ (use Graph 7, } \gamma = 0.7)$$

$$R_n = 0.22, \ M_n = R_n f'_c bh^2 = 0.22(4.0)(16)(18)^2 = 4562 \text{ in.-k}$$

Also from Graph 7, f_s/f_y is between 0.9 and 1.0 which means $\varepsilon_t < \varepsilon_y$ and $\phi = 0.65$ as assumed.

The columns above and below the joint have the same axial load and same cross section, hence the same capacity. If the axial load of the column below had been different from that above, the moment capacity would be different, even with the same cross section. The sum of the nominal column moment capacities at the joint is therefore $2 \cdot 4562 = 9124$ in.-k. The sum of the nominal beam moment capacities is $3595 + 155$ ASCE/ACI Equation 12.8-25 $= 5150$ in.-k. Since $9124 > 1.2 \cdot 5150 = 6180$ in.-k, the "strong-column, weak-beam requirements of ACI 318-08 Chapter 21 for special moment frames are satisfied.

PROBLEMS

21.1 Repeat Example 21.1 if $S_S = 1.8$, $S_1 = 0.6$ and soil site class D. The bottom floor is 16 ft high, and the others are each 14 ft. The value of W for each floor is 400 k, and for the roof, it is 175 k.

21.2 Repeat Example 21.2 assuming a special concrete moment frame.

21.3 Repeat Example 21.2 using the Excel spreadsheet from Chapter 10 to determine the column capacity. (*Ans.* $V_u = 38.7$ k)

21.4 Repeat Example 21.3 using $P_u = 50$ k and ten #8 bars in the column.

A

Tables and Graphs

U. S. CUSTOMARY UNITS

Table A.1 Values of
Modulus of Elasticity for
Normal-Weight Concrete

U.S. customary units

f_c' (psi)	E_c (psi)
3,000	3,160,000
3,500	3,410,000
4,000	3,640,000
4,500	3,870,000
5,000	4,070,000

Source Notes: Tables A.4, A.6, A.14, and A.15, as well as Graph 1, are reprinted from *Design of Concrete Structures* by Winter and Nilson. Copyright © 1972 by McGraw-Hill, Inc., and with permission of the McGraw-Hill Book Company. Graphs 2–13 reprinted from *Design Handbook, Volume 2, Columns* (SP-17A), 1978, with permission of American Concrete Institute.

Tables A.3(a) and A.3(b) are reprinted from *Manual of Standard Practice*, 22nd ed., 1997, second printing, Concrete Reinforcing Steel Institute, Chicago, IL.

Tables A.16 through A.20 are reprinted from Commentary on Building Code Requirements For Reinforced Concrete (ACI 318-77) with permission of the American Concrete Institute.

Table A.2 Designations, Areas, Perimeters, and Weights of Standard Bars

| Bar no. | U.S. customary units | | |
	Diameter (in.)	Cross-sectional area (in.^{2})	Unit weight (lb/ft)
3	0.375	0.11	0.376
4	0.500	0.20	0.668
5	0.625	0.31	1.043
6	0.750	0.44	1.502
7	0.875	0.60	2.044
8	1.000	0.79	2.670
9	1.128	1.00	3.400
10	1.270	1.27	4.303
11	1.410	1.56	5.313
14	1.693	2.25	7.650
18	2.257	4.00	13.600

Table A.3(A) Common Stock Styles of Welded Wire Fabric—U.S. Customary Units

| Style designation | Steel area (sq in. per ft) | | Weight approx. (lb per 100 sq ft) |
	Longit.	Transv.	
Rolls			
6 × 6—W1.4 × W1.4	0.03	0.03	21
6 × 6—W2 × W2	0.04	0.04	29
6 × 6—W2.9 × W2.9	0.06	0.06	42
6 × 6—W4 × W4	0.08	0.08	58
4 × 4—W1.4 × W1.4	0.04	0.04	31
4 × 4—W2 × W2	0.06	0.06	43
4 × 4—W2.9 × W2.9	0.09	0.09	62
4 × 4—W4 × W4	0.12	0.12	86
Sheets			
6 × 6—W2.9 × W2.9	0.06	0.06	42
6 × 6—W4 × W4	0.08	0.08	58
6 × 6—W5.5 × W5.5	0.11	0.11	80
4 × 4—W4 × W4	0.12	0.12	86

Table A.3(B) Sectional Area and Weight of Welded Wire Fabric—U.S. Customary Units

Wire size number[a]		Nominal diameter (inches)	Nominal weight (lb/lin. ft)	Area in sq. in. per ft of width for various spacings						
Smooth	Deformed			Center-to-center spacing						
				2"	3"	4"	6"	8"	10"	12"
W31	D31	0.628	1.054		1.24	0.93	0.62	0.465	0.372	0.31
W28	D28	0.597	0.952		1.12	0.84	0.56	0.42	0.336	0.28
W26	D26	0.575	0.934		1.04	0.78	0.52	0.39	0.312	0.26
W24	D24	0.553	0.816		0.96	0.72	0.48	0.36	0.288	0.24
W22	D22	0.529	0.748		0.88	0.66	0.44	0.33	0.264	0.22
W20	D20	0.505	0.680	1.20	0.80	0.60	0.40	0.30	0.24	0.20
W18	D18	0.479	0.612	1.08	0.72	0.54	0.36	0.27	0.216	0.18
W16	D16	0.451	0.544	0.96	0.64	0.48	0.32	0.24	0.192	0.16
W14	D14	0.422	0.476	0.84	0.56	0.42	0.28	0.21	0.168	0.14
W12	D12	0.391	0.408	0.72	0.48	0.36	0.24	0.18	0.144	0.12
W11	D11	0.374	0.374	0.66	0.44	0.33	0.22	0.165	0.132	0.11
W10	D10	0.357	0.340	0.60	0.40	0.30	0.20	0.15	0.12	0.10
W9.5		0.348	0.323	0.57	0.38	0.285	0.19	0.142	0.114	0.095
W9	D9	0.339	0.306	0.54	0.36	0.27	0.18	0.135	0.108	0.09
W8.5		0.329	0.289	0.51	0.34	0.255	0.17	0.127	0.102	0.085

Table A.3(B) (Continued)

Wire size number[a]		Nominal diameter (inches)	Nominal weight (lb/lin. ft)	Area in sq. in. per ft of width for various spacings						
Smooth	Deformed			Center-to-center spacing						
				2"	3"	4"	6"	8"	10"	12"
W8	D8	0.319	0.272	0.48	0.32	0.24	0.16	0.12	0.096	0.08
W7.5		0.309	0.255	0.45	0.30	0.225	0.15	0.112	0.09	0.075
W7	D7	0.299	0.238	0.42	0.28	0.21	0.14	0.105	0.084	0.07
W6.5		0.288	0.221	0.39	0.26	0.195	0.13	0.097	0.078	0.065
W6	D6	0.276	0.204	0.36	0.24	0.18	0.12	0.09	0.072	0.06
W5.5		0.265	0.187	0.33	0.22	0.165	0.11	0.082	0.066	0.055
W5	D5	0.252	0.170	0.30	0.20	0.15	0.10	0.075	0.06	0.05
W4.5		0.239	0.153	0.27	0.18	0.135	0.09	0.067	0.054	0.045
W4	D4	0.226	0.136	0.24	0.16	0.12	0.08	0.06	0.048	0.04
W3.5		0.211	0.119	0.21	0.14	0.105	0.07	0.052	0.042	0.035
W2.9		0.192	0.099	0.174	0.116	0.087	0.058	0.043	0.035	0.029
W2.5		0.178	0.085	0.15	0.10	0.075	0.05	0.037	0.03	0.025
W2		0.160	0.068	0.12	0.08	0.06	0.04	0.03	0.024	0.02
W1.4		0.134	0.048	0.084	0.056	0.042	0.028	0.021	0.017	0.014

Note: The above listing of smooth and deformed wire sizes represents wires normally selected to manufacture welded wire fabric styles to specific areas of reinforcement. Wire sizes and spacings other than those listed above may be produced provided the quantity required is sufficient to justify manufacture.

[a]The number following the prefix W or the prefix D identifies the cross-sectional area of the wire in hundredths of a square inch. The nominal diameter of a deformed wire is equivalent to the diameter of a smooth wire having the same weight per foot as the deformed wire.

Table A.4 Areas of Groups of Standard Bars (In.^{2})—U.S. Customary Units

Bar no.	Number of bars								
	2	3	4	5	6	7	8	9	10
4	0.39	0.58	0.78	0.98	1.18	1.37	1.57	1.77	1.96
5	0.61	0.91	1.23	1.53	1.84	2.15	2.45	2.76	3.07
6	0.88	1.32	1.77	2.21	2.65	3.09	3.53	3.98	4.42
7	1.20	1.80	2.41	3.01	3.61	4.21	4.81	5.41	6.01
8	1.57	2.35	3.14	3.93	4.71	5.50	6.28	7.07	7.85
9	2.00	3.00	4.00	5.00	6.00	7.00	8.00	9.00	10.00
10	2.53	3.79	5.06	6.33	7.59	8.86	10.12	11.39	12.66
11	3.12	4.68	6.25	7.81	9.37	10.94	12.50	14.06	15.62
14	4.50	6.75	9.00	11.25	13.50	15.75	18.00	20.25	22.50
18	8.00	12.00	16.00	20.00	24.00	28.00	32.00	36.00	40.00

Bar no.	Number of bars								
	11	12	13	14	15	16	17	18	19	20
4	2.16	2.36	2.55	2.75	2.95	3.14	3.34	3.53	3.73	3.93
5	3.37	3.68	3.99	4.30	4.60	4.91	5.22	5.52	5.83	6.14
6	4.86	5.30	5.74	6.19	6.63	7.07	7.51	7.95	8.39	8.84
7	6.61	7.22	7.82	8.42	9.02	9.62	10.22	10.82	11.43	12.03
8	8.64	9.43	10.21	11.00	11.78	12.57	13.35	14.14	14.92	15.71
9	11.00	12.00	13.00	14.00	15.00	16.00	17.00	18.00	19.00	20.00
10	13.92	15.19	16.45	17.72	18.98	20.25	21.52	22.78	24.05	25.31
11	17.19	18.75	20.31	21.87	23.44	25.00	26.56	28.12	29.69	31.25
14	24.75	27.00	29.25	31.50	33.75	36.00	38.25	40.50	42.75	45.00
18	44.00	48.00	52.00	56.00	60.00	64.00	68.00	72.00	76.00	80.00

Table A.5 Minimum Web Width (In.) for Beams with Inside Exposure (1995 ACI Code)[a,b,c]—U.S. Customary Units

Size of bars	Number of bars in single layer of reinforcing							Add for each additional Bar
	2	3	4	5	6	7	8	
#4	6.8	8.3	9.8	11.3	12.8	14.3	15.8	1.50
#5	6.9	8.5	10.2	11.8	13.4	15.0	16.7	1.625
#6	7.0	8.8	10.5	12.3	14.0	15.8	17.5	1.75
#7	7.2	9.0	10.9	12.8	14.7	16.5	18.4	1.875
#8	7.3	9.3	11.3	13.3	15.3	17.3	19.3	2.00
#9	7.6	9.8	12.1	14.3	16.6	18.8	21.1	2.26
#10	7.8	10.4	12.9	15.5	18.0	20.5	23.1	2.54
#11	8.1	10.9	13.8	16.6	19.4	22.2	25.0	2.82
#14	8.9	12.3	15.7	19.0	22.4	25.8	29.2	3.39
#18	10.6	15.1	19.6	24.1	28.6	33.1	37.7	4.51

[a]Minimum beam widths for beams were calculated using #3 stirrups.

[b]Maximum aggregate sizes were assumed not to exceed 3/4 of the clear spacing between the bars (ACI 3.3.2).

[c]The horizontal distance from the center of the outside longitudinal bars to the inside of the stirrups was assumed to equal the larger of 2 times the stirrup diameter (ACI 7.2.2) or half of the longitudinal bar diameter.

Table A.6 Areas of Bars in Slabs (In.^{2}/Ft)—U.S. Customary Units

Spacing (in.)	Bar no.								
	3	4	5	6	7	8	9	10	11
3	0.44	0.78	1.23	1.77	2.40	3.14	4.00	5.06	6.25
$3\frac{1}{2}$	0.38	0.67	1.05	1.51	2.06	2.69	3.43	4.34	5.36
4	0.33	0.59	0.92	1.32	1.80	2.36	3.00	3.80	4.68
$4\frac{1}{2}$	0.29	0.52	0.82	1.18	1.60	2.09	2.67	3.37	4.17
5	0.26	0.47	0.74	1.06	1.44	1.88	2.40	3.04	3.75
$5\frac{1}{2}$	0.24	0.43	0.67	0.96	1.31	1.71	2.18	2.76	3.41
6	0.22	0.39	0.61	0.88	1.20	1.57	2.00	2.53	3.12
$6\frac{1}{2}$	0.20	0.36	0.57	0.82	1.11	1.45	1.85	2.34	2.89
7	0.19	0.34	0.53	0.76	1.03	1.35	1.71	2.17	2.68
$7\frac{1}{2}$	0.18	0.31	0.49	0.71	0.96	1.26	1.60	2.02	2.50
8	0.17	0.29	0.46	0.66	0.90	1.18	1.50	1.89	2.34
9	0.15	0.26	0.41	0.59	0.80	1.05	1.33	1.69	2.08
10	0.13	0.24	0.37	0.53	0.72	0.94	1.20	1.52	1.87
12	0.11	0.20	0.31	0.44	0.60	0.78	1.00	1.27	1.56

Table A.7 Values of ρ Balanced, ρ to Achieve Various ϵ_t Values, and ρ Minimum for Flexure. All Values Are for Tensilely Reinforced Rectangular Sections

f_y	f_c'	3000 psi $\beta_1 = 0.85$	4000 psi $\beta_1 = 0.85$	5000 psi $\beta_1 = 0.80$	6000 psi $\beta_1 = 0.75$
Grade 40 40,000 psi (275.8 MPa)	ρ balanced	0.0371	0.0495	0.0582	0.0655
	ρ when $\epsilon_t = 0.004$	0.0232	0.0310	0.0364	0.0410
	ρ when $\epsilon_t = 0.005$	0.0203	0.0271	0.0319	0.0359
	ρ when $\epsilon_t = 0.0075$	0.0155	0.0206	0.0243	0.0273
	ρ min for flexure	0.0050	0.0050	0.0053	0.0058
Grade 50 50,000 psi (344.8 MPa)	ρ balanced	0.0275	0.0367	0.0432	0.0486
	ρ when $\epsilon_t = 0.004$	0.0186	0.0248	0.0291	0.0328
	ρ when $\epsilon_t = 0.005$	0.0163	0.0217	0.0255	0.0287
	ρ when $\epsilon_t = 0.0075$	0.0124	0.0165	0.0194	0.0219
	ρ min for flexure	0.0040	0.0040	0.0042	0.0046
Grade 60 60,000 psi (413.7 MPa)	ρ balanced	0.0214	0.0285	0.0335	0.0377
	ρ when $\epsilon_t = 0.004$	0.0155	0.0206	0.0243	0.0273
	ρ when $\epsilon_t = 0.005$	0.0136	0.0181	0.0212	0.0239
	ρ when $\epsilon_t = 0.0075$	0.0103	0.0138	0.0162	0.0182
	ρ min for flexure	0.0033	0.0033	0.0035	0.0039
Grade 75 75,000 psi (517.1 MPa)	ρ balanced	0.0155	0.0207	0.0243	0.0274
	ρ when $\epsilon_t = 0.004$	0.0124	0.0165	0.0194	0.0219
	ρ when $\epsilon_t = 0.005$	0.0108	0.0144	0.0170	0.0191
	ρ when $\epsilon_t = 0.0075$	0.0083	0.0110	0.0130	0.0146
	ρ min for flexure	0.0027	0.0027	0.0028	0.0031

$$\frac{M_u}{\phi b d^2} = R_n$$

Table A.8 $f_y = 40{,}000$ PSI; $f_c' = 3000$ PSI—U.S. Customary Units

ρ	$\dfrac{M_u}{\phi bd^2}$	ρ	$\dfrac{M_u}{\phi bd^2}$	ρ	$\dfrac{M_u}{\phi bd^2}$	ρ	$\dfrac{M_u}{\phi bd^2}$
$\rho_{\min}$ for temp. and shrinkage 0.0020	78.74	0.0061	232.3	0.0102	375.3	0.0143	507.6
0.0021	82.62	0.0062	235.9	0.0103	378.6	0.0144	510.7
0.0022	86.48	0.0063	239.5	0.0104	382.0	0.0145	513.8
0.0023	90.34	0.0064	243.1	0.0105	385.3	0.0146	516.9
0.0024	94.19	0.0065	246.7	0.0106	388.6	0.0147	520.0
0.0025	98.04	0.0066	250.3	0.0107	392.0	0.0148	523.1
0.0026	101.9	0.0067	253.9	0.0108	395.3	0.0149	526.1
0.0027	105.7	0.0068	257.4	0.0109	398.6	0.0150	529.2
0.0028	109.5	0.0069	261.0	0.0110	401.9	0.0151	532.2
0.0029	113.4	0.0070	264.6	0.0111	405.2	0.0152	535.3
0.0030	117.2	0.0071	268.1	0.0112	408.5	0.0153	538.3
0.0031	121.0	0.0072	271.7	0.0113	411.8	0.0154	541.4
0.0032	124.8	0.0073	275.2	0.0114	415.1	0.0155	544.4
0.0033	128.6	0.0074	278.8	0.0115	418.4	0.0156	547.4
0.0034	132.4	0.0075	282.3	0.0116	421.7	0.0157	550.4
0.0035	136.2	0.0076	285.8	0.0117	424.9	0.0158	553.4
0.0036	139.9	0.0077	289.3	0.0118	428.2	0.0159	556.4
0.0037	143.7	0.0078	292.9	0.0119	431.4	0.0160	559.4
0.0038	147.5	0.0079	296.4	0.0120	434.7	0.0161	562.4
0.0039	151.2	0.0080	299.9	0.0121	437.9	0.0162	565.4
0.0040	155.0	0.0081	303.4	0.0122	441.2	0.0163	568.4
0.0041	158.7	0.0082	306.8	0.0123	444.4	0.0164	571.4
0.0042	162.5	0.0083	310.3	0.0124	447.6	0.0165	574.3
0.0043	166.2	0.0084	313.8	0.0125	450.8	0.0166	577.3
0.0044	169.9	0.0085	317.3	0.0126	454.0	0.0167	580.2
0.0045	173.6	0.0086	320.7	0.0127	457.2	0.0168	583.2
0.0046	177.4	0.0087	324.2	0.0128	460.4	0.0169	586.1
0.0047	181.1	0.0088	327.6	0.0129	463.6	0.0170	589.1
0.0048	184.8	0.0089	331.1	0.0130	466.8	0.0171	592.0
0.0049	188.5	0.0090	334.5	0.0131	470.0	0.0172	594.9
$\rho_{\min}$ for flexure 0.0050	192.1	0.0091	337.9	0.0132	473.2	0.0173	597.8
0.0051	195.8	0.0092	341.4	0.0133	476.3	0.0174	600.7
0.0052	199.5	0.0093	344.8	0.0134	479.5	0.0175	603.6
0.0053	203.2	0.0094	348.2	0.0135	482.6	0.0176	606.5
0.0054	206.8	0.0095	351.6	0.0136	485.8	0.0177	609.4
0.0055	210.5	0.0096	355.0	0.0137	488.9	0.0178	612.3
0.0056	214.1	0.0097	358.4	0.0138	492.1	0.0179	615.2
0.0057	217.8	0.0098	361.8	0.0139	495.2	0.0180	618.0
0.0058	221.4	0.0099	365.2	0.0140	498.3	0.0181	620.9
0.0059	225.0	0.0100	368.5	0.0141	501.4	0.0182	623.8
0.0060	228.7	0.0101	371.9	0.0142	504.5	0.0183	626.6

Table A.8 (Continued)

ρ	$\dfrac{M_u}{\phi bd^2}$	ρ	$\dfrac{M_u}{\phi bd^2}$	ρ	$\dfrac{M_u}{\phi bd^2}$	ρ	$\dfrac{M_u}{\phi bd^2}$
0.0184	629.5	0.0189	643.6	0.0194	657.6	0.0199	671.4
0.0185	632.3	0.0190	646.4	0.0195	660.3	0.0200	674.1
0.0186	635.1	0.0191	649.2	0.0196	663.1	0.0201	676.9
0.0187	638.0	0.0192	652.0	0.0197	665.9	0.0202	679.6
0.0188	640.8	0.0193	654.8	0.0198	668.6	0.0203	682.3

Table A.9 $f_y = 40{,}000$ PSI; $f'_c = 4000$ PSI—U.S. Customary Units

ρ	$\dfrac{M_u}{\phi bd^2}$	ρ	$\dfrac{M_u}{\phi bd^2}$	ρ	$\dfrac{M_u}{\phi bd^2}$	ρ	$\dfrac{M_u}{\phi bd^2}$
ρ_{min} for temp. and shrinkage 0.0020	79.06	0.0046	179.0	0.0072	275.8	0.0098	369.3
0.0021	82.96	0.0047	182.8	0.0073	279.4	0.0099	372.9
0.0022	86.86	0.0048	186.6	0.0074	283.1	0.0100	376.4
0.0023	90.75	0.0049	190.4	0.0075	286.7	0.0101	379.9
0.0024	94.64	ρ_{min} for flexure 0.0050	194.1	0.0076	290.4	0.0102	383.4
0.0025	98.53	0.0051	197.9	0.0077	294.0	0.0103	387.0
0.0026	102.4	0.0052	201.6	0.0078	297.6	0.0104	390.5
0.0027	106.3	0.0053	205.4	0.0079	301.3	0.0105	394.0
0.0028	110.2	0.0054	209.1	0.0080	304.9	0.0106	397.5
0.0029	114.0	0.0055	212.9	0.0081	308.5	0.0107	401.0
0.0030	117.9	0.0056	216.6	0.0082	312.1	0.0108	404.5
0.0031	121.7	0.0057	220.3	0.0083	315.7	0.0109	408.0
0.0032	125.6	0.0058	224.1	0.0084	319.3	0.0110	411.4
0.0033	129.4	0.0059	227.8	0.0085	322.9	0.0111	414.9
0.0034	133.3	0.0060	231.5	0.0086	326.5	0.0112	418.4
0.0035	137.1	0.0061	235.2	0.0087	330.1	0.0113	421.9
0.0036	141.0	0.0062	238.9	0.0088	333.7	0.0114	425.3
0.0037	144.8	0.0063	242.6	0.0089	337.3	0.0115	428.8
0.0038	148.6	0.0064	246.3	0.0090	340.9	0.0116	432.2
0.0039	152.4	0.0065	250.0	0.0091	344.5	0.0117	435.7
0.0040	156.2	0.0066	253.7	0.0092	348.0	0.0118	439.1
0.0041	160.0	0.0067	257.4	0.0093	351.6	0.0119	442.6
0.0042	163.8	0.0068	261.1	0.0094	355.1	0.0120	446.0
0.0043	167.6	0.0069	264.8	0.0095	358.7	0.0121	449.4
0.0044	171.4	0.0070	268.4	0.0096	362.2	0.0122	452.9
0.0045	175.2	0.0071	272.1	0.0097	365.8	0.0123	456.3

Table A.9 (Continued)

ρ	$\dfrac{M_u}{\phi bd^2}$	ρ	$\dfrac{M_u}{\phi bd^2}$	ρ	$\dfrac{M_u}{\phi bd^2}$	ρ	$\dfrac{M_u}{\phi bd^2}$
0.0124	459.7	0.0161	582.8	0.0198	699.5	0.0235	809.7
0.0125	463.1	0.0162	586.1	0.0199	702.5	0.0236	812.5
0.0126	466.5	0.0163	589.3	0.0200	705.6	0.0237	815.4
0.0127	469.9	0.0164	592.5	0.0201	708.6	0.0238	818.3
0.0128	473.3	0.0165	595.7	0.0202	711.7	0.0239	821.2
0.0129	476.7	0.0166	599.0	0.0203	714.7	0.0240	824.1
0.0130	480.1	0.0167	602.2	0.0204	717.8	0.0241	826.9
0.0131	483.5	0.0168	605.4	0.0205	720.8	0.0242	829.8
0.0132	486.9	0.0169	608.6	0.0206	723.8	0.0243	832.6
0.0133	490.2	0.0170	611.8	0.0207	726.9	0.0244	835.5
0.0134	493.6	0.0171	615.0	0.0208	729.9	0.0245	838.3
0.0135	497.0	0.0172	618.2	0.0209	732.9	0.0246	841.2
0.0136	500.3	0.0173	621.4	0.0210	735.9	0.0247	844.0
0.0137	503.7	0.0174	624.5	0.0211	738.9	0.0248	846.8
0.0138	507.0	0.0175	627.7	0.0212	741.9	0.0249	849.7
0.0139	510.4	0.0176	630.9	0.0213	744.9	0.0250	852.5
0.0140	513.7	0.0177	634.1	0.0214	747.9	0.0251	855.3
0.0141	517.1	0.0178	637.2	0.0215	750.9	0.0252	858.1
0.0142	520.4	0.0179	640.4	0.0216	753.9	0.0253	860.9
0.0143	523.7	0.0180	643.5	0.0217	756.9	0.0254	863.7
0.0144	527.1	0.0181	646.7	0.0218	759.8	0.0255	866.5
0.0145	530.4	0.0182	649.8	0.0219	762.8	0.0256	869.3
0.0146	533.7	0.0183	653.0	0.0220	765.8	0.0257	872.1
0.0147	537.0	0.0184	656.1	0.0221	768.7	0.0258	874.9
0.0148	540.3	0.0185	659.2	0.0222	771.7	0.0259	877.7
0.0149	543.6	0.0186	662.3	0.0223	774.6	0.0260	880.5
0.0150	546.9	0.0187	665.5	0.0224	777.6	0.0261	883.2
0.0151	550.2	0.0188	668.6	0.0225	780.5	0.0262	886.0
0.0152	553.5	0.0189	671.7	0.0226	783.4	0.0263	888.7
0.0153	556.7	0.0190	674.8	0.0227	786.4	0.0264	891.5
0.0154	560.0	0.0191	677.9	0.0228	789.3	0.0265	894.3
0.0155	563.3	0.0192	681.0	0.0229	792.2	0.0266	897.0
0.0156	566.6	0.0193	684.1	0.0230	795.1	0.0267	899.7
0.0157	569.8	0.0194	687.2	0.0231	798.1	0.0268	902.5
0.0158	573.1	0.0195	690.3	0.0232	801.0	0.0269	905.2
0.0159	576.3	0.0196	693.3	0.0233	803.9	0.0270	907.9
0.0160	579.6	0.0197	696.4	0.0234	806.8	0.0271	910.7

Table A.10 $f_y = 50,000$ PSI; $f'_c = 3000$ PSI—U.S. Customary Units

ρ	$\dfrac{M_u}{\phi bd^2}$	ρ	$\dfrac{M_u}{\phi bd^2}$	ρ	$\dfrac{M_u}{\phi bd^2}$	ρ	$\dfrac{M_u}{\phi bd^2}$
ρ_{min} for 0.0020	98.04	0.0056	264.6	0.0092	418.4	0.0128	559.4
temp. and 0.0021	102.8	0.0057	269.0	0.0093	422.5	0.0129	563.2
shrinkage 0.0022	107.6	0.0058	273.5	0.0094	426.6	0.0130	566.9
0.0023	112.4	0.0059	277.9	0.0095	430.6	0.0131	570.6
0.0024	117.2	0.0060	282.3	0.0096	434.7	0.0132	574.3
0.0025	121.9	0.0061	286.7	0.0097	438.7	0.0133	578.0
0.0026	126.7	0.0062	291.1	0.0098	442.8	0.0134	581.7
0.0027	131.4	0.0063	295.5	0.0099	446.8	0.0135	585.4
0.0028	136.2	0.0064	299.9	0.0100	450.8	0.0136	589.1
0.0029	140.9	0.0065	304.2	0.0101	454.8	0.0137	592.7
0.0030	145.6	0.0066	308.6	0.0102	458.8	0.0138	596.4
0.0031	150.3	0.0067	312.9	0.0103	462.8	0.0139	600.0
0.0032	155.0	0.0068	317.3	0.0104	466.8	0.0140	603.6
0.0033	159.7	0.0069	321.6	0.0105	470.8	0.0141	607.2
0.0034	164.3	0.0070	325.9	0.0106	474.8	0.0142	610.9
0.0035	169.0	0.0071	330.2	0.0107	478.7	0.0143	614.5
0.0036	173.6	0.0072	334.5	0.0108	482.6	0.0144	618.0
0.0037	178.3	0.0073	338.8	0.0109	486.6	0.0145	621.6
0.0038	182.9	0.0074	343.1	0.0110	490.5	0.0146	625.2
ρ_{min} for 0.0039	187.5	0.0075	347.3	0.0111	494.4	0.0147	628.7
flexure 0.0040	192.1	0.0076	351.6	0.0112	498.3	0.0148	632.3
0.0041	196.7	0.0077	355.8	0.0113	502.2	0.0149	635.8
0.0042	201.3	0.0078	360.1	0.0114	506.1	0.0150	639.4
0.0043	205.9	0.0079	364.3	0.0115	510.0	0.0151	642.9
0.0044	210.5	0.0080	368.5	0.0116	513.8	0.0152	646.4
0.0045	215.0	0.0081	372.7	0.0117	517.7	0.0153	649.9
0.0046	219.6	0.0082	376.9	0.0118	521.5	0.0154	653.4
0.0047	224.1	0.0083	381.1	0.0119	525.4	0.0155	656.9
0.0048	228.7	0.0084	385.3	0.0120	529.2	0.0156	660.3
0.0049	233.2	0.0085	389.5	0.0121	533.0	0.0157	663.8
0.0050	237.7	0.0086	393.6	0.0122	536.8	0.0158	667.3
0.0051	242.2	0.0087	397.8	0.0123	540.6	0.0159	670.7
0.0052	246.7	0.0088	401.9	0.0124	544.4	0.0160	674.1
0.0053	251.2	0.0089	406.1	0.0125	548.2	0.0161	677.5
0.0054	255.7	0.0090	410.2	0.0126	551.9	0.0162	681.0
0.0055	260.1	0.0091	414.3	0.0127	555.7	0.0163	684.4

Table A.11 $f_y = 50{,}000$ PSI; $f'_c = 4000$ PSI—U.S. Customary Units

ρ	$\dfrac{M_u}{\phi bd^2}$	ρ	$\dfrac{M_u}{\phi bd^2}$	ρ	$\dfrac{M_u}{\phi bd^2}$	ρ	$\dfrac{M_u}{\phi bd^2}$
ρ_{min} for 0.0020	98.53	0.0061	291.3	0.0102	471.6	0.0143	639.6
temp. and 0.0021	103.4	0.0062	295.8	0.0103	475.9	0.0144	643.5
shrinkage 0.0022	108.2	0.0063	300.4	0.0104	480.1	0.0145	647.5
0.0023	113.0	0.0064	304.9	0.0105	484.3	0.0146	651.4
0.0024	117.9	0.0065	309.4	0.0106	488.6	0.0147	655.3
0.0025	122.7	0.0066	313.9	0.0107	492.8	0.0148	659.2
0.0026	127.5	0.0067	318.4	0.0108	497.0	0.0149	663.1
0.0027	132.3	0.0068	322.9	0.0109	501.2	0.0150	667.0
0.0028	137.1	0.0069	327.4	0.0110	505.4	0.0151	670.9
0.0029	141.9	0.0070	331.9	0.0111	509.6	0.0152	674.8
0.0030	146.7	0.0071	336.4	0.0112	513.7	0.0153	678.7
0.0031	151.5	0.0072	340.9	0.0113	517.9	0.0154	682.5
0.0032	156.2	0.0073	345.3	0.0114	522.1	0.0155	686.4
0.0033	161.0	0.0074	349.8	0.0115	526.2	0.0156	690.3
0.0034	165.7	0.0075	354.3	0.0116	530.4	0.0157	694.1
0.0035	170.5	0.0076	358.7	0.0117	534.5	0.0158	697.9
0.0036	175.2	0.0077	363.1	0.0118	538.6	0.0159	701.8
0.0037	180.0	0.0078	367.6	0.0119	542.8	0.0160	705.6
0.0038	184.7	0.0079	372.0	0.0120	546.9	0.0161	709.4
ρ_{min} for 0.0039	189.4	0.0080	376.4	0.0121	551.0	0.0162	713.2
flexure 0.0040	194.1	0.0081	380.8	0.0122	555.1	0.0163	717.0
0.0041	198.8	0.0082	385.2	0.0123	559.2	0.0164	720.8
0.0042	203.5	0.0083	389.6	0.0124	563.3	0.0165	724.6
0.0043	208.2	0.0084	394.0	0.0125	567.4	0.0166	728.4
0.0044	212.9	0.0085	398.4	0.0126	571.4	0.0167	732.1
0.0045	217.5	0.0086	402.7	0.0127	575.5	0.0168	735.9
0.0046	222.2	0.0087	407.1	0.0128	579.6	0.0169	327.4
0.0047	226.9	0.0088	411.4	0.0129	583.6	0.0170	743.4
0.0048	231.5	0.0089	415.8	0.0130	587.7	0.0171	747.2
0.0049	236.1	0.0090	420.1	0.0131	591.7	0.0172	750.9
0.0050	240.8	0.0091	424.5	0.0132	595.7	0.0173	754.6
0.0051	245.4	0.0092	428.8	0.0133	599.8	0.0174	758.3
0.0052	250.0	0.0093	433.1	0.0134	603.8	0.0175	762.1
0.0053	254.6	0.0094	437.4	0.0135	607.8	0.0176	765.8
0.0054	259.2	0.0095	441.7	0.0136	611.8	0.0177	769.5
0.0055	263.8	0.0096	446.0	0.0137	615.8	0.0178	773.2
0.0056	268.4	0.0097	450.3	0.0138	619.8	0.0179	776.8
0.0057	273.0	0.0098	454.6	0.0139	623.7	0.0180	780.5
0.0058	277.6	0.0099	458.9	0.0140	627.7	0.0181	784.2
0.0059	282.2	0.0100	463.1	0.0141	631.7	0.0182	787.8
0.0060	286.7	0.0101	467.4	0.0142	635.6	0.0183	791.5

Table A.11 (Continued)

ρ	$\dfrac{M_u}{\phi bd^2}$	ρ	$\dfrac{M_u}{\phi bd^2}$	ρ	$\dfrac{M_u}{\phi bd^2}$	ρ	$\dfrac{M_u}{\phi bd^2}$
0.0184	795.1	0.0193	827.6	0.0202	859.5	0.0210	887.4
0.0185	798.8	0.0194	831.2	0.0203	863.0	0.0211	890.8
0.0186	802.4	0.0195	834.8	0.0204	866.5	0.0212	894.3
0.0187	806.0	0.0196	838.3	0.0205	870.0	0.0213	897.7
0.0188	809.7	0.0197	841.9	0.0206	873.5	0.0214	901.1
0.0189	813.3	0.0198	845.4	0.0207	877.0	0.0215	904.5
0.0190	816.9	0.0199	849.0	0.0208	880.5	0.0216	907.9
0.0191	820.5	0.0200	852.5	0.0209	883.9	0.0217	911.3
0.0192	824.1	0.0201	856.0				

Table A.12 $f_y = 60{,}000$ PSI; $f_c' = 3000$ PSI—U.S. Customary Units

ρ	$\dfrac{M_u}{\phi bd^2}$	ρ	$\dfrac{M_u}{\phi bd^2}$	ρ	$\dfrac{M_u}{\phi bd^2}$	ρ	$\dfrac{M_u}{\phi bd^2}$
ρ_{min} for 0.0018	105.7	0.0039	223.2	0.0060	334.5	0.0081	439.5
temp. and 0.0019	111.5	0.0040	228.7	0.0061	339.7	0.0082	444.4
shrinkage 0.0020	117.2	0.0041	234.1	0.0062	344.8	0.0083	449.2
0.0021	122.9	0.0042	239.5	0.0063	349.9	0.0084	454.0
0.0022	128.6	0.0043	244.9	0.0064	355.0	0.0085	458.8
0.0023	134.3	0.0044	250.3	0.0065	360.1	0.0086	463.6
0.0024	139.9	0.0045	255.7	0.0066	365.2	0.0087	468.4
0.0025	145.6	0.0046	261.0	0.0067	370.2	0.0088	473.2
0.0026	151.2	0.0047	266.4	0.0068	375.3	0.0089	477.9
0.0027	156.9	0.0048	271.7	0.0069	380.3	0.0090	482.6
0.0028	162.5	0.0049	277.0	0.0070	385.3	0.0091	487.4
0.0029	168.1	0.0050	282.3	0.0071	390.3	0.0092	492.1
0.0030	173.7	0.0051	287.6	0.0072	395.3	0.0093	496.8
0.0031	179.2	0.0052	292.9	0.0073	400.3	0.0094	501.4
0.0032	184.8	0.0053	298.1	0.0074	405.2	0.0095	506.1
ρ_{min} for 0.0033	190.3	0.0054	303.4	0.0075	410.2	0.0096	510.7
flexure 0.0034	195.8	0.0055	308.6	0.0076	415.1	0.0097	515.4
0.0035	201.3	0.0056	313.8	0.0077	420.0	0.0098	520.0
0.0036	206.8	0.0057	319.0	0.0078	424.9	0.0099	524.6
0.0037	212.3	0.0058	324.2	0.0079	429.8	0.0100	529.2
0.0038	217.8	0.0059	329.4	0.0080	434.7	0.0101	533.8

Table A.12 (Continued)

ρ	$\dfrac{M_u}{\phi bd^2}$	ρ	$\dfrac{M_u}{\phi bd^2}$	ρ	$\dfrac{M_u}{\phi bd^2}$	ρ	$\dfrac{M_u}{\phi bd^2}$
0.0102	538.3	0.0111	578.8	0.0120	618.0	0.0129	656.2
0.0103	542.9	0.0112	582.3	0.0121	622.3	0.0130	660.9
0.0104	547.4	0.0113	587.6	0.0122	626.6	0.0131	664.5
0.0105	551.9	0.0114	592.0	0.0123	630.9	0.0132	668.6
0.0106	556.4	0.0115	596.4	0.0124	635.1	0.0133	672.8
0.0107	560.9	0.0116	600.7	0.0125	639.4	0.0134	676.9
0.0108	565.4	0.0117	605.1	0.0126	643.6	0.0135	681.0
0.0109	569.9	0.0118	609.4	0.0127	647.8	0.0136	685.0
0.0110	574.3	0.0119	613.7	0.0128	652.0		

Table A.13 $f_y = 60{,}000$ PSI; $f_c' = 4000$ PSI—U.S. Customary Units

ρ	$\dfrac{M_u}{\phi bd^2}$	ρ	$\dfrac{M_u}{\phi bd^2}$	ρ	$\dfrac{M_u}{\phi bd^2}$	ρ	$\dfrac{M_u}{\phi bd^2}$
ρ_{min} for temp. and shrinkage 0.0018	106.3	0.0041	237.1	0.0064	362.2	0.0087	481.8
0.0019	112.1	0.0042	242.6	0.0065	367.6	0.0088	486.9
0.0020	117.1	0.0043	248.2	0.0066	372.9	0.0089	491.9
0.0021	123.7	0.0044	253.7	0.0067	378.2	0.0090	497.0
0.0022	129.4	0.0045	259.2	0.0068	383.4	0.0091	502.0
0.0023	135.2	0.0046	264.8	0.0069	388.7	0.0092	507.1
0.0024	141.0	0.0047	270.3	0.0070	394.0	0.0093	512.1
0.0025	146.7	0.0048	275.8	0.0071	399.2	0.0094	517.1
0.0026	152.4	0.0049	281.2	0.0072	404.5	0.0095	522.1
0.0027	158.1	0.0050	286.7	0.0073	409.7	0.0096	527.1
0.0028	163.8	0.0051	292.2	0.0074	414.9	0.0097	532.0
0.0029	169.5	0.0052	297.6	0.0075	420.1	0.0098	537.0
0.0030	175.2	0.0053	303.1	0.0076	425.3	0.0099	542.0
0.0031	180.9	0.0054	308.5	0.0077	430.5	0.0100	546.9
0.0032	186.6	0.0055	313.9	0.0078	435.7	0.0101	551.8
ρ_{min} for flexure 0.0033	192.2	0.0056	319.3	0.0079	440.9	0.0102	556.7
0.0034	197.9	0.0057	324.7	0.0080	446.0	0.0103	561.7
0.0035	203.5	0.0058	330.1	0.0081	451.2	0.0104	566.6
0.0036	209.1	0.0059	335.5	0.0082	456.3	0.0105	571.5
0.0037	214.7	0.0060	340.9	0.0083	461.4	0.0106	576.3
0.0038	220.3	0.0061	346.2	0.0084	466.5	0.0107	581.2
0.0039	225.9	0.0062	351.6	0.0085	471.6	0.0108	586.1
0.0040	231.5	0.0063	356.9	0.0086	476.7	0.0109	590.9

Table A.13 (Continued)

ρ	$\dfrac{M_u}{\phi bd^2}$	ρ	$\dfrac{M_u}{\phi bd^2}$	ρ	$\dfrac{M_u}{\phi bd^2}$	ρ	$\dfrac{M_u}{\phi bd^2}$
0.0110	595.7	0.0128	681.0	0.0146	762.8	0.0164	841.2
0.0111	600.6	0.0129	685.6	0.0147	767.2	0.0165	845.4
0.0112	605.4	0.0130	690.3	0.0148	771.7	0.0166	849.7
0.0113	610.2	0.0131	694.9	0.0149	776.1	0.0167	853.9
0.0114	615.0	0.0132	699.5	0.0150	780.5	0.0168	858.1
0.0115	619.8	0.0133	704.1	0.0151	784.9	0.0169	862.3
0.0116	624.5	0.0134	708.6	0.0152	789.3	0.0170	866.5
0.0117	629.3	0.0135	713.2	0.0153	793.7	0.0171	870.7
0.0118	634.1	0.0136	717.8	0.0154	798.1	0.0172	874.9
0.0119	638.8	0.0137	722.3	0.0155	802.4	0.0173	879.1
0.0120	643.5	0.0138	726.9	0.0156	806.8	0.0174	883.2
0.0121	648.2	0.0139	731.4	0.0157	811.1	0.0175	887.4
0.0122	653.0	0.0140	735.9	0.0158	815.4	0.0176	891.5
0.0123	657.7	0.0141	740.4	0.0159	819.7	0.0177	895.6
0.0124	662.3	0.0142	744.9	0.0160	824.1	0.0178	899.7
0.0125	667.0	0.0143	749.4	0.0161	828.3	0.0179	903.9
0.0126	671.7	0.0144	753.9	0.0162	832.6	0.0180	907.9
0.0127	676.3	0.0145	758.3	0.0163	836.9	0.0181	912.0

Table A.14 Size and Pitch of Spirals, ACI Code—U.S. Customary Units

Diameter of column (in.)	Out to out of spiral (in.)	f'_c			
		2500	3000	4000	5000
$f_y = 40{,}000$:					
14, 15	11,12	$\frac{3}{8}$–2	$\frac{3}{8}$–$1\frac{3}{4}$	$\frac{1}{2}$–$2\frac{1}{2}$	$\frac{1}{2}$–$1\frac{3}{4}$
16	13	$\frac{3}{8}$–2	$\frac{3}{8}$–$1\frac{3}{4}$	$\frac{1}{2}$–$2\frac{1}{2}$	$\frac{1}{2}$–2
17–19	14–16	$\frac{3}{8}$–$2\frac{1}{4}$	$\frac{3}{8}$–$1\frac{3}{4}$	$\frac{1}{2}$–$2\frac{1}{2}$	$\frac{1}{2}$–2
20–23	17–20	$\frac{3}{8}$–$2\frac{1}{4}$	$\frac{3}{8}$–$1\frac{3}{4}$	$\frac{1}{2}$–$2\frac{1}{2}$	$\frac{1}{2}$–2
24–30	21–27	$\frac{3}{8}$–$2\frac{1}{4}$	$\frac{3}{8}$–2	$\frac{1}{2}$–$2\frac{1}{2}$	$\frac{1}{2}$–2
$f_y = 60{,}000$:					
14, 15	11, 12	$\frac{1}{4}$–$1\frac{3}{4}$	$\frac{3}{8}$–$2\frac{3}{4}$	$\frac{3}{8}$–2	$\frac{1}{2}$–$2\frac{3}{4}$
16–23	13–20	$\frac{1}{4}$–$1\frac{3}{4}$	$\frac{3}{8}$–$2\frac{3}{4}$	$\frac{3}{8}$–2	$\frac{1}{2}$–3
24–29	21–26	$\frac{1}{4}$–$1\frac{3}{4}$	$\frac{3}{8}$–3	$\frac{3}{8}$–$2\frac{1}{4}$	$\frac{1}{2}$–3
30	17	$\frac{1}{4}$–$1\frac{3}{4}$	$\frac{3}{8}$–3	$\frac{3}{8}$–$2\frac{1}{4}$	$\frac{1}{2}$–$3\frac{1}{4}$

Table A.15 Weights, Areas, and Moments of Inertia of Circular Columns and Moments of Inertia of Column Verticals Arranged in a Circle 5 In. Less Than the Diameter of Column: U.S. Customary Units

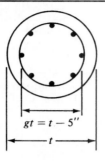

Diameter of column h (in.)	Weight per foot (lb)	Area (in.^{2})	I (in.^{4})	A_s, where $\rho_g = 0.01$*	I_s (in.^{4})[†]
12	118	113	1,018	1.13	6.92
13	138	133	1,402	1.33	10.64
14	160	154	1,886	1.54	15.59
15	184	177	2,485	1.77	22.13
16	210	201	3,217	2.01	30.40
17	237	227	4,100	2.27	40.86
18	265	255	5,153	2.55	53.87
19	295	284	6,397	2.84	69.58
20	327	314	7,854	3.14	88.31
21	361	346	9,547	3.46	110.7
22	396	380	11,500	3.80	137.2
23	433	416	13,740	4.16	168.4
24	471	452	16,290	4.52	203.9
25	511	491	19,170	4.91	245.5
26	553	531	22,430	5.31	292.7
27	597	573	26,090	5.73	346.7
28	642	616	30,170	6.16	407.3
29	688	661	34,720	6.61	475.9
30	736	707	39,760	7.07	552.3

*For other values of ρ_g, multiply the value by 100 ρ_g.

[†]The bars are assumed transformed into a thin-walled cylinder having the same sectional area as the bars. Then $I_s = A_s(\gamma t)^2/8$.

Table A.16 Moment Distribution Constants for Slabs Without Drop Panels[a]

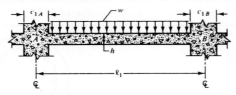

Column dimension		Uniform load FEM = Coef. $(w\ell_2\ell_1^2)$		Stiffness factor[†]		Carryover factor	
$\dfrac{c_{1A}}{\ell_1}$	$\dfrac{c_{1B}}{\ell_1}$	M_{AB}	M_{BA}	k_{AB}	k_{BA}	COF_{AB}	COF_{BA}
	0.00	0.083	0.083	4.00	4.00	0.500	0.500
	0.05	0.083	0.084	4.01	4.04	0.504	0.500
	0.10	0.082	0.086	4.03	4.15	0.513	0.499
	0.15	0.081	0.089	4.07	4.32	0.528	0.498
0.00	0.20	0.079	0.093	4.12	4.56	0.548	0.495
	0.25	0.077	0.097	4.18	4.88	0.573	0.491
	0.30	0.075	0.102	4.25	5.28	0.603	0.485
	0.35	0.073	0.107	4.33	5.78	0.638	0.478
	0.05	0.084	0.084	4.05	4.05	0.503	0.503
	0.10	0.083	0.086	4.07	4.15	0.513	0.503
	0.15	0.081	0.089	4.11	4.33	0.528	0.501
0.05	0.20	0.080	0.092	4.16	4.58	0.548	0.499
	0.25	0.078	0.096	4.22	4.89	0.573	0.494
	0.30	0.076	0.101	4.29	5.30	0.603	0.489
	0.35	0.074	0.107	4.37	5.80	0.638	0.481
	0.10	0.085	0.085	4.18	4.18	0.513	0.513
	0.15	0.083	0.088	4.22	4.36	0.528	0.511
0.10	0.20	0.082	0.091	4.27	4.61	0.548	0.508
	0.25	0.080	0.095	4.34	4.93	0.573	0.504
	0.30	0.078	0.100	4.41	5.34	0.602	0.498
	0.35	0.075	0.105	4.50	5.85	0.637	0.491
	0.15	0.086	0.086	4.40	4.40	0.526	0.526
	0.20	0.084	0.090	4.46	4.65	0.546	0.523
0.15	0.25	0.083	0.094	4.53	4.98	0.571	0.519
	0.30	0.080	0.099	4.61	5.40	0.601	0.513
	0.35	0.078	0.104	4.70	5.92	0.635	0.505
	0.20	0.088	0.088	4.72	4.72	0.543	0.543

Table A.16 (Continued)

Column dimension		Uniform load FEM = Coef. $(w\ell_2\ell_1^2)$		Stiffness factor[†]		Carryover factor	
$\dfrac{c_{1A}}{\ell_1}$	$\dfrac{c_{1B}}{\ell_1}$	M_{AB}	M_{BA}	k_{AB}	k_{BA}	COF_{AB}	COF_{BA}
0.20	0.25	0.086	0.092	4.79	5.05	0.568	0.539
	0.30	0.083	0.097	4.88	5.48	0.597	0.532
	0.35	0.081	0.102	4.99	6.01	0.632	0.524
	0.25	0.090	0.090	5.14	5.14	0.563	0.563
0.25	0.30	0.088	0.095	5.24	5.58	0.592	0.556
	0.35	0.085	0.100	5.36	6.12	0.626	0.548
0.30	0.30	0.092	0.092	5.69	5.69	0.585	0.585
	0.35	0.090	0.097	5.83	6.26	0.619	0.576
0.35	0.35	0.095	0.095	6.42	6.42	0.609	0.609

[a]Applicable when $c_1/\ell_1 = c_2/\ell_2$. For other relationships between these ratios, the constants will be slightly in error.

[†]Stiffness is $K_{AB} = k_{AB}E\dfrac{\ell_2 h^3}{12\ell_1}$ and $K_{BA} = k_{BA}E\dfrac{\ell_2 h^3}{12\ell_1}$

Table A.17 Moment Distribution Constants for Slabs with Drop Panels[a]

Column dimension		Uniform load FEM = Coef. $(w\ell_2\ell_1^2)$		Stiffness factor[†]		Carryover factor	
$\dfrac{c_{1A}}{\ell_1}$	$\dfrac{c_{1B}}{\ell_1}$	M_{AB}	M_{BA}	k_{AB}	k_{BA}	COF_{AB}	COF_{BA}
	0.00	0.088	0.088	4.78	4.78	0.541	0.541
	0.05	0.087	0.089	4.80	4.82	0.545	0.541
	0.10	0.087	0.090	4.83	4.94	0.553	0.541
0.00	0.15	0.085	0.093	4.87	5.12	0.567	0.540
	0.20	0.084	0.096	4.93	5.36	0.585	0.537
	0.25	0.082	0.100	5.00	5.68	0.606	0.534
	0.30	0.080	0.105	5.09	6.07	0.631	0.529
	0.05	0.088	0.088	4.84	4.84	0.545	0.545
	0.10	0.087	0.090	4.87	4.95	0.553	0.544
	0.15	0.085	0.093	4.91	5.13	0.567	0.543
0.05	0.20	0.084	0.096	4.97	5.38	0.584	0.541
	0.25	0.082	0.100	5.05	5.70	0.606	0.537
	0.30	0.080	0.104	5.13	6.09	0.632	0.532
	0.10	0.089	0.089	4.98	4.98	0.553	0.553
	0.15	0.088	0.092	5.03	5.16	0.566	0.551
0.10	0.20	0.086	0.094	5.09	5.42	0.584	0.549
	0.25	0.084	0.099	5.17	5.74	0.606	0.546
	0.30	0.082	0.103	5.26	6.13	0.631	0.541
	0.15	0.090	0.090	5.22	5.22	0.565	0.565
	0.20	0.089	0.094	5.28	5.47	0.583	0.563
0.15	0.25	0.087	0.097	5.37	5.80	0.604	0.559
	0.30	0.085	0.102	5.46	6.21	0.630	0.554
	0.20	0.092	0.092	5.55	5.55	0.580	0.580
0.20	0.25	0.090	0.096	5.64	5.88	0.602	0.577
	0.30	0.088	0.100	5.74	6.30	0.627	0.571
0.25	0.25	0.094	0.094	5.98	5.98	0.598	0.598
	0.30	0.091	0.098	6.10	6.41	0.622	0.593
0.30	0.30	0.095	0.095	6.54	6.54	0.617	0.617

[a]Applicable when $c_1/\ell_1 = c_2/\ell_2$. For other relationships between these ratios, the constants will be slightly in error.

[†]Stiffness is $K_{AB} = k_{AB}E\dfrac{\ell_2 h^3}{12\ell_1}$ and $K_{BA} = k_{BA}E\dfrac{\ell_2 h^3}{12\ell_1}$

Table A.18 Moment Distribution Constants for Slab-Beam Members with Column Capitals

FEM (uniform load w) $= Mw\ell_2(\ell_1)^2$
K (stiffness) $= kE\ell_2h^3/12\ell_1$
Carryover factor $= C$

c_1/ℓ_1	c_1/ℓ_2	M	k	C	c_1/ℓ_1	c_2/ℓ_2	M	k	C
0.00	0.00	0.083	4.000	0.500	0.20	0.00	0.083	4.000	0.500
	0.05	0.083	4.000	0.500		0.05	0.085	4.170	0.511
	0.10	0.083	4.000	0.500		0.10	0.086	4.346	0.522
	0.15	0.083	4.000	0.500		0.15	0.087	4.529	0.532
	0.20	0.083	4.000	0.500		0.20	0.088	4.717	0.543
	0.25	0.083	4.000	0.500		0.25	0.089	4.910	0.554
	0.30	0.083	4.000	0.500		0.30	0.090	5.108	0.564
	0.35	0.083	4.000	0.500		0.35	0.091	5.308	0.574
	0.40	0.083	4.000	0.500		0.40	0.092	5.509	0.584
	0.45	0.083	4.000	0.500		0.45	0.093	5.710	0.593
	0.50	0.083	4.000	0.500		0.50	0.094	5.908	0.602
0.05	0.00	0.083	4.000	0.500	0.25	0.00	0.083	4.000	0.500
	0.05	0.084	4.047	0.503		0.05	0.085	4.204	0.512
	0.10	0.084	4.093	0.507		0.10	0.086	4.420	0.525
	0.15	0.084	4.138	0.510		0.15	0.087	4.648	0.538
	0.20	0.085	4.181	0.513		0.20	0.089	4.887	0.550
	0.25	0.085	4.222	0.516		0.25	0.090	5.138	0.563
	0.30	0.085	4.261	0.518		0.30	0.091	5.401	0.576
	0.35	0.086	4.299	0.521		0.35	0.093	5.672	0.588
	0.40	0.086	4.334	0.523		0.40	0.094	5.952	0.600
	0.45	0.086	4.368	0.526		0.45	0.095	6.238	0.612
	0.50	0.086	4.398	0.528		0.50	0.096	6.527	0.623
0.10	0.00	0.083	4.000	0.500	0.30	0.00	0.083	4.000	0.500
	0.05	0.084	4.091	0.506		0.05	0.085	4.235	0.514
	0.10	0.085	4.182	0.513		0.10	0.086	4.488	0.527
	0.15	0.085	4.272	0.519		0.15	0.088	4.760	0.542
	0.20	0.086	4.362	0.524		0.20	0.089	5.050	0.556
	0.25	0.087	4.449	0.530		0.25	0.091	5.361	0.571
	0.30	0.087	4.535	0.535		0.30	0.092	5.692	0.585
	0.35	0.088	4.618	0.540		0.35	0.094	6.044	0.600
	0.40	0.088	4.698	0.545		0.40	0.095	6.414	0.614
	0.45	0.089	4.774	0.550		0.45	0.096	6.802	0.628
	0.50	0.089	4.846	0.554		0.50	0.098	7.205	0.642
0.15	0.00	0.083	4.000	0.500	0.35	0.00	0.083	4.000	0.500
	0.05	0.084	4.132	0.509		0.05	0.085	4.264	0.514
	0.10	0.085	4.267	0.517		0.10	0.087	4.551	0.529
	0.15	0.086	4.403	0.526		0.15	0.088	4.864	0.545
	0.20	0.087	4.541	0.534		0.20	0.090	5.204	0.560
	0.25	0.088	4.680	0.543		0.25	0.091	5.575	0.576
	0.30	0.089	4.818	0.550		0.30	0.093	5.979	0.593

Table A.18 (Continued)

c_1/ℓ_1	c_1/ℓ_2	M	k	C	c_1/ℓ_1	c_1/ℓ_2	M	k	C
	0.35	0.090	4.955	0.558		0.35	0.095	6.416	0.609
	0.40	0.090	5.090	0.565		0.40	0.096	6.888	0.626
	0.45	0.091	5.222	0.572		0.45	0.098	7.395	0.642
	0.50	0.092	5.349	0.579		0.50	0.099	7.935	0.658
	0.00	0.083	4.000	0.500		0.30	0.094	6.517	0.602
	0.05	0.085	4.289	0.515		0.35	0.096	7.136	0.621
	0.10	0.087	4.607	0.530		0.40	0.098	7.836	0.642
	0.15	0.088	4.959	0.546		0.45	0.100	8.625	0.662
	0.20	0.090	5.348	0.563		0.50	0.101	9.514	0.683
0.40	0.25	0.092	5.778	0.580		0.00	0.083	4.000	0.500
	0.30	0.094	6.255	0.598		0.05	0.085	4.331	0.515
	0.35	0.095	6.782	0.617		0.10	0.087	4.703	0.530
	0.40	0.097	7.365	0.635		0.15	0.088	5.123	0.547
	0.45	0.099	8.007	0.654		0.20	0.090	5.599	0.564
	0.50	0.100	8.710	0.672	0.50	0.25	0.092	6.141	0.583
	0.00	0.083	4.000	0.500		0.30	0.094	6.760	0.603
	0.05	0.085	4.311	0.515		0.35	0.096	7.470	0.624
	0.10	0.087	4.658	0.530		0.40	0.098	8.289	0.645
	0.15	0.088	5.046	0.547		0.45	0.100	9.234	0.667
	0.20	0.090	5.480	0.564		0.50	0.102	10.329	0.690
0.45	0.25	0.092	5.967	0.583					

Table A.19 Moment Distribution Constants for Slab-Beam Members with Column Capitals and Drop Panels

FEM (uniform load w) $= Mw\ell(\ell_1^2)$

K (stiffness) $\quad\quad = kE\ell_2h^3/12\ell_1$

		Constants for $h_2 = 1.25h_1$			Constants for $h_2 = 1.5h_2$		
c_1/ℓ_1	c_2/ℓ_2	M	k	C	M	k	C
	0.00	0.088	4.795	0.542	0.093	5.837	0.589
	0.05	0.088	4.795	0.542	0.093	5.837	0.589
	0.10	0.088	4.795	0.542	0.093	5.837	0.589
0.00	0.15	0.088	4.795	0.542	0.093	5.837	0.589
	0.20	0.088	4.795	0.542	0.093	5.837	0.589
	0.25	0.088	4.795	0.542	0.093	5.837	0.589
	0.30	0.088	4.797	0.542	0.093	5.837	0.589
	0.00	0.088	4.795	0.542	0.093	5.837	0.589
	0.05	0.088	4.846	0.545	0.093	5.890	0.591
	0.10	0.089	4.896	0.548	0.093	5.942	0.594

Table A.19 (Continued)

c_1/ℓ_1	c_2/ℓ_2	Constants for $h_2 = 1.25h_1$			Constants for $h_2 = 1.5h_2$		
		M	k	C	M	k	C
0.05	0.15	0.089	4.944	0.551	0.093	5.993	0.596
	0.20	0.089	4.990	0.553	0.094	6.041	0.598
	0.25	0.089	5.035	0.556	0.094	6.087	0.600
	0.30	0.090	5.077	0.558	0.094	6.131	0.602
	0.00	0.088	4.795	0.542	0.093	5.837	0.589
	0.05	0.088	4.894	0.548	0.093	5.940	0.593
	0.10	0.089	4.992	0.553	0.094	6.042	0.598
0.10	0.15	0.090	5.039	0.559	0.094	6.142	0.602
	0.20	0.090	5.184	0.564	0.094	6.240	0.607
	0.25	0.091	5.278	0.569	0.095	6.335	0.611
	0.30	0.091	5.368	0.573	0.095	6.427	0.615
	0.00	0.088	4.795	0.542	0.093	5.837	0.589
	0.05	0.089	4.938	0.550	0.093	5.986	0.595
	0.10	0.090	5.082	0.558	0.094	6.135	0.602
0.15	0.15	0.090	5.228	0.565	0.095	6.284	0.608
	0.20	0.091	5.374	0.573	0.095	6.432	0.614
	0.25	0.092	5.520	0.580	0.096	6.579	0.620
	0.30	0.092	5.665	0.587	0.096	6.723	0.626
	0.00	0.088	4.795	0.542	0.093	5.837	0.589
	0.05	0.089	4.978	0.552	0.093	6.027	0.597
	0.10	0.090	5.167	0.562	0.094	6.221	0.605
0.20	0.15	0.091	5.361	0.571	0.095	6.418	0.613
	0.20	0.092	5.558	0.581	0.096	6.616	0.621
	0.25	0.093	5.760	0.590	0.096	6.816	0.628
	0.30	0.094	5.962	0.590	0.097	7.015	0.635
	0.00	0.088	4.795	0.542	0.093	5.837	0.589
	0.05	0.089	5.015	0.553	0.094	6.065	0.598
	0.10	0.090	5.245	0.565	0.094	6.300	0.608
0.25	0.15	0.091	5.485	0.576	0.095	6.543	0.617
	0.20	0.092	5.735	0.587	0.096	6.790	0.626
	0.25	0.094	5.994	0.598	0.097	7.043	0.635
	0.30	0.095	6.261	0.600	0.098	7.298	0.644
	0.00	0.088	4.795	0.542	0.093	5.837	0.589
	0.05	0.089	5.048	0.554	0.094	6.099	0.599
	0.10	0.090	5.317	0.567	0.095	6.372	0.610
0.30	0.15	0.092	5.601	0.580	0.096	6.657	0.620
	0.20	0.093	5.902	0.593	0.097	6.953	0.631
	0.25	0.094	6.219	0.605	0.098	7.258	0.641
	0.30	0.095	6.550	0.618	0.099	7.571	0.651

Table A20 Stiffness Factors and Carryover Factors for Columns

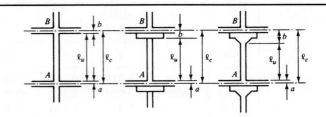

a/b	ℓ_u/ℓ_n	0.95	0.90	0.85	0.80	0.75
0.20	k_{AB}	4.32	4.70	5.33	5.65	6.27
	C_{AB}	0.57	0.64	0.71	0.80	0.89
0.40	k_{AB}	4.40	4.89	5.45	6.15	7.00
	C_{AB}	0.56	0.61	0.68	0.74	0.81
0.60	k_{AB}	4.46	5.02	5.70	6.54	7.58
	C_{AB}	0.55	0.60	0.65	0.70	0.76
0.80	k_{AB}	4.51	5.14	5.90	6.85	8.05
	C_{AB}	0.54	0.58	0.63	0.67	0.72
1.00	k_{AB}	4.55	5.23	6.06	7.11	8.44
	C_{AB}	0.54	0.57	0.61	0.65	0.68
1.20	k_{AB}	4.58	5.30	6.20	7.32	8.77
	C_{AB}	0.53	0.57	0.60	0.63	0.66
1.40	k_{AB}	4.61	5.36	6.31	7.51	9.05
	C_{AB}	0.53	0.56	0.59	0.61	0.64
1.60	k_{AB}	4.63	5.42	6.41	7.66	9.29
	C_{AB}	0.53	0.55	0.58	0.60	0.62
1.80	k_{AB}	4.65	5.46	6.49	7.80	9.50
	C_{AB}	0.53	0.55	0.57	0.59	0.60
2.00	k_{AB}	4.67	5.51	6.56	7.92	9.68
	C_{AB}	0.52	0.54	0.56	0.58	0.59

Notes: 1. Values computed by column analogy method.

2. $k_c = (k_{AB} \text{ from table})\left(\dfrac{EI_0}{\ell_n}\right)$.

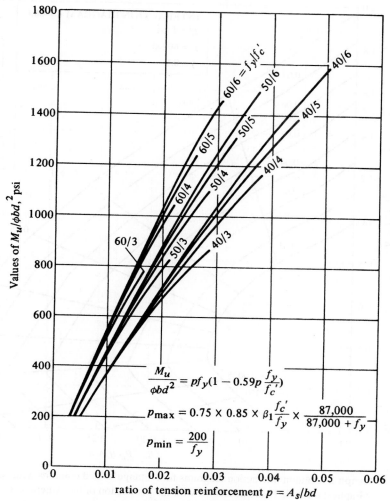

Graph 1 Moment capacity of rectangular sections. (*Note:* The upper ends of the curves shown here for 40 and 50 ksi bars correspond to ρ values for which $\epsilon_t < 0.004$ in the steel.)

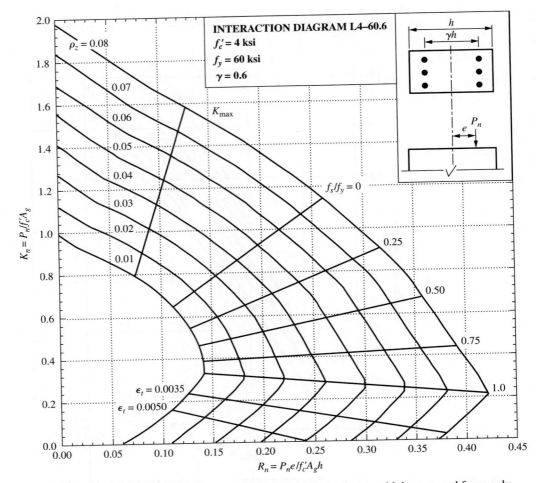

Graph 2 Column interaction diagrams for rectangular tied columns with bars on end faces only. (Graphs 2 through 13 are published with the permission of the American Concrete Institute.)

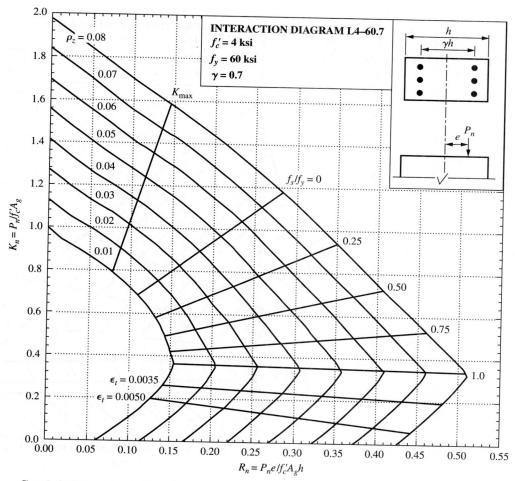

Graph 3 Column interaction diagrams for rectangular tied columns with bars on end faces only.

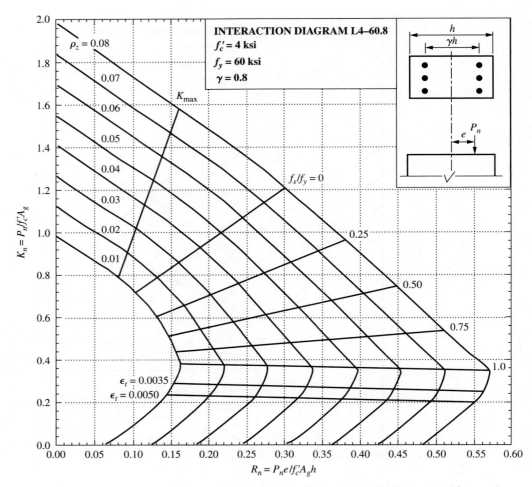

Graph 4 Column interaction diagrams for rectangular tied columns with bars on end faces only.

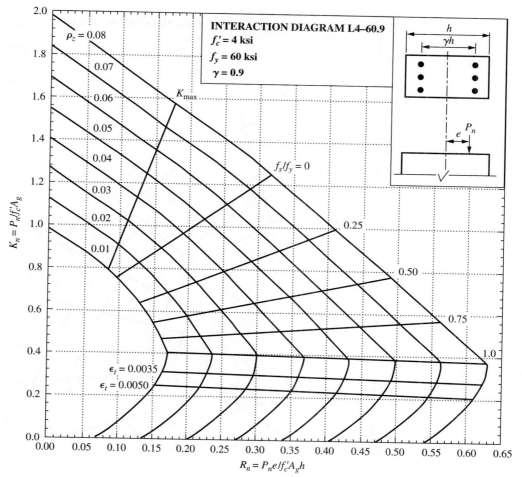

Graph 5 Column interaction diagrams for rectangular tied columns with bars on end faces only.

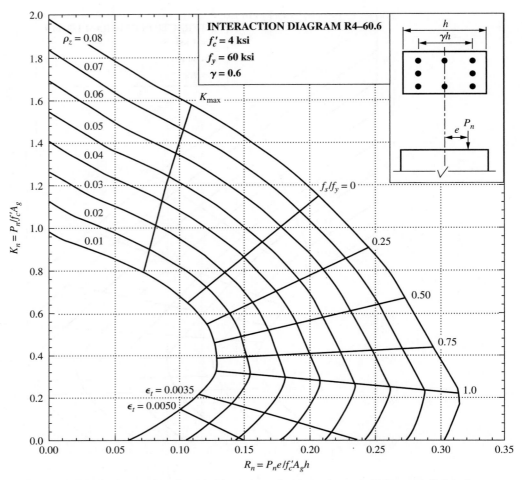

Graph 6 Column interaction diagrams for rectangular tied columns with bars on all four faces.

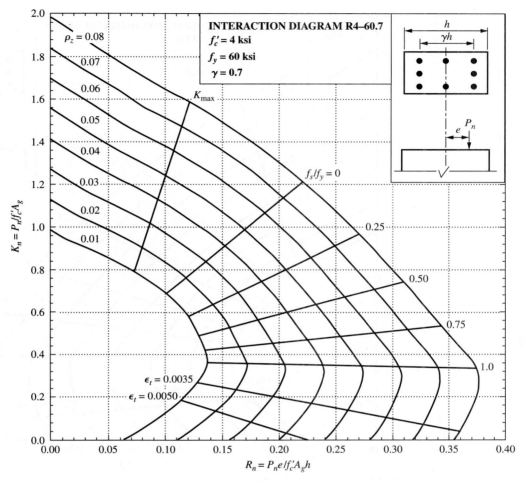

Graph 7 Column interaction diagrams for rectangular tied columns with bars on all four faces.

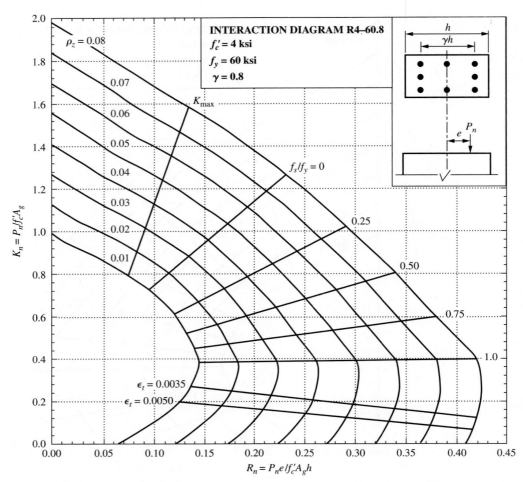

Graph 8 Column interaction diagrams for rectangular tied columns with bars on all four faces.

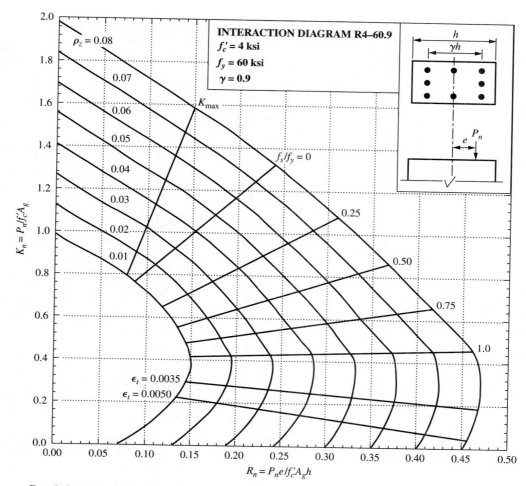

Graph 9 Column interaction diagrams for rectangular tied columns with bars on all four faces.

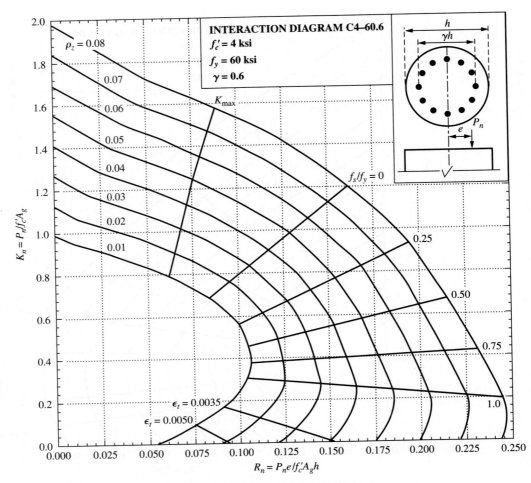

Graph 10 Column interaction diagrams for circular spiral columns.

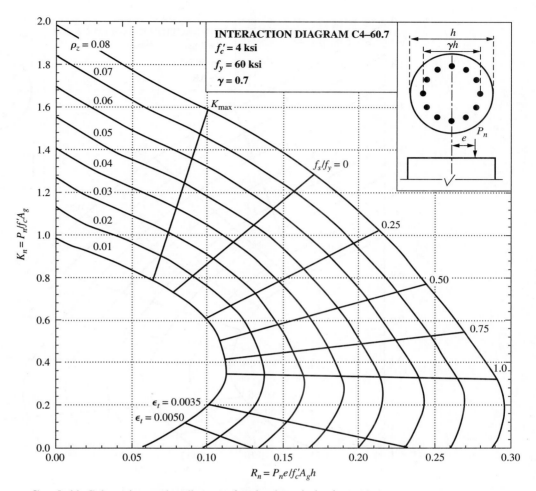

Graph 11 Column interaction diagrams for circular spiral columns.

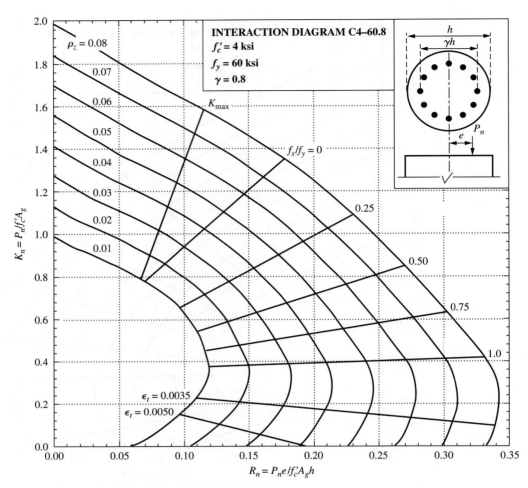

Graph 12   Column interaction diagrams for circular spiral columns.

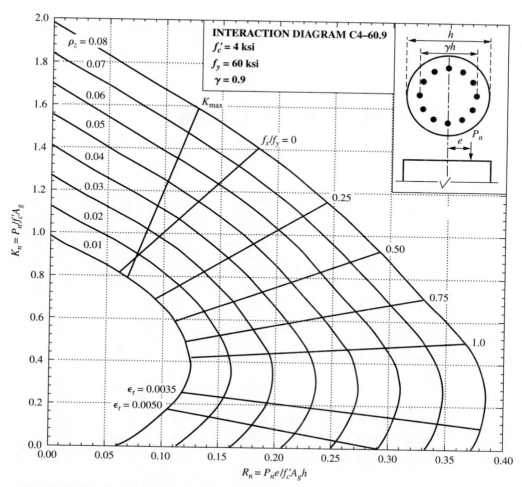

Graph 13 Column interaction diagrams for circular spiral columns.

B

Tables in SI Units

Table B.1 Values of Modulus of
Elasticity for Normal-Weight Concrete

f_c' (MPa)	E_c (MPa)
17	17 450
21	21 500
24	23 000
28	24 900
35	27 800
42	30 450

Table B.2 Designations, Diameters, Areas, Perimeters,
and Masses of Metric Bars

Bar no.	Nominal dimensions		
	Diameter (mm)	Area (mm^2)	Mass (kg/m)
10	9.5	71	0.560
13	12.7	129	0.994
16	15.9	199	1.552
19	19.1	284	2.235
22	22.2	387	3.042
25	25.4	510	3.973
29	28.7	645	5.060
32	32.3	819	6.404
36	35.8	1006	7.907
43	43.0	1452	11.38
57	57.3	2581	20.24

Table B.3 Grades of Reinforcing Bars and Metric Bar Sizes Available for Each

ASTM No.	Steel Grade (MPa)	Bar sizes
A615M Billet	300	#10–#19
	420	#10–#57
	520	#19–#57
A616M Rail	350	#10–#36
	420	#10–#36
A617M Axle	300	#10–#36
	420	#10–#36
A706M Low-alloy	420	#10–#57

Table B.4 Areas of Groups of Standard Metric Bars (mm^2)

Bar Designation	Number of bars								
	2	3	4	5	6	7	8	9	10
#10	142	213	284	355	426	497	568	639	710
#13	258	387	516	645	774	903	1032	1161	1290
#16	398	597	796	995	1194	1393	1592	1791	1990
#19	568	852	1136	1420	1704	1988	2272	2556	2840
#22	774	1161	1548	1935	2322	2709	3096	3483	3870
#25	1020	1530	2040	2550	3060	3570	4080	4590	5100
#29	1290	1935	2580	3225	3870	4515	5160	5805	6450
#32	1638	2457	3276	4095	4914	5733	6552	7371	8190
#36	2012	3018	4024	5030	6036	7042	8048	9054	10 060
#43	2904	4356	5808	7260	8712	10 162	11 616	13 068	14 520
#57	5162	7743	10 324	12 905	15 486	18 067	20 648	23 229	25 810

Table B.4 (Continued)

Bar Designation	Number of bars									
	11	12	13	14	15	16	17	18	19	20
#10	781	852	923	994	1065	1136	1207	1278	1349	1420
#13	1419	1548	1677	1806	1935	2064	2193	2322	2451	2580
#16	2189	2388	2587	2786	2985	3184	3383	3582	3781	3980
#19	3124	3408	3692	3976	4260	4544	4828	5112	5396	5680
#22	4257	4644	5031	5418	5805	6192	6579	6966	7353	7740
#25	5610	6120	6630	7140	7650	8160	8670	9180	9690	10 200
#29	7095	7740	8385	9030	9675	10 320	10 965	11 610	12 255	12 900
#32	9009	9828	10 647	11 466	12 285	13 104	13 913	14 742	15 561	16 380
#36	11 066	12 072	13 078	14 084	15 090	16 096	17 102	18 108	19 114	20 120
#43	15 972	17 424	18 876	20 328	21 780	23 232	24 684	26 136	27 588	29 040
#57	28 391	30 972	33 553	36 134	38 715	41 296	43 877	46 458	49 039	51 620

Table B.5 Minimum Beam Width (mm) for Beams with Inside Exposure (1995 ACI Metric Code)[a,b,c]

Size of Bars	Number of Bars in Single Layer of Reinforcement							Add for each additional bar
	2	3	4	5	6	7	8	
#13	175	213	251	288	326	364	401	37.7
#16	178	219	260	301	342	383	424	40.9
#19	182	226	270	314	358	402	446	44.1
#22	185	232	279	326	373	421	468	47.2
#25	188	239	290	341	391	442	493	50.8
#29	195	252	310	367	424	482	539	57.4
#32	202	267	331	396	460	525	590	64.6
#36	209	281	353	424	496	567	639	71.6
#43	228	314	400	486	572	658	744	86.0
#57	271	386	501	615	730	844	959	114.6

[a]Minimum beam widths for beams were calculated using #10 stirrups.

[b]Maximum aggregate sizes were assumed not to exceed $\frac{3}{4}$ of the clear spacing between the bars (ACI 3.3.2).

[c]The horizontal distance from the center of the outside longitudinal bars to the inside of the stirrups was assumed to equal the larger of 2 times the stirrup diameter (ACI 7.2.2) or half the longitudinal bar diameter.

Table B.6 Areas of Bars in Slabs (mm²/m)

Spacing (mm)	Bar number								
	10	13	16	19	22	25	29	32	36
75	947	1720	2653	3787	5160	6800	8600	10 920	13 413
90	789	1433	2211	3156	4300	5667	7167	9100	11 178
100	710	1290	1990	2840	3870	5100	6450	8190	10 060
115	617	1122	1730	2470	3365	4435	5609	7122	8748
130	546	992	1531	2185	2977	3923	4962	6300	7738
140	507	921	1421	2029	2764	3643	4607	5850	7186
150	473	860	1327	1893	2580	3400	4300	5460	6707
165	430	782	1206	1721	2345	3091	3909	4964	6097
180	394	717	1106	1578	2150	2833	3583	4550	5589
190	374	679	1047	1495	2037	2684	3395	4311	5295
200	355	645	995	1420	1935	2550	3225	4095	5030
225	316	573	884	1262	1720	2267	2867	3640	4471
250	284	516	796	1136	1548	2040	2580	3276	4024
300	237	430	663	947	1290	1700	2150	2730	3353

Table B.7 Values of ρ Balanced, ρ to Achieve Various ϵ_t Values, and ρ Minimum for Flexure. All Values are for Tensilely Reinforced Rectangular Sections

f_y (MPa)	f_c' (MPa)	21 $\beta_1 = 0.85$	28 $\beta_1 = 0.85$	35 $\beta_1 = 0.814$	42 $\beta_1 = 0.764$
	ρ balanced	0.0337	0.0450	0.0538	0.0606
	ρ when $\epsilon_t = 0.004$	0.0217	0.0289	0.0346	0.0390
300	ρ when $\epsilon_t = 0.005$	0.0190	0.0253	0.0303	0.0341
	ρ when $\epsilon_t = 0.075$	0.0144	0.0193	0.0231	0.0260
	ρ min for flexure	0.0047	0.0047	0.0049	0.0054
	ρ balanced	0.0274	0.0365	0.0437	0.0492
	ρ when $\epsilon_t = 0.004$	0.0186	0.0248	0.0297	0.0334
350	ρ when $\epsilon_t = 0.005$	0.0163	0.0217	0.0259	0.0292
	ρ when $\epsilon_t = 0.0075$	0.0124	0.0165	0.0198	0.0223
	ρ min for flexure	0.0040	0.0040	0.0042	0.0046
	ρ balanced	0.0212	0.0283	0.0339	0.0382
	ρ when $\epsilon_t = 0.004$	0.0155	0.0206	0.0247	0.0278
420	ρ when $\epsilon_t = 0.005$	0.0135	0.0181	0.0216	0.0244
	ρ when $\epsilon_t = 0.0075$	0.0103	0.0138	0.0165	0.0186
	ρ min for flexure	0.0033	0.0033	0.0035	0.0039

Table B.7 (Continued)

f_y (MPa)	f'_c (MPa)	21 $\beta_1 = 0.85$	28 $\beta_1 = 0.85$	35 $\beta_1 = 0.814$	42 $\beta_1 = 0.764$
	ρ balanced	0.0156	0.0208	0.0249	0.0281
	ρ when $\epsilon_t = 0.004$	0.0125	0.0167	0.0200	0.0225
520	ρ when $\epsilon_t = 0.005$	0.0109	0.0146	0.0175	0.0197
	ρ when $\epsilon_t = 0.0075$	0.0083	0.0111	0.0133	0.0150
	ρ min for flexure	0.0027	0.0027	0.0028	0.0031

Table B.8 $f_y = 420$ MPa; $f'_c = 21$ MPa—SI Units

	ρ	$\dfrac{M_u}{\phi bd^2}$	ρ	$\dfrac{M_u}{\phi bd^2}$	ρ	$\dfrac{M_u}{\phi bd^2}$	ρ	$\dfrac{M_u}{\phi bd^2}$
ρ_{min} for temp. and shrinkage	0.0018	0.740	0.0048	1.902	0.0078	2.975	0.0107	3.928
	0.0019	0.780	0.0049	1.939	0.0079	3.010	0.0108	3.960
	0.0020	0.820	0.0050	1.976	0.0080	3.044	0.0109	3.991
	0.0021	0.860	0.0051	2.013	0.0081	3.078	0.0110	4.022
	0.0022	0.900	0.0052	2.050	0.0082	3.112	0.0111	4.053
	0.0023	0.940	0.0053	2.087	0.0083	3.146	0.0112	4.084
	0.0024	0.980	0.0054	2.124	0.0084	3.179	0.0113	4.115
	0.0025	1.019	0.0055	2.161	0.0085	3.213	0.0114	4.146
	0.0026	1.059	0.0056	2.197	0.0086	3.247	0.0115	4.177
	0.0027	1.098	0.0057	2.233	0.0087	3.280	0.0116	4.207
	0.0028	1.137	0.0058	2.270	0.0088	3.313	0.0117	4.238
	0.0029	1.176	0.0059	2.306	0.0089	3.347	0.0118	4.268
	0.0030	1.216	0.0060	2.342	0.0090	3.380	0.0119	4.298
	0.0031	1.255	0.0061	2.378	0.0091	3.413	0.0120	4.328
	0.0032	1.293	0.0062	2.414	0.0092	3.446	0.0121	4.359
ρ_{min} flexure	0.0033	1.332	0.0063	2.450	0.0093	3.479	0.0122	4.389
	0.0034	1.371	0.0064	2.486	0.0094	3.511	0.0123	4.418
	0.0035	1.409	0.0065	2.521	0.0095	3.544	0.0124	4.448
	0.0036	1.448	0.0066	2.557	0.0096	3.577	0.0125	4.478
	0.0037	1.486	0.0067	2.592	0.0097	3.609	0.0126	4.508
	0.0038	1.525	0.0068	2.628	0.0098	3.641	0.0127	4.537
	0.0039	1.563	0.0069	2.663	0.0099	3.674	0.0128	4.566
	0.0040	1.601	0.0070	2.698	0.0100	3.706	0.0129	4.596
	0.0041	1.639	0.0071	2.733	0.0101	3.738	0.0130	4.625
	0.0042	1.677	0.0072	2.768	0.0102	3.770	0.0131	4.654
	0.0043	1.715	0.0073	2.803	0.0103	3.802	0.0132	4.683
	0.0044	1.752	0.0074	2.837	0.0104	3.834	0.0133	4.712
	0.0045	1.790	0.0075	2.872	0.0105	3.865	0.0134	4.741
	0.0046	1.827	0.0076	2.907	0.0106	3.897	0.0135	4.769
	0.0047	1.865	0.0077	2.941				

Table B.9 $f_y = 420$ MPa; $f'_c = 28$ MPa—SI Units

	ρ	$\dfrac{M_u}{\phi bd^2}$	ρ	$\dfrac{M_u}{\phi bd^2}$	ρ	$\dfrac{M_u}{\phi bd^2}$	ρ	$\dfrac{M_u}{\phi bd^2}$
ρ_{min} for temp. and shrinkage	0.0018	0.744	0.0059	2.349	0.0100	3.829	0.0141	5.185
	0.0019	0.785	0.0060	2.387	0.0101	3.864	0.0142	5.217
	0.0020	0.825	0.0061	2.424	0.0102	3.898	0.0143	5.248
	0.0021	0.866	0.0062	2.462	0.0103	3.933	0.0144	5.280
	0.0022	0.906	0.0063	2.499	0.0104	3.967	0.0145	5.311
	0.0023	0.946	0.0064	2.536	0.0105	4.001	0.0146	5.342
	0.0024	0.987	0.0065	2.573	0.0106	4.036	0.0147	5.373
	0.0025	1.027	0.0066	2.611	0.0107	4.070	0.0148	5.404
	0.0026	1.067	0.0067	2.648	0.0108	4.104	0.0149	5.435
	0.0027	1.107	0.0068	2.685	0.0109	4.138	0.0150	5.466
	0.0028	1.147	0.0069	2.722	0.0110	4.172	0.0151	5.497
	0.0029	1.187	0.0070	2.758	0.0111	4.205	0.0152	5.528
	0.0030	1.227	0.0071	2.795	0.0112	4.239	0.0153	5.558
	0.0031	1.266	0.0072	2.832	0.0113	4.273	0.0154	5.589
	0.0032	1.306	0.0073	2.869	0.0114	4.306	0.0155	5.620
ρ_{min} flexure	0.0033	1.346	0.0074	2.905	0.0115	4.340	0.0156	5.650
	0.0034	1.385	0.0075	2.942	0.0116	4.373	0.0157	5.681
	0.0035	1.424	0.0076	2.978	0.0117	4.407	0.0158	5.711
	0.0036	1.464	0.0077	3.014	0.0118	4.440	0.0159	5.741
	0.0037	1.503	0.0078	3.051	0.0119	4.473	0.0160	5.771
	0.0038	1.542	0.0079	3.087	0.0120	4.506	0.0161	5.801
	0.0039	1.582	0.0080	3.123	0.0121	4.539	0.0162	5.831
	0.0040	1.621	0.0081	3.159	0.0122	4.572	0.0163	5.861
	0.0041	1.660	0.0082	3.195	0.0123	4.605	0.0164	5.891
	0.0042	1.699	0.0083	3.231	0.0124	4.638	0.0165	5.921
	0.0043	1.737	0.0084	3.267	0.0125	4.671	0.0166	5.951
	0.0044	1.776	0.0085	3.302	0.0126	4.704	0.0167	5.980
	0.0045	1.815	0.0086	3.338	0.0127	4.736	0.0168	6.010
	0.0046	1.854	0.0087	3.374	0.0128	4.769	0.0169	6.040
	0.0047	1.892	0.0088	3.409	0.0129	4.801	0.0170	6.069
	0.0048	1.931	0.0089	3.444	0.0130	4.834	0.0171	6.098
	0.0049	1.969	0.0090	3.480	0.0131	4.866	0.0172	6.128
	0.0050	2.007	0.0091	3.515	0.0132	4.898	0.0173	6.157
	0.0051	2.046	0.0092	3.550	0.0133	4.930	0.0174	6.186
	0.0052	2.084	0.0093	3.585	0.0134	4.963	0.0175	6.215
	0.0053	2.122	0.0094	3.621	0.0135	4.995	0.0176	6.244
	0.0054	2.160	0.0095	3.656	0.0136	5.027	0.0177	6.273
	0.0055	2.198	0.0096	3.690	0.0137	5.058	0.0178	6.302
	0.0056	2.236	0.0097	3.725	0.0138	5.090	0.0179	6.331
	0.0057	2.274	0.0098	3.760	0.0139	5.122	0.0180	6.359
	0.0058	2.311	0.0099	3.795	0.0140	5.154	0.0181	6.388

C

The Strut-and-Tie
Method of Design

C.1   INTRODUCTION

This appendix presents an alternative method for designing reinforced concrete members with force and geometric discontinuities. The method is also very useful for designing deep beams for which the usual assumption of linear strain distribution is not valid. This method of design, commonly referred to as strut-and-tie design is briefly introduced.

C.2   DEEP BEAMS

Section 10.7 of the ACI Code defines a deep beam as a member that

    (a) Is loaded on one face and supported on the opposite face so that compression struts can develop between the load and the supports.

    (b) Has a clear span not more than four times its overall depth or that has regions where concentrated loads are located within two times the member depth from the support.

    Transfer girders are one type of deep beam that occur rather frequently. Such members are used to transfer loads laterally from one or more columns to other columns. Sometimes bearing walls also exhibit deep beam action.

    Deep beams begin to crack at loads ranging from $\frac{1}{3}$ to $\frac{1}{2} P_u$. As a result, elastic analyses are not of much value to us except in one regard: the cracks tell us something about the way the stresses that cause the cracks are distributed. In other words, they provide information as to how the loads will be carried after cracking.

C.3   SHEAR SPAN AND BEHAVIOR REGIONS

The ratio of the shear span of a beam to its effective depth determines how the beam will fail when overloaded. The shear span for a particular beam is shown in Figure C.1 where it is represented by the symbol a. This is the distance from the concentrated load shown to the face of the support. Should the beam be supporting only a uniform load, the shear span is the clear span of the beam.

    When shear spans are long, they are referred to as B regions. These are regions for which the usual beam theory applies—plane sections remain plain before and after bending. The

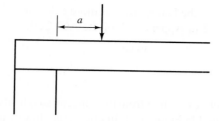

Figure C.1. Shear span.

letter B stands for beam or for Bernoulli (he is the one who presented the linear strain theory for beams).

In some situations the usual beam theory does not apply. When shear spans are short, loads are primarily resisted by arch action rather than beam action. Locations where this occurs are called D regions. The letter D represents discontinuity or disturbance. In such regions plane sections before bending do not remain plane after bending, and the forces obtained with the usual shear and moment diagrams and first-order beam theory are incorrect.

D regions are those parts of members located near concentrated loads and reactions. They also include joints and corbels and other locations where sudden changes in member cross section occur such as where holes are present in members.

According to the St. Venant principle, local disturbances such as those caused by concentrated loads tend to dissipate within a distance approximately equal to the member depth. Figure C.2 shows several typical B and D regions. You should note that the authors used the St. Venant

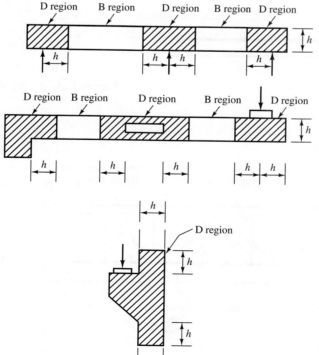

Figure C.2. B and D regions.

principle in this figure to show the extent of the D regions. For more examples the reader should also examine Figures R.A.1.1 and R.A.1.2 in Appendix A of the ACI Code.

C.4   TRUSS ANALOGY

If shear spans are very short, inclined cracks extending from the concentrated loads to the supports tend to develop. This situation is illustrated in Figure C.3. In effect, the flow of horizontal shear from the longitudinal reinforcement to the compression zone has been interrupted. As a result, the behavior of the member has been changed from that of a beam to that of a tied arch where the reinforcing bars act as the ties of an arch.

In Section 8.7 of this text, reference was made to the description of reinforced concrete beams by Ritter-Morsch with the truss analogy method. According to that theory, a reinforced concrete beam with shear reinforcement behaves much like a statically determinate parallel chord truss with pinned joints. The concrete compression block is considered to be the top chord of the fictitious "truss," while the tensile reinforcement is considered to act as the bottom chord. The "truss" web is said to consist of the stirrups acting as vertical tension members, while the portions of the concrete between the diagonal cracks are assumed to act as diagonal compression members. Such a "truss" is shown in Figure C.4, which is a copy of Figure 8.4 presented earlier in this text.

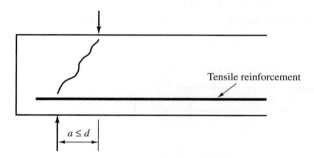

Figure C.3. A very short shear span.

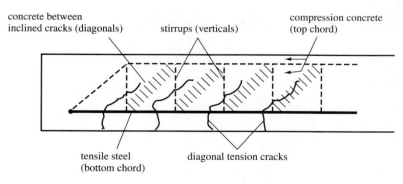

Figure C.4. Truss analogy.

In this figure, the compression concrete and the stirrups are shown with dashed lines. These lines represent the estimated centers of gravity of those forces. On the other hand, the tensile forces are represented with solid lines because those forces clearly act along the reinforcing bar lines.

C.5   DEFINITIONS

A *strut-and-tie model* is a truss model of a D region where the member is represented by an idealized truss of struts and ties.

A *tie* is a tension member in a strut and tie model.

A *strut* is a compression member in a strut and tie model which represents the resultant of the compression field.

A *node* in a strut-and-tie model is the point in a joint where the struts, ties, and concentrated forces at the joint intersect.

The *nodal zone* is the volume of concrete around a node which is assumed to transfer the forces from the struts and ties through the node

C.6   ACI CODE REQUIREMENTS FOR STRUT-AND-TIE DESIGN

Several of the more important Code requirements for strut-and-tie-design are as follows.

Strength of Struts

1. The design strength of a strut, tie, or nodal zone, ϕF_n, must be at least as large as the force in the strut or tie or nodal zone.

$$\phi F_n \geq F_u \qquad \text{(ACI Equation A-1)}$$

In Section 9.3.2 of the ACI Code, ϕ is specified to be 0.75 for strut-and-tie members.

2. The nominal compression strength of a strut which contains no longitudinal reinforcing is to be taken as the smaller value at the two ends of the strut of

$$F_{ns} = f_{cu} A_{cs} \qquad \text{(ACI Equation A-2)}$$

where A_{cs} is the cross-sectional area at one end of a strute taken perpendicular to the axis and f_{ce} is the effective compression strength of the concrete (psi) in a strut or nodal zone. Its value is to be taken as the lesser of (a) and (b) to follow:

(**a**) Effective concrete compression strength in struts

$$f_{ce} = 0.85 \beta_s f_c' \qquad \text{(ACI Equation A-3)}$$

β_s is a factor used to estimate the effect of cracking and confining the reinforcing on the strength of the strut concrete. Values of β_s are given in Appendix Section A.3.2.2 of the ACI Code for different situations. They vary from 0.4 to 0.75, and their meaning and effect are similar to β_1 on the rectangular stress blocks so frequently discussed for beams and columns earlier in this text.

(**b**) Effective concrete compression strength in nodal zones

$$f_{ce} = 0.85 \beta_n f_c' \qquad \text{(ACI Equation A-8)}$$

β_n is a factor used to estimate the effect of the anchorage of ties on the effective compression strength of the nodal zone. Values are specified for different situations in ACI Appendix Section A.5.2 and vary from 0.6 to 1.0, depending on the number of ties and on what bounds the nodal zone.

Strength of Ties

Following the provisions of ACI 318 in their Appendix A-4, the nominal strength of a tie is to be determined with the following expression:

$$F_{nt} = A_{ts}f_y + A_{tp}(f_{se} + \Delta f_p)$$

(ACI Equation A-6)

where A_{ts} equals the area of nonprestressed reinforcing in a tie

f_y = yield strength of the nonprestressed reinforcement
A_{tp} = area of prestressing steel in a tie
f_{se} = effective stress in prestressed reinforcement after losses
Δf_p = increase in stress in prestress steel due to factored loads. The Code in its Section A.4.1 states that it is permissible to use $\Delta f_p = 60,000$ psi for bonded prestressed reinforcement and 10,000 psi for nonbonded prestressed reinforcement. Other values can be used if they can be justified by analysis.

C.7 SELECTING A TRUSS MODEL

When the strut-and-tie method is used for D regions, the results are thought to be more conservative but more realistic than the results obtained with the usual beam theory. To design for a D region of a beam, it is necessary to isolate the region as a free body, determine the forces acting on that body, and then select a system or truss model to transfer the forces through the region.

Once the D region has been identified and its dimensions have been determined, it is assumed to extend a distance h on each side of the discontinuity, or to the face of the support if that value is less than the depth.

The stresses on the boundaries of the region are computed with the usual expression for combined axial load and bending, $P/A \pm Mc/I$. The resulting values must be divided by the capacity reduction factor ϕ for shearing forces (0.75) to obtain the required nominal stresses.

The designer needs to represent the D regions of members, which fail in shear, with some type of model before beginning the design. The model selected for beams with shear reinforcement is the truss model, as it is the best one available at this time.

For this discussion, the beam of Figure C.5 is considered. The internal and external forces acting on this beam, which is assumed to be cracked, are shown. To select a strut-and-tie model

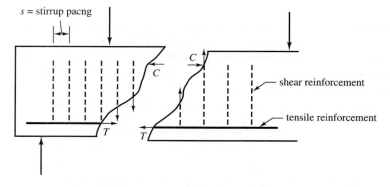

Figure C.5. A beam showing shear and tensile reinforcing.

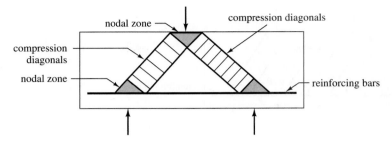

Figure C.6. A short deep beam with the truss model shown.

for such a beam, all the stirrups cut by the imaginary section (see Figure C.5) are lumped into one. In a similar fashion, the concrete parallel to a diagonal is also lumped together in one member.

With the strut-and-tie method, forces are resisted by an idealized internal truss such as the triangular one sketched in Figure C.6. The member and joints of this truss are designed so that they will be able to resist the computed forces. The truss selected must, of course, be smaller than the beam that encloses it, and any reinforcing steel must be given adequate cover. For a first illustration, a short deep beam supporting a concentrated load is shown in Figure C.6.

Various types of nodes are shown in Figure C.7. You should observe that there have to be at least three forces at each joint for equilibrium This is the number of forces necessary for static equilibrium as well as the largest number that can occur in a state of determinate static equilibrium.

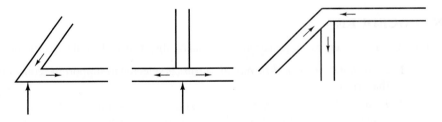

Figure C.7. Various types of truss joints.

If more than three forces meet at a joint when a truss is laid out, the designer will need to make combinations of them in some way so that only three forces are considered to meet at the node. Two possible strut-and-tie models for a deep beam supporting two concentrated loads are shown in Figure C.8. In part (a) of the figure, four forces meet at the location of each concentrated load. As such, we cannot determine all of the forces. An alternative truss is shown in part (b) of the figure in which only three forces meet at each joint.

You can see that the assumptions of the paths of the forces involved in the trusses described might vary quite a bit among different designers As a result, there is no one correct solution for a particular member designed by the strut-and-tie method.

C.8   ANGLES OF STRUTS IN TRUSS MODELS

To lay out the truss, it is necessary to establish the slope of the diagonals (angle θ in Figure C.8 which is measured from the tension chord—the tension reinforcement). According to Schlaich

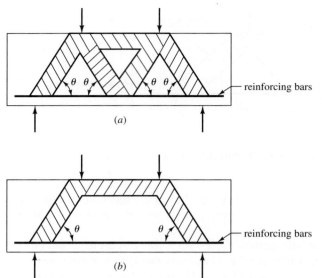

Figure C.8. Two more assumed strut-and-tie trusses.

and Weischede, the angel of stress trajectories varies from about 68° if $l/d \geq 10$ to about 55° if $l/d = 2.0$.[1] A rather common practice, and one that is used in this appendix and is usually satisfactory, is to assume a 2 vertical to 1 horizontal slope for the struts. This will result in a value of $\theta = 63°56'$. The dimensions selected for the truss model must fit into the D region involved, so the angles may need to be adjusted.

C.9   DESIGN PROCEDURE

Following is a step-by-step procedure for using the strut-and-tie design method.

1. *Selection of strut-and-tie model.* A truss is selected to support the concentrated loads, and that truss is analyzed.

2. *Design of vertical stirrups.* A stirrup bar size is assumed, and its strength is assumed to equal its cross-sectional area times it yield stress. The number of stirrups required equals the vertical force divided by the strength of one of the stirrups. The required spacing of these stirrups is determined. If it is too large or too small, a different stirrup size is assumed and the procedure is repeated.

3. *Selection of horizontal reinforcing across beam perpendicular to span.* The Appendix to the Code does not require that reinforcing such as this be used, but it is likely that its use will appreciably reduce cracking. As a result, we can select an amount of steel equal to that listed in ACI Section 11.7.4 for regular deep beam design. There the equation $A_{vh} = 0.0025b_w s_h$ is given, and it is specified that the spacing of such reinforcing not exceed $d/5$ or 12 in.

4. *Computing the strength of struts.* Next ACI Equation A-3 is applied to check needed strut sizes. In actual problems, these struts are the diagonals. As a part of the calculation, the spaces available are compared to the required sizes.

[1]Jörg Schlaich and Dieter Weischede, *Detailing of Concrete Structures*, Bulletin d'Information 150, Comité Euro-International 'du Béton, Paris, March 1982, 163 pp.

5. *Design of ties parallel to beam span.* Horizontal ties parallel to the beam span are needed to resist the horizontal forces in the struts and keep them from cracking. The design strength of such ties is provided by ACI Appendix Equation A-6.

6. *Analysis of nodal zones.* Finally, ACI Appendix Equation A-8 is used to determine the strength of the nodal zones. The reader should note that ACI Appendix Section A.5.2 states that no confinement of the nodal zones is required.

Glossary

Aggregate interlock   Shear or friction resistance by the concrete on opposite sides of a crack in a reinforced concrete member (obviously larger with narrower cracks).

Balanced failure condition   The simultaneous occurrence of the crushing of the compression concrete on one side of a member and the yielding of the tensile steel on the other side.

Bond stresses   The shear-type stresses produced on the surfaces of reinforcing bars as the concrete tries to slip on those bars.

Camber   The construction of a member bent or arched in one direction so that it won't look so bad when the service loads bend it in the opposite direction.

Capacity reduction factors   Factors that take into account the uncertainties of material strengths, approximations in analysis, and variations in dimensions and workmanship. They are multiplied by the nominal or theoretical strengths of members to obtain their permissible strengths.

Cast-in-place concrete   Concrete cast at the building site in its final position.

Column capital   A flaring or enlarging of a column underneath a reinforced concrete slab.

Composite column   A concrete column that is reinforced longitudinally with structural steel shapes.

Compression reinforcement   Reinforcement added to the compression side of beams to increase moment capacity, increase ductility, decrease long-term deflections or to provide hangars for shear reinforcement.

Concrete   A mixture of sand, gravel, crushed rock, or other aggregates held together in a rock-like mass with a paste of cement and water.

Cover   A protective layer of concrete over reinforcing bars to protect them from fire and corrosion.

Cracking moment   The bending moment in a member when the concrete tensile stress equals the modulus of rupture and cracks begin to occur.

Creep or plastic flow   When a concrete member is subjected to sustained compression loads, it will continue to shorten for years. The shortening that occurs after the initial or instantaneous shortening is called *creep* or *plastic flow* and is caused by the squeezing of water from the pores of the concrete.

Dead load   Loads of constant magnitude that remain in one position. Examples: weights of walls, floors, roofs, plumbing, fixtures, structural frames, and so on.

Development length   The length of a reinforcing bar needed to anchor or develop its stress at a critical section.

Doubly reinforced beam   Concrete beams that have both tensile and compression reinforcing.

Drop panels   A thickening of a reinforced concrete slab around a column.

Effective depth   The distance from the compression face of a flexural member to the center of gravity of the tensile reinforcing.

Factored load   A load that has been multiplied by a load factor, thus providing a safety factor.

Flat plate   Solid concrete floor or roof slabs of uniform depths that transfer loads directly to supporting columns without the aid of beams or capitals or drop panels.

Flat slab   Reinforced concrete slab with capitals and/or drop panels.

Formwork   The mold in which semiliquid concrete is placed.

Grade 40 (60) reinforcement   Reinforcement with a minimum yield stress of 40,000 psi (60,000 psi).

Honeycomb   Areas of concrete where there is segregation of the coarse aggregate or rock pockets where the aggregate is not surrounded with mortar. It is caused by the improper handling and placing of the concrete.

Inflection point   A point in a flexural member where the bending moment is zero and where the moment is changing from one sign to the other.

Influence line   A diagram whose ordinates show the magnitude and character of some function of a structure (shear, moment, etc.) as a load of unity moves across the structure.

Interaction curve   A diagram showing the interaction or relationship between two functions of a member, usually axial column load and bending.

L Beam   A T beam at the edge of a reinforced concrete slab which has a flange on only one side.

Lightweight concrete   Concrete where lightweight aggregate (such as zonolite, expanded shales, sawdust, etc.) is used to replace the coarse and/or fine aggregate.

Limit state   A condition at which a structure or some part of that structure ceases to perform its intended function.

Live loads   Loads that change position and magnitude. They move or are moved. Examples: trucks, people, wind, rain, earthquakes, temperature changes, and so on.

Load factor   A factor generally larger than one that is multiplied by a service or working load to provide a factor of safety.

Long columns   *See* Slender columns.

Maximum considered earthquake (MCE)   An extreme earthquake, considered to occur only once every 2,500 years.

Microcrack   A crack too fine to be seen with the naked eye.

Modulus of elasticity   The ratio of stress to strain in elastic materials. The higher its value, the smaller the deformations in a member.

Modulus of rupture   The flexural tensile strength of concrete.

Monolithic concrete   Concrete cast in one piece or in different operations but with proper construction joints.

Nominal strength   The theoretical ultimate strength of a member such as M_n (nominal moment), V_n (nominal shear), and so on.

One-way slab   A slab designed to bend in one direction.

Overreinforced members   Members for which the tensile steel will not yield (nor will cracks and deflections appreciably change) before failure, which will be sudden and without warning due to crushing of the compression concrete.

P-delta moments   *See* Secondary moments.

Plain concrete   Concrete with no reinforcing whatsoever.

Plastic centroid of column   The location of the resultant force produced by the steel and the concrete.

Plastic deformation   Permanent deformation occurring in a member after its yield stress is reached.

Plastic flow   *See* Creep or plastic flow.

Poisson's ratio   The lateral expansion or contraction of a member divided by its longitudinal shortening or lengthening when the member is subjected to tension or compression forces. (Average value for concrete is about 0.16.)

Posttensioned concrete   Prestressed concrete for which the steel is tensioned after the concrete has hardened.

Precast concrete   Concrete cast at a location away from its final position. It may be cast at the building site near its final position but usually is done at a concrete yard.

Prestressing   The imposition of internal stresses into a structure that are of an opposite character to those that will be caused by the service or working loads.

Pretensioned   Prestressed concrete for which the steel is tensioned before the concrete is placed.

Primary moments   Computed moments in a structure which do not account for structure deformations.

Ready-mixed concrete   Concrete that is mixed at a concrete plant and then is transported to the construction site.

Reinforced concrete   A combination of concrete and steel reinforcing wherein the steel provides the tensile strength lacking in the concrete. (The steel reinforcing can also be used to resist compressive forces.)

Secondary moments   Moments caused in a structure by its deformations under load. As a column bends laterally, a moment is caused equal to the axial load times the lateral deformation. It is called a *secondary* or *P-delta moment*.

Seismic Design Category   A classification given to a structure based on its Occupancy Category and the severity of the design earthquake ground motion at the site.

Serviceability   Pertains to the performance of structures under normal service loads and is concerned with such items as deflections, vibrations, cracking, and slipping.

Service loads   The actual loads that are assumed to be applied to a structure when it is in service (also called *working loads*).

Shearheads   Cross-shaped elements such as steel channels and I beams placed in reinforced concrete slabs above columns to increase their shear strength.

Shores   The temporary members (probably wood or metal) that are used for vertical support for formwork into which fresh concrete is placed.

Short columns   Columns with such small slenderness ratios that secondary moments are negligible.

Slender columns (or long columns)   Columns with sufficiently large slenderness ratios that secondary moments appreciably weaken them (to the ACI an appreciable reduction in strength in columns is more than 5%).

Spalling   The breaking off or flaking off of a concrete surface.

Spiral column   A column that has a helical spiral made from bars or heavy wire wrapped continuously around its longitudinal reinforcing bars.

Spirals   Closely spaced wires or bars wrapped in a continuous spiral around the longitudinal bars of a member to hold them in position.

Split-cylinder test   A test used to estimate the tensile strength of concrete.

Stirrups   Vertical reinforcement added to reinforced concrete beams to increase their shear capacity.

Strength design   A method of design where the estimated dead and live loads are multiplied by certain load or safety factors. The resulting so-called *factored loads* are used to proportion the members.

T beam A reinforced concrete beam that incorporates a portion of the slab which it supports.

Tendons Wires, strands, cables, or bars used to prestress concrete.

Tied column A column with a series of closed steel ties wrapped around its longitudinal bars to hold them in place.

Ties Individual pieces of wires or bars wrapped at intervals around the longitudinal bars of a member to hold them in position.

Top bars Horizontal reinforcing bars that have at least 12 in. of fresh concrete placed beneath them.

Transformed area The cross-sectional area of one material theoretically changed into an equivalent area of another material by multiplying it by the ratio of the moduli of elasticity of the two materials. For illustration, an area of steel is changed to an equivalent area of concrete, expressed as.

Two-way slabs Floor or roof slabs supported by columns or walls arranged so that the slab bends in two directions.

Underreinforced member A member that is designed so that the tensile steel will begin to yield (resulting in appreciable deflections and large visible cracks) while the compression concrete has not yet reached its limiting compressive strain. Thus a warning is provided before failure occurs.

Web reinforcement Shear reinforcement in flexural members.

Working loads The actual loads that are assumed to be applied to a structure when it is in service (also called *service loads*).

Working-stress design A method of design where the members of a structure are so proportioned that the estimated dead and live loads do not cause elastically computed stresses to exceed certain specified values. Method is also referred to as *allowable stress design, elastic design*, or *service load design*.

Index

Typical SI Quantities and Units

Quantity	Unit	Symbol	Quantity	Unit	Symbol
Length	meter	m	Stress	pascal (N/m^2)	Pa
Area	square meter	m^2	Moment	newton meter	$N \cdot m$
Volume	cubic meter	m^3	Work	newton meter	Nm
Force	newton	N	Density	kilogram per cubic meter	kg/m^3
Weight	newton per cubic meter	N/m^3	Mass	kilogram	kg

SI Prefixes

Pre/Exname	Symbol	Multiplication factor	
tera	T	$10^{12} =$	1 000 000 000 000
giga	G	$10^9 =$	1 000 000 000
mega	M	$10^6 =$	1 000 000
kilo	k	$10^3 =$	1 000
hecto	h	$10^2 =$	100
deca	da	$10^1 =$	10
deci	d	$10^{-1} =$	0.100
centi	c	$10^{-2} =$	0.010
milli	m	$10^{-3} =$	0.001
micro	μ	$10^{-6} =$	0.000 001
nano	n	$10^{-9} =$	0.000 000 001
pico	p	$10^{-12} =$ 0.000 000 000 001	

Conversion of U.S. Customary to SI Units

Customary U.S. units	SI units
1 in.	25.400 mm = 0.025 400 m
1 in.^{2}	645.16 mm^2 = 6.451 600 m^2 $\times$ 10^{-4}
1 ft	304.800 mm = 0.304 800 m
1 lb	4.448 222 N
1 kip	4 448 222 N = 4.448.222 kN
1 psi	6.894 757 kN/m^2 = 0.006 895 MN/m^2 = 0.006 895 N/mm^2
1 psf	47.880 N/m^2 = 0.047 800 kN/m^2
1 ksi	6.894 757 MN/m^2 = 6.894 757 MPa
1 in.-lb	0.112 985 N $\cdot$ m
1 ft-lb	1.355 818 N $\cdot$ m
1 in.-k	112.985 N $\cdot$ m
1 ft-k	1 355.82 N $\cdot$ m = 1.355 82 kN $\cdot$ m